Progress in Rubber Nanocomposites

Related titles

Dynamic Deformation, Damage and Fracture in Composite Materials and Structures
(ISBN: 978-0-08100-870-6)

Manufacturing of Nanocomposites with Engineering Plastics
(ISBN: 978-1-78242-308-9)

Fillers and Reinforcements for Advanced Nanocomposites
(ISBN: 978-0-08100-079-3)

Woodhead Publishing Series in Composites Science and Engineering: Number 75

Progress in Rubber Nanocomposites

Edited by

Sabu Thomas

Hanna J. Maria

ELSEVIER

AMSTERDAM • BOSTON • HEIDELBERG • LONDON
NEW YORK • OXFORD • PARIS • SAN DIEGO
SAN FRANCISCO • SINGAPORE • SYDNEY • TOKYO
Woodhead Publishing is an imprint of Elsevier

WP WOODHEAD PUBLISHING

Woodhead Publishing is an imprint of Elsevier
The Officers' Mess Business Centre, Royston Road, Duxford, CB22 4QH, United Kingdom
50 Hampshire Street, 5th Floor, Cambridge, MA 02139, United States
The Boulevard, Langford Lane, Kidlington, OX5 1GB, United Kingdom

British Library Cataloguing-in-Publication Data
A catalogue record for this book is available from the British Library

Library of Congress Cataloging-in-Publication Data
A catalog record for this book is available from the Library of Congress

ISBN: 978-0-08-100409-8 (print)
ISBN: 978-0-08-100428-9 (online)

For information on all Woodhead Publishing visit our website at https://www.elsevier.com

Publisher: Matthew Deans
Acquisition Editor: Gwen Jones
Editorial Project Manager: Charlotte Cockle
Production Project Manager: Omer Mukthar
Cover Designer: Greg Harris

Typeset by MPS Limited, Chennai, India

Contents

List of contributors

J. Abraham Mahatma Gandhi University, Kottayam, Kerala, India

S. Agnelli University of Brescia, Brescia, Italy

N.M. Ahmed National Research Center, Giza, Egypt

J. Bahadur Indian Institute of Technology, Roorkee, Uttarakhand, India

V. Cipolletti Pirelli Tyre, Milano, Italy

D. Dasgupta Momentive Performance Materials Pvt. Ltd, Bangalore, Karnataka, India

B. Du Tianjin University, Tianjin, China

M. Dubey Momentive Performance Materials Pvt. Ltd, Bangalore, Karnataka, India

S.H. El-Sabbagh National Research Center, Giza, Egypt

M. Galimberti Politecnico di Milano, Milano, Italy

Y. Ikeda Kyoto Institute of Technology, Matsugasaki, Kyoto, Japan

K. Kanny Durban University of Technology, Durban, South Africa

A. Kato NISSAN ARC, LTD, Yokosuka, Kanagawa, Japan

S. Kohjiya Kyoto University, Umezu, Ukyou-ku, Kyoto, Japan

J. Liu Beijing University of Chemical Technology, Beijing, China

A. Malas The University of Nottingham, Nottingham, United Kingdom

T.P. Mohan Durban University of Technology, Durban, South Africa

R.C.R. Nunes Universidade Federal do Rio de Janeiro, Rio de Janeiro, Brazil

K. Pal Indian Institute of Technology, Roorkee, Uttarakhand, India

V. Panwar Indian Institute of Technology, Roorkee, Uttarakhand, India

T. Phakkeeree Kyoto Institute of Technology, Matsugasaki, Kyoto, Japan

P. Phinyocheep Mahidol University, Ratchthewee, Bangkok, Thailand

K. Sahakaro Prince of Songkla University, Pattani, Thailand

A. Sarkar Momentive Performance Materials Pvt. Ltd, Bangalore, Karnataka, India

A. Saxena Momentive Performance Materials Pvt. Ltd, Bangalore, Karnataka, India

M.M. Selim National Research Center, Giza, Egypt

K.S. Sisanth Mahatma Gandhi University, Kottayam, Kerala, India

B.G. Soares Universidade Federal do Rio de Janeiro, Rio de Janeiro, Brazil

K. Song Massachusetts Institute of Technology, Cambridge, MA, United States

M. Srividhya Momentive Performance Materials Pvt. Ltd, Bangalore, Karnataka, India

M.G. Thomas Mahatma Gandhi University, Kottayam, Kerala, India

S. Thomas Mahatma Gandhi University, Kottayam, Kerala, India

A. Tohsan King Mongkut's University of Technology North Bangkok, Bangsue, Bangkok, Thailand

G.M. Turky National Research Center, Giza, Egypt

D. Wrobel Momentive Performance Materials GmbH, Leverkusen, Germany

L. Zhang Beijing University of Chemical Technology, Beijing, China

Woodhead publishing series in composites science and engineering

1. **Thermoplastic aromatic polymer composites**
   F. N. Cogswell
2. **Design and manufacture of composite structures**
   G. C. Eckold
3. **Handbook of polymer composites for engineers**
   Edited by L. C. Hollaway
4. **Optimisation of composite structures design**
   A. Miravete
5. **Short-fibre polymer composites**
   Edited by S. K. De and J. R. White
6. **Flow-induced alignment in composite materials**
   Edited by T. D. Papthanasiou and D. C. Guell
7. **Thermoset resins for composites**
   Compiled by Technolex
8. **Microstructural characterisation of fibre-reinforced composites**
   Edited by J. Summerscales
9. **Composite materials**
   F. L. Matthews and R. D. Rawlings
10. **3-D textile reinforcements in composite materials**
    Edited by A. Miravete
11. **Pultrusion for engineers**
    Edited by T. Starr
12. **Impact behaviour of fibre-reinforced composite materials and structures**
    Edited by S. R. Reid and G. Zhou
13. **Finite element modelling of composite materials and structures**
    F. L. Matthews, G. A. O. Davies, D. Hitchings and C. Soutis

14. **Mechanical testing of advanced fibre composites**
Edited by G. M. Hodgkinson

15. **Integrated design and manufacture using fibre-reinforced polymeric composites**
Edited by M. J. Owen and I. A. Jones

16. **Fatigue in composites**
Edited by B. Harris

17. **Green composites**
Edited by C. Baillie

18. **Multi-scale modelling of composite material systems**
Edited by C. Soutis and P. W. R. Beaumont

19. **Lightweight ballistic composites**
Edited by A. Bhatnagar

20. **Polymer nanocomposites**
Y-W. Mai and Z-Z. Yu

21. **Properties and performance of natural-fibre composite**
Edited by K. Pickering

22. **Ageing of composites**
Edited by R. Martin

23. **Tribology of natural fiber polymer composites**
N. Chand and M. Fahim

24. **Wood-polymer composites**
Edited by K. O. Niska and M. Sain

25. **Delamination behaviour of composites**
Edited by S. Sridharan

26. **Science and engineering of short fibre reinforced polymer composites**
S-Y. Fu, B. Lauke and Y-M. Mai

27. **Failure analysis and fractography of polymer composites**
E. S. Greenhalgh

28. **Management, recycling and reuse of waste composites**
Edited by V. Goodship

29. **Materials, design and manufacturing for lightweight vehicles**
Edited by P. K. Mallick

30. **Fatigue life prediction of composites and composite structures**
Edited by A. P. Vassilopoulos

31. **Physical properties and applications of polymer nanocomposites**
Edited by S. C. Tjong and Y-W. Mai
32. **Creep and fatigue in polymer matrix composites**
Edited by R. M. Guedes
33. **Interface engineering of natural fibre composites for maximum performance**
Edited by N. E. Zafeiropoulos
34. **Polymer-carbon nanotube composites**
Edited by T. McNally and P. Pötschke
35. **Non-crimp fabric composites: Manufacturing, properties and applications**
Edited by S. V. Lomov
36. **Composite reinforcements for optimum performance**
Edited by P. Boisse
37. **Polymer matrix composites and technology**
R. Wang, S. Zeng and Y. Zeng
38. **Composite joints and connections**
Edited by P. Camanho and L. Tong
39. **Machining technology for composite materials**
Edited by H. Hocheng
40. **Failure mechanisms in polymer matrix composites**
Edited by P. Robinson, E. S. Greenhalgh and S. Pinho
41. **Advances in polymer nanocomposites: Types and applications**
Edited by F. Gao
42. **Manufacturing techniques for polymer matrix composites (PMCs)**
Edited by S. Advani and K-T. Hsiao
43. **Non-destructive evaluation (NDE) of polymer matrix composites: Techniques and applications**
Edited by V. M. Karbhari
44. **Environmentally friendly polymer nanocomposites: Types, processing and properties**
S. S. Ray
45. **Advances in ceramic matrix composites**
Edited by I. M. Low
46. **Ceramic nanocomposites**
Edited by R. Banerjee and I. Manna

47. **Natural fibre composites: Materials, processes and properties**
Edited by A. Hodzic and R. Shanks

48. **Residual stresses in composite materials**
Edited by M. Shokrieh

49. **Health and environmental safety of nanomaterials: Polymer nanocomposites and other materials containing nanoparticles**
Edited by J. Njuguna, K. Pielichowski and H. Zhu

50. **Polymer composites in the aerospace industry**
Edited by P. E. Irving and C. Soutis

51. **Biofiber reinforcement in composite materials**
Edited by O. Faruk and M. Sain

52. **Fatigue and fracture of adhesively-bonded composite joints: Behaviour, simulation and modelling**
Edited by A. P. Vassilopoulos

53. **Fatigue of textile composites**
Edited by V. Carvelli and S. V. Lomov

54. **Wood composites**
Edited by M. P. Ansell

55. **Toughening mechanisms in composite materials**
Edited by Q. Qin and J. Ye

56. **Advances in composites manufacturing and process design**
Edited by P. Boisse

57. **Structural Integrity and Durability of Advanced Composites: Innovative modelling methods and intelligent design**
Edited by P.W.R. Beaumont, C. Soutis and A. Hodzic

58. **Recent Advances in Smart Self-healing Polymers and Composites**
Edited by G. Li and H. Meng

59. **Manufacturing of Nanocomposites with Engineering Plastics**
Edited by V. Mittal

60. **Fillers and Reinforcements for Advanced Nanocomposites**
Edited by Y. Dong, R. Umer and A. Kin-Tak Lau

61. **Biocomposites: Design and Mechanical Performance**
Edited by M. Misra, J. K. Pandey and A. K. Mohanty

62. **Numerical Modelling of Failure in Advanced Composite Materials**
Edited by P.P. Camanho and S. R. Hallett

63. **Marine Applications of Advanced Fibre-reinforced Composites**
Edited by J. Graham-Jones and J. Summerscales

64. **Smart Composite Coatings and Membranes: Transport, Structural, Environmental and Energy Applications**
    Edited by M. F. Montemor
65. **Modelling Damage, Fatigue and Failure of Composite Materials**
    Edited by R. Talreja and J. Varna
66. **Advanced Fibrous Composite Materials for Ballistic Protection**
    Edited by X. Chen
67. **Lightweight Composite Structures in Transport: Design, Manufacturing, Analysis and Performance**
    Edited by J. Njuguna
68. **Structural Health Monitoring (SHM) in Aerospace Structures**
    Edited by F-G. Yuan
69. **Dynamic Deformation and Fracture in Composite Materials and Structures**
    Edited by V. Silberschmidt
70. **Advanced Composite Materials for Aerospace Engineering: Advanced Composite Materials for Aerospace Engineering**
    Edited by S. Rana and R. Fangueiro
71. **Lightweight Ballistic Composites: Military and Law-Enforcement Applications, 2nd Edition**
    Edited by A. Bhatnagar
72. **Handbook of Advances in Braided Composite Materials: Theory, Production, Testing and Applications**
    J. Carey
73. **Novel Fire Retardant Polymers and Composite Materials: Technological Advances and Commercial Applications**
    Edited by D.-Y. Wang
74. **Advanced High Strength Fibre Composites in Construction**
    Edited by M. Fan and F. Fu
75. **Progress in Rubber Nanocomposites**
    Edited by S. Thomas and H. J. Maria

General introduction to rubber compounding

1

K.S. Sisanth, M.G. Thomas, J. Abraham and S. Thomas
Mahatma Gandhi University, Kottayam, Kerala, India

Abbrevations

6PPD	*N*-(1,3-dimethylbutyl)-*N′*-phenyl-*p*-phenylenediamine
ACM	polyacrylate rubber
ACN	acrylonitrile
ADC	azodicarbonamide
ASTM	american standard for testing and materials
BHT	2,6-di-*t*-butyl hydroxyl toluene
BIIR	bromobutyl rubber
BPH	2,2′-methylene-bis-(4-methyl-6-*tert*.butylphenol)
BSH	benzene sulfonylhydrazide
CB	carbon black
CBS	*N*-cyclohexyl-2-benzothiazolsulfenamide
CIIR	chlorobutyl rubber
CLD	caprolactam disulfide
CR	polychloroprene rubber
CSM	chlorosulfonated polyethylene
CTP	cyclohexylthiophthalimide
CV	conventional vulcanization
DBP	dibutyl phthalate
DCBS	*N,N*-dicyclohexyl-2-benzothiazylsulfenamide
DIOP	diisooctylphtalate
DNPT	dinitrosopenta methylene tetra amine
DOP	dioctylphtalate
DOTG	di-*o*-toluylguanidine
DCPD	dicyclopentadiene
DPG	diphenyl guanidine
DPPD	*N,N′*-diphenyl-pphenylenediamine
DPTT	dipentamethylenethiuramtetrasulfide
DTDM	dithiomorpholine
EBC	estate brown crepe
EG	expanded graphite
ENB	ethyldienenorborne
EPDM	ethylene-propylene diamine methylene
EPM	ethylene-propylene rubber
ETMQ	6-Ethoxy-2,2,4-trimethyl-1,2dihydroquinoline
EV	efficient vulcanization
EVA	ethylene-vinyl acetate

Progress in Rubber Nanocomposites. DOI: http://dx.doi.org/10.1016/B978-0-08-100409-8.00001-2

FCM farrel continuos mixers
FEF high abrasion furnace/general purpose furnace
FKM fluorocarbon rubber
FMQ fluro-methyl-polysiloxane
GNP graphite nanoplatelets
HA high ammonia
HD 1,4-hexadiene
HFP hexafluoropropene
IIR isobutylene isoprene rubber
IPPD *N*-isopropyl-*N′*-phenyl-*p*-phenylenediamine
ISAF intermediate super abrasion furnace black
ISNR indian standard natural rubber
LA low ammonia
LCM long continuous mixers
MA medium ammonia
MBS 2-(4-morpholinothio)benzothiazole
MBSS 2-morpholinodithiobenzothiazole
MBT 2-mercaptobenzothiazole
MBTS dibenzothiazyl disulfide
MDI diphenylmethane 4,4′-diisocyanate
MDR moving die rheometer
MQ methyl-polysiloxane
MWNT multiwalled nanotube
NBR nitrile rubber
NFG natural graphite flakes
NSR nigerian standard rubber
ODR oscillating die rheometer
OTOS oxydiethylenethiocarbamylsulfonamide
PAN phenyl-α-naphthylamine
PBN phenyl-ß-naphthylamine
PBR polybutadiene rubber
PLC pale latex crepe
PMQ phenyl-methyl-polysiloxane
PP polypropylene
PRI plasticity retention index
PU polyurethane elastomers
PVC polyvinylchloride
PVMQ phenyl-vinyl-methyl-polysiloxane
Q silicone rubber
RSS ribbed smoke sheet
SAF super abrasion furnace black
SBR styrene-butadiene rubber
SBS styrene-butadiene styrene
SEV semi-efficient vulcanization
SIR standard indonesian rubber
SIS styrene isoprene styrene
SIX sodium isopropyl xanthates
SLR standard lanka rubber
SMR standard malaysian rubber
SPH styrenated phenol

SRF semi reinforcing furnace
SSR specified singapore rubber
SWNT single-walled nanotube
T polysulfide rubber
TBBS *N-tert*-butyl-2 benzothiazolsulfenamide
TDI toluene diisocyanate
TETD tetraethyl-thiuramdisulfide
TFE tetrafluoroethene
T_g glass transition temperature
TMQ 2,2,4-trimethyl-1,2 dihydroquinoline
TMTD tetramethyl-thiuramdisulfide
TMTM tetramethyl-thiurammonosulfide
TPE-A polyether/polyamide block copolymer
TPE-E thermoplastic copolyesters
TPE-U thermoplastic urethane
TSR technically specified rubber
TTR thai tested rubber
VA vinyl acetate
VGC viscosity gravity constant
VMQ vinyl-methyl-polysiloxane
ZBX zincbutylxanthates
ZDBC zinc dibutyldithiocarbamate
ZDMC zinc-dimethyl-dithiocarbamate
ZEDC zinc-diethyl-dithiocarbamate
ZIX zinc isopropyl xanthates
ZMBT zinc-2- mercaptobenzothiazole

1.1 Compounding and its importance

Elastomers such as natural and synthetic rubbers are amorphous polymers and are viscoelastic in nature. These materials have lower mechanical and thermal properties. To overcome this problem various ingredients are added to the rubber matrix to improve its functional properties, this process is termed as compounding. In other words, compounding is the science of selecting and combining elastomers and additives to obtain a uniform mixture that will develop the necessary physical and chemical properties for the finished product [1–3].

The main objectives of rubber compounding are

1. To satisfy the end properties (service requirements)
2. To satisfy better processing requirements
3. To maintain a balance between the properties and price

The processing requirements include proper mixing (incorporation, dispersion, and plasticization) of the compounding ingredients in the elastomeric matrix. The better mixing also gives an idea about the viscosity, scorch safety, and molding characteristics of the rubber compound. The property requirements include hardness, tensile properties (tensile strength, elongation at break, modulus), tear strength, set properties, abrasion resistance, and flex properties.

Classification of compounding ingredients

1. Elastomers
2. Cross-linking agents
3. Accelerators
4. Activators and retarders
5. Antidegradents (antioxidants and antiozonants)
6. Fillers (reinforcing and Non reinforcing)
7. Processing aids (peptizers, plasticizers, softeners, and extenders)
8. Special purpose materials
9. Coloring materials

Elastomer is the key ingredient in the list of compounding ingredients; therefore it is always at the top of the formulation list and is expressed as 100 parts by weight of the total recipe. Rubber chemists generally use the term phr (parts per hundred rubber) for expressing the weight of the compounding ingredients in the formulation.

1.2 Introduction to compounding ingredients

1.2.1 Elastomers

The selection of the elastomer depends on service requirements, cost, and ease of processing. The elastomers are generally two types, natural rubber and synthetic rubbers. Natural rubber is obtained from the rubber tree and it is available in different forms such as sheet, block rubbers, and latex concentrates. The synthetic rubbers are manufactured to overcome the drawback of natural rubber (oil resistance, flame resistance, and weather resistance). Generally used synthetics rubbers are styrene-butadiene rubber (SBR), polybutadiene (BR), nitrile rubbers (NBR), butyl rubbers (IIR), polychloroprene rubber (CR), silicone rubbers (Q), polyurethane, etc.

1.2.1.1 Natural rubber

Natural rubber can be isolated from the rubber tree, the commonly used source, *Hevea brasiliensis*. Natural rubber latex is an aqueous colloidal dispersion of rubber particle and it is extracted by a tapping process. Natural rubber obtained by the biosynthesis pathway consists of *cis*-1,4 polyisoprene units. The rubber particles are spherical in nature and are in the particle size of 0.02–2 microns, specific gravity 0.92, and pH 7. NR latex contains 30–40 wt. % dry rubber, 3–5 wt. % nonrubber constituent, and remaining water. Fresh latex coagulates spontaneously within few hours, but this can be prevented by adding anticoagulants such as ammonia, formalin, etc. Natural rubber is used in different forms such as dry rubber form and liquid form (Fig. 1.1) and (Table 1.1).

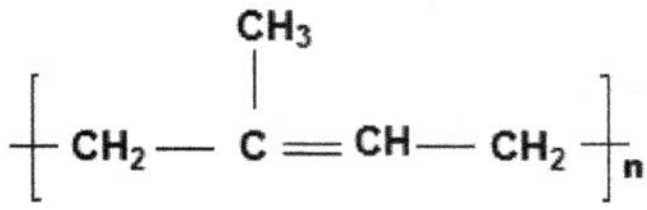

Figure 1.1 Repeating unit of natural rubber.

Table 1.1 Different forms of natural rubber

Latex form	Centrifuged latex (60%) Creamed latex (50–60%)
Dry form	Technically specified rubber (TSR)/block rubber Ribbed smoke sheet (RSS) Pale latex crepe (PLC) Estate brown crepe (EBC) Cuplumps, tree lace

Latex concentrate

Natural rubber latex is obtained from the rubber tree and has a dry rubber content of 30–40%. It is uneconomical to transport the preserved field latex over long distances. The normal procedure is to concentrate it to about 60% of dry rubber contents. The mainly used concentration processes are centrifuging, creaming, evaporation, and electrodecantation methods and among these, the centrifuging method is the most popular method for latex concentration. Indian standard specifications for concentrated latex IS 5430:1981 are given in (Table 1.2).

Table 1.2 Indian standard specifications for concentrated latex IS 5430:1981

Characteristics	Specifications		
	Type HA	Type MA	Type LA
Dry rubber content, % min	60.0	60.0	60.0
Nonrubber solids, % max.	2.0	2.0	2.0
Coagulum content, % max.	0.05	0.05	0.05
Sludge content, % max.	0.10	0.10	0.10
Alkalinity as ammonia, %	0.6 min	Above 0.3 but below 0.6	0.3 max
KOH number, max.	1.0	1.0	1.0
Mechanical stability, *s*, min	475	475	475
Volatile fatty acid number, max.	0.15	0.15	0.15
Copper, ppm on TS, max.	8	8	8
Manganese, ppm on TS, max.	8	8	8

HA-High ammonia, MA-Medium ammonia, LA-Low ammonia.

Technically specified rubber (TSR) and sheet rubber

Technically specified natural rubber is the commercial grade of natural rubber and the grading of rubber is based on the technical specification. This rubber is also known as block rubber. It is prepared by different processes such as decan grana, hevea crumb, dynat, extrusion drying, and granulation process; the hevea crumb process is the most popular method. TSR can be prepared from a latex route and a dry rubber route, however the latex route produces more superior grades. The different grades and types of TSR available in the market include:

- ISNR—Indian Standard Natural Rubber
- SMR—Standard Malaysian Rubber
- SIR—Standard Indonesian Rubber
- SLR—Standard Lanka Rubber
- TTR—Thai Tested Rubber
- SSR—Specified Singapore Rubber
- NSR—Nigerian Standard Rubber

Bureau of Indian Standards formulated specifications for ISNR in 1969 (IS 4588) prescribes 6 grades, mainly ISNR 3 CV, ISNR 3 L, ISNR 5, ISNR 10, ISNR 20, and ISNR 50. The grading of ISNR is based on dirt content; the specification details are shown in (Table 1.3). The sheet rubber is graded according to the "Green Book" standards, the grading of sheet rubber is sorted by visual inspection standard only. The different grades are RSS1X, RSS1, RSS2, RSS3, RSS4, and RSS5; a higher number indicates a more inferior quality.

Properties of natural rubber Glass transition temperature (T_g) of NR is about −70°C and its structure is *cis*-1,4-poryisoprene(2-methyl-1,3-butadiene). NR crystallizes upon stretching; in fact strain-induced crystallization imparts

Table 1.3 Grading of ISNR and its specifications

Characteristics	ISNR 3CV	ISNR 3 L	ISNR 5	ISNR 10	ISNR 20	ISNR 50
Dirt % by mass (max)	0.03	0.03	0.05	0.10	0.20	0.50
Volatile matter, % by mass (max)	0.80	0.80	0.80	0.80	0.80	0.80
Ash, % by mass (max)	0.50	0.50	0.60	0.75	1.00	1.50
Nitrogen, % by mass (max)	0.60	0.60	0.60	0.60	0.60	0.60
Initial plasticity Po (min)	a	30.00	30.00	30.00	30.00	30.00
Plasticity retention index, PRI (min)	60.00	60.00	60.00	50.00	40.00	30.00
Color (livibond scale)	–	6.0	–	–	–	–
Mooney viscosity (1 + 4) at 100°C	60 + 5	–	–	–	–	–
Accelerated storage hardening (max)	8.00	–	–	–	–	–
Color code	Black	Black	Green	Brown	Red	Yellow

[a]As agreed to between the purchaser and the supplier.

outstanding green strength. Natural rubber imparts good tear, tensile, wear, and tack properties. It has a dissipation factor, low heat buildup in dynamic stress and good electrical insulating properties. The main disadvantages are poor weather and ozone resistance, low oil and fuel resistance, and restricted high temperature resistance.

Applications Natural rubber is mainly used in the tyre industry for making different components such as tread, sidewall, and carcass. It also used in the production of conveyor belts and V belts, coatings, latex products, footwear, and adhesives.

1.2.1.2 Synthetic rubbers

The term synthetic rubber refers to rubbery materials produced by chemical synthesis. These materials are homo- or copolymers of diene monomers by emulsion or solution polymerization techniques. Different varieties of synthetic rubbers are available in the market on the basis of polymerization technique and ratio of monomers. Synthetic rubbers are classified into general purpose, special purpose, and speciality purpose synthetic rubbers [4].

General purpose synthetic rubbers

The general purpose synthetic elastomers are used mainly in the tyre industry. The commonly used general purpose synthetic rubbers are Styrene-butadiene rubber (SBR), polybutadiene rubber (PBR) and polyisoprene rubber (IR) etc.

Styrene-butadiene rubber (SBR) SBR is the highest volume synthetic elastomer produced globally. The structure of SBR is given in Fig. 1.2. During the Second World War production was set up on a large scale in US and in Germany. It can be produced by solution and emulsion polymerization of styrene and butadiene. Free radical emulsion polymerization is carried out at two different temperatures: 50°C (hot emulsion grade) and 5−10°C (cold emulsion grade). In the hot process an initiator such as potassium peroxydisulfate is used and in the cold process a redox initiator is used. T_g (−55 to −35°C) depends on the styrene and butadiene ratio, and the density is 0.94 g/cm^3

- *Different grades of SBR*:
  - 1000: Hot emulsion SBR
  - 1500: Cold emulsion SBR
  - 1600: Cold emulsion SBR, masterbatches with 14 phr or less oil
  - 1700: Cold emulsion SBR, oil masterbatches
  - 1800: Cold emulsion SBR, oil-carbon black masterbatches with more than 14 phr oil
  - 1900: Emulsion resin rubber masterbatches.

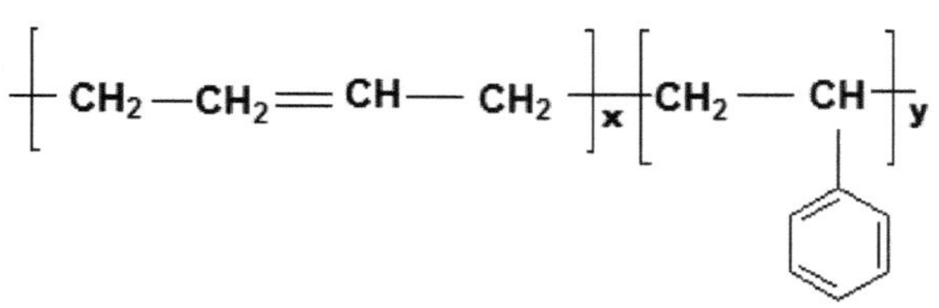

Figure 1.2 Repeating unit of SBR.

- *Trade names*:
  - Ameripol Synpol
  - Stereon
  - Plioflex
  - Krynol
  - Royalene

SBR can also be produced by solution polymerization of styrene and butadiene with alkyl lithium initiator (e.g., butyllithium) in the presence of hydrocarbon solvents such as hexane or cyclohexane. Solution SBR possess narrow molecular weight distribution, higher molecular weight, and higher *cis*-1,4-polybutadiene content than emulsion polymerization SBR. The hot emulsion SBR is more branched polymer than the cold emulsion SBR. Cold SBR shows better abrasion resistance and good dynamic properties. SBR imparts less reversion resistance, higher resistance to abrasion at higher speeds, better resistance to flex fatigue, and it imparts higher filler loading. The main drawbacks of SBR are its inferior mechanical properties (neat form), poor oil resistance, and poor ozone resistance. SBR can be cured by the use of sulfur, sulfur donor systems, and peroxides systems.
Applications Passenger car tyre treads, motor cycle and scooter tyre treads, bead compound for tires, light truck tire treads, conveyor belts, molded shoe soles, microcellular shoe soling, water proof materials, adhesive, molded and extruded goods.

Polybutadiene (PBR/PBD) The structure of PBR is given in (Fig. 1.3). Polybutadiene is the second largest synthetic rubber produced globally. Polybutadiene is the homopolymer of butadiene (C_4H_6) and it is prepared by solution polymerization technique. During the polymerization different microstructures (*cis*-1,4, *trans*-1,4, 1,2 Vinyl) of PBD can be formed. The different microstructures of PBD are shown in Fig. 1.4 and Table 1.4 shows the percentage content of different microstructures while using different catalysts. These microstructures also affect the final properties of polybutadiene. The generally used catalysts are alkyl lithium based, nickel based, cobalt based, titanium based, vanadium based, and neodymium based.

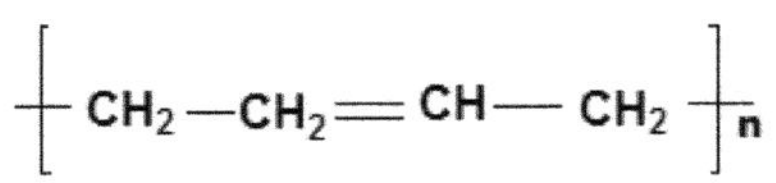

Figure 1.3 Repeating unit of PBR.

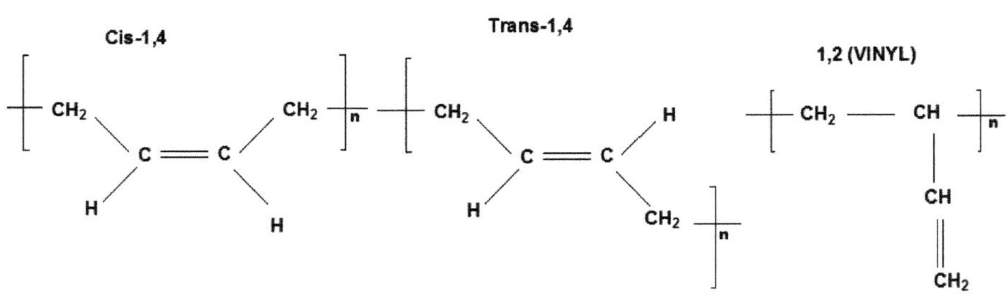

Figure 1.4 Different microstructure of PBD.

Table 1.4 Percentage content of different microstructure using different catalyst

Catalyst system	*Cis* content (%)	*Trans* content (%)	Vinyl content (%)
Neodymium	98	1	1
Nickel based	98	1	1
Cobalt based	97	2	1
Butyllithium based	36	54	10
Titanium based	92	4	4

Properties of PBD The main advantages include better resistance to reversion, high speed abrasion resistance, high resilience, excellent low-temperature flexibility, low heat buildup, and high flex resistance. The main limitations are poor processability, less resistance to hydrocarbon solvents and oils, less resistance to heat and ozone, and low gum strength.

Application It is mainly used in truck tire tread, side walls, car tire tread, light truck tire tread. It is also used in shoe soles, transmission belts, and conveyor belts.

Polyisoprene rubber (IR) Polyisoprene rubber is produced by the solution polymerization of isoprene in the presence of hexane by using the catalyst system. It is also known as synthetic natural rubber, the same empirical formula as natural rubber. But it shows some differences from NR. It is more uniform and lighter in color than NR and possesses less tackiness and low gum strength. The narrower molecular weight distribution of IR exhibits less tendency to strain. The different microstructures formed in the polyisoprene rubber are shown in (Fig. 1.5).

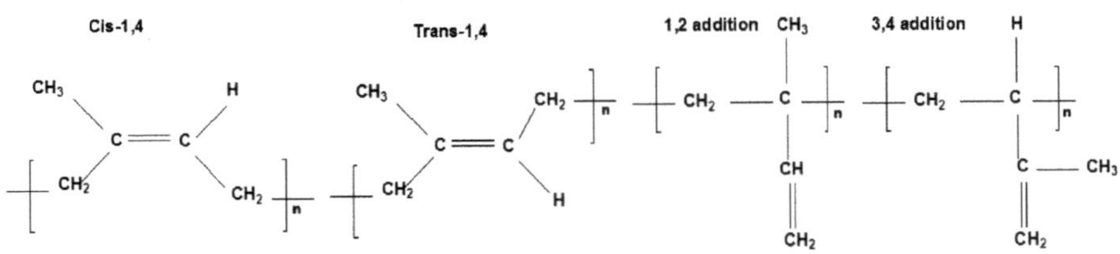

Figure 1.5 Microstructures of IR.

Properties IR Due to the lower amount of nonrubber constituent in IR, they have fewer tendencies for hardening, Lack of strain induced crystallization, and have low tensile strength, abrasion resistance, tack, and modulus. But IR shows better processing parameters, such as better mastication, less mixing time, and lower die swell. IR products impart better compression set properties over NR products. Blending with other rubber can impart good tensile and tear properties.

Applications Applications of IR are similar to natural rubber, mainly used in tyres, conveyor belts, transmission belts, tubes, sports equipment, protective gloves, etc.

Special purpose and speciality synthetic rubbers

In many applications general purpose elastomers are unsuitable due to their insufficient resistance to weather, oils, fire, and elevated temperatures. Special purpose and speciality elastomers have been developed to meet these requirements. The special purpose synthetic rubber includes Butyl Rubbers (IIR- isobutylene isoprene rubber), ethylene-propylene rubber (EPM/EPDM), Nitrile rubber (NBR- Acrylonitrile butadiene rubber), Polychloroprene rubber (CR), and the speciality synthetic rubbers include Silicone rubber (Q rubber), ethylene-vinyl acetate (EVA) copolymer, fluorocarbon rubber (FKM), chlorosulfonated poly-ethylene (CSM), polyurethane elastomers (PU), polyacrylate rubbers (ACM), and polysulfide rubber (T).

Butyl Rubbers (IIR–isobutylene isoprene rubber) IIR are prepared via cationic copolymerization of isoprene (2–3%) with isobutylene (97–98%). The abbreviation IIR stands for isobutylene–isoprene rubber. The monomer unit of IIR is given

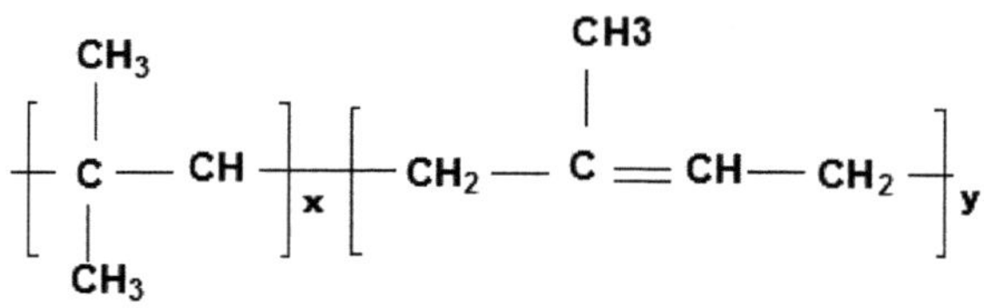

Figure 1.6 Repeating unit of IIR.

in Fig. 1.6. The isoprene part imparts the site for the sulfur cure. Solution polymerization technique is used for the preparation of IIR; the polymerization is kept under the temperature of $-100°C$. The close packing of the isobutylene chain gives the polymer a high degree of impermeability to gases. Chlorobutyl (CIIR) and bromobutyl (BIIR) are the modified types of isobutylene–isoprene rubber containing 1.2% wt of chlorine or bromine, the isoprene unit being the site of halogenation. Introduction of halogen gives greater cure flexibility and enhanced cure compatibility in blends with other diene rubbers. It also confers increased adhesion to other rubbers and metals. The butyl and halobutyl rubbers can be cross-linked with sulfur, dioxime, and resin cure systems. Peroxides cannot be used because they tend to depolymerize the polyisobutylene. Due to the low level of unsaturation in the main chain, more sulfur is required for active thiuram and dithiocarbamate accelerators. Resin curing gives butyl compounds excellent heat stability, this method is used for the production of tyre curing bladders.

Properties

- Low rate of gas permeability
- High long-term thermal stability
- Ozone resistance and weather resistance
- Chemical and moisture resistance
- Vibration damping and high coefficient of friction

Applications The IIR are used in tyre tubes, inner liners for tubeless tyres, in wire and cable applications, tire curing bladders, vibration isolation, radiation resistant gloves, sealants, and sheeting for pond liners and roofing.

Ethylene-propylene rubber (EPM/EPDM) EPM first introduced in 1962, can be divided into two groups: ethylene-propylene rubber (EPM) and ethylene-propylene-diene rubbers (EPDM). Fig. 1.7A shows the repeating unit of EPM. EPM is a copolymer of ethylene and propylene and EPDM is a terpolymer of ethylene, propylene, and

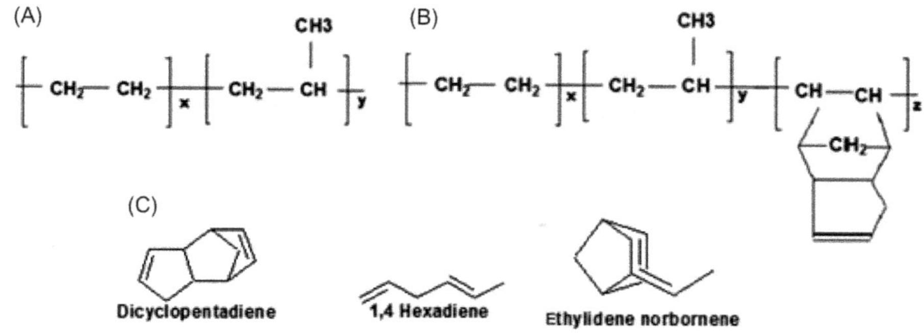

Figure 1.7 (A) Repeating unit EPM, (B) Repeating unit EPDM rubber, and (C) Structure of diene monomers.

diene (Fig. 1.7C). The monomer unit of EPDM is shown in Fig. 1.7B. The most frequently used dienes are dicyclopentadiene (DCPD), ethyldienenorborne (ENB), and 1,4-hexadiene(HD), which offer cross-linking sites for the elastomer. EPM are produced mostly by solution polymerization with Ziegler–Natta type catalysts. EPM rubbers cannot be vulcanized with sulfur because of the absence of unsaturation in the main chain. EPM can be cured with peroxides or radiation. EPDM can be vulcanized with sulfur, peroxide, and radiation.

Properties Both EPM and EPDM rubbers exhibit excellent stability to oxygen, UV light, and also have resistance to ozone and chemicals. It can be loaded with higher amounts of filler and can be cross-linked with sulfur at higher temperature.

Application The EPM/EPDM rubbers are used for wire and cable applications, profile extrusions (window seals and car door seals), washing machine door seals, and agricultural equipment.

Nitrile rubber (NBR) NBR are copolymers of butadiene and acrylonitrile (ACN) which are produced by emulsion polymerization; "hot" and "cold" polymerization methods are used for synthesis. The "hot" polymerized types generally have higher green strength and are slightly harder to process than "cold" copolymers. NBR is available also in the form of Lattices or latex (Fig. 1.8).

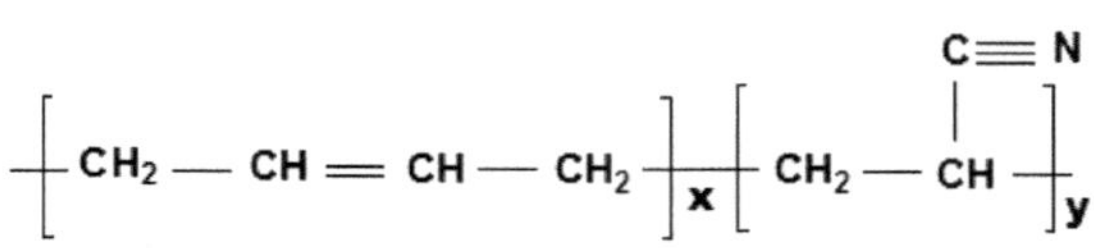

Figure 1.8 Repeating unit NBR.

The effect of ACN in the polymer backbone imparts oil resistance and affects many other properties. Grades of NBR depend on the ACN content; it ranges from 18 to 50%.

Effect of ACN content on the properties

High ACN ⟷ Low ACN

← Resistance to oil and fuel

← Hardness

← Abrasion resistance

← Tensile strength improves

← Gas impermeability

← Heat resistance

← Processing and compatibility with plastics

Resilience improvement →

Plasticizer compatibility →

Low temperature flexibility →

Compression set decreases →

Nitrile rubbers can be cured by sulfur, sulfur donor, and peroxide systems. The solubility of sulfur is less than for other rubbers, so upside down mixing technique is generally practised in NBR compounding. A less sulfur and more accelerator system is generally used. The NBR rubbers have good resistance to oils, so ester-type plasticizers are generally incorporated in NBR compounding. Trade names are Nipol, Krynac, Paracril, Hycar, Perbunan, and Breon.

Applications The excellent oil resistance of NBR makes it useful in oil seals, fuel hoses for fuel stations, hydraulic hoses, cable insulators, roll coverings, conveyor beltings, protective clothing, shoe soles, and plant linings, etc.

Polychloroprene rubber (CR) Polychloroprene rubber is also known as neoprene, (Fig. 1.9) named by Dupont. CR is produced via radical polymerization in aqueous emulsion at temperatures of 10–45°C. The starting material for the production of chloroprene rubber is 2-chloro-1,3-butadiene. The different microstructure formed during the polymerization can be controlled by the polymerization temperature. The higher polymerization temperature is another advantage as it increases the 1,2 vinyl configuration in the chloroprene rubber. During the polymerization, different forms of addition are possible in chloroprene rubber (Fig. 1.10). Chloroprene rubber is also cross-linked with metal oxides systems, based on magnesium oxide/zinc oxide. The most common ratio is magnesium oxide 4.0 phr and zinc oxide 5.0 phr.

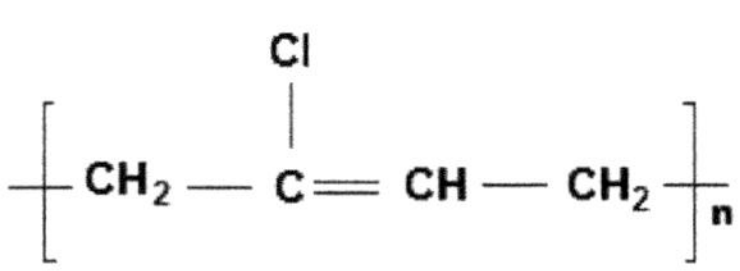

Figure 1.9 Repeating unit CR.

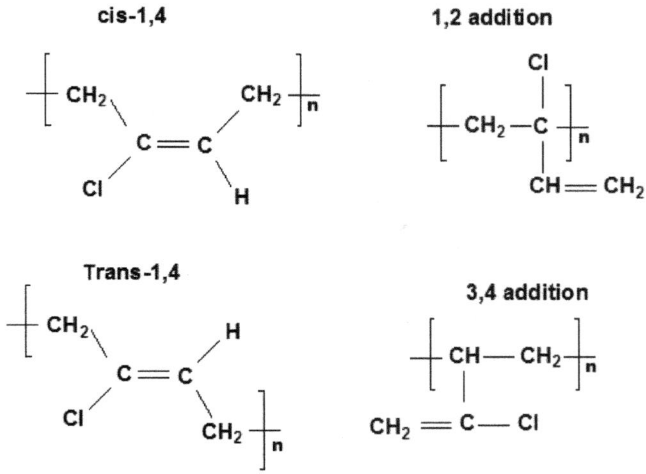

Figure 1.10 Microstructures of CR.

Types of Neoprene rubber

- G type—CR is modified with sulfur and thiruram disulfide
- W type—CR is modified with mercaptan
- T type—highly cross-linked gel form of polychloroprene rubber.

Properties

- The presence of a chlorine atom in the main chain imparts good flame resistance
- Resistance to solvents and chemicals
- Good adhesion to polar substrates such as metals
- Resistance against ozone, sunlight, and weather
- Good low-temperature properties
- CR vulcanizate shows good mechanical properties

Applications

- Wire and cable covering
- Conveyor belts and V belts
- Coated fabric and inflatable hoses
- Adhesives and gaskets
- Vibration isolators

Silicone rubber (Q rubber) Q rubber is an inorganic polymer, the backbone consists of silicone and oxygen atoms (—Si—O—) linkage (Fig. 1.11). There are some pendant groups attached to the polymer chain to form different types of SIR. The SIR are available in different forms, such as block and resinous form, depending on the molecular weight.

Q rubbers are generally polymerized from cyclic oligomers to linear macromolecules. The curing can be carried out at room temperature or elevated temperature. Curing at room temperature occurs with cross-linking agent (e.g., ortho-silicon acid ether) or air, for high temperatures vulcanization peroxides are used (Table 1.5).

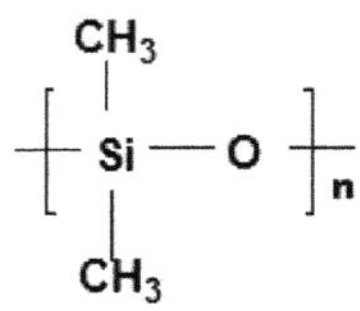

Figure 1.11 General structure of silicone rubber.

Table 1.5 Different types of silicone polymers

Pendant group		Abbreviation	Rubber type
Methyl	CH_3	MQ	Methyl-polysiloxane
Phenyl	C_6H_5	PMQ	Phenyl-methyl-polysiloxane
Vinyl	$CH_2{=}CH$	VMQ	Vinyl-methyl-polysiloxane
Vinyl phenyl	$CH_2{=}CH{-}C_6H_5$	PVMQ	Phenyl-vinyl-methyl-polysiloxane
Trifluoropropyl	$CF_3CH_2CH_2$	FMQ	Fluro-methyl-polysiloxane

Types of silicone polymers

Properties

- Wide operating temperature (−100 to +300°C)
- Excellent resistance to attack by oxygen, ozone, and sunlight
- Nontoxic, odorless, tasteless
- Good electrical insulation properties
- Resistance to low concentration acid, base and salts
- Low toxicity

Limitations of silicones

- Weak mechanical properties with additives
- Weak resistance to steam, acids, and alkalis
- Large shrinkage in molded articles
- Weak oil resistance and solvent resistance
- High cost

Applications Roll coverings, cable insulations, coatings, insulation to power cables and spark plug boots, seals, gaskets, diaphragms, conveyor belts in food industry, pharmaceuticals, and medical components.

Ethylene-vinyl Acetate copolymer (EVA) EVA is a copolymer of EVA (Fig. 1.12). The end property depends on vinyl acetate (VA) content; it varies from 40 to 60 wt %. The elastomeric grade of EVA consists of VA content varying from 40–60 wt %. EVA can be cross-linked by using peroxides and radiation methods. The polymerization technique also varies the VA content in the EVA. EVA shows better blending properties with NR and SBR for better ozone resistance. The main properties are excellent resistance to oxygen, ozone, and light. But EVA shows poor resistance to tear and abrasion. It is mainly used in cable application, medical extrusions, floor materials, coverings, and seals.

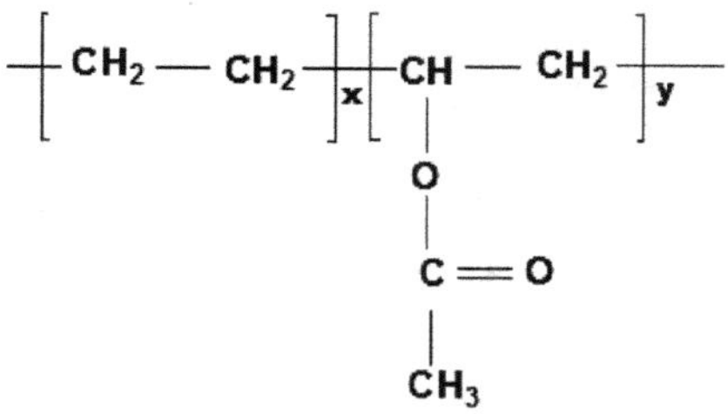

Figure 1.12 Repeating unit of EVA.

Fluorocarbon rubber (FKM) Fluororubbers are co- and terpolymers consisting of at least 75% vinylidene fluoride ($CH_2{=}CF_2$) with the following comonomers

• Hexafluoropropene (HFP)	$CF_2{=}CF{-}CF_3$
• Tetrafluoroethene (TFE)	$CF_2{=}CF_2$
• Perfluoro(methyl vinyl ether)	$CF_2{=}CF{-}O{-}CF_3$

Examples

- Copolymers of vinylidene fluoride and hexafluoropropylene (VF2/HFP)
- Terpolymers of vinylidene fluoride, hexafluoropropylene, and tetrafluoroethylene (VF2/HFP/TFE)
- Terpolymers of vinylidene fluoride, perfluoro(methyl vinyl ether), tetrafluoroethylene (VF2/PVME/TFE)

Properties These rubbers are produced by radical emulsion polymerization using ammonium persulfate as initiator. This rubber has a density of 1.85 g/cm^3 and has a service temperature exceeding 250°C. Vulcanization is done with diamines in combination with basic metal oxides (MgO). The vulcanization is very slow and takes about 24 h at a temperature of 200°C. Fluororubbers allow the highest use temperatures. These products have a very good oil resistance, excellent oxygen, ozone, and weather resistance, and good chemical and solvent resistance. The drawbacks are poor alkali resistance, high price, and relatively poor mechanical properties and cold flexibility.

Applications FKM rubber is mainly used in car and aerospace seals and hoses, fire resistant coverings, gaskets, fuel hoses, O-rings, shaft seals, heat resistant insulators, etc.

Chlorosulfonated polyethylene (CSM) CSM is also known as Hypalon rubber. This rubber is a modification of polyethylene; crystalline thermoplastic material to an amorphous elastomer by chlorosulfonation. CSM grade contains 25 to 43% by weight of chlorine and 0.8 to 1.5% by weight of sulfur. The chemical structure is shown in Fig. 1.13.

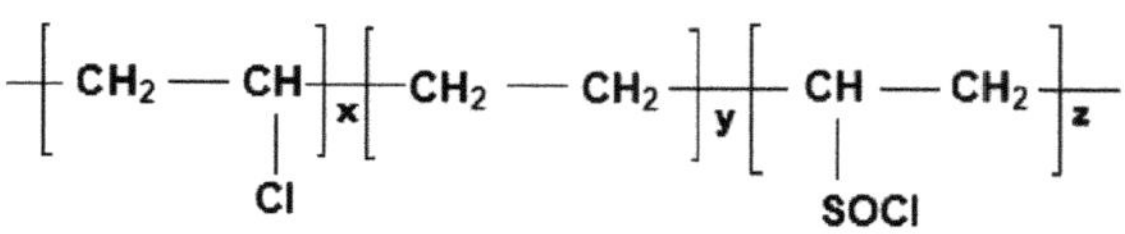

Figure 1.13 Repeating unit of CSM.

It is a polar rubber with excellent chemical resistance, flame resistance, and moderate oil resistance. The unsaturation imparts excellent weather resistance. CSM can be cured with peroxides, metal oxides, and amines. Increasing chlorine content increases oil, fuel, and solvent resistance, but decreases low-temperature flexibility.

Properties of CSM:

- Oxidation and ozone resistance
- Good chemical resistance
- Swelling in some types of oils
- Heat and flame resistance
- Very difficult to process
- High compression set at high temperatures

Applications:

- Cable and wire coverings
- Automotive materials
- Floor materials
- Coated fabrics
- Pond liners
- Hoses

Polyurethane rubbers (PUR) Polyurethanes are prepared by the reaction of the isocyanate group with the hydroxyl group of the alcohol. By varying the reactant amounts and reaction conditions polyurethane can be obtained in different forms, such as millable elastomeric gums, rigid plastics, reactive liquids, and foams. (Fig. 1.14 and Table 1.6).

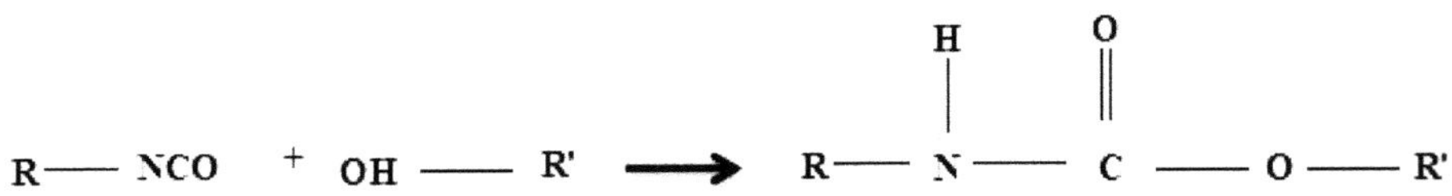

Figure 1.14 Schematics representation of urethane formation.

Table 1.6 Typical isocynate and poly used for the preparation of polyurethane

Typical isocynate	
MDI (diphenylmethane 4,4′-diisocyanate) 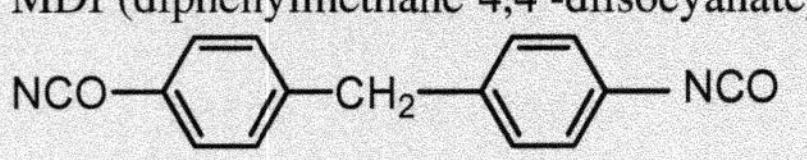	TDI (toluene diisocyanate) 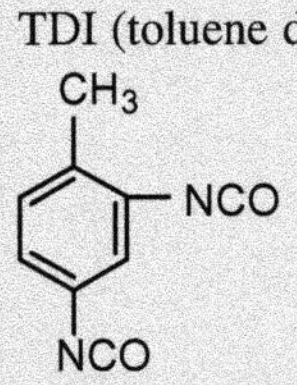
Typical polyol	
Polyethylene adipate (polyester) 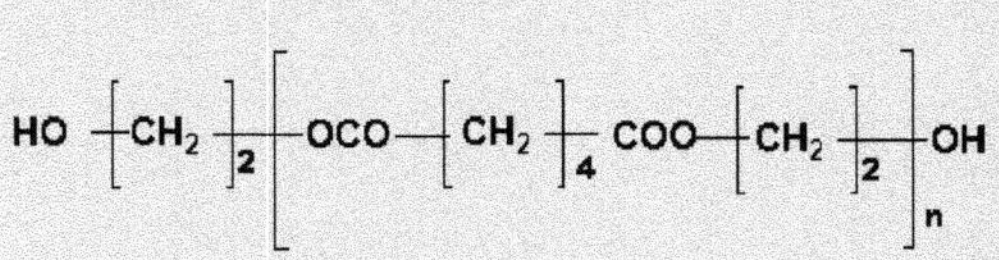	Poly(tetramethylene ether) glycol (polyether) 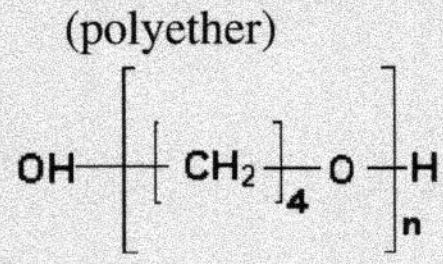

The properties of polyurethane rubbers (PUR) depend on the structure of their chains. Polyester-based PUR usually have better mechanical properties and chemical resistance than polyether-based PUR. Polyether-based PUR have better properties in low temperatures and better hydrolysis resistance. Urethane rubbers can be cured with sulfur or peroxide, and vulcanizates have excellent resistance to weathering, abrasion, and swelling by oil. Some applications are industrial rolls, caster wheels, gaskets, shoe soles, and conveyor belts.

Polyacrylate rubbers (ACM) ACM prepared via a solution or emulsion polymerization process of acrylic esters (ethyl and methyl acrylate, 95%) and reactive cure site monomer (carboxylic acid or chloroethyl vinyl ether, 5%). The general schematic representation of a repeating unit is shown in Fig. 1.15.

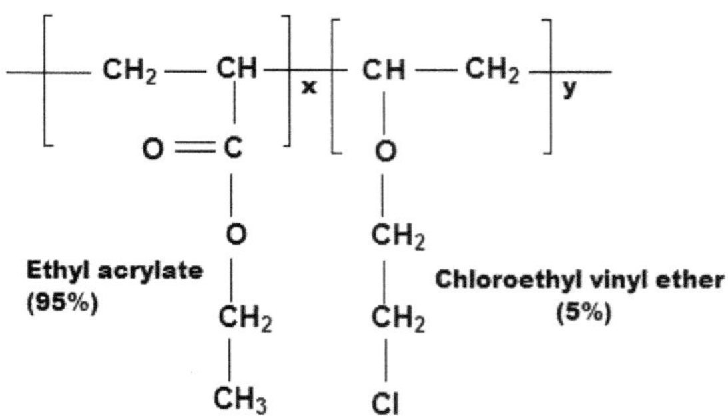

Figure 1.15 Repeating units of ACM.

Butyl acrylate imparts a lower T_g, but poor oil resistance compared to ethyl acrylate, but copolymerization with ACN improves oil resistance. Although acrylate rubbers have excellent weathering resistance, good heat resistance, good oil resistance, and excellent flexing properties, they have poor resistance to alkali and acids and low resistance to hot water. Applications include gaskets, adhesives, O-rings, oil hoses, and transmission seals.

Polysulfide rubber (T rubber) Polysulfide rubbers were first developed in 1920 and are the condensation product of ethylene dichloride and sodium tetrasulfide. Polysulfide rubber is also known as Thiokol. Polysulfide rubbers are formed by the reaction between dihalide and sodium polysulfide. The repeating units are shown in Fig. 1.16 and different types of Thiokol rubbers, sulfur content, and specific gravity are shown in Table 1.7

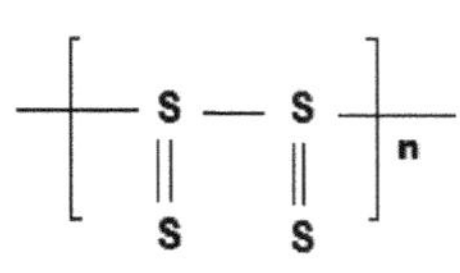

Figure 1.16 Repeating units of polysulfide rubber.

Table 1.7 The different types of thiokol rubbers and their sulfur content, and specific gravity

	Thiokol A	Thiokol B	Thiokol FA	Thiokol ST
Sulfur content	84%	64%	48%	37%
Specific gravity	1.6	1.51	1.34	1.27

Properties

- Excellent oil and solvent resistance
- Good weather and ozone resistance
- Narrow operating temperature range
- Very good low-temperature properties

Applications

- Gaskets
- Paint, oil and fuel hoses
- Permanent putties for fuel tank sealants
- Paint and varnish rolls

1.2.1.3 Thermoplastic elastomers (TPE)

Thermoplastic elastomers (TPE) are a new class of material combining the service properties of elastomers and processing properties of thermoplastic materials. TPE consist of elastomeric soft segment (low T_g) and hard crystallizable segments (high T_g); they are thermodynamically incompatible blends and behave like a single phase. The ratio of hard and soft segments determines the modulus of elasticity, hardness, etc. Advantages of TPE are good properties at low temperatures, excellent abrasion resistance, damping properties, good chemical resistance, and easy processability. TPE are mainly used in the automobile industry, hoses, and sports accessories.

TPE can be divided into

- Block or segmented copolymers
  - Styrene-di/triblock copolymer (SIS, SBS)
  - Thermoplastic urethane elastomers(TPE-U)
  - Thermoplastic copolyesters (TPE-E)
  - Polyether/Polyamide block copolymers (TPE-A)
- Elastomeric blends and thermoplastics
  - NR/PP blends
  - EPDM/PP blends
  - NBR/PVC blends
  - NBR/PP blends

1.2.2 Peptizers

Elastomers exhibit higher viscosity because of high molecular weight so it is very difficult to incorporate fillers and other additives to the matrix. Optimum viscosity level ensures the incorporation of the additives and processing properties, so viscosity control is the key challenging parameter for rubber compounding. The peptizers are used to increase the efficiency of mastication of rubbers, it increases the rate of breakdown of the rubber molecules. Peptizers are added to the rubber at the start of mastication, 0.5 phr or less is used. The main role of peptizers is to accelerate the oxidation rate; they act as an oxidation catalyst and prevent chain recombination.

Examples:

Renacit IV—Zinc salt of pentachlorothiophenol
Pepton 44—Activated dithiobisbenzanilide

1.2.3 Activators

Activators are inorganic and organic chemicals used to activate the action of accelerators. These materials reduce the vulcanization time (cure time) by increasing the rate of vulcanization. Zinc oxide is generally used in combination with a fatty acid (stearic acid) to form an in situ rubber soluble soap–zinc stearate in the rubber matrix. Zinc oxide and steric acid system is the generally used activator system in sulfur cured compounds. The normal dosage of zinc oxide is 2 to 5 phr and stearic acid is 1 to 4 phr. The finer the particle size of zinc oxide, the higher is the effectiveness.

1.2.4 Fillers

Fillers are defined as materials that are added to a polymer formulation to lower the compounding cost or to improve properties. Such materials can be in the form of solid, liquid, or gas. The appropriate selection of these materials, not only helps in reducing the costs but it also helps in improving other properties such as processing and mechanical behaviors. Although these fillers retain their inherent characteristics, very significant differences are often seen depending on the molecular weight, compounding technique, and the presence of other additives in the formulation. Therefore, once the basic property requirements are established, the optimum filler type and the loading level for cost and performance balance must be determined. The recent use of nanofillers in rubber compounding has helped in improving the various properties to a noticeable level. However the structure and other similar properties of the nanofiller play a major role in attaining the properties. Various types of nanofillers based on the dimensions are discussed below [5].

1.2.4.1 One-dimensional nanofiller

These have nano-size (<100 nm) only in one dimension. These are in the form of cylinders and rods. Examples are Carbon nanotubes (CNTs). CNTs have been the main focus of recent research works. CNTs are classified into two categories: Single-walled nanotubes (SWNTs) and multiwalled nanotubes (MWNTs). SWNTs are made up of a rolled graphite sheet capped by hemispheres of fullerenes with diameters ranging from 0.4 to 5.6 nm, lengths of a few micrometers, and an aspect ratio (L/D) of about 1000, while MWNTs consist of two or more concentric cylindrical shells

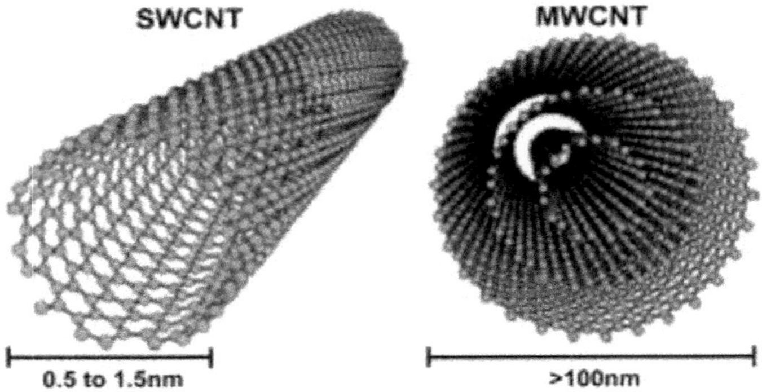

Figure 1.17 Single-walled nanotubes (SWNTs) and multiwalled nanotubes (MWNTs).

of graphite sheets with interlayer separation of about 0.34 nm (Fig. 1.17). SWNTs have a smaller diameter than the diameter of MWNTs, which are about 10–100 nm. CNTs are very expensive compared with other types of nanofillers.

1.2.4.2 Two-dimensional nanofillers

These have nano-size (<100 nm) in two dimensions (Fig. 1.18). These have the shape of plates. Examples are nanoclay, graphene, etc. There are two types of clay structures: 1:1 kaolinite and 2:1 layer silicate. The layered silicate clays or montmorillonite (MMT) are the most commonly used clay for the fabrication of polymer–clay nanocomposites because they are relatively inexpensive and provide exceptional properties because of their high aspect ratio of higher than 1000. The molecular formula of MMT is (Na1/3(Al5/3Mg1/3)$Si_4O_{10}(OH)_2$. It consists of one octahedral layer in between the tetrahedral layers. The tetrahedral sheets are composed of tetrahedrons in which a silicon atom is surrounded by four oxygen atoms. The tetrahedrons are arranged in a hexagonal pattern and form a two-dimensional sheet, while octahedral sheets consist of octahedrons, in which an aluminum atom is surrounded by oxygen atoms. These octahedral and tetrahedral layers form one clay sheet with a thickness of about 0.96 nm and a width of more than 200 nm bonded together by Vander Waals forces. The gap between layers is called the gallery. When some atoms in the crystal structure are replaced with other atoms with different valence, such as Al^{3+} replaced by Mg^{2+}, this leads to negative charges. These charges are balanced by the cations like Na^{+} and K^{+} ions inside the galleries. Thereby the attraction forces become relatively weak and the polymer chains can be intercalated in between the clay layers. Natural graphite flakes are polycrystalline forms of carbon comprising layered planes containing hexagonal arrays of carbon atoms to form an atomically flat stacked material in three dimensions. Covalent bonds bind the carbon atoms in the same plane together, with van der Waals forces between successive layers, separated by 0.337 nm. Because of the very weak van der Waals forces, it is quite easy for small atoms, ions, and molecules to intercalate between the layers to form expanded graphite, graphite nanoplatelets, and graphene.

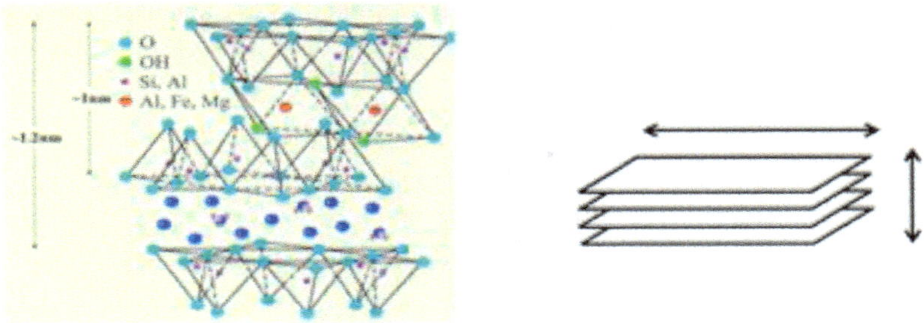

Figure 1.18 Structure of 2D materials montmorillonite.

1.2.4.3 Three-dimensional nanofillers

These have nano-size (<100 nm) in all three dimensions. These are in the form of spherical particles. Inorganic fillers have the form of spherical particles. These nanoparticles offer advantages over one-dimensional and two-dimensional nanoparticles because of their smaller size. Examples of nanoparticles are carbon black (CB) and silica which are used as important components in different applications, e.g., in rubber industries as catalysts and water treatments. Alumina nanoparticles are used as inert fillers in polymers and as catalysts in some reactions (Fig. 1.19).

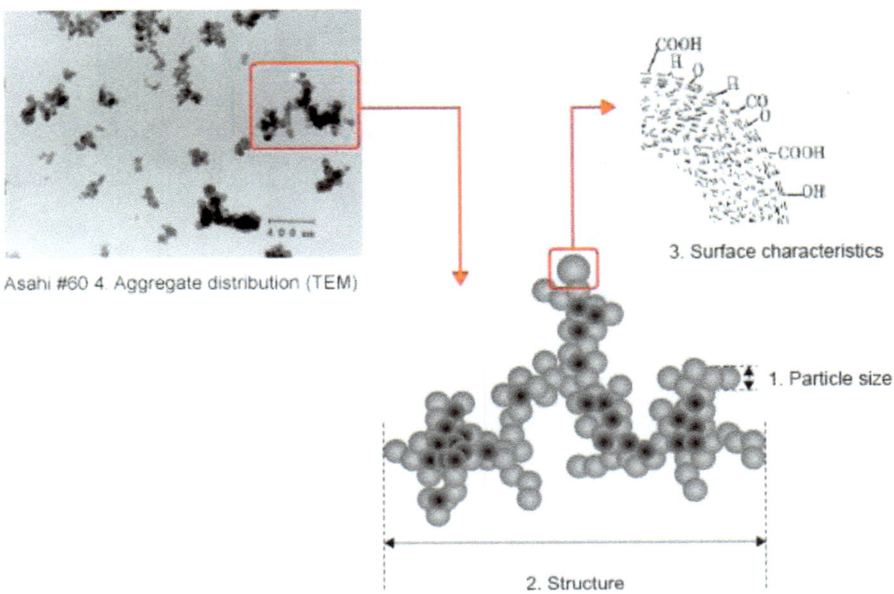

Figure 1.19 Schematic representation of 3D material.

1.2.4.4 Reinforcing and non reinforcing fillers

Fillers are classified into two groups; namely reinforcing fillers and non reinforcing filler. Reinforcing fillers impart good mechanical properties to the vulcanizates, but non reinforcing fillers or inert fillers are generally used to reduce the cost of the compound.

Reinforcing Fillers

The reinforcing fillers, when incorporated in rubber matrix during mixing improve the mechanical properties, such as tensile strength, modulus, abrasion resistance, hardness, and tear resistance. They are classified as black and non black. The reinforcement behavior of filler in polymer matrix depends on the following parameters:

1. Particle size of filler (smaller particle size more reinforcing power)
2. Structure of filler (the higher the structure the higher the reinforcement)
3. Concentration of fillers (optimum filler loading imparts higher reinforcement)
4. Physical and chemical nature of filler
5. Porosity of filler

Examples of reinforcing fillers are CB and precipitated silica. Carbon black is manufactured by four different process: furnace process, thermal process, channel process, and lamp black process. The furnace process is the more popular method for the manufacture of CB. Table 1.8 shows the nomenclature for CBs.

Table 1.8 Nomenclature for carbon blacks

ISO classification	Type	Type name
N110	SAF	Super abrasion furnace
N220	ISAF	Intermediate super abrasion
N330	HAF	High abrasion furnace
N550	FEF	Fast extrusion furnace
N660	FEF	General purpose furnace
N770	SRF	Semireinforcing furnace

Nonreinforcing fillers

Nonreinforcing fillers are also known as inert filler, generally used for reducing the cost of the compound. The size is more than 100 nm, but it alters the specific gravity of the rubber compound. Table 1.9 shows the details of nonreinforcing fillers and their specific gravity values.

1.2.5 Processing aids

These materials are used for improving the processing characteristics of the rubber compounds in mixing, calendering, extrusion, and molding operations. These materials also reduce the cost of the compound. The processing aid consists of plasticizers,

Table 1.9 Details of nonreinforcing fillers and their specific gravity values

Fillers	Specific gravity	Fillers	Specific gravity
China clay	2.50	Calcium oxide	3.4
Calcium carbonate	2.65	Fumed anhydrous silica	2.5
Calcium oxide (lime)	2.19	Precipitated amorphous hydrated silica	2.0
Calcium silicate	2.10	Calcium hydroxide	2.3
China clay (kaolin)	2.60	Magnesium silicate	2.80
Magnesium oxide	2.15 to 3.60	Lead oxide	9.30
Aluminum hydroxide	2.4	Zinc oxide	5.55
Aluminum silicate	2.10	Titanium dioxide (rutile)	4.20

softeners, tackifiers, and extenders. The selection of the processing aid depends on the nature of the polymer, solubility parameter, and processing requirements.

1.2.5.1 Plasticizers

The plasticizers are classified into two forms:

Petroleum-based

The petroleum-based oils are broadly classified as aromatic, paraffinic, or naphthenic, depending on the viscosity gravity constant (VGC) and aniline point. (Table 1.10).

Table 1.10 The VGC constant and aniline point of different petroleum-based oils

Petroleum oils	VGC	Aniline point
Aromatic	0.95–1.00	25–50°C
Naphthenic	0.85–0.9	55–75°C
Paraffinic	0.78–0.82	100–150°C

Ester plasticizers

The ester-type plasticizers are generally used in NBR and CR compounds because they impart good vulcanization properties. The examples for ester-type plasticizers are dibutyl phthalate (DBP), dioctylphtalate (DOP), diisooctylphtalate (DIOP), trixylyl phosphate, and dibutylsebacate.

1.2.5.2 Other processing aids

Table 1.11 shows the generally used softeners and extenders in rubber compounding.

Table 1.11 Generally used softeners and extenders in rubber compounding

Name of processing aid	Example	Use
Factice	Vulcanized vegetable oils	Controls die swell, improves surface quality and preparation of soft compounds
Prevulcanization inhibitors	*N*-Cyclohexylthiophthalimide (CTP)	Retards the cure reaction
Pine product	Pine tar	Improves the tackiness of the compounds
Resins	Coumarone-indene resins	Tackifiers and plasticisers
	High styrene resins	Good processability and high hardness
	Phenol formaldehyde resins	Tackifiers and reinforcing agent
Reclaim rubber	Whole tyre reclaim Butyl reclaim	Controls die swell and reduces the cost

1.2.6 Accelerators

The rate of vulcanization of a rubber compound is controlled by the selection of the accelerator. The accelerators are classified according to the chemical structure and functional action.

Table 1.12 Types of accelerators based on chemical structure and their chemical name

Chemical class	Chemical name	Abbreviation
Thiazole	2-Mercaptobenzothiazole	MBT
	Zinc-2-mercaptobenzothiazole	ZMBT
	Dibenzothiazyl disulfide	MBTS
Guanidines	Diphenyl guanidine	DPG
	Di-*o*-toluylguanidine	DOTG
Sulfenamides	*N*-cyclohexyl-2-benzothiazolsulfenamide	CBS
	N-*tert*- butyl-2 benzothiazolsulfenamide	TBBS
	2-(4-Morpholinothio)benzothiazole	MBS
	N,N-dicyclohexyl-2-benzothiazylsulfenamide	DCBS
Dithocarbamates	Zinc-dimethyl-dithiocarbamate	ZDMC
	Zinc-diethyl-dithiocarbamate	ZEDC
	Zinc dibutyldithiocarbamate	ZDBC
Thiuramsulfides	Tetramethyl-thiurammonosulfide	TMTM
	Tetramethyl-thiuramdisulfide	TMTD
	Tetraethyl-thiuramdisulfide	TETD
Xanthates	Zinc isopropyl xanthates	ZIX
	Sodium isopropyl xanthates	SIX
	Zinc butyl xanthates	ZBX

1.2.6.1 Accelerators based on chemical structures

According to chemical structure, accelerators are classified in to thiazole, guanidines, sulfenamides, dithocarbamates, thiuram sulfides, and xanthates (Table 1.12).

1.2.6.2 Accelerators based on functional action

According to their functional action the accelerators are classified as primary or secondary. Primary accelerators show scorch delay (slow to delayed fast) and secondary accelerators are scorchy (fast curing) (Table 1.13).

Table 1.13 Types of accelerators based on functional action

According to action	Accelerator type
Slow	DPG, DOTG
Delayed	CBS, TBBS, DCBS, MBTS, MBS
Fast	TMTM, TMTD, TETD
Ultrafast	ZIX, ZDMC, ZDEC

1.2.7 Antidegradents

Most of the elastomers contain unsaturated hydrocarbons, which makes them subject to attack by both oxygen, ozone, light, and metal ions. Their properties are thus decreased with an increase in the service period. The antidegradant protects rubber compounds against the attack of oxygen and ozone. Typical loading levels are of the order of 1–4 phr. Antidegradants can be classified as antioxidants or antiozonants.

1.2.7.1 Antioxidants

There are two types of antioxidant used for the rubber compounding, they are amine and phenolic types. Amine types are staining types and phenolic types are nonstaining, used in colored compounds (Table 1.14).

Table 1.14 The common amine-based and phenolic-based antioxidants

Chemical class	Chemical name	Abbreviation
Amine-based antioxidants (staining antioxidants)		
Dihydroquinolines	2,2,4-Trimethyl-1,2 dihydroquinoline	TMQ
	6-Ethoxy-2,2,4-trimethyl-1,2 dihydroquinoline	ETMQ
Naphtylamines	Phenyl-α-naphthylamine	PAN
	Phenyl-ß-naphthylamine	PBN
Phenol-based antioxidants (non staining)		
Monophenols	Styrenated phenol	SPH
	2,6-Di-*t*-butyl hydroxyl toluene	BHT
Bisphenols	2,2′-Methylene-bis-(4-methyl-6-*tert*.butylphenol)	BPH

1.2.7.2 Antiozonants

The various types of antiozonants are given in Table 1.15.

Table 1.15 Different types of antiozonants

Chemical class	Chemical name	Abbreviation
Amine-based antiozonants (staining)		
Paraphenylenediamines	*N*-(1,3-Dimethylbutyl)-*N′*-phenyl-*p*-phenylenediamine *N,N′*-Diphenyl-pphenylenediamine *N*-Isopropyl-*N′*-phenyl-*p*-phenylenediamine	6PPD DPPD IPPD
Phenol-based antiozonants (nonstaining)		
Monophenols Other class	Styrenated phenol Paraffin wax	SPH

1.2.8 Curing agents

The generally used curing agents are sulfur, peroxides, and metal oxides. These materials are used to cross-link the rubber matrix; these cross-links form different bonds, such as carbon–carbon, carbon–sulfur, and sulfur–sulfur linkages. The bond energy also depends on the type of cross-linking formed in the elastomer matrix.

1.2.8.1 Sulfur

The sulfur vulcanization was discovered in 1839 by Charles Goodyear. During vulcanization, the rubber molecule becomes cross-linked and a three-dimensional network structure is formed. During vulcanization rubber loses its tackiness and it is more resistant to solvents, heats, light, etc. Fig. 1.20 shows the schematic representation of sulfur vulcanization.

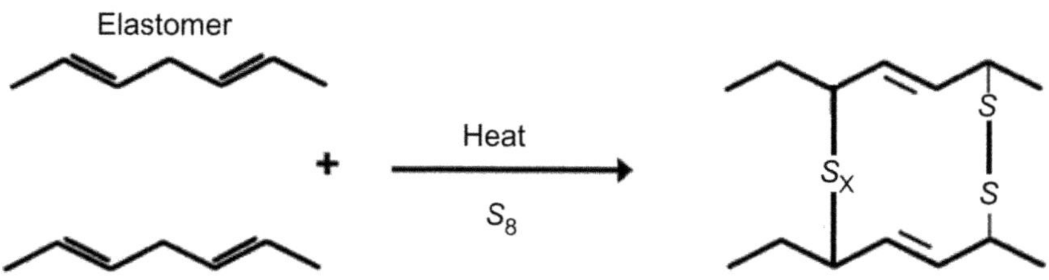

Figure 1.20 Schematic representation of sulfur vulcanization.

Two forms of sulfur used are the rhombic form and amorphous form. The rhombic sulfur is an allotropic form of sulfur, it occurs as an S_8 ring. This is soluble in carbon disulfide and rubber. The amorphous sulfur is also known as insoluble sulfur, it an allotropic form of sulfur, not soluble in carbon disulfide and rubber. At a temperature of 110°C, insoluble sulfur is converted into the rhombic form. Table 1.16 shows the different types of sulfur vulcanization systems.

Table 1.16 Different types of sulfur vulcanization system

	Conventional vulcanization (CV)	Efficient vulcanization (EV)	Semi-efficient vulcanization (SEV)
Sulfur	1.5–3 phr	0–0.5 phr	1.2–1.5 phr
Accelerator	0.5–2 phr	2–6 phr	1.2–1.5 phr
Types of linkage formed	Polysulfidic	Monosulfide	Disulfide
Property improvement	Good tensile properties	Excellent thermal stability	Moderate thermal stability and tensile properties

Sulfur donors (sulfur bearing chemicals)

Organic accelerators and similar compounds that release sulfur at vulcanization temperatures can also be used as vulcanization agents. These materials are added in 3–5 phr and form monosulfidic linkages with excellent heat resistance. The generally used materials are shown in Table 1.17.

Table 1.17 Generally used sulfur donor materials

Sulfur donor	% Sulfur content
Tetramethylthiuram disulfide (TMTD)	13.3
N-oxydiethylene-*N'*-oxydiethylenethiocarbamylsulfonamide (OTOS)	12.9
Dipentamethylenethiuramtetrasulfide (DPTT)	16.6
2-Morpholinodithiobenzothiazole (MBSS)	11.3
Dithiomorpholine (DTDM)	13.6
Caprolactam disulfide (CLD)	11.1
Alkyl phenol disulfide	23.0

1.2.8.2 Peroxides

Organic peroxides are normally used to cross-link saturated and unsaturated rubbers. The peroxide is decomposed forming a free radical on the polymer chain. The peroxide radical forms a carbon–carbon linkage with adjacent polymer chains. A schematic representation of peroxide curing is shown in Fig. 1.21.

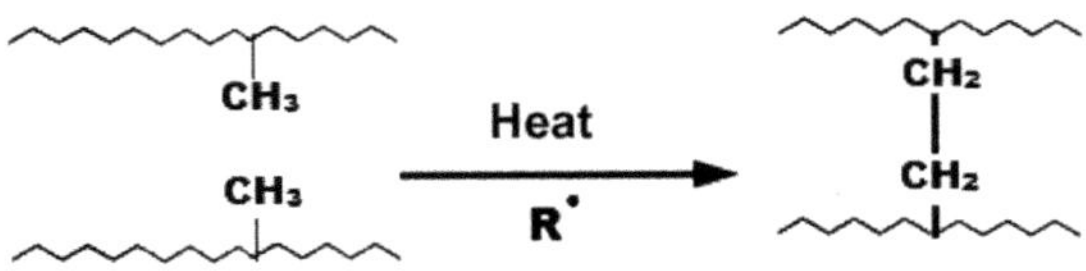

Figure 1.21 Schematic representation of peroxide curing mechanism.

The unsaturated polymers such as NR, SBR, NBR, etc. can also be cross-linked with peroxide, but IIR is degraded by peroxide. The peroxide curing gives superior aging properties to the rubber products. However, peroxide curing is generally higher in cost and requires greater care in storage and processing. Wide varieties of organic peroxides are available and those commonly used are benzoyl peroxide and dicumyl peroxide. The proper choice of a peroxide curing agent depends on stability, activity, intended cure temperature, and effect on processing properties.

1.2.8.3 Resin curing

Phenolic resins are used to cure IIR. This cure system is widely used for the bladders used in curing new tyres, and the curing bags used in the retreading. The low level of unsaturation of IIR requires resin cure activation by halogen-containing materials such as $SnCl_2$ or halogen-containing elastomers such as neoprene. The resin cured butyl vulcanizates have excellent heat resistance.

1.2.8.4 Metal oxides curing

Polychloroprene rubber (Neoprene) and CSM are vulcanized with metal oxides. The commonly used metal oxides are zinc oxide, magnesium oxide, lead oxide, etc. Generally 5 phr of zinc oxide and 4 phr magnesium oxide are used in the formulations. Ethylene thiourea is used as an accelerator for metal oxide curing.

1.2.9 Special purpose additives

1.2.9.1 Blowing agents

These are the materials generally used for the production of cellular/sponge rubber products. There are two type of blowing agents:

1. Inorganic blowing agents—release carbon dioxide gas
   E.g., Sodium bicarbonate, ammonium carbonates, ammonium bicarbonates
2. Organic blowing agents—release nitrogen gas
   E.g., Dinitrosopenta methylene tetra amine (DNPT), Azodicarbonamide (ADC), Benzene sulphonylhydrazide (BSH)

1.2.9.2 Silane coupling agents

Silane coupling agents improve the mechanical properties of silica and silicate containing fillers. A chemical bond is formed between the filler and the rubber matrix. The generally used silane coupling agents are bis-(3-triethoxysilylpropyl) tetrasulfane and 3-thio-cyanatopropyl triethoxysilane. The incorporation of coupling agents improves the cross-link density and tear properties; it reduces compression set, heat buildup and damping behavior.

1.2.9.3 Antistatic agents

In certain rubber articles, such as conveyor belts, textile rollers, trolley wheels, and spinning cots, the buildup of static electricity is undesirable. The generally used antistatic agents are quaternary ammonium salts or ethylene oxide condensate.

1.2.9.4 Flame retardants

These materials are added to rubber compounds for reducing their flammability; this is achieved by a mixture of organic and inorganic materials. The generally used materials are antimony trioxide, chlorinated derivative of paraffin hydrocarbons, zinc borate.

1.3 Rubber processing equipments

1.3.1 Mixing equipments

Mixing is accomplished using two roll mills, internal mixers, and continuous mixers. The resulting compound is then further processed through extrusion, calendaring, etc. The objective of the mixing process is to incorporate compounding ingredients uniformly in the rubber matrix. The mix quality of the rubber compound depends on the processing parameter, such as viscosity, cure characteristics, etc. ASTM D 3182 specifies the mixing equipment, general mixing procedures, vulcanization equipment, and procedures.

1.3.1.1 Two roll mill

A two roll mill consists of two horizontally placed hollow metal rolls rotating towards each other. The distance between the mill rolls can be varied and this gap is known as nip. The speed difference between the rolls is called friction ratio and it allows the shearing action. The back roll moves faster than the front roll; a common friction ratio is 1:1.25. Two roll mill mixing is also known as open mill mixing.

1.3.1.2 Internal mixer

Two different types of internal mixers are generally used—Banbury and intermix. The advantages of internal mixers are they are faster, need fewer operators, need less floor space, etc. The rolls of internal mixers are known as rotors. The various processing parameters in internal mixing are batch size, sequence of loading, ram pressure, rotor speed, mixing time, and temperature of mixing.

1.3.1.3 Continuous mixers

There are mainly two types of continuous mixers—Farrel continuous mixers (FCM) and Long continuous mixers (LCM). They provide processing efficiency, extreme versatility, dependability, and profitability.

FCM – All ingredients can be fed into the mixer separately and liquids can be injected directly into the mixing chamber. Intensive material shear is applied to melt the polymer and to mix all of the ingredients by kneading between the rotors and chamber wall as well as by the rolling action within the material itself.

Some of the most important features are: counter-rotating, non intermeshing rotors at synchronous speed; the unique rotor geometry enables superior dispersive mixing; large rotor tip-to-wall clearance minimizes the effect of wear; large feed opening allows for high filler loadings; and they are energy efficient with low operating costs.

LCM – The LCM design includes a two-stage mixing chamber in combination with 10 L/D long rotors. The primary mixing stage features include: dry blending of the polymer with all other ingredients followed by preheating of dry blend, and subsequently there is a breakdown of larger agglomerates by friction between polymer particles, "the ball mill effect."

Secondary mixing stage features include: the intensive shear between the rotor tips and chamber wall melts the polymer and provides dispersive mixing to incorporate the other ingredients; back mixing via longitudinal cut-back pushes the material back and forth along rotor axes for distributive mixing; uniformity is achieved in the final kneading step by the rolling action between the two rotors. After this step, the material leaves the mixing chamber.

1.3.2 Molding equipments

The process of molding involves both shaping and vulcanization. The basic concept of the rubber molding is that an uncured elastomer rubber compound is placed inside a heated metal mold with pressure to form a cross-linked rubber compound.

1.3.2.1 Compression molding

Compression molding is one of the oldest and most widely used techniques for many products. This process is of low-cost because of the low mold costs and simple process technique. The compression molds are classified into three techniques, they are:

1. Straight mold or flash mold
2. Positive type or plunger type
3. Semi positive

1.3.2.2 Transfer molding

This process involves the transfer of uncured elastomer from one part of the mold to another part within the mold. Transfer molding is costlier than compression molding, but the advantages are the ability to molded into complex shapes, much less flash, and metal encapsulation can be made very easily.

1.3.2.3 Extrusion molding

Extrusion is a process used to create objects of a fixed cross-sectional profile. In this process the materials are subjected to extrusion through a die under controlled temperature and pressure. The extrusion technology is mainly used for making tyre tubes, rubber hoses, and there are two types of extruders

1. Screw type extrusion (Pressure is produced by screw)
2. Ram type extruder (Pressure is produced by ram)

Calendering

A calender consists of two or more opposite moving rolls in a vertical framework. The thickness of the rubber compound is controlled by adjusting the nip gap between the rolls. The calenders are generally used for making sheets of uniform thickness and width. A schematic representation is shown below in (Fig. 1.22).

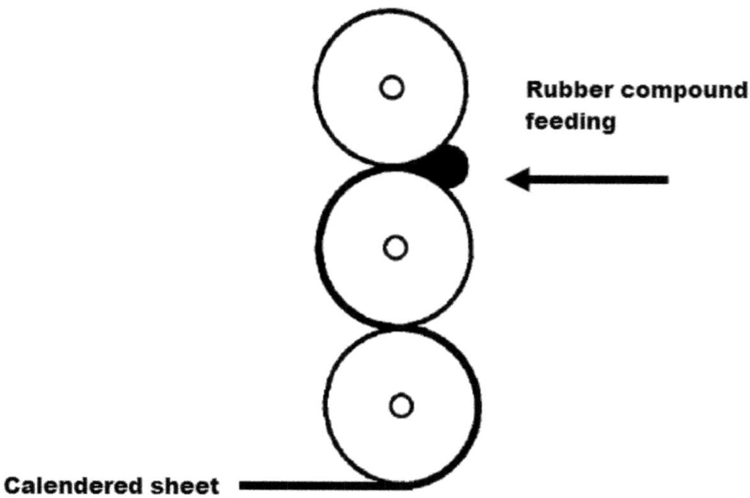

Figure 1.22 Schematic representation of calendering process.

Calender rolls are classified by the number of rolls

1. Two roll calender
2. Three roll calender
3. Four roll calender

1.4 Different vulcanization methods

Vulcanization methods are classified into two (Table 1.18)

- Batch vulcanization: One batch of materials vulcanized at same time
- Continuous vulcanization: Continuous vulcanization of long material

Table 1.18 Different vulcanization methods, procedures and the corresponding products

Methods of vulcanization	Vulcanization procedures	Products manufactured
Batch vulcanization methods	Autoclave or steam pan	Sheeting and foamed goods
	Hot water curing	Rubber lined products
	Hot air ovens	Latex products, postcuring
	Lead curing	Hose
Continues vulcanization	High pressure steam tube	Cables and hose
	Hot air tunnel	Extrusion and latex products
	Fluidized bed curing	Extrudates

1.5 Testing of compounded rubber [6–12]

1.5.1 Processability of rubber compounds

Processability is defined as the behavior and interactions of polymer, filler, oil, and other additives during the various processing stages, such as mixing, extrusion, calendaring, and molding processes. It depends on the cure characteristics and rheological properties of the rubber compound. The main processability parameters are cure time, scorch time, viscosity, extrusion rate, etc.

1.5.2 Cure studies and viscosity

The properties of the uncured rubber compounds are generally studied by Mooney viscometer, Oscillating Disc Rheometer (ODR), and moving die rheometer (MDR). Mooney viscosity measures the stiffness of the uncured compounds, otherwise known as the compound's viscosity. The appropriate standards associated with this test are IS0 289, ASTM D 1646, and DIN 53523-3. The units of measurement are known as Mooney units.

The Mooney viscometer consists of a rotor rotating (@2 rpm) in a closed heated cavity chamber filled with uncured rubber. A shearing action develops between the compound and the rotor, and the resulting torque (resistance of the rubber to the turning rotor) is measured in arbitrary units called Mooney units, which directly relate to torque. The larger the number, the higher the viscosity, e.g., 40 ML 1 + 4 (100°C), where 40 refers to viscosity in Mooney units, *M* indicates Mooney, *L* indicates that a large rotor was used (*S* for small rotor), 1 indicates one minute preheating time, 4 refers to the time in minutes after starting the rotor, and 100°C is test temperature. A typical Mooney graph is shown in Fig. 1.23. The Mooney graph gives an idea about scorch time, cure time, and cure index (Table 1.19).

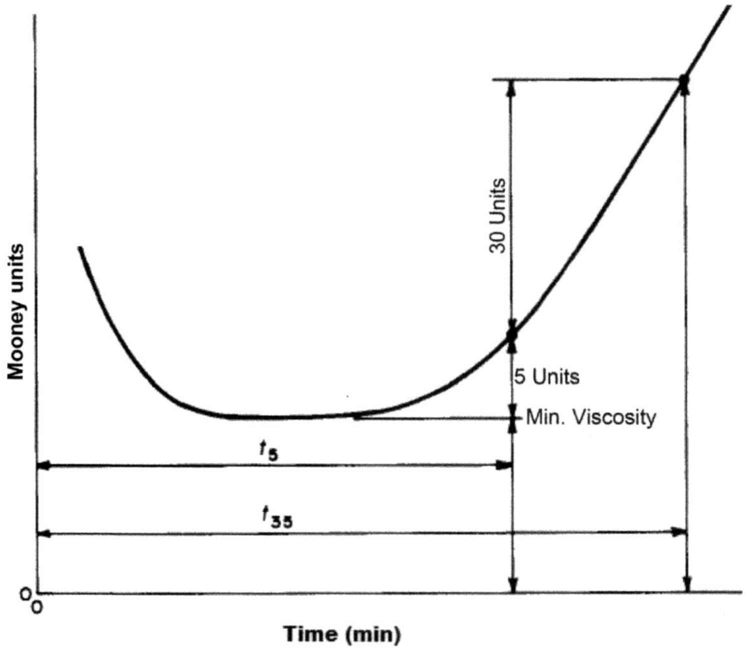

Figure 1.23 Typical Mooney viscometer graph (large rotor).

Table 1.19 Mooney viscometer test report parameters

	Large rotor	Small rotor
Scorch time	t_5	t_3
Cure time	t_{35}	t_{18}
Cure index	$t_{35}-t_5$	$t_{18}-t_3$

ODR are also used for understanding the cure characteristics of the rubber compounds; the appropriate standard is ISO 3417 and ASTM D 2084. The ODR also give the cure time and scorch time; a sample curve is shown in Fig. 1.24. The main limitation of this method is that it is not able to make rheological measurement after scorch time (Table 1.20).

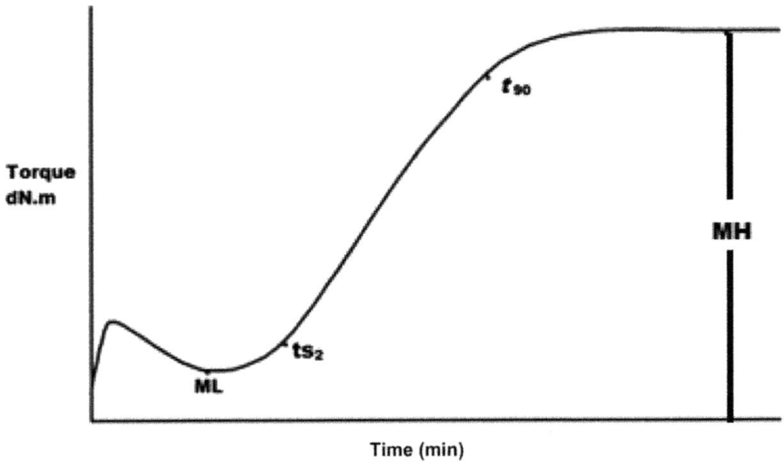

Figure 1.24 ODR cure graph.

Table 1.20 **Rheometer test report parameters**

ML—Minimum torque, dN · m	MH—Maximum torque, dN · m
tS_2—Time to 2 dN · m rise from ML (scorch time) Cure rate index = 100/(cure time—scorch time)	t_{90}—Optimum cure time t_{90} = ML + 0.90(MH−ML)

Rotorless cure meters are used to study the cure properties; the standard methods are IS0 6502 and ASTM D 5289. An example of a rotorless cure meter is the MDR. MDR gives three curves, such as S' (elastic stress), S'' (viscous stress), and tan delta (tan δ). The MDR has a number of advantages over the ODR; the MDR instrument can be operated with different sweeps of frequency, strain, and temperature.

1.5.3 Mechanical properties of compounded rubber

1.5.3.1 Tensile and tear properties

Tensile and tear properties are studied by using a universal testing machine. Tensile properties depend both on the material and the conditions of test (extension rate, temperature, humidity, specimen geometry, pretest conditioning, etc.); therefore materials should be compared only when tested under the same conditions. ASTM D 412 discusses the tensile properties of vulcanized thermoset rubbers and TPE. This testing gives an idea about tensile properties, such as tensile stress, tensile stress at a given elongation, tensile strength, yield point, and ultimate elongation.

Tensile strength is defined as the force per unit area of original cross-section of the sample required to stretch a rubber test piece to its breaking point.

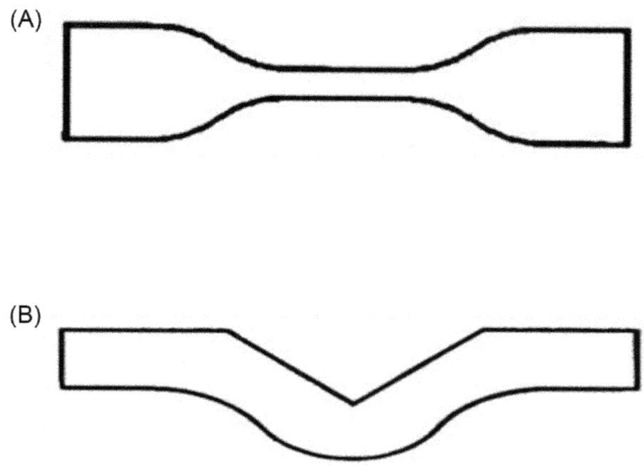

Figure 1.25 (A) Dumbbell shaped test piece for tensile testing. (B) Rubber tear test piece (angular specimen).

Modulus is the amount of force per unit area to stretch the test piece to a given elongation. Elongation is used to describe the ability of the rubber to stretch without breaking. Normally ultimate elongation, expressed as the percentage of original length, is taken at the moment at rupture. Tension tests shall be a uniform rate of grip separation of 500 ± 50 mm/min and standard test temperature shall be $23 \pm 2°C$, Dumbbell (Fig. 1.25A) shaped test specimens are used.

ASTM D 624 refers to the tear testing of rubber samples; this test measures the rubber's ability to resist against the tearing. Several tear test geometries (angular and crescent shaped) are used in tear testing. They are cut from the flat sheet and are stretched in a tensile machine. The test rate of jaw separation shall be 500 ± 50 mm/min and standard test temperature shall be $23 \pm 2°C$. Fig. 1.25B represent the angular tear test specimen.

1.5.3.2 Hardness (ASTM D 2240, ASTM D 1415)

Hardness test is based on measurement of the depth of indentation of a rigid indenter of specified shape and size into the rubber specimen under specified conditions. In ASTM D 2240, durometer hardness is used to determine the relative hardness of soft materials, usually plastic or rubber. The test measures the penetration of a specified indenter into the material under specified conditions of force and time. This test method describes eight types of rubber hardness measurement devices known as durometers: types A, B, C, D, DO, O, OO, and M. Shore A and shore D hardness scales are common, the A scale being used for softer and the D scale being used for harder materials. This test method is not applicable to the hardness testing of coated fabrics. In ASTM D 1415, this test method measures the penetration depth of a rigid ball into the surface of a rubber specimen which is then converted to a hardness scale value (IRHD—International Rubber Hardness Degrees). On the IRHD scale, "0" represents a material having an elastic modulus of zero and 100 represents a material with an infinite elastic modulus. The readings in IRHD are comparable with shore A durometer hardness.

1.5.3.3 Set properties

Compression sets are done using short cylinder or Disc shaped samples, under constant strain or constant stress. It regains after removal of an applied stress or strain measured over a specified time at a specified temperature. It measures the ability of rubber compounds to retain elastic properties after prolonged action of compressive stresses, particularly for the rubber compounds used in machinery mountings, vibration dampers, and seals. ASTM D 395 refers to the methods for the compression set properties of rubber vulcanizates. The standard dimensions used for the specimen are thickness of 12.5 ± 0.5 mm and diameter of 29.0 ± 0.5 mm.

$$\text{Compression set} = \left(\frac{T_o - T_r}{T_o - T_s}\right) \times 100 \tag{1.1}$$

where, T_o is original thickness, T_s is compressed thickness, and T_r is recovered thickness

1.5.3.4 Abrasion resistance (ASTM D 5963)

This test method covers the measurement of abrasion resistance of rubbers (vulcanized thermoset rubbers and TPE) that are subjected to abrasive/frictional wear in actual service. The abrasion resistance is measured by moving a test piece across the surface of an abrasive sheet mounted to a revolving drum, and is expressed as volume loss in mm^3 or abrasion resistance index as a percentage. For volume loss, a smaller number indicates better abrasion resistance, while for the abrasion resistance index, a smaller number denotes poorer abrasion resistance. The abrasion tester consist of a cylindrical drum of diameter 150.0 ± 0.2 mm and a length approximately of 500 mm operating at 40 ± 1 rpm. Test pieces are cylindrical in shape with a diameter of 16 ± 0.2 mm and with a minimum thickness of 6 mm.

$$\text{Calculation of abrasion loss (A)}, \; A = \frac{\Delta m_t \cdot S_0}{d_t \cdot S} \tag{1.2}$$

where A is abrasion loss in mm^3, Δm_t is mass loss of the test piece in mg, d_t is density of the test rubber in mg/m^3, S_0 is "normal abrasiveness" = 200 mg, and S is abrasiveness in mg.

$$\text{ARI} = \left(\frac{\Delta m_1 \cdot d_t}{\Delta m_t \cdot d_1}\right) \times 100 \tag{1.3}$$

where ARI is abrasion resistance index in percent, Δm_1 is mass loss of the standard rubber test piece in mg, t_1 is density of standard rubber in mg/m^3, Δm_t is mass loss of the test rubber piece in mg, and d_t = density of the test rubber in mg/m^3.

1.6 New trends in rubber compounding

1.6.1 Green compounding

Towards the end of the 19th century the introduction of the green tire concept by using silica as the reinforcing filler received wide appreciation [12]. The goal has been to develop a green, innovative technology and to prove to the rubber industry that it could achieve a better rolling resistance by achieving at least equal wet traction without affecting the tread wear. In green tyre technology, the compound mixing procedure has a significant effect. The mixing and processing of silica compounds are comparatively difficult as compared to CB filled compounds. A number of processing aids have been suggested to decrease these difficulties, such as zinc soaps and esters. However, the reduction in Zn content preferred by the manufacturers still demands a newer technique. The scope of improving processability of the compound and to produce a cost-efficient product with enhanced properties still remains a challenge.

1.6.2 REACH regulations

Registration, Evaluation, Authorization, and Restriction of Chemicals (REACH) is a regulation of the European Union. This regulation was put into practise to improve the protection of human health and the environment from the risks that can be caused due to chemicals. It also promotes alternative methods for the hazard assessment of substances in order to reduce the number of tests on animals. In principle, REACH applies to all chemical substances; not only those used in industrial processes but also in our daily lives, for example in cleaning products and paints, as well as in articles such as clothes, furniture, and electrical appliances. Therefore, the regulation has an impact on most companies across the European Union [13]. REACH places the burden of proof on companies. To comply with the regulation, companies must identify and manage the risks linked to the substances they manufacture and market in the European Union. They have to demonstrate to the European chemicals agency how the substance can be safely used, and they must communicate the risk management measures to the users. If the risks cannot be managed, authorities can restrict the use of substances in different ways. In the long run, the most hazardous substances should be substituted with less dangerous ones. This regulation was put into practise on June 1, 2007.

1.7 Conclusion and future outlook

The use of the right rubber compound is important for manufacturing long-lasting and cost-effective products. A lot of developments have occurred in rubber compounding to improve the final performance. However, it is important to know the basic principles of rubber compounding to further develop new novel materials with excellent properties [14–18]. Here, we have given a brief report on the basics of rubber compounding and their various characterisations. The need for the development of an economic and environmental friendly way addressing the global challenges of the 21st century is one major areas that is given prime importance. Progress in the area of green composites which reduces the harmful environmental effects during its life time is one such development that have gained wide appreciaton. Within various compounding strategies available right now one of the fascinating and hopeful areas that gives good scope for the development of materials with excellent properties is the introduction of nano technology in rubber compounding. The future development, while formulating different methodologies, should take into consideration a wide range of potential economic, environmental, and societal implications of the technology.

Reference and further reading

[1] Blow CM, Hepburn C. Rubber technology and manufacture. London; Boston: Butterworth Scientific; 1982.
[2] Hofmann W. New York: Distributed in the USA. by Oxford University Press Rubber technology handbook. New York, NY: Hanser Publishers; 1989.
[3] Morton M, American Chemical Society, Division of Rubber Chemistry. Rubber technology. New York, NY: Van Nostrand Reinhold; 1973.
[4] Dick JS, Annicelli RA. Rubber technology: compounding and testing for performance. Munich: Hanser Publishers, Hanser Gardner Publications; 2009.
[5] Galimberti M, editor. Rubber-clay nanocomposites: science, technology, and applications. New York, NY: John Wiley & Sons; 2011. Available from: http://www.sciencedirect.com/science/article/pii/S0169131714004414.
[6] Annual book of ASTM standards. Section 9, rubber, ASTM International, West Conshohocken, PA: ASTM International; 2005.
[7] Ciesielski A. An introduction to rubber technology. Shawbury, Shrewsbury, Shropshire: Rapra Technology Limited; 1999.
[8] Rubber engineering, Indian Rubber Institute. New York, NY: McGraw-Hill; 2000.
[9] Forrest MJ. Rubber analysis: polymers, compounds and products. Shawbury, UK: Rapra Technology Ltd.; 2001.
[10] Simpson RB. Rubber basics. Shawbury, UK: Rapra Technology Ltd.; 2002.
[11] Brown R. Physical testing of rubber. New York, NY: Springer; 2006.
[12] Scott JR. Physical testing of rubbers. New York, NY: Palmerton Pub. Co.; 1965.
[13] De SK, White JR. Rubber technologist's handbook. Shawbury, Shrewsbury: Rapra Technology Limited; 2001.

[14] Hewitt N. Compounding precipitated silica in elastomers. Norwich, NY: William Andrew Pub.; 2007.
[15] Kemmlein S, Herzke D, Law RJ. Brominated flame retardants in the European chemicals policy of REACH—regulation and determination in materials. J Chromatogr A 2009;1216(3):320–33, (0116).
[16] Tamil Selvan N, Eshwaran SB, Das A, Stöckelhuber KW, Wießner S. Piezoresistive natural rubber-multiwall carbon nanotube nanocomposite for sensor applications. Sensor Actuat A Phys 2016;239:102–13, (20160301).
[17] Ponnamma D, Sadasivuni KK, Wan C, Thomas S, Al-Ali AlMa'adeed M. Flexible and stretchable electronic composites. Springer; 2016.
[18] Sadasivuni KK, Ponnamma D, Kim J, Thomas S. Graphene-based polymer nanocomposites in electronics. Cham [Switzerland]: Springer; 2015.
[19] Galimberti M, Coombs M, Pandini S, Riccò T, Cipolletti V. Delamination of organically modified montmorillonite for reducing the filler networking with carbon black in poly(1,4-cis-isoprene) based nanocomposites. Appl Clay Sci 2015;104:8–17.

Micro- and nano-fillers used in the rubber industry

2

K. Song
Massachusetts Institute of Technology, Cambridge, MA, United States

2.1 Introduction

Particulate fillers are added to rubbers for mechanical reinforcement, electrical or thermal conductivity modification, and ease of processing [1]. Like other composites' preparation and modification mechanisms, several factors influencing the property or processability include [1]:

1. Volume fraction,
2. Particle dimension and geometry,
3. Dispersion quality,
4. Interaction between rubber and fillers or among fillers, and, if applicable,
5. Orientation of fillers for anisotropic properties.

These factors affect one another and achieve synergy if a perfect balance is reached. For example, the dispersion worsens with high filler concentrations, the caused aggregates reduce the interaction at the interface, the bundled structure deteriorates the shear modulus, and additionally, anisotropic fillers are hard to be aligned at high loadings [2,3]. For this reason, the appropriate choice of fillers used in a specific polymer matrix, and optimization of processing techniques and composites properties should be studied for particles of different sizes, geometries, and compatibilities with polymers [1].

Up to now, the most commonly used fillers in the rubber industry are carbon materials (i.e., carbon black, carbon fiber (CF), carbon nanotubes (CNTs), graphite, and graphene), inorganic particles (i.e., nanoclays, polyhedral oligomeric silsesquioxane (POSS), silica, calcium carbonate, talc, zinc oxide, titanium oxide, alumina, halloysite), and biofillers (i.e., cellulose, husk, wood, coir) [1–3]. In the meantime, carbon black is consumed 300% more than all the other particles combined. Following carbon black are kaolin, calcium carbonate, and silica. Among them, carbon black and silica have the longest history in application and are treated as conventional fillers. However, due to the poor dispersion and adhesion, requirements for better dispersion quality on other particles have been attracted great attentions [1–3]. Under this circumstance, nanoscale CNT, graphene, nanoscale calcium carbonate, and other metal oxide ceramic particles become the focus of academic research. These nanoparticles usually have dimensional features between 1 to 100 nm. In this chapter, the review of frequently used rubber matrices and

Progress in Rubber Nanocomposites. DOI: http://dx.doi.org/10.1016/B978-0-08-100409-8.00002-4

characteristic features of various rubber industry-used particles as well as their additive/modification effects on composites are reviewed.

2.2 Rubber category

Rubbers display very special characteristics [4] as shown in Table 2.1, and some examples are:

1. Physically versatile, including large stretch ratio, high resilience, and air/water-proof.
2. Variable friction coefficient, which is high on dry surface and low on wet surface. Based on this property, rubber is useful for power-transmission belts and water environment-used lubricant bearings.
3. Viscoelastic mechanical behavior, which can be measured by creep or stress relaxation tests. Two phenomena are commonly seen. (1) The stress–strain curve under large deformations depending on the maximum loading previously encountered is termed as Mullins effect. (2) The capability to store energy greatly decreases with increased strain, which is termed as Payne effect. This effect is more distinctive in small strain amplitudes and frequently observed in carbon black filled rubber composites. This could be attributed to the deformation-induced breakage of weak physical bonds [3,5].
4. Low thermal conduction, therefore heat builds up under dynamic mechanical conditions and subjects the rubber to thermal fatigue.
5. Low electrical resistance, thus applicable in insulation packaging and protective wearable.

2.2.1 Natural rubber

Natural rubber is abundant in South America, Asia, and Africa. Today the main source of rubber is Asia, with Thailand, Indonesia, and Malaysia being the largest producing countries for around 75% of all natural rubber production. Currently rubber is harvested in the form of latex from certain trees and subsequently refined into rubber-based commercial products. Among the 30 million tonnes of rubber produced every year, around 40% is natural rubber. The remaining is synthetic rubber derived from petrochemical sources. Among these natural rubbers, 70% is used in tires [6].

In the harvesting procedure, the technique of tapping has been used. After peeling the bark skin, latex in the form of colloid is collected in vessels. Uncured rubber has very limited usage. For example, adhesive, insulating, footwear, and so on are very typical applications [6]. After vulcanization, the treated rubber can extend their usage for (1) high end usage of condoms, balloons, medical devices, and supplies, and (2) middle end usage of vehicle tires, windshield wipers, and conveyor belts, as well as (3) low end usage of adhesives, bands, textiles, and erasers.

2.2.2 Styrene-butadiene rubber (SBR)

Styrene-butadiene rubber (SBR) refers to a group of synthetic rubbers derived from styrene and butadiene. These materials have excellent abrasion resistance and aging stability with the protection of additives. The ratio between styrene and butadiene

Table 2.1 **Various types of rubber [4,6,7]**

Rubber type	Repeating unit	Additional remarks
Natural rubber	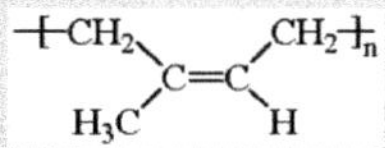	"Frozen" rubber's elasticity is low. Mechanical property depends on environment.
SBR	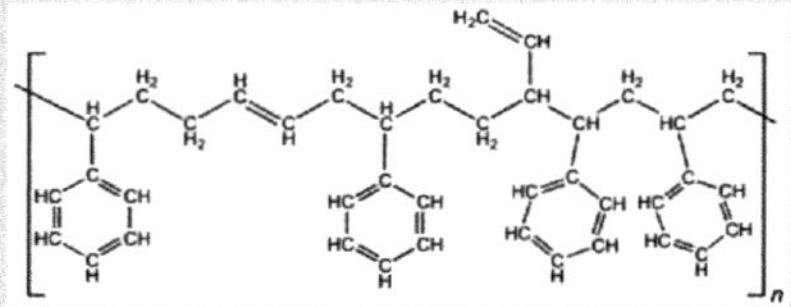	Thermoset rubber should be differentiated from styrene-butadiene block copolymer. Elongation at tear is between 550% and 650%, tensile strength around 20 MPa, glass transition −65 to −50°C, and polydispersity of 2 to 5.
PU	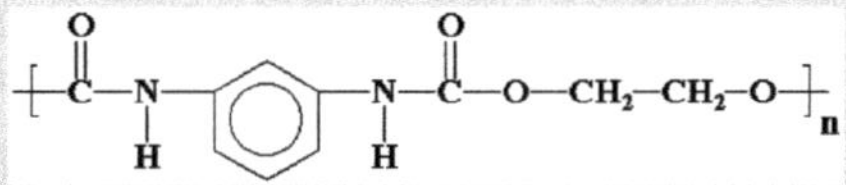	Long and flexible polymer chains contributed by the polyol generates soft or stretchy PU while high amounts of cross-linking give tough or rigid PU. Due to the many choices of isocyanates, polyols, additives, and processing conditions available for PU syntheses, there are a wide range of properties that make PU widely used polymers. These can be divided as flexible, semirigid and rigid categories. PU can also be categorized into low-density foams (0.006 g/cm^{-3}), high-density foams (0.4 g/cm^{-3}), microcellular foams (0.8 g/cm^{-3}), and solid elastomers (1.2 g/cm^{-3}).
SIR	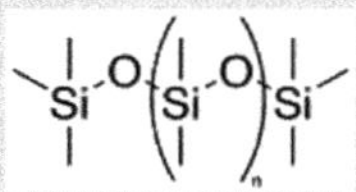	Silicone rubbers are thermally resistant from −100 up to 300°C. Due to the presence of silicone, SIR is not as susceptible as organic rubbers to ozone, UV, heat, or other aging factors. The elongation, creep, cyclic flexing, tear strength, compression, thermal conductivity, fire resistance, and tensile strength are thus superior to organic rubbers in general, especially at extreme temperatures. Elongation 100–1100%, thermal stability between −120 and 300°C, and tensile strength around 10 MPa.

SBR, styrene-butadiene rubber; PU, polyurethane; SIR, Silicone rubber, are synthetic rubbers, in comparison with natural rubber of polyisoprene.

influences the rubber properties. The higher this styrene/butadiene ratio is, the harder the polymer is. The polymerization can be obtained from solution (anionic polymerization) or emulsion (radical polymerization) procedures. The latter is more widely used. The application of SBR often competes with natural rubber. The main applications include pneumatic tires, shoe heels, gum, coatings, and construction materials in buildings [6].

2.2.3 Polyurethane rubber (PU)

Polyurethane rubber (PU) polymers are traditionally and most commonly formed by reactions between di- or polyisocyanate ($R-(N{=}C{=}O)_{n>2}$) and polyol containing two or more hydroxyl groups per molecule ($R'-(OH)_{n>2}$). The reaction is more often than not happening in the presence of a catalyst or by activation with ultraviolet light. Applications of PU include building and construction, transportation, furniture, packaging, textiles, machinery, and footwear.

2.2.4 Silicone rubber (SIR)

Silicone rubber (SIR) is an elastomer composed of silicone containing silicon, carbon, hydrogen, and oxygen, in the form of a polymer. Untreated silicone in gel or liquid form will convert to solid after curing. The curing method can be either vulcanized or catalyzed. The applications of SIR include automobile, cookware, apparel, electronics, medical devices and implants, sealants, lubricants, and toys [6]. For example, polydimentylsiloxane as the basic polymer for SIR can be used for lubricants mainly due to its rheological properties.

2.3 Fillers in the rubber industry

Fillers have been extensively used in the rubber industry to extend service life, reduce cost, and therefore improve service efficiency. The rubber/filler composites display properties between pure rubber and filler, depending on the particle dimensions, dispersion quality, volume, degree of particle orientation (if it exists), and the interaction between matrix and fillers [1]. Rubber fillers have mainly included two categories: (1) Reinforcement fillers [1]. The addition of fillers will modify the mechanical properties including stiffness, strength, toughness, abrasion resistance, antiscratching property, or friction behavior. Carbon black and silica belong to this class; (2) Extending fillers (Table 2.2). The main aim is to extend the desirable properties and make it last longer or more efficiently. TiO_2, MMT, ZnO, and Al_2O_3 are in the latter case. Fillers have played such an important role in the composite modifications and therefore their physical/chemical characteristics and their polymer modification mechanisms have been reviewed here.

Table 2.2 **Generally used rubber fillers and their basic physical properties**

Fibers	Role	Density (g/cm^3)	Diameter (nm)	Tensile Modulus (G a)	Tensile strength (Gpa)	Tensile strain (%)	[a]Price/kg
Carbon fillers							
Carbon black [8,9]	Mechanical, electrical	0.32–0.65	10–400	—	0.010–0.030	—	4
Carbon fibers [1]	Mechanical	1.6–2.2	$>10^3$	600	4	5	20
Carbon nanotbues [1]	Mechanical	0.8–2.0	1–20	1500	63	—	>3000
Graphene [1]	Mechanical, electrical, thermal	2.2	—	1000	—	—	>2000
Inorganic fillers							
Aluminum oxide [10]	Thermal	3.95–4.1	50	375	2.6	—	2
Calcium carbonate [11]	Flame	2.71–2.83	400–3000	69.9	—	—	2
Halloysite [1]	Mechanical	2.6	10–70	600	—	—	1
Kaolin [12]	Mechanical	2.6	—	170	—	—	1
Nanoclay [13]	Mechanical, thermal	2–3	—	Up to 300	—	—	60
POSS [14]	Mechanical, thermal	—	1.5	—	—	—	700
Silica [15]	Thermal	2.2–2.65	30–100	90	1.6	—	100
Talc [16]	Thermal	2.5–2.8	$>10^3$	57.31	—	—	0.8
Titanium oxide [17]	Thermal, permittivity	3.78–4.23	30–40	41.77	53.11	—	4000
Zinc oxide [18]	Mechanical, thermal, electrical	5.61	100	250	—	—	2000
Biofillers							
Cellulose [19,20]	Mechanical	1.5	$>10^3$	150	0.5	20	100
Coir [21]	Mechanical	1.2	$>10^3$	6	0.18	30	100
Husk [22]	Mechanical	1.8–2.1	$>10^3$	22	—	—	0.1
Wood [23]	Mechanical	0.3–1.3	500–2000	13	0.24	—	1

[a]The prices are from Sigma-Aldrich (http://www.sigmaaldrich.com/).

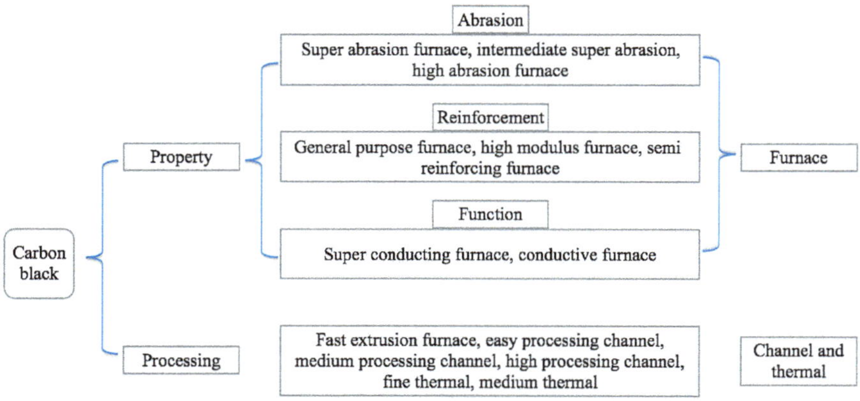

Figure 2.1 Carbon black varieties based on property (i.e., abrasion, mechanical, and functional properties) and processing techniques.

2.3.1 Carbon origin

2.3.1.1 Carbon black

Carbon black is mainly produced by the incomplete combustion of heavy petroleum products including FCC tar, coal tar, ethylene cracking tar, and some vegetable oils. Carbon black has been mainly used in rubber as particles to introduce electrical conductivity, coloring pigment, and mechanical reinforcement [8,9,24,25]. Regarding its components, above 90% elemental carbon combined with oxygen and hydrogen complexes, i.e., carboxylic, quinonic, lactonic, phenolic, ketonic, and others, are located on the surface of blacks. These surface oxygen groups are collectively referred to as volatile content.

Based on the final properties and processing techniques, carbon black can be categorized into different varieties, as shown in Fig. 2.1. For example, based on their properties, carbon black can be used for abrasion, reinforcement, and functional modifications. Based on the syntheses methods, other researchers also differentiate the carbon blacks into (1) furnace, (2) channel, (3) thermal, (4) acetylene, and (5) lamp blacks. The physical properties in the first three categories are listed in Table 2.3. Among these carbon blacks, furnace production is the highest and is mostly used in the rubber industry; channel produced products are dissolvable in solvent and thermal production not only makes carbon black easy to process but also generates high mechanical properties.

2.3.1.2 Carbon fiber

CFs have carbon atoms bonded together in crystals along the fiber axis as the crystal alignment. This crystal alignment and the high aspect ratio give the CF high stiffness, high strength, low density, and high chemical resistance as well as high temperature tolerance [26–28]. The atomic structure consists of sheets of carbon

Table 2.3 Generally used carbon blacks based on their processing, particle size, and content [8, 9, 24, 25]

Category	Furnace	Channel	Thermal
Size (nm)	15–100	15–40	N/A
Processing	Hydrocarbon burning at 1200–1600°C in furnaces	Natural gas burnt on iron plate with limited oxygen	Degradation of hydrocarbons at 800–1000°C
ph	6.5–10	5.5	N/A
Features	High mechanical properties	Oxygen content of 2–10% in form of carboxylic groups, and dissolvable in water	Particle form chains and aggregates, high oxygen content provides active surface activity

atoms arranged in a regular hexagonal pattern (graphene sheets). The difference between CFs and graphite is that in CFs the graphene sheets interlock with one another while in graphite these graphene sheets stack to each other with weak Van der Waals force. This weak adhesion makes graphite soft and easily slides. CFs are relatively more expensive than glass fibers and high-performance polymer fibers. Despite the high cost, they have been widely used in aerospace, luxury automobiles, sports goods, and other civil engineering due to the superior properties.

CFs can be fabricated based on mainly three precursors, namely, polyacrylonitrile (PAN) [27,28], petroleum pitch, and rayon [27]. Around 90% of the CFs produced are from PAN and the other CFs are from rayon or petroleum pitch. The process for making CFs is in part chemical and partially mechanical. The precursor fibers are firstly spun and drawn into long strands, secondly heated in air at intermediate temperature (stabilization around 200–300°C for 30–120 mins), and lastly heated in an inert environment at extremely high temperature (carbonization around 1000–2000°C and graphitization around 3000°C for minutes or hours) [1]. Take PAN-based CF for example. The second stabilization procedure will convert the linear atomic bonding to a more thermally stable ladder bonding. The last carbonization and graphitization treatments lead the carbon atoms in the fiber to align parallel with the CF [1,27,28]. The last heating procedure in the absence of oxygen, termed as carbonization and graphitization, is frequently completed in a furnace with high pressure and controlled heating rates. After this step, the CFs are composed of long and tightly interlocked chains of carbon atoms, with a very limited number of noncarbon atoms remaining.

The obtained CFs, like most fillers in composite engineering, do not interact with polymer matrices well [1]. To give better compatibilities between CFs and polymer matrices, including most rubbers, the CF surfaces are treated slightly with oxygen. The addition of oxygen on one hand provides more active-bonding properties, and on the other hand, this process will etch the CF surface so that the

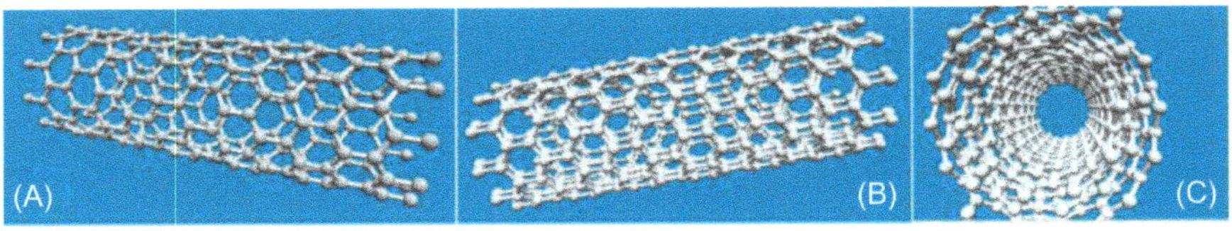

Figure 2.2 Various crystalline carbon nanotubes of (A) single-wall (SWNT), (B) double-wall (DWNT), (C) multiwall (MWNT) showed side-view or cross-section view.

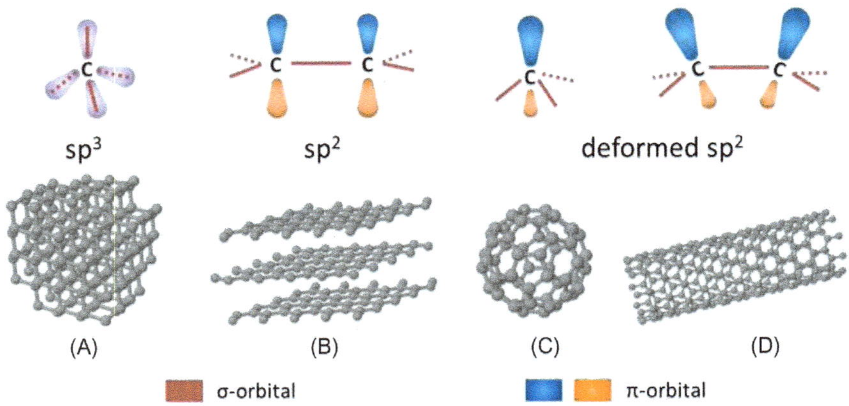

Figure 2.3 Various crystalline carbon forms depending on different hybridization mechanism, (A) sp^3 hybridized carbon in diamond, (B) sp^2 hybridized carbon in graphite, (C) and (D) deformed sp^2 hybridized fullerene and carbon nanotubes (CNTs).

polymer–fiber interface will be more mechanically interactive (i.e., interlock). The CFs have very high modulus of 600 GPa and limited strain of less than 5%, and thus behave mechanically brittle (Table 2.2). To avoid failure during winding, weaving, or knitting, the fiber surface will be treated with ductile coatings [1]. This treatment is called sizing. The coating material is expected to be compatible with the composite matrix. Typical commercial coatings include urethane (which is compatible with rubber matrix), epoxy, nylon, polyester, and some thermoplastics. The coated fibers are finally winded into yarns for composite usage [1].

2.3.1.3 Carbon nanotubes (CNTs)

Apart from carbon black and CF, carbon can also exist in other crystalline or amorphous forms. As seen from Figs. 2.2 and 2.3, diamond is sp^3 hybridized carbon bonded to four other atoms in a tetrahedral manner. Graphite platelets, on the other hand, are carbon atoms bonded together in a planar hexagonal network with a sp^2 hybridization arrangement. Before the discovery of fullerene and CNTs, graphite and diamond were considered the only two allotropic forms of carbon in the solid state.

In 1991, during the study of arc-discharge evaporation synthesis of fullerene, Iijima identified the structure of the multiwall carbon nanotubes (MWNT) at the negative end of the electrode [29]. Shortly thereafter in 1993, the structure of the single-wall carbon nanotubes (SWNT) (Fig. 2.2) were also identified by Iijima and coworkers at NEC corporation [30]. Both SWNT and MWNT are cylindrical nanostructures. The discovery of these new structures ignited a great amount of scientific interest and numerous studies have focused on their synthesis, structure, properties, and applications [31–37].

There are mainly three classes of nanotubes, namely, SWNT, double-wall carbon nanotubes (DWNT), and MWNT (as shown in Fig. 2.2). Synthesis methods include electric arc discharge [38,39], laser ablation [40], and chemical vapor deposition [41]. CNT have excellent mechanical [42], electrical [43,44], and thermal [45,46] behavior, which allow them to exhibit strength to be 200–300 times stronger than steel, density that is lighter than aluminum (2.70 g cm^{-3}), as well as conducting much better than copper (10^5 S/cm at 20°C) (Tables 2.2 and 2.3) [47].

The mechanical properties of CNT are mainly influenced by forces inherent to its CNT molecular structure; in other words, the strong intralayer carbon–carbon bonds (i.e., entirely composed of sp^2 bonds), where both σ-bonding and π-bonding are present, and one weak interlayer interaction force (only for MWNT). Prior to the experimental measurements, CNT properties were analyzed theoretically. *Ab initio* local density calculations determined the CNT stiffness to be $\sim$1.5 TPa [48]. The experimentally measured mechanical properties as listed in Tables 2.2 and 2.3 are in fair agreement with this theoretical prediction. These unique mechanical attributes of CNT make them ideal candidates to substitute or complement conventional fillers used for reinforcing polymer composites [49].

SWNT have excellent electronic properties, exhibiting carrier mobility of $\sim$10,000 $cm^2/(V\ s)$, which is 10 times higher than that of silicon [43]. Electrical current density of CNT are as high as $\sim 4 \times 10^9$ $A^2\ cm^{-2}$, which is three orders higher than that of copper or aluminum [44]. Another potential application for CNT due to its electric capability is the fabrication of computer chips, which may potentially replace the current silicon-based circuits. In addition, CNT also exhibit excellent thermal transport, which is dominated by the intrinsic properties of the strong sp^2 bond/hybridization. This results in efficient heat transfer by lattice vibrations [50], rather than by free carrier contributions as in metals. This gives rise to high K values of $\sim$3000 W/m K.

In summary, CNTs have excellent mechanical stiffness/strength, electrical and thermal conductivities. However, poor dispersion and inefficient interfacial bonding between CNT and polymers limit the full utilization of CNT in reinforcing the composites to their potential. Therefore, both physical and chemical treatments on carbon nanotube have been developed for improved compatibility between polymer and tubes, and will be discussed in later sections in this chapter.

2.3.1.4 Graphene and graphite

Graphite, before the overwhelming research on two-dimensional graphene materials, has a long history in rubber applications. Graphite is a layered material composed of alternating carbon sheets. The weaker van der Waals forces between these layers make intercalation possible. Expanded or decoupled graphene sheets have been produced by an intensive heating method. There is a good registry between graphene and polar/nonpolar polymers due to the presence of −OH or −COOH groups.

In addition, graphite is formed when only three valence electrons bond with neighboring electrons, while leaving one electron to be delocalized among all atoms. The delocalized fourth valence electron is free to move from one orbital to another forming a π-bond network, which results in graphitic sheets with high electrical conductivity. Besides, the technique involved peeling a single layer from graphite using micromechanical cleaving. It has been observed that highly crystalline graphene can exhibit a quantum Hall effect. For this reason, graphene materials are promising candidates for future electronic applications, such as ballistic field-effect transistors. A Nobel Prize was awarded in 2010 for the discovery of graphene and their superior electronic properties.

Similar to CNT, graphene exhibits an in-plane elastic modulus of ~1 TPa and intrinsic strength of 130 GPa [51]. Graphene can also be visualized as an unzipped CNT. These two actually have very similar mechanical modulus and failure strength, thermal and electrical conductivity. A nanoindentation method by atomic force microscopy (AFM) was used to examine the breaking strength of freestanding monolayer graphene membranes. The measured breaking strength was 42 Nm^{-1}, which is considered as the intrinsic strength of a defect-free graphene sheet. Chemically reduced graphite oxide in the form of graphene sheets was also examined for its elastic properties [52]. Stiffness was determined to be 0.25 TPa using the tip-induced deformation experiments. This trend is similar to CNT, where a reduction of inherent properties is observed after chemical treatment of the material.

Another important property for graphene is quantifying the free movement electrons without any scattering or resistance, where electron mobility can reach as high as 20,000 $cm^2/(V\ s)$. This mobility is an order of magnitude higher than the silicon transistors [53]. This movement enables ballistic transport at submicron distances in graphene. At such high speeds, the electrons take less than 0.1 ps to cover the typical distance (30 nm to 1 μm) between the source and drain electrodes in a graphene-based transistor. For this reason, these graphene or graphene platelet properties also make them attractive in multifunctional composites for mechanical reinforcement as well as electrical enhancement. The specific properties of several nanocarbons (nCs) have been discussed in a number of publications. Table 2.4 summarizes the general mechanical, electrical, and thermal properties for these materials.

2.3.2 Inorganic origin

The inorganic fillers frequently used in rubbers are listed below and arranged in an alphabetical order. An attempt to review their characteristics, syntheses, or production methods and applications is made.

2.3.2.1 Calcium carbonate ($CaCO_3$)

$CaCO_3$ is abundant in nature, low cost, highly inert, and appropriate for decent mechanical reinforcement due to the development of nanosized $CaCO_3$ material

Table 2.4 Theoretical and experimental properties of fullerene, CNT, and graphene [42, 45, 46, 48, 51, 54–67]

Properties	Fullerene[a]	SWNT[b]	MWNT[c]	Graphite[d]	Graphene[d]
Dimensions (nm)	~1	Diameter as low as ~1 nm, length up to 5 cm	Diameter 10 to 50 nm	Sheet width several μm	100 nm in-plane dimensions
Density (g cm^{-3})	1.65	0.8	1.8	1.9–2.3	2.26
Elastic modulus (TPa)	—	Theoretically 1.5; experimentally 0.6–1.5	0.4–0.95 (arc-discharge); 0.01–0.45 (CVD)	~1 along the basal plane	~1
Tensile strength (GPa)	—	13–52	11–63 (arc-discharge); 10 (CVD)	—	130
Electrical conductivity (S/cm)	10^5	10^2–10^6	10^3–10^5	4000 (in-plane), 3.3 (c-axis)	4000
Electron mobility (cm^2/(V s))	0.5–6	10^5	10^4–10^5	2×10^4	$>1.5 \times 10^4$
Thermal conductivity (W/m K)	0.4	3000	3500	3000 (in-plane) 6 (along the c-axis)	3000
Thermal expansion (10^{-6}/K)	62	negligible	negligible	−1 (in-plane) 29 (along the c-axis)	−1
Thermal stability (°C)	600	>700 (in air)	>600 (in air), 2800 (in vacuum)	450–650 (in air)	unstable
Specific surface area (m^2/g)	725–1220	>1000	>50	—	2630

[a] C_{60} general properties are listed here [54].
[b] One problem of single-wall carbon nantotubes (SWNT) tensile or bending test in scanning electron microscopy (SEM) or atomic force microscopy (AFM) is the slippage of individual carbon nanotubes (CNT) within one bundle. During this process shear modulus dominates and results in underestimating the effective strengthening of individual SWNT.
[c] CVD-synthesized multiwall carbon nanotubes (MWNT) carry more defects and are type dependent, resulting in significantly reduced mechanical behaviors [67].
[d] Graphene stacks to form graphite thus properties similar at plane dimensions.
Source: Permitted of reprint from thesis [68].

fabrications. Calcium carbonate is mostly found in ground chalk, limestone, and rock in sedimentary earth layers, as well as in shells of marine organisms, snails, pearls, and eggshells. Apart from $CaCO_3$, which is above 98% of the mineral components, $MgCO_3$, Fe_2O_3, Al_2O_3, and SiO_2 are also included. Improvements in mechanical, rheological, and flame retardant properties with $CaCO_3$ have been reported in rubber composites [69–77]. In the rubber industry, natural limestone with an average size of 500–5000 nm and calcium carbonate of around 40 nm are mostly used. In comparison, the precipitated calcium carbonate has larger specific area and thus can better reinforce the rubber matrices.

2.3.2.2 Clay

Clay is a fine-grained natural rock or soil material that combines more than one single clay mineral. Chemically, clays contain Al, Mg, Si, Fe, and other elements. Physically, they are mostly platy (sheetlike structure). Similar to the way graphene forms graphite, single layer metal oxides are bonded together with mainly van der Waals force. Due to this feature, clays have the capability to absorb water, even moisture from air. Clays with moisture content are plastic; after heat treatment, they are hard and stiff. The frequently used clay particles are listed in Table 2.2 and 2.5. In the following section the mostly frequently used halloysite, kaolinite, and montmorillonite (MMT) are reviewed regarding their physical properties and potential reinforcement in rubber materials.

Halloysite nanotubes (HNT) and kaolinite

Halloysite nanotubes (HNT) is composed of alumina-silicate clay with micro hollow and nano tubular structure, and chemically similar to Kaolinite. The neighboring silica and alumina layers curve and form multilayer tubes due to packing disorder. Monolayer HNT ($Al_2Si_2O_5(OH)_4 \cdot nH_2O$) refers to the so-called halloysite-10 Å ($n = 2$). Heating HNT dehydrates it to halloysite-7 Å ($n = 0$ as shown in Fig. 2.4A). A halloysite wall usually contains 15–20 bilayers (aluminum and silicon oxides). Under electron microscopy 3–4 layers were also observed. Chemically, the outer surface of the HNT has properties similar to SiO_2 while the inner cylinder core is related to Al_2O_3 (Fig. 2.4). The charge (zeta potential) behavior of the halloysite particles can be roughly described by superposiiton of mostly negative (at pH 6–7) surface potential of SiO_2 and a small contribution from the positive Al_2O_3 inner surface (at pH 2–8). The physical properties are listed in Table 2.6.

HNT is just one kind of kaolin. Other forms with similar chemical components are dickite, nacrite, and kaolinite. Kaolinite compromises the exact same two layers to halloysite, (i.e., $Al_2O_3 \cdot 2SiO_2 \cdot 2H_2O$), i.e., tetrahedral silica and octahedral alumina. The alumina sheet chemically bonds to one side of the silica sheet and forms the basic crystal lattice. The shrink-swell capacity and cation-exchange capacity in Kaolinite are both low. Hard kaolins have high modulus and soft kaolins have lower modulus, strength, and abrasion resistance due to the presence of coarse grains. The

Table 2.5 **Generally used clay particles and their physical properties**

Category		Chemical formula	Crystal structure	Density (g/cm^3)	Specific area (m^2/g)	Modulus (GPa)
Kaolin	*Kaolinite* [78]	$Al_4Si_4O_{10}(OH)_8$	Triclinic or monoclinic	2.65	50	170
	Halloysite [79]	$Al_2Si_2O_5(OH)_4 \cdot nH_2O$	Monoclinic	2.6	50–100	600
Illite	Illite [80–82]	$(K,H_3O)(Al,Mg,Fe)_2(Si,Al)_4O_{10}[(OH)_2,(H_2O)]$	Monoclinic	2.6–2.9	40–80	62
Smectite	*Montmorillonite*	$Al_2Si_4O_{10}(OH)_2 \cdot nH_2O$	Monoclinic	2–3	60–300	56
	Hectorite [80–82]	$Na_{0.3}(Mg,Li)_3Si_4O_{10}(OH)_2$	Monoclinic	2–3	60–400	90
Palygorskite	Sepiolite [80–82]	$Mg_4Si_6O_{15}(OH)_2 \cdot 6H_2O$	Orthorhombic	0.99–1.28	500	88
	Attapulgite [80–82]	$K_{1-1.5}Al_4(Si_{7-6.5}Al_{1-1.5}O_{25})(OH)_4$	Monoclinic	2.4	300–350	71
Vermiculite	Vermiculite [80–82]	$Mg_3Si_4O_{10}(OH)_2 \cdot nH_2O$	Monoclinic	2.3	70–175	95

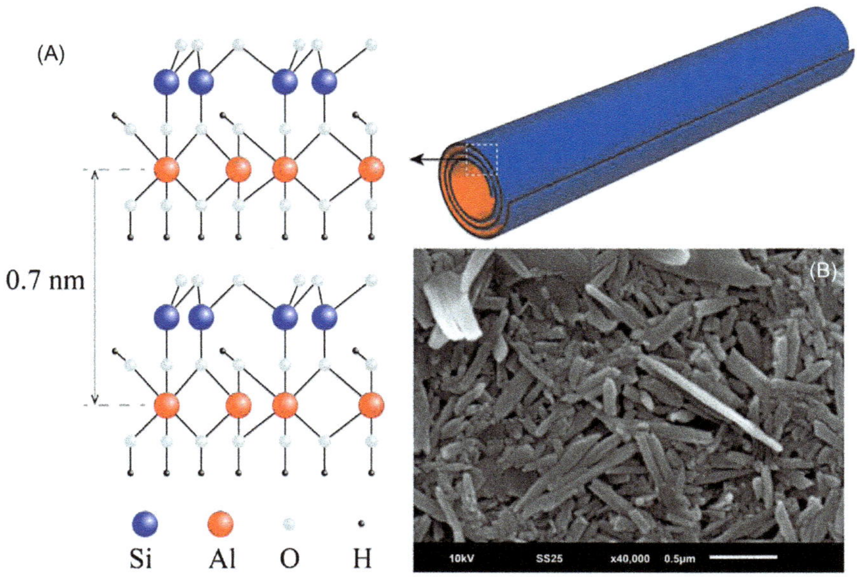

Figure 2.4 (A) Schematic shows the crystal structure, and (B) scanning electron microscopy (SEM) image of halloysite nanotubes (HNT).

Table 2.6 **HNT physical parameters**

Chemical formula	**$Al_2Si_2O_5(OH)_4 \cdot nH_2O$**
Inner diameter	10–40 nm
Out diameter	40–70 nm
Length	0.2–40 μm
Density	2.14–2.59 g cm^{-3}
Elastic modulus [79,83,84]	Up to 600 GPa
Strength	N/A
Specific surface area [85]	50–137 $m^2 \cdot g^{-1}$ (compared to CNT 100–1000 and graphene up to 3630 $m^2 \cdot g^{-1}$)
Pore volume	1.25 $mL \cdot g^{-1}$
Pore space	14–46.8%
Average pore size	7.97–10.02 nm
Crystal system	Monoclinic
Cell parameters [85]	$a = 5.14$ Å, $b = 8.9$ Å, $c = 14.7$ Å, $\beta = 104°$, a:b:c = 0.578:1:1.65, $Z = 2$ (monoclinic)

theoretical Young's modulus was simulated to be around 170 GPa as shown in Table 2.5 [78].

Kaolinite clays will go through a number of phase transformations upon heating. The thermogravimetric data of Kaolinite and polymer materials are plotted in

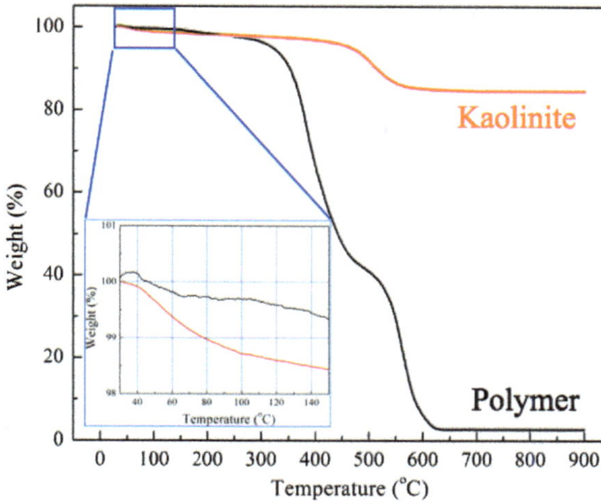

Figure 2.5 Thermal transition of kaolinite as compared to that in polymer material.

Fig. 2.5. The liquid water is removed below 100°C, and remaining hydroxyl groups are expelled from clay up to 500°C. However, the expulsion of water and hydroxyl groups during this procedure is reversible. Moisture in air can be reabsorbed and reintegrate into its fine crystal form. The dehydration from 550°C produces meta form kaolinite and the continual loss is observed up to 900°C. The phase transformation in this procedure is nonreversible. The highly thermal stability of clay is frequently taken advantage of to increase the heating retardance in rubbers.

Montmorillonite (MMT)

MMT consists of an inner octahedral layer sandwiched between two silicate tetrahedral layers as illustrated in Fig. 2.6. The octahedral layer is an alumina layer with aluminum atoms replaced by magnesium. This leaves a negative charge due to the differences in valences of Al and Mg distributed within the platelets that are balanced by positive counter ions. The dimension of a single layer ranges from tens of nanometers to several microns. The interlayer displacing is between 0.7 to 1.8 nm dependent on water moisture intercalation (d-spacing), with each layer thickness of 0.3 nm. The swelling behavior of platelets is benefiting the intercalation of polymer chains that forms soft–hard micro segments. This intends to reinforce the composites efficiently. However, in most cases, due to the polarity between polymer matrix and MMT filer, it is not possible for a lot of macromolecules to enter the spaced layers. Therefore, appropriate treatment to MMT is necessary to increase the interlayer spacing so that more polymer chains can adsorb on the basal surface.

Based on the degree of intercalation, the final composites can be classified as (1) nonintercalated composites and (2) intercalated, and (3) exfoliated composites. For the former case, MMT are in big-size stacks. Delamination and dispersion of the mixture is poor. The formed MMT bundle with the polymer surrounding is termed as a tactoid structure. For the latter two cases, the polymer chains can either enter

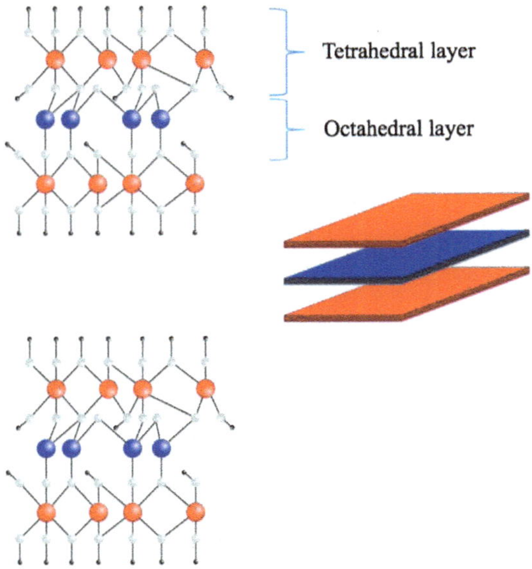

Figure 2.6 Schematic shows the crystal structure of MMT. Between layers there usually are sodium ions that keep a balance due to the valances of Al and Mg ions.

the expanded spacing or adsorb on the single-layered structure. In characterization methods, a shift in wide-angle X-ray diffraction [86], Fourier transformed infrared spectroscopy, Raman, or SANS [87] spectrum can be characterized. Fully exfoliated stacks are hard to achieve, and theoretically, the interlayer spacing characteristic peaks will disappear in the final spectrum. These nanocomposites can be prepared by solution, melt, or in situ polymerization methods [1].

2.3.2.3 Polyhedral oligomeric silsesquioxane (POSS)

POSS particles are nanoscale in size (1–3 nm in diameter), cheap in cost, and flexile regarding their physical and chemical properties. The chemical formula is $(RSiO_{1.5})_n$ (where $n = 4$, 6, 8, 10, and 12, and in most cases 8, as shown in Fig. 2.7). *R* is hydrogen, or alkyl, alkylene, aryl, arylene, or an organo-functional derivative of the akyl, alkylene, aryl, or arylene groups. In each POSS particle, oxygen atoms link 8 silicon atoms. It should be noticed that the POSS particles are often aggregated in the 1–100 μm range, and only dispersed single particles display nano level dimension features. The nanosize POSS is considered as the smallest silica ever obtained. For POSS, it is easy to disperse them in tetrahydrofuran, acetone, methanol, and so on. POSS is unique in that the core is inorganic silicate and the exterior is organic. This microstructure provides mechanical stiffness and thermal stability as well as fire retardant. POSS have attracted a lot of attention due to their function as ceramic precursors and modification effects in nanocomposites. POSS has the capability to control polymer chain movement and simultaneously does not influence the polymer processability and mechanical reinforcement.

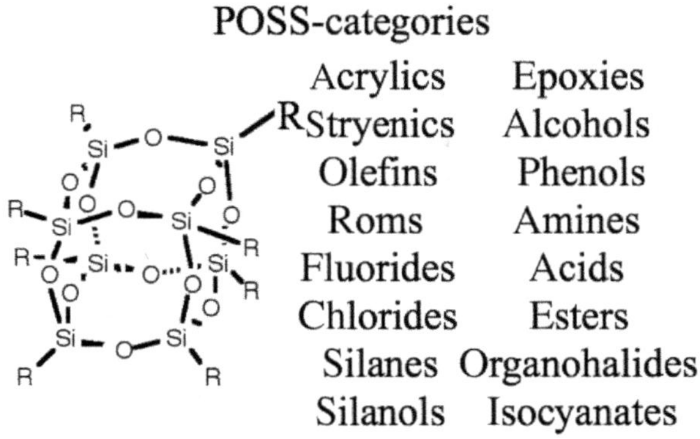

Figure 2.7 POSS molecular structure and reactive groups summarized from [107].

Table 2.7 Generally used silica particles and their physical properties [88]

Category	Size (nm)	Primary size in rubber (nm)	Pore size (nm)	Specific area (m^2/g)	Density (g/cm^3)	Application
Fumed silica	5–50	5–50	N/A	50–600	2.2	Silicon rubber
Arc silica	5–500	50	N/A	25–300	2.2	
Precipitated	5–100	10–80	30	30–800	2.0	Prevulcanized rubber
Xerogels	3–20	50	2–20	250–1000	2.0	
Aerogels	3–20	50	25	250–400	2.0	

2.3.2.4 Silica (SiO_2)

Silica is a naturally occurring spherical particle, with the chemical formula of SiO_2. It is frequently found in sand, quartz, quartzite, perlite, and Tripoli. The form can be either amorphous or crystalline. The surface area for silica is on average more than 100 m^2/g, which defines itself as a "reinforcing particle"—as a comparison, carbon black has a specific area between tens to 100 m^2/g, and is a "semireinforcing particle." An increase of surface area results in higher tensile, abrasion, and tear properties. As shown in Table 2.7, fumed silica, precipitated silica, and ground silica are the three most often used silica particles in rubber-based composites.

Silica with hydroxyl groups, like other clay materials, finds it easy to form bundles between particles due to the hydrogen bonding formation. This will lead to poor dispersion and polymer–filler adhesion. To overcome this problem, surface treatments are necessary. The generally used methods include silane agent inclusion and titanate agent addition. The former is preferable due to wide application and easy processibility.

Generally, the silane formula has four functional molecules (Eq. (2.1)). *R* and *Y* are ethoxy or methoxy groups. The nonhydrolyzed organo-functional groups include amino, methacrylate, or vinyl mercapto groups. These nonhydrolyzed functional groups are able to form effective interactions between polymer and fillers [76–79]. Three (OY) groups are attached to a central silicon atom and these groups hydrolyze to form Si-OH groups, which subsequently adsorb on the silicate or silica surface. *R* links the silicate and a polymer molecule in a continual way. Through the various chemical and physical bonding, the interaction between polymer matrix and reinforcing particles are increased [89]. Eqs. (2.1) to (2.4) also summarize how the coupling agents increase interfacial interactions through hydrolysis (Eq. (2.2)), condensation (Eq. (2.3)), and coupling reactions (Eq. (2.4)) [89]. It can be told that the silane related functional groups form by (1) hydroxyl group generation, and (2) condensation with particle surface by polymerization [89]. As a result, the interfacial interaction efficiency improves and furthermore, the tear, tension, compression, abrasion, bending, and so on will become more resistant [82,89–92].

$$\mathrm{R-Si(OY)_3} \tag{2.1}$$

$$\mathrm{R-Si(OY)_3 + 3H_2O \rightarrow R-Si(OH)_3 + 3Y-OH} \tag{2.2}$$

$$\mathrm{Silica-OH + R-Si-(OH)_3 \rightarrow Silica-O-Si-(OH)_2-R + H_2O} \tag{2.3}$$

$$\begin{aligned}&\mathrm{Silica-O-Si-(OH)_2-R + R_o-Polymer \rightarrow}\\&\mathrm{Silica-O-Si-(OH)_2-R-R_o-Polymer}\end{aligned} \tag{2.4}$$

2.3.2.5 Other less frequently used particles

Alumina trihydrate (ATH, $Al_2O_3 \cdot 3H_2O$, or $Al(OH)_3$)

Alumina trihydrate (ATH) has four polymorphisms, all with aluminum at thecenter and three hydroxyl groups surrounding it. ATH is monoclinic, and has a density of 2.4 g/cm^3. ATH is added to the rubber frequently as an antitracking agent and flame retardant. In the meantime, the addition of ATH can also influence the electrical properties. However, ATH only withstands temperature as high as 200°C. Therefore, the usage of ATH is limited to those polymers processed at lower temperatures (<200°C), and these polymers include some of the rubbers. The addition of ATH in rubbers usually serves as a fire retardant and smoke suppressant.

Barium sulfate ($BaSO_4$)

$BaSO_4$ has an orthorhombic structure and its density is 4.5 g/cm^3. It has been used in foams, floor coverings, high energy or X-ray radiation protections, acid or base corrosion resistance, and absorptivity of light.

Calcium sulfate ($CaSO_4 \cdot nH_2O$)

$CaSO_4$ is orthorhombic in crystal structure, and has a density of 3 g/cm^3. It is a white crystalline solid, odorless, and insoluble in water. If moisture were attached, $CaSO_4 \cdot nH_2O$ changes into a monoclinic structure and decreases the density to 2.3 g/cm^3.

Magnesium carbonate ($MgCO_3$)

$MgCO_3$ is cryptocrytalline, and 3–3.5 g/cm^3 in density. The application of $MgCO_3$ includes the production of magnesium oxide, cosmetics, toothpaste fillers, fire extinguishing and fire proofing fillers, a smoke suppressant in plastics, and a reinforcing agent in rubber. $MgCO_3$ is also usually added into nitrile rubbers (NBR) to improve the dispersion quality of sulfur.

Talc ($Mg_3Si_4O_{10}(OH)_2$)

Talc powder occurs as foliated to fibrous masses, and in a monoclinic or triclinic crystal form. It has a specific density of 2.5–2.8 g/cm^3. The primary application of talc includes fillers in paper, plastic, paint, coatings, rubber, cosmetics, pharmaceuticals, and ceramics.

Titanium dioxide (TiO_2)

Titanium dioxide, also known as titania, is the naturally occurring oxide of titanium, and has the chemical formula of TiO_2. TiO_2 fillers improve the thermal, electrical, and mechanical properties. For TiO_2 the Vickers hardness has been reported to be between 7 and 38 GPa and bulk modulus between 300 and 431 GPa, which is very close to the 446 GPa of diamond [93,94]. TiO_2 have also been synthesized into nanotubes and nanoribbons, with potential applications in catalytic support and photocatalysts. Hollow TiO_2 nanofibers can also be prepared by heating TiO_2-coated CFs to get rid of the carbon core and obtain the hollow nanocrystals.

The applications include pigments in paper and polymers including the use in the rubber industry, ceramics, catalysts, cements, lubricants, concrete manufacturing, pigments, and generation of pure titanium [95]. In addition, the scattering of light on TiO_2 provides the final composites with whiteness and opacity.

Zinc oxide (ZnO)

Zinc oxide has the two main crystal forms of hexagonal and cubic. Its density is 5.61 g/cm^3. The nanostructure of ZnO can be in the form of wires, rods, belts, flowers, or spheres. One dimensional ZnO has been synthesized by a wet chemical method [96], which is cheap, scalable, and environmentally friendly. In addition, the addition of additives, such as polyethylene glycol or polyethylenimine, can be used to tailor the aspect ratio and therefore improve the stress transfer in composite engineering.

ZnO fillers show mechanical reinforcement, semiconductive behavior, and thermal barriers. The hardness is around 4.5 on Mohs scale. ZnO has been widely applied as an additive in plastics, ceramics, glass, cement, paints, adhesives, sealants, pigments, foods, batteries, and so on. Among them, around half of these consumptions are in the rubber industry. Although ZnO can be obtained from the natural mineral zincite, most industrial products are from chemical syntheses [97].

2.3.3 Biofillers

2.3.3.1 Cellulose

Cellulose is a polysaccharide consisting of a linear chain with repeating units from hundreds to thousands (Fig. 2.8A). The chemical formula is $(C_6H_{10}O_5)_n$. Cellulose is abundant in a lot of organic polymers, e.g., 90% in cotton fiber, 50% in wood, 60% in hemp, and 40% in bamboo fibers (BFs) [98–101]. The repeating units in wood are 300 to 1700, while in cotton and other plant fibers they are 800 to 10,000 [102]. Cellulose is a linear polymer chain without coiling or branching. The molecule adopts an extended conformation and acts as a rod-like stiff. The hydroxyl groups on the glucose form hydrogen bonding with another from neighboring chains or even on the same chain, constraining the linear chain firmly and forming the febrile structure. In this linear order and due to the large amounts of hydrogen bonding, cellulose has high strength and stiffness.

The cellulose should be differentiated from hemicellulose. Hemicellulose is also a polysaccharide, but it is derived from sugars in addition to glucose. Another major difference is that hemicellulose consists of shorter chains of between 500 to 3000 repeating sugar units, and these units are branched as shown in Fig. 2.8B [103].

Based on the crystal structure, namely the location of hydrogen bonds between and within cellulose strands, cellulose can be categorized into cellulose I (including I_α and I_β), II, III, and IV. Original cellulose produced by algae or bacteria is I_α and in plants it is I_β. When spun into fibers the cellulose exists in the form of II (Fig. 2.8C). The forms of III and IV can be obtained through chemical treatments but are not frequently used. There are crystalline and amorphous regions in cellulose. The amorphous regions can be eliminated using acid treatment and as a result, the mechanical and thermal properties for cellulose are improved. These treated cellulose crystals have been studied as nanofillers in polymer-based composites [104]. Cellulose has also been widely applicable in paper manufacturing, rayon fiber fabrication, and liquid filtration using cellulose fibers.

2.3.3.2 Husk

The husk or hull in botany is the outer shell or coating on a seed, grain, fruit, or vegetable. The existence of the husk is to protect the seed or fruit during the growing season, because it is made from hard materials of silica and lignin. The main usages of husk have been found in thermal insulating material production and as rubber tire additives (from Goodyear news [105]).

2.3.3.3 Wood—lignin fiber

Wood fibers are usually cellulosic extracted from trees and used mainly in paper manufacturing. Another component in wood is lignins, which are important in the formation of cell walls in wood or bark. Lignins fill the spaces in the cell wall among cellulose, hemicellulose, and pectin components. Lignins are stiffer than

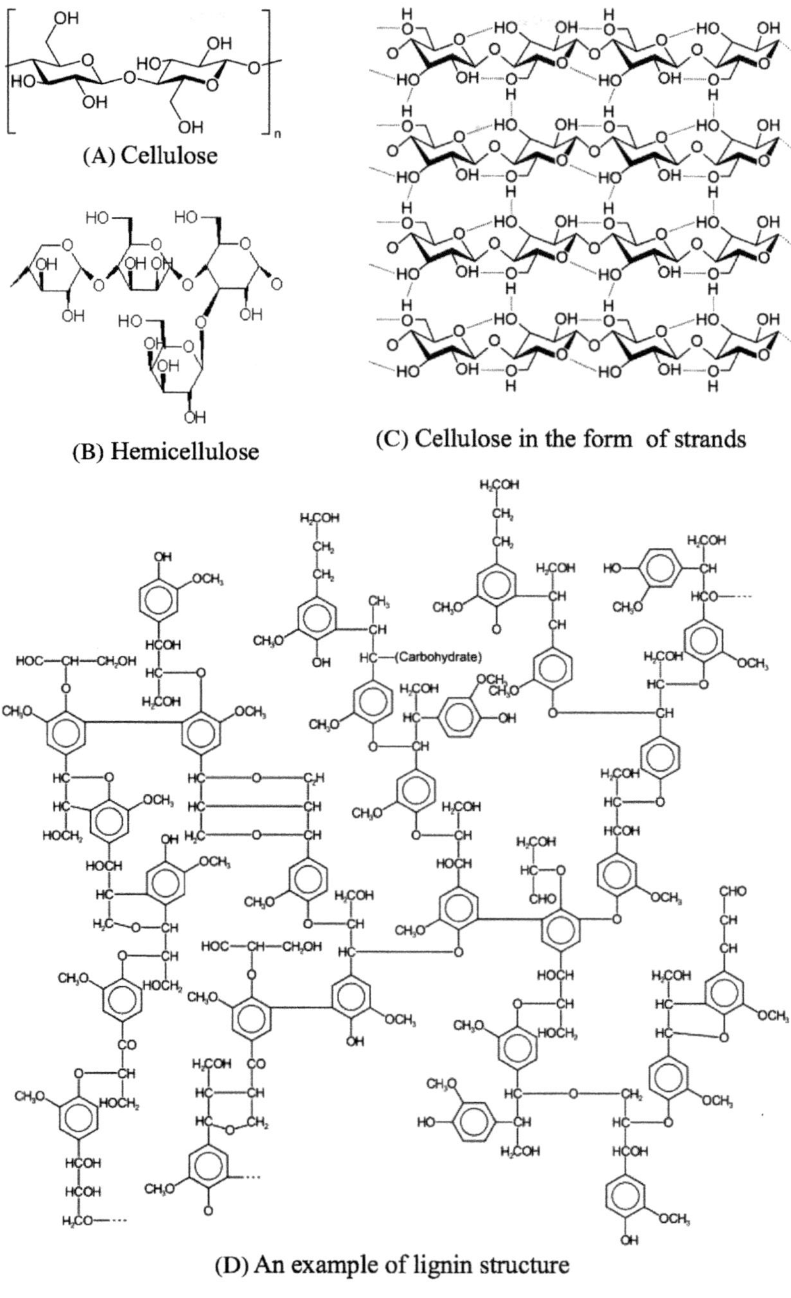

Figure 2.8 Polymer chain conformation in (A) cellulose, (B) hemicellulose, (C) strands of cellulose, and (D) lignin structure.

cellulose and usually they are brittle. The cross-linked structure in lignins is shown in Fig. 2.8D. It can be told that lignin is very aromatic and hydrophobic in nature.

2.3.3.4 Coir—lignocellulosic fiber

Coir is a lignocellulosic fiber obtained from the fibrous mesocarp of coconuts, the fruit of coconut trees (*Cocos nucifera*) cultivated extensively in tropical countries. Coir fibers have been used for making certain traditional products including furnishing materials, rope, farm tools, and so on, which consist only a small percentage of the total coir production. Besides, coir fibers have certain advantages over other natural fibers. For example, they possess high weather resistance due to higher amounts of lignin. Coir fibers absorb water to a much lower extent as compared to most other natural fibers due to their much lower composition of cellulose content. In addition, these fibers can be much more stretched beyond their elastic limit without failure due to the helical arrangement of microfibrils at 45°. On the other hand, the structural parameters and physical properties of coir fibers also accompany disadvantages in the processing techniques. The large and variable diameter, high microfibrillar angle, high lignin and hemicellulose content have been found to relate to poor mechanical reinforcement.

2.4 Impact of particle features on composites properties

Nanoparticles like SWNT, MWNT and graphene exhibit large surface areas of $>10^3\ m^2/g$, $>50\ m^2/g$ [65], and $\sim 2630\ m^2/g$ [66], respectively. Inorganic clays can also reach as high as $100\ m^2/g$. These features can be referred to from Figs. 2.4 to Fig. 2.8. This feature gives rise to the potential for a significant amount of interfacial area, which can be very important for load transfer in the composite material. Despite the fact that the nanoparticles possess the potential for high interfacial interactions, in reality these materials prefer to aggregate. For this reason interfacial interactions and surface area are significantly reduced. Therefore, in rubber-based nanocomposites, good dispersion quality of these particles is important [106].

In ideal filler-dispersions in polymers, isolated/individual tubes, spheres, or sheets with even distribution throughout the matrix will be the main distribution components. However, this is very challenging to achieve especially for high particle concentrations (>10 wt%) [1,87,108,109]. The reasons include (1) entanglement between tubes which occurs in the synthesis process and (2) van der Waals forces between tubes, spheres, and sheets causing agglomerations. If the dispersions are not carefully controlled then the aggregated particles will lead to defective structures in the final composite material. Several different techniques have been used to improve the dispersion of nCs and to optimize the interactions between the polymer and particles. The methods include chemical functionalization, use of surfactants, melt blending, polymer crystallization, in situ polymerization, and mechanical sonication, mixing, or homogenization [110,111].

Rubber-based composites are sometimes produced from solution, which contains polymers, solvent, and particles. For this reason, a popular choice for particle

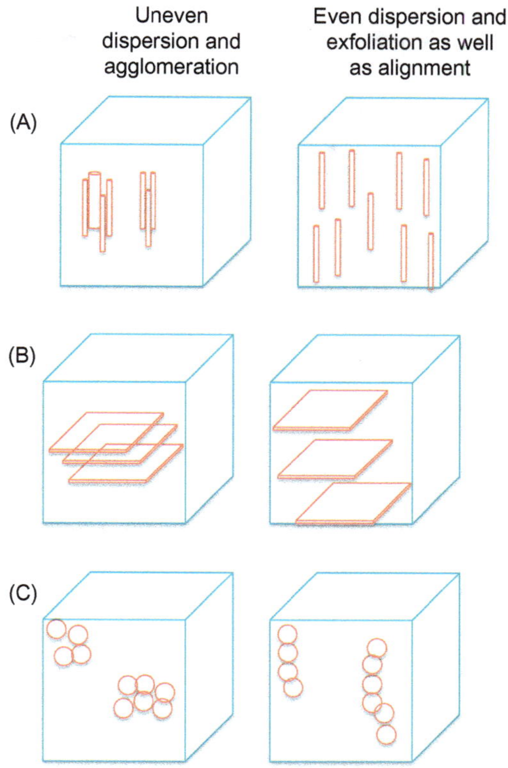

Figure 2.9 Dispersion, exfoliation, and orientation of (A) fibril, (B) platy, and (C) particulate fillers in polymer matrices.

dispersion has been mechanical sonication/agitation/homogenization of the filler in a solvent suitable for subsequent polymer processing. Solution blending involves three major steps: (1) dispersing particles in a suitable solvent, (2) mixing the dispersed particles with the polymers, and (3) evaporating the excess solvent to form the final composite material. For example, studies of dispersion effects on exfoliation and length reduction of CNT have been conducted and reported by several research groups [106,112–116]. While the reduction of the CNT length which occurs during dispersion is not desirable for producing high-performance materials, improving exfoliation has a significant positive impact [117]. Fig. 2.9 shows an example schematic for microstructures containing well-exfoliated particles (i.e., tubes, platelets, and spheres). In general, this is desired to produce a composite with improved mechanical properties.

Polymers are helpful for particle dispersions because they (1) wrap on the carbon surface providing a barrier to prevent reaggregation; (2) crystallize in the presence of nanosized particles, which helps to improve matrix properties [118,119]; and (3) can also be templated by particles at the interphase, which is also important for the composite properties [115]. Since the particle can influence polymer morphology, preservation of its structure during dispersion is another important factor. In particular, for nanoscale fillers, preserving the length can increase the reinforcement efficiency of the fillers in the matrix [86,120,121]. Therefore, in the following section, the effects from particle size, volume, interphase, and orientation will be discussed, respectively.

2.4.1 Particle size, aspect ratio, and surface area—shape parameter

The size of the particles is one the most important factors that influences the reinforcement effects. (1) Large and uneven particles by grinding minerals and coarse filtering usually do not lead to effective mechanical reinforcement. The addition of these particles is aimed to decrease the expensive rubber partition in the mixture. The mechanical reinforcement in most cases remains similar to those composites before particle filling, and the failure strength or strain even drops drastically. (2) On the contrary, small sized particles are found to be very effective in mechanical property improvements, including modulus, strength, tear and abrasion resistance. The reason is due to efficient interfacial interactions and shape parameter-induced stress transfer. The effectiveness of stress transfer from filled particles to polymers depends on particle aspect ratio, specific surface area, and compatibility between matrix and filler phases.

From the perspective of composite mechanics, the length and aspect ratio in transferring stress can be quantified in the length efficiency factor, η_l,

$$\eta_l = 1 - \left[\frac{\tanh(na)}{na}\right] \tag{2.5}$$

$$n = \sqrt{\left[\frac{2G_m}{E_f \ln(2R/d)}\right]} \tag{2.6}$$

$$a = \frac{l}{d} \tag{2.7}$$

where G_m is the shear modulus of polymer matrix, 2R is the distance from the fiber to its nearest neighbor fiber, l and d are the length and diameter of the particle.

Researchers studied the effect of fiber length on the mechanical properties of polymer composites by using all kinds of filled particles. For example, fillers with 40 nm in diameter and 2 μm in length generate an aspect ratio of 50. At fixed fiber concentration and assumed even dispersion, length efficiency is only dependent on aspect ratio as shown in Eq. (2.5). Fig. 2.10 also shows clearly how the length efficiency factor changes with parameter *na*. Based on this curve from Eq. (2.5), polymer shear modulus of 1.7 GPa (assuming a 5 GPa and Poisson ratio of 0.3 in a typical polymer), filler tensile modulus of 150 GPa, and concentration of 10 vol.% generate a value of 0.05 for *n*. A multiplication of aspect ratio of 100 provides a

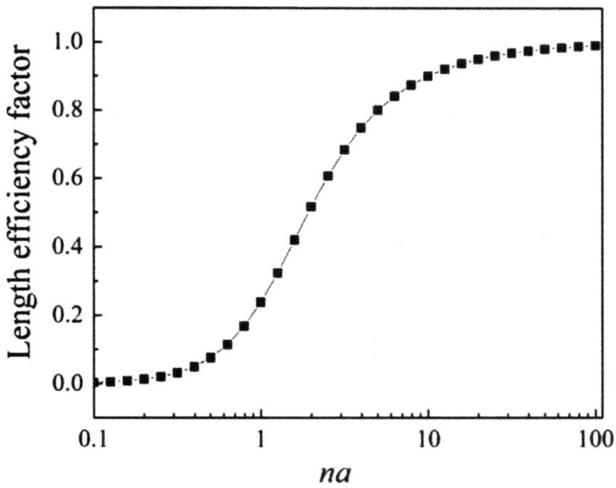

Figure 2.10 Length efficiency factor (η_l) as a function of parameter *na*.

length efficiency of 80%. This is also consistent with the experimental observation that limited length and aspect ratio of fiber will not reinforce the composites.

Another perspective in composite mechanics would relate surface area to stress transfer. A high specific surface area can extremely increase the mechanical properties such as tensile modulus and strength. For specific physical or chemical bonding between polymer and particles, the larger the interaction area the more efficient between these two phases [122]. Reinforcement fillers with specific geometric features have been plotted in Fig. 2.11. Accordingly, the surface area per unit volume (*S/V*) values versus aspect ratio parameters ($a = l/d$) were calculated from Eqs. (2.8) to (2.11). For platy and tubular particles, it is plotted in Fig. 2.12. It can be told that platy and tubular particles have similar reinforcement efficiency due to their large surface area volume ratio. In addition, platy materials intend to reinforce the polymer through the platy direction instead of only the anisotropic direction in tube filled composites.

$$\frac{A_s}{V_s} = \frac{2\pi r^2 + 2\pi rl}{\pi r^2 l} = 2/r + 2/l \tag{2.8}$$

$$\frac{A_s}{V_s} = \frac{4\pi r^2}{4/3\pi r^3} = 3/r \tag{2.9}$$

$$\frac{A_s}{V_s} = \frac{2\pi r^2 + 2\pi rl}{\pi r^2 l} = 2/r + 2/l \tag{2.10}$$

$$\frac{A_s}{V_s} = \frac{2\pi r^2 + 2\pi rl}{\pi r^2 l} = 2/r + 2/l = \left(\frac{2\pi}{V}\right)^{1/3}(a^{-2/3} + 2a^{1/3}) \tag{2.11}$$

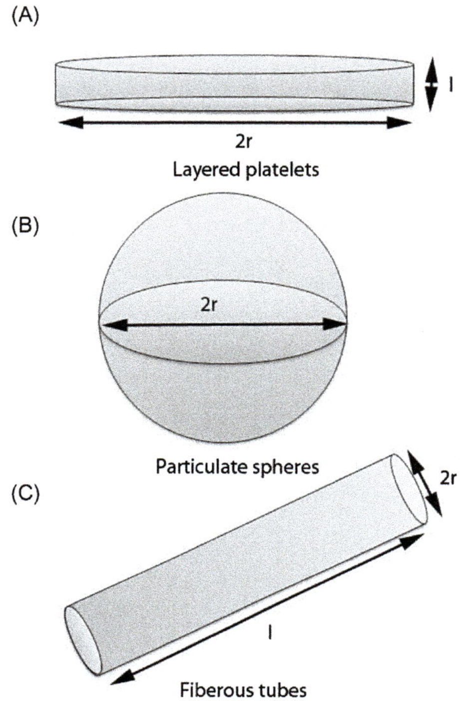

Figure 2.11 Reinforcement particle geometries for (A) platelets, (B) spheres, and (C) tubes.

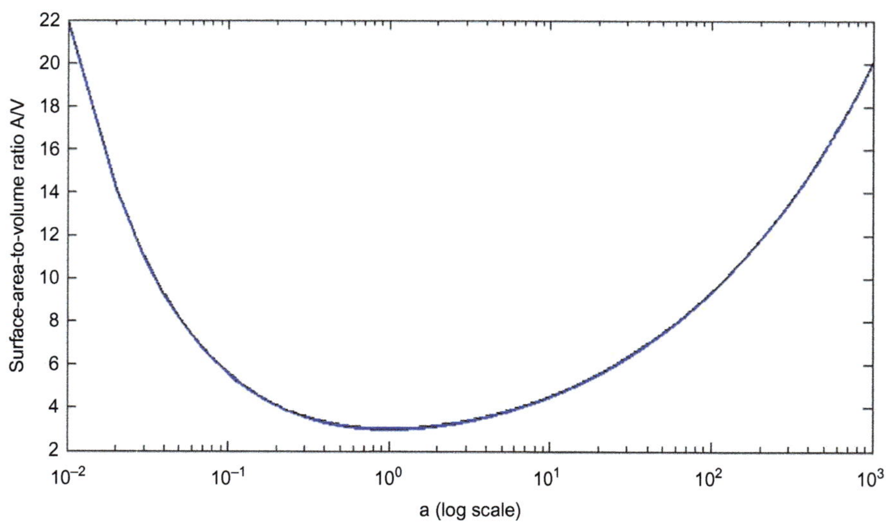

Figure 2.12 Surface-to-volume ratio, *A/V*, of a specific particle with aspect ratio of $a = l/d$, for platy ($a << 1$) and tubular ($a >> 1$) particles.

Particle fillers in rubber composites can be divided into three categories, i.e., two-dimensional layered platelets, one-dimensional tubes, and zero-dimensional spheres. This taxonomy is meaningful since the various dimensions provide different levels of reinforcement efficiencies due to the interaction area plotted in Fig. 2.12. In the rubber industry, layered silicates may interact with polymers in alternative mechanisms, namely, intercalated and exfoliated structures. To achieve the more interactive intercalation between polymers and silicates, compatible agents are usually used. The usage of these agents will reduce the surface energy of the silicates and allow polymer chains to slide between platelets. These agents include amino acid, alkyl ammonium ions, polyetheramines, dihydroimmidazolines, silanes, and so on. In an intercalated structure, well-defined arrangements of rubber polymer chains among the platy sheets will lead to effective reinforcement in rubbers.

One-dimensional tubes are anisotropic and allow stress to transfer more efficiently along tubular direction than the orthogonal direction. CNTs and HNT typically belong to this category. Unfortunately, aggregation is a big problem in these nanoparticles. Therefore, appropriate choices of good solvents and processing techniques are important in producing high-performance rubber composites.

Zero-dimensional spherical articles are not like platy or tubular particles. As shown in Fig. 2.12, the area per unit volume is rather low when the aspect ratio is one. Instead of simply mixing polymers with spheres, very frequently the sol–gel process is used in fabricating organic/inorganic hybrid materials. The composites are formed by a condensation reaction between the functionalized polymer precursors and metal alkoxides. In the procedure, chemical bonding between the filler and polymers are formed so that it is much stronger than the physical bonding. Carbon black and SiO_2 have been mainly used in rubber composites. To achieve effective reinforcement 20–40% concentration is usually applied. Other fillers, such as TiO_2, ZnO, $CaSO_4$, $CaCO_3$, and similar spherical particles have been used in rubber composites.

2.4.2 Percolation threshold

Originally, percolation refers to the movement and filtering of fluids through porous materials. Generally for material science, three percolation thresholds are defined in electrical, rheological, and mechanical properties. An observation of lack of continual increase in material properties beyond the percolation threshold has been widely reported in different polymer system and particle variations. Table 2.8 and Fig. 2.13 list the percolation threshold values in mechanical reinforcement. In the rubber industry, the filler concentration is between 50 and 120 phr, and sometimes even higher.

As previously stated, in particle-reinforced polymers, in addition to tube orientation, dispersion quality and interfacial interactions are another two correlated factors influencing reinforcement efficiency [123]. At a better dispersion quality, exfoliated fillers would intend to interact with more polymer interfacial area and transfer stress more efficiently. However, with the increase of concentrations, particles form aggregates in composites and cause crack initiation, propagation, and failure under mechanical loadings. In particle suspensions, based on the degree of particle interactions, three concentration regimes are defined as dilute, semidilute,

Table 2.8 Percolation thresholds in mechanical reinforcement

References	Polymer	Particles	Percolation (%) in		
			Electrical	Mechanical	Rheological
2005 [126]	SBR	Carbon black		43 vol%	
2006 [127]	SBR	Graphene	0.12 vol%		
2006 [128]	Silicon rubber	Carbon microcoils with 10–30 μm in diameter and 10 μm-1 mm in length	3 wt%	3 wt%	3 wt%
2007 [129]	Methyl vinyl silicone rubber	MWNT as received is about 20–40 nm in diameter and 10–15 μm in length	1.2 vol%		
2007 [130]	Silicone rubber	Graphite nanosheet with 30–80 nm in thickness	0.9 vol%		
2008 [131]	Hydrocarbon rubber	Carbon black	2 vol%		
2009 [132]	Methyl vinyl silicone rubber	Carbon black		3.5 vol%	
2010 [133]	Natural rubber	Bagasse cellulose whiskers with length 84–102 nm and width 4–12 nm		5.15 vol%	
2011 [134]	SBR	MWNT with diameter of 30 nm	0.67 vol%		
2012 [135]	Latex	Graphene	0.62 vol%		
2013 [136]	Polydimethylsilicone (PDMS)	Alkyl-functionalized graphene	0.63 vol%		
2014 [137]	Latex	Graphene	0.25 vol%		
2015 [138]	Natural rubber	MWNT as received is about 20–40 nm in diameter and 10–15 μm in length			3 wt%
2016 [139]	Natural rubber	MWNT as received is about 10–20 nm in diameter and 10–30 μm in length	2 vol%		

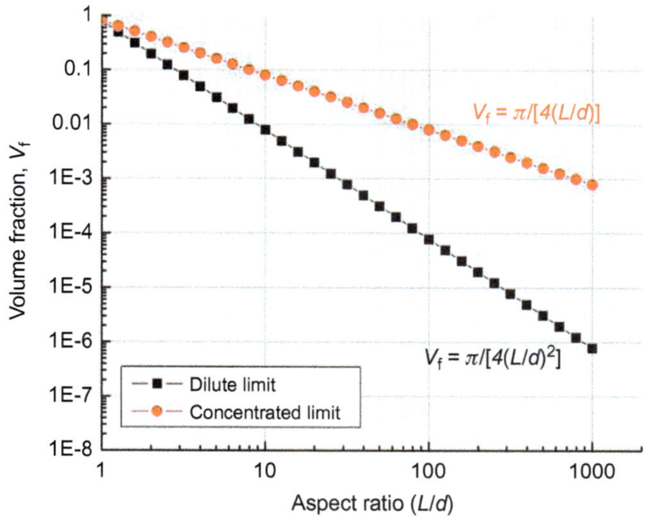

Figure 2.13 Particle aspect ratio and volume fraction based regimes of dilute, semidilute, and concentrated [124,125].

and concentrated [124,125]. These different regimes affect hydrodynamics or rheological properties (i.e., viscosity) in polymer solution or melt preparations, and caution should be taken in composite fabrication procedures.

Rheological properties give information about the deformation and flow of polymer solutions or melts. The use of different fillers in rubber compounds influences these behaviors when compared with unfilled matrices, e.g., the weaker viscoelastic region, wall slippage effects, lower extruding swell with increasing filler content, smoother melt fracture defects with increasing filler content, anisotropic effects in flow, and so on. Therefore great efforts should be devoted to develop polymer-based composites with high mechanical and electrical properties at a low percolation threshold in order to keep an excellent processing of rubber at low cost.

2.4.3 Interfacial interactions

Computational work has also been carried out to predict the interfacial shear stress (IFSS). CNTs filled polymers are provided as examples here. Polymer systems such as polystyrene (PS) [140], epoxy [141], PmPV, and PPA [142] have been calculated using molecular dynamics, where the calculated IFSS was dependent on both the polymer and CNT and ranges from 18 to 186 MPa. Apart from calculations and simulations, direct measurements have also been reported. The techniques and devices (scanning electron [59], transmission electron microscopy [143], AFM [144,145], and scanning probe microscopy [146]) used to measure CNT properties are extended to study interfacial properties in CNT-strengthened polymer composites. Barber's group [144,145] used the "push-in pull-out" method in PE matrices, and reported values of 0.02–0.09 GPa. Compared with the calculated values in the literature (50–100 MPa range) [106,147–149], the measurements are consistent. It is important to note that the IFSS values discussed here are for the noncovalently bonded composites. Larger values are measured by AFM for composites with covalent bonding between the matrix and CNT, and ranged from 0.3 to 0.5 GPa [143,146]. Therefore, in rubber-based particle-filled composites, appropriate physical or chemical methods should be undertaken for improving interfacial interactions.

2.4.4 Particle alignment

Anisotropic particles have been known for their high modulus. For example, HNT can reach up to around 700 GPa at outer diameters of less 50 nm and CNTs can even be as high as 1000 GPa. However, as tubes start aggregating, the accumulating defects and the lack of intertubular registry resulting from diameter differences and helicity variations will lead to a decrease of effective modulus, especially shear modulus. For example, CNTs have been shown to have shear modulus of 6 GPa for 4.5 nm bundles, 2.3 GPa for 9 nm bundles, and 0.7 GPa for 20 nm bundles [42,150,151]. Therefore, a theoretical estimation of the average modulus dependent on bundle size will be necessary to understand the plateau region beyond the percolation threshold. The average effective modulus $<E_x>$ has been calculated using continuum mechanics (Eq. (2.12)) [1]. Modulus along longitudinal direction (E_1), transverse direction (E_2), and Poisson's ratio (ν) are listed in Fig. 2.14. Shear modulus (G_{12}) dependent on bundles size is not available experimentally; here in this

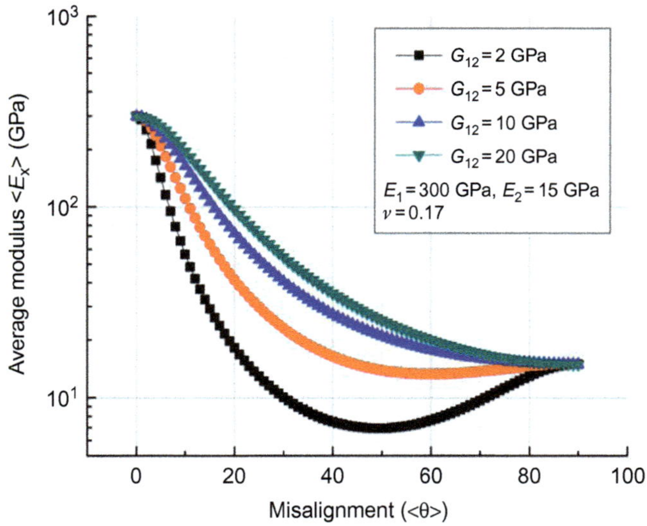

Figure 2.14 Average effective modulus based on nanotube bundle diameter in aggregates. As a comparison, the shear modulus of HNTs with diameters ranging from 200 nm to 220 nm are 1.5 ± 0.26 GPa, while CNTs are 6 GPa for 4.5 nm bundles [42], 2.3 GPa for 9 nm bundles [150], and 0.7 GPa for 20 nm bundles [151]. Considering the average particle diameter of 40 nm, shear moduli were assumed to be 2, 5, 10, and 20 GPa.

research they were taken as 2, 5, 10, and 20 GPa to show the average modulus change with bundle size. It can be seen that with the bundle size increase, the decrease in effective average modulus can be one order of magnitude lower. In addition, for the same bundle size, the average modulus was also found to improve with higher alignment. This is also consistent with the phenomenon as indicated from the trend line in a lot of literature reports [1].

$$\frac{1}{<E_x>} = \frac{<\cos^4\theta>}{E_1} + \frac{<\sin^4\theta>}{E_2} + \left(\frac{1}{G_{12}} - \frac{2\nu_{12}}{E_1}\right) <\cos^2\theta\cos^2\theta>$$
$$= \frac{1}{E_2} + \left(\frac{1}{G_{12}} - \frac{2\nu_{12}}{E_1} - \frac{2}{E_2}\right) <\cos^2\theta> + \left(\frac{1}{E_1} + \frac{1}{E_2} - \frac{1}{G_{12}} + \frac{2\nu_{12}}{E_1}\right) <\cos^4\theta> \tag{2.12}$$

2.5 Summary

The frequently used rubber materials, micro- and nano-fillers used in the rubber-based industry have been summarized. In particular, carbon fillers, inorganic fillers, and biofillers are introduced regarding their physical properties, productions, and applications. The most important several factors including shape parameter, percolation threshold, interfacial interactions, and particle alignment have been analyzed regarding their influence on the final composite performances. In comparison to traditional carbon black and silica, nanostructured fillers have been used in rubber composites. By the help of organic modification, coupling agents, and better dispersion control techniques, the interaction between rubber matrix and particle fillers has been more efficient than before in transferring properties from filler to matrix. On reviewing the rubber industry and the particle fillers, several challenges remain. Appropriate choice and development of particle fillers and rubbers based on the application, optimal exfoliation and more uniform dispersion quality, effective interfacial interactions from physical or chemical modifications, and, especially in anisotropic materials, preferential alignment of fillers will still be the main goals in rubber composite fabrications.

Nomenclature

Materials

BF	Bamboo fiber
BFCP	Bamboo fiber reinforced polymers
CF	Carbon fiber
CFRP	Carbon fiber reinforced polymers
CNTs	Carbon nanotubes
DWNT	Double-wall carbon nanotubes
GF	Glass fiber

GFRP Glass fiber reinforced polymers
GP Graphene platelets
GPRP Graphene platelets reinforced polymers
HNT Halloysite nanotubes
MA Maleic anhydride
MWNT Multiwall carbon nanotubes
MMT Montmorillonite
nCs Nanocarbons
PA Polyamide
PDDA Poly(diallyldimethylammonium chloride)
PDMS Polydimethylsiloxane
PE Polyethylene
PET Polyethylene terephthalate
PHR Parts per hundred parts
PI Polyimide
POSS Polyhedral oligomeric silsesquioxane
PP Polypropylene
PS Polystyrene
PVA Polyvinyl alcohol
PVC Polyvinyl chloride
PU Polyurethane
SWNT Single-wall carbon nantotubes
SBR Styrene-butadiene rubber
SIR Silicone rubber
UHMWPE Ultrahigh molecular weight polyethylene

Processing

AT Annealing with temperature
ATT Annealing with both temperature and tension
LBL Layer-by-Layer

Characterizations

AFM Atomic force microscopy
DMA Dynamic mechanical analysis
DSC Differential scanning calorimetry
DTA Differential thermogravimetic analysis
FTIR Fourier transformed infrared spectroscopy
QCM Quartz crystal microbalance
ROM Rule of mixture
SAXS Small-angle X-ray scattering
SEM Scanning electron microscopy
TEM Transmission electron microscopy
TGA Thermogravimetic analysis
WAXD Wide-angle X-ray diffraction

Units

$g \cdot cm^{-3}$ Gram per unit centimeter cube
vol % Volume percentage

wt %	Weight percentage
μm	Micrometers
nm	Nanometers
MPa	Megapasca
GPa	Gigapasca
J	Joules

Symbols

$\tan\delta$	Dampening parameter
T_c	Crystallization temperature
T_d	Degradation temperature
T_m	Melting temperature
χ_c	Crystallinity degree
H_c	Enthalpy of crystallization
H_m	Enthalpy of melting
E	Tensile/elastic modulus
E'	Storage modulus
E''	Loss modulus
E_f	Tensile/Young's modulus of filled particles
E_m	Tensile/Young's modulus of matrix
σ_f	Tensile/failure stress of filled particles
σ_m	Tensile/failure stress of matrix
σ	Tensile/failure stress (general notation)
σ_y	Yield stress
σ_i	Impact stress
σ_b	Bending stress
ε_t	Tensile/failure strain (general notation)
ε_y	Yield strain (general notation)
V_f	Volume fraction of filled particles
V_m	Volume fraction of matrix

References

[1] Song K, et al. Structural polymer-based carbon nanotube composite fibers: understanding the processing–structure–performance relationship. Materials 2013;6(6):2543–77.

[2] Sengupta R, et al. A short review on rubber/clay nanocomposites with emphasis on mechanical properties. Polym Eng Sci 2007;47(11):1956–74.

[3] Sui G, et al. Preparation and properties of natural rubber composites reinforced with pretreated carbon nanotubes. Polym Adv Technol 2008;19(11):1543–9.

[4] Bhowmick AK, Stephens H. Handbook of elastomers. New York, NY: CRC Press; 2000.

[5] Chandra R, Singh S, Gupta K. Damping studies in fiber-reinforced composites–a review. Compos Struct 1999;46(1):41–51.

[6] Thomas S, Stephen R. Rubber nanocomposites: preparation, properties and applications. John Wiley & Sons; 2010.

[7] Kato A, Ikeda Y, Kohjiya S. Carbon black-filled natural rubber composites: physical chemistry and reinforcing mechanism. Polym Compos 2012;1:515–43.

[8] Heckman F. Microstructure of carbon black. Rubber Chem Technol 1964;37(5):1245–98.

[9] Medalia A. Effect of carbon black on dynamic properties of rubber vulcanizates. Rubber Chem Technol 1978;51(3):437–523.
[10] Callister WD, Rethwisch DG. Materials science and engineering: an introduction, vol. 7. New York, NY: Wiley; 2007.
[11] Kato T, Sugawara A, Hosoda N. Calcium carbonate–organic hybrid materials. Adv Mater 2002;14(12):869–77.
[12] Trckova M, Matlova L, Dvorska L. Kaolin, bentonite, and zeolites as feed supplements for animals: health advantages and risks. A review. Vet Med-UZPI (Czech Republic) 2004;49:389–99.
[13] Lau K-t, Gu C, Hui D. A critical review on nanotube and nanotube/nanoclay related polymer composite materials. Compos Part B Eng 2006;37(6):425–36.
[14] Li G, et al. Polyhedral oligomeric silsesquioxane (POSS) polymers and copolymers: a review. J Inorg Organomet Polym 2001;11(3):123–54.
[15] Muralidharan K, et al. Molecular dynamics studies of brittle fracture in vitreous silica: review and recent progress. J Non-Cryst Solids 2005;351(18):1532–42.
[16] Bailey E, Holloway JR. Experimental determination of elastic properties of talc to 800 C, 0.5 GPa; calculations of the effect on hydrated peridotite, and implications for cold subduction zones. Earth Planet Sci Lett 2000;183(3):487–98.
[17] Kasuga T, et al. Formation of titanium oxide nanotube. Langmuir 1998;14 (12):3160–3.
[18] Wang ZL. Zinc oxide nanostructures: growth, properties and applications. J Phys Condens Matter 2004;16(25):R829.
[19] Iwamoto S, et al. Elastic modulus of single cellulose microfibrils from tunicate measured by atomic force microscopy. Biomacromolecules 2009;10(9):2571–6.
[20] Miyake H, et al. Tensile properties of wet cellulose. Polym J 2000;32(1):29–32.
[21] Bongarde U, Shinde V. Review on natural fiber reinforcement polymer composites. Int J Eng Sci Innov Technol 2014;3(2):431–6.
[22] Chandrasekhar S, et al. Review processing, properties and applications of reactive silica from rice husk—an overview. J Mater Sci 2003;38(15):3159–68.
[23] Bledzki A, Reihmane S, Gassan J. Thermoplastics reinforced with wood fillers: a literature review. Polym Plast Technol Eng 1998;37(4):451–68.
[24] Donnet J-B, Voet A. Carbon black: physics, chemistry, and elastomer reinforcement. New York, NY: M. Dekker; 1976.
[25] Studebaker ML. The chemistry of carbon black and reinforcement. Rubber Chem Technol 1957;30(5):1400–83.
[26] Minus M, Kumar S. The processing, properties, and structure of carbon fibers. Jom 2005;57(2):52–8.
[27] Zhang Y, et al. Low temperature graphitization of interphase polyacrylonitrile (PAN). Carbon NY 2015;91:479–93.
[28] Zhang Y, et al. Tailoring polyacrylonitrile interfacial morphological structure by crystallization in the presence of single-wall carbon nanotubes. ACS Appl Mater Interfaces 2013;5(3):807–14.
[29] Iijima S. Helical microtubules of graphitic carbon. Nature 1991;354(6348):56–8.
[30] Iijima S, Ichihashi T. Single-shell carbon nanotubes of 1-nm diameter. Nature 1993;363(6430):603–5.
[31] Baughman RH, Zakhidov AA, de Heer WA. Carbon nanotubes-the route toward applications. Science 2002;297(5582):787–92.
[32] Dai HJ. Carbon nanotubes: opportunities and challenges. Surf. Sci 2002;500 (1–3):218–41.

[33] Ericson LM, et al. Macroscopic, neat, single-walled carbon nanotube fibers. Science 2004;305(5689):1447–50.
[34] Li Y-L, Kinloch IA, Windle AH. Direct spinning of carbon nanotube fibers from chemical vapor deposition synthesis. Science 2004;304(5668):276–8.
[35] Zhang M, et al. Strong, transparent, multifunctional, carbon nanotube sheets. Science 2005;309(5738):1215–19.
[36] Koziol K, et al. High-performance carbon nanotube fiber. Science 2007;318 (5858):1892–5.
[37] De Volder MFL, et al. Carbon nanotubes: present and future commercial applications. Science 2013;339(6119):535–9.
[38] Bethune DS, et al. Cobalt-catalyzed growth of carbon nanotubes with single-atomic-layerwalls. Nature 1993;363(6430):605–7.
[39] Ebbesen TW, Ajayan PM. Large-scale synthesis of carbon nanotubes. Nature 1992;358 (6383):220–2.
[40] Morales AM, Lieber CM. A laser ablation method for the synthesis of crystalline semiconductor nanowires. Science 1998;279(5348):208–11.
[41] Ren ZF, et al. Synthesis of large arrays of well-aligned carbon nanotubes on glass. Science 1998;282(5391):1105–7.
[42] Salvetat J-P, et al. Elastic and shear moduli of single-walled carbon nanotube ropes. Phys Rev Lett 1999;82(5):944–7.
[43] Zhou X, et al. Band structure, phonon scattering, and the performance limit of single-walled carbon nanotube transistors. Phys Rev Lett 2005;95(14):146805.
[44] Hong S, Myung S. Nanotube electronics: a flexible approach to mobility. Nat Nanotechnol 2007;2(4):207–8.
[45] Kim P, et al. Thermal transport measurements of individual multiwalled nanotubes. Phys Rev Lett 2001;87(21):215502.
[46] Pop E, et al. Thermal conductance of an individual single-wall carbon nanotube above room temperature. Nano Lett 2005;6(1):96–100.
[47] Callister WD, Rethwisch DG. Materials science and engineering: an introduction. 8th ed. New York, NY: John Wiley and Sons; 2010.
[48] Overney G, Zhong W, Tomanek D. Structural rigidity and low frequency vibrational modes of long carbon tubules. Zeitschrift fur Physik D Atoms, Molecules and Clusters 1993;27(1):93–6.
[49] Ajayan PM, Tour JM. Materials science: nanotube composites. Nature 2007;447 (7148):1066–8.
[50] Berman R. Thermal conductivity of glasses at low temperatures. Phys Rev 1949;76 (2):315–16.
[51] Lee C, et al. Measurement of the elastic properties and intrinsic strength of monolayer graphene. Science 2008;321(5887):385–8.
[52] GAmez-Navarro C, Burghard M, Kern K. Elastic properties of chemically derived single graphene sheets. Nano Lett 2008;8(7):2045–9.
[53] Novoselov KS, et al. Two-dimensional atomic crystals. Proc Natl Acad Sci USA 2005;102(30):10451–3.
[54] Dresselhaus MS, Dresselhaus G, Eklund PC. Science of fullerenes and carbon nanotubes: their properties and applications. San Diego, CA: Academic Press; 1996.
[55] Dresselhaus MS. Fifty years in studying carbon-based materials. Physica Scripta 2012;2012(T146):014002.
[56] Tombler TW, et al. Reversible electromechanical characteristics of carbon nanotubes under local-probe manipulation. Nature 2000;405(6788):769–72.

[57] Yu MF, et al. Tensile loading of ropes of single wall carbon nanotubes and their mechanical properties. Phys Rev Lett 2000;84(24):5552–5.
[58] Treacy MMJ, Ebbesen TW, Gibson JM. Exceptionally high young's modulus observed for individual carbon nanotubes. Nature 1996;381(6584):678–80.
[59] Yu MF, et al. Strength and breaking mechanism of multiwalled carbon nanotubes under tensile load. Science 2000;287(5453):637–40.
[60] Wong EW, Sheehan PE, Lieber CM. Nanobeam mechanics: elasticity, strength, and toughness of nanorods and nanotubes. Science 1997;277(5334):1971–5.
[61] Blakslee OL, et al. Elastic constants of compression and annealed pyrolytic graphite. J Appl Phys 1970;41(8):3373–82.
[62] Kelly A, MacMillan NH. Strong solids. 3rd ed New York, NY: Oxford University Press; 1986.
[63] Balandin AA, et al. Superior thermal conductivity of single-layer graphene. Nano Lett 2008;8(3):902–7.
[64] Ma P-C, et al. Dispersion and functionalization of carbon nanotubes for polymer-based nanocomposites: a review. Compos Part A Appl Sci Manuf 2010;41(10):1345–67.
[65] Peigney A, et al. Specific surface area of carbon nanotubes and bundles of carbon nanotubes. Carbon N Y 2001;39:507–14.
[66] Stoller MD, et al. Graphene-based ultracapacitors. Nano Lett 2008;8(10):3498–502.
[67] Xie S, et al. Mechanical and physical properties on carbon nanotube. J Phys Chem Solids 2000;61(7):1153–8.
[68] Song K. Customizing fiber spinning approaches for polymer/nano-carbon composites. Northeastern University; 2014.
[69] Yang L, et al. Toughening and reinforcement of rigid PVC with silicone rubber/nano-$CACO_3$ shell-core structured fillers. J Appl Polym Sci 2006;102(3):2560–7.
[70] Ji B, et al. Application of nano-calcium carbonate in rubber. Dev Appl Mater 2008;5:023.
[71] Le-fan L, Si-dong L, Zhi-fen W. Research development on nano-composite materials of natural rubber. China Elastomerics 2011;3:018.
[72] Peiris, C. and U. Ratnayake, Natural Rubber/Layered Silcate Nanocomposite For Building Applications. 2013.
[73] Cheng G-J, et al. Modification of calcium carbonate by using coupling agent and the influence to the mechanical properties of SBS. Mater Sci Technol 2013;1:008.
[74] Guoqing YJP. Study on surface grafting treatment of superfine CaCO_3 & its uses in rubber. Non-Metallic Mines 2007;1:010.
[75] Chen X-Z, et al. Effect of nano-calcium carbonate/silica composite particles on mechanical properties of styrene-butadiene rubber. China Synth Rubber Ind 2012;35 (3):229–33.
[76] Indrajati, I.N. and I.R. Dewi Statistical Review Of Calcium Nitrate Coagulant On The Natural Rubber Latex Film Strength. In Prosiding Seminar Nasional Kulit, Karet dan Plastik. 2014.
[77] Xin F. Reinforcing effect of modified CaCO_3 on natural rubber. Chinese J Trop Crops 1998;1:002.
[78] Sato H, et al. First-principles studies on the elastic constants of a 1: 1 layered kaolinite mineral. Am Mineral 2005;90(11–12):1824–6.
[79] Lu D, et al. Direct measurements of the young's modulus of a single halloysite nanotube using a transmission electron microscope with a bending stage. J Nanosci Nanotechnol 2011;11(9):7789–93.
[80] Bergaya F, Lagaly G. Handbook of clay science, vol. 5. Newnes; 2013.

[81] Anthony, J., et al., Handbook of Mineralogy, 2003.
[82] Russell J, Wilson M. A handbook of determinative methods in clay mineralogy. New York, NY: Blackie and Son Ltd.; 1987.
[83] Guimarães L, et al. Structural, electronic, and mechanical properties of single-walled halloysite nanotube models. J Phys Chem C 2010;114(26):11358–63.
[84] Lecouvet B, et al. Elastic modulus of halloysite nanotubes. Nanotechnology 2013;24 (10):105704–11.
[85] Joussein E, et al. Halloysite clay minerals—a review. Clay Miner 2005;40 (4):383–426.
[86] Tajaddod N, et al. Exfoliation of boron nitride platelets by enhanced interfacial interaction with polyethylene. Macromol Mater Eng 2016;301(3):315–27.
[87] Song K, et al. Spectral analysis of lamellae evolution and constraining effects aided by nano-carbons: a coupled experimental and simulation study. Polymer (Guildf) 2015;75:187–98.
[88] Rahman IA, Padavettan V. Synthesis of silica nanoparticles by sol-gel: size-dependent properties, surface modification, and applications in silica-polymer nanocomposites—a review. J Nanomater 2012;2012:8.
[89] Plueddemann, E.P., Adhesion through silane coupling agents. 1970.
[90] Park S-J, Cho K-S. Filler–elastomer interactions: influence of silane coupling agent on crosslink density and thermal stability of silica/rubber composites. J Colloid Interface Sci 2003;267(1):86–91.
[91] Zhou H, et al. Fluoroalkyl silane modified silicone rubber/nanoparticle composite: a super durable, robust superhydrophobic fabric coating. Adv Mater 2012;24 (18):2409–12.
[92] Gottenbos B, et al. In vitro and in vivo antimicrobial activity of covalently coupled quaternary ammonium silane coatings on silicone rubber. Biomaterials 2002;23 (6):1417–23.
[93] Dubrovinsky LS, et al. Materials science: the hardest known oxide. Nature 2001;410 (6829):653–4.
[94] Nishio-Hamane D, et al. The stability and equation of state for the cotunnite phase of TiO_2 up to 70 GPa. Phys Chem Mineral 2009;37(3):129–36.
[95] Hernández Battez A, et al. CuO, ZrO2 and ZnO nanoparticles as antiwear additive in oil lubricants. Wear 2008;265(3–4):422–8.
[96] Xu S, Wang ZL. One-dimensional ZnO nanostructures: solution growth and functional properties. Nano Res 2011;4(11):1013–98.
[97] Völz HG, et al. Pigments, inorganic. Ullmann's Encyclopedia of Industrial Chemistry. Wiley-VCH Verlag GmbH & Co. KGaA; 2000.
[98] Siró I, Plackett D. Microfibrillated cellulose and new nanocomposite materials: a review. Cellulose 2010;17(3):459–94.
[99] Petropavlovskii, G. and N. Kotel'nikova, Microcrystalline cellulose (review). Khimiia drevesiny, 1979.
[100] Eichhorn SJ, et al. Review: current international research into cellulose nanofibres and nanocomposites. J Mater Sci 2010;45(1):1–33.
[101] Moezzi A, McDonagh AM, Cortie MB. Zinc oxide particles: synthesis, properties and applications. Chem Eng J 2012;185–186:1–22.
[102] Klemm D, et al. Cellulose: fascinating biopolymer and sustainable raw material. Angewandte Chemie International Edition 2005;44(22):3358–93.
[103] Gibson LJ. The hierarchical structure and mechanics of plant materials. J R Soc Interface 2012;9(76):2749–66.

[104] Pranger L, Tannenbaum R. Biobased nanocomposites prepared by in situ polymerization of furfuryl alcohol with cellulose whiskers or montmorillonite clay. Macromolecules 2008;41(22):8682–7.
[105] http://www.rubbernews.com/article/20150610/NEWS/150619998.
[106] Fu S-Y, et al. The reduction of carbon nanotube (CNT) length during the manufacture of CNT/polymer composites and a method to simultaneously determine the resulting CNT and interfacial strengths. Carbon N Y 2009;47(14):3192–200.
[107] Lichtenhan, J., NanostructuredTM Chemicals: Simplifying "Nano" Technology, in 10th Foresight Conference on Molecular Nanotechnology. 2002.
[108] Song K, et al. Lubrication of poly (vinyl alcohol) chain orientation by carbon nanochips in composite tapes. J Appl Polym Sci 2013;127(4):2977–82.
[109] Song K, Zhang Y, Minus ML. Polymer interphase self-reinforcement and strengthening mechanisms in low-loaded nanocomposite fibers. Macromol Chem Phys 2015;216 (12):1313–20.
[110] Chen G, et al. PMMA/graphite nanosheets composite and its conducting properties. Eur Polym J 2003;39(12):2329–35.
[111] Rafiee MA, et al. Fracture and fatigue in graphene nanocomposites. Small 2010;6 (2):179–83.
[112] Zhang, Y. and M.L. Minus. *Study of Dispersion Effects on Exfoliation and Length Reduction of Single-Walled Carbon Nanotubes towards Fabricating Polymer-based Composites*. in *ACS 2012 Spring meeting*. 2012. San Diego.
[113] Pease LF, et al. Length distribution of single-walled carbon nanotubes in aqueous suspension measured by electrospray differential mobility analysis. Small 2009;5 (24):2894–901.
[114] Green MJ. Analysis and measurement of carbon nanotube dispersions: nanodispersion versus macrodispersion. Polint 2010;59(10):1319–22.
[115] Shokrieh MM, Rafiee R. Investigation of nanotube length effect on the reinforcement efficiency in carbon nanotube based composites. Compos Struct 2010;92 (10):2415–20.
[116] Vichchulada P, et al. Sonication power for length control of single-walled carbon nanotubes in aqueous suspensions used for 2-dimensional network formation. J Phys Chem C 2010;114(29):12490–5.
[117] Hilding J, et al. Dispersion of carbon nanotubes in liquids. J Disper Sci Technol 2003;24(1):1–41.
[118] Chae HG, Minus ML, Kumar S. Oriented and exfoliated single wall carbon nanotubes in polyacrylonitrile. Polymer (Guildf) 2006;47(10):3494–504.
[119] Minus ML, Chae HG, Kumar S. Interfacial crystallization in gel-spun poly(vinyl alcohol)/single-wall carbon nanotube composite fibers. Macromol Chem Physic 2009;210 (21):1799–808.
[120] Meng J, et al. Forming crystalline polymer-nano interphase structures for high-modulus and high-tensile/strength composite fibers. Macromol Mater Eng 2014;299 (2):144–53.
[121] Song K, Zhang Y, Minus ML. Using low concentrations of nano-carbons to induce polymer self-reinforcement of composites for high-performance applications. MRS Online Proceedings Library 2015;1752:137–44.
[122] Coleman JN, et al. Small but strong: a review of the mechanical properties of carbon nanotube-polymer composites. Carbon N Y 2006;44(9):1624–52.
[123] Liu M, et al. Recent advance in research on halloysite nanotubes-polymer nanocomposite. Prog Polym Sci 2014;39(8):1498–525.

[124] Doi M, Edwards SFE. Dynamics of concentrated polymer systems. Part 1.—Brownian motion in the equilibrium state. J Chem Soc, Faraday Transactions 2: Molecular and Chemical Physics 1978;74:1789–801.
[125] Doi M, Edwards SFE. Dynamics of concentrated polymer systems. Part 2.—Molecular motion under flow.. J Chem Soc Faraday Transactions 2: Molecular and Chemical Physics 1978;74:1802–17.
[126] Wan Y, et al. Effect of processing parameters on electrical resistivity and thermo-sensitive properties of carbon-black/styrene–butadiene–rubber composite membranes. Compos Sci Technol 2005;65(11):1769–79.
[127] Stankovich S, et al. Graphene-based composite materials. Nature 2006;442(7100):282–6.
[128] Katsuno T, et al. Observation and analysis of percolation behavior in carbon microcoils/silicone-rubber composite sheets. Appl Phys Lett 2006;88(23):232115.
[129] Jiang M-J, Dang Z-M, Xu H-P. Giant dielectric constant and resistance-pressure sensitivity in carbon nanotubes/rubber nanocomposites with low percolation threshold. Appl Phys Lett 2007;90(4):042914.
[130] Chen L, et al. Silicone rubber/graphite nanosheet electrically conducting nanocomposite with a low percolation threshold. Polym Compos 2007;28(4):493–8.
[131] Bokobza L, et al. Blends of carbon blacks and multiwall carbon nanotubes as reinforcing fillers for hydrocarbon rubbers. J Polym Sci Part B Polym Phys 2008;46(18):1939–51.
[132] He L, Tjong SC. Low percolation threshold of graphene/polymer composites prepared by solvothermal reduction of graphene oxide in the polymer solution. Nanoscale Res Lett 2013;8(1):1–7.
[133] Bras J, et al. Mechanical, barrier, and biodegradability properties of bagasse cellulose whiskers reinforced natural rubber nanocomposites. Ind Crops Products 2010;32(3):627–33.
[134] Tsuchiya K, et al. High electrical performance of carbon nanotubes/rubber composites with low percolation threshold prepared with a rotation–revolution mixing technique. Compos Sci Technol 2011;71(8):1098–104.
[135] Zhan Y, et al. Enhancing electrical conductivity of rubber composites by constructing interconnected network of self-assembled graphene with latex mixing. J Mater Chem 2012;22(21):10464–8.
[136] Hou Y, et al. Positive piezoresistive behavior of electrically conductive alkyl-functionalized graphene/polydimethylsilicone nanocomposites. J Mater Chem C 2013;1(3):515–21.
[137] Tian M, et al. Graphene encapsulated rubber latex composites with high dielectric constant, low dielectric loss and low percolation threshold. J Colloid Interface Sci 2014;430:249–56.
[138] Thankappan Nair S, et al. Selective localisation of multi walled carbon nanotubes in polypropylene/natural rubber blends to reduce the percolation threshold. Composites Sci. Technol. 2015;116:9–17.
[139] Wang L, Dang Z-M. Carbon nanotube composites with high dielectric constant at low percolation threshold. Appl Phys Lett 2005;87(4):042903.
[140] Liao K, Li S. Interfacial characteristics of a carbon nanotube-polystyrene composite system. Appl Phys Lett 2001;79(25):4225–7.
[141] Wong M, et al. Physical interactions at carbon nanotube-polymer interface. Polymer (Guildf) 2003;44(25):7757–64.

[142] Lordi V, Yao N. Molecular mechanics of binding in carbon-nanotube–polymer composites. J Mater Res 2000;15:2770–9.
[143] Wagner HD, et al. Stress-induced fragmentation of multiwall carbon nanotubes in a polymer matrix. Appl Phys Lett 1998;72(2):188–90.
[144] Barber AH, Cohen SR, Wagner HD. Measurement of carbon nanotube--polymer interfacial strength. Appl Phys Lett 2003;82(23):4140–2.
[145] Barber AH, et al. Interfacial fracture energy measurements for multi-walled carbon nanotubes pulled from a polymer matrix. Compos Sci Technol 2004;64(15):2283–9.
[146] Cooper CA, et al. Detachment of nanotubes from a polymer matrix. Appl Phys Lett 2002;81(20):3873–5.
[147] Young K, et al. Strong dependence of mechanical properties on fiber diameter for polymer-nanotube composite fibers: differentiating defect from orientation effects. Acs Nano 2010;4(11):6989–97.
[148] Blighe FM, et al. The effect of nanotube content and orientation on the mechanical properties of polymer–nanotube composite fibers: separating intrinsic reinforcement from orientational effects. Adv Funct Mater 2011;21(2):364–71.
[149] Blond D, et al. Enhancement of modulus, strength, and toughness in poly(methyl methacrylate)-based composites by the incorporation of poly(methyl methacrylate)-functionalized nanotubes. Adv Funct Mater 2006;16(12):1608–14.
[150] Satcurada I, Ito T, Nakamae K. Elastic moduli of the crystal lattices of polymers. J Polym Sci Part C Polym Symp 1967;15(1):75–91.
[151] Popov V, Van Doren V, Balkanski M. Elastic properties of crystals of single-walled carbon nanotubes. Solid State Commun 2000;114(7):395–9.

Mechanism of reinforcement using nanofillers in rubber nanocomposites

3

K. Sahakaro
Prince of Songkla University, Pattani, Thailand

3.1 Introduction

Reinforcement of rubber is essential in order to achieve the required technical and specific properties, especially the strength and the durability of the materials. Micro- and nanofillers have been used to modify the properties of rubber to meet each product's specific requirements. Particulate fillers such as carbon black (CB) and silica are the most commonly and conventionally used fillers for the reinforcement of the rubbers. Their primary particles are in the nanometer range, but they exist in the form of aggregates and agglomerates which have a size in the micrometer range. Due to the small size and increased surface area, nanofillers have emerged as promising fillers for improving the properties of polymers with only low filler loadings. The size, structure, and surface activity of fillers are considered as crucial factors in controlling the reinforcing effect on rubbers. Several types of nanofillers with different geometry, such as nanosilica, carbon nanotubes (CNTs), layered silicates, and graphene, have been intensively investigated.

The mechanical properties of filled-elastomer nanocomposites are influenced by several factors involving filler volume fraction, the filler dispersion degree, temperature, structure and size of the filler, the filler morphology, the filler–matrix compatibility, the nature and strength of the interactions between filler–filler and filler–rubber, and the cross-linking. Sufficient wetting of rubber on the filler surface is required for the improvement of the rubber nanocomposite properties [1], and good adhesion between the nanofillers and the matrix enables an effective stress transfer at the interface. If the interfacial interaction is strong, or if the structure of the polymer at the interface between rubber and filler is very different from the bulk, markedly different properties in the material as a whole can be observed [2]. The large surface-to-mass ratios of nanofillers require a good adsorption of the polymeric material on the particle surface to provide the adhesion levels between the filler and matrix necessary for improved performance. The challenges in preparation of rubber nanocomposites are on obtaining a good filler dispersion and sufficient interactions between the matrix and the filler since they have a strong influence on the mechanical properties of the nanocomposites [3–5]. Compatibilizers have been applied for the nanocomposites in order to reduce the interfacial energy, and to improve the adhesion between the

Progress in Rubber Nanocomposites. DOI: http://dx.doi.org/10.1016/B978-0-08-100409-8.00003-6

different phases; to achieve finer dispersion during mixing; to stabilize the fine dispersion against agglomeration during processing and throughout its service life; and to allow the smooth transfer of stress from one phase to the other [6]. The key area of research dealing with nanocomposite development is therefore to develop a technique to achieve homogeneous dispersion of nanofillers, and to tailor the interfacial interactions. Moreover, the development of nanocomposites aims to achieve some novel properties enabled by the functional properties of nanoparticles.

Reinforcement may be simply defined as the ratio of the elastic modulus of the composite over that of the pure polymer matrix. In general, a more efficient mechanical reinforcement is obtained when a rigid percolating filler network is formed within the soft polymer matrix [7–11] in which an increase of Young's modulus is clearly observed. The reinforcement mechanisms in the rubber nanocomposites are complicated, and are still not well understood due to the complications of a heterogeneous system of multicomponent and cross-linking structure of the rubber that imparts to rubber elasticity in the nanocomposites. Moreover, interactions between filler–filler and filler–rubber as well as the microstructure of the composites all play a concomitant role on the reinforcement. Therefore, the effective enhancement of the properties, such as the elastic modulus, normally cannot be explained by one simple theory. Nowadays, the reinforcement mechanism of the rubber nanocomposites still attracts lots of interest with an emergence of new fillers and different filler/rubber matrix combinations. The interaction between the fillers and the matrix is a key feature that affects the mechanical properties, and has been a center of interest for filler reinforcement mechanism studies. Different reinforcement mechanisms for nanocomposites have been proposed, but up to the present day, there is no finite conclusion with regard to the mechanical reinforcement mechanisms in nanocomposites. So, the topic remains an open question. This chapter aims to underline the main concepts of the reinforcement mechanism as reported by many researchers based on both theoretical and experimental studies using different types of nanofillers in the rubber matrices.

3.2 Reinforcing nanofillers for rubbers

Rubber nanocomposites that are widely investigated employ three groups of fillers as categorized by their geometry, i.e., particulate- or spherical-, tubular-, and layered fillers. Each type of filler has its own characters as briefly described as follows with an emphasis on its reinforcing mechanism.

3.2.1 Particulate or spherical fillers

The two main particulate fillers for rubber reinforcement are CB and silica. Particles with sizes smaller than 100 nm enable reinforcement in rubbers/polymers, and particle structure is a very crucial factor [12]. Reinforcing CB with particle sizes smaller than 100 nm has been utilized in the rubber industry since 1920s. CB is primarily composed of carbon with the presence of only a small amount of other hydrocarbon or oxygen-containing groups on its surface, e.g., phenol, carboxyl, ketone, quinine, lactone, pyrole, and lactone [13]. The primary CB particles are spheres made up of broken quasigraphitic layers whose stacking gives edges with a steplike structure [12]. They exist in various forms of aggregation depending on the production process, and the extent to which aggregates form complex tridimensional objects is referred as "structure." The relatively nonpolar surface of CB makes it compatible with hydrocarbon rubbers. The reinforcing activity is mainly attributed to mechanical interlocking of the rubber onto the CB surface, the chemisorption of the rubber molecules onto the filler surface, and the van der Waal forces between CB and rubber [13]. With respect to surface chemistry, the silica surface is covered by silanol groups (—Si—OH) that are very polar, so considerably chemically active [14]. This feature leads to low compatibility with hydrocarbon rubbers leading to weak filler—rubber interaction [14—16]. For good reinforcement in rubber, silica is practically utilized in combination with the silane coupling agents.

For particulate fillers, particle size, structure, and surface area of the filler have effects on the reinforcement efficiency [15,17—19]. Particle size of the particulate fillers is one of the parameters used for classification. Reinforcing fillers with very small particle size in the range of 10—100 nm have a high surface area that provides more active contact sites with the rubber matrix [20]. The physical properties of filled rubbers are known to correlate with the specific surface area of the filler. Reinforcing or active fillers should give a strong interaction between filler and rubber matrix which depends on the type or form of the functionality on the filler surface, and relates to the affinity for and ability to react with the rubber [21]. "Structure" of fillers formed by aggregation or agglomeration of the primary particles during the production includes shape, density, and size. The higher the structure leads to the better the reinforcement efficiency [17,22]. Four main elements for the reinforcement concept that have been usually adopted to explain the reinforcement in filled rubber compounds are polymer network, hydrodynamic effects, filler—rubber interactions, and filler—filler interactions. For particulate fillers, the study by Payne [23] on the relation between the three-dimensional filler aggregates, and the dependence of the storage modulus on strain led to a better understanding about reinforcement mechanism of the filled rubber. It was demonstrated that the network of aggregates or filler—filler network could easily be ruptured even at small strain. Thereafter, the Payne concept has been widely used to describe the filler—filler interaction as well as filler—rubber interaction [24—26], as shown in Fig. 3.1. For the filled-rubbers, the level of storage modulus depends on four contributions in which three of them (i.e., polymer network, hydrodynamic effects, and filler—rubber interaction) are strain independent, while the filler—filler interaction contribution is strain dependent.

Polymer networks already appear in the unvulcanized rubber in a small amount due to physical cross-links of entangled chains, and some chemical cross-links that occur during polymerization generally via vinyl structures, or occur naturally in the case of natural rubber (NR). After vulcanization, chemical cross-links dominantly contribute to the modulus in which the value is proportional to the cross-link density of the networks.

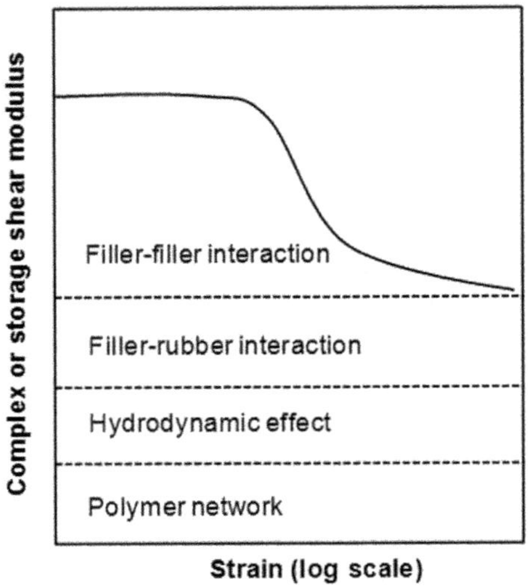

Figure 3.1 Reinforcement effect in particulate filled-rubber based on the Payne concept.

With regard to the strain-independent contribution from hydrodynamic effect, the addition of filler particles increases the modulus of rubber compounds. The simple relationship between the modulus of filled rubber and filler volume fraction can be described by the Einstein–Guth–Gold relation [27] as shown in Eq. (3.1). It should be noted that this equation is only valid for a prediction of the modulus of the elastomers filled with poorly-reinforcing fillers at a small volume fraction of spherical particles ($\emptyset < 0.3$).

$$E_C = E_0(1 + 2.5\phi + 14.1\phi^2) \tag{3.1}$$

where E_C and E_0 represent the Young's moduli of the filled (composites) and unfilled (matrix) elastomers, respectively. This equation is not applicable for well-reinforcing fillers like CB and silica which can enhance the modulus of filled elastomer to a larger extent compared to the value predicted by using Eq. (3.1). The large increase in modulus is thought to be caused by the formation of a three-dimensional network of fillers. Taking into account that the dispersed particles interact at practical concentrations and including the influence of a shape factor (f) for chain-like or nonspherical fillers, the Guth–Smallwood Eq. (3.2) has been proposed as follows [28].

$$E_C = E_0(1 + 0.67f\phi + 1.62f^2\phi^2) \tag{3.2}$$

The ratio of E_C/E_0 is called the strain amplification factor or reinforcement factor.

$$\frac{E_C}{E_0} = 1 + 0.67f\phi + 1.62f^2\phi^2 \tag{3.3}$$

Huber and Vilgis [29] discussed the mechanism of hydrodynamic reinforcement in elastic composites on the principal issues of the Einstein–Smallwood theory. Relative increase of the elastic modulus as a function of filler volume fraction of different elastic composite systems was evaluated based on their theoretical investigations. The result for an elastic matrix filled with spherical core-shell particles with hard core and soft shell (e.g., CB particles with bound rubber) is shown in Fig. 3.2.

Strain-independent contribution to the increase of storage modulus in filled rubber is the filler–filler interactions. This part shows a drastic decrease of the storage modulus when strain amplitude is increased, as shown in Fig. 3.1, due to a breakage of filler–filler network generated by physical bonds, i.e., van der Waals forces and hydrogen bonds, among filler particles and aggregates. The Payne-effect is stronger for silica when compared with that of CB as a consequence of the strong hydrogen bonding between the silica particles [15,30]. Theoretical study on the effect of the filler–filler interactions on the viscoelasticity of filled cross-linked rubber via dissipative particle dynamics (DPD) simulation in systems having dispersed, moderately aggregated, and fully aggregated filler particles also show that the filler–filler interactions have a clear effect on the dynamic shear modulus [31].

Another important contribution to the increase of modulus is from the filler–rubber interactions that create bound rubber. There are different models that have been proposed to describe the filler–rubber interactions initially for CB-filled compounds, as the examples given in Fig. 3.3. Within the composites, there are

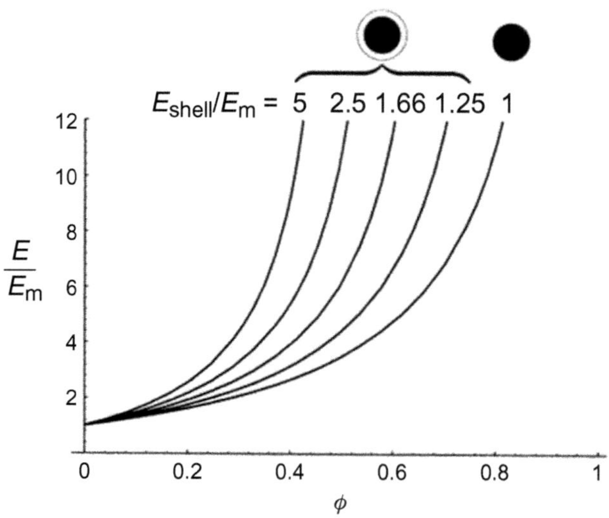

Figure 3.2 Relative increase of the elastic modulus as a function of filler volume fraction for different values of the ratio E_{shell}/E_m. The ratio between shell and core radius is taken as 4/3.

Source: Reprinted with permission from Huber G, Vilgis TA. On the mechanism of hydrodynamic reinforcement in elastic composites. Macromolecules 2002;35:9204–10. Copyright (2002) American Chemical Society.

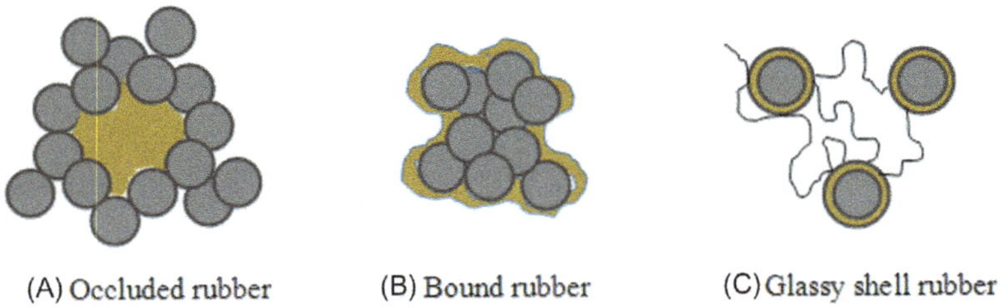

Figure 3.3 Immobilized rubber chains that contribute to modulus of the filled elastomers.

immobilized chains associated with the filler surface. Active fillers can form a filler–filler network by self-association in the rubber matrix, and trap some part of rubber in the void of their structure. These trapped chains in the voids of the filler aggregates are shielded from deformation [32,33]. This immobilized rubber is also called trapped or occluded rubber. Another model suggested that rubber molecules are able to attach on the filler surface either by physical or chemical interactions as well as by chemical reactions, and then the model of shell rubber with an immobilized or glassy state character was proposed [27,34,35]. The glassy rubber shell model has been also reported by O'Brien et al. [36]. These immobilized rubber chains do not contribute to the elastic behavior of the rubber matrix, but act as rigid filler particles. Therefore, the in-rubber structure increases the effective filler loading and hence the strain-independent contribution to the modulus [27,33,34].

Bound rubber is the rubber portion that has strong interactions with the filler surface and cannot be extracted by a good solvent of the polymer. It is often taken as a measure of the filler surface activity [37], and the filler–rubber interactions lead to a three-dimensional structure in uncured filled rubber compounds. Filler–rubber interactions affect on interphase dynamics between filler particles in the filled-rubber including nanocomposites. Bound rubber consists mainly of two polymer phases of different molecular mobility [12,27,32,34,38–40]. The tightly bound phase has molecules that are strongly adhered to the surface of filler, and molecular mobility is strictly constrained, while the loosely bound phase has molecules that are a little more constrained than those in unfilled gum rubber. The two bound layers with different levels of molecular mobility have been studied by advanced techniques such as by nuclear magnetic resonance (NMR) spectroscopy [35,40–44]. The thickness of the tightly bound layer is estimated to be 0.4–1.3 nm.

For silica-silane filled rubber compounds, the silica aggregates form a network by strong hydrogen bonding having the occluded rubber trapped inside. By the use of silane coupling agent, a layer of bound rubber chemically bonded to the silica surface is formed. In addition, some rubber chains are probably also loosely adsorbed on the modified surface which also contributes to bound rubber, but silica–rubber interaction is very weak in this latter case. Under high deformation, the filler network partially breaks open leading to the reduction of occluded rubber, and hence the deformation of the matrix. However, with the aid of coupling agent, the rubber chemically attached to the silica surface and occluded rubber remain grossly immobilized, and therefore still contribute to the modulus [45]. Reinforcement that evolved with mixing time was shown for the silica-filled rubber, and a direct link between the bound rubber content and reinforcement indicators was found, revealing a replacement of strong filler–filler interactions by weak filler–rubber ones [46]. The increase of silane coupling agent increases the chemically bound rubber contents as seen in Fig. 3.4(A), and the increased chemically bound rubber content clearly improves reinforcement index, and reduces loss tangent at 60°C of the silica-reinforced NR compounds as shown in Fig. 3.4(B) as a result of more efficiently stress transfer between the rubber and filler phases [47]. In addition to the use of silane coupling agent, poor interactions between the polar silica surface and nonpolar hydrocarbon rubber have been overcome by various approaches because the silica–rubber interactions have a strong influence on the behaviors and properties of the rubber compounds. Modification of rubbers by introducing polar functional groups that can increase compatibility with silica has been widely investigated. Epoxidized natural rubber (ENR) is one of the successful examples. The epoxy groups of ENR chains and Si–OH groups on the silica surfaces can interact and once the oxirane ring is opened, the covalent bonds can be also formed between the two phases. These interactions and reaction lead to

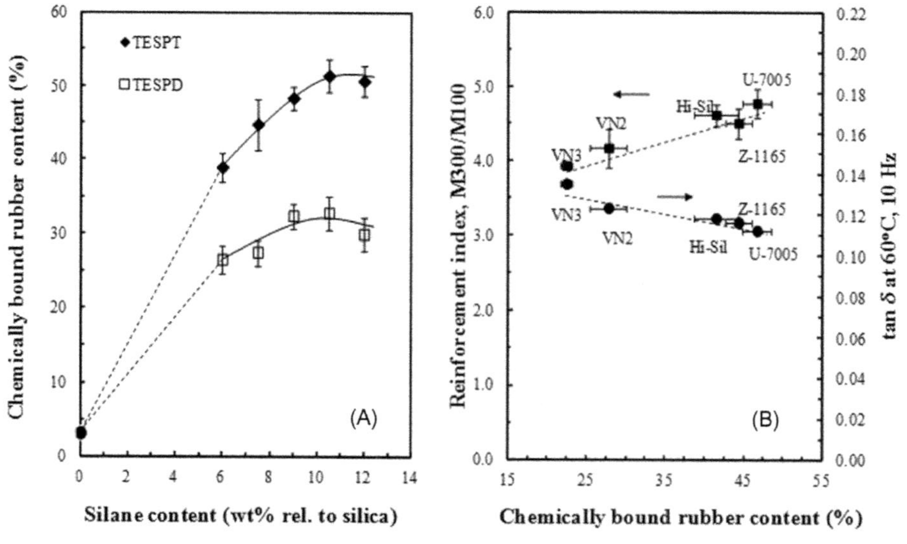

Figure 3.4 Influence of quantity of TESPT and TESPD silane coupling agents on chemically bound rubber content (A); and correlation between reinforcement index and tan δ at 60°C with chemically bound rubber content of the silica-filled NR compounds prepared with different types of silica (B).

Source: Reprinted with permission from Kaewsakul W, Sahakaro K, Dierkes WK, Noordermeer JWM. Optimization of rubber formulation for silica-reinforced natural rubber compounds. Rubber Chem Technol 2013;86(2):313–29. Copyright (2013) Rubber Division, American Chemical Society.

improved dispersion of silica in the rubber matrix and enhanced interfacial adhesion between rubber and silica [48–51]. Modification of the silica surface, such as by grafting with silane coupling agent, has been also reported [52]. In this work, a mobility of solution SBR (S-SBR) rubber chains adsorbed onto filler surface was investigated. By using tube model theory, S-SBR containing modified silica exhibited a combination of increments in topological tube-like constraints and cross-link density in comparison with S-SBR filled with unmodified silica. The modified silica has an increase of the hydrophobic character of silica surface and lessens the extent of secondary network formation by silica particles.

Sternstein and Zhu [53] had found experimentally that the mechanism for reinforcement in nanocomposites is attributed to filler–matrix interactions, rather than filler agglomeration or percolation. Rubber chains that directly adhered to the filler particles and/or in proximity of the filler–rubber interface have different mobility compared to the matrix, depending on their levels or strength of filler–rubber interactions. There are a number of works reported about local segmental dynamics of polymer chains in the vicinity of the filler surface, and gradients of chain mobility proceed from the polymer in direct contact with the filler surface to bulk behavior as well as "bridges" of glassy polymer between filler aggregates where extreme confinement of the polymeric macromolecules is achieved.

The filler–rubber interface model based on stress analysis was proposed for CB-reinforced elastomer by Fukahori [54]. The model consists of double uncross-linked polymer layers of different molecular mobility: the inner glassy hard (GH) layer of thickness 2 nm, and the outer sticky hard (SH) layer of thickness 3–8 nm surrounding a CB particle. The large stress increases with filler content and increasing strain amplitude were interpreted as a result of the strong stress concentration generated around CB particles which transmitted to the whole system. This model is as illustrated in Fig. 3.5.

Fukahori explained that the GH layer increases the effective diameter of a CB particle and its contribution to the stress increase of the system is kept constant over all strain amplitude. When the volume fraction of fine CB is more than 0.20–0.25, the SH layers surrounding two adjacent CB particles are overlapped, and the interconnected SH layers construct the network of the strands of oriented molecules. He proposed that the stress-hardened superstructure produced in the SH layer under large extension supports the large stress concentration and contribute to the large increase in tensile stress.

The presence of glassy layers around the fillers is also a concept for a microscopic model as discussed by Merabia and coworkers [55], as can be seen in Fig. 3.6. This model proposed that strong reinforcement is obtained when glassy layers between fillers overlap. At moderate volume fractions and/or at high temperature, the glassy layers do not overlap, and only the effective filler volume fraction is increased. But at higher filler volume fractions and/or at lower temperature, the glassy layers overlap and the system enters the strongly reinforced regime. It should be noted that this model relies on the validity of interfacial effects regarding the change of dynamics and in particular of the increase of glass transition temperature induced by a solid substrate with strong interactions with the polymer.

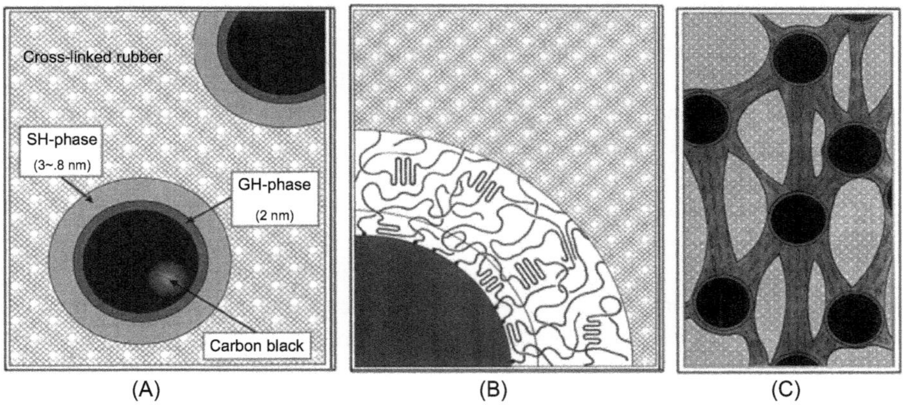

Figure 3.5 Double-layer interface model proposed for carbon black-filled rubber; (A) a double-layer interface model consisting of a glassy hard (GH) layer and a sticky hard (SH) layer; (B) detailed molecular structures in the GH and SH layers as vulcanized; and (C) super network structure of CB particles interconnected by strands of oriented molecules.
Source: Reprinted with permission from Fukahori Y. New progress in the theory and model of carbon black reinforcement of elastomers. J Appl Polym Sci 2005;95(1):60–7. Copyright (2004) John Wiley and Sons.

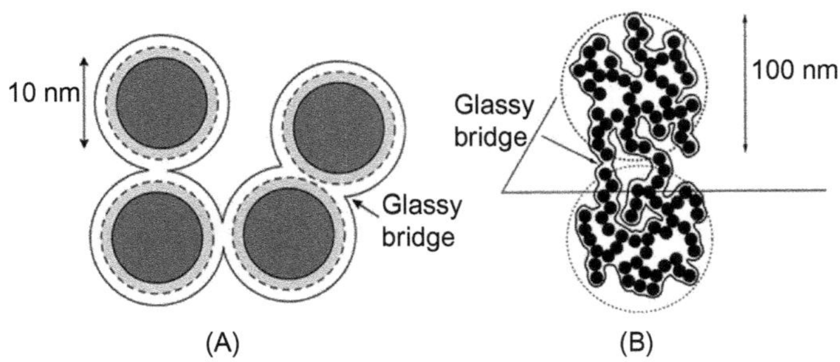

Figure 3.6 (A) Glassy bridges between fillers are formed at high volume fraction or at high temperature; (B) Macroscopic stress is supported by the glassy polymer fraction which bridges two neighboring filler aggregates.
Source: Reprinted with permission from Merabia S, Sotta P, Long DR. A microscopic model for the reinforcement and the nonlinear behavior of filled elastomers and thermoplastic elastomers (Payne and Mullins Effects). Macromolecules 2008;41(21):8252–66. Copyright (2008) John Wiley and Sons.

A study on the mobility of rubber chains around CB and network structure in the rubber matrix of filled ethylene–propylene–diene rubber (EPDM) by low-field proton NMR spectroscopic experiments led to an estimation of the thickness of the immobilized EPDM-CB interfacial layer to be ≥ 0.6 nm [56]. The adsorbed chain fragments restrict chain mobility in the rubbery matrix outside of the interface, and the physical adsorption junctions contributed moderately to the total cross-link density compared to the densities of chemical cross-links and entanglement chains. Based on the use of the dynamic flocculation model to analyze the stress–strain curves, the results showed that a relatively small amount of strongly adsorbed chains impacts the stress–strain properties of the filled-elastomers significantly. The concept that the constrained polymer region is responsible for the reinforcement mechanism is also supported by the work based on an investigation of the viscoelastic behavior and reinforcement mechanism of nano-ZnO reinforced NR nanocomposites [57]. This work discussed that the increase in the storage modulus is related to the amount of nano-ZnO added into the nanocomposites as well as the constrained volume of the rubber chains in the proximity of ZnO nanoparticles. This constrained region around the filler particles that contributes to the enhancement in the storage modulus of the nanocomposites is strongly influenced by the rubber–nanofiller interactions.

Moreover, it has been claimed that the behavior of the elastomer matrix itself is strongly modified by the presence of glassy domains due to a glass transition temperature (T_g) shift of the matrix in the vicinity of fillers [55,58,59]. Chen et al. [60] reported a remarkable shift of T_g of nanosilica-filled NR that was prepared by self-assembly technique to ensure good dispersion. The large increase in T_g is ascribed to be influenced by the restriction of the NR molecular chain movement, as the nanosilica is not just physically bonded with NR, but strongly interacts with NR molecular chains.

Conversely there are a large number of studies that report no significant effect of the presence of filler on the local segmental dynamics of the polymer and the bulk glass transition. A study by Robertson et al. [26] on silica-filled styrene butadiene rubber (SBR) with a filler volume fraction of 0.19 and a presence of silane coupling agent reported that, in contrast to the significant reinforcement effects observed in the strain-dependent shear modulus, the bulk modulus from hydrostatic compression was largely unaltered by the silanes. The coupling agent increases polymer–filler linkages and thus bound rubber, but this bound rubber exhibited glass transition behavior which was similar to the bulk SBR response, as determined by calorimetry and viscoelastic testing. The modification of polymer–filler interface by silane had a strong effect on the nature of the filler network, but it had very little influence on the segmental dynamics of polymer chains proximate to filler particles. A recent report by Huang et al. [61] based on an evaluation of silica and CB-filled NR using several techniques to determine the segmental dynamics and glass transition of the polymer in proximity to the filler surface concluded that a bulk change in the glass transition temperature does not occur upon the introduction of filler particles. Their dielectric spectroscopic measurements showed that the polymer relaxation times of the filled rubbers in the glass-to-rubber transition zone remained essentially equivalent to that of the corresponding unfilled material. The storage moduli of the filled elastomers which are significantly amplified on the rubbery side of the glass transition are discussed in the context of contributions from filler networking. So, there are no finite conclusions that can be drawn yet for this matter.

3.2.2 Tubular filler: carbon nanotubes (CNTs) and nanofibers

CNTs are the most widely used tubular filler in the polymers which have structural properties depending on the diameter, length, and chirality or twist of the nanotubes. CNTs are chemically inert and have good thermal conductivity as well as excellent physical properties due to their high aspect ratio and sp^2 hybridized chemical bonding structure. The single-wall nanotube (SWNT) with a diameter of 1–2 nm consists of a single graphene layer wrapped into a cylindrical shape, and hemispherical caps seal both ends of the tube; each end capped with half of a fullerene molecule [62]. It has a Young's modulus of about 1 TPa [63]. Multiwall nanotube (MWNT) generally exists with diameter of 10–40 nm with length of few micrometers. SWNTs and MWNTs can be synthesized by various techniques, but presently, they are mainly produced by three techniques including arc discharge, laser ablation, and chemical vapor decomposition. The prepared CNTs from all known techniques give mixtures of nanotube chirality, diameters, and lengths along with different amount and type of impurities [64]. For multiwall carbon nanotubes (MWCNTs), the weak van der Waals interaction between layers causes the reduction of the mechanical strength subjected to a uniaxial tensile load in nanocomposites. It has been reported that the inner layers of multiwalled nanotubes cannot effectively take any tensile loads applied at both ends, because the stress transfer ability between the layers of the nanotubes is very weak [65] and only the outmost layer of the nanotubes takes the entire load.

CNT-reinforced polymer and rubber composites attract great attention because of the remarkable properties of CNTs arising from its very high aspect ratio. CNTs enable a very low geometrical percolation threshold and then can offer conductive properties at very low filler volume fraction [66]. The properties of the nanotube-filled polymer composites vary significantly depending on the distributions of the type, diameter, and length of the nanotubes. For effective mechanical reinforcement of CNTs in the polymer matrix, large aspect ratio, good dispersion, alignment, and efficient interfacial stress transfer are required. The main challenge for those who would like to use CNTs to reinforce the elastomer is how to obtain a good and homogeneous dispersion in the matrix. The production of well aligned and well dispersed nanotube/polymer nanocomposites is hardly achieved because agglomeration always takes place during the manufacturing process [67]. To enhance the dispersion of CNTs in the polymer matrix, the most commonly used technique is by chemical modification of CNTs to introduce some functional groups that promote adhesion to the matrix, such as carbonyl and carboxyl, as well as double bonds [68]. These functional groups provide compatibilization with polymer probably via reaction with surface-active agents. Covalent interaction in polymer–CNT nanocomposites occurs when polymer chains are chemically bonded to the nanotubes, giving rise to a stronger adhesion between the polymer matrix and nanotubes that lead to better mechanical performance. Covalent interaction is typically achieved through open-end oxidation or sidewall functionalization [69]. Moreover, surface modified CNTs provide the opportunity for grafting of monomer onto the nanotube in which the grafted polymer helps in compatibilization with a polymer matrix. In addition to surface modification of CNTs, modified rubber polymer can also enhance interactions between them and promote the good dispersion. A comparative study of CNTs addition to three types of NR; i.e., unmodified NR, ENR, and maleated NR (MNR) showed that enhanced interactions of functional groups on CNT surfaces and epoxide groups in ENR molecules help to obtain the homogeneous distribution of CNT particles in the ENR matrix which result in the highest mechanical properties compared to other types of rubber [70]. Another alternative to enhance the bonding strength between the nanotubes and the rubber matrix is to introduce the nanomechanical interlocking of the nanotubes by changing their configuration and/or surface morphology. This mechanical interaction is a noncovalent interaction that only provides a weak interfacial adhesion depending on the polymer type. Different methods of CNT–polymer interaction improvement have been reviewed by Rahmat and Hubert [71].

Due to the geometry of the CNTs with high aspect ratio and relatively inert surface compared to particulate fillers like CB or silica, reinforcement mechanism of CNTs in rubber or polymer is mainly attributed to physical entanglement and networking effect. A study of carboxylated multiwalled CNTs (c-MWCNT) reinforced NR showed a low percolation threshold (<1 wt. %) associated with the formation of an interconnected nanotube network [72]. It was found that the reinforcement occurred mainly at low-strain amplitude which is dominantly contributed by the networking effect, whereas at high strain levels the behavior remained very similar to that of pure NR. The study of NR reinforced by CNTs in comparison with that by CB by Nah et al. [73] showed that the mechanical properties of the NR/CNTs compounds were significantly higher even though the amount of bound rubber that indicates the level of filler–rubber interaction was lower than in the compound with CB. The CNT-filled compounds also showed a greater degree of strain-softening (Payne effect) and stress-softening (Mullins effect) and higher permanent set than the CB-filled compounds, indicating weaker bonding. They suggested that the reinforcement by CNTs is attributed to their large aspect ratio and physical entanglement with rubber molecules, rather than to strong interfacial interaction.

Deng et al. [74] reported the complete stress–strain behaviors of MWNTs/rubber composites under uniaxial tension before rupture, and suggested a novel reinforcement effect of high concentration of MWNTs. Based on their theoretical model developed to understand the reinforcing mechanism and estimate the mechanical properties of MWNTs/rubber composites under large deformation, it was found that the persistence length of tortuous MWNTs and the orientation evolution of MWNTs upon stretching play a key role in the macroscopic mechanical response and the enhancement of the mechanical properties such as the modulus of the rubber composite. It was recommended based on the model prediction that increasing persistence length of MWNTs is a useful way to enhance the modulus of rubber composite, as shown in Fig. 3.7. The model prediction is in line with the experimental results. Their morphological study by using transmission electron microscopy (TEM) of the MWNT-filled NR nanocomposites showed that the MWNTs formed the 3D network structure in the NR matrix. It was discussed that the networks and the interfacial region which formed around MWNTs divided

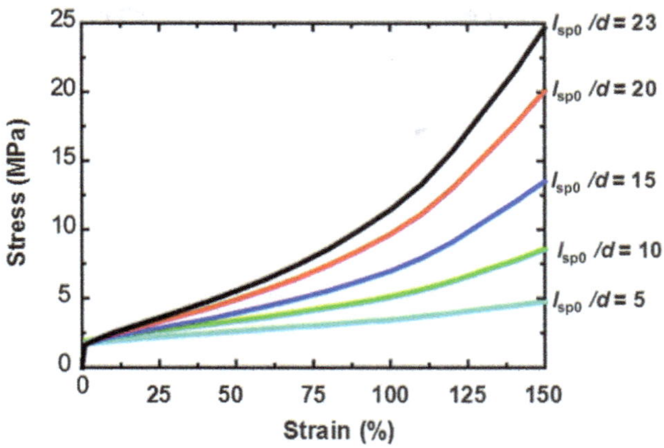

Figure 3.7 Influence of initial MWNTs persistence length (l_{sp0}) on stress–strain response of MWNT/NR composites.
Source: Reprinted (adapted) with permission from Deng F, Ito M, Noguchi T, Wang LF, Ueki H, Niihara K, et al. Elucidation of the reinforcing mechanism in carbon nanotube/rubber nanocomposites. ACS Nano 2011;5(5):3858–66. Copyright (2011) American Chemical Society.

the NR matrix into small units, which is like a cellular structure. Further, the NR matrix was assumed to confine in the cells, and this cellular structure formed by a MWNT network and occluded NR then has a tremendous reinforcement effect on enhancing the storage modulus of NR matrices.

3.2.3 Layered filler: Nanoclay and graphitic nanofillers

3.2.3.1 Nanoclay

Layered silicates consist of two-dimensional (2D) layers of two fused silicate tetrahedral sheets with a dimension of approx. 1 nm thick and 100 nm to 1000 nm long and with an edge-shared octahedral sheet of metal atoms such as Mg or Al [75]. The gap between the two sheets is called the gallery which is normally filled with cations to counterbalance the excess of negative charges. Prior to being used in the polymer, the cations in the layered silicate galleries need to be replaced by organic cations with a long organic chain, such as alkyl ammonium ions, so that the layered silicates can be dispersed in the polymer either in the form of intercalated and/or exfoliated structures depending on the interactions between the polymer and the layered silicates. It has been reported that alkyl ammonium ions with a chain length larger than 8 C-atoms favored the formation of an exfoliated structure, whilst the shorter alkyl chains led to the formation of an intercalated structure [76]. The intercalation of long-chain fatty acids in the interlayer space of organically modified layered silicates was shown to promote the expansion of the interlayer distance and improved the dispersion of organically modified clays (OC) in NR matrix [77,78]. Low crystalline order in the interlayer space of a layered nanofiller such as organically modified clays (OC) leads to easier delamination. Once the nanofillers are evenly dispersed as individual layers, networks are expected to form at very low concentrations in polymer matrices [79]. Preparation techniques and the properties of rubber/nanoclay composites have been extensively reviewed by Sengupta et al. [80]. The intercalation/exfoliation mechanism is influenced by the variability of interlayer spacing due to the swelling of clay layers by intercalation of a wetting liquid, in the form of organo-modifier or the polymer itself. The adsorption mechanism is important for the case of clays having smaller gallery gaps and higher surface energy, and the higher surface energy of the nanoclay facilitates its wetting by the low surface energy polymers. Once the gallery gap is sufficiently created, polymer chains can enter into the interlayer spacing, pushing the silicate layers further apart. Stacks of clay get dissociated from their lattices by means of layer shift parallel to the surface layer, as the polymer adsorbed filler is dragged away under the shear force field [80].

For intrinsically anisotropic fillers such as rods and platelets, Halpin-Tsai equation has been used to describe their reinforcement effect. The Halpin-Tsai model describes explicitly the significant effects of filler geometry alone on the composites' mechanical properties and the model is an effective method to predict the modulus of unidirectional or randomly distributed filler-reinforced nanocomposites [81–83]. For oriented platelets, the reinforcement factor is expressed as Eq. (3.4).

$$\frac{E_C}{E_0} = \frac{(1 + 2f\eta\phi)}{(1 - \eta\phi)} \tag{3.4}$$

where $\eta = \frac{E_f/E_m - 1}{E_f/E_m + 2f}$, f is the aspect ratio, E_f and E_m are the Young's moduli of the filler and pure matrix, respectively. For most of the inorganic fillers including nanoclay, the values of $E_f/E_m \approx 10^5$ and $\eta \cong 1$ [83].

The Halpin-Tsai model has been further modified by Nielsen and Lewis [84,85] by introducing a factor φ, as shown in Eq. (3.5), to consider the maximum volumetric packing fraction of the filler because the original Halpin-Tsai equation predicts lower values than are obtained experimentally at high filler concentration.

$$\frac{E_C}{E_0} = \frac{(1 + 2f\eta\varphi)}{(1 - \eta\varphi)} \tag{3.5}$$

where $\varphi = 1 + \phi[(1 - \phi_m)/\phi_m^2]$ and ϕ_m is the maximum volumetric packing fraction (which is equal to true volume of the filler divided by apparent volume occupied by the filler). The filler apparent volume is related to the polymer chains near the boundaries of the filler that are subjected to restricted mobility and thus contributes to reinforcement effect as if the filler had a greater volume.

For layered fillers, geometry of the platelets plays a crucial role on reinforcement. Buxton and Balazs [86] used the lattice spring model to investigate the effects of filler geometry and intercalation–exfoliation of clay platelets. The simulation results revealed that polymers filled with platelets or fully exfoliated clay have the most significant increase in reinforcement efficiency, whilst incomplete exfoliation of clay platelets leads to a less effective reinforcement. The reinforcement efficiency was attributed to the volume of polymer matrix constrained in the proximity of the particles which also depends on filler–rubber interactions. The concept of immobilized polymer chains in the proximity of the filler surface that influence the reinforcement mechanism is also supported by the work of Rezende et al. [83] on NR–clay nanocomposites. They demonstrated that the improved mechanical properties of the nanocomposites can be modeled by the high aspect ratio of exfoliated clay platelets coupled with immobilized rubber matrix. Moreover, no change in the clay morphology was detected at high deformations, and enhanced mechanical properties were related to ordering of the clay network in the nanocomposite. Ramorino et al. [87] reported that the percolation threshold of the clay for a true clay volume fraction was found at about 0.014 for organo-modified montmorillonite-filled NR. With higher amount of clay than the percolation threshold, filler network is created via confined regions of immobilized rubber joining the filler dispersed particle and contributed to the viscoelastic behavior of the nanocomposites. Also, it has been reported that the filler network in NR–clay nanocomposites with the high level of dispersion is arranged in small groups of finely dispersed clay tactoids (containing ~10 layers) which are largely isolated and separated from each other by ~10–50 nm without forming any rigid filler framework with fractal structure [88]. The inclusion of highly anisotropic nanoclay with large interfacial surfaces in the rubber matrix introduces a mass of physical cross-linking points (entanglements) and the adsorption of rubber chains on the surface of organically modified montmorillonite (OMMT) forms a "denser network domain" due to the strong filler–rubber interaction [88]. However, as like other filler type inclusion, the high elastic modulus of NR–nanoclay nanocomposites at low strain deformation can be mainly attributed to the hydrodynamic effect and the interactions between nanoclay and polymer [89].

It should be noted that the use of nanoclay in the rubbers that have the strain-induced crystallization (SIC) behavior such as NR and synthetic polyisoprene (IR) has attracted interests from the researchers. It was reported that in the presence of OMMT, filled cis-1,4-IR [90] and NR [88,91] showed a development of crystallization at lower strain in the nanocomposites during uniaxial stretching, as well as enhancement of overall crystallization because of the clay orientation and alignment, as shown in Fig. 3.8. The large interfacial surfaces introduced by nanoclay facilitate the overall chain orientation during deformation, resulting in an increase in final crystallite fraction.

3.2.3.2 Graphene

Perfect graphene is not naturally available, but the first report on the preparation and isolation of single graphene layers in Manchester in 2004 attracted an enormous scientific interest due to its exceptional electron transport, mechanical properties, and high surface area. Graphene is an atomically thick, two-dimensional (2D) sheet composed of sp^2 carbon atoms arranged in a honeycomb structure [92]. It has been

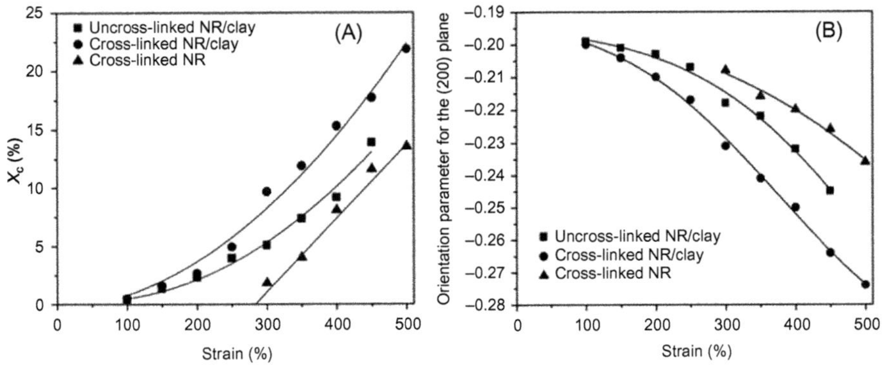

Figure 3.8 Change of crystallinity (X_c) as a function of strain (A); and change of orientation parameters as a function of strain (B) for uncross-linked NR/clay nanocomposite, phenolic resin cross-linked NR/clay nanocomposite and cross-linked neat NR.
Source: Reprinted with permission from Masa A, Iimori S, Saito R, Saito H, Sakai T, Kaesaman A, et al. Strain-induced crystallization behavior of phenolic resin crosslinked natural rubber/clay nanocomposites. J Appl Polym Sci 2015;132(39):42580, http://dx.doi.org/10.1002/app.42580. Copyright (2015) John Wiley and Sons.

reported that single-layer graphene has Young's modulus of 1 TPa and ultimate strength of 130 GPa [93] which is considered to be the strongest material ever measured. Young et al. [94] reviewed that the predicted strength of the monolayer is in the range of 107–121 GPa, which is in very good agreement with the range of values measured experimentally. Graphene is usually developed from a precursor, graphene oxide (GO), and the mechanical properties of GO are inferior to those of graphene due to the disruption of the structure through oxidation and the presence of sp^3 rather than sp^2 bonding. However, the elastic modulus of a single GO sheet is reported to be as high as 250 GPa despite the high concentration of local defects [95,96]. GO is readily available in large quantities, and can be exfoliated and dispersed in a polymer matrix much easier compared to the graphene, so it also attracts interest widely. Moreover, GO has functional groups available, such as epoxide, carbonyl, hydroxyl, and carboxyl groups, allowing the possibility of surface functionalization to interact with various polymers. In contrast, the homogeneous carbon composition of graphene without other heteroatom functionalities makes it more difficult to form strong molecular interactions with polymers and so only weak physical interactions are normally formed. Therefore, the level of adhesion between the graphene and polymer matrix is relatively poor [94].

The interlayer spacing of the graphene is 0.34 nm and the effective thickness of the polymer coils is assumed to be around 2 nm [97]. Because the graphene is atomically thin and the aspect ratios are very large, monolayer graphene can efficiently reinforce the polymers only if the in-plane dimensions of the flakes are sufficiently large ($>30\ \mu m$) [94]. The nanocomposites can be assumed to be made up of parallel graphene flakes separated by thin polymer layer of the same uniform thickness, but the number of graphene layers needed in the flakes for optimum reinforcement is not yet clear. This will depend upon the polymer layer thickness and the efficiency of stress transfer between the graphene layers. It is noted that sliding between the graphene layers can take place easily due to the relatively weak van der Waals force between the individual graphene layers which causes a reduction of reinforcement effect [98].

Both graphene and GO have distinctively high stiffness and strength but the use of them as reinforcing materials for rubber and polymer nanocomposites is really a challenge. The ideal case to obtain highly effective reinforcement is to have good dispersion, the full exfoliation of graphene into single- or few-layer material, and a strong interfacial adhesion between the graphene and polymer to ensure an effective load transfer from polymer to reinforcing fillers [94,99]. A study of graphene-NR nanocomposites by Fu et al. [100] showed that, in the presence of graphene, NR could crystallize at a lower strain compared to unfilled NR, like the case of nanoclay [91] as mentioned earlier, and the strain amplification effects were also observed. The application of the tube model to analyze the chain network structure and determine the network parameters of nanocomposites showed that the addition of graphene has an influence on the molecular network structure and improves the contribution of entanglement to the conformational constraint [100]. The use of ENR promotes interfacial interaction with the surface of GO sheets and enhances the properties of polymeric/inorganic nanocomposites. The morphology of the GO sheets affected the properties, and the disoriented GO resulted in a weaker interfacial interaction with ENR than that of aligned GO [101].

A recent review by Papageorgiou and coworkers [102] for graphene/elastomer nanocomposites concludes that, overall, the inclusion of graphene-based nanofillers can increase the stiffness and strength of elastomers significantly, with the best improvements being obtained with well-exfoliated, uniformly-dispersed fillers. Moreover, the graphene structure, the number of layers, the specific surface and the chemical functionalization of these fillers all play a major role in the final performance of the reinforced elastomers.

3.2.4 Chemical and interface modification on nanofillers

Due to the differences in structure and geometry of each nanofiller type, the reinforcement effects and the final properties of the nanocomposites obtained by the inclusion of different nanofillers are therefore varied. One of the dominant parameters that lead to significant enhancement of the properties of the nanocomposites is a homogeneous and uniform dispersion of the fillers. To increase the filler–polymer interactions and bonding whilst simultaneously decreasing the filler–filler interactions and/or increasing the interlayer distance, nanofillers are commonly modified chemically with functional groups [103]. The increased filler–polymer compatibility and interactions through surface modified particles result in immobilization of the polymer chains in the proximity of filler surface and the occurrence of higher density of physical network or entanglements leading to significantly increased modulus [104].

The filler–rubber interactions either by physical or chemical interactions and/or chemical bonding at the filler surface which result in the attachment of the rubber molecules on the filler surface lead to a change of mobility and dynamics of the rubber chains that finally affect the properties of the materials. For the nanocomposites with fillers of high aspect ratio like CNTs and layered silicate in which the physical entanglement may have stronger influence on the reinforcement than the filler–rubber interaction, the improvement of such interactions still give a positive effect on the reinforcement efficiency. The chemical and interface modifications on nanofillers have been explored by various approaches. Silica surface is commonly treated with silane coupling agent to enhance the interactions with the rubber matrix which can be done either by in situ treatment during mixing or pretreatment of silica before being used [45,47,52]. Recent works on the modifications of the silica surface are such as the grafting of silica with antioxidant 2-mercaptobenzimidazole (MB) that leads to the better dispersion in SBR, the stronger interfacial interaction in SBR/silica nanocomposites, and the remarkable improvement of mechanical properties, compared to those for the γ-chloropropyltriethoxysilane- modified silica/SBR containing equivalent antioxidant component [105]. This work analyzed the large strain behavior of SBR nanocomposites and elucidated that the physical entanglements between the cross-linked rubber chains in the immobilized polymer layer and the grafted MB on the silica surface, i.e., the increased amount of physical cross-linking points between filler and rubber, played an important role in the silica reinforcement of SBR. So, both chemical and physical factors can affect the interfacial interaction in the filled rubber nanocomposites [105]. The immobilized layer prevents the phase separation between filler and rubber, leading to a constrained compatibility and strong filler–rubber interfacial interaction in the nanocomposites. The thickness of the immobilized polymer layer is known to be influenced significantly by the filler dispersion state, filler aspect ratio, and compatibility between filler and rubber [106].

Ionic liquids (ILs) have been applied for interface modification of the nanofillers either in the form of catalyst for the silanization reaction or as a surface modifier for carbon-based fillers. A phosphonium ionic liquid (PIL) was used to promote the silanization reaction between silica and bis (3-triethoxysilylpropyl)-tetrasulfide (TESPT) [107], allowing a possible use of a smaller amount of TESPT that leads to a lower amount of ethanol emission. The interfacial adhesion between silica and the rubber matrix is reported to improve significantly by the addition of a small amount of PIL into a TESPT-modified SBR/silica composite, and the improved interfacial adhesion leads to a more efficient energy transfer between the phases, i.e., less energy loss during dynamic deformation. The IL 1-allyl-3-methylimidazolium chloride was applied to modify CB surface [108], leading to a decreased polymer–filler interaction in the diene nonpolar elastomer, i.e., EPDM and SBR, but an improved polymer–filler coupling in polar rubber, i.e., hydrogenated nitrile rubber, due to attractive dipolar interactions between the polar units of the polymer and the strongly adsorbed IL at the CB surface. The rubber–filler interphase dynamics are affected by different levels of rubber–filler interactions, and strongly confined polymer layers on the filler surface show a glassy-like behavior. The surface modification of the carbon–silica dual phase filler (CSDPF) by ILs also showed that the bound rubber of ILs-CSDPF/NR nanocomposites increased significantly due to the enhanced filler–rubber interactions, resulting in the improved mechanical properties of the vulcanizates [109].

For nanofillers with high aspect ratio like CNTs and layered silicate, chemical modification on the filler surface is also applied in order to promote adhesion between nanofiller and polymer matrix, and to enhance dispersion. CNTs can be surface modified to have some functional groups that can provide compatibilization with the polymer matrix. As reviewed by Rahmat and Hubert [71], there are different methods for interaction improvement, mainly categorized under covalent and noncovalent interactions. The covalent interaction occurs when polymer chains are chemically bonded to the nanotube typically via open-end oxidation or sidewall functionalization. The noncovalent interactions provide a weak interfacial adhesion which is obtained by employing the mechanical techniques such as bridging, increasing specific interfacial area, and wrapping. It was concluded that an optimum CNTs–polymer interaction is a key factor towards reaching the full potential of CNTs in nanocomposites [71], as stronger adhesions between the nanotubes and polymer matrix lead to better mechanical performance. For layered nanofillers, the process of exfoliation needs to be optimized. To achieve exfoliated structure, nanoclay-based composites has been processed using three different modes, including (i) intercalation of difunctional monomers with interactive groups that are capable of exfoliating the clay layers followed by polymerizing a monomer in the exfoliated clay matrix; (ii) intercalation of a surfactant with a long hydrophobic chain that exfoliates the silicate galleries and helps the distribution of the filler in the polymer matrix; and (iii) in situ generation of nanofiller particles in a matrix that subsequently polymerizes to provide a nanocomposite [6]. Through intercalation mode, the tightly bound layers of layered silicate can be separated by firstly exchanging inorganic cations with organic cations attached to large hydrophobic groups to increase the interlamellar gap, then allowing the intercalation process to take place between the galleries that will finally break up the agglomerates into their primary particles [6]. Various functionalized amines, such as stearyl, dodecyl, and hexadecyl, have been employed to modify montmorillonite (MMT) clay for rubber composites in which the intergallery distance is increased by incorporation of the amines [110]. The modification of sodium–montmorillonite (Na-MMT) nanoclay with different amine-based modifying agents having different alkyl chain lengths, i.e., dodecylamine, tetradecylamine, hexadecylamine, and octadecylamine, leads to the enhanced dispersion of nanoclay in NR [111]. It was shown that the longer alkyl chain lengths were more effective in intercalating and expanding the interlayer spacing of clay. Moreover, the use of amine-modifying agent additionally gives a positive effect on the vulcanization reaction of NR, leading to an increased cross-linking degree. The use of sodium dodecyl sulfate results in the increased physical interactions between the rubber and GO nanosheets that lead to an improvement of the mechanical strength of the EPDM/GO nanocomposites [112].

3.3 Mechanism of rubber reinforcement by nanofillers

There are some common features that can be observed in the filled elastomers, such as the filler is usually aggregated and agglomerated to some extent; the Payne effect is associated with the breakdown and reformation of filler agglomerates and aggregates or other filler network structures; percolated structures or filler network is formed at high filler concentrations; and the actual volume fraction of filler or an effective value in the presence of occluded and immobilized rubber that is trapped inside the filler network contributes to reinforcement. Reinforcement mechanisms of filled elastomers have been reviewed by several authors, such as by Kraus [113], Wang [114,115], and Heinrich and Klüppel [116]. One of the mechanisms that is widely discussed related to the reinforcement is the changes in the matrix properties close to or in the proximity of the filler surfaces. The increase of stiffness in nanocomposites is associated to strong filler–matrix interactions, and the effective interaction at the polymer–nanofiller interface is responsible for the modulus reinforcement of rubber nanocomposites [2]. Based on this mechanism, the matrix–filler bonding and debonding mechanisms lead to a nonlinear behavior of the materials [53,117–121]. Another possible mechanism of reinforcement is a formation of filler network above the critical concentration or percolation threshold, which results in a transfer of stress partly to such a network causing an abrupt increase of the material stiffness. For the filler networking concept, the material nonlinearity is attributed to an agglomeration–deagglomeration process of the filler network. Another proposed mechanism is related to the presence of glassy layers around the nanoparticles, which can significantly increase the stiffness of the system.

To obtain more information at the molecular level in the elastomer nanocomposites, computational simulation techniques have been also widely applied for a theoretical study on chain dynamics around filler particles. Liu et al. [122] discussed the progress achieved in the simulation research of three critical topics of elastomer nanocomposites including the dispersion mechanism; the characterization of the nanoscale/microscale structure and dynamics; and the macroscopic viscoelasticity (the Payne effect), mechanical reinforcement, and thermal conductivity. With regard to the second topic, there are several aspects involved, including the change of molecular chain configuration close to the nanofiller surface, the strength of interfacial interactions and/or bonding that determines the stress transfer efficiency, the interfacial segmental dynamics that leads to the shift of the glass transition temperature, the formation of the filler network and its structure, and the chemical cross-links generated by the vulcanization process. The structural changes in unfilled and filled elastomers during uniaxial deformation were studied by using a large-scale dissipative dynamics simulation by Gavrilov et al. [123] and the results revealed that the subchains connected to the filler particles are deformed significantly more, whereas part of the subchains which are not connected to the filler particles are deformed slightly more than in the unfilled matrix. A study by applying the tube model on equilibrium statistical mechanics showed that, by the addition of nanoparticles, a highly ordered structure with a huge amount of entanglements is formed wherein the polymer is nanoscopically confined. These physical entanglements exhibit freedom of movement under stretching, but in a lower volume because of confinement [124]. Based on a theoretical investigation via an idealized model of a polymer network and filler, Liu et al. [125] reported that the reinforcement mechanism stems from the nanoparticle-induced chain alignment and orientation, as well as the limited extensibility of chain bridges formed between neighboring nanoparticles at large deformation. By introducing enough chemical coupling between the rubber and the filler, an upturn in the modulus at large deformation is observed in the Mooney-Rivlin plot, attributed to the limited chain extensibility at large deformation.

3.3.1 Percolation phenomena

One of the reinforcement mechanisms that has been used to describe the large enhancement of modulus or electrical properties is percolation phenomenon [126]. The formation of filler network, above the critical concentration or percolation threshold, results in the abrupt increase of stiffness, or modulus, or conductivity of the nanocomposites. This percolation phenomenon depends mainly on the interparticle distance of the reinforcing fillers dispersed in the rubber matrix, which is strongly related to the filler size, loading level, and dispersion state. It is generally observed that the modulus starts to deviate from the pure hydrodynamic prediction and increases quite abruptly above a critical volume fraction, which depends on the type of fillers. However, under large strain amplitude, breaking down of the filler network occurs, leading to the nonlinear drop of the modulus [23,127,128]. Phenomena such as the Payne effect are well documented and are usually attributed to particle network formation [23,129]. Nanoparticle aggregation and network formation are also known to have significant effects on polymer rheology [130].

Huber and Vilgis [131] determined the percolation threshold value in CB filled rubbers based on a model which predicts that the excess modulus $(E\text{-}E_0)/E_0$, where E and E_0 are the initial modulus of filled and unfilled rubber, respectively, scales linearly with the filler volume fraction below the percolation threshold, and with a power law with exponent 4 for higher amounts of filler. Wang et al. [132] studied the variation of the tensile strength of elastomers filled with nanofillers (CB, zinc oxide) with varying filler content within a low loading range. They showed that as the loading of nanofillers increases, the tensile strength of rubber nanocomposites (SBR, EPDM) firstly increases slowly, then increases abruptly, and finally levels off. Moreover, a shift of the percolation point to the higher the filler content was observed when the filler particle size was increased. This work proposed the concept of a critical particle–particle distance to explain the observed percolation phenomenon. Furthermore, they suggested that rubber strengthening through nanoparticles is attributed to the formation of stretched straight polymer chains between neighbor particles, induced by the slippage of adsorbed polymer chains on the filler surface during tension. Therefore, even though chemical bonding and the filler–rubber interfacial interactions are crucial for rubber reinforcement, partial physical absorption of rubber chains on the filler surface is also necessary to facilitate the formation of stretched straight chains. Le Strat et al. [133] showed that at higher volume fraction of silica in cross-linked poly(dimethylsiloxane) nanocomposites, a percolation phenomenon occurred with strong interactions leading to a high reinforcement level which is driven by interparticle contacts and aggregate morphology. Above the percolation threshold, polymer chains are trapped inside the lattice spacing of an infinite percolation network. So, a certain amount of these segments are "occluded" within the filler network voids, and partially shielded from the mechanical solicitation. In this case, strong interparticle interactions allow the transmitting of mechanical forces through the filler network. It was discussed that the percolation threshold, Φc, is governed by the shape of the aggregates, and above Φc, the level of the nanocomposite mechanical reinforcement is only governed by the rigidity of the network.

3.3.2 Reinforcing mechanism under small deformation

Linear viscoelastic response can develop at small strain, but nonlinear behavior of the filled compounds can be observed at a certain strain threshold. Either static or dynamic measurements can be applied to study this behavior. The reinforcement of filler in the rubber matrix may be simply described by the equations of Guth–Gold and Guth–Smallwoood as previously shown in Eqs. (3.1) and (3.2). However, it should be noted that these relationships failed to predict the high level of reinforcement observed at low filler concentration.

Gerspacher et al. [134] proposed that a reinforcement at low strains ($<6\%$) is due to the formation of a filler network that interpenetrates the matrix polymer network. Below this strain level, energy dissipation was attributed to the breakdown and reformation of the filler network, while energy storage occurred in the polymer network. They proposed a new mechanism for the strain above the threshold value, i.e., under this condition the filler network is separated into "subnetworks," each of which is composed of many aggregate particles, and there is no longer a presence of "percolated structure within the polymer matrix." If the strain is further increased, a further subdivision into smaller subnetworks occurs. It can be viewed such that the filler network and percolated structure is present at low strain, but network breakdown occurs at higher strain. This is in line with the concept of the Payne effect originally proposed for particulate fillers which demonstrated that the network of aggregates or filler–filler network could easily be ruptured even at small strain causing a reduction of storage modulus [23].

However, different points of view to the Payne effect have been addressed such as by Maier and Goritz [118] and Sternstein and coworkers [53,119,135] who propose that the reduction of modulus when strain is increased is due to the release of polymer chains from the filler surface, and that removal of the strain leads to chain readsorption and recovery in the modulus. Later, Sternstein et al. [136] further suggest that the initial high reinforcement at very low strains is due to the presence of "trapped entanglements" or "filler-induced conformational restriction" in the vicinity of the filler surface. The latter is caused not only by its own polymer–surface bonds but also by other chains that are themselves bound to the same particle. This work proposed filler-induced conformation restriction (FICR) model based on the physics of the polymer matrix in the presence of filler, i.e., "sticky chains on solid filler surfaces have reduced and restricted conformational freedom leading to enhanced mean bonding lifetimes determined not only by the filler–matrix bond strength and temperature, but also by the frequency and magnitude of any externally imposed stress or strain." Based on their investigation using untreated fumed silica (NT200)-filled poly (vinyl acetate) at various filler loading and the neat polymer, the results showed that the filler addition increases the storage shear modulus, and promotes longer terminal relaxation times relative to those for the neat polymer. Furthermore, based on the FICR model, within the rubbery plateau region, reinforcement varies exponentially with filler concentration, and can be much larger than that predicted from theories proposed by Einstein and Guth–Gold equation.

3.3.3 Reinforcing mechanism under large deformation

At sufficiently high strains, the nonlinear viscoelastic properties are observed. The reduced elastic modulus above the threshold strain is termed as the Payne effect which is influenced by the filler network's reversible breakdown and rearrangement. At high strain, rubber chains between cross-links are extended, and stress-hardening is usually observed prior to the rupture. The presence of fillers causes an increase of modulus at high stains as well as at low strain. It has been generally considered that at higher strain, beyond filler network breakup, the reinforcement mechanism is primarily local strain amplification. At large strain, the stress is remarkably increased at a given strain while the hysteretic loss during deformation cycle is also increased. Based on studies by a combination of three different techniques: the uniaxial stretching mechanical test; in situ X-ray scattering measurements for chain segment orientation; and cross-link density analysis by NMR, it was reported that the mechanical response (i.e., an increase of shear modulus) at medium/large strains is essentially driven by strain amplification effects, while, in the linear regime, there is a strong additional reinforcement which is not related to the properties of the elastomer matrix [137]. Herewith, the strain amplification factor is expressed as the average strain of the composites over that in the pure matrix. The stress–strain investigation of MWCNTs-filled SBR nanocomposites by varying loading of the nanofillers also showed that as the content of MWCNTs increases, the stress values increase due to the reinforcement or a "strain amplification" effect in the presence of rigid filler [138]. At low concentrations the tubes exist more or less individually while at higher contents the coils of the tubes overlap. The coil–coil overlap may be thought of as analogous to the polymer chain entanglements.

In addition to filler–rubber interaction, the influence of nanoparticles on the rubber network or cross-links is also of importance to explain the improved properties of the rubber nanocomposites. Lopez-Manchado et al. [124] evaluated the influence of inorganic nanoparticles on cross-linking mechanism of elastomers by applying the tube model on equilibrium statistical mechanics, and the results showed that the addition of nanoparticles led to a formation of highly ordered structure with a huge amount of entanglements, wherein the polymer is nanoscopically confined. These physical links exhibit freedom of movement under stretching, but in a lower volume because of confinement. Therefore, network molecular parameters such as lateral tube dimensions or average molecular mass of the chains decrease in presence of nanoparticles. Herewith, it should be noted that the tube model takes into account the importance of topological constraints or entanglements along the contour of strands. Each individual chain is effectively confined in a determined volume with tube-like geometry, and within this tube, the polymer is free to explore all possible configurations parallel and perpendicular to the axis of the tube [139].

An example of the work based on the studies on the nanocomposites comprising of NR/5 phr Saponite (layered silicate) organically modified by octadecylammonium salt in comparison with the NR/5 phr CB (N234) [124], showed that the addition of nanofiller to NR significantly enhances the modulus. At low deformations, the elastic response of the material is contributed by both chemical cross-links and topological constraints or entanglements. When the network chains are stretched, the entanglement slippage takes place leading to a decrease in reduced stress. At higher elongations, the entanglements behave as effective cross-links by which its network density is increased by the addition of fillers, and the increase is more pronounced in the case of nanofiller. At low loading of filler (5 phr), only a few rubber chains remain free in nanoparticles-filled nanocomposite with layered silicate of a high aspect ratio under optimum dispersion, whereas the majority of the rubber chains remain free in the CB-filled elastomer in which the movement is restricted mainly in a zone at the filler/elastomer interface. Based on the tube model study, it was revealed that the rubber is nanoscopically confined forming a highly ordered and entangled structure due to the formation of intense nanoparticle/elastomer interactions, and this structure is responsible for the high performance of these materials even at low filler percentages in the compound.

The complications in nanofiller-reinforced elastomer are inevitably partly caused by cross-links. Under large strain, elastomer chains are subjected to very high local strains and/or an increased density of effectively active constraints (cross-links and entanglements) in the vicinity of filler particles, due to trapping effects associated with adsorption of chains on the filler surfaces [136,140,141]. A significant fraction of the chains might contribute to modulus enhancement at large strain amplitude due to finite chain extensibility [53]. Desorption/absorption of polymer chains at filler surface has been also applied to explain nonlinear behaviors at high strain [118]. Therefore, the response of the nanocomposites under load has contributions from the matrix, filler, and interfaces.

3.4 Conclusions

Reinforcement of rubber by nanofillers attracts numerous interests because of their potential to give not only superior mechanical properties, but also some functional properties to the rubber. The rubber nanocomposites are heterogeneous systems, in which the interfaces, interactions between the filler–filler and filler–rubber, as well as the dynamics of the rubber chains that are in contact or in proximity of the filler surface have a role on the reinforcement mechanism. The reinforcement efficiency of the most commonly used particulate fillers mainly depends on particle size and specific area, structure, filler–rubber and filler–filler interactions, as well as dispersion level. The Payne concept has been widely used to describe the reinforcement effect in the presence of particulate fillers in which the strain-dependent contribution from filler–filler interactions, and the strain-independent filler–rubber interactions contribute greatly to the increase of modulus. In association to this filler–rubber interaction, several models including bound rubber and immobilized or constrained rubber chains have been proposed to describe the reinforcement mechanism. For tubular fillers with high aspect ratio that enables a very low percolation threshold, reinforcement efficiency depends on the distributions of the type, diameter, and length of the nanotubes. In this system in which the reinforcement mechanism of the nanotubes in rubber is mainly attributed to physical entanglement and networking effect; large aspect ratio, good dispersion, alignment, and efficient interfacial stress transfer are required. For the layered filler, such as nanoclay, delamination to form exfoliated or intercalated structures in the rubber matrix is crucial. The significant increase in reinforcement efficiency will be achieved if the layers are fully exfoliated and load can be effectively transferred between the rubber and filler. The reinforcement efficiency will be also affected by the volume of polymer matrix constrained in the proximity of the particles which depends on filler–rubber interactions. Physical entanglements of the rubber chains that adsorb on the large interfacial area of nanofillers also contribute to a number of cross-link junctions and thus the modulus of the materials.

Despite their different geometry, the addition of nanofillers into the rubber matrix have some common effects, i.e., there are the changes in the matrix properties close or in the proximity to the filler surfaces; the increase of stiffness in nanocomposites is associated to strong filler–matrix interactions, and hence effective load transfer at the interface; filler network is formed above the critical concentration or percolation threshold; and there is the presence of constrained polymer chains or glassy layers around the nanoparticles. At small deformation where linear viscoelastic behavior is observed, filler networks play a major role on reinforcement. There are two main different views with regard to the cause of the reduction in modulus with increasing strain; i.e., a breakdown and reformation of filler network, and the release of polymer chains from the filler surface (i.e., deadsorption vs readsorption of the polymer chains). Whereas, under large deformation where the materials have nonlinear response, the strain amplification primarily effects the reinforcement, and desorption/absorption of polymer chains at the filler surface has been also applied to explain nonlinear behaviors at high strain.

So far, even though there is a lot of information available with good progress made concerning to a better understanding of reinforcement mechanism, not a single model is capable of convincingly describing the reinforcement mechanism yet. Further development of either new nanofillers or new polymer–nanofiller composites, and both experimental and theoretical studies, will lead to the emergence of new information that will help us to better understand the reinforcement process. Hopefully, with help from recent advanced technologies of characterization tools together with theoretical predictions, the precise reinforcement mechanism at the nanoscale in the nanocomposites will be eventually elucidated.

References

[1] Bhattacharya M, Maiti M, Bhowmick AK. Influence of different nanofillers and their dispersion methods on the properties of natural rubber nanocomposites. Rubber Chem Technol 2008;81:782–808.

[2] Maiti M, Bhattacharya M, Bhowmick AK. Elastomer nanocomposites. Rubber Chem Technol 2008;81:384–469.

[3] Tjong SC. Structural and mechanical properties of polymer nanocomposites. Mat Sci Eng R 2006;53(3–4):73–197.

[4] Jancar J, Douglas JF, Starr FW, Kumar SK, Cassagnau P, Lesser AJ, et al. Current issues in research on structure/property relationships in polymer nanocomposites. Polymer 2010;51(15):3321–43.

[5] Montes H, Chaussée T, Papon A, Lequeux F, Guy L. Particles in model filled rubber: dispersion and mechanical properties. Eur Phys J E 2010;31(3):263–8.

[6] Kroshefsky RD, Price JL, Mangaraj D. Role of compatibilization in polymer nanocomposites. Rubber Chem Technol 2009;82:340–68.

[7] Chabert E, Bornert M, Bourgeat-Lami E, Cavaillé JY, Dendievel R, Gauthier C, et al. Filler-filler interactions and viscoelastic behavior of polymer nanocomposites. Mat Sci Eng A 2004;381:320–30.

[8] Dalmas F, Cavaillé JY, Gauthier C, Chazeau L, Dendievel R. Viscoelastic behavior and electrical properties of flexible nanofiber filled polymer nanocomposites. Influence of processing conditions. Compos Sci Technol 2007;67(5):829–39.

[9] Oberdisse J, El Harrak A, Carrot G, Jestin J, Boué F. Structure and rheological properties of soft-hard nanocomposites: influence of aggregation and interfacial modification. Polymer 2005;46(17):6695–705.

[10] Tatou M, Genix AC, Imaz A, Forcada J, Banc A, Schweins R, et al. Reinforcement and polymer mobility in silica-latex nanocomposites with controlled aggregation. Macromolecules 2011;44(22):9029–39.

[11] Zhu Z, Thompson T, Wang SQ, von Meerwall ED, Halasa A. Investigating linear and nonlinear viscoelastic behavior using model silica-particle-filled polybutadiene. Macromolecules 2005;38(21):8816–24.

[12] Leblanc JL. Rubber-filler interactions and rheological properties in filled compounds. Prog Polym Sci 2002;27:627–87.

[13] Wolff S. Chemical aspects of rubber reinforcement by fillers. Rubber Chem Technol 1996;69(3):325–46.

[14] Meon W, Blume A, Luginsland HD, Uhrlandt S. Silica and silane. In: Rodgers B, editor. Rubber compounding chemistry and applications. New York, NY: Marcel Dekker Inc; 2004.

[15] Medalia AI. Filler aggregates and their effect on reinforcement. Rubber Chem Technol 1975;47(2):411–33.

[16] Wagner MP. Reinforcing silicas and silicates. Rubber Chem Technol 1976;49(3): 703–74.

[17] Li Y, Wang MJ, Zhang T, Zhang F, Fu X. Study on dispersion morphology of silica in rubber. Rubber Chem Technol 1994;67(4):693–9.

[18] Sircar AK, Voet A. Immobilization of elastomer at the carbon black particle surface. Rubber Chem Technol 1970;43(5):973–80.

[19] Wolff S, Wang AJ, Tan EH. Filler-elastomer interactions. Part VII. Study on bound rubber. Rubber Chem Technol 1993;66(2):163–77.

[20] Magee RW. Evaluation of the external surface area of carbon black by nitrogen adsorption. Rubber Chem Technol 1995;68(4):590–600.

[21] Wang MJ, Wolff S. Filler-elastomer interactions. Part V. Investigation of the surface energies of silane-modified silicas. Rubber Chem Technol 1992;65:715–35.

[22] Blume A. Analytical properties of silica–a key for understanding silica reinforcement. Kautsch Gummi Kunstst 2000;53(6):338.

[23] Payne AR. A note on conductivity and modulus of carbon black-loaded rubbers. J Appl Polym Sci 1965;9(3):1073.

[24] Coran AY, Donnet JB. A note on conductivity and modulus of carbon black-loaded rubbers. Rubber Chem Technol 1992;65(5):1016.

[25] Donnet JB. Black and white fillers and tire compound. Rubber Chem Technol 1998;71:323–41.

[26] Robertson CG, Lin CJ, Bogoslovov RB, Rackaitis P, Quinn JD, Roland CM. Flocculation, reinforcement and glass transition effects in silica-filled styrene butadiene rubber. Rubber Chem Technol 2011;84(4):507–19.

[27] Pliskin I, Tokita N. Bound rubber in elastomers: analysis of elastomer-filler interaction and its effect on viscosity and modulus of composite systems. J Appl Polym Sci 1972;16:473–92.

[28] Wolff S, Donnet JB. Characterization of fillers in vulcanizates according to the Einstein-Guth-Gold equation. Rubber Chem Technol 1990;63(1):32–45.

[29] Huber G, Vilgis TA. On the mechanism of hydrodynamic reinforcement in elastic composites. Macromolecules 2002;35:9204–10.

[30] Payne AR, Whittaker RE. Low strain dynamic properties of filled rubbers. Rubber Chem Technol 1971;44(2):440–78.

[31] Raos G, Moreno M, Elli S. Computational experiments on filled rubber viscoelasticity: what is the role of particle-particle interactions? Macromolecules 2006;39(19): 6744–51.

[32] Medalia AIJ. Morphology of aggregates. 6. Effective volume of aggregates of carbon black from electron microscopy. Application to vehicle absorption and to die swell of filled rubber. Colloid Interface Sci 1970;32(1):115.

[33] Kraus G. A carbon black structure-concentration equivalence principle. Application to stress-strain relationships of filled rubbers. Rubber Chem Technol 1971;44(1):199–213.

[34] Smit PPA. Glass transition in carbon black reinforced rubber. Rubber Chem Technol 1968;41(5):1194–202.

[35] Litvinov VM, Steeman PAM. EPDM-carbon black interactions and the reinforcement mechanisms, as studied by low resolution H-1 NMR. Macromolecules 1999;32(25): 8476–90.

[36] O'Brien J, Cashell E, Wardell GE, McBrierty VJ. NMR investigation of interaction between carbon black and cis-polybutadiene. Macromolecules 1976;9(4):653–60.

[37] Dannenberg EM. Bound rubber and carbon black reinforcement. Rubber Chem Technol 1986;59(3):512–24.

[38] Brennan JJ, Jermyn TE, Boonstra BB. Carbon black-polymer interaction: a measure of reinforcement. J Appl Polym Sci 1964;8(6):2687–706.

[39] Kraus G. Reinforcement of elastomers by carbon black. Rubber Chem Technol 1978;51 (2):297–321.
[40] Nishi T. Effect of solvent and carbon-black species on rubber-carbon black interactions studied by pulsed NMR. J Polym Sci Part B Polym Phys 1974;12(4):685–93.
[41] Kaufman S, Slichter WP, Davies DD. Nuclear magnetic resonance study of rubber-carbon black interactions. J Polym Sci Part B Polym Phys 1971;9(5):829–40.
[42] Kenny JC, McBrierty VJ, Rigbi Z, Douglass DC. Carbon black filled natural rubber. 1. Structural investigations. Macromolecules 1991;24(2):436–43.
[43] Lüchow H, Breier E, Gronski W. Characterization of polymer adsorption on disordered filler surfaces by transversal ^{1}H NMR relaxation. Rubber Chem Technol 1997;70 (5):747–58.
[44] Yatsuyanagi F, Kaidou H, Ito M. Relationship between viscoelastic properties and characteristics of filler-gel in filled rubber system. Rubber Chem Technol 1999;72 (4):657–72.
[45] Luginsland HD, Frohlich J, Wehmeier A. Influence of different silanes on the reinforcement of silica-filled rubber compounds. Rubber Chem Technol 2002;75(4): 563–70.
[46] Majeste JC, Vincent F. A kinetic model for silica-filled rubber reinforcement. J Rheol 2015;59(2):405–27.
[47] Kaewsakul W, Sahakaro K, Dierkes WK, Noordermeer JWM. Optimization of rubber formulation for silica-reinforced natural rubber compounds. Rubber Chem Technol 2013;86(2):313–29.
[48] Kaewsakul W, Sahakaro K, Dierkes WK, Noordermeer JWM. Cooperative effects of epoxide functional groups on natural rubber and silane coupling agents on reinforcing efficiency of silica. Rubber Chem Technol 2014;87(2):291–310.
[49] Saramolee P, Sahakaro K, Lopattananon N, Dierkes WK, Noordermeer JWM. Comparative properties of silica- and carbon black-reinforced natural rubber in the presence of epoxidized low molecular weight polymer. Rubber Chem Technol 2014;87 (2):320–39.
[50] Sengloyluan K, Sahakaro K, Dierkes WK, Noordermeer JWM. Silica-reinforced tire tread compounds compatibilized by using epoxidized natural rubber. Eur Polym J 2014;51:69–79.
[51] Xu TW, Jia ZX, Luo YF, Jia DM, Peng Z. Interfacial interaction between the epoxidized natural rubber and silica in natural rubber/silica composites. Appl Surf Sci 2015;328:306–13.
[52] Qu LL, Yu GZ, Wang LL, Li CQ, Zhao QS, Li J. Effect of filler-elastomer interactions on the mechanical and nonlinear viscoelastic behaviors of chemically modified silica-reinforced solution-polymerized styrene butadiene rubber. J Appl Polym Sci 2012;126 (1):116–26.
[53] Sternstein SS, Zhu AJ. Reinforcement mechanism of nanofilled polymer melts as elucidated by nonlinear viscoelastic behavior. Macromolecules 2002;35(19):7262–73.
[54] Fukahori Y. New progress in the theory and model of carbon black reinforcement of elastomers. J Appl Polym Sci 2005;95(1):60–7.
[55] Merabia S, Sotta P, Long DR. A microscopic model for the reinforcement and the nonlinear behavior of filled elastomers and thermoplastic elastomers (Payne and Mullins Effects). Macromolecules 2008;41(21):8252–66.
[56] Litvinov VM, Orza RA, Kluppel M, van Duin M, Magusin PCMM. Rubber-filler interactions and network structure in relation to stress-strain behavior of vulcanized, carbon black filled EPDM. Macromolecules 2011;44(12):4887–900.

[57] Bindu P, Thomas S. Viscoelastic behavior and reinforcement mechanism in rubber nanocomposites in the vicinity of spherical nanoparticles. J Phys Chem B 2013;117 (41):12632–48.
[58] Berriot J, Montes H, Lequeux F, Long DR, Sotta P. Evidence for the shift of the glass transition near the particles in silica-filled elastomers. Macromolecules 2002;35(26):9756–62.
[59] Dequidt A, Long DR, Sotta P, Sanséau O. Mechanical properties of thin confined polymer films close to the glass transition in the linear regime of deformation: theory and simulations. Eur Phys J E 2012;35:61–83.
[60] Chen Y, Peng Z, Kong LX, Huang MF, Li PW. Natural rubber nanocomposite reinforced with nano silica. Polym Eng Sci 2008;48(9):1674–7.
[61] Huang ML, Tunnicliffe LB, Thomas AG, Busfield JJC. The glass transition, segmental relaxations and viscoelastic behaviour of particulate-reinforced natural rubber. Eur Polym J 2015;67:232–41.
[62] Dresselhaus MS, Lin YM, Rabin O, Jorio A, SouzaFilho AG, Pimenta MA, et al. Nanowires and nanotubes. Mater Sci Eng C 2003;23(1–2):129–40.
[63] Lu JP. Elastic properties of single and multilayered nanotubes. J Phys Chem Solids 1997;58(11):1649–52.
[64] Moniruzzaman M, Winey KI. Polymer nanocomposites containing carbon nanotubes. Macromolecules 2006;39(16):5194–205.
[65] Lau KT, Chipara M, Ling HY, Hui D. On the effective elastic moduli of carbon nanotubes for nanocomposite structures. Compos Part B 2004;35(2):95–101.
[66] Chazeau L, Gauthier C, Chenal JM. Mechanical properties of rubber nanocomposites: how, why ... and then? In: Thomas S, Stephen R, editors. Rubber nanocomposites: preparation, properties and applications. Singapore: John Wiley & Sons (Asia); 2000. p. 291–330.
[67] Mukhopadhyay K, Dwivedi CD, Mathur GN. Conversion of carbon nanotubes to carbon fibers by sonication. Carbon 2002;40(8):1373–6.
[68] Thorstenson ET, Ren Z, Chou TW. Advances in the science and technology of carbon nanotubes and their composites: a review. Compos Sci Technol 2001;61(13): 1899–912.
[69] Zhu J, Kim JD, Peng HQ, Margrave JL, Khabashesku VN, Barrara EV. Improving the dispersion and integration of single-walled carbon nanotubes in epoxy composites through functionalization. Nano Lett 2003;3(8):1107–13.
[70] Nakaramontri Y, Nakason C, Kummerlowe C, Vennemann N. Influence of modified natural rubber on properties of natural rubber-carbon nanotube composites. Rubber Chem Technol 2015;88(2):199–218.
[71] Rahmat M, Hubert P. Carbon nanotube–polymer interactions in nanocomposites: a review. Compos Sci Technol 2011;2(1):72–84.
[72] Bhattacharyya S, Sinturel C, Bahloul O, Saboungi ML, Thomas S, Salvetat JP. Improving reinforcement of natural rubber by networking of activated carbon nanotubes. Carbon 2008;46(7):1037–45.
[73] Nah C, Lim JY, Cho BH, Hong CK, Gent AN. Reinforcing rubber with carbon nanotubes. J Appl Polym Sci 2010;118(3):1574–81.
[74] Deng F, Ito M, Noguchi T, Wang LF, Ueki H, Niihara K, et al. Elucidation of the reinforcing mechanism in carbon nanotube/rubber nanocomposites. ACS Nano 2011;5(5): 3858–66.
[75] Stephen R, Thomas S. Nanocomposites: state of the art, new challenges and opportunities. In: Thomas S, Stephen R, editors. Ruber nanocomposites: preparation, properties and applications. Singapore: John Wiley & Sons (Asia); 2010. p. 1–20.

[76] Lan T, Kaviratna PD, Pinnavaia TJ. Mechanism of clay tactoid exfoliation in epoxy-clay nanocomposites. Chem Mater 1995;7(11):2144–50.
[77] Rooj S, Das A, Stöckelhuber KW, Mukhopadhyay N, Bhattacharyya AR, Jehnichen D, et al. Pre-intercalation of long chain fatty acid in the interlayer space of layered silicates and preparation of montmorillonite/natural rubber nanocomposites. Appl Clay Sci 2012;67–68:50–6.
[78] Rooj S, Das A, Stöckelhuber KW, Wang DY, Galiatsatos V, Heinrich G. Understanding the reinforcing behavior of expanded clay particles in natural rubber compounds. Soft Matter 2013;9(14):3798–808.
[79] Galimberti M, Cipolletti V, Musto S, Cioppa S, Peli G, Mauro M, et al. Recent advancements in rubber nanocomposites. Rubber Chem Technol 2014;87(3):417–42.
[80] Sengupta R, Chakraborty S, Bandyopadhyay S, Dasgupta S, Mukhopadhyay R, Auddy K, et al. A short review on rubber/clay nanocomposites with emphasis on mechanical properties. Polym Eng Sci 2007;47(11):1956–74.
[81] Halpin JC. Stiffness and expansion estimates for oriented short fiber composites. J Compos Mater 1969;3:732–44.
[82] Halpin JC, Kardos JL. Halpin-Tsai equations–review. Polym Eng Sci 1976;16(5):344–52.
[83] Rezende CA, Bragança FC, Doi TR, Lee LT, Galembeck F, Boué F. Natural rubber-clay nanocomposites: mechanical and structural properties. Polymer 2010;51(16):3644–52.
[84] Nielsen LE. Generalized equation for elastic moduli or composite materials. J Appl Phys 1970;41(11):4626–7.
[85] Lewis TB, Nielsen LE. Dynamic mechanical properties of particulate-filled composites. J Appl Polym Sci 1970;14(6):1449–61.
[86] Buxton GA, Balazs AC. Lattice spring model of filled polymers and nanocomposites. J Chem Phys 2002;117(16):7649–58.
[87] Ramorino G, Bignotti F, Pandini S, Riccò T. Mechanical reinforcement in natural rubber/organoclay nanocomposites. Compos Sci Technol 2009;69(7–8):1206–11.
[88] Carretero-González J, Retsos H, Verdejo R, Toki S, Hsiao BS, Giannelis EP, et al. Effect of nanoclay on natural rubber microstructure. Macromolecules 2008;41(18):6763–72.
[89] Vu YT, Mark JE, Pham HL, Engelhardt M. Clay nanolayer reinforcement of cis-1, 4-polyisoprene and epoxidized natural rubber. J Appl Polym Sci 2001;82(6):1391–403.
[90] Fu X, Huang GS, Xie ZT, Xing W. New insights into reinforcement mechanism of nanoclay-filled isoprene rubber during uniaxial deformation by in situ synchrotron X-ray diffraction. RSC Adv 2015;5(32):25171–82.
[91] Masa A, Iimori S, Saito R, Saito H, Sakai T, Kaesaman A, et al. Strain-induced crystallization behavior of phenolic resin crosslinked natural rubber/clay nanocomposites. J Appl Polym Sci 2015;132(39):42580 <http://dx.doi.org/10.1002/app.42580>.
[92] Kim HW, Abdala AA, Macosko CW. Graphene/polymer nanocomposites. Macromolecules 2010;43(16):6515–30.
[93] Lee C, Wei X, Kysar JW, Hone J. Measurement of the elastic properties and intrinsic strength of monolayer grapheme. Science 2008;321(5887):385–8.
[94] Young RJ, Kinloch IA, Gong L, Novoselov KS. The mechanics of graphene nanocomposites: a review. Compos Sci Technol 2012;72(12):1459–76.
[95] Gómez-Navarro C, Burghard M, Kern K. Elastic properties of chemically derived single graphene sheets. Nano Lett 2008;8(7):2045–9.

[96] Suk JW, Piner RD, An J, Ruoff RS. Mechanical properties of mono-layer graphene oxide. ACS Nano 2010;4(11):6557–64.
[97] Allen MJ, Tung VC, Kaner RB. Honeycomb carbon: a review of graphene. Chem Rev 2010;110(1):132–45.
[98] Gong L, Young RJ, Kinloch IA, Riaz I, Jalil R, Novoselov KS. Optimizing the reinforcement of polymer-based nanocomposites by graphene. ACS Nano 2012;6(3): 2086–95.
[99] Wu X, Lin TF, Tang ZH, Guo BC, Huang GS. Natural rubber/graphene oxide composites: effect of sheet size on mechanical properties and strain-induced crystallization behavior. eXPRESS Polym Lett 2015;9(8):672–85.
[100] Fu DH, Zhan YH, Yan N, Xia HS. A comparative investigation on strain induced crystallization for graphene and carbon nanotubes filled natural rubber composites. eXPRESS Polym Lett 2015;9(7):597–607.
[101] She XD, He CZ, Peng Z, Kong LX. Molecular-level dispersion of graphene into epoxidized natural rubber: morphology, interfacial interaction and mechanical reinforcement. Polymer 2014;55(26):803–10.
[102] Papageorgiou DG, Kinloch IA, Young RJ. Graphene/elastomer nanocomposites. Carbon 2015;95:460–84.
[103] Basu D, Das A, Stockelhuber KW, Wagenknecht U, Heinrich G. Advances in layered double hydroxide (ldh)-based elastomer composites. Prog Polym Sci 2014;39(3): 594–626.
[104] Chen X, Gug J, Sobkowicz MJ. Role of polymer/filler interactions in the linear viscoelasticity of poly (butylene succinate)/fumed silica nanocomposite. Compos Sci Technol 2014;95:8–15.
[105] Zhong B, Jia Z, Hu D, Luo Y, Jia D. Reinforcement and reinforcing mechanism of styrene–butadiene rubber by antioxidant-modified silica. Compos Part A 2015;78: 303–10.
[106] Tadiello L, D'Arienzo M, Di Credico B, Hanel T, Matejka L, Mauri M, et al. The filler–rubber interface in styrene butadiene nanocomposites with anisotropic silica particles: morphology and dynamic properties. Soft Matter 2015;11(20):4022–33.
[107] Tang ZH, Huang J, Wu XH, Guo BC, Zhang LQ, Liu F. Interface engineering toward promoting silanization by ionic liquid for high-performance rubber/silica composites. Ind Eng Chem Res 2015;54(43):10747–56.
[108] Fleck F, Froltsov V, Kluppel M. Polymer-filler interphase dynamics and reinforcement of elastomer nanocomposites. Soft Mater 2014;12:S121–34.
[109] Wang J, Jia H, Ding L, Xiong X, Gong X. The mechanism of carbon–silica dual phase filler modified by ionic liquid and its reinforcing on natural rubber. Polym Compos 2015. Available from: http://dx.doi.org/10.1002/pc.
[110] Sadhu S, Bhowmick AK. Effect of chain length of amine and nature and loading of clay on styrene-butadiene rubber-clay nanocomposites. Rubber Chem Technol 2003;76(4):860–75.
[111] Sookyung U, Nakason C, Thaijaroen W, Vennemann N. Influence of modifying agents of organoclay on properties of nanocomposites based on natural rubber. Polym Test 2014;33:48–56.
[112] Allahbakhsh A, Mazinani S. Influences of sodium dodecyl sulfate on vulcanization kinetics and mechanical performance of EPDM/graphene oxide nanocomposites. RSC Adv 2015;5(58):46694–704.
[113] Krause G. Reinforcement of elastomers. New York, NY: John Wiley & Sons; 1965.

[114] Wang MJ. Effect of polymer-filler and filler-filler interactions on dynamic properties of filled vulcanizates. Rubber Chem Technol 1998;71(3):520−89.
[115] Wang MJ. The role of filler networking in dynamic properties of filled rubber. Rubber Chem Technol 1999;72(2):430−48.
[116] Heinrich G, Kluppel M. Recent advances in the theory of filler networking in elastomers. Adv Polym Sci 2002;160:1−44.
[117] Funt JM. Dynamic testing and reinforcement of rubber. Rubber Chem Technol 1988;61(5):842−65.
[118] Maier PG, Goritz D. Molecular interpretation of the Payne effect. Kautsch Gummi Kunstst 1996;49(1):18−21.
[119] Chazeau L, Brown JD, Yanyo LC, Sternstein SS. Modulus recovery kinetics end other insights into the Payne effect for filled elastomers. Polym Compos 2000;21(2):202−22.
[120] Montes H, Lequeux F, Berriot J. Influence of the glass transition temperature gradient on the nonlinear viscoelastic behavior in reinforced elastomers. Macromolecules 2003;36(21):8107−18.
[121] Gauthier C, Reynaud E, Vassoilee R, Ladouce-Stelandre L. Analysis of the nonlinear viscoelastic behaviour of silica filled styrene butadiene rubber. Polymer 2004;45(8):2761−71.
[122] Liu J, Zhang LQ, Cao DP, Shen JX, Gao YY. Computational simulation of elastomer nanocomposites: current progress and future challenges. Rubber Chem Technol 2012;85(3):450−81.
[123] Gavrilov AA, Chertovich AV, Khalatur PG, Khokhlov AR. Study of the mechanisms of filler reinforcement in elastomer nanocomposites. Macromolecules 2014;47(15):5400−8.
[124] Lopez-Manchado MA, Valentin JL, Carretero J, Barroso F, Arroyo M. Rubber network in elastomer nanocomposites. Eur Polym J 2007;43(10):4143−50.
[125] Liu J, Wu SZ, Zhang LQ, Wang WC, Cao DP. Molecular dynamics simulation for insight into microscopic mechanism of polymer reinforcement. Phys Chem Chem Phys 2011;13(2):518−29.
[126] Zhu Z, Thompson T, Wang SQ, von Meerwall ED, Halasa A. Investigating linear and nonlinear viscoelastic behavior using model silica-particle-filled polybutadiene. Macromolecules 2005;38(21):8816−24.
[127] Heinrich G, Klüppel M, Vilgis TA. Reinforcement of elastomers. Curr Opin Solid State Mater Sci 2002;6(3):195−203.
[128] Klüppel M, Schuster RH, Heinrich G. Structure and properties of reinforcing fractal filler networks in elastomers. Rubber Chem Technol 1997;70(2):243−55.
[129] Payne AR. The dynamic properties of carbon black-loaded natural rubber vulcanizates. Part I. J Appl Polym Sci 1962;6(19):53−7.
[130] Cassagnau P. Melt rheology of organoclay and fumed silica nanocomposites. Polymer 2008;49(9):2183−96.
[131] Huber G, Vilgis TA. Universal properties of filled rubbers: mechanisms for reinforcement on different length scales. Kautsch Gummi Kunstst 1999;52(2):102−7.
[132] Wang ZH, Liu J, Wu SZ, Wang WC, Zhang LQ. Novel percolation phenomena and mechanism of strengthening elastomers by nanofillers. Phys Chem Chem Phys 2010;12(12):3014−30.
[133] Le Strat D, Dalmas F, Randriamahefa S, Jestin J, Wintgens V. Mechanical reinforcement in model elastomer nanocomposites with tuned microstructure and interactions. Polymer 2013;54(5):1466−79.

[134] Gerspacher M, O'Farrell CP, Yang HH. A proposed mechanism for the reinforcement of elastomers in the rubbery plateau by carbon black. Kautsch Gummi Kunstst 1994;47(5):349–53.
[135] Zhu AJ, Sternstein SS. Nonlinear viscoelasticity of nanofilled polymers: interface, chain statistics and properties recovery kinetics. Compos Sci Technol 2003;63(8): 1113–26.
[136] Sternstein SS, Amanuel S, Shofner ML. Reinforcement mechanisms in nanofilled polymer melts and elastomers. Rubber Chem Technol 2010;83(2):181–98.
[137] Pérez-Aparicio R, Vieyres A, Albouy PA, Sanséau O, Vanel L, Long DR, et al. Reinforcement in natural rubber elastomer nanocomposites: breakdown of entropic elasticity. Macromolecules 2013;46(22):8964–72.
[138] Peddini SK, Bosnyak CP, Henderson NM, Ellison CJ, Paul DR. Nanocomposites from styrene-butadiene rubber (SBR) and multiwall carbon nanotubes (MWCNT) part 2: mechanical properties. Polymer 2015;56:443–51.
[139] Edwards SF. Theory of rubber elasticity. Br Polym J 1977;9(2):140–3.
[140] Bogoslovov RB, Roland CM, Ellis AR, Randall AM, Robertson CG. Effect of silica nanoparticles on the local segmental dynamics in poly(vinyl acetate). Macromolecules 2008;41(4):1289–96.
[141] Fukahori Y. Generalized concept of the reinforcement of elastomers. Part 1: carbon-black reinforcement of rubbers. Rubber Chem Technol 2007;80(4):701–25.

Interphase characterization in rubber nanocomposites

4

K. Song
Massachusetts Institute of Technology, Cambridge, MA, United States

4.1 Introduction

Generally, for polymer composites filled with well-dispersed nano-size particles, four parameters, (i) volume fractions, (ii) particle dimensions, and (iii) orientations (if applicable especially for anisotropic materials), as well as (iv) polymer/filler interactions, determine the final bulk mechanical properties [1]. A balance of these parameters is needed to achieve stiffer and stronger properties in composites. For example, with an increase in nanoparticle concentrations, the load carrying capability improved; however, at the same time dispersion quality and interactions between polymer and fillers deteriorated due to less interfacial contact area. Aggregations induce localized stress concentrations, which led to defect initialization and crack propagation until failure. In such cases, it will limit the consistent increases in modulus and hardness parameters at high nanoparticle loadings. Under specific loadings and particulate filler dispersion and orientations, interactions between polymer and particles will mainly be determinant on the final property [2]. Both physical and chemical treatments have been applied to form stronger linkages, and the interfacial registry and their characterization as well as their influence on composite mechanics will be reviewed in this work. Before the overview of the characterization of the interphase/interface regions, it is necessary to consider the formation mechanisms of these bonding types.

4.1.1 Interdiffusion

Interdiffusion among molecular chains to form interphase regions is commonly seen in polymer materials (Fig. 4.1A). This can happen either within one polymer or between two polymers. The chain ends intend to entangle with one another and simultaneously promote the adhesion strength. One example is the semicrystalline polymer fibers. The entanglement between chains is also the reason why polymer fibers are formed: (i) during processing, molecular segments slide relative to each other and improve the mechanical properties, (ii) at the same time, the fiber will not break due to the entanglement, and (iii) the final fibrillar structure contains both extended and entangled polymer chains (Fig. 4.2). In rubber-based composites, the entanglements due to cross-linking are all through the composites and therefore the shared interface regions make up a great portion.

Progress in Rubber Nanocomposites. DOI: http://dx.doi.org/10.1016/B978-0-08-100409-8.00004-8

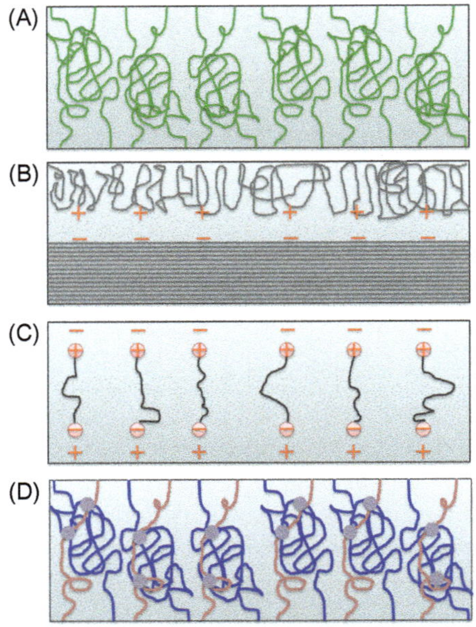

Figure 4.1 Interfacial formation mechanism of (A) polymer entanglement [3], (B) adsorption or wetting attraction including van der Waals [4], (C) polyelectrolyte or hydrogen bonding [5], (D) chemical reaction (i.e., cross-linking, and block-copolymerization or in situ polymerization as indicated by the chemical bonding, which are purple dots in the schematic) [6].

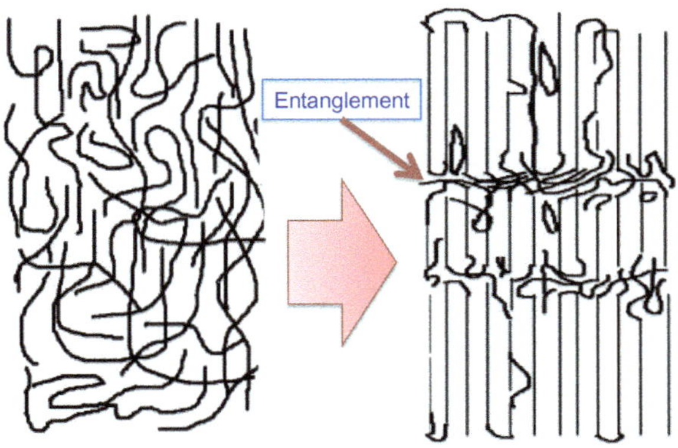

Figure 4.2 Interfusion of polymer chains caused entanglements.

4.1.2 Adsorption or wetting caused by van der Waals force

Adsorption or wetting happens when two surfaces are close enough and adhere to each other, primarily due to van der Waals force (Fig. 4.1B). The occurrence of wetting can be treated with thermodynamics (Fig. 4.3).

Figure 4.3 Contact angle θ and surface energy γ for a liquid drop on a solid surface based on thermodynamics model.

The surface energy or work of adhesion, W_a, can be expressed as Eq. (4.1). Based on the equation, wetting is favorable if interfacial surface energy, γ_{SL}, is small as compared to the sum surface energies between solid/vapor and liquid/vapor.

$$W_a = \gamma_{SV} + \gamma_{LV} - \gamma_{SL} \tag{4.1}$$

$$\gamma_{SV} = \gamma_{LV} \cos\theta + \gamma_{SL} \tag{4.2}$$

where S, L, and V stand for solid, liquid, and vapor, respectively.

To simplify the stable state, equilibrium along the horizontal direction can be established. The equilibrium wetting or contact angle, θ, is dictated by Young Equation, as shown in Eq. (4.2). Complete wetting can happen if solid surface energy, γ_{SV}, is greater than liquid surface energy, γ_{LV}, plus interfacial energy, γ_{SL}. Usually interfacial energy is smaller than either liquid or solid surface energy. However, for some polymer materials, such as polyethylene with surface energy of 31 mJ m^{-2} it is hard to wet the surface of glass (560 mJ m^{-2}) or graphite (70 mJ m^{-2}).

4.1.3 Electrolyte or hydrogen bonding

Surfaces carrying positive or negative electrical charges can be tailored for sustained adhesive force (Fig. 4.1C). In polymer composite fabrication, layer-by-layer (LBL) is an example demonstrating this mechanism (Fig. 4.4). During LBL deposition, a suitable growth substrate (usually charged) is dipped back and forth between dilute baths of positively and negatively charged polyelectrolyte solutions. During each dip a small amount of polyelectrolyte is adsorbed and the surface charge is reversed, allowing the gradual and controlled build-up of electrostatically cross-linked films of polycation–polyanion layers. Scientists have demonstrated thickness control of such films down to the single-nanometer scale. LBL films can also be constructed by substituting charged species such as nanoparticles or clay platelets in place of or in addition to one of the polyelectrolytes. LBL deposition has also been accomplished using hydrogen bonding instead of electrostatics [7].

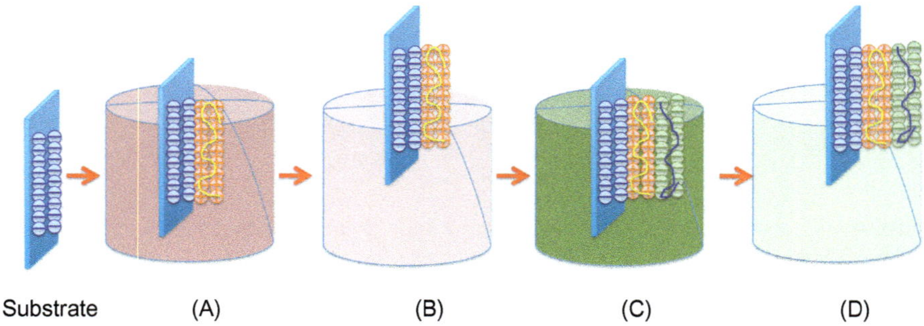

Figure 4.4 Layer-by-layer deposition of polyeletrolyte of alternative positive and negative charges. The washing procedure eliminates extra undesirable polymer layers. (A) Polycation, (B) Wash, (C) Polyation, (D) Wash.

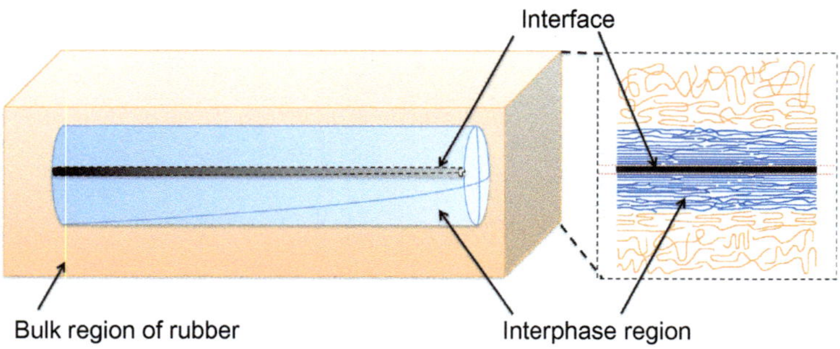

Figure 4.5 Schematic shows the bulk region of rubber, interface immediate to rubber and filler, and also interphase region close to filled particles.

4.1.4 Chemical reaction

Chemical reactions can happen at the interface or in the interphase (Fig. 4.1D). These chemical reactions can be cross-linking induced by an agent in rubber materials or polymerization in thermoplastics.

4.1.5 Interface and interphase (namely interfacial region)

Nanoscale particles are reported to induce template or crystallized polymers in the vicinity of an interface [8–13]. These newly formed regions are called interphase regions. The interphases have been found to exhibit properties more enhanced than the bulk polymer in the composite [14–16]. In mechanical and functional property estimations, the consideration of these interphases is significant to guarantee precision in theoretical property predictions or correlations [17–20]. The interface, interphase, and bulk regions are plotted in Fig. 4.5. It can be said that the interphase is the volume region between nanoparticles and polymer matrices. More often than not, the interphase properties are different from either the bulk or the filler phase [8,10,11,21].

4.1.6 Factors influencing interfacial interactions

Nanoscale particles are found to induce the most effective reinforcement in rubber and other polymer-based composites. This is often attributed to (i) interaction area per unit volume or weight, which can be more than 1000 times difference between nano- and microscale particles, (ii) surface chemistry. Inorganic particles and organic polymers are like water and oil. To increase the compatibility, physical or chemical treatments are used. Take carbon black (CB) for example. Physical treatment of heating at extreme temperature at around 2000°C can get rid of the functional groups on the surface through a carbonization procedure. It has been found in many research results that this treatment can greatly reduce the surface activity and as a result of this treatment mechanical, tribological, and thermal properties decreased. As a comparison, before heat treatment CB was found to bond with polymers due to the presence of carboxyl and other functional groups. Therefore, it is important to characterize the interfacial interactions both qualitatively and quantitatively.

4.2 Interphase characterization in rubber composites

Effective interfacial interactions are the basic requirements for achieving optimal reinforcement and low hysteresis in rubber matrices. Particles including CB, silica, layered silicates, talk, biofillers, carbon nanotubes (CNTs), and graphene have been extensively studied in both academia and industry. Weather interphase can be formed in the vicinity of the interface and how active the interface is will depend on particle features as well as physical or chemical treatments. For example, CNTs are observed to initiate polymer crystallizations based on the template, and clay can be intercalated so that polymers can fill among the many single sheets. Although most of the current literature is about CB and layered silicates filled rubber composites, this chapter will try to explain from a more general perspective, which is useful in rubber materials and practical in some other polymer composites. The following section provides information about the developments in interphase and interfacial interaction characterizations.

4.2.1 Mechanical characterization

As mentioned for polymer/nano–filler composites, mechanical enhancement could be achieved through three aspects: (i) well-dispersed fillers; (ii) preferential alignment of polymer chains/nano-fillers in line with mechanical loading direction; and (iii) effective interfacial interactions between matrix and fillers. To better understand how the interphase influences the final mechanical properties and to obtain their qualitative and/or quantitative characterization, a review on static mechanical tests and dynamic mechanical behavior, as well as atomic force measurements regarding their measurement mechanisms is provided here.

4.2.1.1 Static mechanical test (tension, compression, bending, indentation)

An understanding of composite mechanics will help estimation of interphase existence and their volume in rubber composites. A simple calculation method is based on the modified Rule-of-Mixture (ROM, Eq. (4.3)), namely, the Cox-Krenchel method (Eq. (4.4)), by taking filler length and orientation factor into consideration,

$$E_c = E_m V_m + E_f V_f \tag{4.3}$$

$$E_c = E_m V_m + \eta_o \eta_l E_f V_f \tag{4.4}$$

where E and V stand for the modulus and volume fraction for control polymer (i.e., E_m and V_m) and fibers (i.e., E_f and V_f). Here orientation factor, η_l, and length efficiency factor, η_o, were defined based on Krenchel's method [22] and shear lagtheory [23],

$$\eta_o = \frac{\int_0^{\frac{1}{2}\pi} I(\varphi)\cos^4\varphi d\varphi}{\int_0^{\frac{1}{2}\pi} I(\varphi) d\varphi} \tag{4.5}$$

It is often assumed that the distribution of rods, $I(\varphi)$ as a function of φ, can be described by a Gaussian or Lorentzian distribution. η_o is 0 for perpendicular distribution to loading axis, 1 for unidirectional along loading direction and 0.325 for randomly distributed particles. $I(\varphi)$ have been measured in polarized microscope, Raman [24,25], and X-ray diffraction [26–28], as well as scattering [29] techniques.

Researchers have studied effects of nanofiller loading variations and also orientation effects on mechanical properties in polymer composites. The relationship is also plotted in Fig. 4.6, assuming a polymer modulus of 5 GPa, nanofiller modulus of 40 GPa, and volume fraction/orientation factor of a range between 0 and 1. It can be observed that the optimal design of advanced composites is to improve both parameters of orientation and volume fractions.

For length efficiency factor,

$$\eta_l = 1 - \left[\frac{\tan h(na)}{na}\right] \tag{4.6}$$

$$n = \sqrt{\left[\frac{2G_m}{E_f \ln(2R/d)}\right]} \tag{4.7}$$

$$a = \frac{l}{d} \tag{4.8}$$

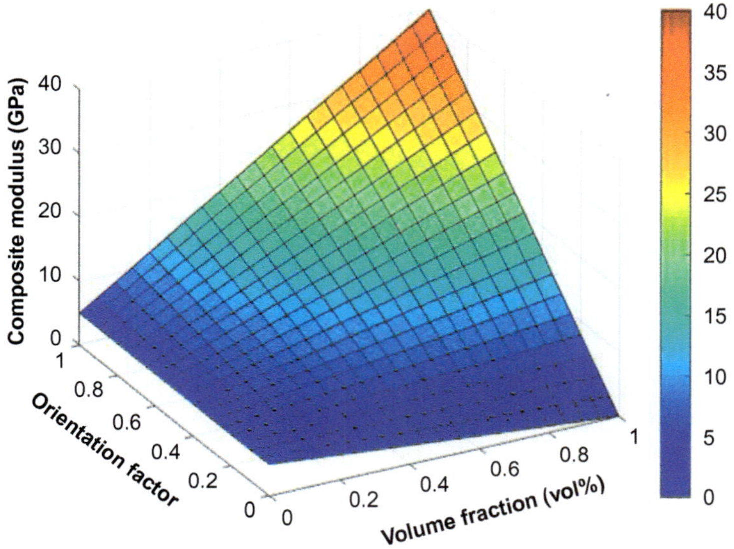

Figure 4.6 Composite modulus as a function of parameters of both orientation factor and volume fraction.

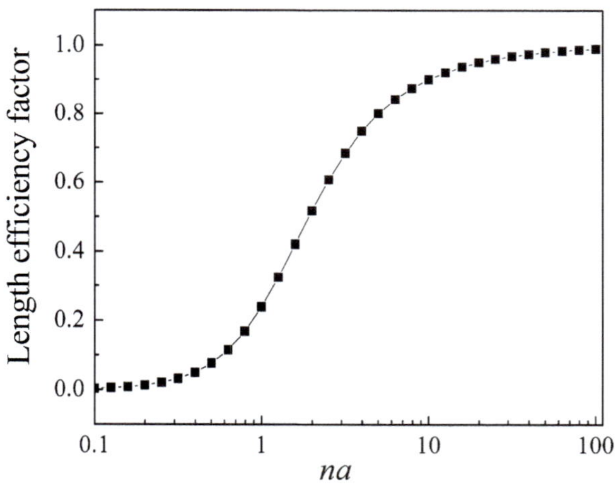

Figure 4.7 Length efficiency factor (η_l) as a function of parameter *na*.

where G_m is the shear modulus of polymer matrix, 2R is the distance from the fiber to its nearest neighbor fiber, and l and d are the length and diameter of the particle, respectively.

Researchers studied the effect of fiber length on the mechanical properties in polymer composites. At fixed fiber concentration with assumed even dispersion, length efficiency is only dependent on aspect ratio. Fig. 4.7 also shows clearly how the length efficiency factor changes with parameter *na*. Based on this curve from Eq. (4.6), e.g., shear modulus of 100 MPa, tensile modulus of 40 GPa, and concentration of 48 vol.% generate a value of 0.05 for n. A multiplication of aspect ratio of 20 provides a length efficiency of 100%. This is also consistent with the experimental observation that a lower aspect ratio of nanofiller will not reinforce the composites [13].

Strength can be treated in a similar way. When the thickness of polymer wrapping was not observed, the simplified form of strength in composites will become,

$$\sigma_c = \sigma_m V_m + \eta_o \eta_l \sigma_f V_f \tag{4.9}$$

$$\sigma_f = \frac{l_f}{2R} \sigma_{shear} \tag{4.10}$$

Another modified ROM based on the equation above accounts for the influence from length and orientation factors, as:

$$\sigma_c = \sigma_m V_m + \eta_o \eta_l \sigma_f V_f \approx \eta_o \eta_l \sigma_f V_f \tag{4.11}$$

This equation holds when the product between polymer strength and volume fraction is much lower than that from the filler contributions. Inserting the length efficiency factor, the equation above could be modified into [25]:

$$\sigma_c = \sigma_m V_m + \eta_o \eta_l \sigma_f V_f \approx \eta_o \eta_l \sigma_f V_f \approx \eta_o \eta_l \left(\frac{l_f}{2R} \sigma_{shear} \right) V_f \tag{4.12}$$

Likewise, if orientation efficiency factor and dimensional parameters were known, the relationship between composite strength and various loadings will be obtained, which is also the interfacial shear strength (IFSS) or the polymer shear yield strength, whichever is lower.

In a similar way, not only the shear strength of the interfacial phase, but also the modulus of this interphase could be determined. Double rule of mixture is the model developed to calculate the modulus taking into consideration the interfacial stiffness contributions. As shown in the equation [30]:

$$E_c = E_m V_m + \eta_o \eta_l E_f V_f + \left[\left(\frac{b}{r} \right)^2 + \frac{2b}{r} \right] V_f (E_c - E_a) \tag{4.13}$$

where the first two terms (same as Eq. 4.4) in the equation above give the standard ROM for nanoparticles dispersed in an amorphous polymer matrix. The last term in the equation describes the correction to the composite Young's modulus introduced by the presence of the ordered polymer phase. Differentiation of the equation could derive [30]:

$$\frac{dE_c}{dV_f} = (\eta_o \eta_l E_f - E_m) + \left[\left(\frac{b}{r} \right)^2 + \frac{2b}{r} \right] (E_c - E_a) \tag{4.14}$$

Normalized modulus and crystallinity combined with known amorphous polymer stiffness and orientation efficiency factor will provide the corresponding crystalline polymer stiffness and efficient particle stiffness contribution. Based on this model, the modulus of the polymer crystalline interphase in the polyvinyl alcohol (PVA)/CNT composite system was computed to be 46 ± 5.5 GPa, a much larger value than 1.92 ± 0.33 GPa, which is the amorphous polymer stiffness [30]. Although this model was derived from semicrystalline polymers, it is also applicable in any interphase including polymer composites.

In cases where composites with interphase regions are produced but not many different volume fractions are available, another modified ROM without considering the orientation factor or length factor will be [12]:

$$E_c = E_m V_m + E_i V_i + E_f V_f \tag{4.15}$$

where m, i, and f stand for matrix, interphase, and filler. If available, length efficiency factor (Eq. (4.6)) and orientation factor (Eq. (4.5)) can also be combined in this Eq. (4.15) for more precise estimation.

4.2.1.2 Rheology

The inclusion of nanoparticles influences the rheological properties of rubber compounds. The increased concentration of nanoparticles in rubbers improves the viscosity values based on the Guth–Gold equation [31],

$$\eta_c = \eta_m(1 + 2.5v_f + 14.1v_f^2) \tag{4.16}$$

where v_f is the volume fraction of nanoparticles. Similarly, it is found that Young's modulus in CB filled gum was governed by this rule,

$$E_c = E_m(1 + 2.5v_f + 14.1v_f^2) \tag{4.17}$$

When nano-sized particles are included, shape parameters, f, considering higher specific area than that in CB-based rubber materials were included in the equation,

$$E_c = E_m(1 + 0.67fv_f + 1.67f^2v_f^2) \tag{4.18}$$

In most cases, this shape parameter is taken as an aspect ratio. This equation is more valid for predicting modulus when polymer chains adsorb on nanoparticle surface.

4.2.1.3 Viscoelasticity

Particle filled rubber materials possess rubbery features, including viscoelasticity. This makes a clear evaluation of the rubber composite mechanics very difficult. The discussions above mainly focus on linear mechanical behavior usually within small strains. Other nonlinear mechanical behavior includes Mullins effect (stress softening under large-strain deformation) and the Payne effect (strain-dependent dynamic modulus within small strain).

The Payne effect refers to the observation that with the increase of deformation, the stiffness decreases. Stress–strain curves show lower slopes with strain increase, and more straightforwardly dynamic mechanical measurements display a lower plateau of storage modulus as deformation continues. In composites, the rubber molecules adsorb on the particle surface and become deadhesive under larger strains, therefore, a smaller slope in stress–strain curves are observed.

Mullins effect refers to the phenomenon that under large deformations, stiffness reaches the maximum for the first test compared to those in the following test at the same or even lower strain. This is associated with cross-linking bond breakage, strongly stretched molecules, and polymer chain shortening. Similar to the Payne effect, if glassy rubber forms on the particle surface, under the first large deformation this glassy layer will be irreversibly broken and lead to smaller stiffness in the following tests. It is important to consider the influence of these factors in analyzing the composite reinforcement and the role interfacial interaction plays.

4.2.1.4 Dynamic mechanical analysis (DMA) of tension, compression, and bending modes

Dynamic mechanical analysis (DMA) provides information regarding rheological/viscoelastic properties of polymeric materials as a function of temperature or frequency. The storage, (E'), and the loss (E'') dynamic moduli, as well as dampening parameter (tan δ) are monitored during the temperature sweep test. In contrast to traditional thermal analysis techniques for polymers (e.g., differential scanning calorimetry (DSC), thermalgravimetric analysis (TGA), and differential thermal analysis (TMA)), DMA provides more sensitive characterization of transitions associated with chain movements, molecular relaxations, and their relationship to the interactions between the matrix and filled fibers. The dynamic parameters were ominously influenced by the increase in fiber length, aspect ratio, and orientation as well as loading, but not in a geometric progression. Thus DMA is important toward understanding the characteristics for the confined polymer regions as well as other microstructural parameters such as morphology and the nature of constituents in polymer composites. The general dynamic mechanical properties of several polymer and polymer-based composite materials (i.e., PVA [9,12,13,19,21,26,32,33], epoxy [34], Poly(methyl methacrylate) (PMMA) [35], carbon fiber (CF) [35], primitive path (PP) [36]) have been previously investigated in detail, and reinforced effects of filled particles have been examined. The DMA analyses greatly depend on the composites compositions (i.e., fiber content, concentration, compatibilizer, plasticizer, additive, orientation) and the test mode (i.e., tension, compression, temperature sweep, frequency sweep).

To understand the dynamic mechanisms, physical meanings of measured parameters should be understood. The storage modulus relates to Young's' modulus.

This data is associated with stiffness of a material and determines how stiff or flexible the material is. It also reflects the material's capability to store energy applied on it for future mechanical response. Loss modulus relates to the viscous response to mechanical loadings and reflects the material's capability to dissipate energy. In some studies, researchers believe that the loss modulus is sensitive to internal segment motions, transitions, relaxations, morphology variations, or phase generations. Here in this section, the dampening parameter, the ratio between storage modulus and loss modulus, will be focused upon in relation to the microstructure in polymer/nanofiller composites. Usually the higher the dampening parameter in specific materials with unit phase, the greater the content of nonelastic components; however in multiphase composites, this data may be dealt with cautiously because of its complicated suggestions of phases.

The matrix–fiber interface (interfacial region) condition strongly affects the mechanical properties of composites and quite often its dampening level too. Interfacial bonding in fiber-reinforced composites can be considered to be weak, ideal, or strong. The ideal interface plays the role of transferring loads and does not contribute to dampening. Kubat et al. [37] proposed a parameter A (relative dampening at interface to that of matrix) and interfacial dampening ($\tan\delta$) to describe poor adhesion in composites, and Ziegel et al. [38] derived a parameter B to depict strong adhesion that does not contribute to dampening in composites, as shown in Eqs. (4.19) and (4.20). The property derived composition and interphase models will be presented in this section.

$$A = \frac{1}{1 - V_f}\frac{\tan\delta_c}{\tan\delta_m} - 1 \tag{4.19}$$

$$B = \left(\frac{\tan\delta_m}{\tan\delta_c} - 1\right)/1.5V_f \tag{4.20}$$

where $\tan\delta_m$ and $\tan\delta_c$ stand for dampening parameters in control and composites, respectively, and V_f means the fiber concentration.

Two-phase model

The interfacial interaction between the filler and the polymer matrix in a two-phase model can be studied by correlating the dampening parameter and interaction efficiency. The theoretical estimation has been observed experimentally.

$$\frac{\tan\delta_{\text{control}}}{\tan\delta_{\text{composite}}} = 1 + 1.5V_f B \tag{4.21}$$

$$B = \left(1 + \frac{\Delta R}{R_0}\right)^3 \tag{4.22}$$

where the term $B = \left(1 + \frac{\Delta R}{R_0}\right)^3$ is the relative increase in particle diameter for spheres, V_f represents the volume fraction of the fillers, B is the intercalation parameter describing the interfacial adhesion between the fillers and the polymer matrix quantitatively. A larger value of B means a better interaction between the two phases. For particles without tubular shape or spherical dimensions, or even where the particle dimensions are unknown, studying the interaction parameter will be more convenient.

This phenomenon has also been taken advantage of by creating hydroxylation functional groups on fibers (i.e., halloysite tubes, CNTs, boron nitride tubes, glass fibers, CFs, as well as natural fibers) to further reinforce the epoxy nanocomposites [34]. The calculated interaction parameters for composites without and with hydroxylation are listed in Table 4.1. The low and negative value in epoxy/p-HNTs (halloysite nanotubes) indicates a weakly physical interaction between the two phases. However, there is a higher B value in epoxy/h-HNTs, which is ascribed to the hydroxylation group density on the external surface of h-HNTs and this leads to better stiffening and toughening effects. To see application of this method, rubber composites including various filler particles and alternative rubber matrices are discussed as follows.

Carbon fillers Carbon nanofiller reinforced composites are applicable in this analysis. Dampening parameters and interaction parameters were recalculated and are listed in Table 4.1. Traditional particles were also compared to nanoparticles regarding their interaction efficiency with rubbers. CB and single-wall carbon nantotube (SWNT) were incorporated into rubber, both of which increased the glass transition temperatures and lowered the dampening parameters as shown in Table 4.1 [39]. As compared to CB, SWNT showed a higher value of B suggesting a better interaction between rubber and SWNT.

Graphene related materials including graphite and graphene, like CNTs, are also hot research areas due to their nanostructure-induced high performance in rubber composites. In research, graphene oxide (GO)/styrene-butadiene rubber (SBR) composites with complete exfoliation of GO sheets were prepared by aqueous-phase mixing of GO colloid with SBR latex, followed by cocoagulation procedures [40]. The results demonstrated that GO/SBR composites with 2.0 vol.% GO is comparable with those of the SBR composites reinforced with 13.1 vol% CB, with a low mass density and a good gas barrier ability to boot. Compared to most traditional fillers with dampening parameters ranging from 0 to 2, these GO-compounded rubber composites showed a value as high as 56 [40]. This was attributed to the effective interaction due to fully exfoliated GO sheets.

In another study based on an SBR matrix, multiwall carbon nanotube (MWNT), graphene platelets, and expanded graphene were compared regarding their reinforcement efficiency [41]. Graphene showed higher interaction parameters than in SWNT-filled SBR. Between graphene sheets and expanded graphene, better interfacial interactions were found in the latter case, when both are 20 phr.

Inorganic fillers Inorganic clay particles are known for their low cost, effective reinforcement, and compatibility with rubber matrices after appropriate treatments.

Table 4.1 **Dampening parameter of various particle-filled composites**

Samples	Concentration (wt%)	Dampening parameter (tan δ)	*B*
Pure epoxy [34]	0	0.82	N/A
Epoxy/p-HNTs	4.8	0.84	−0.65
Epoxy/h-HNTs	4.8	0.76	1.78
Carbon fillers			
NR [39]	0	1.7	
NR/carbon black	10	1.5	0.89
NR/SWNT	10	1.35	1.73
NR/carbon black	40	1.1	1.21
SBR [40]	0	1.6	
SBR/GO	0.2	1.4	47.62
SBR/GO	0.4	1.2	55.56
SBR/GO	1.2	0.8	55.56
SBR/GO	2.0	0.7	42.86
SBR [41]	0	1.75	
SBR/MWNT	10	1.55	0.86
SBR/graphene platelets	20	1.3	2.31
SBR/expanded graphene	20	1.1	3.94
Inorganic fillers			
SBR [42]	0	1.5	
SBR/MMT	1.86	1.4	2.56
SBR/MMT	4.52	1.3	2.27
SBR/MMT	8.7	1.2	1.92
SBR/MMT	12.45	1.0	2.68
SBR/MMT	15.94	0.8	3.66
Fluoroelastomer [43]	0	1.11	
Fluoroelastomer/HNT	5	1.1	0.22
Fluoroelastomer/HNT	10	0.9	1.48
Fluoroelastomer/HNT	15	0.9	0.99
Fluoroelastomer/HNT	20	1.1	0.00
Latex [44]	0	2.3	
Latex/silica	10	2.1	0.63
Latex/silica	20	1.8	0.93
Latex/silica	30	1.6	0.97
Latex/silica/silane	30	1.5	1.19
Bio-fillers			
SBR [45]	0	1.25	
SBR/cellulose	3	1.2	5.56
SBR/cellulose	5	1.2	3.33

(*Continued*)

Table 4.1 **(Continued)**

Samples	Concentration (wt%)	Dampening parameter (tan δ)	*B*
SBR/cellulose	10	1.05	2.86
SBR/cellulose	15	1.0	2.22
NR [46]	0	2.0	
NR/husk	10	2.0	0.00
NR/husk	20	2.0	0.00
NR/husk	30	1.7	0.39
NR/husk	40	1.6	0.42
NR/husk	50	1.6	0.33
NR@10 Hz [47]	*0*	*0.075*	*N/A*
NR/Coir fiber@10 Hz	*30*	*0.170*	*− 1.24*
NR@1 Hz	*0*	*0.15*	*N/A*
NR/Coir fiber@1 Hz	*60*	*0.19*	*− 0.23*
NR@10 Hz	*0*	*0.075*	*N/A*
NR/Coir fiber@10 Hz	*60*	*0.26*	*− 0.47*
NR@1 Hz	*0*	*0.15*	*N/A*
NR/Coir fiber@1 Hz	*10*	*0.11*	*0.40*
NR/Coir fiber@1 Hz	*20*	*0.12*	*0.28*
NR/Coir fiber@1 Hz	*60*	*0.20*	*− 0.28*
NR/Coir fiber_B@0.1 Hz	30	0.25	−0.89
NR/Coir fiber_S@0.1 Hz	30	0.13	0.34
NR/Coir fiber_J@0.1 Hz	30	0.22	−0.71
NR/Coir fiber_T@0.1 Hz	30	0.32	−1.18
NR [48]	0	0.17	
NR/NiZn Ferrite	4	0.16	1.04
NR/NiZn Ferrite	8	0.14	1.79
NR/NiZn Ferrite	12	0.13	1.71
Hybrid composites including more than one filler			
SBR [49]		2.0	
SBR/nanoclay	5	1.7	2.35
SBR/nanoclay	10	1.5	2.22
SBR/nanoclay	15	1.5	1.48
SBR/nanoclay	20	1.2	2.22
SBR/nanoclay/carbon black	5	1.2	8.89
SBR/nanoclay/carbon black	10	1.1	5.45
SBR/nanoclay/carbon black	15	1.0	4.44
SBR/nanoclay/carbon black	20	0.8	5.00
NR [50]	0	2.1	
NR/sisal fiber/oil palm fiber	0.16	0.965	3.369
NR/sisal fiber/oil palm fiber	0.15	0.95	3.607
NR/sisal fiber/oil palm fiber	0.15	0.7834	4.17
NR/sisal fiber/oil palm fiber	0.15	0.75	4.28

Note: p-HNTs were prepared using H_2O_2 to remove organic impurities while h-HNTs were further treated using NaOH to form hydroxylation functional groups on HNT surfaces. The pure and composite films were prepared using the cast mold method [34]. Similarly, coir fibers were treated using no chemicals (*B*), sodium hydroxide/NR solution/toluene diisocyanate solution (*J*), resorcinol-formaldehyde-latex (*S*), and bleaching solutions of sodium silicate and hydrogen peroxide (*T*) in water solutions [47].

Montmorillonite (MMT) was incorporated into NBR through a melt intercalation process. The increases in tensile tests and dynamic mechanical measurements are consistent, which were due to the nanoscale effects after intercalation and strong interactions between the NBR matrix and the clay interface [42].

HNT, a tubular structure resembling CNT, has the same chemical composition to kaolin and possesses high stiffness and hardness values. A bis-phenol curing system was used to fabricate the HNT-included elastomer composites. Both mechanical and dynamic mechanical properties were improved [43]. The interaction parameters were found to increase with high concentrations up to 15 wt%. Higher loadings lowered this value due to the reduced curing efficiency of the bis-phenol curing system. As a result, both the interaction at the interface and the physical properties became inferior [43].

Silica particles, another traditionally used filler in the rubber industry, have been extensively studied. For example, both precipitated silica and fused silica have been mixed with latex for mechanical and dynamical studies [44]. The continual increase up to 30 phr was observed and attributed to a breakdown of filler transient networks and loss of trapped rubber chains in the filler structure. During dynamic tests, the higher loading generated a denser network and thus increased the energy dissipation capability. In addition, silane treatment provided a stronger filler–polymer bonding and therefore decreased the dampening parameter more than that without surface treatment at the same loading of silica [44].

Biofillers Biofillers have been used in rubber matrices for reinforcement and cost-efficiency. Cellulose is one kind of stiff and strong filler potentially used in the composite industry. It has shown a depressing effect in dampening parameter peaks in composites [45]. Concentrations up to 15 phr from rubber matrix displayed a continuous decrease in the dampening parameter, showing a good dispersion quality and effective interaction efficiency. Rice husk is one material containing some cellulose materials and can be directly put into rubber-based materials [46]. The interaction efficiency was also compared to commercial particles of CB and silica. The results showed similar levels of dampening parameters.

It can be also shown [47] that the untreated coir fiber showed better interaction with rubbers when concentration was below 30 wt%, beyond which the interaction parameters all went down to negative values. Also, a lower concentration of 10 wt % displayed higher interaction efficiency, mainly due to better distribution of nanofillers within the matrix, as compared to composites of 20 wt% loadings. Chemical treatments showed different effects on mechanical performance. An appropriate coupling agent (i.e., resorcinol–formaldehyde-latex (RFL)) was found to increase the rubber/coir fiber compatibility and therefore increased its storage modulus, loss modulus, and thermal stabilities.

Other micro and nanoparticles Magnetic nanoparticles have been attractive in electrical data storage devices and their assistance in retarding the energy dissipation has been studied in rubber composites. NiZn Ferrite nanoparticles decreased the dampening parameters in natural rubber [48]. This was attributed to the lower degree of their mobility in polymeric chains. Higher nanoparticle concentrations further decreased the peak of dampening parameters. However, the interaction parameter B was not necessarily increasing proportionally with concentrations. This could be due to the possible aggregates that leads to a less effective contact area between polymers and particles. As a result, the higher loading composites displayed better capacity in energy dissipation but interaction parameter B decreased a little bit from 1.79 (8% loading) to 1.71 (12% loading) [48].

Hybrid composites containing more than one particle Usage of more than one kind of particle in rubber has been studied extensively due to the many choices of various fillers. The different dimensional features at the micro- and nanoscale generate synergistic effects and will improve mechanical as well as other properties. In one study, CB, belonging to traditional rubber fillers, was incorporated together with nanoclay particles [49]. The NBR/nanoclay black composites were also fabricated as comparisons. The increase of nanoclay consistently decreased the dampening parameter peaks. It was also found the combination of CB with nanoclay with a CB concentration of 20 phr showed much higher interaction parameters than those without CB. Other electron microscopy and spectra analyses also confirmed the unique intercalated clay structure and properties brought up by these two particles [49].

Biofibers of two or more kinds can also be used together. Natural rubber was reinforced with sisal and oil palm fibers and subjected to DMA to determine the dynamic properties. The fibers were subjected to alkali treatment of different concentrations of namely 0.5, 1, 2, and 4% and the dynamic properties were also examined. The interfacial strength, proportional to the interaction parameters, increases with alkali treatment and maximum bond strength was found in fibers treated with 4% NaOH [50].

Three-phase model

PVA macromolecules have been found to possess a template growth on nanofiller surfaces [14,15,17,18, 51−53]. This newly formed interphase plays the role of stress-transfer between polymer matrix and nanofillers and can be assessed quantitatively by DMA. This fractional constrained region of PVA can be computed in the following manner. According to the linear viscoelastic behavior, the relationship between the energy loss fraction, W, and the dampening parameter, tan δ, can be given in Eq. (4.23). It is well known that loss of energy during dynamic tests on pure polymer fibers (i.e., control) mainly arises due to the presence of less constrained and amorphous/low-ordered regions. This phenomenon can be analyzed by examining the tan δ peak associated with the glass transition region for the polymer material. This peak amplitude diminishes and/or shifts as ordering in the polymer material increases. By assuming a proportional relationship between W and the fractions of amorphous/low-ordered regions (V_a) (Eq. (4.24)), the more constrained/ordered regions (V_c) can be calculated using Eqs. (4.24) and (4.25). The baseline portions pertaining to the amorphous and constrained regions are determined using the DMA tan δ data from the control fibers. Therefore, Eqs. (4.24) and (4.25) use the loss fraction from the control fiber ($W_{control}$), where $V_{a\_control}$ is the volume fraction of the unconstrained (lower ordered regions), $V_{c\_control}$ is the volume fraction of the more constrained regions (crystalline regions), and α is the proportionality constant.

$$W = \frac{\pi E''}{\pi E'' + E'} = \frac{\pi \tan \delta}{\pi \tan \delta + 1} \quad (4.23)$$

$$V_{a\_control} = \alpha W_{control} \quad (4.24)$$

$$V_{c\_control} = (1 - \alpha) W_{control} \quad (4.25)$$

It is recognized that for the nanocomposite fibers the ordered portion consists of both crystalline ($V_{c\_composite}$) and interphase (V_{m2}) regions. Therefore, a modification of the constrained regions in composites is included in Eqs. (4.26) and (4.27), where $W_{control}$ and $W_{composite}$ represent the energy loss fraction of the control and composite fiber samples, respectively. These loss fractions are measured from peak-amplitude of the glass-transition peak. Eq. (4.27) is derived from Eqs. (4.23)–(4.26) to determine the ratio $V_{m2}/V_{c\_control}$. This provides an indication for the presence of interphase regions in the composite samples, which go beyond the general semicrystalline polymer matrix structures present in the control fibers.

$$V_{c\_composite} = V_{c\_control} - V_{m2} = (1 - \alpha) W_{composite} \quad (4.26)$$

$$\frac{V_{m2}}{V_{c\_control}} = 1 - \frac{W_{composite}}{W_{control}} \quad (4.27)$$

Fig. 4.8 shows a plot for $V_{m2}/V_{c\_control}$ as a function of filler concentration, where all composite fibers show an increasing presence of interphase regions as compared to the control fibers. Comparison between the ROM analysis and these DMA measurements (Fig. 4.8B) indicate that the composite fibers exhibit regions of self-reinforcement (interphase), which enhance stress-transfer mechanisms in the materials.

The three-phase model described above has actually been studied theoretically before [54], the interphase is referred to as a distinct third phase between the fiber and bulk matrix with its own viscoelastic properties. Parametric studies indicate that the fiber aspect ratio, the elastic modulus, and the dampening properties of the interphase materials appear to be dominant parameters which significantly influence the dynamic stiffness and viscoelastic dampening properties of the three-phase composite system. The main drawback of the model is that the input data for the interphase, i.e., geometric and material properties of interphase, have been interpolated and no experimental values are available [54].

The two-phase model and three-phase model have also been studied theoretically in micromechanics models [55]. For various longitudinal, transverse, and shear loss factors in composite materials (epoxy/glass fiber composites), the percentage contribution of dampening from fiber, matrix, and interphase are predicted. The three-phase model suggested that in unidirectional laminate, poor adhesion

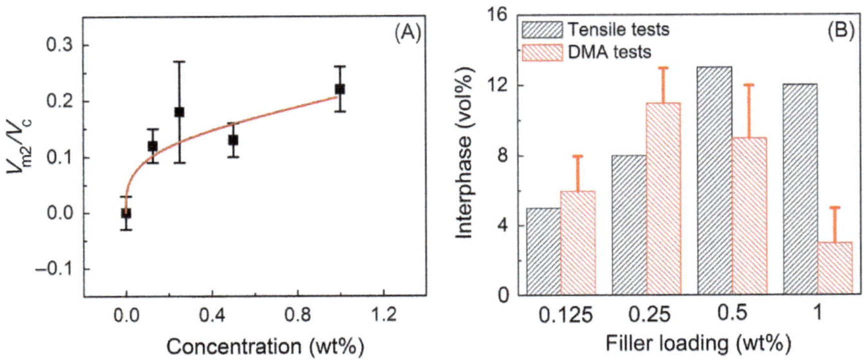

Figure 4.8 (A) Plot shows $V_{m2}/V_{c\_control}$ versus the filler loading (wt.%). (B) Plot of interphase volume fraction (V_{m2}) predictions for all composites using DMA analysis as compared with measurements from static tensile tests. Interphase growth is shown to increase with filler concentration.

Source: Permission obtained from Wiley Song K, Zhang Y, Minus ML. Polymer interphase self-reinforcement and strengthening mechanisms in low-loaded nanocomposite fibers. Macromol Chem Phys (2015);216:1313–20. http://dx.doi.org/0.1002/macp.201500011.

between matrix and fiber would lead to low elastic stiffness and high dampening. The method mentioned above derived the interphase in highly interphase occupied composites; however, for a normal quality interphase, more sensitive and accurate dampening measurement methods are needed to identify the contribution from interphase to the overall dampening parameter.

4.2.1.5 Atomic force microscopy (AFM)

The nature of the interfacial bonding affects the elastic and fracture properties of the composite in a number of ways. The interaction adhesion between polymers and nanofillers can be characterized using atomic force microscopy (AFM). Subsequently, the fractured surfaces with varying degrees of topography can be visualized by in situ scanning electron microscopy (SEM) or scanning probe microscopy (SPM) imaging. Bonding interactions between most particles and polymers were observed with pullout of fillers, in other words, shear debonding and sliding. These measurements have little or no attempt to change the normal stress across the interface but more often than not only use simple variants of the shear lag theory to interpret the measured data. Up to now, the shear-strength as a single fiber was pulled out from polymers is a standard for evaluating the interaction between these two phases. Research has shown that the shear strength ranges between polymers (i.e., epoxy, PMMA) and CNTs from tens to 632 MPa [56], and boron nitride from 125 to 728 MPa [57]. These measurements of shear strength also correspond to the normal stress levels in the composites. Although the single-fiber pullout test is simple for interpreting the preliminary failure mode in composites, caution should be taken when relating shear stress to normal stress, mainly due to the following reasons:

1. Roughness is a factor that cannot be excluded for promoting shear stress but not necessarily improving normal stress, which is exactly the one being measured in the fiber pullout test.
2. Manually mimicking fiber pullout by a "push-pull" method may be different from the real situation in composites, where at least the distribution of particles and exfoliation quality also matter in stress allocations.
3. Industrial processing techniques also determine several interfacial topographies that may be different from bench-scale mimicking.

For a single-fiber pullout test, the schematic and mechanics model is given in Fig. 4.9. The debonding shear stress can be approximately calculated as follows. The shear stress at radius of r_1 and r_2 at a given axial distance dx are equal (Eq. (4.28)).

$$2\pi r_1 \tau_1 dx = 2\pi r_2 \tau_2 dx \tag{4.28}$$

Similarly any shear stress along the radial direction, at radius of ρ can be related to shear stress at an interfacial region (Eq. (4.29)).

$$\tau = \tau_i \frac{r}{\rho} \tag{4.29}$$

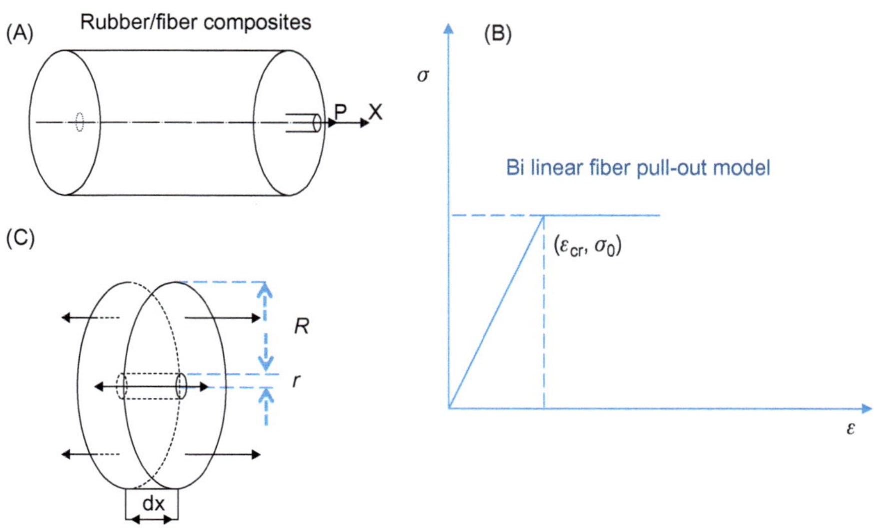

Figure 4.9 Schematic model and stress distribution as during the single-fiber pull-out.

The displacement of fiber, u, along fiber axis, derived at the movement along radial direction, ρ, defines the shear strain (Eq. (4.30)). Integration on both sides will generate Eqs. (4.32) and (4.33).

$$\frac{du}{d\rho} = \gamma = \frac{\tau}{G_\mathrm{m}} = \frac{\tau_\mathrm{i}\frac{r}{\rho}}{G_\mathrm{m}} = \frac{\tau_\mathrm{i}\frac{r}{\rho}}{\frac{E_\mathrm{m}}{2(1+\upsilon_\mathrm{m})}} = \frac{2(1+\upsilon_\mathrm{m})\tau_\mathrm{i}}{E_\mathrm{m}}\frac{r}{\rho} \tag{4.30}$$

$$\int_{u_\mathrm{r}}^{u_R} 1du = \frac{2(1+\upsilon_\mathrm{m})\tau_i r}{E_\mathrm{m}} \frac{\int_r^R 1d\rho}{\rho} \tag{4.31}$$

$$u_R - u_\mathrm{r} = \frac{2(1+\upsilon_\mathrm{m})\tau_\mathrm{i} r}{E_\mathrm{m}} \ln\frac{R}{r} \tag{4.32}$$

$$\tau_\mathrm{i} = \frac{(u_\mathrm{R} - u_\mathrm{r})\cdot E_\mathrm{m}}{2(1+\upsilon_\mathrm{m})r\,\ln\frac{R}{r}} \tag{4.33}$$

The interfacial shear stress keeps a balance with the tensile stress on fiber cross-section, therefore,

$$2\pi r dx \tau_\mathrm{i} = -\,\pi r^2 d\sigma_\mathrm{f} \tag{4.34}$$

$$\frac{d\sigma_\mathrm{f}}{dx} = \frac{-\,2\tau_\mathrm{i}}{r} \tag{4.35}$$

Substituting Eq. (4.35) into Eq. (4.34) then,

$$\frac{d\sigma_\mathrm{f}}{dx} = \frac{-2}{r}\cdot\frac{(u_\mathrm{R} - u_\mathrm{r})\cdot E_\mathrm{m}}{2(1+\upsilon_\mathrm{m})r\,\ln\frac{R}{r}} = \frac{-\,(u_\mathrm{R} - u_\mathrm{r})\cdot E_\mathrm{m}}{(1+\upsilon_\mathrm{m})r^2\cdot\ln\frac{R}{r}} \tag{4.36}$$

When the fibers are initiating the debonding procedure, the displacement conditions along the fiber axis are as follows. This corresponds to perfect bonding condition, i.e., shear strain at the interface is the same for both matrix and fiber. The shear strain on the infinity to fiber is zero as an unstrained state.

$$\frac{du_\mathrm{r}}{dx} = \varepsilon_\mathrm{f} \tag{4.37}$$

$$\frac{du_\mathrm{R}}{dx} = 0 \tag{4.38}$$

Derivations of Eq. (4.28) on both sides to x gives,

$$\frac{d^2\sigma_f}{dx^2} = \frac{n^2}{r^2}\sigma_f \tag{4.39}$$

where,

$$n^2 = \frac{E_m}{E_f(l + v_m)\ln\frac{R}{r}} \tag{4.40}$$

The stress boundary conditions for Eq. (4.28) will be,

$$\sigma_f(0) = \sigma_0 \tag{4.41}$$

$$\sigma_f(L) = 0 \tag{4.42}$$

Applying these stress boundary conditions in Eq. (4.28),

$$\sigma_f = \sigma_0 \left\{ \frac{\sin h[n(L-x)/r]}{\sin h(nL/r)} \right\} \tag{4.43}$$

substituting this Eq. (4.43) to (4.28), the shear stress can be obtained,

$$\begin{aligned} \tau_i &= -\frac{r}{2}\frac{d\sigma_f}{dx} \\ &= \frac{n\sigma_0}{2}\cos h\left[\frac{n(L-x)}{r}\right]\cos ech\left(\frac{nL}{r}\right) \end{aligned} \tag{4.44}$$

Especially when $x = 0$ the debonding shear stress can be deduced as,

$$\tau_i = \frac{n\sigma_0}{2}\cos th\left[\frac{n(L/r)}{r}\right] \tag{4.45}$$

Based on this derivation, and assuming,

1. a polymer with modulus $E_m = 5$ GPa and Poisson's ratio $\nu_m = 0.3$, fiber modulus $E_f = 500$ GPa, length of 500 nm, diameter of 10 nm, as well as a variant of volume fraction from 2.25 vol% to 76.31 vol%,
2. the shear stress of the composite will be one tenth of the composite modulus, and,
3. perfect alignment and bonding of fiber to polymer matrix,

the obtained shear stress at interface as a function of parameter n was plotted in Fig. 4.10. It can be observed that the interfacial shear stress measurement from experiments based on a number of polymers and fibers are within this range (Table 4.2).

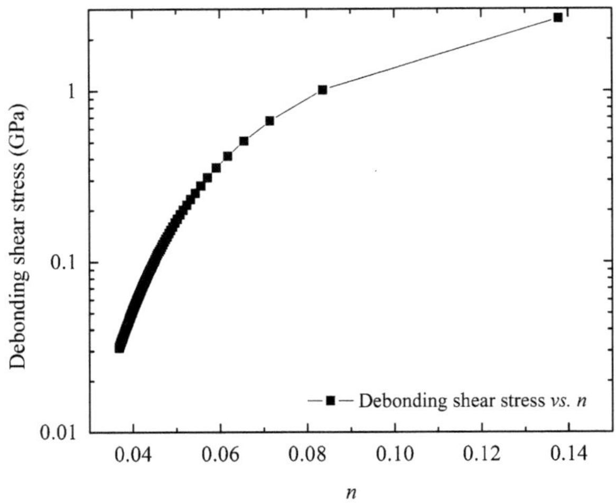

Figure 4.10 Debonding shear stress as a function of dimensionless constant, *n*, with a shear stress range from 2.64 GPa ($n = 0.14$) to 31 MPa ($n = 0.04$).

Similar to the determination of nanotube strength (Table 4.2), nanoparticle–polymer interfacial strength is of great importance since both of them are significant factors in determining the mechanical stiffness/strength/toughness/antidelamination/antibending of the bulk polymer composites. There are many kinds of nanoparticles and here CNTs are taken as an example. Due to the difficulties in manipulating nanometer-sized particles, the nanotubes have been determined previously via complicated methods using expensive equipment or specially designed devices. As shown in Table 4.2, the measured strength of CNT varies according to the CNT type, quality, and different measuring methods.

Actually before the proved feasibility of instrumental measurement, a body of computational work was carried out to predict the IFSS. Polymer systems such as PS [58], epoxy [59], PmPV, PPA [60] etc. were calculated using molecular dynamics, resulting in calculate polymer and CNT dependent IFSS ranges from 18 to 186 MPa. Later, Raman spectroscopy was used to detect the shift of the D' band (around 2662 cm^{-1}) in frequency under tension. The relative ratio between the shift under special strain and modulus has been studied [61].

The techniques and devices (SEM [62], transmission electron microscopy (TEM) [63], AFM [64,65], SPM [66]) used to tune CNT can thus be extended to the case of interfacial area in CNT-strengthened polymer composites. However, another challenge in characterizing the specialized interfacial area is that in most cases, the embedded CNT in polymers cannot be easily extracted after processing due to the presence/thickness of the polymer. Large amounts of fracture tests will be required before a satisfying sample is obtained. A trial of manually embedding a single tube or tube bundles has been conducted; subsequent pullout tests are feasible to measure the shear-strength directly. Nonetheless, the real interaction between CNT and polymer bulk in the real interface may be different due to the transformed

Table 4.2 Reported interfacial tensile strength of CNT and shear strength properties of polymer/CNT composite fibers

Author	Sample	Characterization	Strength (GPa)	Remark
Nanofillers				
Yu et al. [62]	MWNT	EM	11–63	Stress-strain measurements on individual arc-MWNT inside an EM
Demczyk et al. [71]	MWNT	TEM	150	Pulling and bending test on individual acr-MWNT in situ TEM; force-displacement curves give tensile strength
Xie et al. [72]	MWNT	Specially designed stress–strain puller	3.6	CVD synthesized MWNT were pulled using puller with special design
Salvetat et al. [73]	SWNT	AFM	N/A	Small diameter SWNT bundles was conducted by AFM using bending; shear modulus between 0.7–6.5 GPa and elastic modulus 0.1–1.3 TPa were obtained
Polymer/nanofiller interphase				
Wagner et al. [63]	Acr-discharge synthesized Single WMNT/polymer	TEM	Up to 0.5	Examined the fragmentation of WMNT in polymer showed one order of magnitude larger stress transfer than that from traditional CF in polymer composites
Cooper et al. [66]	MWNT/epoxy	SPM	0.04–0.38	Carbon nanotubes bridging across holes in an epoxy matrix have been drawn out using the tip of a scanning probe microscope while recording the forces involved.
	SWNT ropes/epoxy		~0.35	

(*Continued*)

Table 4.2 (Continued)

Author	Sample	Characterization	Strength (GPa)	Remark
Barber et al. [64]	MWNT/polyethylene–butene matrix	AFM	0.047	Reproducible nanopullout experiments measured remarkably high separation stress, indicated the effective strengthening, especially in close vicinity of the carbon nanotube.
Barber et al. [65]	MWNT/polyethylene–butene matrix	AFM	0.02–0.09	WMNT were mounted on AFM tip before pushing in polymer; on cooling the CNT were embedded into polymer and subsequent pull-out test were conducted
Fu et al. [69]	MWNT/epoxy	Tensile tester	0.044	Tensile test on composite with different MWNT loadings were conducted
Coleman et al. [24,25]	SWNT/PVA	Tensile tester	~0.040	Tensile test on composite fibers with different SWNT loadings were conducted, and theoretical calculations were done
Coleman et al. [70]	CNT/PMMA	Tensile tester	~0.088	Tensile test derived shear stress

or modified structure of CNT. As mentioned previously, polymer wrapping may be present on the CNT surface, be it in various tubular, blob or kebab forms [11], and thus various shear behaviors.

The large value of 0.5 GPa of polymer/CNTIFSS, shown in Table 4.2, estimated by Weigner et al. [63] was attributed to covalent bonding between CNT and polymer introduced by photoinduced reactions. The functional moieties on CNT typically provide better interfacial load transfer via bonding and/entanglement with the polymer. This has also been proven by molecular simulations; the shear strength was reported to have a ten times increase even when as low as 0.3% grafting density of polymer matrix is formed on CNT [67].

Cooper et al. [66] used AFM as the tool to manipulate CNT from epoxy, and observed 0.35 GPa for tubes with short embedded lengths, but 0.04–0.09 GPa with longer embedded lengths. Also, it was shown that the IFSS decreased with an increase in the embedded length of CNT (as low as 40 MPa) when the embedded length of CNT was more than 500 nm.

Barber's group [64,65] have used the "push-in pull-out" trick in Polyehtylene (PE), and reported values between 0.02–0.09 GPa. In addition, the same group found that the average interfacial stress required to remove a single MWNT from the PE–butene matrix is 47 MPa, the stuck depth of MWNT was around 10 nm. This strength is around 10 times larger than the adhesion level between the same type of polymer and CFs; this example also shows the importance of filler size on the interfacial strength: the larger the MWNT outside diameter, the smaller the IFSS.

As mentioned in the interphase section, the polymer wrapping on the CNT surface will have a different morphology closer to the CNT than the bulk. Thus it is reasonable that the polymer chains close to the interphase behave differently than the bulk too. This is due to the fact that CNT diameter is of similar magnitude to the radius of gyration for the polymer. In the report corroborating an interfacial region of nonbulk polymer, a "sheathing layer" of polycarbonate on pulled out MWNT was imaged using AFM, which provided the evidence of significant interfacial interaction between MWNT and polymer [68].

Combining tensile tests and theoretical calculations, Fu and Coleman et al. [24,25,69,70] separately reported close values, lying in the region of 50–100 MPa. Noteworthy is that the IFSS are for the noncovalently bonded composites. Thus it becomes obviously significant to understand the interfacial adhesion mechanism at the molecular level to further optimize the interface in composite fiber systems. Theoretical calculations including molecular dynamics simulation and elasticity calculations become necessary. Liao et al. [58] found that the CNT/polymer adhesion exists without the presence of atomic bonding, instead coming from electrostatic and van der Waals interactions and the stress/deformation arising from the mismatch in the coefficients of thermal expansion between polymer and CNT.

4.2.2 Thermal characterization

Polymer wrapping thickness could be derived from mechanical properties as discussed above, and it can be calculated from thermal tests including DSC and thermogravimetric analyzer.

4.2.2.1 Thermogravimetric analysis (TGA)

Thermogravimetric analysis, or TGA, is one method of TMA in which changes in physical and chemical properties of materials are measured as a function of increasing temperature (with constant heating rate), or as a function of time (with constant temperature and/or constant mass loss). TGA provides information including both secondary transitions (i.e., vaporization, absorption, adsorption, desorption, molecular group movement, loss of volatiles or moisture) and chemical reaction (i.e., decomposition, oxidation, reduction).

In polymer-based composites, TGA in most cases have been used for studying thermal stability. Take PDMS for example. Pure PDMS shows thermal degradation at around $-69°C$, while the silica and titania filled PDMS showed lower and lower thermal transition with increased concentrations. This delay in degradation is due to the interaction between hydroxyl groups and PDMS polymer chain formed hydrogen bonding [74].

4.2.2.2 Differential scanning calorimetry (DSC)

DSC is another thermoanalytical technique in which the difference in the amount of heat required to increase the temperature of a sample and reference is measured as a function of temperature. Both sample and reference pan go through the same heating history and the reference should have a well-defined heat capacity over the range of temperatures being scanned. The basic principle is simple, i.e., for the sample to go through a phase transition, more or less heat will be absorbed or released to maintain the same temperature of sample to the reference pan. Physical changes including crystallization, melting, solidification, oxidation, and even secondary transitions of functional group movements can be detected.

In rubber-based composites, the adsorption of molecular chains on particles is detectable by DSC. For example, pure PDMS has a crystallization temperature T_m around $-68°C$. When silica or titania was included in PDMS, both the T_m and crystallization degree decreased [75]. At silica concentrations of 10%, the melting temperature decreased to $-43°C$. This phenomenon has been observed in a lot of polymer systems. The observation was attributed to the lower crystallization capability when polymer chains are cross-linked on the particle surface [76].

Interphase can form around particles. Polymer wrapping thickness could be computed by the crystallinity degree and volume fraction fitted relationship using the equation [30],

$$\left[\frac{\pi(b+r)^2 - \pi r^2}{\pi r^2}\right] = \frac{d\chi}{dV_f} \tag{4.46}$$

$$\frac{b^2 + 2br}{r^2} - \frac{d\chi}{dV_f} = 0 \tag{4.47}$$

where b is the thickness of interphase, r is the radius of tubular particles, V_f is the volume fraction of particles, and χ is the crystallinity degree in each differently concentrated samples. It is straightforward that the interphase dimensions around a nanotube can be obtained based on the crystallization degree in composites.

Rubbers including natural rubber and polydimethyl siloxane are known to be semicrystalline polymers and crystallize either under room temperature or under strain. The glass transition temperature is $-25°C$ for natural rubber and $-70°C$ for PDMS. Strain-induced crystallization is actually helpful for composite reinforcement. This is due to the local stress or strain concentration in the vicinity of nanoparticles. When nanoparticles are included in the composites, cross-linking density is influenced. For example, PDMS has been studied regarding the crystallization behavior when silica [77] and CB [78] are incorporated. In both composites, the polymer displayed reduced capability in crystallization.

In composites, with the addition of nanoparticles the strain at the crystallization onset decreases. In composites, nanoparticles have been considered as new effective cross-links. Effective cross-linking density has been defined as the sum of chemical cross-links, entanglements, and polymer chains adsorption on nanoparticle surfaces. One theory describes the evolution of crystallization and the crystal size variations [79]. When the entanglements are below a criterion, the nuclei number would increase with the effective cross-link density increasing at the same time. The crystal size decreases. When the entanglements are above a criterion, the limited polymer movements will constrain the crystal nucleation as well as crystal growth.

4.2.3 Electron microscopy

Interphase formation and interfacial modifications have been reported to influence composites' properties. By investigating mechanical or thermal properties, the interface compositions or interfacial interactions can be derived. As compared to these methods, electron microscopy (EM), including SEM and TEM, have been extensively used in interfacial observations. These microscopic observations are straightforward and very useful in analyzing the interfacial interactions. It is noteworthy that the in situ mechanical test using SEM, SPM, and TEM have been discussed in Table 4.2 and will not be included in the following section.

4.2.3.1 Scanning electron microscopy (SEM)

SEM provides images of a sample by scanning it with a focused beam of electrons. These electrons interact with atoms in the sample and by analyzing the information of electron beam position and intensity generates an image of topography. A specimen is often prepared and observed in vacuum, wet, and controlled temperature conditions. Additionally, a specimen must be electrically conductive on the surface and electrically grounded to prevent accumulation of electrons during the scanning test. Specifically for rubber-based samples, an ultrathin coating of electrically conductive materials is usually deposited on the sample surface by either sputter coating or high vacuum evaporation.

In rubber-based composites, fractography is frequently used in SEM for the observation of fracture and failure surfaces of a specimen. The sample needs to be small enough to be put in an SEM chamber, and mounted on a specimen holder in some cases, when cryo-cutting is a requirement due to the viscoelasticity. Polymer wrapping on the particle surface can be observed if the composites fail in cohesive regions.

SEM is very useful in observing different morphologies, especially the fractured mode, including "sword-scabbard" or "core-shell" [80], "shish-kebab" [81], and "bridging effect" [82], however, the specific interphase based on particle–interphase interfacial features, dispersion of individual nanoparticles, and possible reinforcement effects will be more revealed in TEM images.

4.2.3.2 Transmission electron microscopy (TEM)

TEM is another microscopy technique in which a beam of electrons is transmitted through an ultrathin sample, and interacts with the specimen as it passes through it. The resolution of TEM is higher than that of SEM. The sample prepared for TEM is thin so that electrons can pass through. To prepare thin enough sample cuttings from bulk materials, microtoming is frequently used in rubber materials. Most rubber-based materials are soft and need to be cryo-cut below their glass-transition temperature. The ideal sample thickness is less than 100 nm.

Bright-field TEM imaging is useful for nanoparticle dispersion information. In bright-field TEM, the intensity of transmitted beam is influenced by the sample thickness and density. As a result, the polymer matrix and particle-involved composite are observed with different contrast due to their density and crystal structure differences. The morphological observations were useful to understand different schemes of material failures.

4.2.3.3 Scanning transmission electron microscopy (STEM)

Scanning transmission electron microscopy does bright-field and dark-field imaging similarly to TEM as the electron beam is transmitted though an ultrathin specimen and interacts with the ultrathin specimen. At the same time, the electron beam focuses on a narrow area of the thin specimen and collects the image as does the SEM.

4.2.4 Spectral analysis

In platy particle filled polymer composites, there are mainly two types of morphology, namely, intercalation and exfoliation. Generally, several strategies have been used for intercalated nanocomposites successfully, namely,

1. Exfoliation-adsorption. For exfoliation-adsorption, most hydrophilic monomers, prepolymers and water-soluble polymers have been used. These molecules can be intercalated in the untreated MMT and produce the final intercalated composites [83].
2. In situ intercalative polymerization. For in situ polymerization, the layered structure is swollen within a liquid monomer. Nylon [83], polycaprolactone [83], and epoxy [83]

materials have been incorporated in this method. This method is the most efficient one to generate intercalated structures. The prerequisite is to find the appropriate clay, organic modifier, and good solvent.

3. Melt intercalation. In the melt state, the presence of polymer chains allows the separation between silicate layers [83]. However, even when the molecules are compatible with the layer surface, the intercalation efficiency is not high. After all, the electrostatic attraction between the silicate layers is strong.
4. Template synthesis. The polymer chains act as a template for layer formations in aqueous polymer gel media.

Exfoliated morphology has been well recognized as the best morphology for high performance in polymer composites with rather low particle concentrations. Once the platy particles are intercalated by polymer chains, the original interlayer spacing distances will become larger than that before intercalation. On the contrary, when the nanocomposites are precipitated from solution through coagulation (e.g., sinking in a large amount of ethanol), the polymer may move out. At this moment when the polymer chains are expelled, some grafted molecules might also be taken out of the space by the movement and lead to the decrease of *d*-spacing. Spectral analyses will be useful in analyzing this delicate spacing change.

4.2.4.1 Wide-angle X-ray diffraction (WAXD)

Wide-angle X-ray diffraction measurements are important for providing information about the clay layer dispersion quality and exfoliation degree. Before exfoliation, the layered sheets display peaks between 1° and 24° for 2θ, corresponding to a *d*-spacing of 0.34 to 2.2 nm. After exfoliation, these layered structures would disappear and thus the distinct features in the low 2θ ranges will be lost too. Examples including graphite and MMT particles are given in Table 4.3.

For montmorillonite-included composites, Octylamine $CH_3(CH_2)_7NH_2$ (C8-clay), Dodecylamine amine $CH_3(CH_2)_{11}NH_2$ (C12-clay), and Octadecylamine $CH_3(CH_2)_{17}NH_2$ (C18-clay), modified particles displayed (001) peak at different locations [42]. Before treatment, the *d*-spacing is around 1.23 nm; C8-clay showed a *d*-spacing of 1.36 nm, C12-clay displayed a *d*-spacing of 1.52 nm, and C18-clay even decreased this number to 2.18 nm. When natural rubber was mixed for nanocomposites fabrications, this (001) characteristic peak disappears, suggesting the fully intercalated silicate layers.

4.2.4.2 Fourier transform infrared spectroscopy (FTIR)

Fourier transform infrared spectroscopy (FTIR) is the technique used to obtain an infrared spectrum of absorption or emission of a solid, liquid and gas. FTIR can be used to collects high spectral resolution data over a wide range, usually between 5000 and 400 cm^{-1} for mid-IR region wavelength, and between 10,000 and 4000 cm^{-1} for near-IR region wavelength. For a typical FTIR, the resolution is 4 cm^{-1}. A straightforward way for this absorption measurement is to shine a monochromatic light beam at a sample and measure how much of it is absorbed. Repeating corresponding test parameters can control the number of measurements

Table 4.3 **WAXD parameters before and after modification**

Particles	Peak	Modifier	2 θ	Polymers	$^{a}d_1$ (nm)	$^{b}d_2$ (nm)
Graphite [84]	001	—	26.6	—	0.335	—
		Ultrasonic treatment	26.6	—	—	0.335
		SDS	2.55	—	—	3.47
		SDS	1.02	SBR	—	8.67
MMT [42]	001	—	7.16	—	0.62	—
		Octylamine $CH_3(CH_2)_7NH_2$	6.5	—	—	1.36
		Dodecylamine amine $CH_3(CH_2)_{11}NH_2$	—	—	—	1.52
		Octadecylamine $CH_3(CH_2)_{17}NH_2$	—	—	—	2.18
MMT [42]	001	—	7.16	—	—	—
		Octylamine $CH_3(CH_2)_7NH_2$	3.92	Natural rubber	0.62	2.25
MMT [85]	001	Hexadecyl trimethyl ammonium bromide	4.48	—	1.97	—
		—	2.96	PC	—	2.98
		—	2.64	PS	—	3.34
		—	2.48	EPDM	—	3.56
MMT [86]	001	Primary amine	4	—	2.10	—
		Primary amine	—	EPDM	—	2.32
		Curative	—	EPDM	—	3.08
		Vulcanization	—	EPDM	—	3.30
		Quaternary amine	—	—	2.50	—
		Quaternary amine	—	EPDM-MA	—	4.20
		Vulcanization	—	EPDM-MA	—	4.69

^{a}d-spacing before intercalation.
^{b}d-spacing after intercalation. ethylene/propylene/diene rubber, EPDM.

for each different wavelength. In FTIR, light of many frequencies can be measured simultaneously and this process can be repeated many times. This gives FTIR an advantage, i.e., it results in a higher signal-to-noise ratio for a specific scanning time.

FTIR can be used for microscopy and imaging of rubber and particle interfacial modifications. For example, the physically-absorbed water can be characterized through peaks between 3800 and 3200 cm^{-1} (i.e., —OH stretching vibrations). More conveniently, when this vibration combines with the bending mode of water molecules, it generates bands from 5050 to 5350 cm^{-1} range. In PDMS-based composites, both silica [87] and titania [88] have been well studied regarding this IR absorption.

4.2.4.3 *Nuclear magnetic resonance (NMR)*

Nuclear magnetic resonance (NMR) is a physical phenomenon in which nuclei in a magnetic field absorb and reemit electromagnetic radiation. NMR spectroscopy can be used to study molecular physics, crystals, and noncrystalline materials. By studying the peaks from NMR, atoms within a molecule or a collection of molecules of the same type can be distinguished. Specifically in particle-filled elastomer composites, the transverse nuclear magnetic relaxation time, T_2, is affected by physical and chemical cross-links on the particle surface. For example, PDMS is said to adsorb on the silica surface with a thickness of several nanometers. This layer of molecules is more restricted than those further away from the particle surface [77]. This can be evidenced by ^{1}H NRM relaxation methods using spin-echo techniques [89].

4.3 Interfacial modification

Polymer matrix and particle fillers are opposite in wetting behavior, i.e., polymers are hydrophobic, and most particles are hydrophilic. The compatibility is often modified for improved interfacial interactions and increased mechanical or thermal performance, and several examples are given below.

4.3.1 *Coupling agents*

One of these effective modification methods is coupling agents. The primary function of the coupling agent is to provide a strong chemical link between the oxide groups on the filler surface and the polymer chains through chemical reactions. Silane coupling agents, with chemical formula of $R{-}Si{-}X_3$, is commonly used for interfacial treatment in composites. As it is known that filled particles intend to absorb water from air or the surrounding environment, and the absorbed water content greatly reduces the wettability of the fibers (e.g., glass fibers will reduce from 500 mJ m^{-2} to 15 mJ m^{-2} in surface tension once wetted). Silane, on the other hand, will compete with water molecules to form hydrogen bonds with the hydroxyl groups bound to the fiber surface. Once the fillers are dry, the water will be driven off. Simultaneously, condensation takes place in silane/silane and silane/filler as well as silane/polymer. Because of this, a polysiloxane layer forms on the fiber surface, protruding the silane R- groups to the surrounding environment in a form of R/polymer chain entanglement or cross-linking. This bonding form depends on the characteristics of polymers. In situ polymerization of polymers in the presence of silane tends to form silane/polymer cross-linking and a mixture of silane/thermoplastics will frequently lead to a polymer chain entanglement.

It is also noteworthy that the absence of interface modification will be beneficial for toughening mechanisms in composites. When the interface debonds easily, the crack-deflection is promoted and friction sliding is propagated unless a whole fiber is pulled out. This procedure is expected to absorb a substantial amount of energy and thus increase the toughness of the whole composite.

4.3.2 Polymer coating

Chemical modifications are always benefiting the mechanical reinforcement. The decreases in mechanical properties have been found and attributed to the foreign chemicals induced during the chemical reactions. However, if a polymer matrix is coated around the particles, no further chemical modification will be then necessary. The presence of polymer chains on particle surfaces at the same time will repel particles from forming big chunks and improve the dispersion quality [13]. Besides, CNT has evidenced an increase in compatibility with polymer compared with CFs after oxidation [90]. SBR matrix was adhesive to the CNT surfaces in a stronger bonding and enhanced the mechanical properties in composites [91]. In another study [92], an improvement in the properties was observed when the surfaces of CNT were modified with polymer chains, which act as a bridge between the fibers and polymer matrix.

4.3.3 Layered sheets intercalation

Layered sheets included in rubber composites are mostly polar and hydrophobic. Polymers of nylon and epoxy are effective in intercalating some pristine silicates but for other polymers it may be not effective. The modification of these layered sheets involves usage of salts such as alkyl ammonium, phosphonium, pyridinium, imidazolium, and so on. The inclusion of these salts makes the surface of sheets more hydrophilic and simultaneously increases the interlayer spacing. In a general procedure, dilute dispersions of layered particles are prepared through sonication, mechanical mixing, and centrifugation. Surfactant is then added for the ion exchange to finish. The reaction usually takes place at elevated temperature until organosilicate products are filtered and collected. For example, MMT is modified with aminosilanes along with surfactants and increased interlayer spacing is found. The mixture with rubber afterwards will allow polymer chains to expand these sheets and improve the mechanical performance [93].

4.3.4 Natural fiber modification

The low cost, easy availability, low density, high specific mechanical properties, less wear or corrosion in processing techniques, and environmentally friendly characteristics of some natural fiber reinforcements in polymers have been the primary benefits for their commercialization. Despite these advantages, the use of natural fibers has been limited, mainly due to the poor adhesion with the polymer matrix as compared to their synthetic counterparts. Various natural fiber surface treatments have been reported and summarized. The interfacial interaction improvements due to these surface treatments have been achieved by both physical and chemical means. For example, sodium hydroxide solutions of various concentrations on coir fiber [47] were examined for optimal properties. The formation of a large number of voids on fibers promotes better mechanical anchorage between fiber and rubber matrix. It is also applicable to layered silicates and able to reduce the particulate surface energy. Bleaching with sodium silicate and hydrogen peroxide in water, toluene diisocyanate soaking, and RFL treatments were also compared for their surface treatment efficiency differences and their influence in mechanical properties.

4.4 Summary

Nanoparticles–polymer interactions can determine the mechanical, thermal, and other functional properties in composites. Either physical coating or chemical cross-linking can improve interfacial interaction efficiency. Therefore, the characterization of interfacial interactions is important. Mechanical and thermal properties, by incorporating interphase regions, can be greatly increased. These measurements can also determine the volume fraction of the newly formed interphase or interfacial interaction area. In addition, more straightforward methods such as electron microscope and spectral analyses are significant supplementary methods. In sum, this chapter reviews the currently and frequently used characterization means for interphase and/or interfacial interaction efficiency measurements.

Nomenclature

Materials

CNTs Carbon nanotubes
DWNTs Double-wall carbon nanotubes
GP Graphene platelets
GPRP Graphene platelets reinforced polymers
HNTs Halloysite nanotubes
MA Maleic anhydride
MWNTs Multiwall carbon nanotubes
MMT Montmorillonite
POSS Polyhedral oligomeric silsesquioxane
PVA Polyvinyl alcohol
PVC Polyvinyl chloride
PU Polyurethane
SWNTs Single-wall carbon nanotubes

Characterizations

AFM Atomic force microscopy
DMA Dynamic mechanical analysis
DSC Differential scanning calorimetry
DTA Differential thermogravimetric analysis

FTIR Fourier transformed infrared spectroscopy
ROM Rule of mixture
SAXS Small-angle X-ray scattering
SEM Scanning electron microscopy
TEM Transmission electron microscopy
TGA Thermogravimetric analysis
WAXD Wide-angle X-ray diffraction

References

[1] Halpin JC, Tsai SW. Environmental Factors in Composite Materials Design. U.S. Air Force Tech. Rep. AFML TR; 1967. p. 67–423.
[2] Ma P-C, et al. Dispersion and functionalization of carbon nanotubes for polymer-based nanocomposites: a review. Compos Part A Appl Sci Manuf 2010;41(10):1345–67.
[3] Porter RS, Johnson JF. The entanglement concept in polymer systems. Chem Rev 1966;66(1):1–27.
[4] Good RJ. Contact angle, wetting, and adhesion: a critical review. J Adhes Sci Technol 1992;6(12):1269–302.
[5] Mortimer DA. Synthetic polyelectrolytes—a review. Polym Int 1991;25(1):29–41.
[6] Watkinson A, Wilson D. Chemical reaction fouling: a review. Exp Thermal Fluid Sci 1997;14(4):361–74.
[7] Lee H, et al. Strategies for hydrogen bonding based layer-by-layer assembly of poly (vinyl alcohol) with weak polyacids. Macromolecules 2011;45(1):347–55.
[8] Tajaddod N, et al. Exfoliation of boron nitride platelets by enhanced interfacial interaction with polyethylene. Macromol Mater Eng 2016;301(3):315–27.
[9] Song K, Zhang Y, Minus ML. Polymer interphase self-reinforcement and strengthening mechanisms in low-loaded nanocomposite fibers. Macromol Chem Phys 2015;216(12):1313–20.
[10] Meng J, et al. Forming crystalline polymer-nano interphase structures for high-modulus and high-tensile/strength composite fibers. Macromol Mater Eng 2014;299(2):144–53.
[11] Zhang Y, et al. Tailoring polyacrylonitrile interfacial morphological structure by crystallization in the presence of single-wall carbon nanotubes. ACS Appl Mater Interfaces 2013;5(3):807–14.
[12] Song K, et al. Lubrication of poly (vinyl alcohol) chain orientation by carbon nano-chips in composite tapes. J Appl Polym Sci 2013;127(4):2977–82.
[13] Song K, et al. Structural polymer-based carbon nanotube composite fibers: understanding the processing–structure–performance relationship. Materials 2013;6(6):2543–77.
[14] Minus ML, et al. Development of nanoscale interfacial morphology in polymer-based nano-carbon hybrid structures for stress transfer improvement. in Society of Plastics Engineers - EUROTEC 2011 Conference Proceedings; 2011.
[15] Minus ML, et al. Interfacial structure development in polymer nano-carbon composite fibers. in International SAMPE Technical Conference; 2012.
[16] Song K., Zhang Y., and Minus ML. Gel-spinning of poly (vinyl alcohol)/carbon nano--chips composite tapes. in abstracts of papers of the american chemical society; 2012. Amer Chemical Soc 1155 16th St, NW, Washington, DC 20036 USA.
[17] Minus, M., et al. Using nano-carbon templates to control polymer matrix micro-structure formation and properties in the composite. In: SPE automotive composites conference & exhibition (ACCE). Detroit; 2013.

[18] Song K, Minus M. Inducing polymer interphase formation in the composite at low nano-carbon loading. 17th Annual nanotech and tech connect world conference and expo. Washington, D.C: CRC Press; 2014.

[19] Song K, Zhang Y, Minus ML. Using low concentrations of nano-carbons to induce polymer self-reinforcement of composites for high-performance applications. MRS Online Proc Libr 2015;1752:137–44.

[20] Song K, et al. Application of the electron density correlation function for structural analysis of x-ray scattering/diffraction information from polymer-based nano-composites. MRS Online Proc Libr 2015;1754. p. null-null.

[21] Song K. Customizing fiber spinning approaches for polymer/nano-carbon composites. Northeastern University; 2014.

[22] Cox H. The elasticity and strength of paper and other fibrous materials. Br J Appl Phys 1952;3(3):72.

[23] McCrum NG, Buckley CP, Bucknall CB. Principles of polymer engineering. 2nd ed. USA: Oxford University Press; 1997. p. 276–8.

[24] Young K, et al. Strong dependence of mechanical properties on fiber diameter for polymer-nanotube composite fibers: differentiating defect from orientation effects. ACS Nano 2010;4(11):6989–97.

[25] Blighe FM, et al. The effect of nanotube content and orientation on the mechanical properties of polymer–nanotube composite fibers: separating intrinsic reinforcement from orientational effects. Adv Funct Mater 2011;21(2):364–71.

[26] Song K, et al. Spectral analysis of lamellae evolution and constraining effects aided by nano-carbons: a coupled experimental and simulation study. Polymer (Guildf) 2015;75:187–98.

[27] Pichot V, et al. Structural and mechanical properties of single-wall carbon nanotube fibers. Phys Rev B 2006;74(24) 245416-8.

[28] Chen M, et al. Characterization of single-walled carbon nanotube fibers and correlation with stretch alignment. MRS proceedings. Cambridge University Press; 2004.

[29] Zhou W, et al. Single wall carbon nanotube fibers extruded from super-acid suspensions: preferred orientation, electrical, and thermal transport. J Appl Phys 2004;95(2):649–55.

[30] Coleman JN, et al. Reinforcement of polymers with carbon nanotubes. The role of an ordered polymer interfacial region. Experiment and modeling. Polymer (Guildf) 2006;47(26):8556–61.

[31] Fukahori Y, et al. Modified guth-gold equation for carbon black-filled rubbers. Rubber Chem Technol 2013;86(2):218–32.

[32] Park J-S, Park J-W, Ruckenstein E. On the *viscoelastic properties of poly*(vinyl alcohol) and chemically cros*slinked poly(vinyl alcohol).* J Appl Polym Sci 2001;82(7):1816–23.

[33] Nishio Y, Manley RSJ. Cellulose-poly(vinyl alcohol) blends prepared from solutions in N,N-dimethylacetamide-lithium chloride. Macromolecules 1988;21(5):1270–7.

[34] Zeng SS, et al. Facile hydroxylation of halloysite nanotubes for epoxy nanocomposite applications. Polymer (Guildf) 2014;55(25):6519–28.

[35] Saha N, Banerjee AN, Mitra B. Dynamic mechanical study on unidirectional polyethylene fibers-PMMA and glass fibers-PMMA composite laminates. J Appl Polym Sci 1996;60(5):657–62.

[36] Wielage B, et al. Processing of natural-fibre reinforced polymers and the resulting dynamic–mechanical properties. J Mater Proc Technol 2003;139(1–3):140–6.

[37] Kubat J, Rigdahl M, Welander M. Characterization of interfacial interactions in high density polyethylene filled with glass spheres using dynamic-mechanical analysis. J Appl Polym Sci 1990;39(7):1527–39.

[38] Ziegel K, Romanov A. Modulus reinforcement in elastomer composites. I. Inorganic fillers. J Appl Polym Sci 1973;17(4):1119–31.
[39] López-Manchado M, et al. Dynamic mechanical and Raman spectroscopy studies on interaction between single-walled carbon nanotubes and natural rubber. J Appl Polym Sci 2004;92(5):3394–400.
[40] Mao Y, et al. High performance graphene oxide based rubber composites. Sci Rep 2013;3.
[41] Das A, et al. Rubber composites based on graphene nanoplatelets, expanded graphite, carbon nanotubes and their combination: a comparative study. Compos Sci Technol 2012;72(16):1961–7.
[42] Kim JT, Oh TS, Lee DH. Preparation and characteristics of nitrile rubber (NBR) nanocomposites based on organophilic layered clay. Polym Int 2003;52(7):1058–63.
[43] Rooj S, Das A, Heinrich G. Tube-like natural halloysite/fluoroelastomer nanocomposites with simultaneous enhanced mechanical, dynamic mechanical and thermal properties. Eur Polym J 2011;47(9):1746–55.
[44] Prasertsri S, Rattanasom N. Fumed and precipitated silica reinforced natural rubber composites prepared from latex system: mechanical and dynamic properties. Polym Test 2012;31(5):593–605.
[45] Cao X, et al. Preparation and properties of carboxylated styrene-butadiene rubber/cellulose nanocrystals composites. Carbohydr Polym 2013;92(1):69–76.
[46] Da Costa H, et al. Mechanical and dynamic mechanical properties of rice husk ash–filled natural rubber compounds. J Appl Polym Sci 2002;83(11):2331–46.
[47] Geethamma VG, et al. Dynamic mechanical behavior of short coir fiber reinforced natural rubber composites. Compos Part A Appl Sci Manuf 2005;36 (11):1499–506.
[48] Flaifel MH, et al. Thermal conductivity and dynamic mechanical analysis of NiZn ferrite nanoparticles filled thermoplastic natural rubber nanocomposite. Compos Part B: Eng 2013;52:334–9.
[49] Praveen S, et al. Synergistic effect of carbon black and nanoclay fillers in styrene butadiene rubber matrix: development of dual structure. Compos Part A Appl Sci Manuf 2009;40(3):309–16.
[50] Jacob M, et al. Dynamical mechanical analysis of sisal/oil palm hybrid fiber-reinforced natural rubber composites. Polym Compos 2006;27(6):671–80.
[51] Minus ML, Chae HG, Kumar S. Single wall carbon nanotube templated oriented crystallization of poly(vinyl alcohol). Polymer (Guildf) 2006;47(11):3705–10.
[52] Minus ML, Chae HG, Kumar S. Interfacial crystallization in gel-spun poly(vinyl alcohol)/single-wall carbon nanotube composite fibers. Macromol Chem Phys 2009;210(21):1799–808.
[53] Minus ML, Chae HG, Kumar S. Observations on solution crystallization of poly(vinyl alcohol) in the presence of single-wall carbon nanotubes. Macromol Rapid Comm 2010;31(3):310–16.
[54] Chaturvedi SK, Tzeng GY. Micromechanical modeling of material damping in discontinuous fiber three-phase polymer composites. Compos Eng 1991;1(1):49–60.
[55] Vantomme J. A parametric study of material damping in fibre-reinforced plastics. Composites 1995;26(2):147–53.
[56] Chen X, et al. Quantitative nanomechanical characterization of the van der Waals interfaces between carbon nanotubes and epoxy. Carbon NY 2015;82:214–28.
[57] Chen X, et al. Mechanical strength of boron nitride nanotube-polymer interfaces. Appl Phys Lett 2015;107(25):253105.

[58] Liao K, Li S. Interfacial characteristics of a carbon nanotube-polystyrene composite system. Appl Phys Lett 2001;79(25):4225–7.
[59] Wong M, et al. Physical interactions at carbon nanotube-polymer interface. Polymer (Guildf) 2003;44(25):7757–64.
[60] Lordi V, Yao N. Molecular mechanics of binding in carbon-nanotube–polymer composites. J Mater Res 2000;15:2770–9.
[61] Cooper CA, Young RJ, Halsall M. Investigation into the deformation of carbon nanotubes and their composites through the use of Raman spectroscopy. Compos Part A Appl Sci Manuf 2001;32(3–4):401–11.
[62] Yu M-F, et al. Strength and breaking mechanism of multiwalled carbon nanotubes under tensile load. Science 2000;287(5453):637–40.
[63] Wagner HD, et al. Stress-induced fragmentation of multiwall carbon nanotubes in a polymer matrix. Appl Phys Lett 1998;72(2):188–90.
[64] Barber AH, Cohen SR, Wagner HD. Measurement of carbon nanotube--polymer interfacial strength. Appl Phys Lett 2003;82(23):4140–2.
[65] Barber AH, et al. Interfacial fracture energy measurements for multi-walled carbon nanotubes pulled from a polymer matrix. Compos Sci Technol 2004;64(15):2283–9.
[66] Cooper CA, et al. Detachment of nanotubes from a polymer matrix. Appl Phys Lett 2002;81(20):3873–5.
[67] Frankland SJV, et al. Molecular simulation of the influence of chemical cross-links on the shear strength of carbon nanotube–polymer interfaces. J Phys Chem B 2002;106 (12):3046–8.
[68] Ding W, et al. Direct observation of polymer sheathing in carbon nanotube–polycarbonate composites. Nano Lett 2003;3(11):1593–7.
[69] Fu S-Y, et al. The reduction of carbon nanotube (CNT) length during the manufacture of CNT/polymer composites and a method to simultaneously determine the resulting CNT and interfacial strengths. Carbon N Y 2009;47(14):3192–200.
[70] Blond D, et al. Enhancement of modulus, strength, and toughness in Poly(methyl methacrylate)-based composites by the incorporation of Poly(methyl methacrylate)-functionalized nanotubes. Adv Funct Mater 2006;16(12):1608–14.
[71] Demczyk BG, et al. Direct mechanical measurement of the tensile strength and elastic modulus of multiwalled carbon nanotubes. Microsc Microanal 2006;12(Suppl. S02):934–5.
[72] Xie S, et al. Mechanical and physical properties on carbon nanotube. J Phys Chem Solids 2000;61(7):1153–8.
[73] Salvetat J-P, et al. Elastic and shear moduli of single-walled carbon nanotube ropes. Phys Rev Lett 1999;82(5):944–7.
[74] Wen J, Mark JE. Precipitation of silica-titania mixed-oxide fillers into poly (dimethylsiloxane) networks. Rubber Chem Technol 1994;67(5):806–19.
[75] Aranguren MI. Crystallization of polydimethylsiloxane: effect of silica filler and curing. Polymer (Guildf) 1998;39(20):4897–903.
[76] Bokobza L, Diop AL., Reinforcement of Silicone Rubbers by Sol-Gel In Situ Generated Filler Particles. Rubber Nanocomposites: Preparation, Properties and Applications; 2010. p. 63.
[77] Dewimille L, Bresson B, Bokobza L. Synthesis, structure and morphology of poly (dimethylsiloxane) networks filled with in situ generated silica particles. Polymer (Guildf) 2005;46(12):4135–43.
[78] Chenal J-M, et al. New insights into the cold crystallization of filled natural rubber. J Polym Sci Part B Polym Phys 2007;45(8):955–62.

[79] Chenal J-M, et al. Parameters governing strain induced crystallization in filled natural rubber. Polymer (Guildf) 2007;48(23):6893–901.
[80] Li T, et al. Polylactide, nanoclay, and core–shell rubber composites. Polym Eng Sci 2006;46(10):1419–27.
[81] Yang J, et al. Direct formation of nanohybrid shish-kebab in the injection molded bar of polyethylene/multiwalled carbon nanotubes composite. Macromolecules 2009;42(18):7016–23.
[82] Tian M, et al. Surface modification of fibrillar silicate and its reinforcing mechanism on FS/rubber composites. Compos Sci Technol 2005;65(7):1129–38.
[83] Komarneni S. Feature article. Nanocomposites. J Mater Chem 1992;2(12):1219–30.
[84] Yang J, et al. Improved mechanical and functional properties of elastomer/graphite nanocomposites prepared by latex compounding. Acta Mater 2007;55(18):6372–82.
[85] Ma J, et al. A new approach to polymer/montmorillonite nanocomposites. Polymer (Guildf) 2003;44(16):4619–24.
[86] Gatos KG, Karger-Kocsis J. Effects of primary and quaternary amine intercalants on the organoclay dispersion in a sulfur-cured EPDM rubber. Polymer (Guildf) 2005;46(9):3069–76.
[87] Legrand A, et al. Hydroxyls of silica powders. Adv Colloid Interface Sci 1990;33 (2):91–330.
[88] Legrand AP. The surface properties of silicas. Wiley; 1998.
[89] Litvinov V, Barthel H, Weis J. Structure of a PDMS layer grafted onto a silica surface studied by means of DSC and solid-state NMR. Macromolecules 2002;35 (11):4356–64.
[90] Tsubokawa N. Preparation and properties of polymer-grafted carbon nanotubes and nanofibers. Polym J 2005;37(9):637–55.
[91] Girun N, et al. Multi-wall carbon nanotubes/styrene butadiene rubber (SBR) nanocomposite. Fuller Nanotub Carbon Nanostruct 2007;15(3):207–14.
[92] Kong H, Gao C, Yan D. Controlled functionalization of multiwalled carbon nanotubes by in situ atom transfer radical polymerization. J Am Chem Soc 2004;126(2):412–13.
[93] Varghese S, Karger-Kocsis J, Gatos K. Melt compounded epoxidized natural rubber/ layered silicate nanocomposites: structure-properties relationships. Polymer (Guildf) 2003;44(14):3977–83.

Rubber nanocomposites with nanoclay as the filler

5

K. Kanny and T.P. Mohan
Durban University of Technology, Durban, South Africa

5.1 Introduction

Filler is defined as finely divided particles that are used to improve several properties and the performance of the host matrix; however, this depends upon the application. Composite materials consisting of the polymeric matrix reinforced with inorganic fillers have content of nanocomposites where filler size is of nanometric scale and of the characteristic length scale of the polymer. This decreased size, however, increases specific surface area. The commonly used fillers include carbon black, calcium carbonate, and silica [1]. The results of such a method have been recognized by observing a significant improvement in the physical properties of the composites consisting of a very small amount of nanofiller [2]. However, the important advantages of nanofillers are not only to reinforce the rubber matrix but also to improve a number of other properties such as barrier properties, electrical, membrane properties, polymer blend compatibility, and flammability resistance [3].

Nanoparticle additives, such as carbon nanotubes, silicone nanopowder, nano silver, and nanoclay (the latter being the topic of interest in this chapter), are widely used in various industries such as cable coatings, adhesives, inks, pharmaceutical, and automotive. One of the most common forms of clay is montmorillonite (MMT),which is smetic clay of particle thickness of 1 nm and length of a few 100 nm. It has been shown in several studies that the tensile strength, tensile modulus, flexural modulus, flexural strength, and thermal stability are some of the properties that increase as a result of nanoclay being added to a polymer matrix [4].

Nanoclay technology is an interesting research area due to clays being cost-effective and available easily from natural sources. Furthermore, there are several methods of infusing the nanoparticles in a polymer matrix and these methods do not require much effort. For exfoliation to take place with clay, two considerations are crucial: steric repulsion of chains to push the platelets away from one another, and polymer–clay compatibility for polymer chain penetration into the clay galleries [5]. Compatibility does not guarantee exfoliation, but strong polymer–clay combination can encourage intercalation without exfoliation. Common techniques that are introduced to prepare polymer–clay nanocomposites, are improvements of the clay by exchange of the interlayer cation with an organic cation or by chemical improvements to form organophilic clay for compatibility and shearing while milling of clay–polymer melts for exfoliation. By using water as an exfoliating agent

Progress in Rubber Nanocomposites. DOI: http://dx.doi.org/10.1016/B978-0-08-100409-8.00005-X

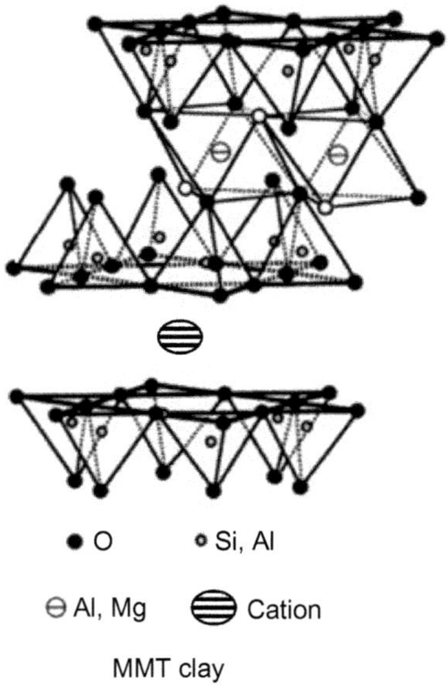

Figure 5.1 Schematic diagram showing layered structure of montmorillonite clay.

chemical improvement or organocation exchange is prevented. Furthermore, the physical chemistry of the particles is controlled by colloidal and solution properties. The octahedral sheets in MMT can be seen in Fig. 5.1. Besides MMT, other clay minerals have been used; such as synthetic fluoro-hectorite, synthetic micas, hectorite, and sepiolite. MMT comes from the 2:1 family of phyllosilicate clay minerals; where every crystal layer consists of a silica tetrahedral layer sandwiched between two octahedral sheets of aluminum and magnesium hydroxides.

5.2 Nanoclay structure, chemical modification, and characterization

Cations and water molecules are intercalated into the interlayer region of MMT clay. However, the gap between two successive layers is too small to accept polymer molecules into it. Organic modification is one of the methods that can be used to solve the problem by increasing the interlayer distance between clay nanolayers with the introduction of the ammonium compounds with a long hydrophobic tail as shown in Fig. 5.2. However, organic modification is not an accurate step to achieve intercalated and exfoliated structures in the rubber matrix. This does not always result in markedly improved mechanical properties of the rubber. This may be caused by the incapability of a considerable part of the layered material to separate into separate layers. This also leads to a lack of

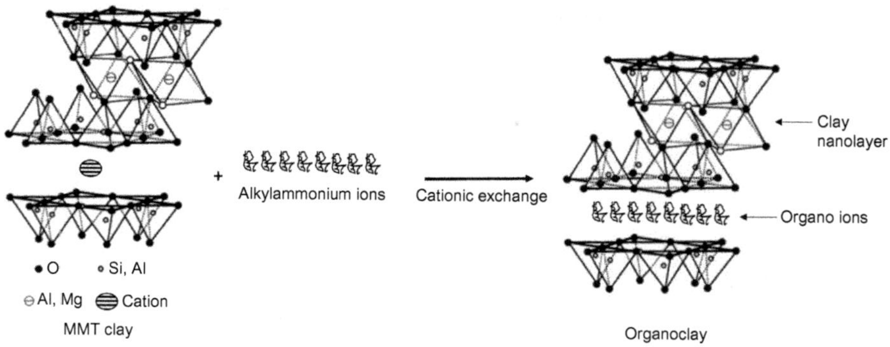

Figure 5.2 The process of formation of organoclay.

similarity between organic polymers and the inorganic layered silicates. Compared with thermoplastics, the dispersion of organoclay in rubber is more difficult due to the amorphous nature, low surface energy, and high viscosity of the rubber polymers being the vital causes. The reinforcing efficiency of organoclay also increases with a higher mixing temperature and shearing forces between rubber and clay during processing.

The sol–gel process is a chemical modification method adopted to clay for organomodification. Such methods are used primarily for the fabrication of materials starting from a chemical solution which acts as the precursor for an integrated network (or gel) of either distinct particles or network clays. Clay may form a sol (quick clay) if it is washed adequately to remove the counter ions. Quick clay may be gelled if enough counter ions are added, so that the colloidal particles combine. By adsorption of charged species on the surface on the particles, repulsion between the particles will increase and agglomeration will be prevented. The surface of a particle is covered by ionic groups, which determines the surface potential [6]. The synthesis of sol–gel MMT depends on procedure time and treatments, which can have an effect on the synthesis rates of sol–gel acids on MMT surfaces [7]. Organoclay preparation involves the following procedure. A small amount of sodium MMT is dispersed in half a liter of distilled water for 24 h at room temperature, using a magnetic stirrer. Using an aqueous solution of salt, the amount of surfactant added was equivalent to the (cationic exchange capacity) of clay, this cation exchange reaction occurs rapidly. This results in an organoclay suspension, which is then mixed for an additional 12 h. This suspension is afterwards filtered using various types of filter paper. The resulting organoclay was dispersed into 50 mL of fresh distilled water and mixed further for 4 h. It is later dried at 60°C under vacuum for 24 h. The final mixture is after that ground to form a fine powder. The abovementioned procedure is one of the methods used to make organoclay; however, various similar procedures are used to produce different types of organoclays [8]. In Fig. 5.2 the cation is removed and is replaced with an organo ion.

5.3 Type of rubbers and their characteristic properties

5.3.1 Nitrile rubber

Nitrile rubber (NBR) is commonly considered the workhorse of the industrial and automotive rubber products industries. NBR is actually a complex family of unsaturated copolymers of acrylonitrile and butadiene. By selecting an elastomer with the appropriate acrylonitrile content in balance with other properties, the NBR rubber is used in a wide variety of application areas, such as oil, fuel, and chemical resistance. In the automotive area, NBR is used in fuel and oil handling hose, seals and grommets, and water handling applications. With a temperature range of −40°C to +125°C, NBR materials can withstand severe automotive service environments. On the industrial side, NBR finds uses in roll covers, hydraulic hoses, conveyor belting, graphic arts, oil field packers, and seals for all kinds of plumbing and appliance applications. Like most unsaturated thermoset elastomers, NBR requires formulating with added ingredients, and further processing to make useful articles. Additional ingredients typically include reinforcement fillers, plasticizers, protectants, and vulcanization packages. Processing includes mixing, preforming to required shape, and application to substrates, extrusion, and vulcanization to make the finished rubber article. Mixing and processing are typically performed on open mills, internal mixers, and extruders. Finished products are found in industry as injection or transfer molded products, such as extruded hose or tubing, calendered sheet goods (floor mats and industrial belting) [9].

5.3.2 Chloroprene rubber

Chloroprene is the common name for the organic compound 2-chlorobutanol-1, 3-diene. This colorless liquid is the monomer in the production of the polymer polychloroprene, a type of synthetic rubber. Chloroprene rubber (CR) possesses the following properties: good chemical resistance; high resistance to aging and ozone; low risk of glazing; better wear resistance than natural rubber (NR); higher cost than NR [10]. Chloroprene elastomers are rigid, usually soft, but always elastically deformable under load. After stress, elastomers return to their initial shape. When heated, elastomers do not become plastic (and are not deformable), and will not melt. At high temperatures, decomposition takes place and the material is destroyed. The elasticity of an elastomer material is defined by the degree of crosslinking, which is called vulcanization or curing. Vulcanized elastomers are also called "rubber."

5.3.3 Styrene rubber

Poly(styrene-butadiene-styrene), or SBS, is a hard rubber that is used for products like the soles of shoes, tire treads, and other places where durability is important. It is a type of copolymer called a block copolymer. Its backbone chain is made up of three segments. The first is a long chain of polystyrene. The middle is a long chain of polybutadiene, and the last segment is another long section of polystyrene [11].

Polystyrene is a tough hard plastic, and this gives SBS its durability. Polybutadiene is rubbery, and this gives SBS its rubber-like properties. In addition, the polystyrene chains tend to clump together. When one styrene group from one SBS molecule joins one clump, and the other polystyrene chain of the same SBS molecule joins another clump, the different clumps become tied together with rubbery polybutadiene chains. This gives the material the ability to retain its shape after being stretched. SBS is made with an advanced chemistry called living anionic polymerization.

SBS is also a type of unique material called a thermoplastic elastomer. These are materials that behave like elastomeric rubbers at room temperature, but when heated, can be processed like plastics. Most types of rubber are difficult to process because they are cross-linked. However, SBS and other thermoplastic elastomers manage to be rubbery without being cross-linked, making them easy to process into nifty useful shapes [12].

5.4 Preparation of rubber nanoclay composites

Fig. 5.3 shows the schematic of NBR nanoclay composite formation. The various dispersion methods have different effects to the properties of the processed rubber–clay nanocomposites. Many physical properties are greatly dependent on the degree of nanoparticles dispersion in matrices. If there is no chemical bonding between the polymer and the particle, negative results will be seen in the

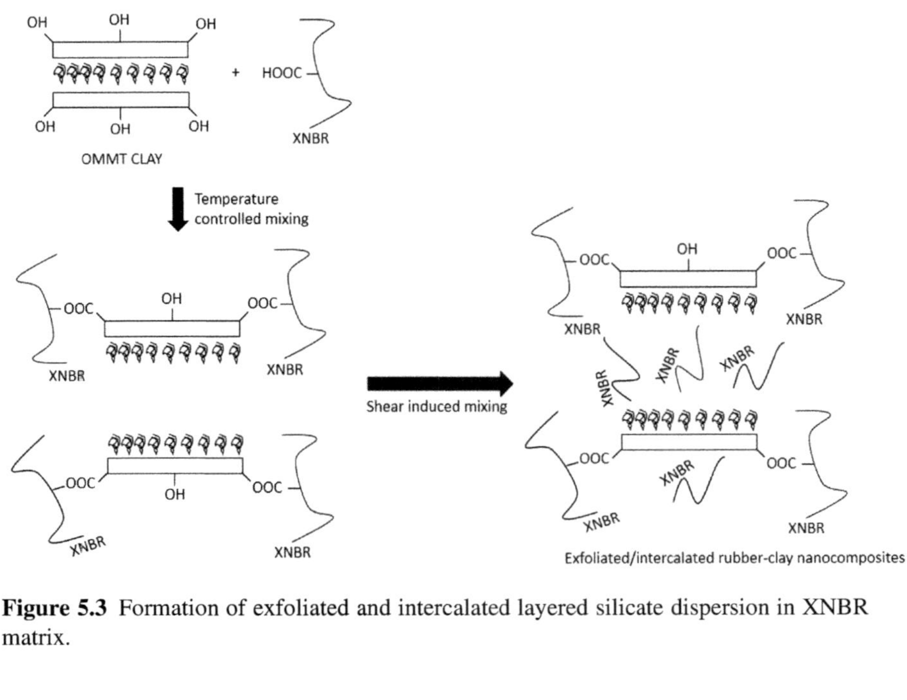

Figure 5.3 Formation of exfoliated and intercalated layered silicate dispersion in XNBR matrix.

mechanical properties of the composites, this is because the strong interfacial bonding can effectively transfer the load to the reinforcement. Therefore it is crucial that the correct manufacturing technique and dispersion methods are used for any particular type of rubber so as to ensure optimal bonding takes place with the filler.

5.4.1 Nitrile rubber with organically modified montmorillonite (OMMT)

NBR is a copolymer of acrylonitrile and butadiene and is mainly made by emulsion polymerization. By selecting appropriate acrylonitrile content, one can tailor the diverse properties in order to use NBR for a variety of applications, such as roll covers, conveyor belting, oil field packers, hydraulic hoses, and seals for all kinds of plumbing and appliances, and also for oil, fuel, and chemically resistant materials.

A variety of loadings of organically modified montmorillonites (OMMTs) were firstly added into low temperature-grade hydrogenated acrylonitrile butadiene rubber (LTG-HNBR) with the clay/rubber mass ratio of 0/100, 5/100, 10/100, 15/100, and 20/100 by a mixing roller for 3 min at room temperature. Thereafter, they were then kneaded to prepare LTG-HNBR blends on a two roll mill at 90°C for 10 min. On the other hand, a comparison of processing method was made so that some LTG-HNBR blends with the mass ratio of 10/100 were continued to be mixed for 10 min at room temperature. Other compounding and cross-linking additives (including 6.0 phr DCP, 3.0 phr ZnO, 3.0 phr MgO, 3.0 phr antioxidant, 2.0 phr stearic acid, and 1.5 phr triallyl isocyanurate) were added. They were continued to be kneaded at room temperature for about 10 min. Thereafter, the blends were hot-pressed under the pressure of 15 MPa and vulcanized at 160°C for 20 min, then naturally cooled down to room temperature. Finally, the vulcanized specimens were further put into an oven at the temperature of 150°C for 4 h sequentially to vulcanize entirely and avoid the aging effect on their short-term properties [13,14].

5.4.2 Chloroprene rubber

CR is well recognized for its high gum vulcanizate strength arising from strain-induced crystallization. The uncured rubber also shows storage hardening due to slow crystallization. It has excellent physical properties, weather resistance, thermal resistance, and can retain properties at low temperatures. The effect of two chemically opposite type of nanofillers, namely MMT and layered double hydroxide (LDH), both in pristine and organically modified forms Organically modified layered double hydroxide (OLDH and OMMT), influences the mechanical properties and crystallization behavior of CR.

The compounding of CR with fillers and vulcanization ingredients was prepared by a laboratory size open two roll mixing mill (Polymix 110 L, size: 203 × 102 mm Servitech GmbH, Wustermark, Germany). At first, necessary amounts of magnesium oxide (MgO) were mixed with masticated CR at 80°C. After addition of MgO, fillers and other ingredients like accelerator, stearic acid were compounded sequentially and finally ZnO was added after cooling down the mixing mill to 30°C. The total compounding cycle was finished within 15 min and the friction ratio of this mixing mill was 1:1.2 during the mixing tenure. After mixing the rubber with the above procedure the compounded sample, thus obtained, was subjected to a curing study to get the optimum curing time. This study was carried out with the help of a moving die rheometer at 160°C. The stocks were cured under pressure at 160°C to the optimum cure (t90) [15,16].

5.4.3 Styrene rubber

Natural rubber–styrene butadiene rubber type rubber is often used in automotive tires, using nanoclays additives. Compound mixing was carried out in two stages, in the first stage uncured rubber was mixed with crucial additives and the mixing was carried out in a laboratory manual controlled temperature control unit. Then a master batch, which consisted of uncured rubber and additives, was mixed at 60 rpm screw speed for 6 min and the batches were dumped at 145°C. After suitable maturation of the master batches (28 h), it was combined with the nanoclay and curatives at different dosages. The mixing was done at 40 rpm for 4 min and the batches were dumped at 100–110°C. Thereafter, final mixing and 24 h maturation was done and then molding was started. The compounds were sheeted out at the required thickness using laboratory mill. Moldings were done in a 22 in. × 24 in. steam heated automatic press at 140°C at a pressure of 10 MPa for 40 min. The molded pieces were cooled in water and conditioned at 23 ± 2°C and 50% relative humidity for 24 h before testing. Also, a two roll mixer was also adopted for rubber with 3 wt.% nanoclay composites [17,18].

5.4.4 Rubber blends

The compatibility between polychloroprene (CR) and ethylene propylene diene monomer (EPDM) rubber by using nanoclay as a compatibilizer and a typical mixing procedure was followed to get the CR/EPDM blend compounds. At first the clay was treated with stearic acid in a mortar and placed in the oven at 100°C for 30 min. After heating, the mass was ground with a pestle in its hot state and again the vessel was kept at 100°C. This hot material was cooled to room temperature and again it was ground. This compound was used as filler. The EPDM rubber was mixed with zinc oxide (ZnO) and the stearic acid modified clay was incorporated into it in a two roll mixing mill for 5 min. This mix, thus obtained, was added with premasticated CR and finally the curatives (accelerator and sulfur) were incorporated to the rubber compounds. For compounds without any clay (gum compounds), the stearic acid was added after the addition of ZnO to the EPDM. Finally the rubber compound was compression molded at 160°C to prepare samples with approximately 2 mm thickness. The curing time used in preparation of the vulcanization was calculated previously by making use of a moving die rheometer (Scarabeaus-SIS-V50). A detailed investigation on the sulfur curing activity of stearic acid

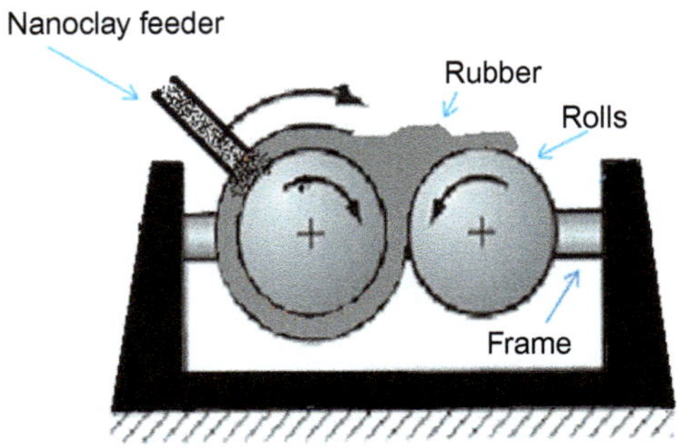

Figure 5.4 Two roll mixing mill machine.

modified clay has been done and reported elsewhere [19,20]. Prestrain of magnitude 40% was chosen to hold the sample with the size of $10 \times 2 \times 35$ mm^3.

Wetting experiments (Wilhelmy method and capillary penetration method) were performed, using the dynamic contact angle meter and tensiometer DCAT 21 (DataPhysics Instruments GmbH, Filderstadt, Germany) [21,22].

5.5 Manufacturing techniques

5.5.1 *Two roll mill mixing*

This process is mixing that is carried out between two contrary rotors as shown in Fig. 5.4. The wicks of both of the rotors are arranged in a parallel manner and possess different speeds. Due to the difference in tangential speed and nip construction, the rubber compounds within the machine experience pressure energy and displacement during the process of mixing. Pressure energy increases the displacement energy, and displacement energy breaks the rubber molecule bonding. The breakage of rubber molecules allows carbon and other chemicals to be uniformly mixed with rubber, which is desirable so that the rubber produced is isotropic in nature [23].

Two roll mixing mills are widely used for mixing and kneading rubber blends of chloroprene, styrene, and other types of rubbers. Nanoclay and other rubber constituents are released from a feeder within the framework.

5.5.2 *Latex compounding*

NR-based nanocomposites with the 10 wt% natural and synthetic layered silicates are produced via the latex compounding method. The process begins when raw latex, which is derived from tropical rubber trees, is mixed with curing agents in a steel tank. A cooling system outside the mixing tank keeps the compound at constant temperatures to avoid variations. The resultant compound is fed using the material handling system so it can be mixed with nanoclay powder as shown in Fig. 5.5. Pressing and molding into desired shapes then follows and compounds are

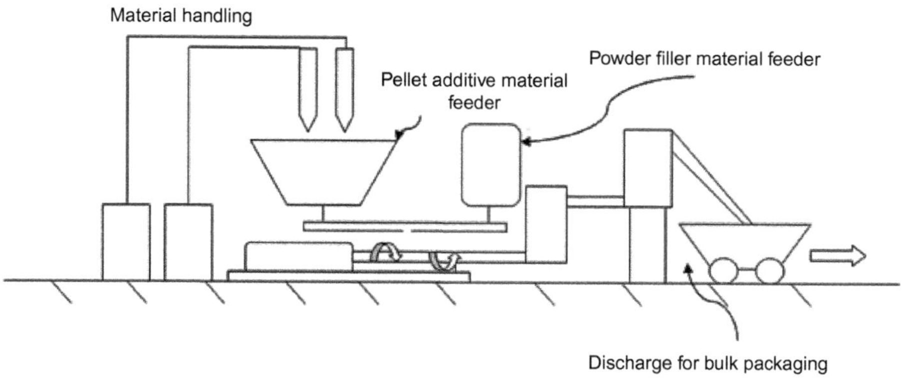

Figure 5.5 Stages involved in latex compounding.

taken for packaging. The nanocomposites are prepared by compounding the dispersions of clays and other latex chemicals necessary for vulcanization. Layered silicate rubber nanocomposites outperformed the reference material in all aspects. This is attributed to the intercalation of the silicates and to the formation of a skeleton silicate network in the NR matrix [24].

5.5.3 In situ polymerization

Fig. 5.6 shows in situ polymer chemistry; in situ polymerization means "in the polymerization mixture." There are numerous unstable molecules which must be synthesized in situ (i.e., in the reaction mixture but cannot be isolated on their own) for use in various processes, of which one of these processes is manufacturing of nanorubber.

One example of this method is protein nanogels made by the in situ polymerization method. In such a type of in situ polymerization, a chemical encapsulation technique is involved that is very similar to interfacial coating. The distinguishing characteristic of in situ polymerization is that no reactants are included among the core material. All polymerization occurs in the continuous phase, rather than on both sides of the interface between the continuous phase and the core material [25].

5.5.4 Freeze drying

Pretreatment includes any method of treating the product before freezing. This should include concentrating the product and addition of components to increase stability, preserve appearance, decreasing vapor pressure solvent, or increasing the surface area. In many instances the decision to pretreat a product is based on theoretical knowledge of freeze drying and its requirements.

Fig. 5.7 shows the steps followed in freeze drying. In a lab, this is often done by placing the material in a freeze-drying flask and rotating the flask in a bath, called a shell freezer, which is cooled by mechanical refrigeration, dry ice and methanol, or

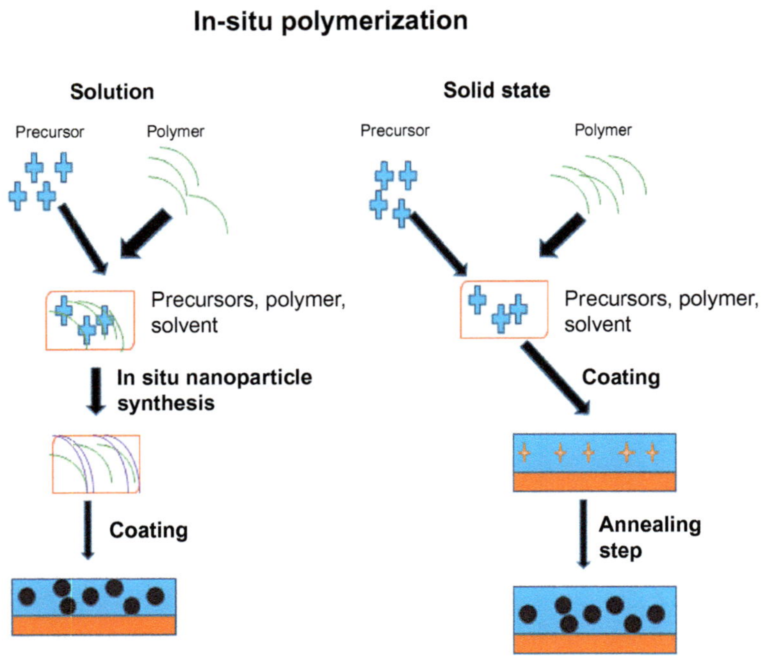

Figure 5.6 Schematic approach of in situ polymerization.

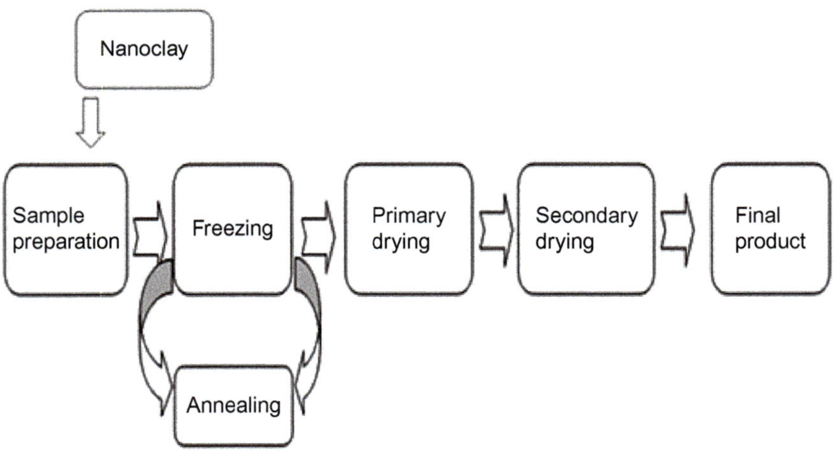

Figure 5.7 Freeze drying cycle of rubber–nanoclay composites.

liquid nitrogen on a larger scale. In this step, it is important to cool the material below its triple point, the lowest temperature at which the solid and liquid phases of the material can coexist. This ensures that sublimation rather than melting will occur in the following steps. Larger crystals are easier to freeze dry. To produce larger crystals, the product should be frozen slowly, resulting in the destruction of more

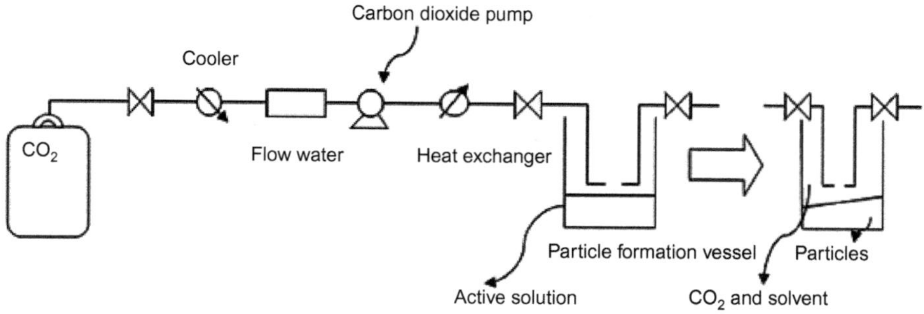

Figure 5.8 Schematic presentation for supercritical carbon dioxide assisted preparation.

cells, which can lead to an increasingly poor texture and nutritive content. In this case, the freezing is done rapidly, in order to lower the material to below its eutectic point quickly, thus avoiding the formation of ice crystals. Usually, the freezing temperatures are between −50 and −80°C. The freezing phase is the most critical in the whole freeze-drying process, because the product can be spoiled if improperly done. However, amorphous materials do not have a eutectic point, but they do have a critical point below which the product must be maintained to prevent melt-back or collapse during primary and secondary drying. During the primary drying phase, the pressure is lowered (to the range of a few millibars), and enough heat is supplied to the material for the ice to sublime [26]. The secondary drying phase aims to remove unfrozen water molecules, since the ice was removed in the primary drying phase.

5.5.5 Supercritical CO_2 assisted preparation

A supercritical carbon dioxide assisted preparation process is shown in Fig. 5.8. There is a huge interest in the design and development of advanced polymeric materials using clean and sustainable technologies. During this process, the carbon dioxide is pumped to a tank (extractor) containing the raw materials that are to be prepared. The carbon dioxide in a gaseous state extracts the desired particles from the raw material and is further directed to a separation vessel. The useful particles remain on the bottom of the vessel for extraction while the carbon dioxide and the solvent are recycled. The preparation of membranes using these technologies encounters numerous advantages due to the tunable properties of supercritical carbon dioxide [27].

5.6 Nanocomposite structure and characterization of structure and morphology

5.6.1 Transmission electron microscopy (TEM)

In order to understand the dispersion of nanoclays in the rubber matrix, their morphologies are checked by TEM. Fig. 5.9 shows TEM images of the LTG-HNBR

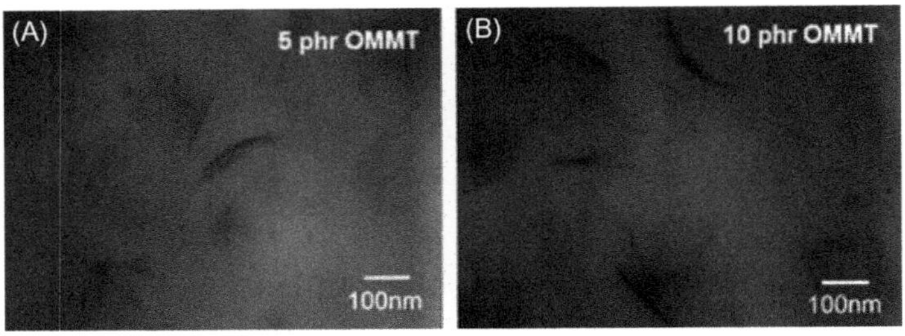

Figure 5.9 Representative TEM images of the rubber nanocomposites including different OMMT amounts: (A) 5 phr and (B) 10 phr [28].

nanocomposites with 5 phr and 10 phr clays, respectively. There are two phase domains in the composites, i.e., matrix and dispersed Organo-MMT (OMMT) phase. It can be seen that OMMT clays are partially exfoliated and intercalated in the rubber matrix. It was observed that there were more intercalated structures (thick layer-bundles) of clays in 10 phr OMMTs than in 5 phr OMMTs. The OMMT tends to agglomerate in the matrix when their amounts are above 10 phr. It is noticeable that the aggregations or agglomerations of plate-like OMMTs would suppress the interactions of clay networks with the matrix [28].

To accurately describe the dispersion of clay particles on the properties of a rubber matrix, a digital analysis for the clay particles can also be conducted. Most attention is paid to the nanocomposites with 5 phr and 10 phr OMMTs because other measurements, such as SEM observations, are difficult to make a clear distinction for fine clay-structures and other contents. The TEM images of polymer compounding with clays can be digitized. Thus the length, thickness, and aspect ratio of the traced particles is simultaneously analyzed by various applicable software (such as Image-Pro Plus). Fig. 5.10 shows a series of histograms of particle length and aspect ratio for HNBR nanocomposites containing various amounts of OMMTs. All of the features show broad distributions. On the other hand, special defined sizes of dispersed clays are also calculated on base of imaging analysis (i.e., the number and weight averages of the length, thickness, and aspect ratio). The statistical results of particle analysis on nanocomposites containing 5 phr and 10 phr OMMTs are listed in Table 5.1. The average particle lengths for the nanocomposite with 10 phr OMMTs are a bit larger than that for the 5 phr nanocomposite. The situation parallels the average particle thickness. However, a different trend is seen in their average aspect ratios, both for the number average and the weight average. This suggests that a small amount of OMMT can more easily exfoliate than intercalate because the two roll mill can provide stronger shear stress on the matrix to peel fewer clays. Furthermore, the higher the amounts of OMMTs in the matrix are, the closer to the original clays their contour features are (a large length and thickness), implying the development of bundled or agglomerated clay-platelets with the increase of OMMT amounts. It was shown that all data on the aspect ratio is roughly close between the nanocomposites with 5 phr and 10 phr OMMTs. This means that there would be major improvements on basic properties of the composite with 10 phr OMMTs owing to the increase of clay amounts.

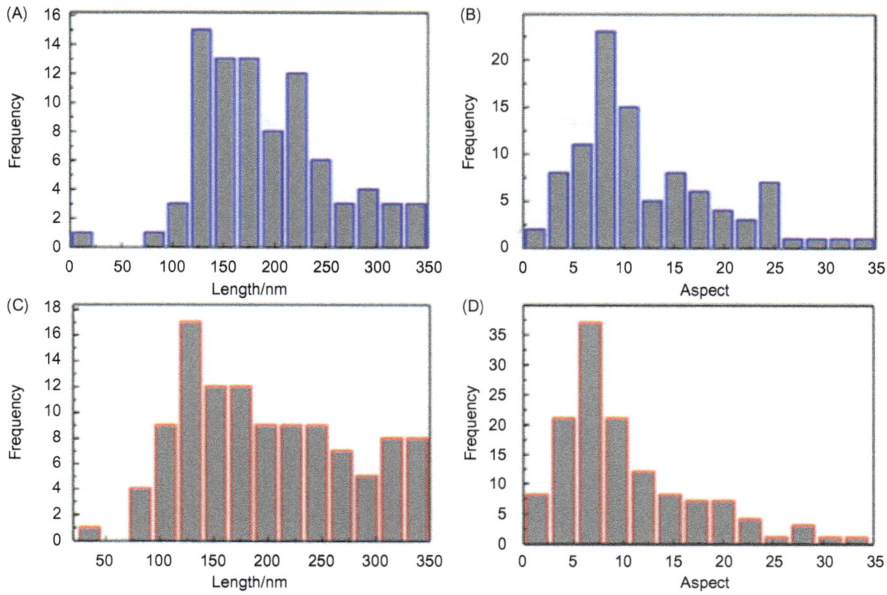

Figure 5.10 Histograms of particle length and aspect ratio for rubber nanocomposites with different OMMT amounts: (A and B) 5 phr; (C and D) 10 phr [28].

5.6.2 Small angle X-ray scattering (SAXS)

Small angle X-ray scattering (SAXS) provides supplementary microstructure information about the distribution of clays in the rubber matrix to TEM observations. SAXS scattering patterns of rubber, original OMMT, and related rubber nanocomposites are shown in Fig. 5.11. Fig. 5.11(A) shows 2D scattering striations of organoclay and its nanocomposites. The scattering patterns become more distinct for nanocomposites than the original OMMT, which is due to the higher aspect ratio of peeled clays in the rubber matrix [29]. In agreement with TEM results, the multi-peak structures in the intensity-scattering vector (q) plots show that OMMTs mainly present intercalated ones in LTG-HNBR matrix, as shown in Fig. 5.11(B), however no peaks give the idea of the amorphous state of neat rubber. For nanocomposites, the d-spacing of the clays is projected from the scattering patterns according to Bragg's law. It is seen that the d-spacing ((0 0 1) primary peak) enlarges the interlayer space of OMMTs from 2.22 nm to 3.21 nm upon 5 phr when mixed with rubber matrix. Further increases of, respectively, 3.41 nm, 3.42 nm, and 3.45 nm are observed when more OMMTs (10 phr, 15 phr, and 20 phr, correspondingly) are added into the matrix. Furthermore, the scattering intensity is also improved owing to the increase of clay amounts. Fig. 5.11(C) shows the OMMT content dependence of d-spacing ((0 0 1) peak) and corresponding scattering angle 2θ. The angle of 2θ can be expressed by using the vector q as per Eq. (5.1)

Table 5.1 The statistical results of particle analysis for different nanocomposites [28]

The amounts of OMMT (phr)	Total number of particles	Number average particle length (l_n, nm)	Number average particle thickness (t_n, nm)	Weight average particle length (l_w, nm)	Weight average particle thickness (t_w, nm)	Number average particle aspect ratio $(l/t)_n$	l_n/t_n	Weight average particle aspect ratio $(i/t)_w$	l_w/t_w
5	96	190.9	30.5	212.3	45.1	12.2	6.3	16.4	4.7
10	124	199.6	36.5	228.9	52	10.1	5.5	14.6	4.4

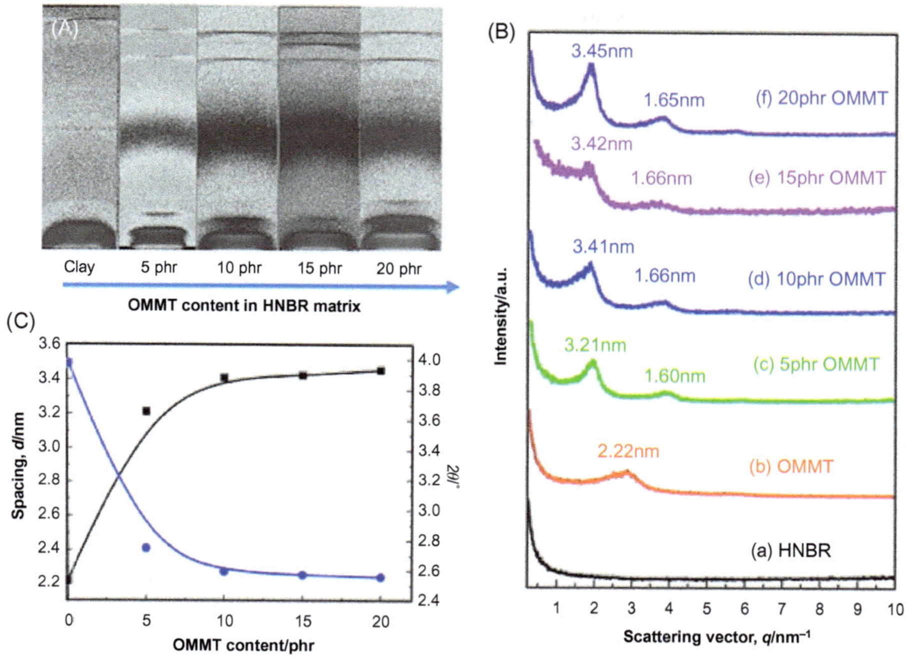

Figure 5.11 SAXS scattering patterns of neat rubber and corresponding nanocomposites: (A) 2D SAXS patterns; (B) the plots of intensity versus scattering vector q; (C) the OMMT content dependence of d-spacing (the black square) and 2θ at (0 0 1) peak (the blue circle) [28].

$$2\theta = 2 \times \arcsin\left(\frac{\lambda q}{4\pi}\right) \tag{5.1}$$

5.6.3 Cure characteristics

Fig. 5.12 shows the % cure versus period of the time for the nanocomposites. It shows that the cure characteristics of base rubber changes when the nanoclays are filled. Table 5.2 shows the time required to achieve 90% cure in nanocomposites series. Base rubber takes 12.68 min to cure and it is continuously reduced when the clay concentration is increased in the base rubber. Maximum decrease is observed for 3 wt.% nanoclay filled rubber and the 90% cure is at 9.71 min, a 30% faster curing than rubber. The faster curing may be attributed to the cross-link reactions of onium ions of clay and matrix rubber [30,31]. The fast curing at 3 wt.% clay rubber result in intercalated structure since the rubber polymer may not have enough time and volume to separate the clay nanolayers to exfoliate.

A schematic illustration of nanocomposite formation, based on the curing mechanism, from initial to final stage is depicted in Fig. 5.13. Fig. 5.13(A) shows

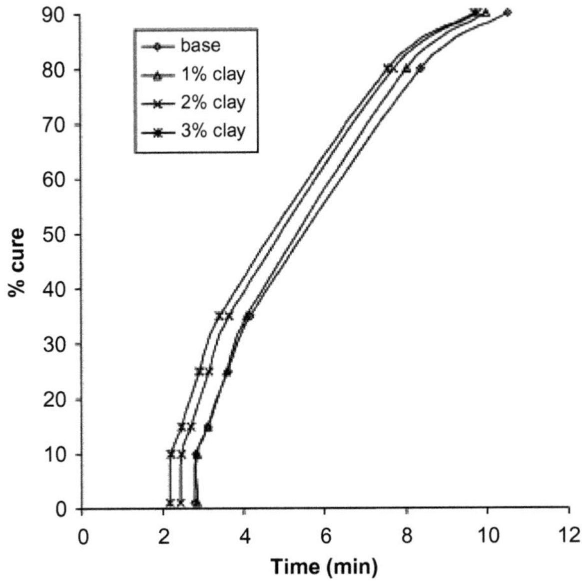

Figure 5.12 % cure versus time for nanocomposite series [30].

Table 5.2 Time for achieving 90% cure of nanocomposites series [30]

Clay wt. %	Time to get 90% cure, min
0	12.68
1	10.01
2	9.81
3	9.71

the initial stage of nanocomposite formation. In this stage a phase separated matrix and nanoclay exist in the composite structure. In the second stage, rubber polymers enter into the interlayer spacing of nanoclays and enlarges the interlayer distance Fig. 5.13(B). After this, more matrix polymer enters into the interlayer spacing of nanoclay and separates the nanolayers further and eventually forms the random dispersion of clay's nanolayers in the polymer matrix Fig. 5.13(C). This random dispersion of nanolayers in matrix is called an exfoliated structure. In a special case, if nanolayers of clay were not able to separate and maintain their original existence of interlayer spacing of nanoclay in polymer matrix, then the composite is called conventionally clay filled particle microcomposites [13]. The cure kinetics, nanoclay concentration, and mixing process affect the final nanocomposite structure.

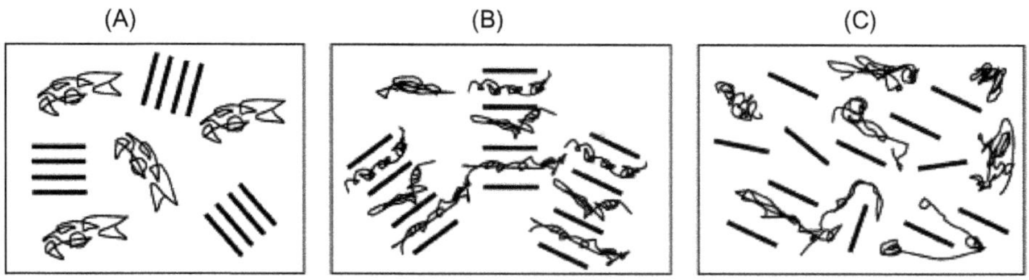

Figure 5.13 Diagram of rubber-clay nanocomposite formation during (A) phase separated rubber-clay mixing, (B) intercalation and (C) exfoliation [30].

Table 5.3 The mechanical properties of neat rubber, its nanocomposites, and a compared case with 10 phr carbon black (N330) [19]

Property	OMMT content (phr)					N330 content (phr)
	0	5	10	15	20	10
Tensile strength (MPa)	1.03	1.94	9.61	10.29	11.12	2.4
Elongation at break (%)	89	100	268	257	261	158
Permanent set (%)	3.0	2.2	10.2	11.5	14.4	5.2
Tearing strength (KN/m)	8.25	11.3	19	19.5	22.3	13.7
Swelling ratio (at 100°C for 72 h, in No.10 oil)	1.42	1.40	1.39	1.38	1.38	1.41

5.7 Properties of nanocomposites

5.7.1 Mechanical properties

Table 5.3 lists the mechanical properties of LTG-HNBR nanocomposites combined with various amounts of OMMTs. It is found that tensile strength of nanocomposites increases dramatically with the increment of OMMT content. For example, it has grown more than 10 times for the case with 20 phr OMMTs (11.1 MPa) in contrast with the neat one. Afterwards, a tendency of the tensile strength to increase runs slowly with the increments of OMMT amounts, such as the 20 phr one. Improved hardness, the elongation at break, and tearing strength of rubber nanocomposites are in concordance with tensile results. Their permanent set is also raised by the enhancements of the viscosity from networks of OMMT fillers. When comparing the organoclay with carbon black as filler for LTG-HNBR, the reinforcement of the organoclay is more noticeable. An example is when high abrasion furnace black (N330) with the amount of 10 phr was added into the matrix. As expected, its tensile strength is only 2.4 MPa which is far less than that with the

Table 5.4 Hardness and tear properties of nanocomposite series [34]

Clay, wt.%	Shore A hardness	Cold tear strength, N/mm
0	60	208.6
1	65	214.5
2	67	219.4
3	70	187.3

Table 5.5 Effect of processing condition on properties of 3 wt% nanoclay composites [34]

Process condition	Tensile strength, MPa	Modulus at 100% elongation, MPa	Elongation at break, %	Cold tear strength, N/mm
Banbury	17.112	3.640	380	187.3
Two roll	17.610	3.630	377	186.1

same clay content. Nah et al. [23] brought forward the model to explain the reinforcement of clays on polymer nanocomposites by the increase of the crack path around these silicate layers ("zig-zag" route). Nanoclay is capable of dissipating input energy to withstand greater stress than those with carbon black. This is true for the polyurethane [32], carboxylated hydrogenated nitrile butadiene rubber matrix composites [33], and equally valid in our nanocomposites.

Table 5.4 shows the hardness and cold tear strength of rubber–nanoclay composites. It is observed that hardness of rubber is 60 Shore A and the hardness is continuously increased as the clay concentration increases in the polymer matrix. Maximum increase in hardness is obtained at 3 wt. % nanoclay concentration of 70 Shore A. The increase in hardness of nanocomposites suggests that the resistance to indentation is greater when nanoclays are filling in the rubber polymer matrix.

The cold tear strength of base rubber is 208.6 N/mm. The cold tear strength increased to a maximum value at 2 wt. % nanoclay, and at 3 wt. % reduces the cold tear strength. The decrease in these properties at 3 wt. % is due to the agglomeration. The agglomerated particle is the weak area under the applied load. The crack initiation or propagation is more likely in the agglomeration zone than in the well dispersed nanoclay zone. To overcome this effect, the processing technique was changed to reduce agglomeration in 3 wt. % clay filled composites. The 3 wt. % nanoclay filled composites was processed by a two roll mixer and the results are shown in Table 5.5. The results show that the property of 3 wt. % nanoclay filled composites is almost constant in both the processing conditions. Hence the results suggest that the agglomeration of nanoclays is an inevitable phenomenon in the composites at higher concentration of clays and causes reduced properties irrespective of processing conditions and may be improved by use of coupling agents [34].

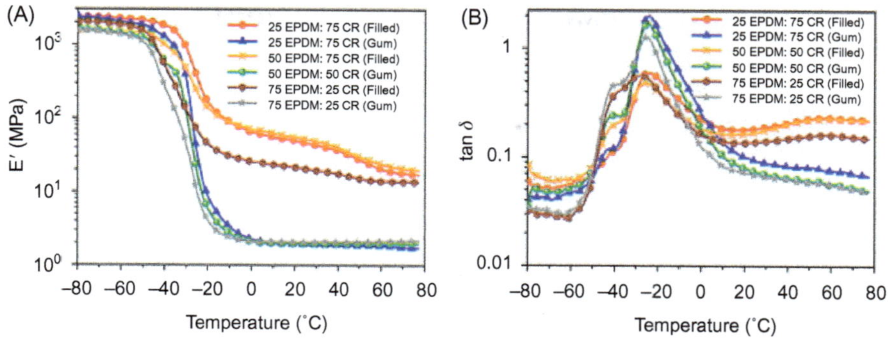

Figure 5.14 (A) Storage modulus (E') versus temperature and (B) tan δ versus temperature plots of CR/EPDM blends in absence and in presence of nanoclay [35].

5.7.2 Dynamic mechanical analysis (DMA)

Dynamic mechanical analysis has been done to understand the dynamic response of the blend after the addition of clay. The dependency of storage modulus obtained from oscillatory tension deformation as a function of temperature is given in Fig. 5.14. All samples show a steep decrease of E' value at the temperature range between $T = -50$ and $-20°C$ followed by a rubbery plateau (Fig. 5.14A). The increase of modulus values at room temperature by the addition of the clay was observed. The storage modulus increases from 2 MPa to 54 MPa with the addition of only 10 phr clay in the EPDM/CR matrix at a ratio of 25/75. A good state of exfoliation of the clay in the CR matrix reinforced the rubber blend very strongly. Sufficiently strong filler–filler interactions as well as the compatibility between CR and EPDM play major roles in giving rise to a very highly reinforced rubber matrix.

The glass transition temperature (T_g) of a rubber or polymer depends on the structure and cooperative mobility of the chain segments. In the case of partially compatible blends, the T_gs of the blend components are expected to be shifted towards each other as compared with the pure components. The T_gs remain largely unaltered for a completely incompatible blend. Fig. 5.14B shows the tan δ dependencies on the temperature for the CR/EPDM blend systems, and the glass transition temperature of the corresponding rubbers are shown in Table 5.6. The two pure vulcanized rubbers show a T_g at -37 and $-24°C$, and even after incorporation of the clay the values remain unaltered. However, those rubbers remaining in the blend system give rise to different T_g as compared with their virgin rubbers. It is clear from Table 5.6 that the T_gs are approaching more closely when the blend matrix is filled with 10 phr clay. This indicates that the clay has a strong capability to offer compatibility between the two hetero-rubbers. A shift of 9°C was observed for 75/25 blend of CR/EPDM. An opposite type of behavior was found for a heterogeneous rubber blend as revealed by the situation when the clay interferes with the compatibility between chloroprene and carboxylated nitrile rubber (XNBR) during the self-cross-linking reaction.

Table 5.6 Glass transition temperature of CR/EPDM rubber blend

Rubber (EPDM: CR)	Glass transition temperature, T_g (°C)					
	Without clay			With clay		
	EPDM	CR	Separation between two peaks	EPDM	CR	Separation between two peaks
100:0	−37	*a*	–	−37	*a*	–
75:25	−44	−25	19	−42	−26	16
50:50	−42	−25	17	−41	−27	14
25:75	−40	−24	16	−35	−28	7
0:100	*a*	−24	–	*a*	−24	–

The curing recipe used in this study is given as follows: 5 phr zinc oxide, 10 phr stearic acid, 7 phr ZDMC, and 1 phr sulfur. Here "phr" stands for parts per hundred of rubber. "*a*" stands for absent for that particular case.

It is also important to note that even at higher temperatures ($T = 40-60°C$) a second relaxation process is observed for the blends containing clay. This relaxation process is flat and broad. Such a process at the same temperature range has been reported by us for a CR–clay composite. In this work, CR chain segments are partially crystallized in the presence of organically modified clay. Very recently [36], such a process has been explained by considering the fact that, during the measurement of the dynamic-mechanical properties, the anisotropic fillers are arranged in a way that they have to reorient by following the applied mechanical field, and additional energy dissipation sets in resulting in the process occurring at higher temperatures [37]. Previous studies also found that melt-like thermal transition of alkyl chains of the surfactant molecules of the modified clay could show a high temperature relaxation peak. Due to the order to disorder-like transition at this temperature the alkyl chains of the surfactant behave like a liquid state, and thus plasticize the rubber chains resulting in an extra relaxation at high temperature [38].

5.7.3 Rheological properties

In order to display the thermal stability or thermal resistant properties, the thermogravimetric analysis (TGA) curves of neat rubber and corresponding composite containing 10 phr OMMTs are shown in Fig. 5.15. The mass of neat HNBR is only 6.4% after the decomposition, its onset decomposition temperature is 372°C, and the ending decomposition temperature is 498°C. In contrast, the residual mass of nanocomposite is 20.1%, its onset and ending decomposition temperatures are 392 and 508°C, respectively. The addition of OMMTs improves the thermal stability of rubber bulk just as described in the OMMT/EPDM system. Moreover, the decomposition of the surfactants on the organoclays is still not seen in the plots owing to their strong interactions with matrix. The improved thermal stability of

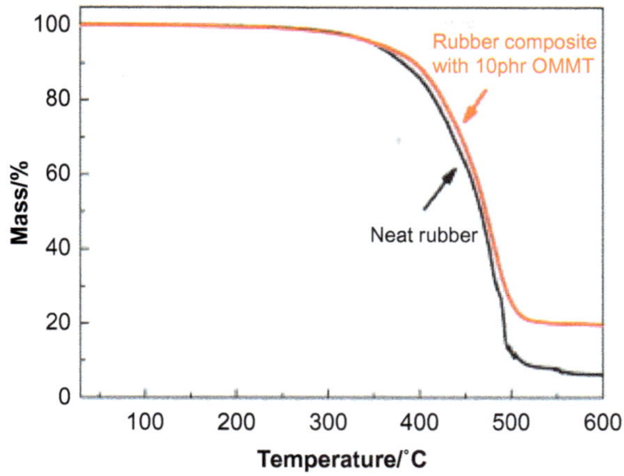

Figure 5.15 The T_g curves of rubber and its nanocomposites containing 10 phr OMMTs [28].

nanocomposites is attributed to the action of clay layers as a superior insulator, a mass transport barrier to the volatile products generated, or the presence of inorganic phases such as SiO_2, Al_2O_3, and MgO in clay particles during decomposition. Inspired by the swelling properties, the slowing down of the escape of the volatile products in nanocomposites is also from the labyrinth paths of the silicate layers in the polymer matrix.

5.7.4 Swelling

The degree of swelling of LTG-HNBR composites at equilibrium is an important property to evaluate the number of filler–rubber attachments and their ability of medium-resistance, especially for oil. To study the equilibrium of the swelling, different soaking times (24 h, 48 h, 72 h, and 120 h, respectively) at the temperature of 100°C were examined. Without the consideration of possible anisotropic swelling caused by the structures of OMMTs [39], an average swelling volume is measured by Archimedes' principle. It was found that the swelling ratios of all specimens was constant (an equilibrium swelling) after they were soaked for 72 h. The nanocomposites have better barrier properties for hydraulic oil by comparison with the neat rubber one. It is well explained that the dispersion of nanofillers increases the tortuosity of the path to diffuse oil molecules during the permeation [40]. As the clays do not swell, the equilibrium swelling ratio of the rubber phase Q_r from the equilibrium swelling ratio Q of the composite and from the filler volume fraction Φ_{OM} [41], is calculated as per Eq. 5.2.

$$Q_r = (Q - \Phi_{OM})/(1 - \Phi_{OM}) \quad (5.2)$$

The relationship between the swelling ratio Q_r of the rubber and clay fraction Φ_{OM} is shown in Fig. 5.16. Note that the swelling of rubber for all nanocomposites

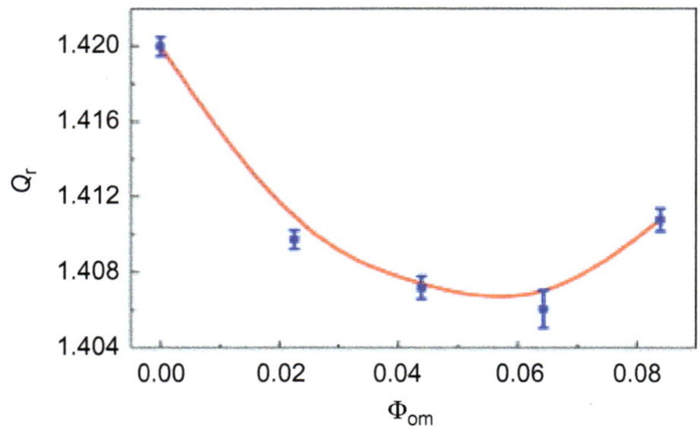

Figure 5.16 The plot of the swelling ratio Q_r of the rubber versus clay fraction Φ_{OM} for the rubber nanocomposites in hydraulic oil [28].

is less than the neat one. The swelling of rubber reaches a low point when the OMMT content is between 10 phr and 15 phr. Bokobza et al. investigated the NR reinforced by clays and proposed that the extent of restricted swelling for the rubber matrix reflects the interactions of organoclays with them. So this strengthens the fact that the interaction of OMMTs with the rubber matrix at such content is very strong. It is also stressed that the dispersion and morphology state of fillers heavily affects the barrier property of the bulk for oil liquid.

5.8 Conclusion and applications

In this chapter nanoclay was used as a filler mixed with different types of rubber matrices. Firstly, for the LTG-HNBR nanocomposites with organoclays, the OMMTs form partially intercalated and exfoliated morphologies in the rubber matrix. It is evident that there are strong clay-networks in the matrix where the addition of 10 phr is the optimum quantity to exploit it. In addition, the mechanical properties of LTG-HNBR are significantly improved by the good dispersion of clays. Whereas, polychloroprene rubber shows significant variations in properties when small amounts of fillers are incorporated. However, after addition of this unmodified filler the mechanical properties were not changed significantly. Modification of the surface of the clay by cationic species improved the physical and dynamic properties to a remarkable degree by the virtue of exfoliation–intercalation processes. However, in the case of nanoclay as an effective compatibilizer as well as a reinforcing agent in the blend of CR and EPDM rubber, it is observed that the nanoclay decreases the interfacial energy between the phases by accumulating at the interfaces, permits a finer intercalation–exfoliation process during mixing and vulcanization, and provides a strong reinforcement to the rubber blends. Also observed was that an exfoliated nanocomposite structure up to 2 wt% clay and at 3 wt% nanoclay intercalated structure is obtained in the composite.

Marked improvements in modulus, hardness, and thermal mass decomposition are observed in this nanocomposite series and in particularly at 2 wt. % nanoclay filled rubber composites. The prospects of using nanoclay in the commonly used rubber materials is encouraging, and this can be further taken into the high end rubber applications, such as aerospace and sporting products, which can realize the improved viscoelastic and thermal properties. Overall rubber nanocomposites can be used widely in the automotive industry as well as in aerospace applications.

References

[1] Das A, Wang D-Y, Stöckelhuber KW, Jurk R, Fritzsche J, Kluppel M, et al. Rubber–clay nanocomposites: some recent results. Adv Polym Sci 2011;239:85–166.
[2] Zachariaha AK, Chandrab AK, Mohammed PK, Parameswaranpillaic J, Thomas S. Experiments and modeling of non-linear viscoelastic responses in natural rubber and chlorobutyl rubber nanocomposites. Appl Clay Sci 2016;123:1–10.
[3] Rezende CA, Bragança FC, Doi TR, Lee L-T, Galembeck F, Boué F, et al. Natural rubber-clay nanocomposites: mechanical and structural properties. Polymer (Guildf) 2010;51:3644–52.
[4] Sengupta R, Chakraborty S, Bandyopadhyay S, Dasgupta S, Mukhopadhyay R, Auddy K, et al. A short review on rubber/clay nanocomposites with emphasis on mechanical properties. Polym Eng Sci 2007;47:1956–74.
[5] Yaragallaa S, Meera AP, Kalarikkala N, Thomas S. Chemistry associated with natural rubber–graphene nanocomposites and its effect on physical and structural properties. Ind Crop Prod 2015;74:792–802.
[6] Cerisuelo JP, Alonso J, Aucejo S, Gavara R, Hernández-Muñoz P. Modifications induced by the addition of a nanoclay in the functional and active properties of an EVOH film containing carvacrol for food packaging. J Memb Sci 2012; 423–424:247–56.
[7] Ganjaee Sari M, Ramezanzadeh B, Shahbazi M, Pakdel AS. Influence of nanoclay particles modification by polyester-amide hyperbranched polymer on the corrosion protective performance of the epoxy nanocomposite. Corros Sci 2015;92:162–72.
[8] Singla P, Mehta R, Upadhyay SN. Clay modification by the use of organic cations. Green Sustain Chem 2012;2:21–5.
[9] Misman MA, Azura AR, Hamid ZAA. The physical and degradation properties of starch-graft-acrylonitrile/carboxylated nitrile butadiene rubber latex films. Carbohydr Polym 2015;128:1–10.
[10] Ahmed K. An investigation on chloroprene-compatibilized acrylonitrile butadiene rubber/high density polyethylene blends. J Adv Res 2015;6(6):811–17.
[11] Usha Devi KS, Ponnamma D, Causin V, Maria HJ, Thomas S. Enhanced morphology and mechanical characteristics of clay/styrene butadiene rubber nanocomposites. Appl Clay Sci 2015;114:568–76.
[12] Zhang Y, Liu Q, Zhang S, Zhang Y, Cheng H. Gas barrier properties and mechanism of kaolin/styrene-butadiene rubber nanocomposites. Appl Clay Sci 2015;111:37–43.
[13] Balachandran M, Devanathan S, Muraleekrishnan R, Bhagawan SS. Optimizing properties of nanoclay-nitrile rubber (NBR) composites using face centre composite design. Mater Design 2012;35:854–62.

[14] LeBaron PC, Wang Z, Pinnavaia TJ. Polymer-layered silicate nanocomposites: an overview. Appl Clay Sci 1999;15:11–29.
[15] Das A, Costa FR, Wagenknecht U, Heinrich G. Nanocomposite based on chloroprene rubber, effect of chemical nature and organic modification of nanoclay on the vulcanization properties. Eur Polym J 2008;44:3456–65.
[16] Siriwong C, Sae-Oui P, Sirisinha C. Comparison of coupling effectiveness among amino-, chloro-, and mercapto silanes in chloroprene rubber. Polym Test 2014;38:64–72.
[17] Li Y, Wu Y, Luo Y, Chan TW, Zhang L, Wu S. A combined experimental and molecular dynamics simulation study on the structures and properties of three types of styrene butadiene rubber. Mater Today Commun 2015;4:35–41.
[18] Wongthong P, Nakason C, Pan Q, Rempel GL, Kiatkamjornwong S. Styrene-assisted grafting of maleic anhydride onto daproteinized natural rubber. Eur Polym J 2014;59:144–55.
[19] Das A, Stöckelhuber KW, Jurk R, Jehnichen D, Heinrich G. A general approach to rubber–montmorillonite nanocomposites: intercalation of stearic acid. Appl Clay Sci 2011;51:117–25.
[20] Conzatti L, Stagnaro P, Colucci G, Bongiovanni R, Priola A, Lostritto A, et al. The clay mineral modifier as the key to steer the properties of rubber nanocomposites. Appl Clay Sci 2012;61:14–21.
[21] Stöckelhuber KW, Das A, Jurk R, Heinrich G. Contribution of physico-chemical properties of interfaces on dispersibility, adhesion and flocculation of filler particles in rubber. Polymer (Guildf) 2010;51:1954.
[22] Nah C, Ryu H, Kim W, Chang Y. Preparation and properties of acrylonitrile-butadiene copolymer hybrid nanocomposites with organoclays. Polym Int 2003;52:1359–64.
[23] Yu S, Adams M, Gururajan B, Reynolds G, Roberts R, Wu C-Y. The effects of lubrication on roll compaction, ribbon milling and tabletting. Chem Eng Sci 2013;86:9–18.
[24] Xing W, Tang M, Wu J, Huang G, Li H, Lei Z, et al. Multifunctional properties of graphene/rubber nanocomposites fabricated by a modified latex compounding method. Compos Sci Technol 2014;9:67–74.
[25] Cromer BM, Scheel S, Luinstra GA, Coughlin EB, Lesser AJ. In-situ polymization of isotactic polypropylene-nanographite nanocomposites. Polymer (Guildf) 2015;80: 275–81.
[26] Wais U, Jackson AW, Zuo Y, Xiang Y, He T, Zhang H. Drug nanoparticles by emulsion-freeze drying via the employment of branched block copolymer nanoparticles. J Control Release 2016;222:141–50.
[27] Yang H, Wang N, Ren Y, Cai L, Chen Z, Xu Q. Supercritical CO_2 assisted preparation of 3D graphene-pyrrole/carbon nanotubes/polyaniline nanoarchitectures for efficient supercapacitor electrodes. Mater Lett 2015;139:471–4.
[28] Zhang J, Wang L, Zhao Y. Improving performance of low-temperature hydrogenated acrylonitile butadiene rubber nanocomposites by using nanoclay. Mater Design 2013;50:322–31.
[29] Gatos KG, Karger-Kocsis J. Effect of the aspect ratio of silicate platelets on the mechanical and barrier properties of hydrogenated acrylonitrile butadiene rubber (HNBR)/layered silicate nanocomposites. Eur Polym J 2007;43:1097–104.
[30] Mohan TP, Kanny K. Effect of nanoclay reinforcement on structure; thermal and mechanical properties of natural rubber-stryene butadiene rubber (NR-SBR). J Ind Eng Chem 2011;17:264–70.

[31] Mathewa G, Rhee JM, Lee YS, Park DH, Nah C. Cure kinetics of ethylene acrylate rubber/clay nanocomposites. J Ind Eng Chem 2008;14:60.

[32] Das A, Stöckelhuber KW, Heinrich G. Influence of layered silicate on the self-crosslinking of polychloroprene and carboxylated nitrile rubber. Macromol Chem Phys 2009;210(2):189–99.

[33] Das A, Mahaling RN, Stockethuber KW, Heinrich G. Reinforcement and migration of nanoclay in polychloroprenelethylene-propylene-diene-monomer rubber blends. Compos Sci Technol 2011;71:276–81.

[34] Bensadoun F, Kchit N, Billotte C, Bickerton S, Trochu F, Ruiz E. A study of nanoclay reinforcement of biocomposites made by liquid composite molding. Int J Polym Sci 2011;2011. Article ID 964193.

[35] Ganter M, Gronski W, Reichert P, Mülhaupt R. Rubber nanocomposites: morphology and mechanical properties of BR and SBR vulcanizates reinforced by organophilic layered silicates. Rubber Chem Technol 2001;74:221.

[36] Lu YL, Li Z, Yu ZZ, Tian M, Zhang LQ, Mai YW, et al. Compos Sci Technol 2007;67:2903.

[37] Azeez AA, Rhee KY, Park SJ, Hui D. Epoxy clay nanocomposites–processing, properties and applications: a review. Compos Part B Eng 2013;45:308–20.

[38] Herrmann W, Uhl C, Heinrich G, Jehnichen D. Analysis of HNBR-montmorillonite nanocomposites morphology, orientation and macroscopic properties. Polym Bull 2006;57:395–405.

[39] Kim J, Oh T, Lee D. Morphology and rheological properties of nanocomposites based on nitrile rubber and organophilic layered silicates. Polym Int 2003;52:1203–8.

[40] Bokobza L, Chauvin J-P. Reinforcement of natural rubber: use of in situ generated silica and nanofibres of sepiolite. Polymer (Guildf) 2005;46:4144–51.

[41] Joly S, Garnaud G, Ollitrault R, Bokobza L. Organically modified layered silicates as reinforcing fillers for natural rubber. Chem Mater 2002;14:4202–8.

Rubber nanocomposites with graphene as the nanofiller

6

A. Malas
The University of Nottingham, Nottingham, United Kingdom

6.1 Introduction

Polymer nanocomposites are described as the materials with more than one phase, structurally, where at least one dimension (generally of the dispersed phase) is in the nanometers range. Recently the development of polymer nanocomposites materials by introducing nano-sized fillers in the polymer matrices has evolved into an appealing area of research. The combination of polymer with nanofiller particulates is very interesting not only due to the reinforcing effect of nanofiller to polymer but also for the achievement of good electrical properties depending on the morphology of the nanocomposites.

Rubber nanocomposites composed of nanofillers are now essentially considered as important materials based on the fact that the physicomechanical, electrical, and other properties can be customized by incorporating nanofillers to fit the requirements in adaptable industrial applications. Different fabrication methods have been adopted to prepare rubber nanocomposites as compared to the method of preparation of thermoplastics polymer nanocomposites. The vulcanization of rubber nanocomposite in the presence of cross-linking agent, accelerator, and activator is one of the important steps in the preparation method to cure it at a specific temperature and time so that it could be useful material for different applications. Carbonaceous nanofillers, such as carbon black (CB), naturally occurring graphite, expanded graphite (EG), graphene oxide (GO), reduced GO, graphene, carbon nanotubes (CNTs), fullerenes, and carbon nanofibers (CNFs), are mostly used to prepare rubber nanocomposites for different applications. Over more than 100 years, CB has been extensively used to reinforce the different rubber matrices [1–3]. There are also so many research works based CNTs/rubber nanocomposites. Nevertheless, graphene is now extensively used due to its extraordinary physical characteristics compared to other nanofillers [4–15].

Graphite is the most familiar and naturally available allotrope of carbon. It contains planar (sp^2 hybridized) atomic carbon layers which are bound together by weak van der Waals forces of attraction. The two-dimensional (2D), single layer, and honeycomb-like lattice structure of graphite is familiar as graphene. Graphene has high modulus of elasticity ($\sim$1 TPa), high electron mobility, high thermal conductivity ($\sim$5000 W/mK), optical transparency, and large aspect ratio [16–21].

Progress in Rubber Nanocomposites. DOI: http://dx.doi.org/10.1016/B978-0-08-100409-8.00006-1

Due to the small interlayer spacing ($\sim$0.335 nm) of graphite, it is very difficult for the polymer chains to get into the gallery spacing which indeed decrease the state of dispersion of graphite flakes [22]. Therefore, several methods have been employed to increase the interlayer spacing of graphite which can be increased by more than 1 nm in a few circumstances, relying on the size of the intercalant. Graphite intercalation compounds (GICs) are the first intercalated compounds prepared by inserting potassium ions into the gallery space of graphite. The intercalants are stabilized between the intergallery spaces of graphite through ionic or polar interactions without disturbing the graphene framework. GICs could also be formed with other alkali metal ions as well as with anions like bisulfate, halogens, or nitrate. Interlayer spacing of graphite can also be increased by consecutive chemical and heat treatment to obtain EG [23]. The intergallery spacing of graphite can also be highly increased by vigorously oxidizing it by using several oxidants to get GO which will be discussed in the next section of this chapter.

In the recent past, graphene has been utilized as effective nanofillers for the fabrication of rubber nanocomposites with improved physical, mechanical, and electrical properties. One of the difficulties during the preparation of graphene/rubber nanocomposites is the aggregation of graphene nanosheets of high surface area due to van der Waals force of attraction which result in a decrease in the overall properties of the final materials. Aggregations of graphene flakes can be reduced by attaching polymers to its surface or by functionalizing it with some polar groups. The aggregation of functionalized or grafted graphene sheets is generally reduced by the hydrophobic or hydrophilic groups by strong polar–polar interactions or by their bulky size [24–26].

In this chapter, importance will be given to the methods utilized for the synthesis of graphene as well as fabrication of graphene filled rubber nanocomposites, along with the physical, mechanical, and electrical properties obtained and different applications.

6.2 Graphite, graphene oxide, reduced graphene oxide, and graphene

The structure of graphite, which is one of the important allotropes of carbon, is shown in Fig. 6.1. A description of graphite has already been included in the introduction part of this chapter. Although graphite is a three-dimensional carbonaceous material built with millions of sheets of graphene, graphite oxide is slightly distinctive. Many oxygenated functional groups are imported in the graphite flakes by the oxidation of graphite using a mixture of strong oxidizing agents, which indeed increases the interlayer spacing as well as makes the material hydrophilic. The hydrophilic nature of GO enables it to be exfoliated in water using ultrasonication and finally generates single or few layer functionalized graphene (FG), familiar as GO. So, graphite oxide is a multilayer system while GO is a few layers or sometimes single layer.

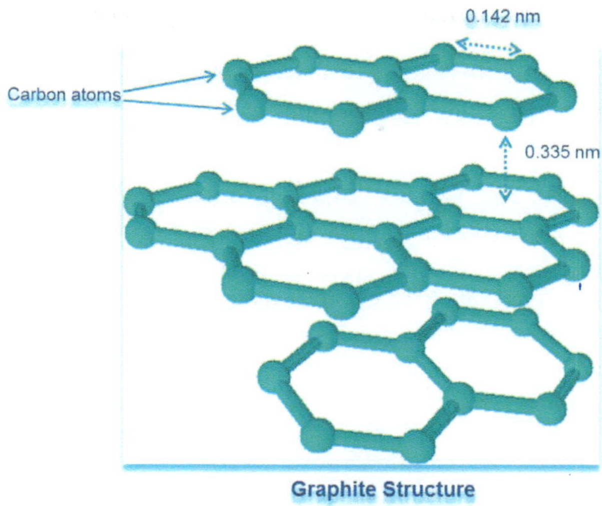

Figure 6.1 Three-dimensional structure of Graphite.

Due to the presence of different functional groups in GO, it is easily dispersed in water and other organic solvents as well. GO is prone to disperse finely in different polar polymer matrices due to the same reason. Regarding electrical conductivity, GO exhibits moderate electrical conductivity due to the breakdown of the planar network of graphite after the vigorous oxidation.

To recover the electrical conductivity as well as the graphitic character, the functional groups of GO are generally reduced by treating it with a chemical reducing agent or by very high temperature heat treatment under inert atmosphere. Depending on the adaptation of different reduction methods, the reduced graphene oxides (rGOs) are generally classified as chemically reduced GO (CRGO) and thermally reduced graphene oxide (TRGO). Features of graphene have already been described in the introduction.

6.2.1 Synthesis of graphene oxide and reduced graphene oxide

Generally, GO was synthesized by the Brodie's method [27], which was reported over 150 years ago. In this method, he oxidized graphite by treating it with potassium chlorate and fuming nitric acid, and a ratio of C/O/H of 2.2/1/0.8 was found out. Later the chemical method used to prepare GO by Brodie was further modified by Staudenmaier [28], Hummers [29], and others, as well as by some variation of these methods [30]. They had synthesized GO by treating graphite with a mixture of concentrated sulfuric acid (H_2SO_4), sodium nitrate ($NaNO_3$), and potassium permanganate ($KMnO_4$). They also obtained identical oxidation (C: O ratio 2:1) and observed oxygen-containing functional groups in the synthesized GO [21,31]. The primary benefit of the graphite oxide pathway for the synthesis of GO, as well as the logic as to why it has been utilized widely to fabricate polymer nanocomposites, is its ability to be uniformly dispersed (stable dispersion) to single GO sheets in various solvents. Therefore, we could obtain a homogeneous dispersion of GO in a soluble matrix in the common solvent. Furthermore, these methods could be utilized to scale up the synthesis of GO. In spite of its certain benefits, it has some crucial disadvantages like low conductivity due to the disruption of the conjugated structure after vigorous oxidation and inferior thermal stability. The schematic of the synthesis of GO is represented in Fig. 6.2A.

The poor electrical and thermal properties of GO can be reestablished to a certain extent, by treating it with chemical reducing agents or by thermal reduction under inert atmosphere. Sodium borohydride [32], hydrazine monohydrate [33],

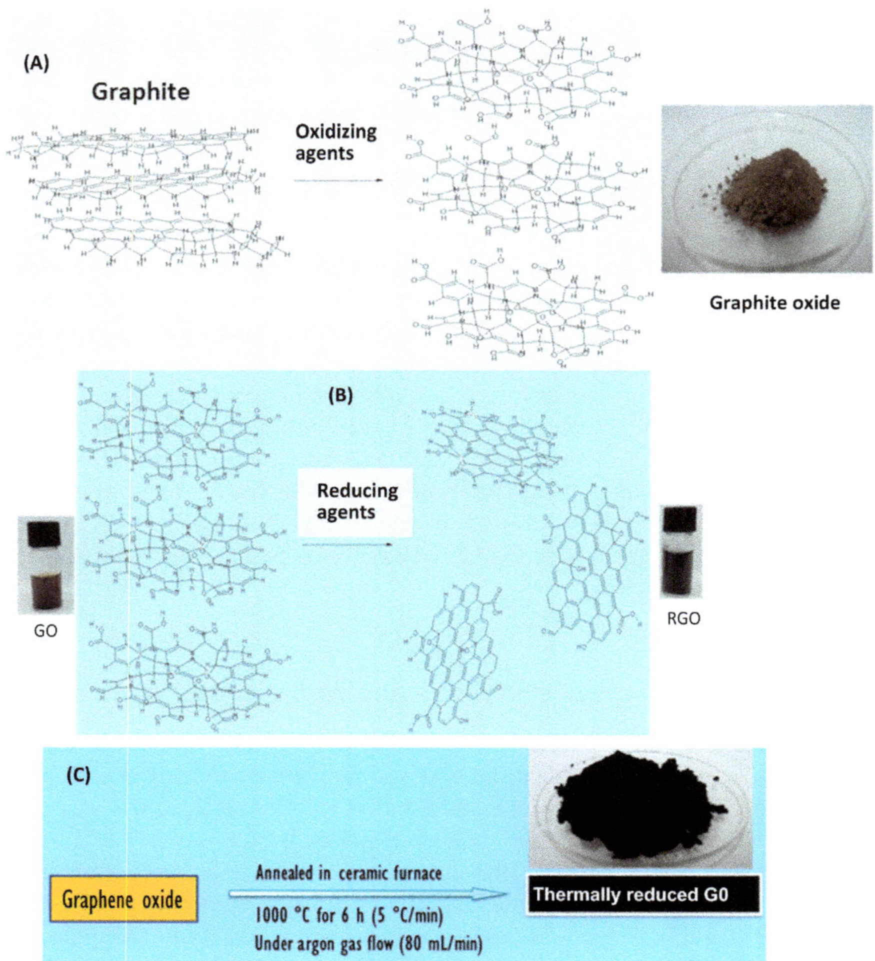

Figure 6.2 Schematic of the synthesis of graphite oxide (A), chemically reduced graphene oxide-CRGO (B) and thermally reduced graphene oxide-TRGO (C).

L-ascorbic acid [34], lithium aluminum hydride ($LiALH_4$) [35], hydroxyl amine (NH_2OH) [36], or hydrogen iodide (HI) [37] could act as effective chemical reducing agents to restore the properties in GO by partially removing the oxygenated functionalities as well as by recovering some of the conjugated double-bonded structures [38]. Fig. 6.2B represents the schematic of the synthesis of RGO chemically. Rapid heating of dry graphite oxide under inert gas and high temperature (1000°C) is generally carried out by the thermal reduction method [39].

6.2.2 Synthesis of graphene

Due to the exceptional properties of graphene, much interest is paid to the graphene nowadays, principally in the field of nanocomposites, which indeed has led to the development of several methods for the synthesis of defect free, highly graphitic, and low cost graphene in large amounts. The physical, mechanical, and electrical properties of graphene are highly dependent upon the synthesis method adopted. Some of the familiar approaches to produce high quality graphene will be discussed in the next few sections of this chapter.

6.2.2.1 Chemical vapor deposition (CVD)

Amongst all the methods adopted to synthesize graphene, chemical vapor deposition (CVD) on transition metal substrates turned into the most favorable approach. It can produce graphene with high aspect ratio. During the course of the CVD method, carbonaceous gases are charged into the reactor as well as the gases going through the hot zone, where hydrocarbon gases break down to carbon radicals at the transition metal surface and, after that, build single-layer and few-layers graphene. During the course of the process, the metal substrate acts as a catalyst by lowering the energy barrier of the reaction as well as driving the graphene deposition mechanism, which indeed influences the quality of graphene [40]. Single layer and few-layer graphene (FLG) have been synthesized by CVD from carbonaceous gases on catalytic metal surfaces or by surface segregation of C-containing gas dissolved in the bulk of those metals. The synthesis of graphene by CVD on Nickel (Ni) substrate was initially developed in 2008 [41]. Deposition of graphene and FLG has been studied using a variety of transition metal substrates [42–46]. Nowadays, copper (Cu) is widely used as a substrate. A variety of carbonaceous feedstocks can also be used as the source of graphene [47]. The first major problem besides synthesizing high quality graphene on a metal substrate using CVD is to separate or exfoliate the graphene from the substrate, as it is not simple to separate without destroying the structure or influencing the properties of the graphene. At the same time, the CVD method is costly because of the high energy requirements and the metal substrates are generally regenerated during each synthesis process. To scale-up this procedure, a roll-to-roll synthesis process has been adopted recently through which graphene is grown on Cu-coated rolls and after that shifted to a polymer film [42]. Graphene synthesized utilizing the CVD method is an auspicious material that could be usable in several applications like sensors or electronic devices, coatings, actuation, etc. and this method could transform as well as advance the production of graphene filled/reinforced nanocomposites.

6.2.2.2 Thermal exfoliation

Thermal exfoliation methods are also able to produce near to complete exfoliated single-layer graphene. These methods are in comparison to much early efforts at thermally expanding and exfoliating graphite, where only a partially exfoliated structure was achieved [48–51]. This method has several advantages in comparison with mechanical exfoliation, such as fastness, solvent free, etc. Thermal exfoliation takes place within seconds at high temperature [52,53]. Dry graphene is required to prepare electrodes in lithium ion batteries, where graphite oxide is mainly used as the starting material; thermal exfoliation results in simultaneous exfoliation and reduction of graphite oxide. The oxygenated functionalities of graphitic layers are decomposed and produce gases during rapid heating, and the gases create pressure between adjacent graphitic sheets. Exfoliation occurs when this pressure surpasses the van der Waals force of attractions between the graphene sheets of graphite oxide. Graphite oxide, EG, and GICs are generally used rather than pure graphite as starting materials for thermal exfoliation due to the presence of functionalities in them. 10 nm thick graphite nanoplatelets can be synthesized by the rapid heating of acid intercalated graphite in a microwave [54]. In particular, high temperatures (1050°C), high pressure (700 $m^2\ g^{-1}$), and fast heating rates (>2000°C/min) are necessary to fully exfoliate graphite flakes [55,56]. By this method few- and many-layer graphene can be synthesized in bulk amounts that still exhibit a few interesting features of SLG. Due to that reason, graphene prepared by thermal exfoliation is most popularly utilized for the fabrication of nanocomposite based on polymer.

6.2.2.3 Mechanical exfoliation in solution

The stacked graphite materials become exfoliated after the application of mechanical forces to solutions of stacked graphite materials by mean of stirring, shaking, or ultrasonication. Some methods also use selective solvents that facilitate the exfoliation by interacting favorably with the layered materials, reducing the energy consumption and reducing reaggregation of exfoliated sheets as well. Several intercalants (ions), surfactants, and functionalization of graphene sheets of graphite are also significantly used to avoid the van der Waals interactions and to keep exfoliated graphene flakes dispersed in solution. The exfoliation of graphite materials in organic solvents like N-methyl-pyrrolidone [57], dimethyl-formamide [58], dimethyl sulfoxide [59], tetrahydrofuran [60], or in different kind of solvents (e.g., per fluorinated aromatic [61]) goes through in the following steps: dispersion in a selective solvent, exfoliation, and purification [62]. In this method, the required energy to exfoliate the graphene sheets in the solution is provided by sonication and the energy gained from the solvent–graphene interactions may facilitate the process as well. Hence, it is necessary for this process to choose a suitable solvent with surface tension similar to or higher than the graphene–graphene interaction energy.

As well as using organic solvents for this method, aqueous-surfactant suspensions have also been found to effectively exfoliate graphite, as explored in the research work of Lotya and coworkers [63]. Aqueous solutions of inorganic salts like NaCl and $CuCl_2$ have also been efficient to exfoliate the graphite sheets [64]. The yield of graphene sheets in this method is quite on the lower side. Mechanical exfoliation of graphite in solvent is convenient for the synthesis of graphene that can be used as nanofillers directly to prepare nanocomposites by the solution intercalation method in the same solvent. However, this method is not environmentally friendly and is expensive as well.

6.2.2.4 Other methods

Graphite oxide exhibits an identical layered structure to graphite, but it contains a huge amount of oxygen-containing groups, which results in an increase in the interlayer spacing and makes the atomic-thick layers hydrophilic as well. Therefore, graphite oxide layers can easily be exfoliated in water by mild ultrasonication. The exfoliated sheets containing only one or few layers of carbon atoms like graphene are known as GO. The oxygenated functionalities of GO can be (most of the functional groups) reduced by treating it with different chemical reducing agents to graphene-like sheets with the recovery (partial) of the conjugated structure. The rGO flakes are generally treated as one kind of chemical derivative of graphene. Reduced graphene oxides (rGOs) have also been designated by some other names such as FG, chemically modified graphene, chemically converted graphene, or reduced graphene [65]. The main target of any reduction method is to synthesize graphene-like structures with desired properties identical to the pristine graphene achieved from direct mechanical exfoliation (i.e., the "Scotch tape method") of individual layers of graphite. To date, many attempts have been made, but the final target is still difficult to achieve. Reducing GO by using different reducing agents like $NaBH_4$, hydrazine monohydrate ($N_2H_4.H_2O$), NH_2OH, etc. is a very scalable method, but the rGOs produced often exhibit moderate surface area and electronic conductivity.

6.2.3 Different characterizations

6.2.3.1 X-ray diffraction (XRD)

X-ray diffraction (XRD) experiments are generally carried out to measure the *d*-spacing (intergallery spacing) and crystallinity of graphite as well as its derivatives. This method can also be used to investigate the intercalation, exfoliation of graphite, and the development of RGO, as well as graphene. It can be seen from Fig. 6.3 that a sharp reflection of graphite under normal measurement conditions (CuKα radiation and $\lambda = 0.154$ nm) is found at $2\theta \sim 26°$, corresponding to the *d*-spacing of ~ 0.335 nm. After vigorous oxidation to graphite oxide, the main peak of graphite shifted towards a lower angle $2\theta \sim 11°$, corresponding to *d*-spacing of ~ 0.818 nm (intercalation) and the peak intensity also reduced (exfoliation). The transition of graphite to graphite oxide is also known as hydrophobic to hydrophilic

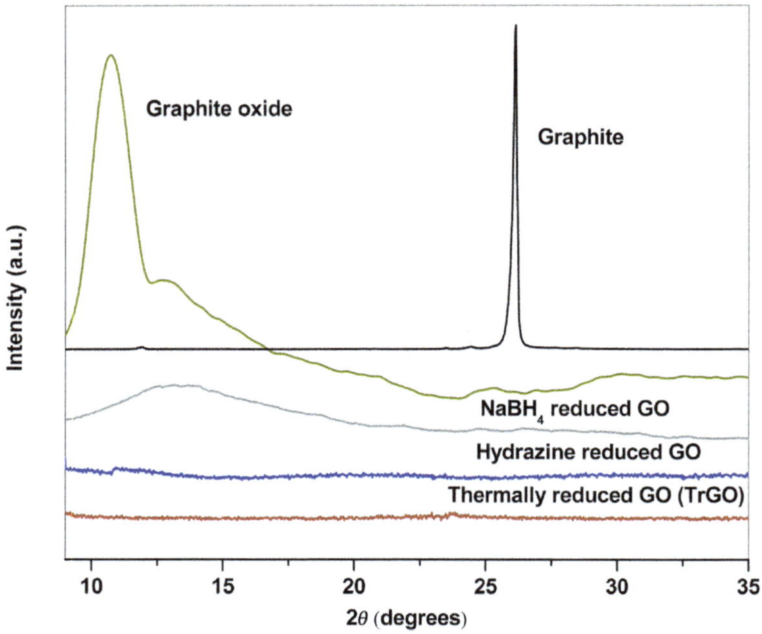

Figure 6.3 XRD patterns of graphite oxide (synthesized using improved method) [30] and reduced graphene oxides-rGOs (synthesized using different reducing agents).

(the presence of different oxygenated functionalities) transition. After reduction of different functional groups of GO by various reducing agents, the primary peak of graphite oxide was almost diminished (Fig. 6.3), which indicates the exfoliated nature of rGOs and also hydrophilic to hydrophobic transition. The main peak of graphite becomes broader with the decreasing number of layers and ultimately disappears in XRD measurements for monolayer graphene [66]. Generally, XRD measurement is done on bulk graphite materials; hence XRD can give only a comparative estimation concerning the average number of layers.

6.2.3.2 Raman spectroscopy

Raman spectrophotometer is one of the most effective tools that is significantly used in investigations on graphene to distinguish single-layer, bilayer, and few-layer of graphene, as well as to measure the defects and graphitic character (for GO and rGOs). Even monolayer graphene can provide a strong resonance Raman scattering [67,68]. The three most intense characteristic peaks of graphene and graphite are noticed at $\sim 1350\ cm^{-1}$ (the D band), the G band at $\sim 1580\ cm^{-1}$, and the 2D band at $\sim 2700\ cm^{-1}$. The precise information about the number of layers of graphene can be observed from Raman spectroscopic study [69]. Liu and coworkers [70] as well as Dresselhaus et al. [71] have demonstrated that the typical 2D band emerges and exhibits variations for graphene with different numbers of layers.

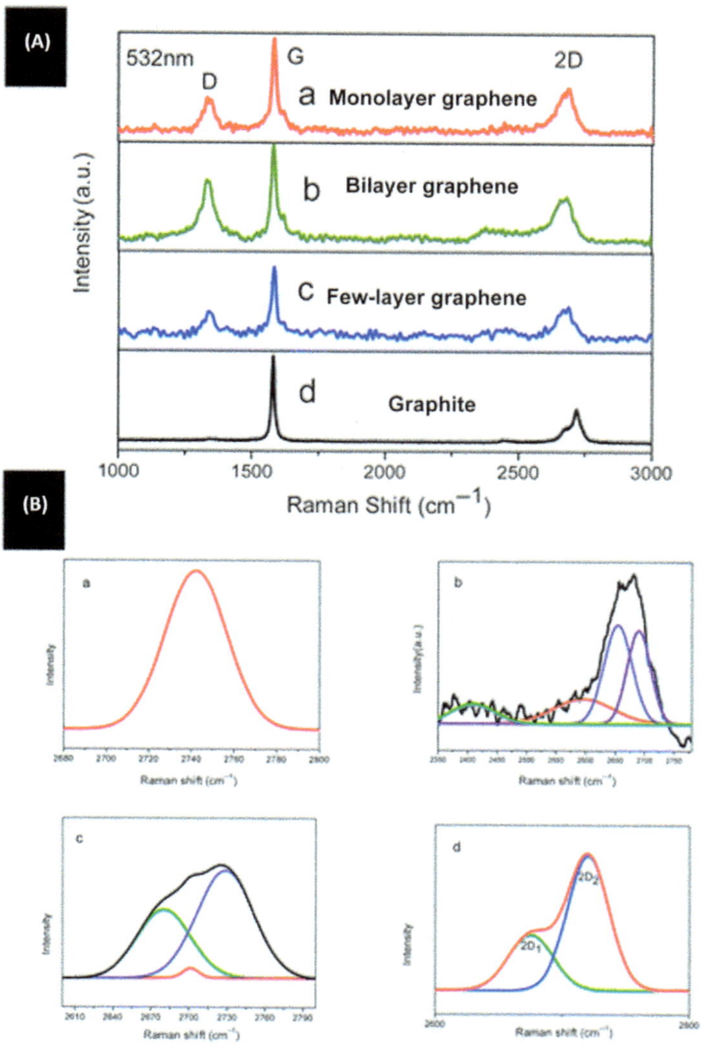

Figure 6.4 Raman spectra of graphene sheets with different numbers of layers (A), (B) fitted curves of the 2D band of Raman spectra using an excitation laser wavelength of 532 nm (a) monolayer graphene, (b) bilayer graphene, (c) few-layer graphene, (d) graphite.
Source: Reproduced with permission from Liu K, Liu L, Luo Y, Jia D. One-step synthesis of metal nanoparticle decorated graphene by liquid phase exfoliation. J Mater Chem 2012;22 (38):20342–52. Copyright 2012, Royal Society of Chemistry.

The 2D peak is generally upshifted and broadened with the increase of the number of layers of graphene. The presence of the D band in the Raman spectra of graphite materials (Figs. 6.4 and 6.5) is connected to the presence of defects and edges [72]. The ratio of the D and G bands' intensity (I_d/I_g) signifies the extent of defects. I_d/I_g ratio for FGs (GO and rGOs) prepared by bulk synthesis is generally higher than

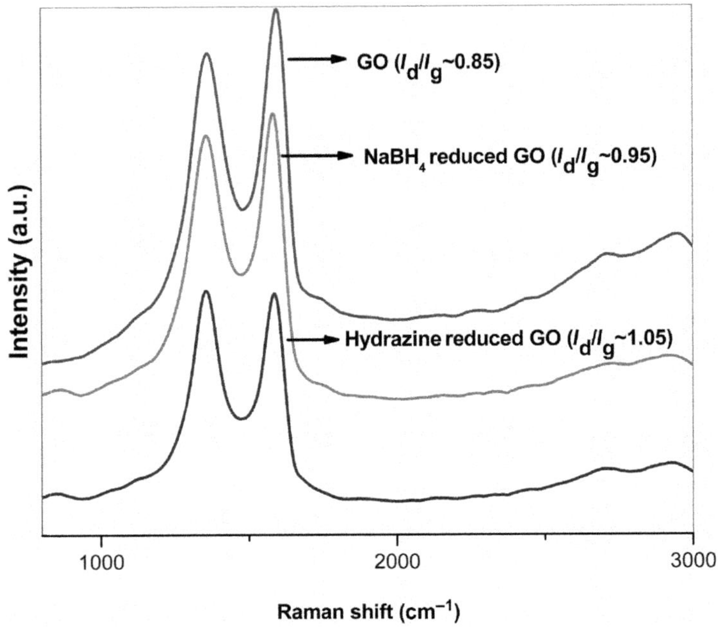

Figure 6.5 Raman spectra of GO and different reduced graphene oxides (rGOs).

that of graphite flakes, which is due to the formation of edges as well as defects in the graphene layers during exfoliation by vigorous oxidation.

6.2.3.3 X-ray photoelectron spectroscopy (XPS)

X-ray photoelectron spectroscopy (XPS) technique is generally used to examine the presence of different functional groups (ratio of different elements quantitatively) on GO as well as the degree of reduction of GO to graphene by different reduction processes. It can be seen from Fig. 6.6 that the high-resolution XPS spectrum [73] of C1s from GO (solid line) has been deconvoluted into four smaller peaks (dashed lines), those are attributed to the C—C (sp^2 bonded carbon, corresponding to the binding energy of ~284.8 eV), C—O (epoxy/hydroxyl, ~286.9 eV), C=O (carbonyl, ~287.9 eV), and O—C=O (carboxyl, ~289.4 eV), suggesting a significant extent of oxidation of graphene [74–76]. After the reduction of GO by different procedures, a significant decrease in the peak intensity of oxygen-containing functional groups is generally observed based on the C1s XPS spectrum of reduced graphene sample [34,77,78], and the C=O peak is almost diminished, implying the partial elimination of the oxygenated functionalities [79]. Additionally, the degree of oxidation of graphite and degree of reduction of GO also can be evaluated by calculating the ratio of carbon to oxygen of the different samples from XPS analysis.

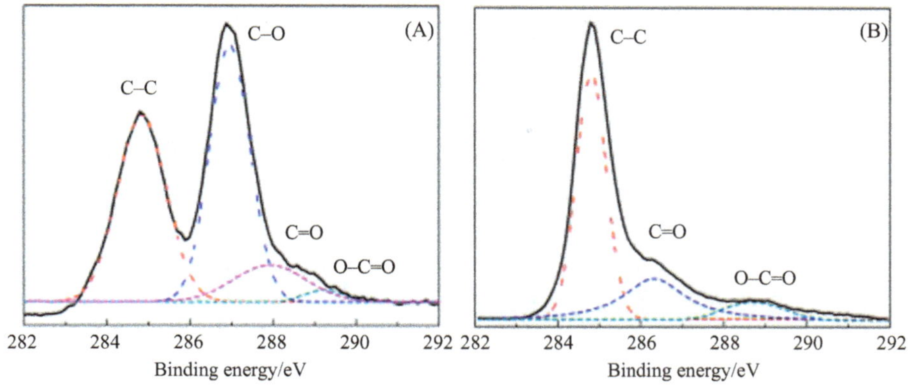

Figure 6.6 High-resolution XPS spectra of C1s from GO (A) and graphene (B).
Source: Reproduced with permission from Ding J, Yan W, Sun S, Bao J, Gao C. Fabrication of graphene/CaIn 2 O 4 composites with enhanced photocatalytic activity from water under visible light irradiation. Int J Hydrogen Energy 2014;39(1):119–26. Copyright 2014, Elsevier.

6.2.3.4 Microscopic analysis

As a result of absorbance ($\sim$2.3%) of visible light by a single atomic layer of graphene sheet, individual sheets of graphene nanomaterials can be observed directly in an optical microscope [80]. High-resolution transmission electron microscopy (HR-TEM) technique is also capable to locate the sheets of graphene with various number of atomic carbon layers [81,82]. Novoselov and coworkers observed the thickness of monolayer graphene to be 0.4 nm using atomic force microscopy (AFM) [83]. From then on, many researchers working with graphene materials have been using AFM to measure the thickness of the graphene sheets (Fig. 6.7). To investigate the wrinkles, ripples, and other different structure of graphene materials, scanning electron microscopy (SEM) and scanning tunneling microscopy have also been employed [84,85].

6.2.4 Novel properties

6.2.4.1 Electrical property

Majority of experimental investigations on graphene target on the utilization of its electronic properties. The fascinating electrical properties of graphene are being used in a number of applications in nanoelectronics. Graphene is a zero-overlap semimetal (with both electrons and holes as charge carriers) with very high electrical conductivity which is one of the most valuable features. The electronic properties of graphene are governed by the bonding and antibonding (the valance and conduction bands) electrons of the pi orbitals of the conjugated double bonds in its structure. The higher electron mobility of graphene has been reported by several researchers [16,18,86]. The electron mobility of graphene (15,000 $cm^2\ V^{-1}\ s^{-1}$) can be enhanced significantly by removing the impurities to the maximum extent.

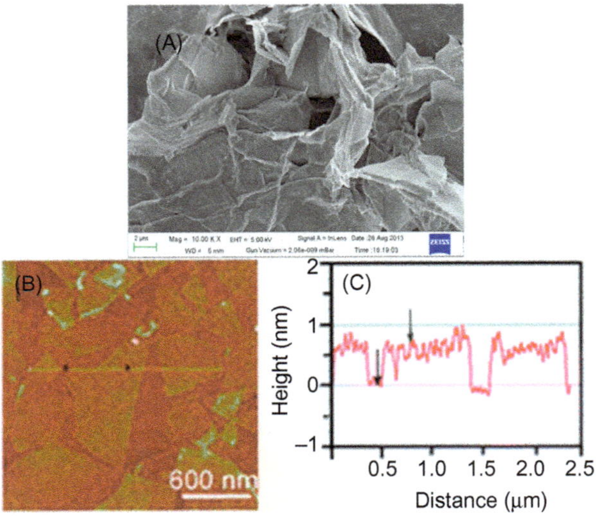

Figure 6.7 FESEM image $NaBH_4$ reduced graphene oxide (A), and (B) tapping mode AFM image of the L-AA reduced GO sheets on mica substrate, (C) the height profile of the AFM image, (Fig. 6.7B,C)
Source: Reproduced with permission from Zhang J, Yang H, Shen G, Cheng P, Zhang J, Guo S. Reduction of graphene oxide via l-ascorbic acid. Chem Commun 2010;46(7):1112–4. Copyright 2010, Royal Society of Chemistry.

6.2.4.2 Thermal property

One of the important features of graphene is its superior thermal conductivity. Graphene exhibits a very high in-plane thermal conductivity which is greatly influenced by atomic defects, interfacial interactions, and edges. The in-plane thermal conductivity of graphene at ambient temperature is amongst the best of any known material, about 2000–4000 W m^{-1} K^{-1} for freely suspended samples [87–89]. Heat flow along the c axis of graphene and graphite is strongly restricted by weak interplane van der Waals force of interactions.

An optothermal Raman investigation reported that the thermal conductivity of freely suspended FLG decreases with the increase in the number of layers, reaching the value of bulk graphite beyond 3 or 4 layers of graphene [90]. The decrease in thermal conductivity of graphene beyond a certain number of layers is due to the dissimilarities in the phonon scattering with the increasing number of layers through which more phase-space conditions become feasible for phonon scattering [88].

6.2.4.3 Mechanical property

Another outstanding property of graphene is its inherent mechanical strength. Graphene is the strongest material ever invented, with an ultimate tensile strength of 130 GPa, compared to 0.4 GPa for A36 structural steel, or 0.37 GPa for Kevlar fiber, which is due to the strength of its 0.142 nm long carbon bonds. Graphene is not only exceptionally strong but also very lightweight (0.77 mg/m^2) compared to the same dimensions of paper (1 m^2 of paper is about 1000 times heavier). The modulus of elasticity determined by nanoindentation experiments upon a monolayer graphene lying across a hole is 1000 ± 100 GPa [20,91]. Many other experimental efforts have been used as well to determine the elastic modulus of graphene. Presently, a free-standing indentation experiment based on AFM is the most direct way to get the elastic modulus of graphene. Lee and coworkers [20] reported the elastic modulus of 1.02 TPa for single-layer graphene and Frank et al. [92] achieved a value of 0.5 TPa for a few layered graphene sheets ($n < 5$). Zhang and coworkers [93] also calculated the elastic modulus of monolayer (~0.89 TPa), bilayer (~0.39 TPa), and multiple layer graphene flakes using an orchestrated nanoindenter, and they observed that the elastic modulus of graphene decreases with the increase in the layer number. Lee et al. [94] observed a very high elastic modulus through measuring the strain enforced by creating a pressure difference across graphene membranes using Raman spectroscopy, and the approximated elastic moduli of single layer and bilayer graphene are 2.4 and 2.0 TPa, respectively. It is generally said that a single layer of graphene (weight ~1 gm) is plentiful in every dimension and sufficient to cover a whole football ground.

6.3 Graphene–rubber nanocomposites

6.3.1 Fabrication methods

To synthesize high performance conventional composite materials as well as to achieve uniform and thermodynamically balanced dispersion of the filler particulates in the polymer matrices, there are two major obstructions to be solved, i.e., aggregation and weak polymer–filler interfacial adhesion. Deaggregation of the filler association, which consists of tens, hundreds, or even millions of filler particulates, accumulated with a profoundly active interparticle force [95], and strong polymer–filler interfacial adhesion are necessary for better mechanical interconnection between the matrix and the filler. The above two requisites are also very much essential to achieve improvement in the properties of polymer-based nanocomposites. With respect to the dimension of the different nanofillers there are some extra obstructions that should be solved for nanocomposites formation, such as entropic responses of the polymers in nanoscopic restrictions between two-dimensional filler particulates; deagglomeration of convoluted one dimensional filler assemblage like CNTs; and conquering the considerably faster kinetics of nanofiller reaggregation. To overcome all the above hurdles we aspire to focus on these challenges by detailing the most common synthetic routes for nanocomposites formation [96]. The conditions that have been discussed above to achieve improved physical and mechanical properties of the polymer nanocomposites can be applicable for most of the other nanofillers.

6.3.1.1 Solution intercalation/latex blending

This method comprises mixing of a fix polymer solution with the layered graphene fillers. It depends on a solvent system where the polymer is soluble and the layered graphene FG sheets are uniformly dispersed in the same solution through high-speed shear mixing or ultrasonication or stirring. The polymer and the layered graphene solutions in the common solvent are mixed together. The polymer chains are generally intercalated into the interlayer spacing of graphene flakes by forcing out the solvent molecule from the intergallery space. The intercalated structure persists even after solvent removal resulting in polymer-layered graphene nanocomposites. The class of the solvent is vital in this method to initiate the inclusion of the polymer chains into the intergallery space of layered nanofiller and also the polarity of the solvent is a driving factor for intercalation [97,98]. The solution intercalation/latex blending method is most frequently used in academic research work for the development of graphene/ elastomer composites [5,99–148].

In latex blending, the elastomers are used in the form of latex. This method is indeed environmentally friendly due to the absence of any toxic organic solvent. In order to achieve significant interactions and interfacial adhesions between the polymer matrix and fillers, a number of steps have generally been inducted into this method. Those steps are latex mixing/cocoagulation or, cocoagulation which can enhance the extent of dispersion. The faster coagulation process of the emulsion can restrict the graphene sheets to restack with each other in the emulsion kinetically [104–110,115,118,134]. Although there are a number of advantages of the solution mixing method, the elimination of the solvents used during the nanocomposites preparation remains a serious concern for the utilization of this method on a large scale. The high cost of the solvents and the difficulties related to their proper disposal have an adverse effect upon the scale-up and therefore prospective adoption of this method by industry. Furthermore, the quantity/quality of solvents, mixing time/speed, sonication, etc. could significantly influence the properties of the final product.

6.3.1.2 Melt blending

In this method, different nanofillers are dispersed in the molten rubber matrix (beyond the softening point of the polymer) and the nanocomposites are fabricated by shear mixing. The melt intercalation technique is better among the other two methods, i.e., in situ intercalative polymerization and solution intercalation, and is favored by the industry due to the absence of organic solvents as well as its compatibility with the industrial processes like injection molding and extrusion [149–160]. There are polymers which are not suitable to be processed in the in situ intercalative polymerization or solution intercalation; the melt intercalation method admits the use of such polymers for the preparation of nanocomposites. The chemical modification of different fillers could be done in this method. In this method different compatibilizers can be introduced to achieve better filler–matrix interactions. This technique can also be significantly used for both polar and nonpolar rubbers [161]. The disadvantages of this method are as follows. As a result of the elevated temperatures that are required during the addition of inorganic fillers into the rubber matrix, the final product is sometimes prone to degradation. In this method, the effective dispersion of different fillers in rubber matrix is sometimes hindered by the use of higher filler fractions, along with the high viscosity of the rubber matrix [162]. High shear forces are generally applied to overcome the high viscosity of the matrix which results in damage in the graphene or GO flakes i.e., lowering the aspect ratio of graphene. Graphene/rubber nanocomposites prepared by this method exhibit poor dispersion of graphene sheets in comparison with the other popular methods used to prepare nanocomposites such as solution intercalation/latex blending and in situ intercalative polymerization [100]. Although it is difficult to achieve fine dispersion of filler through the melt intercalation method, improvement in some of the properties of composites prepared by this method have been disclosed in the literature.

6.3.1.3 In situ polymerization

In situ intercalative polymerization has been significantly used in the past to prepare layered silicates/polymer nanocomposites where the clay layers are dispersed in the monomer solution followed by polymerization [163,164]. The silicate sheets are inflated inside the liquid monomer or a monomer solution resulting in the creation of polymer in the interlayer space of layered silicates after polymerization of the monomer. Polymerization occurs by either radiation or heat and also can occur by the diffusion of an appropriate initiator or catalyst by cation exchange inside the intergallery space before the swelling step [165]. This technique for the synthesis of nanocomposites has been utilized for a long time [97]. Zeng and coworkers successfully synthesized polystyrene-layered silicate and polyaniline-layered silicate nanocomposites by this technique and they observed a highly ordered form of a single chain of polyaniline alternatively stacked with the natural clay layers [166]. This method has also been utilized by a few researchers to prepare graphene–rubber nanocomposites due to the similarity in structure between clay and graphene [100,167,168]. This method has some disadvantages. It will work only for low viscous polymers. As a result of attachment of polymer chains with graphene sheets during in situ polymerization, the connection between the graphene sheets are obstructed, which indeed lowers the electrical properties of the nanocomposites. But due to the good interaction between rubber and graphene sheets, mechanical, thermal, and barrier properties of the nanocomposites are generally improved.

6.3.2 Characterizations

6.3.2.1 X-ray diffraction (XRD)

XRD technique is generally adopted by the researchers in order to observe partially the extent of dispersion of graphene sheets or functionalized graphene sheets (FGSs) in different rubber matrices. Wu and coworkers fabricated natural rubber (NR)/GO nanocomposites by the latex blending method and observed no diffraction peak of GO in the XRD patterns of the nanocomposites, which indicated an exfoliated structure of the nanocomposites [169]. Malas et al. prepared GO filled rubber nanocomposites based on polychloroprene and chlorosulfonated polyethylene rubbers using

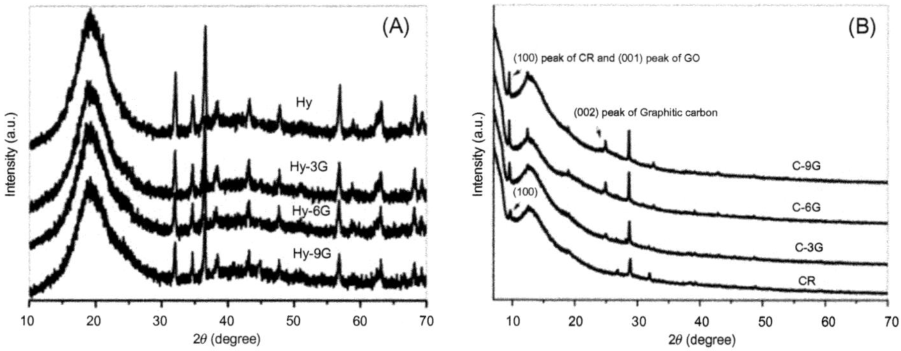

Figure 6.8 XRD patterns of (A) GO/Hypalon (chlorosulfonated polyethylene) and (B) GO/Polychloroprene rubber (CR) nanocomposites with 3, 6, and 9 wt% loading. *Source*: Reproduced with permission from Malas A, Das CK. Effect of graphene oxide on the physical, mechanical and thermo-mechanical properties of neoprene and chlorosulfonated polyethylene vulcanizates. Compos Part B Eng 2015;79:639–48. Copyright 2015, Elsevier.

both solution/melt intercalation methods. They also observed the absence of the main diffraction peak of GO in the XRD patterns of the nanocomposites (Fig. 6.8) [112]. Bai et al. [126] observed exfoliated structure (absence of 001 peak of GO in the diffraction pattern) of nanocomposites based on hydrogenated carboxylated nitrile-butadiene rubber reinforced with exfoliated GO. Similar results were noticed by Xiong et al. [130] for bromobutyl rubber based nanocomposites reinforced with ionic-liquid-modified GO. Along with the microstructure (intercalated and exfoliated structure) of the graphene/rubber nanocomposites, the degree of crystallinity of the nanocomposites, which indeed plays a major part in governing the physical and mechanical features of the material, can also be investigated using XRD.

6.3.2.2 Cure behavior

The cross-linking density of different rubber nanocomposites can be verified by measuring the cure characteristics (torque vs time experiment) of the samples [105,111,112,116,134,150,158,162]. Minimum torque (ML), maximum torque (MH), and torque difference (ΔS) are considered as a measure of viscosity, stock modulus, and degree of cross-linking of the rubber composite, respectively [170–172]. The cure curves are generally shifted towards a shorter time side with the increase in the addition of filler content, which signifies the increase in speed of the vulcanization process. The MH and ML values of rubber nanocomposites may be increased simultaneously. Zhan et al. [162] observed enhancement in the MH value and in the mechanical properties of exfoliated graphene (GE)/NR composites prepared by ultrasonically-assisted latex mixing along with in situ reduction of GO. Frasca et al. reported an increase in the MH value for multilayer graphene (MLG)/chlorine-isobutene-isoprene rubber (CIIR) and they also observed an increase in the cross-linking density for the nanocomposites [173].

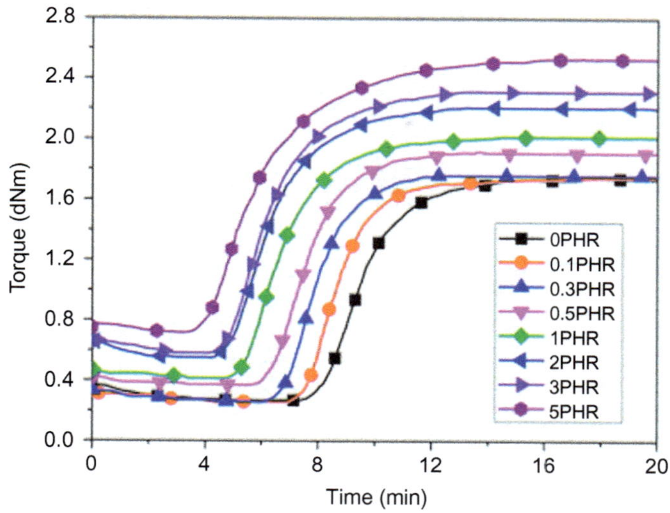

Figure 6.9 Curing curves of NR and GE/NR nanocomposites with different graphene loadings.
Source: Reproduced with permission from Wu J, Xing W, Huang G, Li H, Tang M, Wu S, et al. Vulcanization kinetics of graphene/natural rubber nanocomposites. Polymer 2013;54 (13): 3314–23. Copyright 2013, Elsevier.

Wu and coworkers prepared graphene filled (with different loadings) rubber nanocomposites and cured those nanocomposites with sulfur as a cross-linking system. They studied the vulcanization kinetics of the nanocomposites and observed that the curing curves are systematically shifted towards the shorter time with the increase in graphene content, indicating that the vulcanization process of NR is markedly accelerated. Simultaneously, the minimum and maximum torque values of the nanocomposites were increased in the presence of graphene (Figs. 6.9 and 6.10). They also concluded that the induction period of the vulcanization process is highly restrained at low graphene loading and then shows a moderate decrease, at the same time the vulcanization rate is increased at first and afterwards shows a depression [114].

6.3.2.3 Microscopy

Microscopic study of polymer nanocomposites are generally carried out to investigate the morphology as well as to observe the state of dispersion of the filler in the polymer matrix [174]. Hernandez et al. [153] prepared NR based composites consisting of FGSs using melt blending, and used transmission electron microscopy (TEM) to observe the state of dispersion of graphene sheets. They observed a fine dispersion of graphene sheets in the NR matrix (Fig. 6.11A and B). Xing and coworkers [146] fabricated GE/SBR nanocomposites by latex blending and they noticed that the exfoliated GE nanosheets are uniformly dispersed in the SBR matrix and

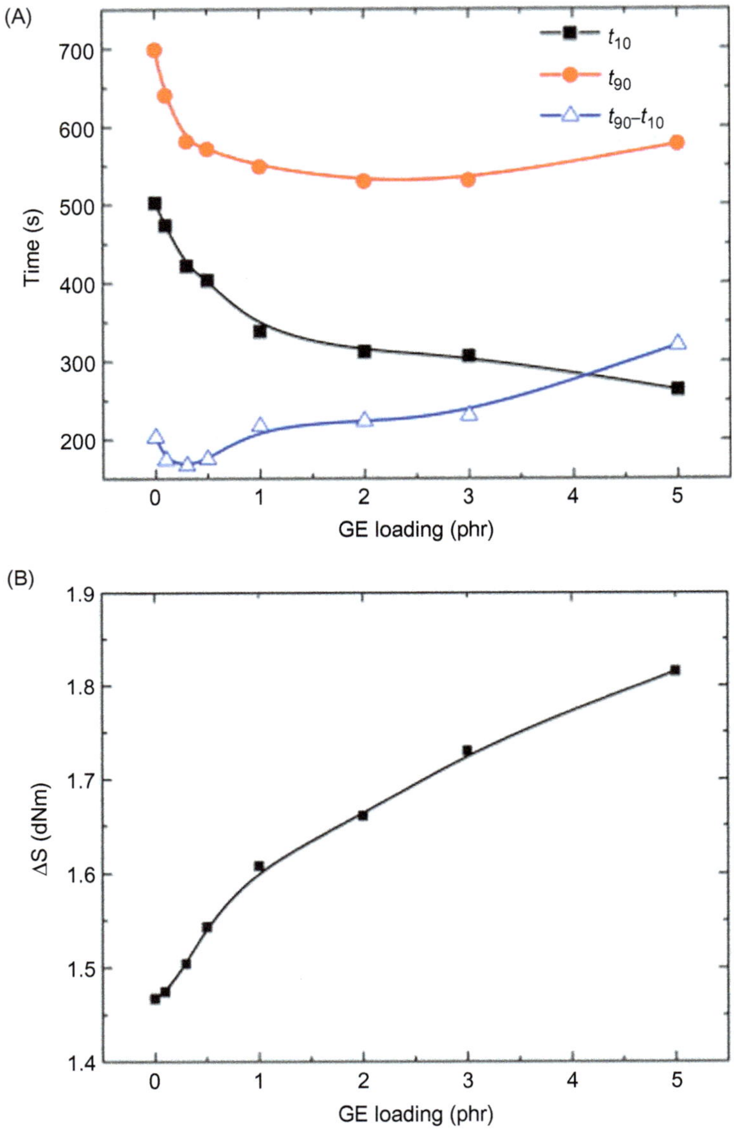

Figure 6.10 Effect of graphene loading on the vulcanization parameters determined from the curing curves. (A) Scorch time, optimum cure time, and the difference between scorch time and optimum time; (B) the difference between minimum and maximum torque (ΔS).
Source: Reproduced with permission from Wu J, Xing W, Huang G, Li H, Tang M, Wu S, et al. Vulcanization kinetics of graphene/natural rubber nanocomposites. Polymer 2013;54 (13): 3314–23. Copyright 2013, Elsevier.

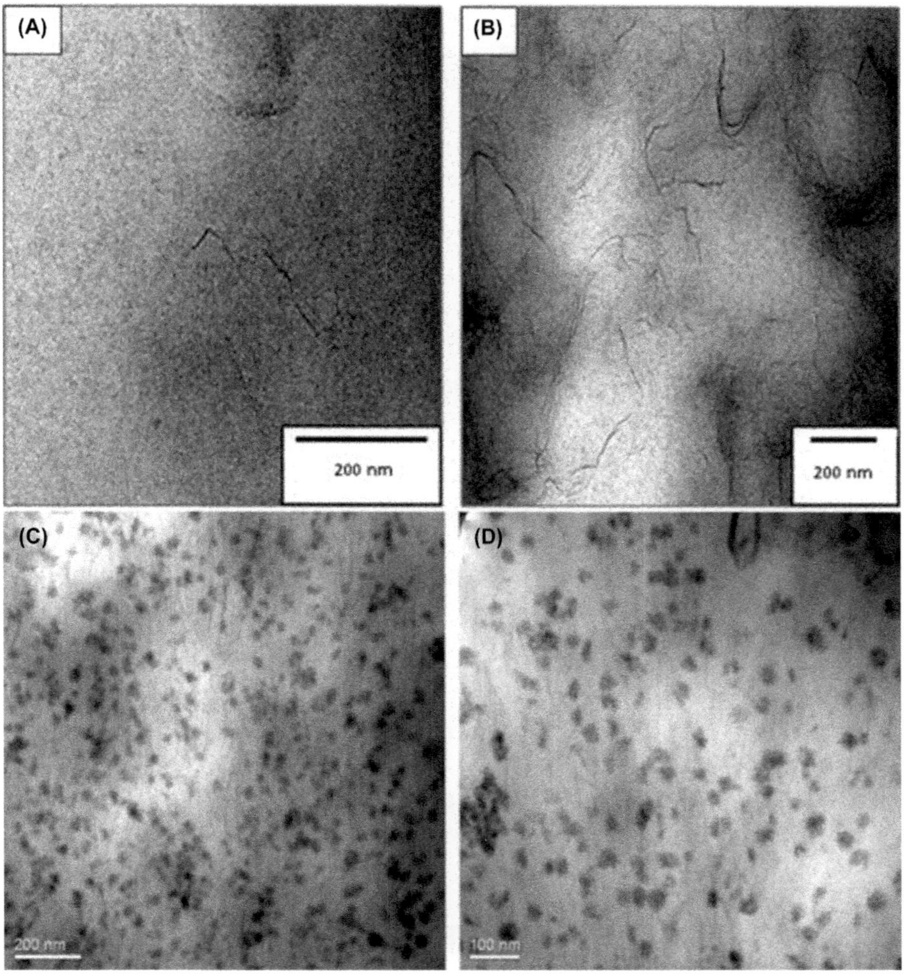

Figure 6.11 TEM images of NR/FGS nanocomposites, with FGS content as a variable: (A) 0.1 phr and (B) 0.5 phr. Reproduced with permission from [153] Copyright 2012, Elsevier. (C) and (D) TEM images of the 7GE/SBR nanocomposite at different magnifications. *Source*: Reproduced with permission from Xing W, Tang M, Wu J, Huang G, Li H, Lei Z, et al. Multifunctional properties of graphene/rubber nanocomposites fabricated by a modified latex compounding method. Compos Sci Technol 2014;99:67–74. Copyright 2014, Elsevier.

also observed that the GE nanosheets are composed of a single layer or a few stacked layers at the highest filler loading. No severe graphene platelet tactoids are formed (Fig. 6.11C and D). Potts et al. [106] prepared RGO/NR nanocomposites using latex cocoagulation procedure and they used TEM to investigate the dispersion of the nanocomposites. For an uncured nanocomposites sample, they observed a nonuniform particle density over the cross-section, with submicrometer regions void of particles surrounded by areas of high particle density. Different electron microscopy (scanning and transmission) techniques are considerably used to observe and also to draw conclusions regarding the state of dispersion of graphene filler particulates for most polymer nanocomposites. TEM has been becoming preferable for this purpose as it supplies information on the internal composition of the nanocomposite at the highest magnification compared to SEM [175]. SEM can also turn out to be advantageous for surface morphology characterization and, specifically, for the observation of the state of dispersion of filler from tensile or cryo-fractured surfaces.

6.3.2.4 Contact angle measurement

Contact angle tests of the polymer nanocomposites are generally conducted to investigate the surface hydrophobicity. Evaluation of the interactions between fillers and the polymer matrix (interfacial energy difference) can usually be done by measuring the contact angles of certain liquids on the polymer nanocomposites. K. J. Berean and coworkers synthesized polymer nanocomposite membrane based on polydimethylsiloxane (PDMS) filled with graphene nanosheets. They observed that the addition of graphene into the PDMS matrix has made the membrane more hydrophobic compared to neat PDMS membrane (Fig. 6.12) [176]. Chen et al. [123] measured contact angles of GO/ethylene-propylene-diene rubber (EPDM) and GO/EPDM/petroleum resin (PR) blends nanocomposites. They observed a significant change in the contact angle for the GO/EPDM nanocomposites due to the approximately matched surface energy of GO (43.8 mJ m^{-2}) and EPDM (34.0 mJ m^{-2}) which indeed improves the dispersion of GO in EPDM matrix. They also did not notice any obvious changes of the contact angles of EPDM/PR matrices after adding GO. As a result of higher interfacial energy difference between GO and EPDM/PR blends compared to GO and EPDM, the dispersion of GO in EPDM matrix was

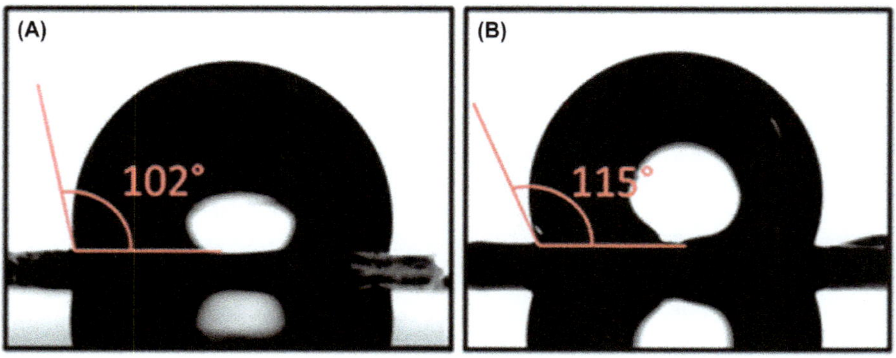

Figure 6.12 (A) Water droplets contact angle for PDMS. (B) Water droplets contact angle for 0.5 wt% graphene–PDMS.
Source: Reprinted (adapted) with permission from Berean KJ, Ou JZ, Nour M, Field MR, Alsaif MM, Wang, Y, et al. Enhanced gas permeation through graphene nanocomposites. J Phys Chem C 2015;119(24):13700–12 Copyright 2015, American Chemical Society.

better than that in EPDM/PR blends. Hence, contact angle measurement can be also used to investigate the influence of the filler on the conversion of a composite to hydrophobic or hydrophilic along with the state of dispersion of fillers.

6.3.3 Mechanical properties

Over the last few decades, different inorganic nano-sized fillers have generally been incorporated into the rubber matrices in order to improve the properties of rubber matrices successfully. The nanofillers can introduce their high modulus and mechanical strength to the bulk rubber materials and obstruct the spreading of cracks during the breaking of the rubber nanocomposites. Nanocomposites can be categorized into three classes based on the number of dimensions of the incorporated filler particulates in the nanometer range. CB, silica, aluminum oxide (Al_2O_3), titanium dioxide (TiO_2), and zinc oxide (ZnO) are such dispersed nanofillers having three dimensions in the nanometer range [177,178]. Cellulose whiskers, CNTs, and CNFs exhibit two dimensions in the nanometer range [179–181], whereas layered silicates, EG, graphene sheets, and layered double hydroxides (LDHs) flakes consist of only one dimension in the nanometer range [164,182–184]. Although the dimension and the geometries of the filler particulates are decisive in the enhancement of the physicochemical properties of rubber composites, the state of dispersion of filler particulates in the matrix also plays a most crucial role in controlling the properties of these nanocomposites. The rubber composites exhibit inferior physical and mechanical properties when the filler particles form aggregates, bundles, and stacks in the matrix, since these kinds of filler structures act as defects in the composites. Between all the different fillers, none of them have the certain exceptional features that single layer and defect-free graphene can introduce to the polymer matrices, such as superior Young's modulus of ~ 1 TPa (in plane), tensile strength of ~ 130 GPa, and the higher elongation at break (25%) [20,103]. Graphene sheets dispersed in the rubber matrix in the wrinkled state (wrinkled graphene) indeed act as a crack propagation barrier in the materials and also the same state of graphene improves the mechanical interlocking between the matrix and the nanofiller [185]. Young's modulus of most of the rubbers can be enhanced considerably by dispersing graphene uniformly in it which indeed makes graphene–rubber nanocomposites useful for various applications.

6.3.3.1 Tensile features

So many researchers have already investigated as well as are now studying the effect of graphene fillers on the mechanical properties of a number of rubber vulcanizates (natural and synthetic) [5,99,100,102–108,111,113–116,118–120,122,125,126,128,131,133–135, 137,140,142,143,146–148,150,151,153,155,158–160,162,167,168]. For most investigations, researchers used GO, rGOs, and FGs as potential nanofillers rather than graphene to reinforce the different rubber matrices. They observed, in the majority number of cases, a significant improvement in the tensile strength and Young's modulus of the rubber nanocomposites. Regarding the elongation at break of graphene filled rubber nanocomposites, researchers observed (few times) an increase at low graphene loading as well as observed a decrease at high graphene loading due to the formation of stacked graphene sheets. Generally, % elongation at break of the rubber nanocomposites is decreased after the incorporation of filler into the matrix.

Xing and coworkers [146] prepared (GE/SBR) nanocomposites with versatile properties as well as enhanced mechanical properties. GE/SBR nanocomposites exhibit a significant increase in the tensile strength with increasing GE loading. They observed 260% increase in the tensile strength and a 140% improvement in the strain at break for 0.3 phr GE loaded nanocomposites, indicating the promising reinforcing effect of GE. They achieved 11 times higher tensile strength for the rubber nanocomposites compared to neat polymer for 7 phr GE loaded SBR composites as well as noticed the same strain at break for the nanocomposite as compared to neat SBR. They also compared the reinforcement effect of GE with the traditional fillers. They noticed that the tensile strength of 7GE/SBR nanocomposite is a little lower than that of 40 phr carbon black (CB-N 330) filled SBR composite, but slightly higher than those of 30 phr of CB (N330) or 40 phr of fumed nanosilica loaded SBR composite.

Hernandez et al. [153] fabricated FGSs filled NR nanocomposites using a laboratory scale two roll mixing mill and studied the effect of FGSs on the mechanical properties of NR. They observed a reduction of tensile strength of the nanocomposite at low FGS concentration (0.1 phr). They explained it as a result of distortion of the cross-linking density of the natural network of NR in the presence of very small amount FGSs. They noticed an enhancement in the elastic modulus and tensile strength as well without negatively influencing the elastic properties of the material with the increase in loading of FGSs.

Wu et al. [111] prepared NR based nanocomposites filled with surface functionalized GO (bis(triethoxysilylpropyl)tetrasulfide (BTESPT) functionalized GO). SGO acted as multifunctional nanofillers for rubber vulcanizates. They noticed that the tensile strength of the NR based nanocomposites increases significantly with increasing filler content, but the effect subsequently subsides when the SGO and GO contents exceeds 0.3 wt% and 0.1 wt%, respectively (Fig. 6.13). This phenomenon is very similar to the percolation theory that is generally utilized to interpret the impact strength of rubbery particle-toughened plastics [60] and the electrical conductivity of conductive filler-loaded polymer nanocomposites [33]. They observed a higher reinforcing effect of SGO compared to GO. The tensile strength of the GO/NR nanocomposites (0.1 wt% loading) was increased by 33% in comparison with that of neat NR. The enhancement in the tensile strength of 0.1 wt% GO/NR composites was much lower than that of the corresponding SGO/NR nanocomposites. They explained it as a result of the superior reinforcing effect of SGO. They reported that SGO played a role of a multifunctional nanofiller which acted as a reinforcing filler cum cross-linking agent due to the presence of polysulfide groups in the modifier of GO. Frasca et al. [173] prepared MLG ($\sim$10 layer of graphene sheets) filled CIIR nanocomposites using a two roll mill as well as using both solution mixing and melt mixing (two roll mixing mill). They noticed significant improvement in the mechanical properties for the nanocomposites prepared

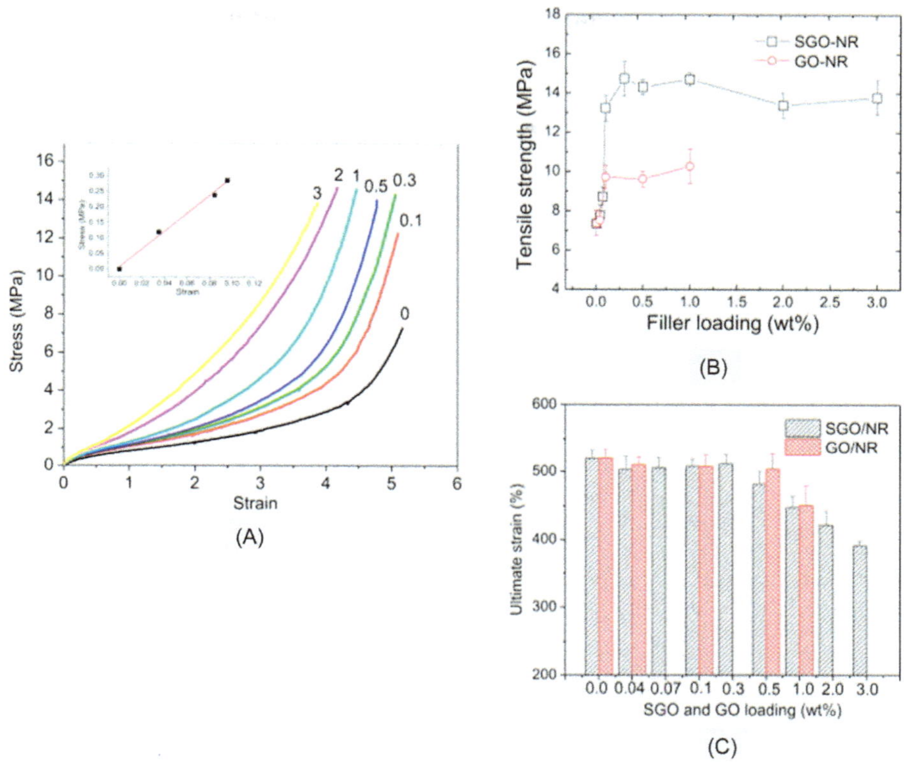

Figure 6.13 (A) Representative stress–strain curves of unfilled NR and SGO/NR nanocomposites. The numbers in the figure indicate the weight fractions of SGO in the nanocomposites, (Inset indicates linear fitting to the data below 10% of the stress–strain curves.) tensile strength (B) and ultimate strain (C) of SGO/NR and GO/NR nanocomposites with various nanofiller loading.
Source: Reproduced with permission from Wu J, Huang G, Li H, Wu S, Liu Y, Zheng J. Enhanced mechanical and gas barrier properties of rubber nanocomposites with surface functionalized graphene oxide at low content. Polymer 2013;54(7):1930–7. Copyright 2013, Elsevier.

using solution intercalation followed by melt mixing method compared to the nanocomposites prepared by only two roll milling (melt mixing). They observed a 38% increase in Young's moduli of the nanocomposites mixed in the solution intercalation method.

Schopp and coworkers [186] did a comparative investigation on styrene-butadiene rubber (SBR) composites filled with conventional and novel carbon-based fillers. At the same time, they studied the effect of filler type, content, and dispersion process on the physicochemical properties of SBR vulcanizates. They prepared FG/SBR nanocomposites based on TRGO and CRGO using two aqueous dispersion blend strategies. They characterized the mechanical properties of those

Table 6.1 Mechanical properties of SBR composites

Sample	Elongation at break (%)	Tensile strength (MPa)	Stress at 50% strain (MPa)	Stress at 300% strain (MPa)	Hardness (Shore A)
SBR	324 ± 43	5.1 ± 1.1	0.8 ± 0.1	4.7 ± 0.7	53.3 ± 0.4
SBR/25 phr RCB	365 ± 12	8.2 ± 0.3	1.2 ± 0.1	6.7 ± 0.2	62.7 ± 0.9
SBR/25 phr CNT	260 ± 10	9.2 ± 0.4	2.1 ± 0.1	[a]	69.7 ± 0.6
SBR/25 phr EG40	334 ± 16	4.6 ± 0.1	1.1 ± 0.1	4.2 ± 0.2	59.8 ± 0.7
SBR/25 phr MLG350	91 ± 26	9.9 ± 2.2	6.5 ± 0.4	[a]	83.9 ± 0.6
SBR/25 phr CRGO	295 ± 9	12.6 ± 0.4	2.8 ± 0.1	12.4	74.7 ± 1.2
SBR/25 phr TRGO	88 ± 7	17.5 ± 0.6	10.9 ± 0.4	[a]	87.1 ± 0.4

[a]Breaking occurs below 300% strain.
Source: Reproduced with permission from Schopp S, Thomann R, Ratzsch KF, Kerling S, Altstädt V, Mülhaupt R. Functionalized graphene and carbon materials as components of styrene-butadiene rubber nanocomposites prepared by aqueous dispersion blending. Macromol Mater Eng 2014;299(3):319–29. Copyright 2014, Wiley.

nanocomposites containing different carbon-based fillers by stress–strain tests and Shore A hardness measurements. The mechanical properties of SBR containing 25 phr carbon filler as a function of the filler type are represented in Table 6.1. The elongation at break and tensile strength (normalized values relative to the neat SBR) are displayed in Fig. 6.14A. They observed that the incorporation of all these carbon fillers introduce higher tensile strength for the different composites. Tensile strength of the different nanocomposites increases with the increase in filler loading. Depending on the higher aspect ratio and uniform dispersion of the different carbon fillers in the SBR matrix, the reinforcing effect of those fillers was improved. The following are the filler ranking regarding their higher specific surface area: EG40 < RCB < CNT < MLG350 < CRGO < TRGO. They achieved highest tensile strength (increase of 240%) for SBR/25 phr TRGO containing rubber composites. The elongation at break of neat rubber is generally reduced after incorporation of carbonaceous filler into it. They observed a little change in the elongation at break value for RCB, CNT, and EG40 loaded composites and substantially lower elongation at break for MLG350 and TRGO. They also observed a simultaneous increase in elongation at break and tensile strength for the CRGO filled SBR composites. They observed most pronounced improvement (highest) in Shore A hardness for MLG350/SBR composites, and ~60% (TRGO/SBR), 40% (CRGO/SBR), 30% (CNT/SBR), 20% (RCB/SBR), and 10% (EG40/SBR) increase in Shore A hardness for the different rubber nanocomposites (Fig. 6.14B).

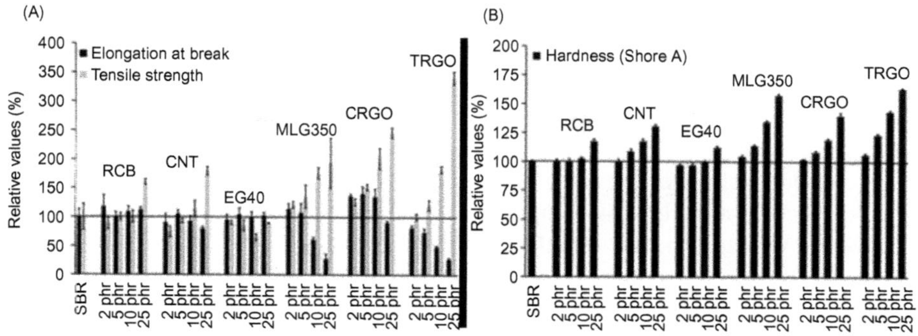

Figure 6.14 (A) Elongation at break and tensile strength for SBR/C filler composites relative to neat SBR and (B) Shore A hardness of SBR/C filler composites.
Source: Reproduced with permission from Schopp S, Thomann R, Ratzsch KF, Kerling S, Altstädt V, Mülhaupt R. Functionalized graphene and carbon materials as components of styrene-butadiene rubber nanocomposites prepared by aqueous dispersion blending. Macromol Mater Eng 2014;299(3):319–29. Copyright 2014, Wiley.

Potts and coworkers widely investigated the reinforcement effect of RGO on the different properties of NR vulcanizates [106,187]. They prepared the nanocomposites using both the solution treatment and two roll mixing mill. They observed an exceptional improvement in the multifunctional properties of the nanocomposites prepared by the coagulation of RGO colloidal suspensions and natural rubber latex [106]. In a different investigation [187], they studied the effect of thermally exfoliated graphene oxide (TEGO) on the physicochemical properties of NR vulcanizates. They used a conventional rubber melt blending method to disperse TEGO in NR using a two roll mill. They noticed that the modulus at 100% elongation (M100) and at 300% elongation (M300) of the TEGO/NR nanocomposites decreased compared to neat NR at 2 phr and 3 phr filler loading as well as observed a small increase over neat NR with 4 and 5 phr TEGO loading. They observed 38% higher value of M100 for 2 phr L-TEGO (Latex premixed TEGO)/NR nanocomposites compared to M100 of the 5 phr TEGO/NR nanocomposites. They observed higher tensile strength values for L-TEGO/NR nanocomposites compared to TEGO/NR nanocomposites due to the better dispersion of L-TEGO filler in the rubber matrix.

Tang and coworkers [188] have also studied the influence of graphene surface chemistry on the dispersion of graphene and interfacial adhesion in SBR/graphene composites and on the resultant properties of the composites as well. They used in situ chemical reduction method to produce rGO itself in the SBR latex during composite preparation and observed that uniform dispersion as well as good interfacial adhesion of rGO (G) affects the final properties of the rubber composite. According to them, the determinant factor to achieve good dispersion of graphene is the COx fraction of in situ chemically rGO (G). They noticed a significant improvement in the modulus and tensile strength for both GO and graphene containing rubber composites. SBR/G composites showed superior performance compared to the SBR/GO composites. They also noticed that the moduli and tensile strength of the nanocomposites were gradually improved as the COx fraction of graphene decreased, which was due to the improved interfacial adhesion and uniform dispersion of graphene (at 3 phr loading). The moduli and tensile strength of SBR/G3 were increased by about 240% and 420%, respectively, compared to the neat SBR. The moduli and strength for SBR/GN3-T95 were improved by about 530% and 490%, respectively, which can be attributed to the improved interfacial adhesions, good dispersion, and high N-doping level.

Overall, we can conclude that the stiffness and strength of various elastomers are generally improved by significant extents after the addition of graphene-based nanofillers. The substantial improvements in the respective mechanical properties of the nanocomposites can only be achieved with highly exfoliated or uniformly-dispersed graphene in the rubber matrix. The desired properties of graphene containing rubber nanocomposites can be obtained by using a combination of the specific preparation methods and surface functionalization of the graphene. The amount of graphene filler loading also can control the desired properties. The Young's modulus of rigid polymers is of the order of 10^9 Pa in comparison with only 10^6 Pa for rubber vulcanizates and the modulus of the graphene based fillers ($\sim 10^{12}$ Pa) which is much higher than the modulus of the all the rubber matrices. For a certain % loading fraction of graphene, the influence upon the rubber modulus after the incorporation of graphene will be much more evident compared to rigid polymers like thermoplastics.

6.3.3.2 Dynamic-mechanical thermal analysis (DMTA)

The response of a given material to an oscillatory deformation (application of sinusoidal stress) as a function of temperature is generally measured by dynamic mechanical analyzer (DMA). It gives important messages regarding both the viscoelastic characteristics of rubber nanocomposites and molecular structure, by exploring the mechanical relaxations of the rubber matrix. DMA results consist of three important parameters such as storage modulus (G′), loss modulus (G″), and loss tangent (tan δ). Tan δ is valuable in determining the existence of molecular mobility transitions like glass transition temperature (T_g). Rubber nanocomposites represent themselves technologically as an important class of materials due to their thermo-mechanical and dynamic-mechanical properties. Specifically, rubbers vulcanizates at the transition region from glassy to rubbery state have excellent capability for vibration damping. The damping characteristics of rubber and its composites are generally determined by the peak height and area under the loss tangent (tan δ) curve as well as the value of the loss modulus at the respective temperature of use. Relaxation phenomena originating in rubber nanocomposites involve inputs from both the rubber matrix and the dispersed nanofiller, and in the majority of circumstances it reflects the interactions between nanofillers and rubber matrix. Relaxation phenomena mainly arise due to the phase changes, molecular dynamic effects, and interfacial effects. Many investigations have been carried out to study the viscoelastic properties of rubber nanocomposites containing graphene/FG as nanofiller using DMA [103–106,109,118,124–126,133,140,142,150,155,156,159,160,162,168,189].

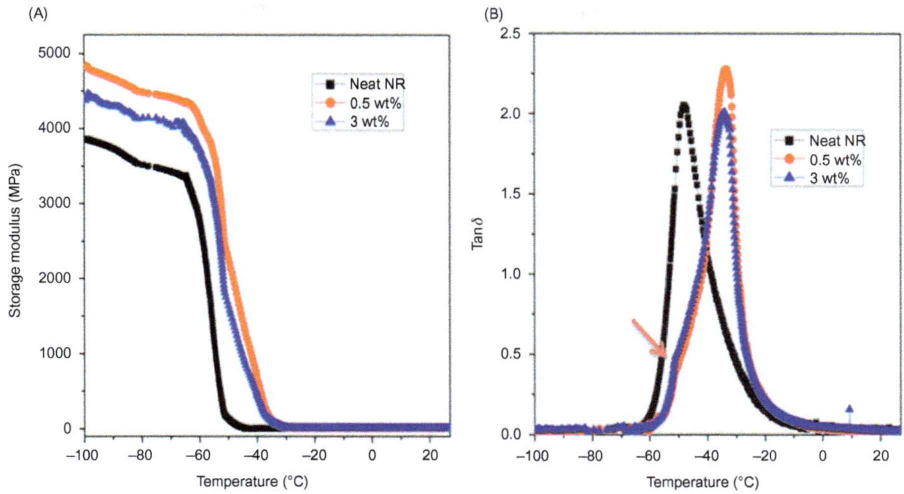

Figure 6.15 DMA plots of NR-thermally reduced graphene oxide (TRGO) composites (A) storage modulus (B) damping factor (tan δ).
Source: Reproduced with permission from Yaragalla S, Meera AP, Kalarikkal N, Thomas S. Chemistry associated with natural rubber–graphene nanocomposites and its effect on physical and structural properties. Ind Crops Prod 2015;74:792–802. Copyright 2015, Elsevier.

Yaragalla and coworkers [190] prepared TRGO containing NR based nanocomposites by mechanical melt mixing method. They noticed an improvement in the storage modulus for the TRG filled NR composites. The storage modulus of the nanocomposites (3 wt% loading of TRG) was increased by 25% compared to neat NR (Fig. 6.15). They observed additional relaxation in the tan δ curves (formation of the hump) for the TRG/NR composites which are due to the formation of strong thick interface between TRG and NR (Fig. 6.15B). As a result of strong $\pi-\pi$ interactions associated with π bonds of TRG and double bonds of NR, the glass transition temperature was significantly increased (15°C) due to the restriction of NR chain mobility by the TRG sheets.

Araby et al. [103] measured the storage modulus, loss modulus, and loss factor of SBR and its nanocomposites containing GnPs (Fig. 6.16). They observed a 4°C increase in the glass transition temperature (T_g) for the SBR based nanocomposites at 10.5 vol% GnPs loading. However, a decrease in the glass transition temperature can sometimes also be observed for graphene/rubber nanocomposites, which is due to the plasticization or a dilution effect that reduces the physical entanglements in the rubber composites [104]. They also noticed a significant enhancement in storage modulus in both glassy and rubbery regions. For 10.5 vol% GnPs containing nanocomposites, the storage modulus was increased by three times (from 900 to 2609 MPa) in the glassy region, and more than seven times (from 2.4 to 18 MPa) in the rubbery region compared to neat SBR. This significant increase in storage modulus in rubbery region indicates good interactions and interfacial adhesion between SBR chains and GnPs. Loss modulus was also increased in the rubbery region than

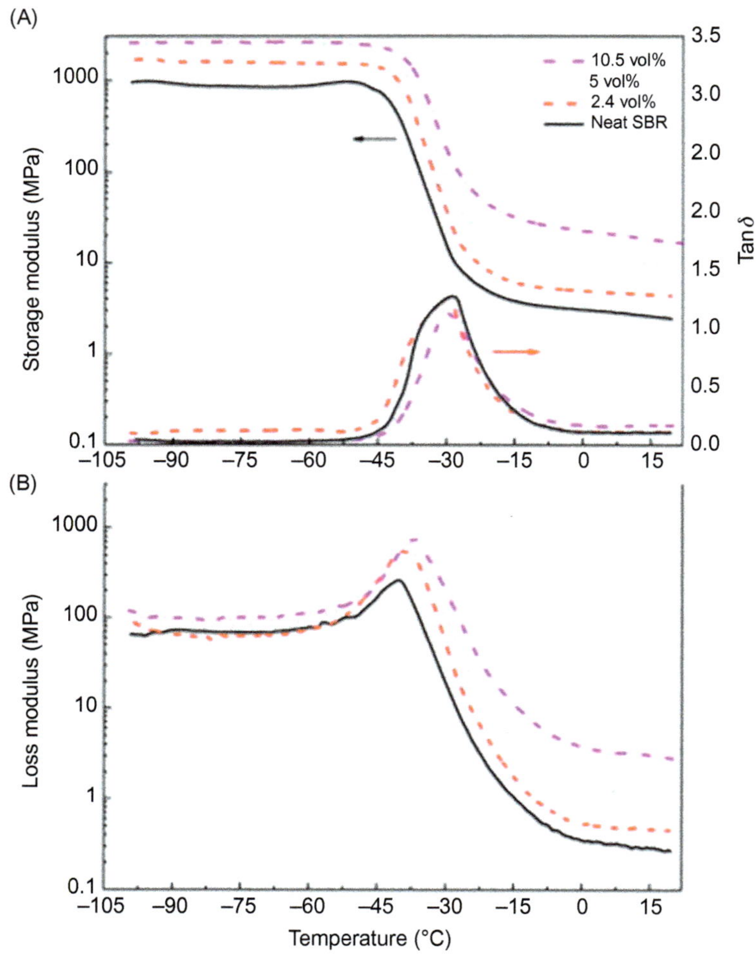

Figure 6.16 Dynamic mechanical properties of neat SBR and graphene nanoplatelets (GnPs) filled SBR nanocomposites (A) and (B) storage modulus, tan δ, and loss modulus as a function of temperature.

Source: Reproduced with permission from Araby S, Meng Q, Zhang L, Kang H, Majewski P, Tang Y, et al. Electrically and thermally conductive elastomer/graphene nanocomposites by solution mixing. Polymer 2014;55(1):201–10. Copyright 2014, Elsevier.

that in glassy region. Potts et al. [106] reported an approximately 20-fold increase in storage modulus for a NR filled with GO by solution treatment in the rubbery region.

Typically, DMA experiments of graphene containing rubber vulcanizates report improvements in the storage modulus as well as the glass transition temperature of the matrix as a result of uniform dispersion of graphene and good interactions/interfacial adhesion between graphene (high aspect ratio) and rubber matrix which cause restriction in the mobility of the macromolecular segments of rubber at high temperatures.

6.3.4 Thermal behavior

6.3.4.1 Thermal degradation behavior

A thermogravimetric analyzer (TGA) generally measures the amount and rate of change in the mass of a sample as a function of time or temperature in a regulated environment. The measurements are mainly carried out to investigate the oxidative and/or thermal degradation behavior of materials as well as their compositional features. This characterization method can analyze materials that show mass loss or gain as a result of thermal degradation, oxidation, or loss of volatile substances. It is very useful for the study of polymeric materials, including thermoplastics, thermosets, elastomers, and polymeric nanocomposites, since it determines the working temperature at which the materials can be used for certain applications. There are a number of literatures where the influence of the incorporation of graphene into the rubber matrices on the thermal stability of the base rubber vulcanizates have been studied [104,116,124,125,130,133,139−141,143,148,149,155,156,162,168,191]. The thermal degradation behavior of graphene filled rubber nanocomposites generally relies upon the nature of graphene, the fabrication methods involved, and the state dispersion of graphene, as well as the chemical functionalities of graphene.

The improvement in the thermal stability of graphene filled rubber vulcanizates can be attributed to the heat shielding effect of uniformly-dispersed graphene flakes in the rubber matrix, where the graphene sheets act as an efficient heat absorbers by dissipating more heat than the rubber matrix and do not allow heat to build-up within the host matrix, and thereby inhibit oxidation at the early stage of degradation. A mass transport barrier is generally formed by the exfoliated graphene sheets to the volatile products produced during the decomposition of the hybrid nanocomposites.

Gan et al. [143] observed that the thermal stability of silicone rubber (SR) was significantly increased after the addition of graphene nanoribbon (GNR) into the rubber matrix (Fig. 6.17A). They also noticed that the onset degradation temperature of the SR/GNR nanocomposites was increased slightly compared to the neat SR. All the GNR containing SR nanocomposites along with neat SR decomposed with a one-step process, which indicates that the GNR did not destroy the network of the SR. They noticed that the inclusion of GNR flakes also led to a noticeable increase in the temperature of the maximum mass loss rate of the rubber matrix (Fig. 6.17B).

The use of different surfactants during the fabrication of rubber nanocomposites by solution blending methods to disperse graphene uniformly in the common solvent [124], or the chemical functionalization of graphene with different organic moieties [150], which are very prone to thermal degradation compared to the rubber matrix, may trigger untimely degradation of the rubber nanocomposites since they decompose at comparatively much lower temperatures than the host rubber.

6.3.4.2 Thermal conductivity

Exceptional thermal transport features of graphene dispersion have the capability for thermal management in miniaturized electronic devices [192−194], for heat actuated and shape memory polymers [195]. The thermal conductivity of polymers

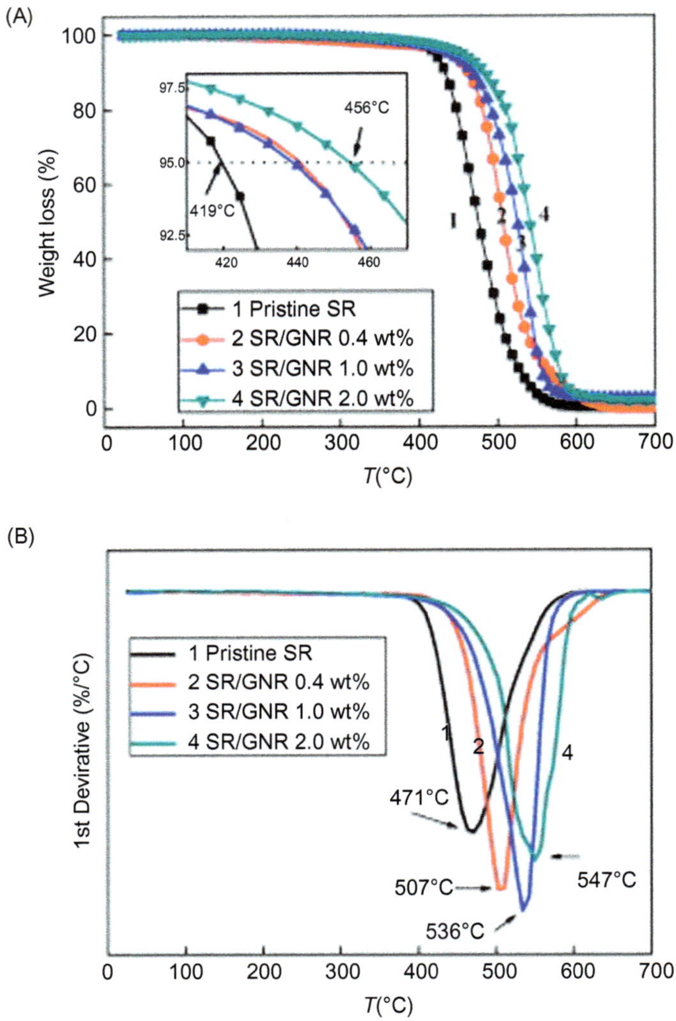

Figure 6.17 (A) TGA curves and corresponding (B) DTG curves of the pristine SR and the SR/GNR nanocomposites.
Source: Reproduced with permission from Gan L, Shang S, Yuen CWM, Jiang SX, Luo NM. Facile preparation of graphene nanoribbon filled silicone rubber nanocomposite with improved thermal and mechanical properties. Compos Part B Eng 2015;69:237–42. Copyright 2015, Elsevier.

ranges from 0.1 to 1 W/ (m.K) and the thermal conductivity of graphene is ~5000 W/ (m.K) at room temperature. Since the thermal energy is generally transported in the mode of lattice vibration (phonons), significant thermal resistance at the filler–polymer and filler–filler [196] interfaces will be imparted by the poor thermal coupling in vibration modes [197]. Thermal conductivity of rubber vulcanizates is generally increased after the incorporation of graphene in the rubber matrix. Graphene nanosheets containing rubber nanocomposites exhibit lower interfacial resistance, higher thermal conductivity, as well as a greater heat capacity compared to the neat rubber due to its sheet-like morphology. The lattice vibrations (phonons) also are mainly responsible for the conduction of thermal energy in rubber vulcanizates; hence the phonon scattering or acoustic impedance mismatch must be lowered to achieve higher thermal conductivity for rubber composites. Uniformly-dispersed graphene nanosheets can form strong bonds with the rubber matrix which indeed lowers the acoustic impedance mismatch. The formation of interconnecting networks of graphene in the rubber matrix can also help to increase the thermal conductivity of the nanocomposites. But the formation of graphene aggregates and the poor thermal coupling in vibration modes between the graphene and rubber or between the graphene sheets make the composites poor in thermal conductivity.

Most of the investigations summarized the effect of graphene sheets addition on the thermal conductivity of graphene/rubber composites. They generally observed an increase in thermal conductivity due to the presence of the highly thermally conducting graphene [103,106,114,117,130,149,160,162,191,199]. Song and coworkers [198] synthesized rubber nanocomposites containing uniform and stable dispersion of low defect graphene flakes (l-GFs), which are prepared using aqueous-phase mixing of exfoliated l-GFs with SBR latex. They observed thermal conductivity of 0.3445 W mK^{-1} for the 5 phr l-GFs filled SBR composites, which was significantly higher than those of the other composites (Fig. 6.18A). They also observed high quality dispersion of l-GFs in rubber matrix which indeed increase the filler performance compared to the GO/SBR composites. The increased phonon scattering at the defects of GO may also influence the thermal conductivity of the GO filled nanocomposites [200]. They also reported temperature dependence of thermal conductivity for different filler containing SBR composites at temperatures (T) ranging from 100 to 300 K (Fig. 6.18B). They noticed that the thermal conductivity of the l-GFs/SBR composites (for other composites as well) decreases near linearly with decreases in temperature up to 150 K, and the T-dependence weakens below 150 K. This linear temperature-dependent behavior is consistent with the Umklapp phonon scattering (characteristic for crystalline materials) [88]. Potts et al. [106] reported higher thermal conductivity for the solution treated reduced graphene (RG) filled NR nanocomposites compared to nanocomposites produced by melt blending in a two roll mill. They observed an interconnecting network for the solution treated RG/NR composites that resulted in the improvement in the thermal conductivity of the nanocomposites.

Song and coworkers [201] fabricated graphene nanoplatelets (GnPs) /SR composites with the assistance of the Flacktek SpeedMixer. They observed a gradual increase in the thermal conductivity of composite with the increase in GnPs loading (0 to 8 wt%). They reported 53.1% increase in thermal conductivity for the 8 wt% GnPs loaded composite compared to neat SR which was due to the uniform dispersion of GnPs in the rubber matrix. Significant interfacial adhesion between GnPs and SR helped to decrease the interfacial resistance between filler and rubber which

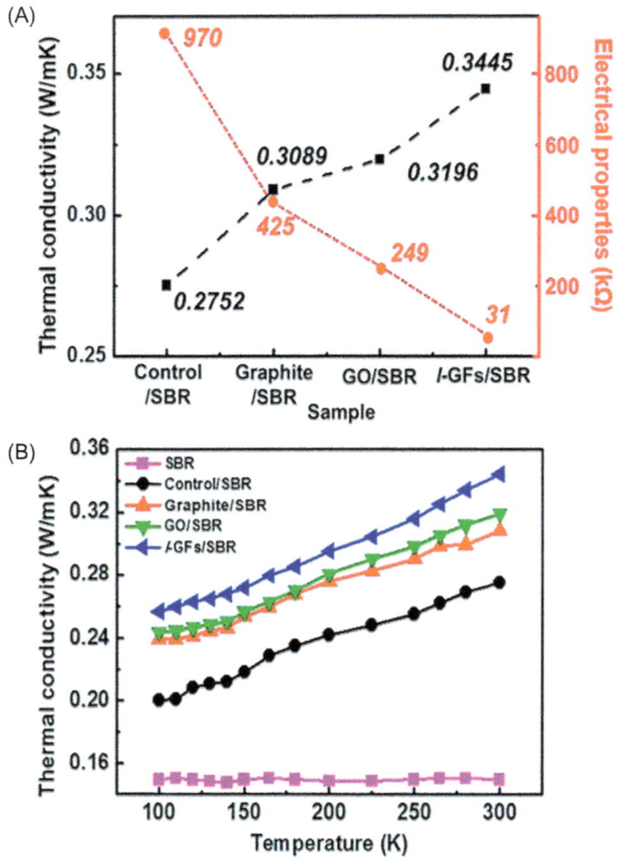

Figure 6.18 Thermal and electrical properties of the SBR composites. (A) Thermal and electrical conductivities. (B) Temperature dependence of the thermal conductivity for different SBR composites.
Source: Reproduced with permission from Song SH, Kim JM, Park KH, Lee DJ, Kwon OS, Kim J, et al. High performance graphene embedded rubber composites. RSC Adv 2015;5 (99): 81707–12. Copyright 2015, Royal Society of Chemistry.

indeed generates thermal conductive pathways and networks. More thermal conductive pathways were achieved and the progressions of thermal conductive networks were increased with the increase in GnPs loading.

6.3.5 Gas barrier properties

Development of significant barrier properties (gas and liquid) in rubber nanocomposites by controlling their microstructure could be one of the most important achievements in order to make these materials marketable. Rubber vulcanizates normally exhibit poor barrier properties. The barrier properties of rubber matrices can be significantly improved by incorporation of impermeable lamellar fillers such as clay tactoids and graphene materials with high aspect ratio to change the path of diffusion of gas/liquid molecules. The major concern is to achieve a uniform dispersion and exfoliation of the filler platelets in the rubber matrix to produce a tortuous diffusion pathway for the gas molecules to obtain increased barrier properties. In most of the research based on rubber nanocomposites, the barrier properties are investigated by using gas. The utilization of graphene as effective barrier materials is very essential to improve the barrier properties of rubber matrix significantly, since the defect-free monolayer graphene are impermeable to all gases [202]. Many researchers have tried to develop graphene filled rubber nanocomposites with high barrier properties but have failed to improve a number of times due to the presence of a large free volume in nanocomposites and, therefore, achieved poor gas barrier properties [5,99,100,102,107,108,110,111,134,140,145,148,203].

Yan et al. [145] measured the oxygen permeability of neat NR and rGO containing NR composites consist of segregated and not-segregated morphologies at 0% and 50% relative humidity (RH) at 38°C (Fig. 6.19). They noticed that the oxygen permeability decreases when increasing the content of rGO for both the segregated and not-segregated structure of the nanocomposite films. They reported that the oxygen permeability of rubber nanocomposites with segregated morphology decreased from 5×10^{-7} to 3.2×10^{-7} cm^3 (STP) cm/cm^{-2} s^{-1} atm^{-1} after the addition of only 1 phr rGO sheets at 0% RH. The oxygen permeability further decreased up to 2×10^{-7} cm^3 (STP) cm/cm^{-2} s^{-1} atm^{-1} with the incorporation of 6 phr of rGO. They noted different effects imparted by the two different morphologies on the gas permeation behavior, i.e., the nanocomposites with segregated morphology exhibited oxygen permeability significantly lower than that of not-segregated nanocomposites at all the filler contents which is attributed to the formation of a percolating and interconnected rGO that resulted in a severe "tortuous pathway" for the gas diffusion. They also measured water vapor permeability of the different nanocomposites and noticed a decrease in the water vapor permeability for the nanocomposites with the increase in rGO loading, and the films with segregated morphology exhibited water vapor permeability lower than that of not-segregated nanocomposites.

Wu et al. [111] measured air permeability of the BTESPT functionalized GO (SGO)/NR nanocomposites (P) at 25°C and they observed that the value of P/P_0 (air permeability of unfilled NR (P_0)) decreased significantly to a value of 52% for 0.3 wt% SGO loading, after which it slowly decreased. They noticed that 0.3 wt% of SGO is equivalent to 16.7 wt% of nanoclay in improving the gas barrier property of NR [204], which may be due to the uniform dispersion of the SGO flakes with high aspect ratio that resulted in a decreased cross-section and a tortuous path for gas diffusion. They also observed superior gas barrier properties of SGO filled composites as compared to TRGO filled NR composites [205], which may be due to the structural difference between SGO (flat sheets) and TRGO (wrinkled). There is a report stating that the diameter of the TRGO flakes is decreased to ~500 nm independent of the size of GO (starting material for synthesizing TRGO), thus the specific surface area of TRGO is smaller than that of SGO [206] which results in the inferior gas barrier property of TRGO filled NR.

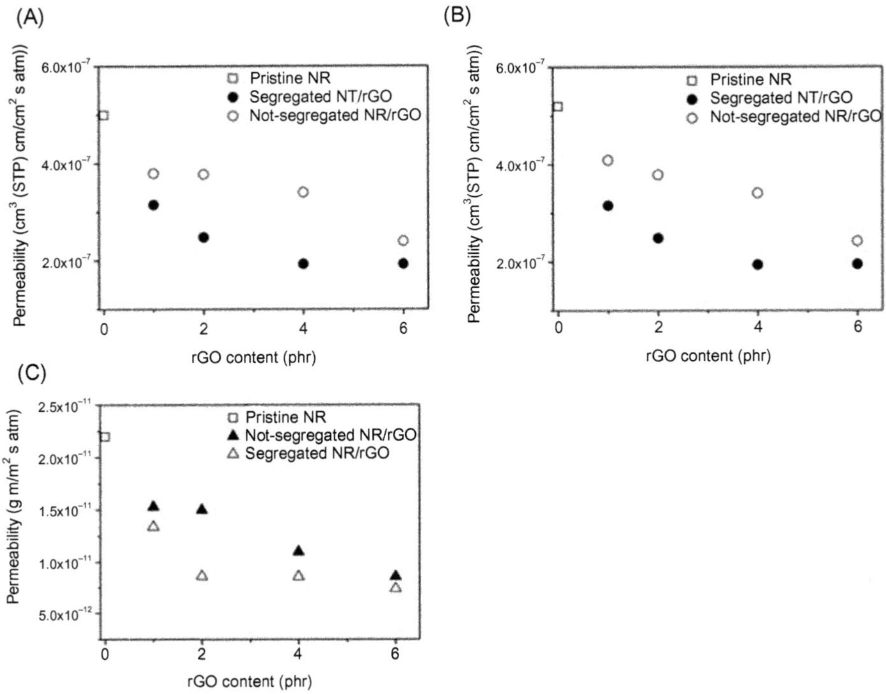

Figure 6.19 Gas permeability of rGO/NR film as a function of rGO content: oxygen permeability measured at 38°C: (A) 0% Relative Humidity (RH), (B) 50% RH, and (C) water vapor permeability of rGO/NR film measured at 38°C.
Source: Reproduced with permission from Yan N, Buonocore G, Lavorgna M, Kaciulis S, Balijepalli SK, Zhan Y, et al. The role of reduced graphene oxide on chemical, mechanical and barrier properties of natural rubber composites. Compos Sci Technol 2014;102:74–81. Copyright 2014, Elsevier.

Song et al. [198] observed an increase in the gas barrier property for l-GFs/SBR nanocomposites due to the uniform dispersion of l-GFs and also due to the exfoliated structure of the nanocomposites. They achieved the lowest oxygen transmission rate value for the l-GFs/rubber composites of 6.87×10^{-14} m^4 sec^{-1} N^{-1}. The l-GFs/SBR composites exhibited a 35% increase in the barrier property compared to the CB containing SBR vulcanizates, and a 22% increase compared to the GO/ rubber nanocomposites, which can be attributed to the uniform dispersion low defect l-GFs nanosheets in SBR matrix.

6.3.6 Electrical properties

6.3.6.1 Dielectric properties

Dielectric spectroscopy has been demonstrated to be notably useful in the investigation of polymer dynamics at interfaces as well as in confining geometries due to their increased sensitivity with decreasing domain size and vast dynamic range, with an exceptional range of temperature and frequency. Exploring the effects of nano-confinement on the segmental relaxation dynamics of the polymer, such as by controlling shifts of transition temperatures, alterations in thermodynamic parameters, and the cooperativity of the relaxation mechanism, broadband dielectric spectroscopic analysis disclosed the effect of entropic (e.g., isolation in chain-end, density deviations, or disentanglement) and enthalpic (hydrogen bonding interactions in the interfaces) influences on chain mobility [207,208]. Dielectric spectroscopy technique has also been used to investigate the dynamics of the macromolecular chains of rubber in various graphene filled rubber nanocomposites [99,101,111,116,121,122,138,152–154,209]. The dielectric properties (dielectric constant, dielectric permittivity, etc.) of graphene containing polymer nanocomposites are significantly influenced by the aspect ratio of the graphene as well as its interfacial adhesion with the polymer matrix [210]. Maxwell-Wagner-Sillars mechanism for dielectric characteristics interpreted polymer–filler interfacial bonding as comparable to the donor–acceptor complexes. During the application of electric field (flow of current) to the polymer nanocomposite, the charges are generally accumulated at the interfaces between two dielectric materials with dissimilar relaxation times [99].

Mensah et al. [116] synthesized GO/acrylonitrile–butadiene rubber (NBR) nanocomposites using solution mixing. They measured the real part of dielectric permittivity (ε') and the loss factor at different frequency (frequency sweep). They observed a considerable increase in the real part of permittivity values for the GO loaded NBR composites at a lower frequency as compared to neat NBR. But they also noticed a gradual decrease in dielectric permittivity at higher frequencies along with the reduction of their respective loss factor with an increase in frequency. The increase in the permittivity values for the nanocomposites at lower frequency can be attributed to slow changes of the electric field which provide enough time for permanent and induced dipoles to align themselves towards the direction of the applied field, leading to increased polarization. The increase in dielectric permittivity (ε') at low frequencies can be associated with the increased conductivity, and/or electrode polarization, and/or interfacial polarization.

Singh et al. [121] prepared RGO (thermally exfoliated) filled NBR based nanocomposites by the solution mixing method. They observed an increase in both the real and imaginary part of permittivity for the nanocomposites with the increase in the RGO loading (Fig. 6.20). The real part of permittivity (relative) was increased from 4.4 to 6.8 and the imaginary part of permittivity (relative) was increased from 0.8 to 2.8 at 9.6 GHz with the increase in RGO loading from 2% to 10%. They also found a significant increase in the loss tangent values in the range of 0.22–0.18 for 2 wt% loading of RGO, 0.37–0.32 for 4 wt% loading and 0.44–0.39 for 10 wt% loading over the frequency range of 4–12 GHz.

The general criteria which basically influence the dielectric features of graphene containing rubber nanocomposites are the state of dispersion, the significant reinforcement of graphene to the rubber matrix, the thickness of the nanocomposite samples, and the amount of filler loading.

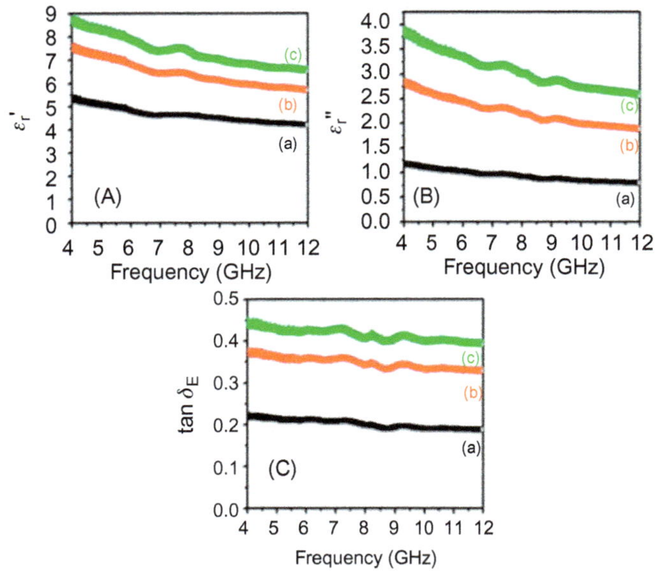

Figure 6.20 (A) Real part of permittivity (ε_r'), (B) imaginary part of permittivity (ε_r''), (C) tangent loss for RGO/NBR composites. (a) 2 wt%, (b) 4 wt%, and (c) 10 wt%.
Source: Reproduced with permission from Singh VK, Shukla A, Patra MK, Saini L, Jani RK, Vadera SR, et al. Microwave absorbing properties of a thermally reduced graphene oxide/ nitrile butadiene rubber composite. Carbon 2012;50(6):2202–8. Copyright 2012, Elsevier.

6.3.6.2 Electrical conductivity

Due to high electrical conductivity of graphene based materials, these materials find use in several potential applications, such as in devices and other electronics. It has been reported that stacked RGO sheets ("Paper"-like materials) show electrical conductivity up to 35,100 S/m [211]. As a result of the incorporation of such kinds of highly conducting fillers into the insulating polymers, the bulk electrical conductivity of the respective polymer is generally increased by many orders of magnitude. In order to transform an insulating matrix to an electrically conductive one, it is essential for the % loading of the conducting filler to be above a certain value which is known as the electrical percolation threshold, where a conductive filler–filler network is developed [212]. The increase in electrical conductivity of polymer nanocomposite after achieving the electrical percolation threshold as a function of filler content could be modeled by a simple power-law equation [213].

$$\sigma_c = \sigma_f * t(\phi - \phi_c)$$

where σ_c is the nanocomposite conductivity, σ_f is the filler conductivity, ϕ is the filler volume fraction, ϕ_c is the percolation threshold, and t is a scaling exponent. There is no strict requirement of filler–filler direct contact for current flow; preferably, electronic conduction in composite films can occur through the tunneling between delicate polymer layers encompassing the filler particulates. When conducting filler concentrations are low and also uniformly dispersed in the insulating polymer matrix, electron tunneling resistance between the conducting filler junctions controls the electron transport in polymer composites [214]. The state of dispersion of different fillers in the polymer matrix can be assessed qualitatively from the percolation threshold. The electrical conductivity of graphene containing rubber nanocomposites not only depends on the state of dispersion and aspect ratio of graphene, but also the functionalization of the graphene can play a role to improve the electrical conductivity of the nanocomposites by inducing good interactions and interfacial adhesions with the rubber matrix. The arrangement of the graphene sheets in the matrix is also crucial because the percolation threshold is generally decreased in the case of random orientation of graphene and increased for parallel orientation. Many researchers also have studied and developed graphene containing rubber nanocomposites with high electrical conductivity, as well as with very low percolation threshold in some cases [5,100,103,106,107,116–120,122,124,125, 138–140,148,153,155,160,209].

Matos and coworkers [124] prepared GO and rGO filled NR based nanocomposites by the latex blending method. They used cetyltrimethylammonium bromide surfactant to uniformly disperse rGO in NR latex as well. They observed higher electrical conductivity (lower electrical resistivity) for the rGO filled nanocomposites. They achieved a very low percolation threshold (between 0.1 and 0.5 wt%) due to the better contact between the graphene sheets (Fig. 6.21A). They reported a decrease in the electrical resistivity of four orders of magnitude (10^7 to 10^3 Ω cm) for the 2 wt% of rGO loaded rubber composites compared to neat rubber.

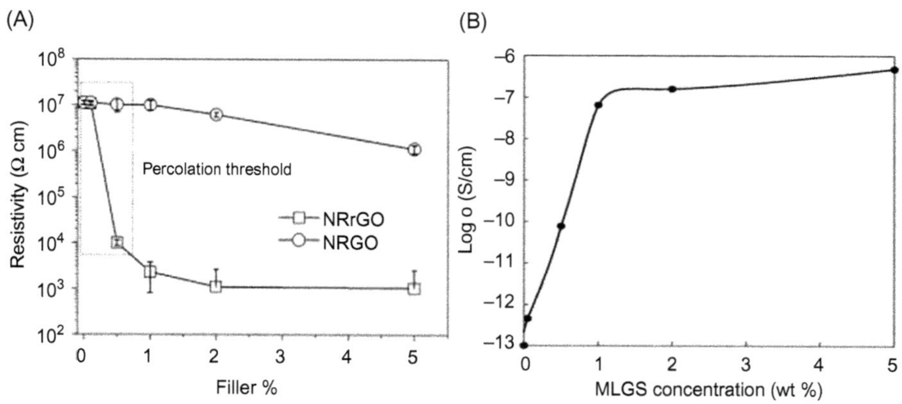

Figure 6.21 (A) Resistivity as a function of filler loading for NR nanocomposites that contain rGO and GO. Reproduced with permission from [124] Copyright 2014, Elsevier. (B) Electrical conductivity as a function of MLGS concentration (wt%) in SBR.
Source: Reproduced with permission from Kim JS, Yun JH, Kim I, Shim SE. Electrical properties of graphene/SBR nanocomposite prepared by latex heterocoagulation process at room temperature. J Ind Eng Chem 2011;17(2):325–30. Copyright 2011, Elsevier.

Kim et al. [139] prepared multilayered graphene sheets (MLGS)/SBR nanocomposites by the heterocoagulation method (latex blending). They reported that the electrical conductivity of rubber nanocomposites was increased from 4.52×10^{-13} to 4.56×10^{-7} S/cm with the increase in MLGS loading from 0.1 to 5 wt%, which was due to the high aspect ratio and electrical conductivity of MLGS (Fig. 6.21B). They also achieved a low percolation threshold in between 0.5–1.0 wt% loading of MLGS which can be attributed to the effective mixing of MLGS sheets in the rubber latex using the heterocoagulation method without damaging the fillers or the filler network. They also explained this significant increase in the electrical conductivity by the segregated network concept [215].

Araby et al. [103] prepared graphene nanoplatelets (GnPs) filled SBR based nanocomposites by the solution mixing method. They observed a small decrease in the electric volume resistivity for the GnPs/SBR nanocomposites before 5 vol% loading, and also observed a significant drop inbetween 5–7 vol% loading of GnPs. It was explained that GnPs at low volume fractions cannot connect with each other to form a percolated network which is required for electron mobility. After the incorporation of more GnPs in the rubber matrix, the distance between graphene sheets becomes smaller and at a certain volume %, a conductive network forms in the matrix providing pathways for electron mobility. They achieved a percolation threshold for the rubber nanocomposites between 5 and 6.5 vol%.

Luo et al. [118] formed three-dimensional (3D) interconnected graphene structures in NR matrix using self-assembly integrating latex compounding technology by exploiting electrostatic adsorption between poly(diallyldimethylammonium chloride) modified graphene (positively charged) (PDDA-graphene) and NR latex particles (negatively charged) as the driving force (Fig. 6.22A). They achieved a 3D network structure of graphene sheets in the rubber matrix which empowered the

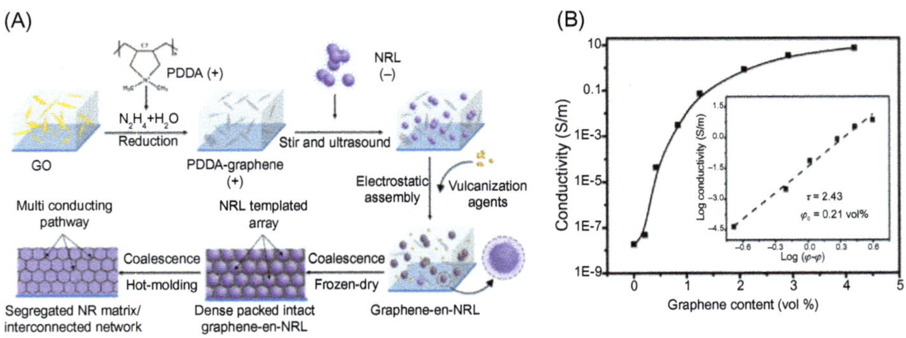

Figure 6.22 (A) Schematic description of the multistep process for the preparation of graphene-filled NR nanocomposites and (B) Electrical conductivity of PDDA-graphene/NR nanocomposites as a function of graphene content. Inserted figure shows the double-logarithmic plot of electrical conductivity vs. ($\varphi - \varphi_0$).
Source: Reproduced with permission from Luo Y, Zhao P, Yang Q, He D, Kong L, Peng Z. Fabrication of conductive elastic nanocomposites via framing intact interconnected graphene networks. Compos Sci Technol 2014;100:143–51. Copyright 2014, Elsevier.

NR based nanocomposites with exceptional electrical conductivity of 7.31 S/m with 4.16 vol% graphene loading as well as an extremely low percolation threshold of 0.21 vol% (Fig. 6.22B).

6.4 Conclusions and prospects

Graphene is a multifunctional nanofiller which can play a significant role in increasing the mechanical, dynamic mechanical, electrical, thermal, and gas barrier properties of different rubber vulcanizates at very low levels of loading. Graphene can introduce a better gas barrier property along with stiffness to the rubber matrix compared to layered nanoclay. It can also be employed as better nanofillers compared to CNTs to improve the electrical and thermal conductivity of the rubber vulcanizates. Highly exfoliated graphene sheets (single layer or very few numbers of layers) with high aspect ratio along with fewer defects can be produced using the CVD (Bottom-up strategy) method, but in very low amounts. To synthesize graphene in bulk quantity, chemical reduction and thermal reduction of chemically modified graphene precursors, mainly GO, are generally employed to obtain CRGO and TRGO, which can act as effective reinforcing nanofillers for the various rubber matrices.

Though both CRGO and TRGO contain certain functional groups on their surface, they exhibit significant electrical and thermal conductivity. Due to the presence of functional groups on the surface of CRGO and TRGO, it is easier to disperse those fillers into solvents directly, which is advantageous for the solution blending with soluble rubbers. In an alternative way, polymer nanocomposites can be synthesized via in situ polymerization of monomers solution containing dispersed graphene sheets. But the solvent blending (solution mixing) method is still advantageous over in situ polymerization regarding the improvements in properties of nanocomposites. Rubber nanocomposites can also be fabricated by melt compounding, which is industrially favorable method due to its low cost and environmentally friendly nature. But the disadvantage of this method is poor dispersion of different nanofillers in the rubber matrix, which indeed results in inferior physicochemical properties of the final product.

Microscopic analysis, such as TEM, is generally used to observe the state of dispersion of graphene sheets in polymers. The state of dispersion of graphene can also be derived partially from the electrical percolation threshold of the polymer nanocomposites, since the electrical percolation threshold is inversely proportional to the aspect ratio of randomly oriented graphene nanosheets. The parallel location of graphene sheets to the surface of polymer films can increase the gas barrier property of polymer composites. Stiffness (modulus) of the nanocomposites can be significantly increased by graphene for the nanocomposites based on rubber vulcanizates rather than for glassy polymer because of the higher stiffness contrast between rubber matrix and graphene.

The structure/surface of the filler is sometimes modified to improve the interactions and interfacial adhesions between filler and rubber matrix which can affect the physicochemical properties of the end product. Some selective properties of the rubber nanocomposites may improve by modifying the structure of the filler or the matrix, but at the same time may deteriorate other properties as well by doing this. It is well established that the random distribution of graphene in the polymer matrix can improve the electrical conductivity; the formation of an aligned network of graphene sheets can increase the gas barrier or mechanical properties in fixed directions. The fabrication methods of the rubber nanocomposites need to be tailored each time to achieve the desired property for the intended application, since it is hardly possible to enhance all the desired properties of the rubber nanocomposite together.

Until now, it has been shown that the addition of graphene based fillers into various rubber matrices can lead to a considerable improvement in the mechanical, electrical, dynamic-mechanical, thermal, and barrier properties of the rubber nanocomposites. A number of fabrication strategies were assessed in order to get good dispersion of the fillers and to achieve better filler–filler/filler–matrix interactions, at the same time various characterization techniques were considered. In general as a concluding remark, we can insist that the utilization of graphene based fillers to prepare this unique type of materials can improve some of the disadvantages of rubber vulcanizates and yield superior rubber nanocomposites which then can be utilized in a number of applications such as tire-tread, automotive, aerospace, tribological, biomedical, sensing devices, actuator, nanoelectronics, etc.

References

[1] Donnet JB, Bansal RC, Wang MJ, editors. Carbon black. New York, NY: Marcel Dekker; 1993.

[2] Heinrich G, Klüppel M, Vilgis TA. Reinforcement of elastomers. Curr Opin Solid State Mater Sci 2002;6(3):195–203.

[3] Ciesielski A. An introduction to rubber technology. iSmithers Rapra Publishing; 1999.

[4] Dittrich B, Wartig KA, Hofmann D, Mülhaupt R, Schartel B. Carbon black, multiwall carbon nanotubes, expanded graphite and functionalized graphene flame retarded polypropylene nanocomposites. Polym Adv Technol 2013;24(10):916–26.

[5] Ozbas B, O'Neill CD, Register RA, Aksay IA, Prud'homme RK, Adamson DH. Multifunctional elastomer nanocomposites with functionalized graphene single sheets. J Polym Sci Part B Polym Phys 2012;50(13):910–16.

[6] Mao Y, Wen S, Chen Y, Zhang F, Panine P, Chan TW, et al. High performance graphene oxide based rubber composites. Sci Rep 2013;3.

[7] He C, She X, Peng Z, Zhong J, Liao S, Gong W, et al. Graphene networks and their influence on free-volume properties of graphene–epoxidized natural rubber composites with a segregated structure: rheological and positron annihilation studies. Phys Chem Chem Phys 2015;17(18):12175–84.

[8] Seyedin MZ, Razal JM, Innis PC, Jalili R, Wallace GG. Achieving outstanding mechanical performance in reinforced elastomeric composite fibers using large sheets of graphene oxide. Adv Funct Mater 2015;25(1):94–104.

[9] Yan N, Xia H, Zhan Y, Fei G. New insights into fatigue crack growth in graphene-filled natural rubber composites by microfocus hard-X-ray beamline radiation. Macromol Mater Eng 2013;298(1):38–44.

[10] Zaman I, Kuan HC, Meng Q, Michelmore A, Kawashima N, Pitt T, et al. A facile approach to chemically modified graphene and its polymer nanocomposites. Adv Funct Mater 2012;22(13):2735–43.
[11] Yangwu Z, Shanthi M, Weiwei C, Xuesong L, Ji Won S, Jeffrey RP, et al. Graphene and graphene oxide-synthesis, properties and application. Adv Mater 2012;22:3906–24.
[12] Yang J, Tian M, Jia QX, Shi JH, Zhang LQ, Lim SH, et al. Improved mechanical and functional properties of elastomer/graphite nanocomposites prepared by latex compounding. Acta Mater 2007;55(18):6372–82.
[13] Kim H, Abdala AA, Macosko CW. Graphene/polymer nanocomposites. Macromolecules 2010;43(16):6515–30.
[14] Kalaitzidou K, Fukushima H, Drzal LT. Multifunctional polypropylene composites produced by incorporation of exfoliated graphite nanoplatelets. Carbon 2007;45 (7):1446–52.
[15] Compton OC, Kim S, Pierre C, Torkelson JM, Nguyen ST. Crumpled graphene nanosheets as highly effective barrier property enhancers. Adv Mater 2010;22(42):4759–63.
[16] Geim AK, Novoselov KS. The rise of graphene. Nat Mater 2007;6(3):183–91.
[17] Singh V, Joung D, Zhai L, Das S, Khondaker SI, Seal S. Graphene based materials: past, present and future. Prog Mater Sci 2011;56(8):1178–271.
[18] Novoselov KSA, Geim AK, Morozov S, Jiang D, Katsnelson M, Grigorieva I, et al. Two-dimensional gas of massless Dirac fermions in graphene. Nature 2005;438 (7065):197–200.
[19] Balandin AA, Ghosh S, Bao W, Calizo I, Teweldebrhan D, Miao F, et al. Superior thermal conductivity of single-layer graphene. Nano Lett 2008;8(3):902–7.
[20] Lee C, Wei X, Kysar JW, Hone J. Measurement of the elastic properties and intrinsic strength of monolayer graphene. Science 2008;321(5887):385–8.
[21] Dreyer DR, Park S, Bielawski CW, Ruoff RS. The chemistry of graphene oxide. Chem Soc Rev 2010;39(1):228–40.
[22] Malas A, Hatui G, Pal P, Das CK. Synergistic effect of expanded graphite/carbon black on the physical and thermo-mechanical properties of ethylene propylene diene terpolymer. Polym Plast Technol Eng 2014;53(7):716–24.
[23] Malas A, Pal P, Giri S, Mandal A, Das CK. Synthesis and characterizations of modified expanded graphite/emulsion styrene butadiene rubber nanocomposites: mechanical, dynamic mechanical and morphological properties. Compos Part B Eng 2014;58:267–74.
[24] Li D, Mueller MB, Gilje S, Kaner RB, Wallace GG. Processable aqueous dispersions of graphene nanosheets. Nat Nanotechnol 2008;3(2):101–5.
[25] Shan C, Yang H, Han D, Zhang Q, Ivaska A, Niu L. Water-soluble graphene covalently functionalized by biocompatible poly-L-lysine. Langmuir 2009;25(20):12030–3.
[26] Si Y, Samulski ET. Synthesis of water soluble graphene. Nano Lett 2008;8(6):1679–82.
[27] Brodie BC. Sur le poids atomique du graphite. Ann Chim Phys 1860;59(466):e472.
[28] Staudenmaier L. Verfahren zur darstellung der graphitsäure. Ber Dtsch Chem Ges 1898;31(2):1481–7.
[29] Hummers Jr WS, Offeman RE. Preparation of graphitic oxide. J Am Chem Soc 1958;80(6) 1339–1339
[30] Marcano DC, Kosynkin DV, Berlin JM, Sinitskii A, Sun Z, Slesarev A, et al. Improved synthesis of graphene oxide. ACS Nano 2010;4(8):4806–14.
[31] Zhu Y, Murali S, Cai W, Li X, Suk JW, Potts JR, et al. Graphene and graphene oxide: synthesis, properties, and applications. Adv Mater 2010;22(35):3906–24.

[32] Shin HJ, Kim KK, Benayad A, Yoon SM, Park HK, Jung IS, et al. Efficient reduction of graphite oxide by sodium borohydride and its effect on electrical conductance. Adv Funct Mater 2009;19(12):1987–92.
[33] Stankovich S, Dikin DA, Dommett GH, Kohlhaas KM, Zimney EJ, Stach EA, et al. Graphene-based composite materials. Nature 2006;442(7100):282–6.
[34] Zhang J, Yang H, Shen G, Cheng P, Zhang J, Guo S. Reduction of graphene oxide via L-ascorbic acid. Chem Commun 2010;46(7):1112–14.
[35] Ambrosi A, Chua CK, Bonanni A, Pumera M. Lithium aluminum hydride as reducing agent for chemically reduced graphene oxides. Chem Mater 2012;24(12):2292–8.
[36] Zhou X, Zhang J, Wu H, Yang H, Zhang J, Guo S. Reducing graphene oxide via hydroxylamine: a simple and efficient route to graphene. J Phys Chem C 2011;115 (24):11957–61.
[37] Cataldo F, Ursini O, Angelini G. Graphite oxide and graphene nanoribbons reduction with hydrogen iodide. Fuller Nanotub Carbon Nanostruct 2011;19(5):461–8.
[38] Robinson JT, Perkins FK, Snow ES, Wei Z, Sheehan PE. Reduced graphene oxide molecular sensors. Nano Lett 2008;8(10):3137–40.
[39] Song NJ, Chen CM, Lu C, Liu Z, Kong QQ, Cai R. Thermally reduced graphene oxide films as flexible lateral heat spreaders. J Mater Chem A 2014;2(39):16563–8.
[40] Zhang Y, Zhang L, Zhou C. Review of chemical vapor deposition of graphene and related applications. Acc Chem Res 2013;46(10):2329–39.
[41] Yu Q, Lian J, Siriponglert S, Li H, Chen YP, Pei SS. Graphene segregated on Ni surfaces and transferred to insulators. Appl Phys Lett 2008;93(11):113103.
[42] Li X, Cai W, An J, Kim S, Nah J, Yang D, et al. Large-area synthesis of high-quality and uniform graphene films on copper foils. Science 2009;324(5932):1312–14.
[43] Sutter PW, Flege JI, Sutter EA. Epitaxial graphene on ruthenium. Nat Mater 2008;7 (5):406–11.
[44] Coraux J, N'Diaye AT, Busse C, Michely T. Structural coherency of graphene on Ir (111). Nano Lett 2008;8(2):565–70.
[45] Sutter P, Sadowski JT, Sutter E. Graphene on Pt (111): growth and substrate interaction. Phys Rev B 2009;80(24):245411.
[46] Varykhalov A, Rader O. Graphene grown on Co (0001) films and islands: electronic structure and its precise magnetization dependence. Phys Rev B 2009;80(3):035437.
[47] Ruan G, Sun Z, Peng Z, Tour JM. Growth of graphene from food, insects, and waste. ACS Nano 2011;5(9):7601–7.
[48] Chen G, Wu D, Weng W, Wu C. Exfoliation of graphite flake and its nanocomposites. Carbon 2003;41(3):619–21.
[49] Chung DDL. Exfoliation of graphite. J Mater Sci 1987;22(12):4190–8.
[50] Chung DDL. Review graphite. J Mater Sci 2002;37(8):1475–89.
[51] Viculis LM, Mack JJ, Mayer OM, Hahn HT, Kaner RB. Intercalation and exfoliation routes to graphite nanoplatelets. J Mater Chem 2005;15(9):974–8.
[52] Schniepp HC, Li JL, McAllister MJ, Sai H, Herrera-Alonso M, Adamson DH, et al. Functionalized single graphene sheets derived from splitting graphite oxide. J Phys Chem B 2006;110(17):8535–9.
[53] Huh SH. Thermal reduction of graphene oxide. InTech Open Access Publisher; 2011.
[54] Xiang J, Drzal LT. Thermal conductivity of exfoliated graphite nanoplatelet paper. Carbon 2011;49(3):773–8.
[55] Chen W, Yan L, Bangal PR. Preparation of graphene by the rapid and mild thermal reduction of graphene oxide induced by microwaves. Carbon 2010;48(4):1146–52.

[56] Zhang HB, Wang JW, Yan Q, Zheng WG, Chen C, Yu ZZ. Vacuum-assisted synthesis of graphene from thermal exfoliation and reduction of graphite oxide. J Mater Chem 2011;21(14):5392–7.
[57] Hernandez Y, Nicolosi V, Lotya M, Blighe FM, Sun Z, De S, et al. High-yield production of graphene by liquid-phase exfoliation of graphite. Nat Nanotechnol 2008;3(9):563–8.
[58] Blake P, Brimicombe PD, Nair RR, Booth TJ, Jiang D, Schedin F, et al. Graphene-based liquid crystal device. Nano Lett 2008;8(6):1704–8.
[59] Park S, An J, Jung I, Piner RD, An SJ, Li X, et al. Colloidal suspensions of highly reduced graphene oxide in a wide variety of organic solvents. Nano Lett 2009;9(4):1593–7.
[60] Paredes JI, Villar-Rodil S, Martinez-Alonso A, Tascon JMD. Graphene oxide dispersions in organic solvents. Langmuir 2008;24(19):10560–4.
[61] Bourlinos AB, Georgakilas V, Zboril R, Steriotis TA, Stubos AK. Liquid-phase exfoliation of graphite towards solubilized graphenes. Small 2009;5(16):1841–5.
[62] Ciesielski A, Samorì P. Graphene via sonication assisted liquid-phase exfoliation. Chem Soc Rev 2014;43(1):381–98.
[63] Lotya M, Hernandez Y, King PJ, Smith RJ, Nicolosi V, Karlsson LS, et al. Liquid phase production of graphene by exfoliation of graphite in surfactant/water solutions. J Am Chem Soc 2009;131(10):3611–20.
[64] Niu L, Li M, Tao X, Xie Z, Zhou X, Raju AP, et al. Salt-assisted direct exfoliation of graphite into high-quality, large-size, few-layer graphene sheets. Nanoscale 2013;5(16):7202–8.
[65] Eda G, Chhowalla M. Chemically derived graphene oxide: towards large-area thin-film electronics and optoelectronics. Adv Mater 2010;22(22):2392–415.
[66] Khanra P, Kuila T, Bae SH, Kim NH, Lee JH. Electrochemically exfoliated graphene using 9-anthracene carboxylic acid for supercapacitor application. J Mater Chem 2012;22(46):24403–10.
[67] Graf D, Molitor F, Ensslin K, Stampfer C, Jungen A, Hierold C, et al. Spatially resolved Raman spectroscopy of single-and few-layer graphene. Nano Lett 2007;7(2):238–42.
[68] Ferrari AC, Meyer JC, Scardaci V, Casiraghi C, Lazzeri M, Mauri F, et al. Raman spectrum of graphene and graphene layers. Phys Rev Lett 2006;97(18):187401.
[69] Young RJ, Kinloch IA. Graphene and graphene-based nanocomposites. Nanosci Nanostruct Chem 2012;1:145.
[70] Liu K, Liu L, Luo Y, Jia D. One-step synthesis of metal nanoparticle decorated graphene by liquid phase exfoliation. J Mater Chem 2012;22(38):20342–52.
[71] Malard LM, Pimenta MA, Dresselhaus G, Dresselhaus MS. Raman spectroscopy in graphene. Phys Rep 2009;473(5):51–87.
[72] Zhang W, Zhang Y, Tian Y, Yang Z, Xiao Q, Guo X, et al. Insight into the capacitive properties of reduced graphene oxide. ACS Appl Mater Interfaces 2014;6(4):2248–54.
[73] Ding J, Yan W, Sun S, Bao J, Gao C. Fabrication of graphene/CaIn 2 O 4 composites with enhanced photocatalytic activity from water under visible light irradiation. Int J Hydrogen Energy 2014;39(1):119–26.
[74] An X, Jimmy CY, Wang Y, Hu Y, Yu X, Zhang G. WO 3 nanorods/graphene nanocomposites for high-efficiency visible-light-driven photocatalysis and NO 2 gas sensing. J Mater Chem 2012;22(17):8525–31.
[75] Gu L, Wang J, Cheng H, Zhao Y, Liu L, Han X. One-step preparation of graphene-supported anatase TiO_2 with exposed {001} facets and mechanism of enhanced photocatalytic properties. ACS Appl Mater Interfaces 2013;5(8):3085–93.

[76] Stankovich S, Dikin DA, Piner RD, Kohlhaas KA, Kleinhammes A, Jia Y, et al. Synthesis of graphene-based nanosheets via chemical reduction of exfoliated graphite oxide. Carbon 2007;45(7):1558–65.
[77] Pham VH, Cuong TV, Hur SH, Oh E, Kim EJ, Shin EW, et al. Chemical functionalization of graphene sheets by solvothermal reduction of a graphene oxide suspension in N-methyl-2-pyrrolidone. J Mater Chem 2011;21(10):3371–7.
[78] Pei S, Cheng HM. The reduction of graphene oxide. Carbon 2012;50(9):3210–28.
[79] Li Q, Guo B, Yu J, Ran J, Zhang B, Yan H, et al. Highly efficient visible-light-driven photocatalytic hydrogen production of CdS-cluster-decorated graphene nanosheets. J Am Chem Soc 2011;133(28):10878–84.
[80] Nair RR, Blake P, Grigorenko AN, Novoselov KS, Booth TJ, Stauber T, et al. Fine structure constant defines visual transparency of graphene. Science 2008;320(5881) 1308–1308.
[81] Blake P, Hill EW, Neto AC, Novoselov KS, Jiang D, Yang R, et al. Making graphene visible. Appl Phys Lett 2007;91(6):063124.
[82] Ni ZH, Wang HM, Kasim J, Fan HM, Yu T, Wu YH, et al. Graphene thickness determination using reflection and contrast spectroscopy. Nano Lett 2007;7(9):2758–63.
[83] Novoselov KS, Geim AK, Morozov SV, Jiang D, Zhang Y, Dubonos SA, et al. Electric field effect in atomically thin carbon films. Science 2004;306(5696):666–9.
[84] Tung VC, Allen MJ, Yang Y, Kaner RB. High-throughput solution processing of large-scale graphene. Nat Nanotechnol 2009;4(1):25–9.
[85] Decker R, Wang Y, Brar VW, Regan W, Tsai HZ, Wu Q, et al. Local electronic properties of graphene on a BN substrate via scanning tunneling microscopy. Nano Lett 2011;11(6):2291–5.
[86] Zhang Y, Tan YW, Stormer HL, Kim P. Experimental observation of the quantum Hall effect and Berry's phase in graphene. Nature 2005;438(7065):201–4.
[87] Chen S, Moore AL, Cai W, Suk JW, An J, Mishra C, et al. Raman measurements of thermal transport in suspended monolayer graphene of variable sizes in vacuum and gaseous environments. ACS Nano 2010;5(1):321–8.
[88] Balandin AA. Thermal properties of graphene and nanostructured carbon materials. Nat Mater 2011;10(8):569–81.
[89] Chen S, Wu Q, Mishra C, Kang J, Zhang H, Cho K, et al. Thermal conductivity of isotopically modified graphene. Nat Mater 2012;11(3):203–7.
[90] Ghosh S, Bao W, Nika DL, Subrina S, Pokatilov EP, Lau CN, et al. Dimensional crossover of thermal transport in few-layer graphene. Nat Mater 2010;9(7):555–8.
[91] Wei Y, Wang B, Wu J, Yang R, Dunn ML. Bending rigidity and Gaussian bending stiffness of single-layered graphene. Nano Lett 2012;13(1):26–30.
[92] Frank IW, Tanenbaum DM, Van der Zande AM, McEuen PL. Mechanical properties of suspended graphene sheets. J Vac Sci Technol B 2007;25(6):2558–61.
[93] Zhang Y, Pan C. Measurements of mechanical properties and number of layers of graphene from nano-indentation. Diam Relat Mater 2012;24:1–5.
[94] Lee JU, Yoon D, Cheong H. Estimation of young's modulus of graphene by Raman spectroscopy. Nano Lett 2012;12(9):4444–8.
[95] Israelachvili J. Intermolecular and surface forces. San Diego, CA: Academic Press; 1991.
[96] Malas A. Elastomeric nanocomposites based on organoclays, expanded graphite and graphene oxide. Kharagpur India, Indian Institute of Technology Kharagpur (PhD thesis); 2014.
[97] Theng BKG. Formation and properties of clay-polymer complexes. Amsterdam /Oxford/New York: Elsevier Scientific Publishing Company; 1979. p. 362.

[98] Ganter M, Gronski W, Reichert P, Mulhaupt R. Rubber nanocomposites: morphology and mechanical properties of BR and SBR vulcanizates reinforced by organophilic layered silicates. Rubber Chem Technol 2001;74(2):221–35.
[99] Kumar SK, Castro M, Saiter A, Delbreilh L, Feller JF, Thomas S, et al. Development of poly (isobutylene-co-isoprene)/reduced graphene oxide nanocomposites for barrier, dielectric and sensingapplications. Mater Lett 2013;96:109–12.
[100] Kim H, Miura Y, Macosko CW. Graphene/polyurethane nanocomposites for improved gas barrier and electrical conductivity. Chem Mater 2010;22(11):3441–50.
[101] Wu S, Tang Z, Guo B, Zhang L, Jia D. Effects of interfacial interaction on chain dynamics of rubber/graphene oxide hybrids: a dielectric relaxation spectroscopy study. RSC Adv 2013;3(34):14549–59.
[102] Sadasivuni KK, Saiter A, Gautier N, Thomas S, Grohens Y. Effect of molecular interactions on the performance of poly (isobutylene-co-isoprene)/graphene and clay nanocomposites. Colloid Polym Sci 2013;291(7):1729–40.
[103] Araby S, Meng Q, Zhang L, Kang H, Majewski P, Tang Y, et al. Electrically and thermally conductive elastomer/graphene nanocomposites by solution mixing. Polymer 2014;55(1):201–10.
[104] Chen B, Ma N, Bai X, Zhang H, Zhang Y. Effects of graphene oxide on surface energy, mechanical, damping and thermal properties of ethylene-propylene-diene rubber/petroleum resin blends. RSC Adv 2012;2(11):4683–9.
[105] Tang Z, Wu X, Guo B, Zhang L, Jia D. Preparation of butadiene–styrene–vinyl pyridine rubber–graphene oxide hybrids through co-coagulation process and in situ interface tailoring. J Mater Chem 2012;22(15):7492–501.
[106] Potts JR, Shankar O, Du L, Ruoff RS. Processing–morphology–property relationships and composite theory analysis of reduced graphene oxide/natural rubber nanocomposites. Macromolecules 2012;45(15):6045–55.
[107] Zhan Y, Lavorgna M, Buonocore G, Xia H. Enhancing electrical conductivity of rubber composites by constructing interconnected network of self-assembled graphene with latex mixing. J Mater Chem 2012;22(21):10464–8.
[108] Kang H, Zuo K, Wang Z, Zhang L, Liu L, Guo B. Using a green method to develop graphene oxide/elastomers nanocomposites with combination of high barrier and mechanical performance. Compos Sci Technol 2014;92:1–8.
[109] Khajehpour M, Sadeghi S, Yazdi AZ, Sundararaj U. Tuning the curing behavior of fluoroelastomer (FKM) by incorporation of nitrogen doped graphene nanoribbons (CNx-GNRs). Polymer 2014;55(24):6293–302.
[110] Scherillo G, Lavorgna M, Buonocore GG, Zhan YH, Xia HS, Mensitieri G, et al. Tailoring assembly of reduced graphene oxide nanosheets to control gas barrier properties of natural rubber nanocomposites. ACS Appl Mater Interfaces 2014;6(4):2230–4.
[111] Wu J, Huang G, Li H, Wu S, Liu Y, Zheng J. Enhanced mechanical and gas barrier properties of rubber nanocomposites with surface functionalized graphene oxide at low content. Polymer 2013;54(7):1930–7.
[112] Malas A, Das CK. Effect of graphene oxide on the physical, mechanical and thermo-mechanical properties of neoprene and chlorosulfonated polyethylene vulcanizates. Compos Part B Eng 2015;79:639–48.
[113] Moghaddam SZ, Sabury S, Sharif F. Dispersion of rGO in polymeric matrices by thermodynamically favorable self-assembly of GO at oil–water interfaces. RSC Adv 2014;4(17):8711–19.
[114] Wu J, Xing W, Huang G, Li H, Tang M, Wu S, et al. Vulcanization kinetics of graphene/natural rubber nanocomposites. Polymer 2013;54(13):3314–23.

[115] Yan N, Xia H, Wu J, Zhan Y, Fei G, Chen C. Compatibilization of natural rubber/high density polyethylene thermoplastic vulcanizate with graphene oxide through ultrasonically assisted latex mixing. J Appl Polym Sci 2013;127(2):933–41.
[116] Mensah B, Kim S, Arepalli S, Nah C. A study of graphene oxide-reinforced rubber nanocomposite. J Appl Polym Sci 2014;131:16.
[117] Hu H, Zhao L, Liu J, Liu Y, Cheng J, Luo J, et al. Enhanced dispersion of carbon nanotube in silicone rubber assisted by graphene. Polymer 2012;53(15):3378–85.
[118] Luo Y, Zhao P, Yang Q, He D, Kong L, Peng Z. Fabrication of conductive elastic nanocomposites via framing intact interconnected graphene networks. Compos Sci Technol 2014;100:143–51.
[119] Beckert F, Trenkle S, Thomann R, Mülhaupt R. Mechanochemical route to functionalized graphene and carbon nanofillers for graphene/SBR nanocomposites. Macromol Mater Eng 2014;299(12):1513–20.
[120] Yang H, Liu P, Zhang T, Duan Y, Zhang J. Fabrication of natural rubber nanocomposites with high graphene contents via vacuum-assisted self-assembly. RSC Adv 2014;4(53):27687–90.
[121] Singh VK, Shukla A, Patra MK, Saini L, Jani RK, Vadera SR, et al. Microwave absorbing properties of a thermally reduced graphene oxide/nitrile butadiene rubber composite. Carbon 2012;50(6):2202–8.
[122] Liu S, Tian M, Yan B, Yao Y, Zhang L, Nishi T, et al. High performance dielectric elastomers by partially reduced graphene oxide and disruption of hydrogen bonding of polyurethanes. Polymer 2015;56:375–84.
[123] Bokobza L, Bruneel JL, Couzi M. Raman spectroscopy as a tool for the analysis of carbon-based materials (highly oriented pyrolitic graphite, multilayer graphene and multiwall carbon nanotubes) and of some of their elastomeric composites. Vib Spectrosc 2014;74:57–63.
[124] Matos CF, Galembeck F, Zarbin AJ. Multifunctional and environmentally friendly nanocomposites between natural rubber and graphene or graphene oxide. Carbon 2014;78:469–79.
[125] Barrett JS, Abdala AA, Srienc F. Poly (hydroxyalkanoate) elastomers and their graphene nanocomposites. Macromolecules 2014;47(12):3926–41.
[126] Bai X, Wan C, Zhang Y, Zhai Y. Reinforcement of hydrogenated carboxylated nitrile–butadiene rubber with exfoliated graphene oxide. Carbon 2011;49(5):1608–13.
[127] Ponnamma D, Sadasivuni KK, Strankowski M, Guo Q, Thomas S. Synergistic effect of multi walled carbon nanotubes and reduced graphene oxides in natural rubber for sensing application. Soft Matter 2013;9(43):10343–53.
[128] Ozbas B, Toki S, Hsiao BS, Chu B, Register RA, Aksay IA, et al. Strain-induced crystallization and mechanical properties of functionalized graphene sheet-filled natural rubber. J Polym Sci Part B Polym Phys 2012;50(10):718–23.
[129] Pradhan B, Srivastava SK. Synergistic effect of three-dimensional multi-walled carbon nanotube–graphene nanofiller in enhancing the mechanical and thermal properties of high-performance silicone rubber. Polym Int 2014;63(7):1219–28.
[130] Xiong X, Wang J, Jia H, Fang E, Ding L. Structure, thermal conductivity, and thermal stability of bromobutyl rubber nanocomposites with ionic liquid modified graphene oxide. Polym Degrad Stabil 2013;98(11):2208–14.
[131] Nawaz K, Khan U, Ul-Haq N, May P, O'Neill A, Coleman JN. Observation of mechanical percolation in functionalized graphene oxide/elastomer composites. Carbon 2012;50(12):4489–94.

[132] Raghu AV, Lee YR, Jeong HM, Shin CM. Preparation and physical properties of waterborne polyurethane/functionalized graphene sheet nanocomposites. Macromol Chem Phys 2008;209(24):2487–93.
[133] Chen Z, Lu H. Constructing sacrificial bonds and hidden lengths for ductile graphene/polyurethane elastomers with improved strength and toughness. J Mater Chem 2012;22(25):12479–90.
[134] Wei J, Jacob S, Qiu J. Graphene oxide-integrated high-temperature durable fluoroelastomer for petroleum oil sealing. Compos Sci Technol 2014;92:126–33.
[135] Khan U, May P, O'Neill A, Coleman JN. Development of stiff, strong, yet tough composites by the addition of solvent exfoliated graphene to polyurethane. Carbon 2010;48(14):4035–41.
[136] Loomis J, Fan X, Khosravi F, Xu P, Fletcher M, Cohn RW, et al. Graphene/elastomer composite-based photo-thermal nanopositioners. Sci Rep 2013;3.
[137] Choi JT, Kim DH, Ryu KS, Lee HI, Jeong HM, Shin CM, et al. Functionalized graphene sheet/polyurethane nanocomposites: effect of particle size on physical properties. Macromol Res 2011;19(8):809–14.
[138] Tian M, Zhang J, Zhang L, Liu S, Zan X, Nishi T, et al. Graphene encapsulated rubber latex composites with high dielectric constant, low dielectric loss and low percolation threshold. J Colloid Interface Sci 2014;430:249–56.
[139] Kim JS, Yun JH, Kim I, Shim SE. Electrical properties of graphene/SBR nanocomposite prepared by latex heterocoagulation process at room temperature. J Ind Eng Chem 2011;17(2):325–30.
[140] Xing W, Wu J, Huang G, Li H, Tang M, Fu X. Enhanced mechanical properties of graphene/natural rubber nanocomposites at low content. Polym Int 2014;63(9):1674–81.
[141] Lian H, Li S, Liu K, Xu L, Wang K, Guo W. Study on modified graphene/butyl rubber nanocomposites. I. Preparation and characterization. Polym Eng Sci 2011;51(11):2254–60.
[142] Liu X, Kuang W, Guo B. Preparation of rubber/graphene oxide composites with in-situ interfacial design. Polymer 2015;56:553–62.
[143] Gan L, Shang S, Yuen CWM, Jiang SX, Luo NM. Facile preparation of graphene nanoribbon filled silicone rubber nanocomposite with improved thermal and mechanical properties. Compos Part B Eng 2015;69:237–42.
[144] Li F, Yan N, Zhan Y, Fei G, Xia H. Probing the reinforcing mechanism of graphene and graphene oxide in natural rubber. J Appl Polym Sci 2013;129(4):2342–51.
[145] Yan N, Buonocore G, Lavorgna M, Kaciulis S, Balijepalli SK, Zhan Y, et al. The role of reduced graphene oxide on chemical, mechanical and barrier properties of natural rubber composites. Compos Sci Technol 2014;102:74–81.
[146] Xing W, Tang M, Wu J, Huang G, Li H, Lei Z, et al. Multifunctional properties of graphene/rubber nanocomposites fabricated by a modified latex compounding method. Compos Sci Technol 2014;99:67–74.
[147] Dong B, Liu C, Zhang L, Wu Y. Preparation, fracture, and fatigue of exfoliated graphene oxide/natural rubber composites. RSC Adv 2015;5(22):17140–8.
[148] Stanier DC, Patil AJ, Sriwong C, Rahatekar SS, Ciambella J. The reinforcement effect of exfoliated graphene oxide nanoplatelets on the mechanical and viscoelastic properties of natural rubber. Compos Sci Technol 2014;95:59–66.
[149] Dao TD, Lee HI, Jeong HM. Alumina-coated graphene nanosheet and its composite of acrylic rubber. J Colloid Interface Sci 2014;416:38–43.

[150] Wei J, Qiu J. Allyl-functionalization enhanced thermally stable graphene/fluoroelastomer nanocomposites. Polymer 2014;55(16):3818–24.
[151] Allahbakhsh A, Mazinani S, Kalaee MR, Sharif F. Cure kinetics and chemorheology of EPDM/graphene oxide nanocomposites. Thermochim Acta 2013;563:22–32.
[152] Al-Hartomy OA, Al-Ghamdi A, Al-Salamy F, Dishovsky N, Shtarkova R, Iliev V, et al. Effect of carbon nanotubes and graphene nanoplatelets on the dielectric and microwave properties of natural rubber composites. Adv Compos Mater 2013;22(5):361–76.
[153] Hernández M, del Mar Bernal M, Verdejo R, Ezquerra TA, López-Manchado MA. Overall performance of natural rubber/graphene nanocomposites. Compos Sci Technol 2012;73:40–6.
[154] Al-Hartomy OA, Al-Ghamdi A, Dishovsky N, Shtarkova R, Iliev V, Mutlay I, et al. Dielectric and microwave properties of natural rubber based nanocomposites containing graphene. Mater Sci Appl 2012;3:453–9.
[155] Yan D, Zhang HB, Jia Y, Hu J, Qi XY, Zhang Z, et al. Improved electrical conductivity of polyamide 12/graphene nanocomposites with maleated polyethylene-octene rubber prepared by melt compounding. ACS Appl Mater Interfaces 2012;4(9):4740–5.
[156] Valentini L, Bolognini A, Alvino A, Bon SB, Martin-Gallego M, Lopez-Manchado MA. Pyroshock testing on graphene based EPDM nanocomposites. Compos Part B Eng 2014;60:479–84.
[157] Das A, Boldt R, Jurk R, Jehnichen D, Fischer D, Stöckelhuber KW, et al. Nano-scale morphological analysis of graphene–rubber composites using 3D transmission electron microscopy. RSC Adv 2014;4(18):9300–7.
[158] Varghese TV, Kumar HA, Anitha S, Ratheesh S, Rajeev RS, Rao VL. Reinforcement of acrylonitrile butadiene rubber using pristine few layer graphene and its hybrid fillers. Carbon 2013;61:476–86.
[159] Das A, Kasaliwal GR, Jurk R, Boldt R, Fischer D, Stöckelhuber KW, et al. Rubber composites based on graphene nanoplatelets, expanded graphite, carbon nanotubes and their combination: a comparative study. Compos Sci Technol 2012;72(16):1961–7.
[160] Araby S, Zaman I, Meng Q, Kawashima N, Michelmore A, Kuan HC, et al. Melt compounding with graphene to develop functional, high-performance elastomers. Nanotechnology 2013;24(16):165601.
[161] Sadasivuni KK, Ponnamma D, Thomas S, Grohens Y. Evolution from graphite to graphene elastomer composites. Prog Polym Sci 2014;39(4):749–80.
[162] Zhan Y, Wu J, Xia H, Yan N, Fei G, Yuan G. Dispersion and exfoliation of graphene in rubber by an ultrasonically-assisted latex mixing and in situ reduction process. Macromol Mater Eng 2011;296(7):590–602.
[163] Ma J, Xu J, Ren JH, Yu ZZ, Mai YW. A new approach to polymer/montmorillonite nanocomposites. Polymer 2003;44(16):4619–24.
[164] Alexandre M, Dubois P. Polymer-layered silicate nanocomposites: preparation, properties and uses of a new class of materials. Mater Sci Eng R Rep 2000;28(1):1–63.
[165] Ray SS, Okamoto M. Polymer/layered silicate nanocomposites: a review from preparation to processing. Prog Polym Sci 2003;28(11):1539–641.
[166] Zeng QH, Wang DZ, Yu AB, Lu GQ. Synthesis of polymer–montmorillonite nanocomposites by in situ intercalative polymerization. Nanotechnology 2002;13(5):549.
[167] Paszkiewicz S, Szymczyk A, Špitalský Z, Mosnáček J, Kwiatkowski K, Rosłaniec Z. Structure and properties of nanocomposites based on PTT-block-PTMO copolymer and graphene oxide prepared by in situ polymerization. Eur Polym J 2014;50:69–77.

[168] Lee YR, Raghu AV, Jeong HM, Kim BK. Properties of waterborne polyurethane/functionalized graphene sheet nanocomposites prepared by an in situ method. Macromol Chem Phys 2009;210(15):1247–54.
[169] Wu X, Lin TF, Tang ZH, Guo BC, Huang GS. Natural rubber/graphene oxide composites: effect of sheet size on mechanical properties and strain-induced crystallization behavior. Express Polym Lett 2015;9:8.
[170] Teh PL, Ishak ZM, Hashim AS, Karger-Kocsis J, Ishiaku US. Effects of epoxidized natural rubber as a compatibilizer in melt compounded natural rubber–organoclay nanocomposites. Eur Polym J 2004;40(11):2513–21.
[171] Ismail H, Chia HH. The effects of multifunctional additive and vulcanization systems on silica filled epoxidized natural rubber compounds. Eur Polym J 1998;34 (12):1857–63.
[172] Malas A, Pal P, Das CK. Effect of expanded graphite and modified graphite flakes on the physical and thermo-mechanical properties of styrene butadiene rubber/polybutadiene rubber (SBR/BR) blends. Mater Des 2014;55:664–73.
[173] Frasca D, Schulze D, Wachtendorf V, Morys M, Schartel B. Multilayer graphene/chlorine-isobutene-isoprene rubber nanocomposites: the effect of dispersion. Polym Adv Technol 2016;. Available from: http://dx.doi.org/10.1002/pat.3740.
[174] Malas A, Das CK. Development of modified expanded graphite-filled solution polymerized styrene butadiene rubber vulcanizates in the presence and absence of carbon black. Polym Eng Sci 2014;54(1):33–41.
[175] Papageorgiou DG, Kinloch IA, Young RJ. Graphene/elastomer nanocomposites. Carbon 2015;95:460–84.
[176] Berean KJ, Ou JZ, Nour M, Field MR, Alsaif MM, Wang Y, et al. Enhanced gas permeation through graphene nanocomposites. J Phys Chem C 2015;119(24):13700–12.
[177] Jana SC, Jain S. Dispersion of nanofillers in high performance polymers using reactive solvents as processing aids. Polymer 2001;42(16):6897–905.
[178] Saujanya C, Radhakrishnan S. Structure development and crystallization behaviour of PP/nanoparticulate composite. Polymer 2001;42(16):6723–31.
[179] Calvert P. Potential applications of nanotubes. In: Ebbesen TW, editor. Carbon nanotubes. Boca Raton, FL: CRC press; 1997.
[180] Favier V, Cavaille JY, Canova GR, Shrivastava SC. Mechanical percolation in cellulose whisker nanocomposites. Polym Eng Sci 1997;37(10):1732–9.
[181] Coleman JN, Khan U, Blau WJ, Gun'ko YK. Small but strong: a review of the mechanical properties of carbon nanotube–polymer composites. Carbon 2006;44 (9):1624–52.
[182] Theng BKG. The chemistry of clay-organic reactions. New York, NY: Wiley; 1974.
[183] Chen GH, Wu DJ, Weng WG, Yan WL. Preparation of polymer/graphite conducting nanocomposite by intercalation polymerization. J Appl Polym Sci 2001;82 (10):2506–13.
[184] Pan YX, Yu ZZ, Ou YC, Hu GH. A new process of fabricating electrically conducting nylon 6/graphite nanocomposites via intercalation polymerization. J Polym Sci Part B Polym Phys 2000;38(12):1626–33.
[185] Wakabayashi K, Pierre C, Dikin DA, Ruoff RS, Ramanathan T, Brinson LC, et al. Polymer-graphite nanocomposites: effective dispersion and major property enhancement via solid-state shear pulverization. Macromolecules 2008;41(6):1905–8.
[186] Schopp S, Thomann R, Ratzsch KF, Kerling S, Altstädt V, Mülhaupt R. Functionalized graphene and carbon materials as components of styrene-butadiene

rubber nanocomposites prepared by aqueous dispersion blending. Macromol Mater Eng 2014;299(3):319–29.
[187] Potts JR, Shankar O, Murali S, Du L, Ruoff RS. Latex and two-roll mill processing of thermally-exfoliated graphite oxide/natural rubber nanocomposites. Compos Sci Technol 2013;74:166–72.
[188] Tang Z, Zhang L, Feng W, Guo B, Liu F, Jia D. Rational design of graphene surface chemistry for high-performance rubber/graphene composites. Macromolecules 2014;47(24):8663–73.
[189] Li C, Feng C, Peng Z, Gong W, Kong L. Ammonium-assisted green fabrication of graphene/natural rubber latex composite. Polym Compos 2013;34(1):88–95.
[190] Yaragalla S, Meera AP, Kalarikkal N, Thomas S. Chemistry associated with natural rubber–graphene nanocomposites and its effect on physical and structural properties. Ind Crops Prod 2015;74:792–802.
[191] AngeláRodriguez-Perez M, de Saja J, AngeláLopez-Manchado M. Functionalized graphene sheet filled silicone foam nanocomposites. J Mater Chem 2008;18(19):2221–6.
[192] Wang S, Tambraparni M, Qiu J, Tipton J, Dean D. Thermal expansion of graphene composites. Macromolecules 2009;42(14):5251–5.
[193] Prasher RS, Chang JY, Sauciuc I, Narasimhan S, Chau D, Chrysler G, et al. Nano and micro technology-based next-generation package-level cooling solutions. Intel Technol J 2005;9(4).
[194] Biercuk MJ, Llaguno MC, Radosavljevic M, Hyun JK, Johnson AT, Fischer JE. Carbon nanotube composites for thermal management. Appl Phys Lett 2002;80 (15):2767–9.
[195] Liang J, Xu Y, Huang Y, Zhang L, Wang Y, Ma Y, et al. Infrared-triggered actuators from graphene-based nanocomposites. J Phys Chem C 2009;113(22):9921–7.
[196] Zhong H, Lukes JR. Interfacial thermal resistance between carbon nanotubes: molecular dynamics simulations and analytical thermal modeling. Phys Rev B 2006;74 (12):125403.
[197] Pollack GL. Kapitza resistance. Rev Mod Phys 1969;41(1):48.
[198] Song SH, Kim JM, Park KH, Lee DJ, Kwon OS, Kim J, et al. High performance graphene embedded rubber composites. RSC Adv 2015;5(99):81707–12.
[199] Lin Y, Liu K, Chen Y, Liu L. Influence of graphene functionalized with zinc dimethacrylate on the mechanical and thermal properties of natural rubber nanocomposites. Polym Compos 2015;36(10):1775–85.
[200] Shenogin S, Bodapati A, Xue L, Ozisik R, Keblinski P. Effect of chemical functionalization on thermal transport of carbon nanotube composites. Appl Phys Lett 2004;85 (12):2229–31.
[201] Song Y, Yu J, Yu L, Alam FE, Dai W, Li C, et al. Enhancing the thermal, electrical, and mechanical properties of silicone rubber by addition of graphene nanoplatelets. Mater Des 2015;88:950–7.
[202] Bunch JS, Verbridge SS, Alden JS, Van Der Zande AM, Parpia JM, Craighead HG, et al. Impermeable atomic membranes from graphene sheets. Nano Lett 2008;8 (8):2458–62.
[203] Bhattacharya M, Biswas S, Bhowmick AK. Permeation characteristics and modeling of barrier properties of multifunctional rubber nanocomposites. Polymer 2011;52 (7):1562–76.
[204] Wu YP, Wang YQ, Zhang HF, Wang YZ, Yu DS, Zhang LQ, et al. Rubber–pristine clay nanocomposites prepared by co-coagulating rubber latex and clay aqueous suspension. Compos Sci Technol 2005;65(7):1195–202.

[205] Prud'Homme RK, Ozbas B, Aksay I, Register R, Adamson D. Functional graphene rubber nanocomposites. In: Patent US, editor. United States patent, vol. US7745528B2. US: The Trustees of Princeton University; 2010. p. 1–80.

[206] McAllister MJ, Li JL, Adamson DH, Schniepp HC, Abdala AA, Liu J, et al. Single sheet functionalized graphene by oxidation and thermal expansion of graphite. Chem Mater 2007;19(18):4396–404.

[207] Priestley RD, Rittigstein P, Broadbelt LJ, Fukao K, Torkelson JM. Evidence for the molecular-scale origin of the suppression of physical ageing in confined polymer: fluorescence and dielectric spectroscopy studies of polymer–silica nanocomposites. J Phys Condens Matter 2007;19(20):205120.

[208] Schönhals A, Goering H, Schick C, Frick B, Zorn R. Glass transition of polymers confined to nanoporous glasses. Colloid Polym Sci 2004;282(8):882–91.

[209] Wang Z, Nelson JK, Hillborg H, Zhao S, Schadler LS. Nonlinear conductivity and dielectric response of graphene oxide filled silicone rubber nanocomposites. In: Electrical insulation and dielectric phenomena (CEIDP), 2012 Annual Report Conference on (pp. 40–43). IEEE; 2012.

[210] Nelson JK. Dielectric polymer nanocomposites. New York, NY: Springer; 2010. p. 1–338.

[211] Chen H, Müller MB, Gilmore KJ, Wallace GG, Li D. Mechanically strong, electrically conductive, and biocompatible graphene paper. Adv Mater 2008;20(18):3557–61.

[212] Balogun YA, Buchanan RC. Enhanced percolative properties from partial solubility dispersion of filler phase in conducting polymer composites (CPCs). Compos Sci Technol 2010;70(6):892–900.

[213] Potts JR, Dreyer DR, Bielawski CW, Ruoff RS. Graphene-based polymer nanocomposites. Polymer 2011;52(1):5–25.

[214] Toker D, Azulay D, Shimoni N, Balberg I, Millo O. Tunneling and percolation in metal-insulator composite materials. Phys Rev B 2003;68(4):041403.

[215] Grunlan JC, Mehrabi AR, Bannon MV, Bahr JL. Water-based single-walled-nanotube-filled polymer composite with an exceptionally low percolation threshold. Adv Mater 2004;16(2):150–3.

Rubber nanocomposites with polyhedral oligomeric silsesquioxanes (POSS) as the nanofiller

7

D. Dasgupta[1], *M. Srividhya*[1], *A. Sarkar*[1], *M. Dubey*[1], *D. Wrobel*[2] and *A. Saxena*[1]
[1]Momentive Performance Materials Pvt. Ltd, Bangalore, Karnataka, India, [2]Momentive Performance Materials GmbH, Leverkusen, Germany

7.1 Introduction

In the physical sense, the term "Rubber" refers to materials characterized by low elastic modulus and high "elongation at break." However, such materials are also named as "elastomers." It is worth noting here that, often, a term "rubbery phase" is being used to represent an entangled polymer residing at any temperature higher than the glass transition temperature but lower than that inducing melt flow and exhibiting a steady viscoelastic behavior over a range of temperature and frequency.

Typically, for a material to behave as rubber requires macromolecular chains to be partially interconnected by chemical bonds and to the most extent conformationally entangled. However, a cross-linked elastomer in its own, does not qualify to the stringent performance requirements for many real-world applications and must be homogeneously blended with comparatively rigid components offering high surface to volume ratio known as "reinforcing fillers" [1–3]. In the majority of cases, at least one dimension of the filler resides in the nanometer length scale and such fillers are to be referred to as nanofillers. Nanofillers in such rubber nanocomposites primarily impart mechanical reinforcement to the polymer base.

The reinforcement of an elastomeric matrix caused by nanofiller was rationalized by comparing the elastomer with a colloidal suspension of spherical particles (of their van der Waals radii) in a given viscous fluid [4a], therefore, the viscosity (η) of the dispersion can be written as

$$\eta = \eta_o[1 + 2.5\phi] \tag{7.1}$$

where η_0 refers to the viscosity of the pure fluid (dispersion medium) and ϕ designates the volume fraction of the system occupied by the colloidal spheres (dispersed

Progress in Rubber Nanocomposites. DOI: http://dx.doi.org/10.1016/B978-0-08-100409-8.00007-3

phase). In 1944 Smallwood[4b] showed that adding filler particles to an elastomeric medium results in a reinforcement modulus E which can be expressed as

$$E = E_{\mathrm{m}}[1 + 2.5\phi] \tag{7.2}$$

where E_{m} is the elastic modulus of the unfilled rubber matrix and ϕ refers to the volume fraction occupied by the filler. Such reinforcement is termed as hydrodynamic effect. However, this equation holds true only for low deformations and essentially based on few approximations like: (i) low filler volume fraction; (ii) spherical aspect of the filler; and (iii) noncompressible filler particles. In particular, the approximation of spherical morphology of filler particles does not represent the real scenario, as in the case of most of the conventional nanofillers (i.e., carbon black, fumed silica (FS)), they tend to aggregate to yield oblate aspects and when such aggregates undergo higher order association, they may yield agglomerates having fractal morphologies. To account for such variations, Guth [5] proposed a modified equation (i.e., Eq. (7.3)) by adding nonlinearity to Eq. (7.2) which essentially considers the average aspect ratio "F," defined as the ratio of its longest length to its perpendicular breadth where ϕ is the volume fraction of the rubber nanocomposite occupied by filler particles. Hydrodynamic reinforcement, thus, plays a crucial role in the mesoscale level.

$$E = E_{\mathrm{m}}[1 + 0.67\phi F + 1.62\phi^2 F^2] \tag{7.3}$$

Nevertheless, an easily stretchable cross-linked unfilled rubber essentially suffers from low modulus. The entangled polymer chains continue to get straightened along the elongation axis and when it attains the most stretched conformation, a further strain causes bond rupture and the critical strain value is referred to as elongation at break. A reinforced nanocomposite provides resistance to elongation through polymer–filler interaction, thereby promoting effective stress propagation. This results in pushing the "elongation at break" to the higher elongation value. Such reinforcement in rubber is quantified as tensile strength.

The kind of effect filler particles impart during elastomer straining can be described by Fig. 7.1. The energy required reaching point B (the maximum possible elongation) from point A (at equilibrium conformation under no stress) and to break

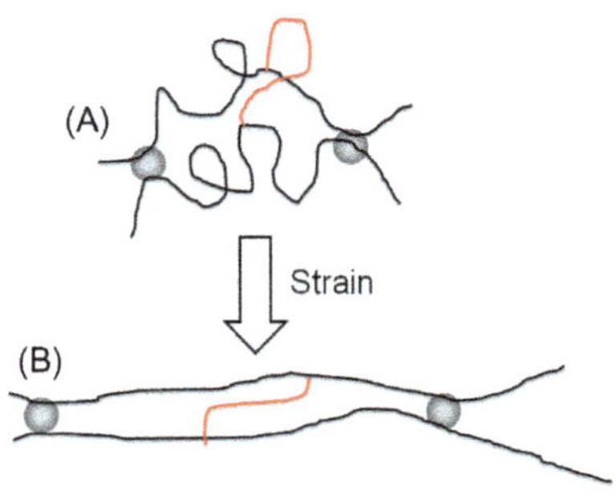

Figure 7.1 Sketch highlighting the effect of tensile stress on elastomer network.

polymer–polymer as well as polymer–filler interfaces accounts for tensile modulus. The deformation essentially includes conformational changes along the polymer chains.

Notably, nanofiller reinforcement also improves tear resistance of the elastomer, which is the resistance towards crack propagation or crack initiation which might generate under tension. Reinforcing filler having small size, high surface area and well-constructed polymer–filler interface effectively provides high tear resistance. Nanofillers also enhance the abrasion resistance as well as the "fatigue life" of the rubber.

Within the scope of the present chapter, we intend to confine our discussion into organic and inorganic synthetic polymers and their composites exhibiting characteristic elastomeric properties.

While carbon black reinforces natural or synthetic rubbers like Poly-isoprene, styrene butadiene rubber efficiently, inorganic nanofillers, such as nanoscopic silica, titania, and nanoclays, constitute the majority of the inorganic elastomeric composites [6–13]. The choice of reinforcing fillers used to compound an elastomer is typically governed by the factors like chemical compatibility with the polymer matrix, filler surface functionality, thermal stability, refractive index, etc. There are some fillers that promote preferential chemical bond formation between polymer and filler surfaces at their molecular volume, thereby contributing to a significant extent of the mechanical reinforcement. One of such nanofillers includes the polyhedral oligomeric silsesquioxane (POSS) class of materials. Having a predefined size and functionality, POSS offers compatibilization to a wide variety of polymers.

7.2 Structure & Synthesis of POSS

POSS are one among the unique members of the silsesquioxane family distinguished by their existence in the form of discrete cage-like structures. Historically, the name "silsesquioxane" was originated from the combination of three syllables "sil = silicon," "sesqui = one and half" and "oxane = oxygen" that is usually applied to all those siloxane structures having an empirical formula $RSiO_{1.5}$, where R can be hydrogen atom or a wide range of organo-functional groups [14–17]. It can be easily envisaged from their empirical formula that they have a composition that lies between the highly rigid silica structure (SiO_2) and highly flexible polyorganosiloxane structure $(R_2SiO)_n$. In fact, it has been shown in the past that by modifying the nature of the organic groups and optimizing the conditions of preparation to some extent, one could obtain the appropriate silsesquioxane-like structure that can provide multiple product benefits including improved processability, flexibility, thermal and oxidative stability, etc. those were otherwise not achievable from either silica or polyorganosiloxane materials [18]. Recently, silsesquioxane-based materials have found potential utility in a variety of applications, including as a component in polymer nanocomposites [19–24], catalysts [25–31], models for silica surfaces [32–34], low-k dielectrics [35–37], antimicrobial agents [38,39], emitting layers in organic light-emitting diodes [40], and coatings [41,42]. Broadly, there are four basic structures of silsesquioxane: (1) random; (2) ladder; (3) open-cage; and (4) closed-cage (see: Fig. 7.2) [43–45].

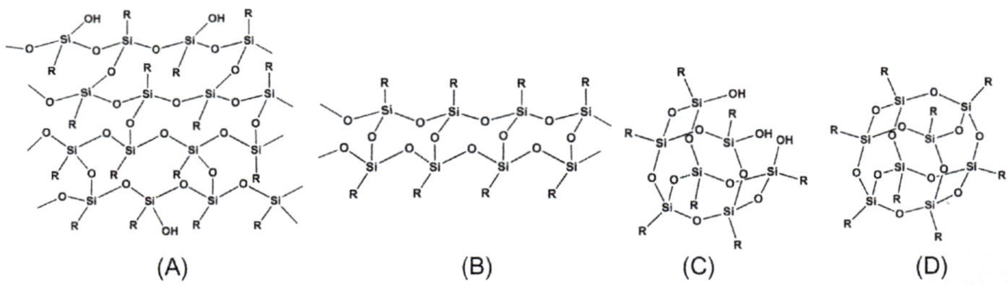

Figure 7.2 Some schematic structures of silsesquioxanes. (A) random; (B) ladder; (C) open-cage; and (D) closed-cage.

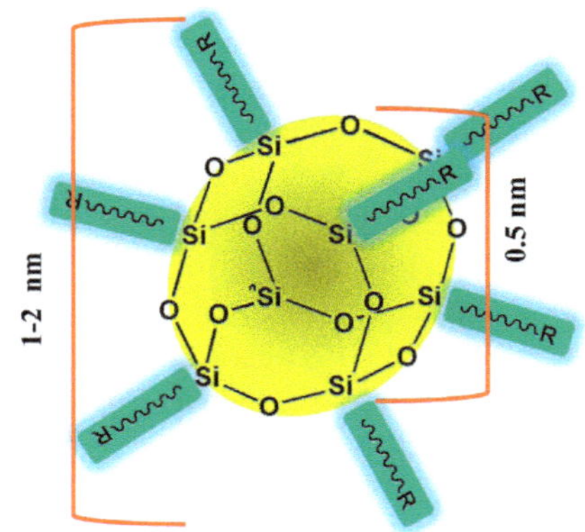

Figure 7.3 Typical dimensions of a silsesquioxane molecule.

Among these, the latter two categories are called POSS due their cage-like core structures. In the closed-caged silsesquioxane category, there also exists a number of discrete molecules with the general formula $[RSiO_{1.5}]_n$ where *n* is commonly 6, 8, 10, or 12. However, the majority of research and development in this area focuses particularly on the synthesis and studies of octameric polyhedral silsesquioxane $[RSiO_{1.5}]_8$, also referred to as "T_8 POSS" (*T* refers to the fact that each Si atom is being connected to three oxygen atoms). Some of the important features that make T_8 POSS very unique and differentiated are its nanoscopic size, highly symmetric core structure, and the distribution of the eight pendant arms from the eight corners of the cage structure in a three dimensional arrangement. The T_8 POSS consists of a rigid silica core (body diagonal = 0.53 nm) with eight organic functional groups anchored to the vertices of the silica core. The silica core and the organic moieties together create a spherical molecule periphery with an average diameter of 1–2 nm (Fig. 7.3). Therefore, these molecules can often be viewed as nano-sized silica particles whose surface is uniformly grafted with organic moieties. Also, these molecules often belong to a highly symmetric (T_h) point group. The molecules with such a high symmetry can increase the probability of minimizing defects in the composites, as any molecular misalignment occurs within the composition, these molecules can be realigned easily (with less energy required) with adjacent molecules. This unique combination of properties of octameric silsesquioxane makes these molecules an ideal building block for the design and construction of novel nanocomposite materials.

Although, T_8 POSS is the most prevalent closed cage structure, the existence of Q_8 (Q refers to Si atoms being connected to four oxygen atoms) structure with the stoichiometry $[R_8Si_8O_{20}]$ is also known. The molecule $[(RO)_8Si_8O_{12}]$ has the same cube structure but alkoxy (OR) instead of R linked to silicon. However, a polyanionic version of Q-POSS structure has been reported to form when high surface area silica is treated with a solution of tetra-alkylammonium hydroxide, yielding Q_6, Q_8, or Q_{10} polyanionic cage structures of the composition $[R_4N^+]_n[^-OSiO_{1.5}]_n$ [46].

7.3 Synthesis

The silsesquioxanes are synthesized using a common and simplistic reaction scheme (7.i), where a trifunctional organosilane with a general formula $RSiX_3$, where X is the alkoxy or halide group, is hydrolytically condensed in presence of a stoichiometric amount of water.

$$nRSiX_3 + 3n/2H_2O \rightarrow (RSiO_{3/2})_n + 3nHX \qquad (7.i)$$

It should be noted here, that there is no known general and straightforward synthetic route that can selectively give any specific structure of silsesquioxane in a quantitative yield. However, as mentioned earlier, by varying certain reaction parameters, it can be possible to tune the reaction up to a level where it produces a specific type of silsesquioxane in majority. It has been established in the past, that the silsesquioxane formation takes place in two distinct steps that occur rather simultaneously during the hydrolytic condensation reaction [47]. The trifunctional organosilane is first hydrolyzed to produce silanols. These silanols are then condensed with one another to form oligosiloxane intermediates. These oligosiloxane intermediates contain unreacted hydroxyl groups and further undergo inter/intramolecular condensation and eventually a rearrangement reaction to give different types of silsesquioxanes. The intermolecular condensation pathway favors the formation of random/ladder structures whereas the intramolecular condensation pathway (owing to entropic advantage) mainly favors the formation of POSS structures. Anything that minimizes the intermolecular interaction among oligosiloxane intermediates would effectively help the formulation of POSS structures. The reaction carried out in neat or highly concentrated solutions favors such intermolecular interactions, thereby increasing the formation of silsesquioxane polymers. Therefore, high yield POSS syntheses are generally achieved by performing the hydrolytic condensation in dilute conditions.

The condensation as well as purity of the synthesized POSS can be well characterized by ^{29}Si NMR, where, the signal pertaining to tri-alkoxysilane (~ -35 ppm) will tend to disappear as the reaction progresses and a new signal corresponding to

T3 structure (where the Si atom is connected to three other Si atoms through Si—O—Si bonds) appears at around −67 ppm.

Typically, nonreactive "*R*" peripheral functionalities are used to achieve solubility, stability, or resin compatibility, whereas one or more reactive groups promote further modification, grafting, or polymerization [44]. Functional POSS structures also impart curable functionality (i.e., acrylate) without increasing organic volatile content. The POSS surface functional groups play a crucial role in controlling properties of POSS-containing hybrids. If it is monofunctional, it can be included in the side chain, if bifunctional it can be copolymerized in the polymer matrix and if it is multifunctional, then it can be used for multipoint cross-linking.

7.4 Properties of POSS-Nanocomposites

POSS can be incorporated in polymer/rubber composites in two ways, namely physical blending and chemically attaching to the polymer backbone or as rigid cross-linking points in the network. [48] Composites, where, POSS is physically compounded most often suffer from microphase separation at comparatively higher loadings. Joshi et al. studied the viscoelastic properties of methyl POSS-high density polyethylene (HDPE) blend and have interestingly shown that, upon adding up to 1 wt% of POSS, there is a decrease in complex viscosity which indicates a significant change of polymer chain entanglement induced by POSS; whereas, a steady increase in viscosity was observed above 1 wt% of POSS Loading in HDPE [49].

Wu et al. [50a] have investigated the morphology and the rheological properties of polystyrene-isobutyl POSS random copolymer and observed that the isobutyl POSS segments remained dispersed with molecular dimensions in the polymer even at a significantly higher loading of POSS in copolymer composition. It was also found that, with an increase in the relative fraction of Isobutyl POSS over styrene, a gradual decrease in the glass transition temperature was observed which suggests a phenomenon converse to the "free volume theory" of polymer glass transition. The fact that the rubbery plateau modulus decreases systematically with the POSS content refers to significant dilation of the polymer chain "radius of reptation tube" thereby yielding decrease in entanglement density [50b].

Kopesky et al. found that below 1 vol%, untethered isobutyl POSS blended in poly(methyl methacrylate) (PMMA) effectively decreases the zero shear viscosity. It is assumed in such cases that POSS being molecularly sized can occupy interchain positions, yielding de-entanglement of chains. Kopesky and coworkers have also studied the blend of POSS with PMMA—Isobutyl POSS random copolymer and observed strong reinforcement over addition of POSS occurring concurrently with a significant retardation of chain relaxation, even higher as compared to PMMA—POSS copolymer alone. The observation has been substantiated with the evidence of distinct POSS crystallization [51].

Seurer et al. [52] studied the POSS aggregation behavior in ethylene—propylene—POSS terpolymer where the T_8 POSS structure is introduced as pendent groups to the ethylene-propylene copolymer backbone. The peripheral organic functionality of the POSS structure was altered between isobutyl, ethyl, and phenyl moieties. The chain–chain aggregation of the copolymers was studied through wide angle X-ray diffraction (WAXD) to probe the POSS crystalline domains. The WAXD experiments carried out on isobutyl POSS terpolymer revealed a weak diffraction signal with a broadening of peak when compared to the WAXD profile of the corresponding POSS monomer which typically crystallizes in a three-dimensional hexagonal array. Interestingly, when the POSS periphery were functionalized by phenyl functionality, the copolymer was found to contain POSS crystallites characterized by a strong diffraction peak at $2\theta = 8$ similar to what was observed for the corresponding Phenyl substituted POSS macromer. The enhanced crystallization of POSS domains might have been resulted by the improved symmetry around the POSS segments and/or unfavorable interactions of peripheral phenyl groups with the ethylene-propylene copolymer matrix. The authors also extended their scientific inquisitiveness towards ethyl functionalized POSS and observed again a poorly defined POSS crystallites formation. It has been further demonstrated that the phenyl POSS aggregation in the corresponding terpolymer imparted the thermoplastic elastomeric properties, wherein, the POSS crystallites act as glassy cross-linking domains between the rubbery polyethylene–polypropylene soft segments. A copolymer with 21 wt% of POSS attained an elongation at break of value 720% characterizing elastomeric properties. However, a further increase in POSS composition led to loss of elasticity. For ethyl and isobutyl POSS containing copolymers, the POSS segments, as they remain in molecular states, merely crystallize and only plasticize the thermoplastic matrix.

Heeley et al. [53] have investigated the physical blending of T_8 POSS functionalized with paraffinic arms and low density polyethylene. The authors characterized the Polyethylene (PE)/T_8C8 (octyl functionalized POSS) blends by small angle X-ray scattering, which indicates a preferential self-aggregation of alkylated T_8 POSS in the PE matrix. However, with increasing the alkyl arm lengths to dodecyl and further to octa-decyl the POSS aggregation was seen to be diminishing most likely due to improved compatibility with PE matrix.

Zhang et al. [54] have designed a pendent POSS containing polysiloxane copolymer with terminal hydrides as cross-linker to vinyl PDMS, thereby reinforcing the addition cured rubber which would otherwise have used pendent polyhydrosiloxanes as cross-linkers.

Saxena and coworkers [55] have investigated the effect of modified POSS loading in fumed silica (FS) reinforced polysiloxane elastomeric composite prepared by addition curing to liquid silicone rubber (LSR). The FS surface was pretreated by methyl-tri-methoxy silane (MTMS) or phenyl-tri-methoxy silane (PTMS) and concurrently the POSS peripheral functionality was tuned as methyl or phenyl moieties. POSS aggregate formation is observed when octa-methyl POSS (OM) is blended in 2.5 wt% loading in LSR reinforced with about 24 wt% of treated FS (Fig. 7.4). The LSR sheets were cryo-microtomed into electron beam transparent sections and observed under the transmission electron microscope. As can be envisaged, the lighter contrast refers to the elastomer matrix, whereas the dark contrast could either be the silica nanoparticle or the POSS. When OM is compounded in both

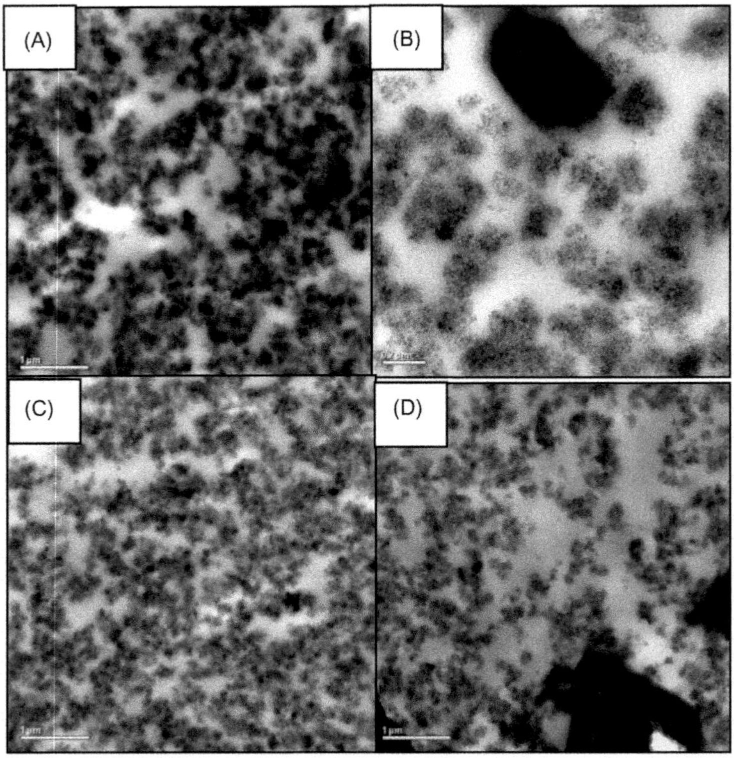

Figure 7.4 Transmission electron micrographs of (A) LSR filled with MTMS modified fumed silica; (B) LSR filled with MTMS modified fumed silica blended with octa-methyl POSS; (C) LSR filled with PTMS modified fumed silica; and (D) LSR filled with PTMS modified fumed silica blended with octa-methyl POSS.
Source: Image by Momentive.

MTMS and PTMS modified silica composites, large dark objects are seen to be dispersed throughout the elastomeric bulk which could possibly be the crystallized POSS as they impart higher electron density due to systematic packing and high density. However, electron microscopic images cannot be considered to be a perfect representation of the global composite morphology and must be supported by other bulk analytical techniques such as rheological characterizations.

Dynamic rheological investigations were carried out with the same set of reinforced LSR formulations in the uncured state. A plot of angular frequency (ω) versus storage modulus (G') provided an understanding on the filled polymer morphology (A frequency invariant G' indicates percolated network structure), whereas the higher magnitude of G' represents the superior reinforcement.

The frequency profiles (Fig. 7.5) of the same set of uncured rubber composites suggest a 10-fold higher storage modulus for MTMS modified composite as compared to that containing PTMS treated silica nanofiller in a comparable proportion.

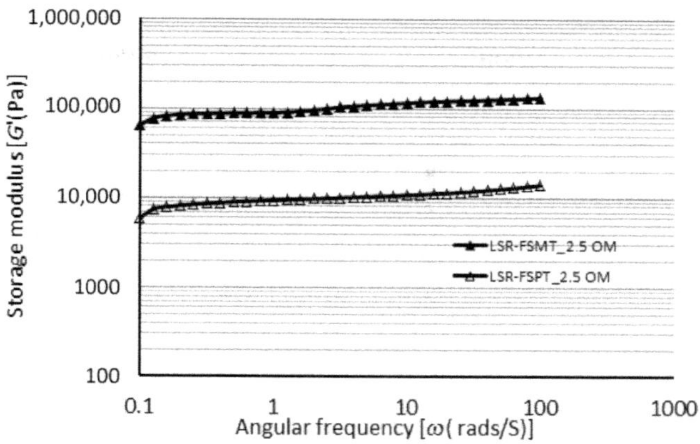

Figure 7.5 Frequency sweep rheology profiles for uncured PDMS composite of the indicated formulation. LSR-FSMT-2.5 OM represents uncured LSR polymer filled with MTMS modified fumed silica (FS) and blended with 2.5 wt% of octa-methyl POSS (OM); LSR-FSPT-2.5 OM represents uncured LSR polymer filled with PTMS modified fumed silica (FS) and blended with 2.5 wt% of octa-methyl POSS (OM). [for both the samples, the fumed silica loadings remain same].
Source: Image by Momentive.

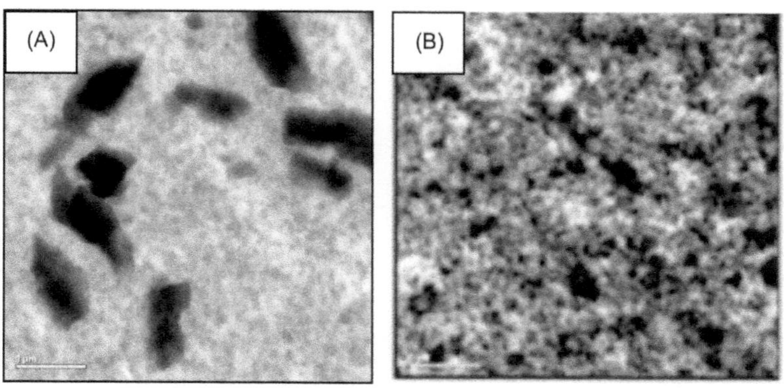

Figure 7.6 Transmission electron microscopic images for LSR filled with MTMS modified fumed silica at different loadings of octa-methyl POSS viz (A) 5 wt% (B) 10 wt%.
Source: Image by Momentive.

Such a result can happen due to preferential POSS agglomeration triggered by differentiating incompatibility. It has also been reported that the processing shear force helps to break apart the POSS crystallites at relatively higher loading, whereas at low loading POSS creates secluded crystalline domains isolated by polymers, where most of the shear force dies down (Fig. 7.6).

Mark and coworkers have investigated the effect of isobutyl functionalized T_8 POSS loading in poly(dimethylsiloxane) curable resin (polymer) composition. Wide angle X-ray data suggested a preferential POSS dissolution at higher blending temperature and recrystallization of POSS upon cooling. It has also been reported that the peroxide vulcanization of the PDMS/POSS blend dramatically declines the rubber cross-link density and hence the mechanical properties of the cured silicone rubber in reference to unfilled vulcanized rubber [56].

The observation has been rationalized by contemplating a side reaction (as elaborated in Fig. 7.7), where an active radical chain at the vinyl function gets chemically

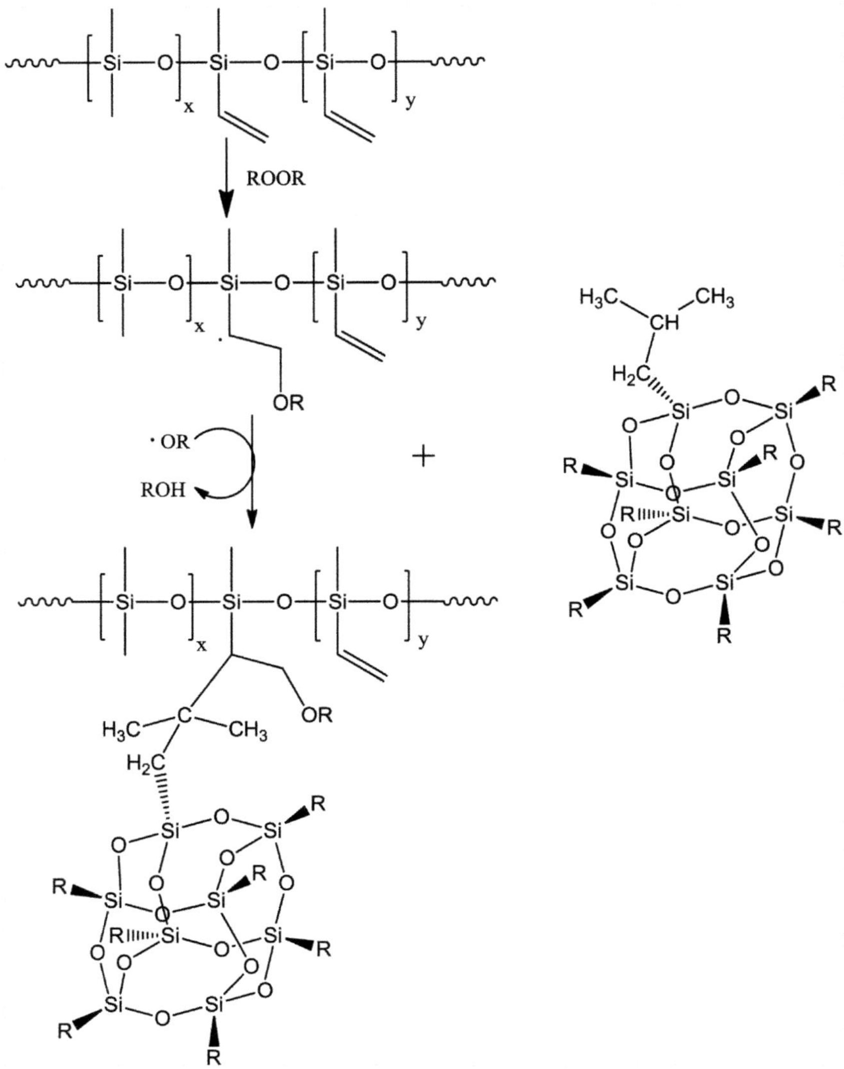

Figure 7.7 A schematic representation of side reactions occurring during peroxide triggered vulcanization of polysiloxane elastomeric composite containing isobutyl POSS (All the substituents designated as *R* refer to isobutyl groups) [57].

tethered with a POSS unit instead of connecting to a neighboring PDMS vinyl radical, rendering the radical macromer unavailable for polymer chain cross-linking.

Lee and Chang [57] have investigated melt blending followed by peroxide vulcanization of ethylene propylene diene (EPDM) with acrylisobutyl POSS. Solid state ^{29}Si NMR study confirmed the chemical grafting of acrylisobutyl POSS onto the EPDM rubber matrix while, microcompositional analysis by energy dispersive X-ray spectroscopy showed that the Si-elemental mapping was uniform throughout the bulk rubber nanocomposite, emphasizing the molecular level blending which led to significant improvement of tensile modulus, tear strength, as well as the elongation at break of the reinforced elastomer.

The converse results in the case of polysiloxane rubber [56] versus EPDM could be understood by looking at the radical promoted vulcanization mechanism of EPDM.

As can be seen in Fig. 7.8, the EPDM macroradical forms at the tertiary carbon, which can either find another similar macroradical to yield an interchain cross-link or can involve unsaturation to create a branch cross-linking point [58]. Thus, even if the acrylated POSS gets grafted to the polymer, vulcanization can still proceed unlike a peroxide curable vinyl polysiloxane which does not form additional macroradical anywhere in the siloxanes backbone and the cross-linking

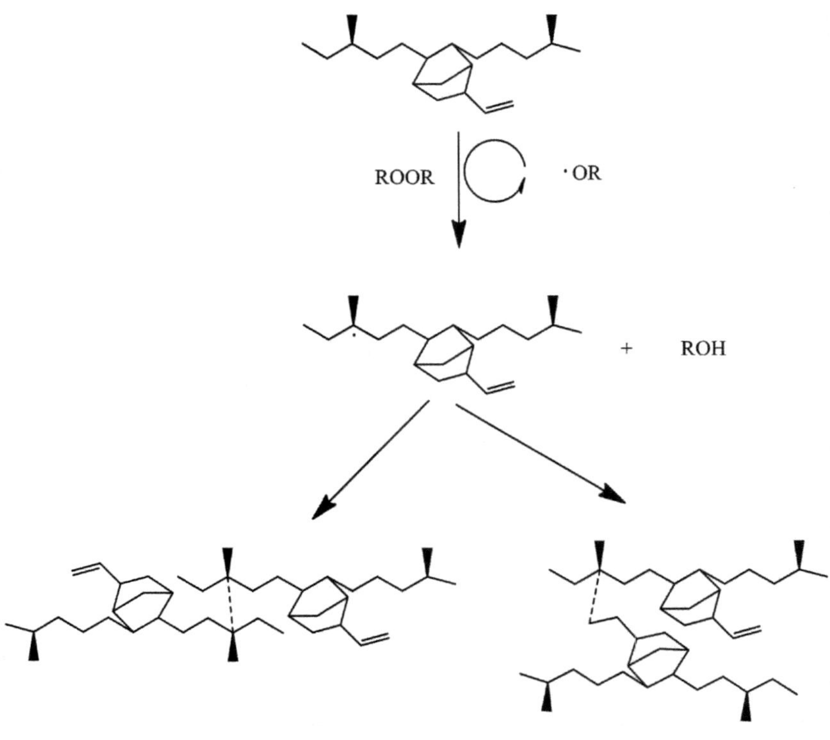

Figure 7.8 Schematic representations of peroxide vulcanization of EPDM [58].

only proceeds through vinyl groups. However, in case of EPDM vulcanization, there should always be a trade-off between cross-link density and POSS grafting density.

Zhang et al. [59] systematically studied the effect of hepta-phenyl POSS cage structures grafted onto the polysiloxane backbone on the PDMS crystallization in silicone rubber nanocomposite. It has been reported that up to 3 wt% of POSS grafting facilitates the PDMS crystallization due to nucleation effect enhancing the crystallization rate by 2.5 times that of pure polydimethylsiloxane. An increase in POSS loading leads to preferential POSS crystallization which imparts a detrimental effect to PDMS crystallization. It is worth noting here that PDMS segment crystallization is of particular interest to polysiloxane-based adhesives, sealants, or elastomers, where a change in the extent of crystallization and/or the rate would translate to properties like adhesion failure, elastic hardening, etc., particularly for subambient temperature applications.

7.5 POSS as Structure Directing Component

Apart from reinforcing thermoplastic and elastomeric composites, the POSS class of building blocks has been extensively used as structure directing agents to attempt diverse applications. An "ABA" triblock copolymer has been synthesized using controlled radical polymerization [60]. At some particular relative degree of polymerization, the polymer yields defined microphase segregated structures comprising of polybutyl acrylate cylindrical phases arranged in poly (POSS-methacrylate) matrix. Such a design strategy can be extended to fabricate organic–inorganic hybrid materials wherein a segregated organic phase can be removed by selective etching processes like by calcination, plasma radiation, etc. to yield mesoporous materials [61].

Wells et al. [62] studied the effect of POSS structures in silicone matrices for high performance applications where there is exposure of atomic oxygen. POSS T_8 cage structures are particularly efficient for protecting material surfaces from erosion caused by atomic oxygen. Polydimethyl-siloxanes upon atomic oxygen attack initiate shrinkage and cracks by the loss of methyl groups from the PDMS and gradual silica formation. POSS demonstrated promising results over conventional PDMS upon oxidative erosion. During the radical oxidation exposure, the Si–R bond preferentially gets cleaved leading to lateral fusion of POSS cages forming an adherent, amorphous silica coating. The process is commonly referred to as "glassification" because the resulting amorphous silica layer has the same chemical composition as glass. Unlike glass, this amorphous layer is flexible having a typical thickness of about 50–100 nanometers (Fig. 7.9).

Wang et al. [63] leveraged octa-phenyl POSS as a precursor to fabricate organic–inorganic porous structures through Friedel-Craft and Scholl coupling reaction. The resulting resins yield microporous or mesoporous structures offering high specific surface area. Tan and coworkers reported porous network formation by coupling tetrabromo-phenyl silane with octa-vinyl POSS in defined stoichiometry using Heck reaction [64]. Such organic–inorganic microporous hybrids are also reported to be used for hydrogen storage applications. Bhowmick et al. reported the strategic use of oligomeric polysilsesquioxane to exfoliate "Graphene" nanosheets through POSS grafting onto graphitic surface. [65]

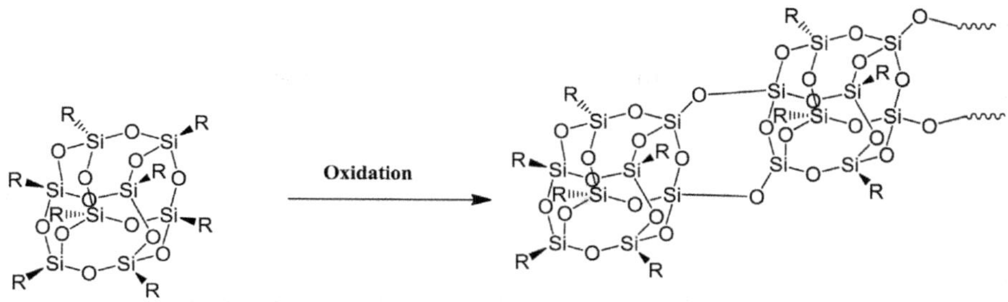

Figure 7.9 Schematic representations of oxidation reaction on POSS leading to glassification.

7.6 Summary

The chapter provides a comprehensive overview of the structure–property research results accomplished on POSS, which combine the most beneficial properties of both organic and inorganic systems, by far, the smallest reinforcing nanofiller available for elastomeric composites. Such a highly thermally stable oligosiloxane framework provides the hydrodynamic reinforcement while the tailormade peripheral organic functionality offers the preferential interactions with a variety of polymeric matrices leading to compatible nanocomposite formation. The current chapter also revisits the explorations of the tailormade POSS structures used to fabricate rubber nanocomposites. The beneficial role of POSS-like molecular fillers as reinforcement agents and/or as structure directing building blocks has been elucidated. In most of the POSS tethered polymers, the matrix–filler incompatibility and the preferential crystallization of POSS segments were used as tunable linkages to design elastomeric materials. The peripheral organic functionality of POSS were reported to play a key role in controlling the POSS dispersion in fumed silica-based addition cured LSR. The radical grafting of organofunctionalized POSS were reported to reinforce the composite during EPDM based rubber vulcanization, while causing detrimental effects to radically curable polysiloxane rubber composite. POSS-based molecular units were also utilized effectively as precursors for developing functional structures like radiation protective coatings, nanoporous organic–inorganic framework for gas storage, catalysis, or low dielectric composites.

Acknowledgment

The authors gratefully acknowledge Dr. Hans Rafael Winkelbach and Dr. Sabine Couderc for valuable discussions. The authors also acknowledge Dr. Jinendra Sethia for fruitful discussions.

References

[1] Treloar LRG. The physics of rubber elasticity. 3rd ed. Oxford: Clarendon Press; 1975.

[2] (a) Paul DR, Mark JE. Fillers for polysiloxane ("silicone") elastomers. Prog Polym Sci 2010;35:893–901.

(b) Mark JE, Erman B. Rubber-like elasticity: a molecular primer. 2nd ed. Cambridge: Cambridge University Press; 2007.
[3] Bhowmick AK, Stephens HL. Handbook of elastomers. New York, NY: Marcel Dekker; 2001.
[4] (a) Einstein A. Anna Phys Leipzigm 1905;17:549.
(b) Smallwood HM. J Appl Phys 1944;15:758.
[5] (a) Guth E. Theory of filler reinforcement. J Appl Phys 1945;16:20.
(b) Guth E, Gold O. On the hydrodynamical theory of the viscosity of suspensions. Phys Rev 1938;53:322.
[6] Kraus G. Reinforcement of elastomers by carbon black. Adv Polym Sci 1971;8:155.
[7] Thomas S, Stephen R. Rubber nanocomposites: preparation, properties, and applications. India; 2010. ISBN 978-0-470-82345-3.
[8] Saxena A, Dasgupta D, Bhat S, Tiwari S, Samantara L, Wrobel D. Beneficial role of "sol–gel" synthesized mesoscale silica networks on the performance of liquid silicone elastomer. J Appl Polym Sci 2014;131(7):1–7.
[9] Das A, Jurk R, Stoeckelhuber KW, Majumder PS, Engelhardt T, Fritzsche J, et al. Processing and properties of nanocomposites based on layered silicate and carboxylated nitrile rubber. J Macromol Sci Chem 2009;46:7–15.
[10] Sadhu S, Bhowmick AK. Preparation and properties of styrene–butadiene rubber based nanocomposites: the influence of the structural and processing parameters. J Appl Polym Sci 2004;92:698–709.
[11] Takeuchi H, Cohen C. Reinforcement of poly(dimethylsiloxane) elastomers by chain-end anchoring to clay particles. Macromolecules 1999;32(20):6792–9.
[12] Schmidt DF, Giannelis EP. Silicate dispersion and mechanical reinforcement in polysiloxane/layered silicate nanocomposites. Chem Mater 2010;22(1):167–74.
[13] Pradhan B, Srivastava SK, Saxena A. Mechanical and thermal properties of silane grafted organomodified montmorillonite reinforced silicone rubber nanocomposites. J Nanosci Nanotechnol 2012;12(12):8975–84.
[14] Baney RH, Itoh M, Sakakibara A, Suzuki T. Silsesquioxanes. Chem Rev 1995;95(5):1409–30.
[15] Provatas A, Matisons JG. Synthesis and applications of silsesquioxanes. Trends Polym Sci 1997;5:327–33.
[16] Frye CL, Collins WT. Oligomeric silsesquioxanes, $(HSiO_{3/2})_n$. J Am Chem Soc 1970;92:5586–8.
[17] Ramirez SM, Diaz YJ, Campos R, Stone RL, Haddad TS, Mabry JM. Incompletely condensed fluoroalkyl silsesquioxanes and derivatives: precursors for low surface energy materials. J Am Chem Soc 2011;133(50):20084–7.
[18] Suenaga K, Nasa T Cosmetic product U.S. patent application publication no. 2011/0040062A1.
[19] Zhang C, Babonneau F, Bonhomme C, Laine RM, Soles CL, Hristov HA, et al. Highly porous polyhedral silsesquioxane polymers. Synthesis and characterization. J Am Chem Soc 1998;120:8380–91.
[20] Sellinger A, Laine RM. Silsesquioxanes as synthetic platforms. 3. Photocurable, liquid epoxides as inorganic/organic hybrid precursors. Chem Mater 1996;8:1592–3.
[21] Choi J, Kim SG, Laine RM. Organic/inorganic hybrid epoxy nanocomposites from aminophenylsilsesquioxanes. Macromolecules 2004;37:99–109.
[22] Choi J, Yee AF, Laine RM. Organic/inorganic hybrid composites form cubic silsesquioxanes. Epoxy resins of octa(dimethylsiloxyethylcyclohexyl-epoxide) silsesquioxane. Macromolecules 2003;36(15):5666–82.

[23] Takamura N, Viculis L, Zhang C, Laine RM. Completely discontinuous organic/inorganic hybrid nanocomposites by self-curing of nanobuilding blocks constructed from reactions of $[HMe_2SiOSiO_{1.5}]_8$ with vinylcyclohexene. Polym Int 2007;56:1378–91.
[24] Asuncion MZ, Laine RM. Silsesquioxane barrier materials. Macromolecules 2007;40 (3):555–62.
[25] Feher FJ, Blanski RL. Olefin polymerization by vanadium-containing silsesquioxanes: synthesis of a dialkyl-oxo-vanadium (V) complex that initiates ethylene polymerization. J Am Chem Soc 1992;114:5886–7.
[26] Feher FJ, Budzichowski TA. Silasesquioxanes as ligands in inorganic and organometallic chemistry. Polyhedron 1995;14:3239–53.
[27] Ropartz L, Morris RE, Schwarz GP, Foster DF, Cole-Hamilton DJ. Dendrimer-bound tertiary phosphines for alkene hydroformylation. Inorg Chem Commun 2000; 3:714–17.
[28] Riollet V, Quadrelli EA, Copéret C, Basset J-M, Andersen RA, Köhler K, et al. Grafting of $[Mn(CH_2{}^tBu)_2(tmeda)]$ on silica and comparison with its reaction with a silsesquioxane. Chem Eur J 2005;11:7358–65.
[29] Maschmeyer T, Klunduk MC, Martin CM, Shephard DS, Thomas JM, Johnson BFG. Modeling the active sites of heterogeneous titanium-centred epoxidation catalysts with soluble silsesquioxane analogues. Chem Comm 1997;1847–8.
[30] Duchateau R, Abbenhuis HCL, van Santen RA, Meetsma A, Thiele SK-H, van Tol MFH. Half-sandwich titanium complexes stabilized by a novel silsesquioxane ligand: soluble model systems for silica-grafted olefin polymerization catalysts. Organometallics 1998;17:5222–4.
[31] Solans-Monfort X, Filhol J-S, Copéret C, Eisenstein O. Structure, spectroscopic and electronic properties of a well defined silica supported olefin metathesis catalyst, $[(\equiv SiO)Re(\equiv CR)(=CHR)(CH2R)]$, through DFT periodic calculations: silica is just a large siloxy ligand. New J Chem 2006;30:842–50.
[32] Feher FJ, Newman DA, Walzer JF. Silsesquioxanes as models for silica surfaces. J Am Chem Soc 1989;111:1741–8.
[33] Feher FJ, Budzichowski TA, Blanski RL, Weller KJ, Ziller JW. Facile syntheses of new incompletely condensed polyhedral oligosilsesquioxanes: $[(c\text{-}C_5H_9)_7Si_7O_9(OH)_3]$, $[(c\text{-}C_7H_{13})_7Si_7O_9(OH)_3]$, and $[(c\text{-}C_7H_{13})_6Si_6O_7(OH)_4]$. Organometallics 1991;10: 2526–8.
[34] Contreras-Torres FF, Basiuk VA. Imidazo[1,2-a]pyrazine-3,6-diones derived from α-amino acids: a theoretical mechanistic study of their formation via pyrolysis and silica-catalyzed process. J Phys Chem A 2006;110:7431–40.
[35] Leu CM, Reddy M, Wei K-H, Shu C-F. Synthesis and dielectric properties of polyimide-chain-end tethered polyhedral oligomeric silsesquioxane nanocomposites. Chem Mater 2003;15:2261–5.
[36] Leu C-M, Chang Y-T, Wei K-H. Synthesis and dielectric properties of polyimide-tethered polyhedral oligomeric silsesquioxane (POSS) nanocomposites via POSS-diamine. Macromolecules 2003;36(24):9122–7.
[37] Liu Y-L, Lee H-C. Preparation and properties of polyhedral oligosilsesquioxane tethered aromatic polyamide nanocomposites through Michael addition between maleimide-containing polyamides and an amino-functionalized polyhedral oligosilsesquioxane. J Polym Sci Part A Polym Chem 2006;44:4632–43.
[38] Chojnowski J, Fortuniak W, Rosciszewski P, Werel W, Lukasiak J, Kamysz W, et al. Polysilsesquioxanes and oligosilsesquioxanes substituted by alkylammonium salts as antibacterial biocides. J Inorg Organomet Polym Mater 2006;16:219–30.

[39] Majumdar P, Lee E, Gubbins N, Stafslien SJ, Daniels J, Thorson CJ, et al. Synthesis and antimicrobial activity of quaternary ammonium- functionalized POSS (Q-POSS) and polysiloxane coatings containing Q-POSS. Polymer 2009;50:1124–33.
[40] Chan KL, Sonar P, Sellinger A. Cubic silsesquioxanes for use in solution processable organic light emitting diodes (OLED). J Mater Chem 2009;19:9103–20.
[41] Gromilov SA, Basova V T, Emel'yanov DY, Kuzmin V A, Prokhorova SA. Layer arrangement in the structure of octakis-(trimethylsiloxy)octasilsesquioxane and dodecakis (trimethylsiloxy)-octasilsesquioxane. J Struct Chem 2004;45:471–5.
[42] Gromilov SA, Emel'yanov DY, Kuzmin AV, Prokhorova SA. Structural organization of layers in octakis-(trimethylsiloxy)octasilsesquioxane. J Struct Chem 2003;44:704–6.
[43] Kuo SW, Chang FC. POSS related polymer nanocomposites. Prog Polym Sci 2011; 36:1649–96.
[44] Cordes DB, Lickiss PD, Rataboul F. Recent developments in the chemistry of cubicpolyhedral oligosilsesquioxanes. Chem Rev 2010;110:2081–173.
[45] Wu J, Mather PT. POSS polymers: physical properties and biomaterials applications. Polym Rev 2009;49:25–63.
[46] Laine RM. Nanobuilding blocks based on the $[OSiO_{1.5}]_x$ (x = 6, 8, 10) octasilsesquioxanes. J Mater Chem 2005;15:3725–44.
[47] (a) Lickiss PD, Rataboul F. Fully condensed polyhedral oligosilsesquioxanes (POSS): from synthesis to application. Adv Organomet Chem 2008;57:1–116.
(b) Harrison PG. Silicate cages: precursors to new materials. J Organomet Chem 1997;542:141–83.
(c) Wallace WE, Guttman CM, Antonucci JM. Polymeric silsesquioxanes: degree- of-intramolecular-condensation measured by mass spectrometry. Polymer 2000;41: 2219–26.
[48] Sahoo S, Bhowmick AK. Polyhedral Oligomeric Silsesquioxane (POSS) Nanoparticles as New Crosslinking Agent for Functionalized Rubber. Rubber Chem. and Technol. 2007;80(5):826–37.
[49] Joshi M, Butola BS. Studies on nonisothermal crystallization of HDPE/POSS nanocomposites. Polymer 2004;45:4953–68.
[50] (a) Wu J, Haddad TS, Kim GM, Mather PT. Rheological behavior of entangled polystyrene – polyhedral oligosilsesquioxane (POSS) copolymers. Macromolecules 2007;40:544–54.
(b) de Gennes PG. Reptation of a polymer chain in the presence of fixed obstacles. J Chem Phys 1971;55:572–9.
[51] Kopesky ET, Haddad TS, Cohen RE, McKinley GH. Thermomechanical properties of poly(methyl methacrylate)s containing tethered and untethered polyhedral oligomeric silsesquioxanes. Macromolecules 2004;37:8992–9004.
[52] Seurer B, Coughlin EB. Ethylene–propylene–silsesquioxane thermoplastic elastomers. Macromol Chem Phys 2008;209:1198–209.
[53] Heeley EL, Hughes DJ, Taylor PG, Bassindale AR. Morphology and crystallization kinetics of polyethylene/long alkyl-chain substituted polyhedral oligomeric silsesquioxanes (POSS) nanocomposite blends: a SAXS/WAXS study. Eur Polym J 2014;51: 45–56.
[54] Zhang Y, Mao Y, Chen D, Wu W, Yi S, Moa S. Synthesis and characterization of addition-type silicone rubbers (ASR) using a novel cross linking agent PH prepared by vinyl-POSS and PMHS. Polym Degrad Stabil 2013;98(4):916–25.
[55] Joshi V, Srividhya M, Dubey M, Ghosh AK, Saxena A. Effect of functionalization on dispersion of POSS-silicone rubber nanocomposites. J Appl Polym Sci 2013;130(1):92–9.

[56] Liu L, Tian M, Zhang W, Zhang L, Mark JE. Crystallization and morphology study of polyhedral oligomeric silsesquioxane (POSS)/polysiloxane elastomer composites prepared by melt blending. Polymer 2007;48:3201–12.
[57] Lee KS, Chang YW. Peroxide vulcanized EPDM rubber/polyhedral oligomeric silsesquioxane nanocomposites: vulcanization behavior, mechanical properties, and thermal stability. Polym Eng Sci 2015;55(12):2814–20.
[58] Baldwin FP, Borzel P, Cohen CA, Makoswki HS, van de Castle JF. Rubber Chem Technol 1970;43:522.
[59] Zhang D, Liu Y, Shi Y, Huang G. Effect of polyhedral oligomeric silsesquioxane (POSS) on crystallization behaviors of POSS/polydimethylsiloxane rubber nanocomposites. RSC Adv 2014;4:6275–83.
[60] Pyun J, Matyjaszewski K, Wu J, Kim GM, Chun SB, Mather PT. ABA triblock copolymers containing polyhedral oligomeric silsesquioxane pendant groups: synthesis and unique properties. Polymer 2003;44:2739–50.
[61] Lee HD, Park S, Gu W, Russell TP. Highly ordered nanoporous template from triblock copolymer. ACS Nano 2011;5(2):1207–14.
[62] Wells B, Brandhorst H, Isaacs-Smith T, Lichtenhan JD, Fu BX. Proceedings of AIAA IECEC07 conference June 25–29.
[63] Wang S, Tan L, Zhang C, Hussain I, Tan B. Novel POSS-based organic–inorganic hybrid porous materials by low cost strategies. J Mater Chem A 2015;3:6542–8.
[64] Wang D, Yang W, Li L, Zhao X, Feng S, Liu H. Hybrid networks constructed from tetrahedral silicon-centered precursors and cubic POSS-based building blocks via Heck reaction: porosity, gas sorption, and luminescence. J Mater Chem A 2013;1:13549–58.
[65] Mondal T, Bhowmick AK, Krishnamoorti R. Butyl lithium assisted direct grafting of polyoligomeric silsesquioxane onto graphene. RSC Adv. 2014;4:8649–56.

Rubber nanocomposites with new core-shell metal oxides as nanofillers

8

S.H. El-Sabbagh, N.M. Ahmed, G.M. Turky and M.M. Selim
National Research Center, Giza, Egypt

8.1 Introduction

Rubber products are used in many applications throughout the world. Recently, polymer blending was in need of an appropriate process to acquire materials with convenient effectiveness and properties. Some of these polymers are immiscible; and compatibilizers are the required ingredient to overcome the weak interfacial adhesion, which leads to inferior mechanical properties [1−5]. However, compatibilizers may form an interface between the immiscible blend components, so that imposed stresses can be transferred between the phases via the covalent bonds along the copolymer backbone. Styrene butadiene rubber (SBR), which is a nonpolar rubber, possesses good mechanical properties, better ozone resistance, weatherability, and abrasion resistance than natural rubber. Acrylonitrile-butadiene rubber (NBR) (polar rubber) has good solvent resistance compared to SBR, but its mechanical characteristics are much poorer [6,7].

Blending of SBR/NBR blends leads to an incompatible one [5], and it is well known that the physicomechanical properties of SBR and NBR are very weak in the absence of a reinforcing filler. Consequently, the physicomechanical properties of their blends are also inferior, which originates principally from the incompatibility of both components with each other in view of the fact that NBR is more polar than SBR [4]. Nanocomposites based on polymer blends are of major importance due to their superior mechanical properties and well-balanced physical characters including cost [4].

Inorganic fillers are used to improve physical and mechanical performance such as color, rigidity, dimensional stability, toughness, and transparency. Recently, there has been an increasing interest in the use of nanometal oxides as reinforcing fillers [1]. In other words, polymers reinforced with nanometal oxides open up new pathways for engineering flexible composites that exhibit better mechanical and chemical properties. Polymers reinforced with nanometal oxides become popular every day because of their improved mechanical strength, flexibility in design, as well as better chemical, electrical, and optical properties. In the rubber industry, a variety of fillers are used to improve and modify the physical properties of elastomeric materials. The addition of rigid filler particles, even in small amounts, to an elasto-

Progress in Rubber Nanocomposites. DOI: http://dx.doi.org/10.1016/B978-0-08-100409-8.00008-5

mer strongly influences its response to mechanical stimuli; nanofillers are the ultimate choice because of their ability to modify properties at very low levels of incorporation [2,3].

Titanium dioxide is used in rubber compounding as a colorant to impart whiteness to any desired end product, it is used in sidewalls of automobile tires. It is chemically inert and slightly basic with a pH of 7–8. The free-chalking, weathering properties of anatase titanium-dioxide provide a (self-cleaning) white appearance in outdoor applications. It can be also used as a vulcanizing agent and a cure-modifier, and can be employed as functional filler, cross-linking agent, and heat stabilizer [8]. Its main defect is the absorption of light energy which creates free radicals in the compound that ultimately leads to the degradation of the polymer system [9].

Silica fume is a by-product that is formed from the smelting process in the ferrosilicon industry which produces very pure SiO_2 vapors; the reduction of high-purity quartz to silicon at temperatures up to 2000°C produces SiO_2 vapors, leading to the oxidation and condensation in low temperature zones to tiny particles consisting of very pure noncrystalline silica. This silica is a fine lightweight fluffy amorphous powder; its performance proves its high capability as a reinforcing material for the manufacturing of highly stable rubber products, such as tires. The significant enhancement of tensile strength was achieved due to the physical entanglements of rubber molecules with the networks formed by silica particles. The openings formed in the networked silica additionally contribute to improve its dispersion by penetration of rubber molecules into it [10].

Mixed structured or hybrid particles are now attracting more and more investigatory interest. These particles are constructed of different materials with different chemical compositions, which show ultimately distinctive properties of varied materials together; especially, manipulating the surface functions to meet diverse application requirements. In general, the properties of these structured particles are different from their individual constituents [11]. The precipitation of different layers is carried out for different reasons, such as surface modification, increased stability, dispersibility, and to control the release of the material in the bulk.

Coupling agents are designed to increase the interaction between the filler and the rubber to produce better dispersion of filler in rubber. It is very common to use a silane coupling agent in combination with silica or titanium dioxide in the automobile tire industry [12]. On the other hand, use of silane coupling agents in silica- or titanium dioxide-filled rubber compounds causes improvements in modulus, compression set, heat build-up, and abrasion resistance. The addition of coupling agents significantly improves the silica dispersion [13–15], but the low electrical conductivity still remains a problem. For various applications, the silica-filled rubber with sufficient electrical conductivity would be advantageous for its antistatic properties.

Broadband Dielectric Spectroscopy (BDS) is considered to be a very important technique for addressing the dynamics of polymeric systems; by using this tool the relaxation dynamics at the molecular scales, the charge transport leading to the conductivity, and the interfacial polarization in many systems of heterogeneous structured polymers [16–20] can be separated. It seems to be important to establish the space charge polarization and to study its relaxation mechanism.

This chapter describes a trial to compensate for the advantages and disadvantages of silica and titania by applying new hybrid or core-shell pigments, which is a new trend of combining more than one component by chemically deposited them on each other with specific ratios. The core is always a cheap material, which is silica fume waste in this case comprising 90% of the whole pigment composition, and the shell contains nanotitanium dioxide which is used in rubber and plastic applications but is expensive and has fatal disadvantages. By applying these new core-shell nanotitania/silica (TS) pigments new properties will be exhibited by rubber composites to get over the disadvantage of the degradation of polymer matrix caused by titanium particles, since it will be present in low concentration. Also employing waste material as major component reduces the cost of rubber products and helps to keep the environment safe.

In this chapter silica fume waste will be covered with layers of different thickness of nanotitanium dioxide to study their effect on the different mechanical, rheological, morphological, and dielectric properties of SBR/NBR 50/50 rubber blend. Also, the effect of these layers on the rubber composite properties and the different loadings in each composite to reach the optimum concentration of their use will be studied. In order to get information about the molecular mobility and the interfacial polarization in heterogeneous structured materials in relation to the morphology of the investigated samples, BDS will be employed in the frequency range of 0.1 to 10 MHz and at temperatures ranging from −50 up to 150°C. The experimental results will be reported by means of the complex parameters: permittivity, ε^*, conductivity, σ^*, and electric modulus, M^*, and their dependence on the frequency and temperature.

8.2 Experimental

- SBR (1502) with a specific gravity of 0.945, styrene content 23.5% and Mooney viscosity M_L (1 + 4) of 48–58 at 100°C was purchased from Transportation & Engineering Company (TRENCO)-Alexandria.
- NBR under the, tradename perbunan, was supplied from Bayer AG, Germany with acrylonitrile content of 34 ± 1%, density 0.99 g/cm^3, and Mooney viscosity M_L (1 + 4) of 65 ± 7 at 100°C.
- Stearic acid and zinc oxide (ZnO) with specific gravity 0.9–0.97 and 5.55–5.61, respectively, at 15°C were used as activators.
- High abrasion furnace (HAF) carbon black (N-330), with specific gravity 1.78–1.82, and particle size of 40 nm.
- Poly (ethylene glycol), average molecular weight (Mn) 950–1050 was purchased from SIGMA-ALDRICH.
- N-cyclohexyl-2-benzothiazole sulfonamide, pale gray powder with specific gravity of 1.27–1.31 at room temperature (25 ± 1°C), melting point 95–100°C was used as an accelerator.

Table 8.1 **Chemical composition of silica fume**

Concentration of main constituents	Wt., %
SiO_2	97.98
TiO_2	–
Al_2O_3	0.39
Fe_2O_3 Total	1.04
MnO	0.08
CuO	0.02
MgO	0.83
CaO	0.32
SrO	0.003
Na_2O	0.53
K_2O	0.75
P_2O_5	0.07
SO_3	0.005
L.O.I	4.93

- Elemental sulfur, a fine pale yellow powder with a specific gravity of 2.04−2.06 at room temperature, was used as the vulcanizing agent.
- Polymerized 2, 2, 4-trimethyl-1, 2-dihydroquinoline was used as an antioxidant.
- Silica fume waste was obtained from ferrosilicon factories in Aswan, Egypt. Its chemical composition is represented in Table 8.1, the waste was collected from the chimneys of the factory and this may be the cause of its gray color as it contains a very high percentage of carbon resulting from the smelting of this alloy production.
- Titanium tetrachloride was from Lobachemie, India and its assay is 99.5%.

8.2.1 Preparation of TiO_2/SiO_2 core-shell pigments

Titanium tetrachloride was added in three concentrations of 5−10 mL to 100 mL hydrochloric acid. Silica was impregnated in these three solutions and left for some time to assure complete covering while stirring. Ammonia solution was then added drop-wise to these impregnated silicas to adjust their pH until complete precipitation of the pigments. The formed paste is then filtered through a buchner system and washed very well. This paste was then calcined at 650−850°C. The preparation of the pigments is represented in the following scheme:

$$TiCl_4/SiO_2 \xrightarrow[NH_4OH]{HCl} TiOH_4/SiO_2 + NH_4Cl \xrightarrow[500-750^\circ C]{\Delta} TiO_2/SiO_2 \tag{8.1}$$

Two concentrations of titanium dioxide were deposited on the silica surface; these two concentrations were determined using X-ray fluorescence (XRF) technique. Table 8.2 includes these results which show that TiO_2 concentrations are ascending from TS1 to T_{S2}.

Table 8.2 XRF analysis results of the prepared TiO_2/SiO_2 pigments

Concentration of main constituents (wt. %)	Silica fume	$TiO_2/SiO_2$1 (TS1)	$TiO_2/SiO_2$2 (T_{S2})
SiO_2	97.98	95.61	94.78
TiO_2	–	10.19	12.42
Al_2O_3	0.39	0.35	0.33
Fe_2O_3	1.04	1.27	1.23
MgO	0.08	0.60	0.57
CaO	0.02	0.33	0.32
Na_2O	0.83	0.17	0.18
K_2O	0.32	0.98	0.86
P_2O_5	0.003	0.06	0.06
SO_3	0.53	0.005	0.007
L.O.I	0.75	0.53	0.76

8.2.2 Preparation of rubber compounds

Rubber was premixed with all compounding ingredients according to ASTM D3182-07 (2012). Mixing was carried out on a laboratory two roll mill. The speed of the slow roll was 24 rev/min, with a gear ratio of 1:1.4. The compounded rubbers were left overnight before vulcanization. Then, the vulcanization was carried out at 152 ± 1°C in an electrically heated press under a pressure of about 4 MPa to obtain vulcanized rubber sheets of 2 mm thickness. The rubber formulations are given in Table 8.3.

Table 8.3 Formulations of rubber mixes

Ingredients, phr[a]	N0	N1	N2	N3	N4	S1	S2	S3	S4
TS1	–	3	6	9	12	–	–	–	–
T_{S2}	–	–	–	–	–	3	6	9	12

Base recipe: NBR/SBR, 100; Stearic acid, 2; Zinc Oxide, 5; HAF, 20; Poly (ethylene glycol) PEG, 1; silane coupling agent, 1; CBS, 1; TMQ, 1; Sulphur, 1.5.
[a]Part per hundred parts of rubber.

8.2.3 Bound rubber determination

Bound rubber content was determined via keeping the core-shell pigments/rubber master batches (without curatives) at room temperature for 7 days for conditioning, after this period the compound was cut into small pieces; about 0.5 g was precisely weighed and wrapped with filter paper. The wrap was immersed in 100 mL toluene at room temperature. The solvent was renewed on the 4th day and after 7 days; the samples were taken out from the solvent and dried to a constant weight in vacuum. The bound rubber percentage (BR%) was calculated by the following equation:

$$BR\% = \frac{(M_B - M_F - M_D)}{M_B} \times 100\% \qquad (8.2)$$

where M_B is the weight of uncured mix before immersion in the proper solvent, M_F is the weight of filler in the uncured mix, and M_D is the weight of rubber dissolved in the solvent (toluene).

8.2.4 Methods of testing

8.2.4.1 Rheometric characteristics

M_L (minimum torque), M_H (maximum torque), T_{c90} (optimum cure time), T_{S2} (scorch time), and cure rate index were determined according to ASTM D2084-11 using a Monsanto (Akron, OH, USA) oscillating disk rheometer, model 100. The measurements were carried out at 152 ± 1°C.

8.2.4.2 Mechanical properties

The tensile strength, stress at 100% strain, and elongation at break of the obtained vulcanizates were tested according to ASTM D412-06a (2013) with a Zwick testing machine. Five measurements were performed for each sample and the average was taken.

8.2.4.3 Swelling study

Swelling tests were carried out by soaking a specific weight of rubber samples in toluene at room temperature for 24 h. The equilibrium swelling in toluene (Q%) was determined according to the standard method (ASTM D471-06). The equilibrium swelling Q% can be calculated according to the following equation:

$$Q = [(w_s - w_d)/w_d] \times 100\% \tag{8.3}$$

where w_s is the weight of the swelled specimen and w_d is the weight of the dried specimen.

8.2.4.4 Cross-linking density

The cross-linking density (ν)mol/cm^3 of NBR/SBR was determined on the basis of solvent-swelling measurements (toluene solvent for 24 h at 25 ± 1°C) using the Flory–Rehner equation [21,22].

$$\nu = 0.5/M_c \tag{8.4}$$

where M_c is the molecular weight between cross-links (g/mol).

$$M_C = \left[\frac{-\rho V_S\left(V_r^{1/3} - \frac{2V_r}{f}\right)}{[\ln(1 - V_r) + V_r + \mu V_r^2]}\right] \tag{8.5}$$

where ρ is the density of rubbers in g/cm^3, V_s is the molar volume of the solvent (toluene), f is the network functionality, μ the interaction parameter of rubbers, and

V_r is the volume fraction of swollen rubber that can be obtained from the mass, density of rubber, and solvent.

8.2.4.5 Hardness

Hardness was measured using a Shore A durometer according to ASTM D2240.

8.2.4.6 Strain energy determination

Strain energy values were obtained by plotting stress–strain curves for vulcanized rubber using unnotched samples, and integrating the area under the curve up to the used particular extension. Simpson's rule [23,24] was applied to calculate the strain energy, and then the calculated strain energies were plotted against the corresponding strain. This curve was used to obtain the strain energies for particular extensions.

8.2.4.7 Scanning electron microscopy (SEM/EDAX)

Scanning electron microscopy (SEM) was performed using a JXA-840A electron probe microanalyzer (Jeol, Japan), this analysis was used in this study for two purposes, first for the determination of pigment particles morphology, and the second is to determine the dispersion of the pigment particles in the rubber matrix. The test was done by breaking the rubber specimen in liquid nitrogen; then covering the cross-section surface with a very thin layer of gold to avoid electrostatic charging during examination. An energy-dispersive was used in this work to determine the elements deposited on silica surface to confirm the preparation of core-shell particles.

8.2.4.8 Transmission electron microscope (TEM)

Various pigments were examined using a JEOL JEM 2100 transmission electron microscope (TEM) (Japan) with a microanalyzer electron probe. This instrument was used in determining the particle shapes and sizes of the prepared pigments.

8.2.4.9 X-ray fluorescence (XRF)

The different concentrations of each element in the prepared pigments were determined using an Axios sequential WDX-ray fluorescence (XRF) spectrometer, PANalytical 2005, USA.

8.2.4.10 FTIR spectroscopy

A JASCO FTIR-6100E Fourier transform infrared spectrometer (Japan) was operated in the absorption mode in a wavelength ranging from 4000–400 cm^{-1} after mixing with potassium bromide and pressed in the form of disks. The spectra were collected at a resolution of 4 cm^{-1}.

8.2.4.11 Dielectric measurements

All presented dielectric and electrical data were obtained by a high resolution ALPHA analyzer with an active sample holder head (Novocontrol, Montabaur,

Germany) and an active sample head. This technique is used to measure the complex dielectric and electrical functions and their dependence on frequency from 10^{-1} Hz to 10^7 Hz and temperature ranging from -50 to 150°C. Samples were prepared in a parallel plate geometry. Therefore, the samples were mounted between the two gold-plated electrodes (20 mm) of the sample holder. Isothermal frequency scans were carried out, and the temperature was controlled by a Quatro Novocontrolcryo-system with a temperature stability of 0.1 K.

8.3 Results and discussion

8.3.1 Characterization of the prepared pigments

Figs. 8.1 and 8.2 show the morphology of the prepared pigments using SEM and TEM. From the featured photos it can be seen that silica fume possesses spherical particle shapes, while titanium dioxide possesses a platelet structure. The new pigments possess the two shapes altered on each other, i.e., the spheres of silica carry on their surfaces and around the spheres the tiny plates of nanotitanium dioxide that appear as a roughness on the sphere surfaces, as can be seen in Fig. 8.1. SEM micrographs shown in Fig. 8.2, also confirmed TEM photos by showing tiny plates

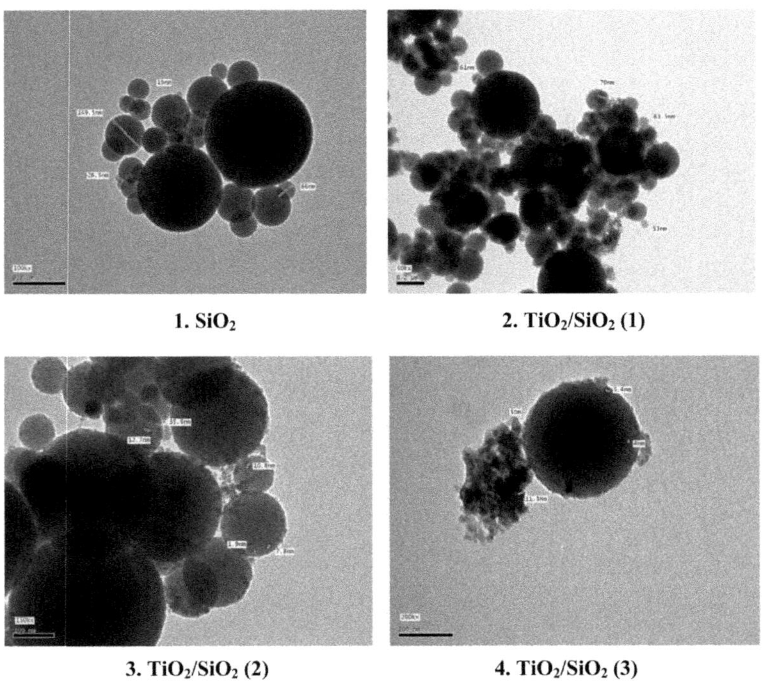

Figure 8.1 TEM micrographs of silica fume and covered silica with titanium dioxide at magnification of 12×10^3.

Figure 8.2 SEM micrographs of silica fume and covered silica with titanium dioxide at magnification of 12×10^3.

on silica spherical surfaces. From these two figures, it can be detected that the prepared pigment particles are formed with two shapes and two particle sizes. Both SEM and TEM showed the size of titanium dioxide in the shell to be in the nanoscale indicating the formation of nanotitanium dioxide on micronized silica fume. It is interesting to point out that the component which is in direct contact with the matrix is nano-TiO_2, but although its small particles are spreading on the silica surface there are voids between these particles allowing the silica also to have contact with the matrix, i.e., both components exert their effect on the matrix.

In general, the plate structure of pigments (nano-TiO_2) helps in the slippage of rubber chains from the surface of pigment particles leading to improved elasticity. This elasticity is attributed to the plasticizing and the conformational effects on the polymer at the pigment–matrix interface, indicating that the pigment particles are in a stacked condition and polymer chains are diffused inside the pigment [25].

Fig. 8.3 shows the EDX analysis of prepared pigments. EDX or energy-dispersive X-ray analysis can detect the elements on the surface up to one micron depth. As can be seen from the chart, titanium was detected revealing its presence on silica surfaces, and since titanium dioxide has more than one energy level it appears in more than one place in the chart, while silica has only one energy level, and thus it has only one peak.

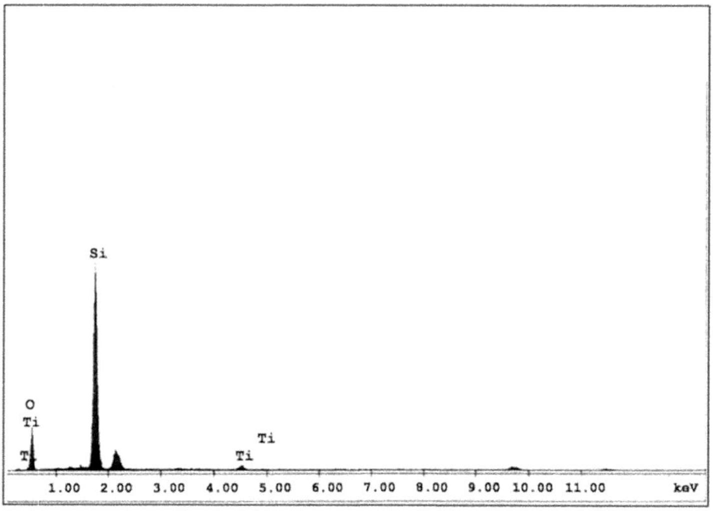

Element	Wt. %	At. %	K-Ratio	Z	A	F
O K	43.47	57.83	0.1068	1.0219	0.2401	1.0007
Si K	54.42	41.23	0.4036	0.9861	0.7519	1.0002
Ti K	2.11	0.94	0.0169	0.8790	0.9091	1.0000
Total	100.00	100.00				

Figure 8.3 EDAX analyses data for TiO_2/SiO_2 pigments.

According to the results given in Table 8.2 representing XRF results, it was obvious that nano-TiO_2 was not present in the original silica fume, but it is now present in low ratios in the three prepared pigments indicating that only a very thin upper layer was formed and this layer was the cause of the properties exhibited by the new core-shell pigments rather than its individual components.

Infrared spectrometry is used to study the chemical interaction between the TiO_2/SiO_2 core-shell pigments and NBR/SBR blends. The absorption bands of core-shell nano-TiO_2/SiO_2 in the IR spectra are represented in Fig. 8.4. This figure shows broad bands at 1631 and 1099 cm^{-1} which are attributed to the stretching vibrations of titanium dioxide, while the intense sharp bands at 806 and 464 cm^{-1} are characteristic for the stretching vibrations of the silica.

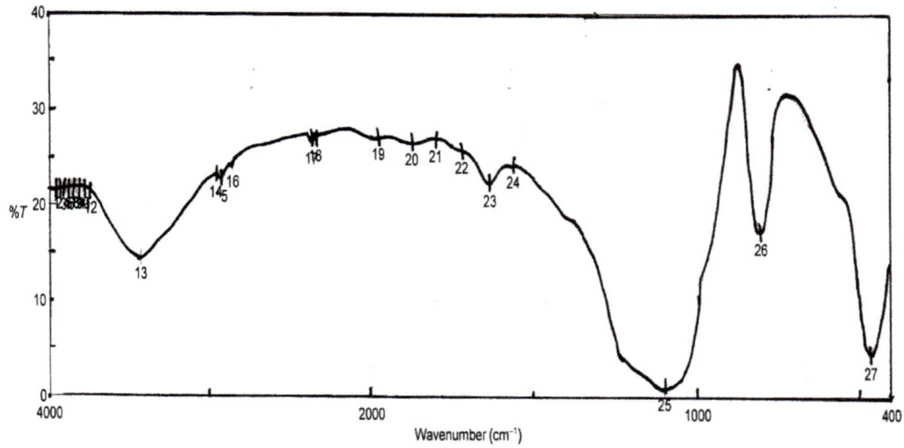

Figure 8.4 FT-IR of TiO_2/SiO_2 core-shell pigments.

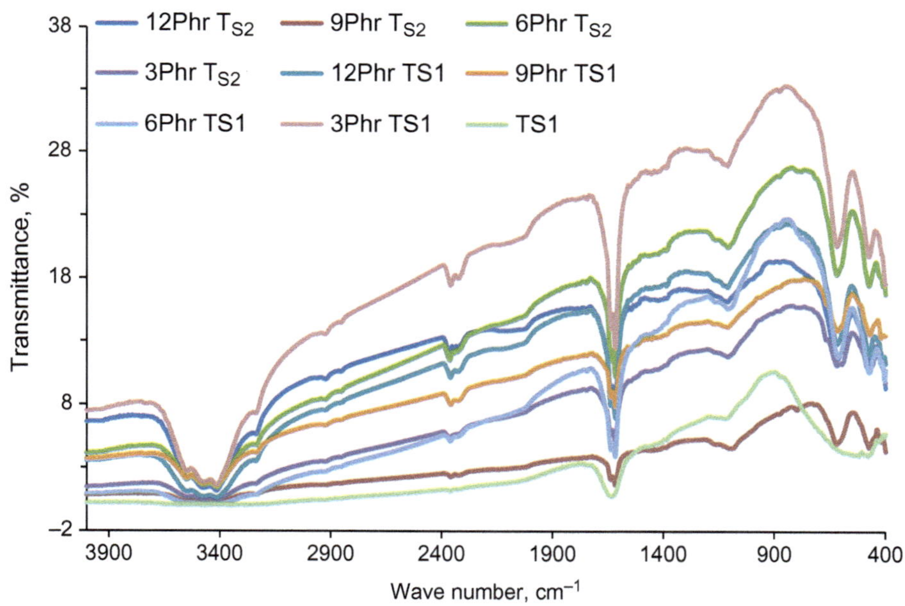

Figure 8.5 FT-IR of SBR/NBR rubber blends loaded with different concentrations of TiO_2/SiO_2.

The IR spectrum of the rubber blends/TS composites shown in Fig. 8.5, revealed the presence of the characteristic bands of both TiO_2/SiO_2 and NBR/SBR with a $5 - 15\ cm^{-1}$ shift from the typical IR spectrum of their parents Fig. 8.5.

Fig. 8.5 shows a typical IR spectrum of NBR/SBR, and it is clear that the band appearing at $873\ cm^{-1}$ can be attributed to the presence of the CH bending frequency of styrene. As a result, the band at $873\ cm^{-1}$ in all vulcanizates (Fig. 8.5) could be one of the styrene moieties present in SBR. Bands at 3.542−3.413 cm^{-1} are attributed to the stretching vibrations of OH groups in the composites. The bands at 2.927 and 2.925 cm^{-1} are assigned to CH_2 stretching of NBR/SBR matrix along with those in all vulcanizates containing TiO_2/SiO_2, while bands at 1.637−1.617 cm^{-1} are attributed to the double bonds in polybutadiene and other unsaturated components. The presence of nitrile rubber can be confirmed from the nitrile peaks represented in bands at 2225 cm^{-1}. Also, NBR/SBR vulcanizates containing the investigated core-shell pigments revealed a shift in the wavelengths of the characteristic bands as well as a change in their intensities, and thus it can be deduced that this shift is attributed to the chemical binding of these fillers (TiO_2/SiO_2) to the NBR/SBR rubber matrix.

8.3.2 Rheological properties

The physical properties of pigmented NBR/SBR rubber blends determines the quality of rubber products, therefore, it is necessary to study the rheological behavior of NBR/SBR rubber blends loaded with nano-TiO_2/SiO_2 (TS). M_H depends on both the extent of cross-linking and the reinforcement of NBR/SBR/TS rubber composites; it has been observed that the incorporation of different concentrations of nano-TiO_2/SiO_2 into the NBR/SBR matrix increases the (M_H) value in both cases of NBR/SBR/TS1 and NBR/SBR/T_{S2} composites that are represented in Fig. 8.6. This increase was an indirect hint for improved interfacial adhesion between NBR/SBR rubber blends and core-shell pigments in the presence of a coupling agent, and also this was associated with the increase in the rigidity of the blend system due to pigment loading, which reduces the mobility of polymer chains in the matrix [3]. It is also evident that the M_L increased with the increase in the loadings of core-shell nano-TiO_2/SiO_2 and passes through a maximum at 9phr loading of TS1, then after that a sharp decrease occurred at 12phr as in Fig. 8.6. It was reported earlier that M_L corresponds to pigment−pigment interagglomeration [26,27], and since incorporation of the silane coupling agent into the pigmented rubber composites reduces the pigment−pigment networking factor, high M_L values were obtained. This is believed to be due to the strong polymer−pigment interaction through ion molecule interaction and hydrogen bonding. The lower the value of M_H, the weaker the pigment−pigment interaction, and thus lower viscosity of the compound. On the other hand the incorporation of 12phr TS in NBR/SBR rubber blends in the presence of a silane coupling agent reduced the pigment−pigment interaction, leading to a lower viscosity, which enhances the processability of NBR/SBR/TS composites [28,29].

The difference between maximum (M_H) and minimum (M_L) torque (ΔM) is taken as a rough estimate of the cross-link density [3,30], and it can be noticed that the cross-link density in the case of composites containing T_{S2} is higher than those containing TS1; these values are given in Table 8.4. Theoretically, the torque difference ($M_H - M_L$) represents the shear dynamic modulus, which is indirectly related to the cross-link density of the rubber compound [31]. The total cross-link density is contributed to both the sulfide and physical cross-links [32,33]; while in the case of adding a silane coupling agent to NBR/SBR/nano-TiO_2/SiO_2 rubber composites,

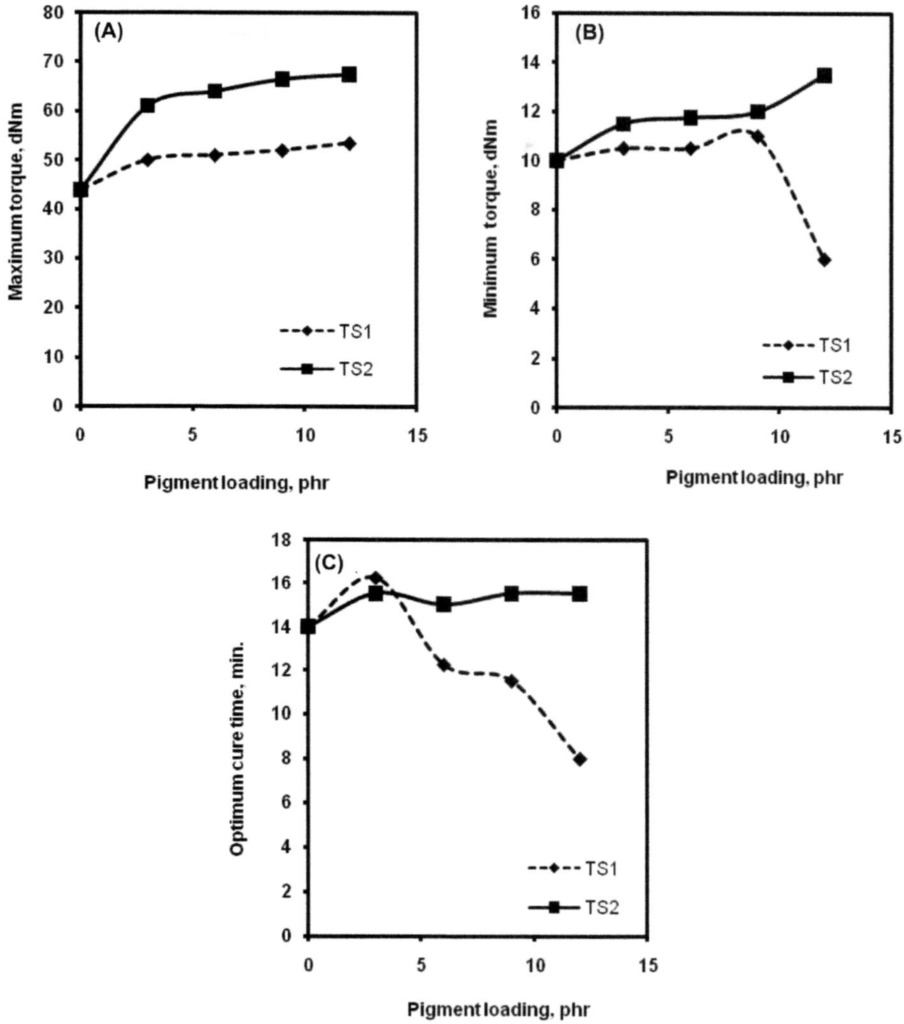

Figure 8.6 The relation between filler loading with (A) maximum torque, (B) minimum torque, (C) optimum cure time.

the torque difference of vulcanizates increased. This is clearly attributed to the action of the coupling agent as a pigment dispersant, which is due to the silane coupling agent accelerating the sulfur reaction and enhancing the state of the sulfide cross-links. The silane coupling agent increased the degree of silica dispersion and formed the additional physical cross-links caused by NBR/SBR/silica attachments.

Fig. 8.6C shows that cure time (T_{c90}) was increased for initial pigment loading at 2phr of TS1 and then there was a decrease in the case of additional loading. It was seen that the higher loading of these pigments (12phr of T_{S2}) leads to a lower cure time compared to that of an unpigmented blend and also to NBR/SBR/TS

Table 8.4 Rheometric characteristics of NBR/SBR/TS composites at 152°C

Properties/ no. of sample	N0	N1	N2	N3	N4	S1	S2	S3	S4
ΔM, dNm	33.75	39.5	40.5	41	47	49.5	52.25	54.5	54
T_{S2}, min.	1.25	1.25	1.125	1.188	1.375	1	1.25	1.25	1.375
CRI, min^{-1}	7.84	6.66	8.98	9.697	15.09	6.8996	7.27	7.018	7.079
α_f	—	0.1704	0.2	0.215	0.422	0.466	0.548	0.615	0.6

ΔM is the difference between maximum torque and minimum torque; T_{S2}—scorch time; T_{c90}—optimum cure time; CRI—cure rate index; α_f-specific constants for the fillers.

composites containing other loadings. The silane coupling agent may be considered as a cocuring agent since the polar parts of this additive reacted with the silanol groups of silica to transform the silica from hydrophilic into hydrophobic by interacting relatively less with the zinc oxide which is an ingredient in the formulations. Also, this decrease was attributed to the amine content of the silane coupling agent which is one of the accelerator activators that increases the cure rate [34]. Amine may also be utilized to overcome cure retardation problems in silica reinforcement [28,35]. In contrast, the incorporation of T_{S2} in NBR/SBR rubber composites resulted in a significant increase in the cure time and it almost followed the same trend, this may be attributed to the increase of TiO_2 concentration in the shell causing an increase in the cure time (T_{c90}).

The ratio between the increases in torque of the loaded composites to that of the gum was found to be directly proportional to pigment loading. The slope of the linear plot showing the relative torque increased as a function of prepared nano-TS loading and was defined by Wolff [36,37] as α_f. These different terms are expressed in Eq. (8.6) as follows;

$$\alpha = \frac{\Delta M_{\max(\text{pigment})} - \Delta M_{\max(\text{gum})}}{\Delta M_{\max(\text{gum})}} \tag{8.6}$$

where $\Delta M_{\max}$ (pigment), and $\Delta M_{\max}$ (gum) represent the changes in M_H during vulcanization for the composite and rubber, respectively.

The values of reinforcement factor α_f are listed in Table 8.5, it is clear that α_f values increased with increasing core-shell concentration except for rubber blends containing 12phr of T_{S2} where α_f decreased with increasing pigment loading. Both $\Delta M_{\max}$ and α value increased according to the reinforcement ability of the pigments. The highest value of α_f represents the strength of pigments and consequently their reinforcing effect in rubber. Therefore, α_f of rubber composites containing different concentration of TS core-shell pigments can be arranged in the following ascending order:

NBR/SBR/nano-T_{S2} > NBR/SBR/nano-TS1

Table 8.5 **Physicomechanical properties of NBR/SBR/TS composites**

Properties/no. of sample	N0	N1	N2	N3	N4	S1	S2	S3	S4
Strain energy, MJ/m^3	0.556	0.717	0.916	0.937	0.799	0.612	0.621	0.615	0.504
Young's modulus, N/mm^2	0.011	0.002	0.027	0.04	0.03	0.015	0.040	0.026	0.016
Mol. Wt. between cross-linking, M_C g/mol	2168	1988	1835	1430	1480	2106	2067	1911	1854
Cross-link density, $\nu \times 10^4$	2.32	2.51	2.72	3.49	3.38	2.37	2.42	2.62	2.67
C_1, MPa	−0.55	−0.33	−1.27	−2.69	−1.03	1.81	−2.07	−1.95	−1.41
C_2, MPa	2.09	1.59	3.36	5.15	2.53	4.07	4.59	4.43	3.49

8.3.3 Dispersion of core-shell pigments in rubber matrix

The degree of TS pigments dispersion in NBR/SBR phase in the presence of a silane coupling agent can be determined quantitatively by Eq. (8.7) [28,38,39].

$$L = \eta_r - m_r \tag{8.7}$$

Where η_r is the relative viscosity and m_r is the relative modulus, and they can be determined from rheometric data by using the following relation:

$$\eta_r = \frac{M^f_{min}}{M^0_{min}} \tag{8.8}$$

and

$$m_r = \frac{M^f_{max}}{M^0_{max}} \tag{8.9}$$

Where M denotes the torque and superscripts "f" and "0" are related to the loaded and unloaded polymer, respectively. From the data shown in Table 8.4, it can be seen that 12phr TS1 caused a lowering in the L value. The higher the TS1 loading the lower the L value, and thus better core-shell (T_{S2}) dispersion in NBR/SBR matrix. The comparison between the values of TS1 to that of T_{S2} is presented in Table 8.5, and it can be deduced that a similar loading of TS1 caused lower L values of NBR/SBR/TS1 composites, This can be attributed to the additional function of the TS1-silane coupling agent as an internal plasticizer which caused the reduction of the pigment–pigment interaction leading to better pigment dispersion compared to that of T_{S2}. This explanation was in agreement with the data in Table 8.4. It was noticed that, at similar loading the M_L of TS1 was lower than that of T_{S2}.

8.3.4 Bound rubber and filler-to-rubber interactions in core shell (TiO_2/SiO_2) compounds

The bound rubber was formed through the physical adsorption of rubber molecules on the pigment aggregates during the mixing process as described by Schuater [40]. In this process, the rubber molecules can penetrate the interior of an aggregate, leading to physical entrapment of the free rubber molecules. Otherwise, bound rubber is formed in pigmented rubber compounds by physical adsorption, chemisorptions, and mechanical interaction [41]. Formation of bound rubber depends on factors such as the polarity, microstructure of the polymer, and structure and surface activity of pigment or filler [42]. The pigment–rubber interaction affects the level of the bound rubber content. Fig. 8.7 shows that the addition of TS1 increased the bound rubber due to the enhancement between Si and the NBR/SBR rubber interfacial interaction. Also, the bound rubber formed by the SBR/NBR/TS1-silane

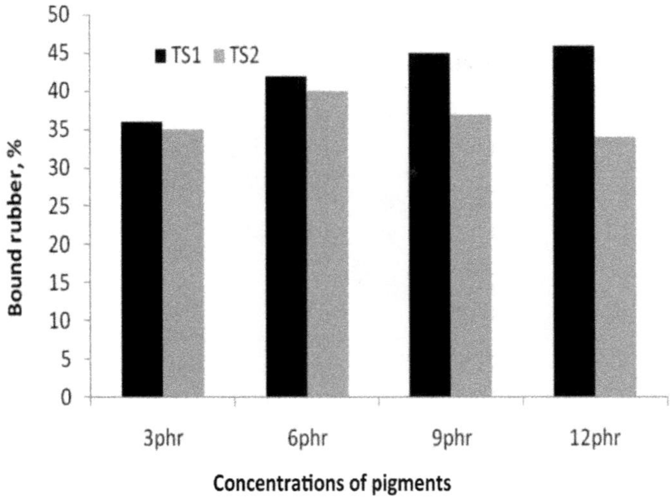

Figure 8.7 Variation of bound rubber versus different concentrations of core shell nano-TS/NBR/SBR.

coupling agent compound is chemically attached, which is obviously due to the hydrophobic nature of the silica surface as a result of silanization due to the presence of the silane coupling agent. The increase in concentrations of the core-shell in the presence of the coupling agent consequently resulted in more pigment-to-rubber interaction (12phr TS1). In the presence of silane coupling agent, on addition of T_{S2} in NBR/SBR rubber blends, it was evident that bound rubber increased up to 6phr, and then a slight decrease with further addition was detected. This may be due to the improvement of carbon black dispersion in the rubber matrix which tends to decrease the formation of bound rubber.

8.3.5 Mechanical properties

The plots of stress–strain for the blend vulcanizates resulting from tensile experiments are shown in Fig. 8.8A,B. It was found that these curves are sensitive to the concentration of the core-shell pigment loadings (TS) and consequently the different concentrations of TiO_2 in the shell. Also, it was seen that for a given strain all pigmented systems exhibited higher stresses than that of unpigmented ones; the same observation was reported by Helaly et al. [43]. Comparatively higher values of failure stress were obtained for composites loaded with 6phr T_{S2} followed by those containing 12phr TS1, and then those loaded by 9phr TS1. It was also noticed that the stiffness of the samples increased with increasing concentration of the core-shell pigment loadings. Otherwise, the stress–strain curves of NBR/SBR/TS showed typical behavior for an elastomer and can be divided into three stages:

- The first stage for small strain, in it the stress is increased slowly due to the unwinding of entangled chains.

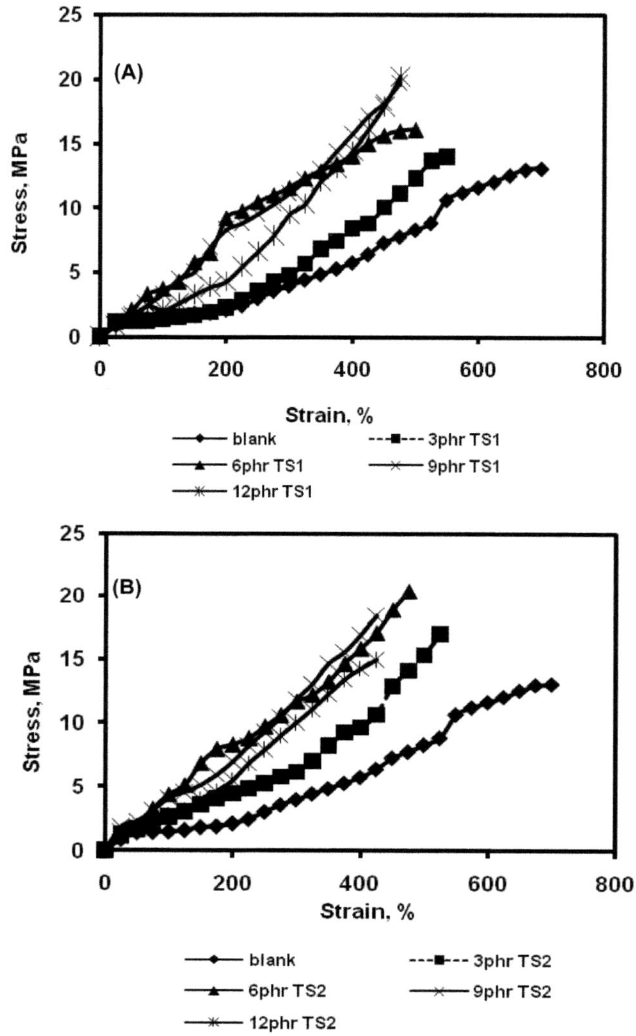

Figure 8.8 Stress–Strain curve of NBR/SBR loaded with different concentrations of: (A) nano-TS1; (B) nano-T_{S2}.

- The second stage for medium strain and in it the stress increased with uniform rate due to stretching in the chains.
- The third stage for high strain and it was observed that the stress was increased in it due to high stretching of the long chains.

The other contribution that arises from incorporation of core-shell TS pigments into NBR/SBR rubber blends in the presence of a silane coupling agent was the formation of pigment–rubber interaction, leading to additional physical cross-links in the network structure of the blends. Consequently, the energy absorbed per unit volume (W) is expected to increase in the deformed rubber blends, and thus the energy absorbed can be written as;

$$W = \int \sigma(\varepsilon)d\varepsilon \tag{8.10}$$

where σ is stress as a function of strain ε; therefore, the greater the area under the stress–strain curve is an indication of a higher energy absorption capacity. Unpigmented NBR/SBR exhibited the lowest area under the stress–strain curves and consequently, they have the lowest absorption energy. On the other hand, addition of 12phr TS1 exhibited the highest absorption energy other than that of 6phr T_{S2}, these findings are in good agreement with the data presented in Table 8.5 which shows the increase in the strain energy due to the incorporation of core-shell pigments (TS) in NBR/SBR rubber blends with the presence of a silane coupling agent. It is well known that the physicomechanical properties of NBR and SBR rubbers are relatively poor in the absence of reinforcing pigments or fillers, and consequently, the physicomechanical properties of their blends, which originate principally from the incompatibility of both rubbers with each other, are due to the higher polarity of NBR than SBR.

Fig. 8.9 represents the tensile strength of composites loaded with core-shell TS1 which increases with increasing core-shell content, but for the composites loaded with T_{S2} pigments the tensile strength first increased up to 6phr then decreased with further loading. The tensile strength enhancement was attributed to the ability of the silane coupling agent to improve the NBR/SBR/TS interactions. The silica surface was saturated with hydrophilic silanol groups, which were relatively incompatible with NBR/SBR, and their interaction with NBR/SBR was relatively low. The silane coupling agent [27] may chemically modify the surface of silica, and transform it to hydrophobic silica, this modified silica was more compatible with NBR/SBR and as a result the improved NBR/SBR/TS interaction led to the formation of physical cross-links. These further cross-links contributed to the total cross-link density [28,33]. The greater the physical cross-links, the higher the reinforcing efficiency of the pigment, and thus higher tensile strength can be detected.

Also, it can be noticed from Fig. 8.9 that the increase in tensile properties takes place with a slight decrease in the elongation at break % with the increment of core-shell pigment content. This may be attributed to the addition of core-shell pigments resulting in rendering the rubber blend vulcanizates less elastic, i.e., the brittleness increased [44]. Fig. 8.9 shows that increasing core-shell loadings leads to the decrease in elongation at break and the reduction in the volume fraction ($\mathcal{E}_R$) of rubber. This also can be rendered to the traveling of cracks through the weaker interfacial regions and the composite fractures at lower values of elongation [45] which are typical characteristics for inorganic filled composite [46].

The hardness values are plotted in Fig. 8.9, representing the relation between NBR/SBR rubber blends with different pigment loadings. From this figure it can be seen that the addition of TS pigments showed a marked increase in hardness reflecting an improved stiffness for NBR/SBR/TS (rubber–pigment) composites. The

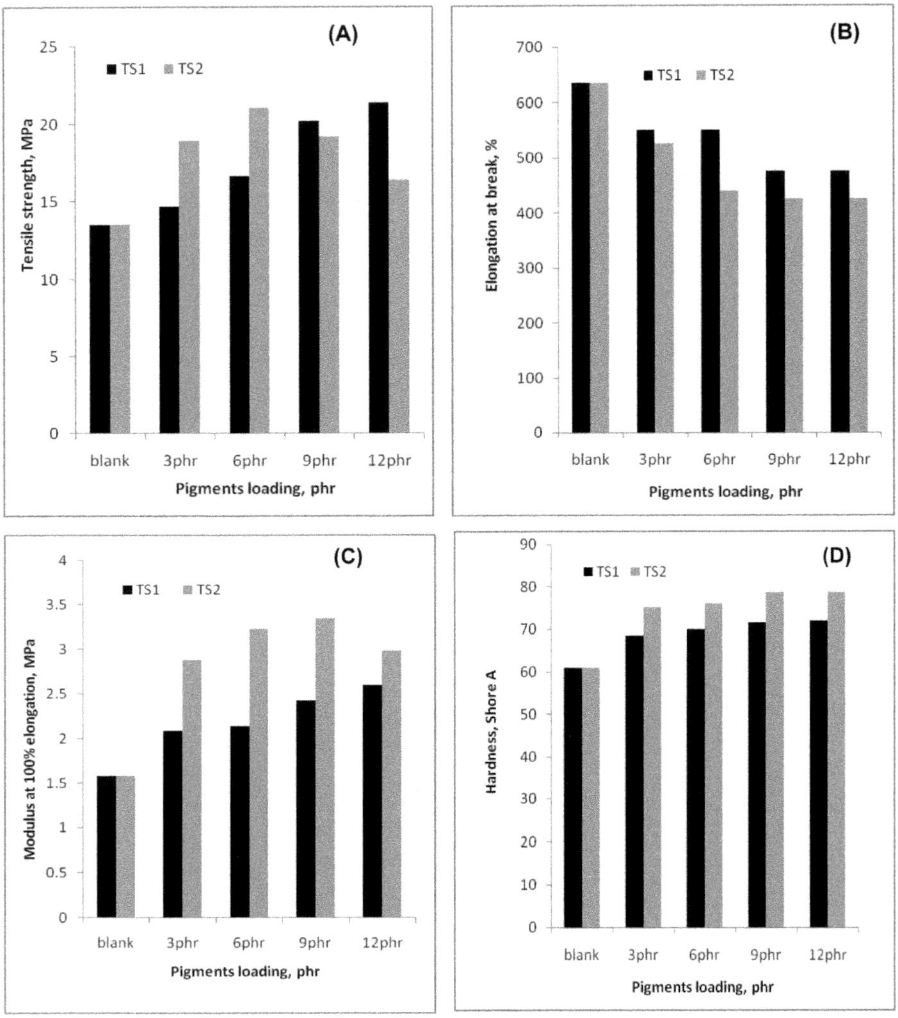

Figure 8.9 Variation of (A) tensile strength, (B) elongation at break, (C) modulus at 100% elongation, and (D) hardness versus different concentrations of nano-TS/NBR/SBR composites.

presence of these stiff pigments and the entanglement of polymer chains rendered the rubber blend composites harder than the respective pigmented blends [47]. On the other hand, the presence of TS pigments within the rubber matrix obviously increased the rigidity of the composites. Moreover, NBR/SBR blends mixed with T_{S2} in the presence of a silane coupling agent showed higher Young's modulus than composites containing TS1. The result showed that T_{S2} particles in NBR/SBR/T_{S2} composites exhibited enhanced stiffness in the blends better than that of NBR/SBR/TS1 composites [8].

Good abrasion resistance is observed for NBR/SBR/TS composites and it increases with the pigment loading. Hardness Shore A of a material is usually found to increase with the addition of pigments. Compatibilized NBR/SBR blends exhibit higher hardness compared to that of the unpigmented ones.

In order to investigate the influence of core-shell nano-TiO_2/SiO_2 content on the mechanical properties of the (50/50) NBR/SBR rubber blend, the affine deformation based on the Mooney-Rivlin model was selected to visualize the mechanism of the interaction of pigment or filler with rubber blends.

Treloar [48] established the framework to understand rubber elasticity at small deformations. The "affine" deformation of a network of Gaussian chains can be equivalently understood from the perspective of thermodynamic elasticity [49] or strain invariants, such as storable elastic energy [50].

The strain invariant approach of Mooney-Rivlin [1,23,51–55] is extensively utilized for filled or pigmented rubbers providing a straightforward approach to examine the deviation of a complex elastomeric system from ideality. The relationship between the stress–strain is expressed as:

$$\frac{\sigma}{2(\lambda - \lambda^{-2})} = C_1 + C_2\lambda^{-1} \tag{8.11}$$

where σ is tensile stress, λ is the strain, and C_1 and C_2 are constants reflecting characteristic of the networks. C_2 represents the nonGaussian aspects of the network and parameters of rubber blend composites, while C_1 is a quantity pertained to the ideal elastic behavior [8], and C_2 expresses the departure from it. From Fig. 8.10, it is worth pointing out that in the range of investigated deformation, no upturn in the modulus is evidenced for the unpigmented composites which do not exhibit any strain-induced crystallization but a large increase in stress was obtained for NBR/SBR/TS composites revealing once more the existence of interaction between the metal oxide and the organic phase. On the other hand, the strain at which the upturn occurs decreased with the pigment loading, this result suggested the presence of some interaction between nano-TiO_2/SiO_2 pigments and the NBR/SBR chain. As the content of metal oxide increases, the C_2 value increases Table 8.5. The Mooney-Rivlin constant C_2 was observed [51,56] to be associated with intermolecular forces and flexibility between polymer chains. This value increases with an increase in proportion of metal oxide (such as the pigment–pigment interaction). Among the blend components the higher C_2 value is noted for NBR/SBR/TS showing the presence of higher chain entanglement; higher chain entanglement shows better molecular level mixing. Thus the observed mechanical properties variation can be exactly correlated with the Mooney-Rivlin equation.

8.4 Swelling properties

The equilibrium swelling of NBR/SBR rubber blends composites are shown in Fig. 8.11. It can be observed that in the case of NBR/SBR/TS composites, the

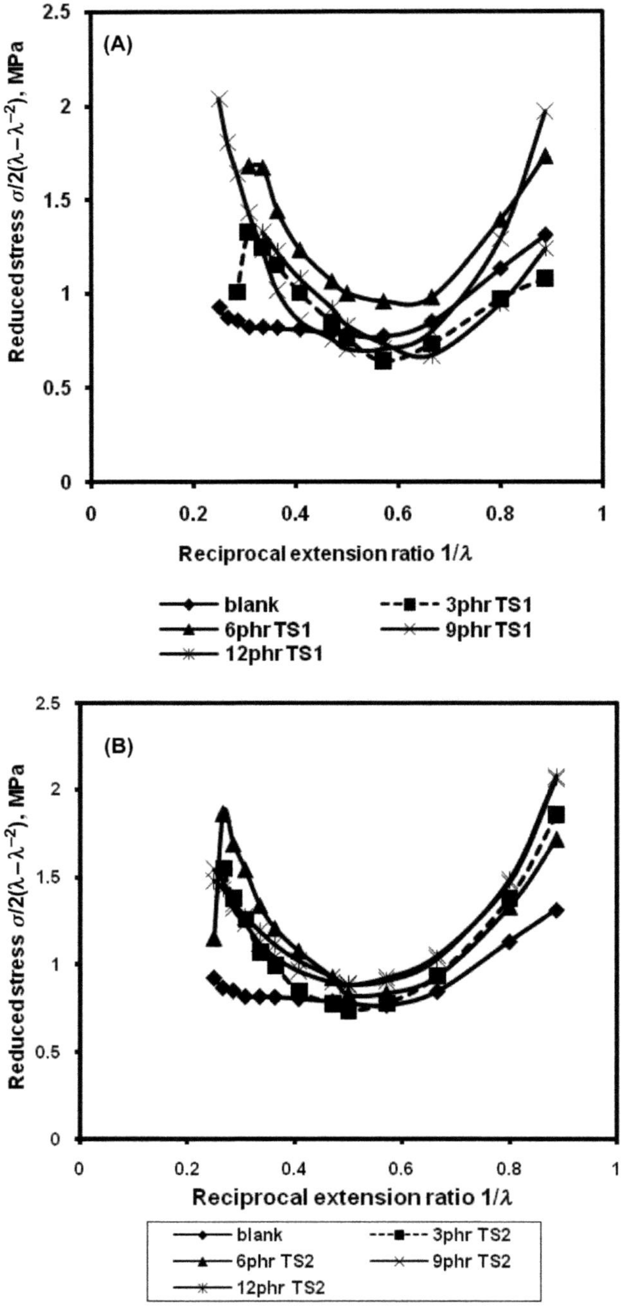

Figure 8.10 Mooney-Rivlin for NBR/SBR blends loaded with different concentrations of (A) TS1 and (B) T_{S2}.

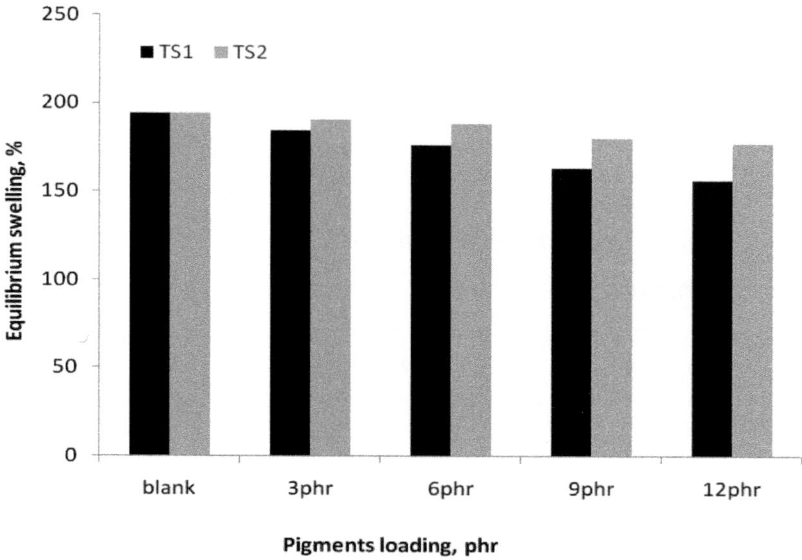

Figure 8.11 Variation of equilibrium swelling versus different concentrations of nano-TS/NBR/SBR composites.

equilibrium swelling decreased with increasing TS pigment loadings which may be attributed to the fact that TiO_2 particles do not swell in toluene, whereas in T_{S2} the titanium concentration increased and therefore the equilibrium swelling is nearly the same level for all composites containing T_{S2} [45]. If an enhanced bonding between TS pigments and NBR/SBR rubber matrix existed, a stronger cross-link system can be formed and the extent of cross-links in pigmented composites can be reflected from the cross-link density. The toluene uptake was considerably decreased after the addition of TS pigments to rubber vulcanizate, this can be due to the presence of TS pigments in NBR/SBR blends matrices which restrict the penetration of solvents into the intermolecular spaces, and hence decrease the swelling percentage [24,25]. On the other hand, the equilibrium swelling was utilized to determine the cross-linking density of NBR/SBR rubber composites. At equilibrium, the chemical forces tending to dissolve the rubber in the solvent are balanced by restraining forces exerted by the rubber network.

On the other hand, the swelling and M_C values of rubber blends loaded with TS pigments decreased with the increase in core-shell pigment content, which can be attributed to the increase in the cross-linking density and thus the increase of the nonswellable part (TiO_2 and HAF) in the rubber matrix.

Assuming that swelling could be completely restricted at the rubber–pigment interface, this may be due to the good adhesion between them. Also, it is worth mentioning that the soluble fraction values of rubber loaded with different concentrations of the prepared core-shell pigment are higher than that of unpigmented rubber vulcanizates, which can be attributed to the nature of the used pigment and the partial solubility of these pigments in organic solvents.

The above results were confirmed with the cross-linking density determined from the equilibrium swelling according to the Flory–Rehner relation [57]. Table 8.5 shows that the values of ν were increased in the case of using T_{S2} pigments due to the presence of additional physical and chemical cross-links, as was observed from the swelling and values of M_C.

8.4.1 Morphology of rubber composites

8.4.1.1 Transmission electron microscopy (TEM)

Fig 8.12A shows a TEM image for neat NBR/SBR. It can be seen that the extended chain reflects the flexibility of NBR/SBR rubber. Also, the same figure is used to observe the dispersion of core-shell (nano-TiO_2/silica) particles in NBR/SBR

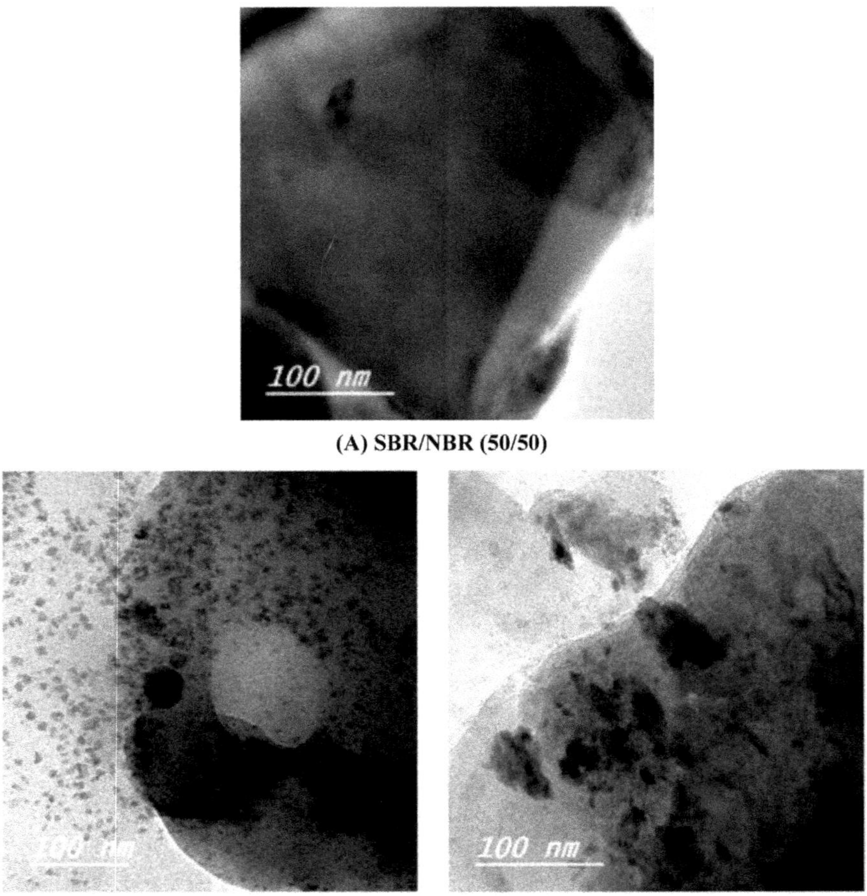

(A) SBR/NBR (50/50)

(B) TS1/SBR/NBR **(C) TS2/SBR/NBR**

Figure 8.12 TEM micrographs of (A) neat SBR/NBR, (B) TS1/SBR/NBR and (C) T_{S2}/SBR/NBR at magnification $10x10^9$.

composites, showed that the introduction of SiO_2/TiO_2 into rubber was found to be homogeneously distributed particles among the rubber body Fig. 8.12B. The increase in the amount of SiO_2/TiO_2 in rubber led to the formation of agglomerates within the rubber, Fig. 8.12C. This means that the homogeneous distribution of the filler is limited by the amount of the particles added.

8.4.1.2 Scanning electron microscopy (SEM)

SEM has been used to study the effect of core-shell (TS) pigments as compatibilizers on the morphology of NBR/SBR rubber blends. Fig. 8.13A shows that the morphology of unpigmented 50/50 NBR/SBR rubber blends exhibited two phases and lamellae were obtained resembling cavities which are irregularly distributed all over the surfaces. While in the case of NBR/SBR blend loaded with 6phr TS1, it can be seen that dispersion of the core-shell particles in the rubber phase resulted in a reduction in the number of cavities and cracks, while TS1 particles were distributed in scattered planes and seem to have smoother surfaces with a reduction in the matrix deformation as illustrated in Fig. 8.13B. In other words, the addition of 6phr

Figure 8.13 Scanning electron microscopy for NBR/SBR blends loaded with 6phr of pigment magnification of 1000 ×.

T_{S2} (Fig. 8.13C) has improved the NBR/SBR blend morphology and consequently the compatibility of the blends due to the reduction of the interfacial tension between NBR and SBR. These results confirmed the results obtained from TEM analysis.

8.4.2 Dielectric measurements dielectric measurements

In this section, three selective samples were set for dielectric measurements; these samples are unpigmented (50/50) NBR/SBR (blank), NBR/SBR/12phr TS1 (S9), and NBR/SBR/3phr TS1 (S6). Fig. 8.14 represents the dielectric loss ε'' of S6 versus frequency and temperature as a representative example.

It was detected that conventional BDS revealed one relaxation process; the relaxation peak tends to be broad and shifted towards lower frequencies with decreasing temperatures. Results are presented as a three-dimensional representation in Fig. 8.14. The peak is overtailed by a conductivity contribution at lower frequencies/higher temperatures [58].

Fig. 8.15 illustrates graphically the frequency dependence of the dielectric loss ε'' (upper) and the imaginary part of electric modulus M'' (lower) at zero Celsius for the three samples. The complex electrical modulus representation $M^*(\nu) = 1/\varepsilon^*(\nu)$ is also adopted to analyze the data which physically corresponds to the relaxation of the electric field in the material if the electric displacement remains constant. It is noticed from the figure that the conductivity contribution is represented as a semilinear increase with decreasing frequency shown at lower frequencies in the dielectric loss representation which become sharp and definite relaxation peak in the M'' representation [16,17]. The relaxation time of the conductivity was estimated from the frequency at the maximum peak position, where there is no remarkable effect of the core-shell loading on the conductivity relaxation time. The conductivity contribution in both representations is followed by the segmental dynamics that is related to the glass transition and usually called α-relaxation.

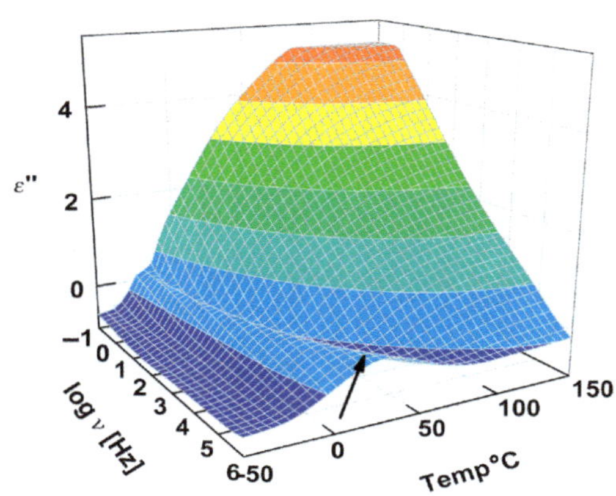

Figure 8.14 Three-dimensional representation of the dielectric loss ε'' of S6 versus frequency and temperature.

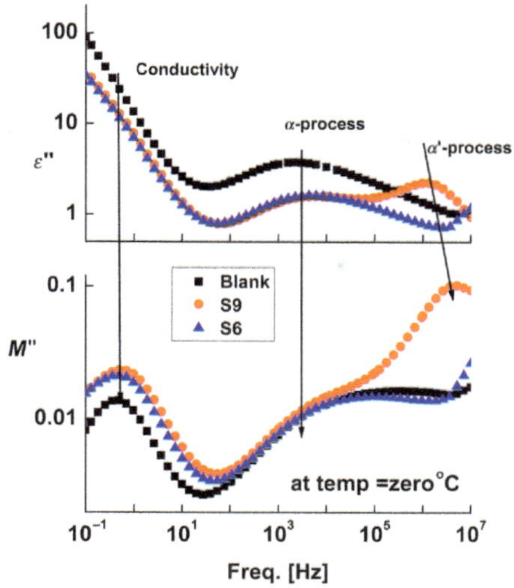

Figure 8.15 Frequency dependence of ε'' and M'' for the three samples at 0°C.

It is clear from Fig. 8.15 that α-relaxation (dynamic glass transition) of the blend is not affected by the addition of the core-shell in both loadings in S6 and S9 (containing 3phr and 12phr, respectively). This is in good agreement with that reported recently in [59], which considered this view of the framework of the free volume and the glass transition that is in principle a cooperative process, where a characteristic correlation length is involved and the change in free volume is too small to affect the glass transition [17,59,60].

Surprisingly a new relaxation peak was developed at higher frequencies in the case of the S9 sample, which can be explained by the fact that there is an extension of the interfacial region around the core-shell pigments in the matrix. The role of the interface regions between the nano- or even microparticle and the polymer matrix has attracted high attention in the literature. This interface region (interphase) is directly influenced by an adjacent nanoparticle surface and can play a decisive role in the performance of the nanocomposite material, especially for the transport of small penetrating molecules. This is also in good agreement with the mechanical properties which revealed that S9 loaded with 12phr TS1 exhibited the highest absorption energy other than that of S6 loaded with 3phr TS1 [61].

Fig. 8.16 showed the frequency dependence of the conductivity at different temperatures for the S9 sample as a representative example. Three ranges of temperature were suggested here; the first range is the lower one ranging between −50 and −25°C Fig. 8.16. It is clear from the figure that the sample behaves as a perfect insulator at very low temperatures, this is because the conductivity linearly decreases with decreasing frequency without a remarkable effect in the temperature range from −50 to −40°C, and then the alternating current (AC) conductivity reaches order of 10^{-16} S/cm at frequency 0.1 Hz. The main significance of performing the conductivity experiments at low temperatures is to find out if there is

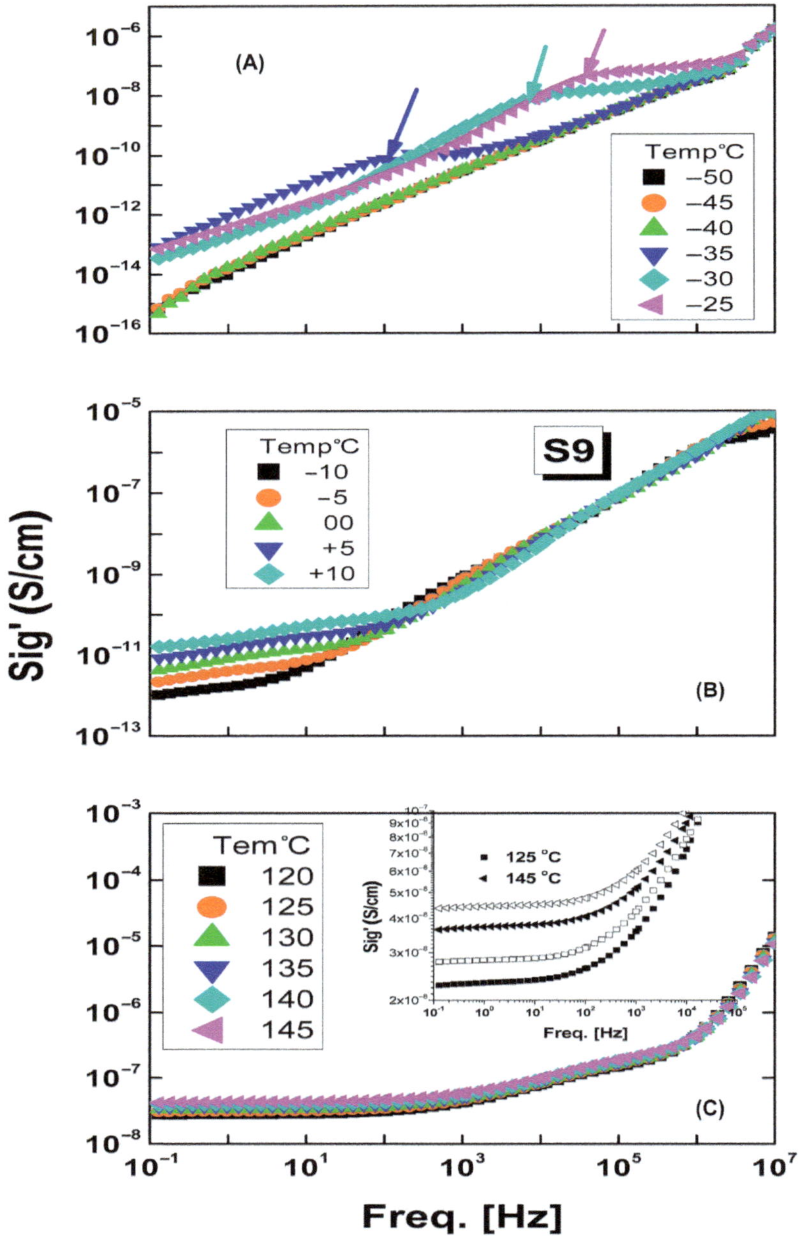

Figure 8.16 AC conductivity against frequency for TS1/ NBR/SBR (S9) at three different temperature ranges. (A) the lower temperature range from − 50 up to − 25, (B) Intermediate range of temperatures, from − 10 up to + 10, (C) The higher temperature range from 120 up to 145.

deviation from linearity as a function of frequency or not. It was expected that the direct current (DC) conductivity is rather low, but increasing temperature from -35 to $-25°C$ led to a shift of an anomalous peak towards a higher frequency, this is due to the mobility of adsorbed rubber chains on the shell (TiO_2) surface at the organic/inorganic interface. In accordance with the previous results, there is another relaxation peak at a higher frequency and it can be also attributed to the interfacial polarization. This can be explained on the basis of the relationship that was suggested recently by Ortiz-Serna et al. [61] in their two-phase model. The model suggested that there are two regions with different mobility at the interface of the rubber segmental group.

At the intermediate range of temperatures (-10 up to $+10°C$), Fig. 8.16(B) shows that at higher frequencies down to some characteristic frequencies, the conductivity followed the power law ($\sigma'(\omega) = A\omega^s$), and there is no effect of temperature on the conductivity values nor on the *s* parameter, which is the slope of linearity. Further decrease from the characteristic frequency is less dependent on frequency but more dependent on temperature which was clearly noticed, leading to an increase of conductivity from 1.6×10^{-12} at $-10°C$ to 2.6×10^{-11} at $+10°C$ (at spot point frequency 1 Hz). The conductivity values were not considered to be DC conductivity since it is still dependent on the frequency.

The frequency dependence of AC-conductivity at higher temperatures (ranging from 120–145°C) is shown in Fig. 8.16C. At lower frequencies, the results obtained exhibit a DC conductivity plateau, σ_{dc}, and there was no indication of the electrode polarization at such high temperatures. It is clear that the effect of temperature on the σ_{dc} is significantly low (ranging between 2.8 and 3.8×10^{-8}). The inset of Fig. 8.16 (C) is just a rescaling of the relation. It can be concluded that the increase of conductivity slows as temperature increases, which means that the conductivity process follows the Vogel-Fulcher-Tammann (VFT) equation just like the segmental motion which is a cooperative process and is related to the glass transition temperature.

The dielectric spectra can be also presented by the complex electric modulus $M^*(\omega)$ is defined as the inverse of complex permittivity $\varepsilon^*(\omega)$.

$$\varepsilon^*(\omega) = \varepsilon'(\omega) - i\varepsilon''(\omega) = \frac{1}{M^*} = \frac{\sigma^*}{i\omega\varepsilon_o}$$

In the modulus representation, the conductivity contribution is converted into a peak. Also, several dielectric active processes can be detected as peaks in the modulus representation. The modulus representation of the blank showed the conductivity as a peak in Fig. 8.17. The conductivity peak shifted towards higher frequencies with increasing temperature from $50 - 70°C$, whereas at higher temperatures the peaks are coinciding (shown in the inset of the figure). The temperature dependence of the frequency was represented in Fig. 8.18 at maximum peak positions ν_{max} following, once again, the VFT-equation. Also, the maximum value of the imaginary part M''_{max} of the modulus slightly increased with increasing temperature (ranging between $0.02 - 0.03$) as shown in the inset of Fig. 8.18 [61]. The principal parameters derived from conductivity/frequency plot are DC conductivity (σ_{dc}) and the critical frequency (ν_c). They are related roughly to the microstructure of the blend

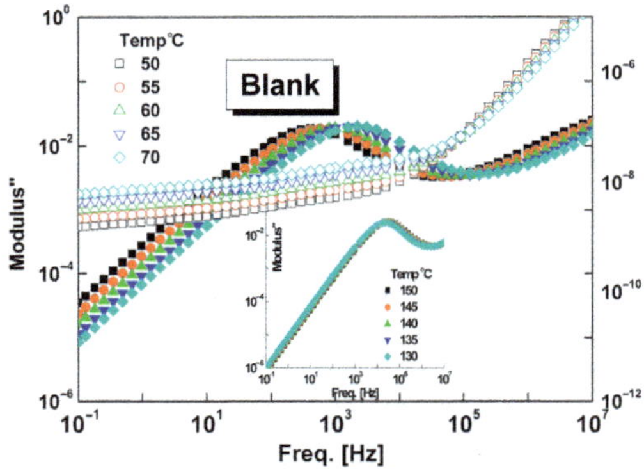

Figure 8.17 The imaginary part of the electric modulus as function of NBR/SBR blend frequency at temperatures ranging between 50–70°C.

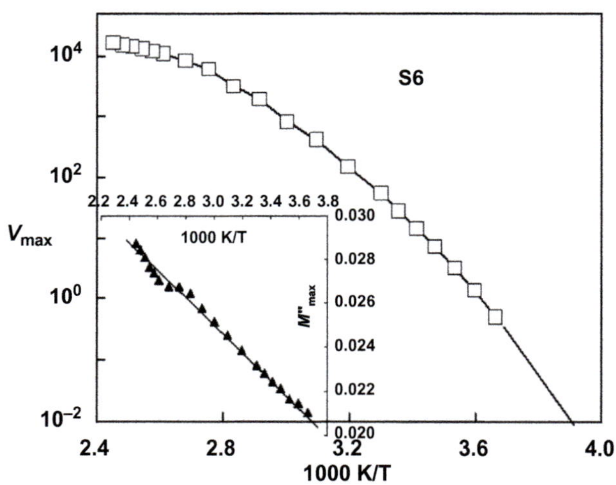

Figure 8.18 The temperature dependence of the frequency at the maximum peak position of TS1/SBR/NBR (S6). The inset shows the effect of temperature on the maximum value of the imaginary modulus M''_{max}.

and the microdynamics of the charge carriers. The additives to the rubber during the vulcanization process, especially carbon black, are responsible for producing the free charge carriers in the tested samples. It is clear from the experimental results that no enhancement of conductivity was detected due to the filler, which reflects the absence of any conductive paths inside the rubber blends.

After all, it can be concluded that the mechanism of cooperation between segmental motion, conductivity, and glass transition temperature is still questionable.

A schematic sketch which explains the idea of the whole chapter is shown in Fig. 8.19.

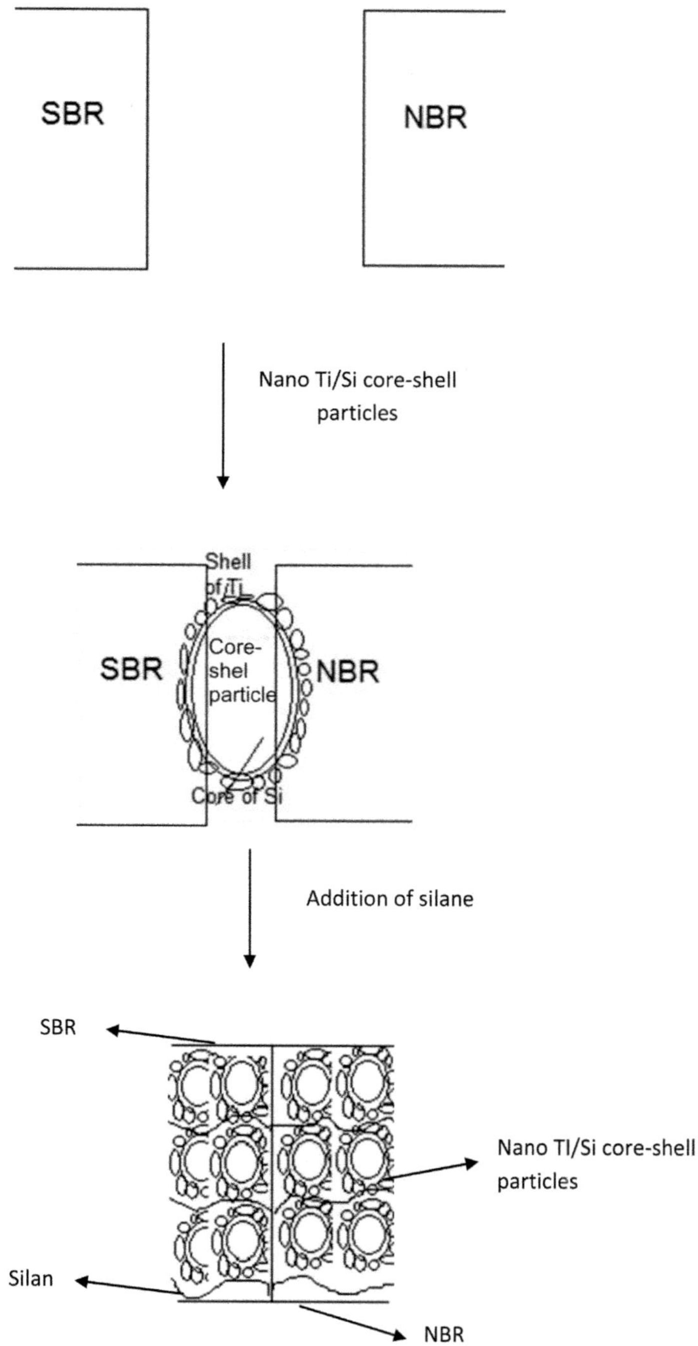

Figure 8.19 Schematic diagram explaining the compatibility caused by the new pigments in the presence of a silane coupling agent.

8.5 Conclusions

Interaction of nano-TiO_2/SiO_2/siliane coupling agent on NBR/SBR blends showed that a reduction in optimum cure time was observed for 12phr TS1 and also for 6phr T_{S2}; this clearly indicated that the efficiency of the pigment as reinforcing filler increased the rate of curing in blends containing 12phr TS1. As the pigment loading increases, mechanical properties such as tensile strength, modulus, hardness, and Young's modulus were found to be also increased, while elongation at break was decreased. The improvement in mechanical properties was supported by the increased value of cross-link density. Also from the stress−strain measurements, it was concluded that mechanical properties of NBR/SBR were improved by addition of 12 and 6phr of nano-TS1 and nano-T_{S2} in the presence of a silane coupling agent as a compatibilizer owing to the good compatibility between NBR and SBR. TS1 and T_{S2} also affected the values of C_1 and C_2 as calculated using the Mooney-Rivlin equation as well as Young's modulus.

Also, three processes were investigated using BDS which are conductivity, interfacial polarization, and segmental motions. The three processes followed the VFT empirical equation assuring that there is some kind of correlation between them. In the case of a rubber blend loaded with 12phr TS1 (S9) there was a further absorption peak at higher frequency in agreement with the mechanical properties, which revealed that this sample exhibited the highest absorption energy of the whole group.

References

[1] El-Sabbagh SH, Tawfik ME, Helaly FM. Preparation and characterization of blends containing natural and some synthetic rubbers with synthesized aromatic polyester. J Appl Polym Sci 2008;109:2823−35.

[2] Abou-Helal MO, El-Sabbagh SH. A study on the compatibility of NR/EPDM blends using electrical and mechanical techniques. J Elasto Plast 2005;37:319−46.

[3] Essawy AH, Tawfik ME, Khalil AM, El-Sabbagh SH. Systematic organophilization of montmorillonite: the impact thereof on the rheometric and mechanical characteristics of NBR and SBR based nanocomposites. J Polym Eng Sci 2014;54:942−8.

[4] Essawy HA, Tawfik ME, El-Sabbagh SH. Rubber nanocomposites based on compatibilized NBR/SBR blends using a series of amphiphilic montmorillonites. J Elasto Plast 2014;46:113−31.

[5] Essawy HA, Khalil AM, Tawfik ME, El-Sabbagh SH. Compatibilization of NBR/SBR blends using amphiphilic montmorillonites: a dynamic mechanical thermal study. J Elasto Plast 2014;46:514−52.

[6] Noriman NZ, Ismail H, Rashid AA. Curing characteristics, mechanical and morphological properties of styrene butadiene rubber/virgin acrylonitrile- butadiene rubber (SBR/vNBR) and styrene butadiene rubber/recycled acrylonitrile-butadiene rubber (SBR/NBRr) blends. Polym Plast Technol Eng 2008;47:1016−102.

[7] Noriman NZ, Ismail H, Rashid AA. Characterization of styrene butadiene rubber/recycled acrylonitrile-butadiene rubber (SBR/NBRr) blends: the effects of epoxidized natural rubber (ENR-50) as a compatibilizer. Polym Test 2010;29:200−8.

[8] Nguyen VG, Linsmeier Ch, Jang MK, Kim MY, Nah JW, Souzth K, et al. Preparation of polypropylene/TiO_2 nanocomposites using modified TiO_2 nanoparticles. KGK 2012;65:63−7.
[9] Ahmed NM, Tawfik H. The role of silica fume pigments in corrosion protection of steel surfaces. Chapter 4, corrosion protection, ISBN 979-953-307-263-9, published: January 25, 2012 under CCBY 3.0 license, in subject Materials Science. Available from: http://dx.doi.org/10.5772/1843, INTECH Ltd.
[10] Ahmed NM, Tawfik H. Study on the anticorrosive behavior of new hygiene structured pigment based on waste core and nano shell in alkyd paints. Chapter 15, developments in corrosion protection, ISBN 978-953-51-1223-5, Published: February 20, 2014 under CCBY 3.0 license, in subject Materials Science. Available from: http://dx.doi.org/10.5772/57243, INTECH Ltd.
[11] Ahmed NM, Selim MM. Tailored ferrites-kaolin anticorrosive hybrid pigments in solvent-based paints for protection of cold-rolled steel. Pig Res Technol 2010;39:101−11.
[12] Ward AA, Bishai AM, Hanna FF, Yehia AA, Stoll B, Von Soden W, et al. Dielectric behaviour of silica filled vinyl solution-SBR in the presence of silane coupling agent. KGK 2006;654−8.
[13] Poh BT, Ng CC. Effect of silane coupling agents on the mooney scorch time of silica-filled natural rubber compound. Euro Polym J 1998;34:975−9.
[14] Ten Brinke JW, Debnath SC, Reuvekamp LAEM, Noordermee JWM. Mechanistic aspects of the role of coupling agents in silica−rubber composites. Compos Sci Technol 2003;63:1165−74.
[15] Heinrich G, Vilgis T. Why silica technology needs S-SBR in high performance tires? The physics of confined polymers in filled rubbers. KGK 2008;61:368−76.
[16] Darwish WM, Abd El-Ghaffar MA, Turky GM, Darwish WM, Abd El-Ghaffar MA, Turky GM. Synthesis and electric modulus formalism of novel metal-phthalocyanine bridged polymers. J Inorg Organomet Polym Mater 2014;24:858−64.
[17] Darwish WM, Turky GM. Dielectric response in the first silicon phthalocyanine network polymer. J Inorg Organomet Polym Mater 2014;24:347−54.
[18] Ismaila MN, Turky GM. Effect of fillers and vulcanizing systems on the physico-mechanical and electrical properties of EPDM vulcanizates. J Polym Plast Technol Eng 2001;40.
[19] Füllbrandt M, Purv J, Purohit PJ, Schönhals A. Macromolecules 2013;46:4626−32.
[20] Tsangaris GM, Psarras GC, Kouloumbi N. Electric modulus and interfacial polarization incomposite polymeric systems. J Mater Sci 1998;33:2027−37.
[21] Flory PJ, Rehner J. Statistical mechanics of cross-linked polymer network, I: rubber like elasticity. J Chem Phys 1943;11:512−20.
[22] Khalaf AI, Ward AA, Abd El-Kader AE, El-Sabbagh SH. Effect of selected vegetable oils on the properties of acrylonitrile- butadiene rubber vulcanizates. Polimery 2015;60:905−18.
[23] Rivlin RS, Thomas AG. Rupture of rubber. I. Characteristic energy for tearing. J Polym Sci 1953;10:291. Available from: http://dx.doi.org/10.1002/pol.1953.120100303.
[24] Ahmed NM, El-Sabbagh SH. The influence of hybrid phosphate−alumina pigments on properties of ethylene-propylene-diene rubber composites. Mater Des 2011;32: 303−21.
[25] El-Sabbagh SH, Mahmoud DS, Zawrah MF, Ahmed NM, Sabaa MW. Investigation on the properties of rubber composites containing modified clay. Pigm Res Technol 2015; 44:131−42.

[26] Manna AK, De PP, Tripathy DK, De SK, Peiffer DG. Bonding between precipitated silica and epoxidized natural rubber in the presence of silane coupling agent. J Appl Polym Sci 1999;74:389–98.
[27] Das A, Naskar N, Kumar Basu D. Thiophosphoryl disulfides as crosslinking agents for chloroprene rubbe. J Appl Polym Sci 2004;91:1913–19.
[28] Surya I, Ismail H, Azura AR. The comparison of alkanolamide and silane coupling agent on the properties of silica-filled natural rubber (SMR-L) compounds. Polym Test 2014;40:24–32.
[29] Ismail H, Nordin R, Noor A. Cure characteristics, tensile properties and swelling behaviour of recycled rubber powder-filled natural rubber compounds. Polym Test 2002;21:565–9.
[30] Arroyo M, Lopez-Monchado M, Herrero B. Organo-montmorillonite as substitute of carbon black in natural rubber compounds. Polymer 2003;44:2447–53.
[31] Surya I, Ismail H, Azura AR. Alkanolamide as an accelerator, filler-dispersant and a plasticizer in silica-filled natural rubber compounds. Polym Test 2013;32:1313–21.
[32] Kraus G. John Wiley & Sons Inc Reinforcement of elastomers. New York, NY: Intrescience; 1965
[33] Polmanteer K, Lentz C. Reinforcement studies-effect of silica structure on properties and crosslink density. Rubb Chem Technol 1975;48:795–809.
[34] Long H, editor. Basic compounding and processing of rubber, rubber division. Ohio, USA: American Chemical Society Inc. The University of Akron; 1985.
[35] Stephens HL. The compounding and vulcanization of rubber. In: Morton M, editor. Rubber technology. New York, NY: Van Nostrand Reinhold; 1987. p. 20–58.
[36] Wolff S, Wang MJ. Filler–elastomer interactions. Part IV. The effect of the surface energies of fillers on elastomer reinforcement. Rubb Chem Technol 1992;65:329–42.
[37] El-Sabbagh SH, Ahmed NA, Ward AA. Effect of kaolin–metal oxides core–shell pigments on the properties of styrene-butadiene rubber composites. Mater Des 2012;40:343–55.
[38] Lee B. Reinforcement of uncured and cured rubber composites and its relationship to dispersive mixing-an interpretation of cure meter rheographs of carbon black loaded SBR and cis-polybutadiene compounds. Rubb Chem Technol 1979;52:1019–29.
[39] Pal P, De S. Effect of reinforcing silica on vulcanization, network structure, and technical properties of natural rubber. Rubb Chem Technol 1982;55:1370–88.
[40] Schuater RH. The challenge a head-new polymer filler systems. Rubber World 2001;224:24–8.
[41] Zafarmehrabian R, Gangali ST, Ghoreishy MHR, Davallu M. The effects of silica/carbon black ratio on the dynamic properties of the tread compounds in truck tires. E-J Chem 2012;9:1102–12.
[42] Tangudom P, Thongsang S, Sombatsompop N. Cure and mechanical properties and abrasive wear behavior of natural rubber, styrene–butadiene rubber and their blends reinforced with silica hybrid fillers. Mater Des 2014;53:856–64.
[43] Helaly M, El Sabbagh SH, El Kinawy OS, El Sawy SM. Effect of synthesized zinc stearate on the properties of natural rubber vulcanizates in the absence and presence of some filler. Mater Des 2011;32:2835–43.
[44] Boonstra BB. In: Blow CM, editor. Rubber technology and manufacture. London: Butterworth and Co; 1971. p. 229.
[45] My AF, Mohd SR, Ishak ZA, Omar AK. Rice husk ash as a filler in polypropylene: effect of wax and silane coupling agents. Plastic Rubber Compos Appl 1994; 21:225–35.

[46] Wolff S, Wang MJ. Rubber Chem Technol 1996;69:325.
[47] Ahmad SJ, Yudong H, Li W. Effect of the clay exfoliation on structure and physical properties. Iran Polym J 2004;13:415–25.
[48] Treloar LRG. The physics of rubber elasticity. 3rd ed. Oxford: Clarendon Press; 1976.
[49] Krihbaum WR, Roe R-J. Survey of the theory of rubber like elasticity. Rubber Chem Technol 1965;38:1039–69.
[50] Ferry JD. Viscoelastic properties of polymers. 3rd ed. New York, NY: John Wiley & Sons Inc; 1980.
[51] Rivlin RS, Saunders Phil Trans DW. R Soc Lond 1951;A 243:251.
[52] Hamza SS, El-Sabbagh SH, Shokr F. Elastic behavior of NR/IIR rubber blend loaded with different compatibilizers. Int J Polym Mater 2008;57:203–15.
[53] El-Sabbagh SH, Ahmed NM, Selim MM. Preparation and characteristion of high performance rubber vulcanizates loaded with modified aluminium rubber vulcanizates loaded with modified aluminium oxide. Pig Res Technol 2006;35:119–30.
[54] Ward A, El-Sabbagh SH, Abd El-Gaffar MA. Studies on the dielectric and physical properties of phosphate pigment/rubber composites. KGK 2013;40:29–40.
[55] Ramesan MT. Effect of silica on uncompatibilized and compatibilized styrene butadiene rubber and nitrile rubber blends. Int J Polym Mater 2011;60:1130–46.
[56] Ramesan MT, Rosamma A, Khanh NV. Studies on the cure and mechanical properties of blends of natural rubber with dichlorocarbene modified styrene-butadiene rubber and chloroprene rubber. React. Funct. Polym 2005;62:41–50.
[57] Choudhary S. Effect of mechanical and electromagnetic irradiation on dielectric properties of solid polymer electrolytes. IJPAP 2015;53:110–18.
[58] Ahmed K, Sirajuddin Nizami S, Riza NZ. Reinforcement of natural rubber hybride composites based on marable sludge/silica and marable sludge/rice husk derived silica. J Adv Res 2014;5:165–73.
[59] Ismail MN, Turky GM. Effect of fillers and vulcanizing systems on the physico-mechanical and electrical properties of EPDM vulcanizates. Polym Plast Technol Eng 2001;40:1.
[60] Salgueiro W, Marzocca A, Somoza A, Consolati G, Gerveny S, Quasso F, et al. Dependence of the network structure of cured styrene–butadiene rubber on the sulphur content. Polymer 2004;45(23):6037–44.
[61] Ortiz-Serna P, Carsí M, Redondo-Foj M, Sanchis MJ. Electrical conductivity of natural rubber–cellulose II nanocomposites. J Non-Cryst Solids 2014;405:180–7.

Rubber nanocomposites with metal oxides as nanofillers

9

B.G. Soares
Universidade Federal do Rio de Janeiro, Rio de Janeiro, Brazil

9.1 Introduction

Rubbers (elastomers) are soft thermosetting polymeric materials which present subambient glass-rubber transition temperature, excellent elastic properties and flexibility. They have found important applications in different fields, such as automobile, aerospace, biomedical, etc. The incorporation of rigid fillers in elastomer compounds has been a common practice for improving mechanical properties (reinforcement), as well as thermal, dynamic-mechanical, electrical, and magnetic properties, among others [1]. The reinforcement effect of the rigid fillers may be attributed to the filler–matrix interactions, which depend on the structure of both filler and matrix components. Traditionally carbon black and silica are the most popular reinforcing agents in the rubber industry. Because of the low affinity between silica and apolar rubbers, a small amount of functionalized silane is usually used as a coupling agent for improving the silica–rubber interactions and enhancing the filler distribution [2].

Metal oxides also play a prominent position in the rubber industry, with zinc oxide (ZnO) being considered one of the most important. ZnO, together with stearic acid, is recognized as an efficient activator during the accelerated sulfur vulcanization of unsaturated rubber [3,4]. ZnO and magnesium oxide (MgO) are also successfully employed as curing agents for carboxylated-based rubbers [5,6] and polychloroprene (CR) [7] and may act as a reactive compatibilizing agent for rubber blends containing carboxylic groups in both components [8–11] or between carboxylated nitrile rubber and chlorosulfonated polyethylene [12]. Besides influencing the curing process of rubbers, such divalent metal oxides can be used as reinforcing fillers [7]. Some metal oxides also impart specific properties to the rubber vulcanizates. For example, magnesium hydroxide, $Mg(OH)_2$, is known as an efficient flame retardant [13]; aluminum oxide, Al_2O_3, enhances the dielectric properties of silicone rubbers and makes them important materials for outdoor high voltage insulation application [14]; iron oxide, Fe_3O_4, and the corresponding ferrites are very important for the development of microwave absorbers materials [15]; titanium dioxide (TiO_2) acts as a photocatalyst [16,17], imparts good antimicrobial activity [17], and also enhances the dielectric properties of rubber materials [18]; and so on.

In all filled rubbery materials, the filler distribution and filler particle size are important parameters for attaining outstanding properties. In this context,

Progress in Rubber Nanocomposites. DOI: http://dx.doi.org/10.1016/B978-0-08-100409-8.00009-7

nanometric particles have been attracting considerable attention due to the excellent properties that can be achieved by using lower amounts of filler compared with micrometric particles [19]. The success of nanoparticles is mainly devoted to their high surface area, which results in a higher interfacial area and increasing filler–matrix interaction.

The field of nanocomposite science and technology has experienced a growing interest and has been the subject of several books and reviews [20–22]. Different nanometric particles have been compounded with rubbers for different purposes. Most of them involve layered silicate, carbon-based nanofillers, or silica. For example, layered silicates have been used to improve reinforcement of rubbers and also to impart barrier properties [23]. Carbon-based nanofillers, including carbon nanotubes, carbon nanofibers, and graphene have been compounded with rubber to impart special characteristics such as better thermal, electrical, and mechanical properties [24–28]. Also, nanosilica and the corresponding organically functionalized silica constitute important materials for improving tear strength, abrasion resistance, and dynamic-mechanical properties [29,30]. The present chapter deals with the use of metal oxide nanoparticles on the development of rubber-based nanocomposites. As will be discussed in the following sections, the metal oxide-based nanocomposites find various important applications.

9.2 Preparation of metal oxide nanoparticles

The main metal oxides used to compound with rubber materials include ZnO, iron oxide, zirconium oxide, titanium oxide (TiO_2), Al_2O_3, and MgO. Different methods have been reported for the preparation of metal oxide nanoparticles and depend on the nature of the metal oxide. The chemical approach has been of particular interest because of the mild conditions usually required for their preparation and the possibility of attaining different morphologies (spherical, nanorods, needles, nanowires, nanoplates, etc.) by changing the synthetic parameters. Moreover, some of these techniques can be adapted for preparing organic–inorganic hybrid materials, which is very interesting for promoting the compatibilization with polymer matrices. Some of the most versatile techniques include the precipitation process from the corresponding salt solutions and also the sol–gel synthesis.

The sol–gel process is generally based on the hydrolysis and condensation of a metal alkoxide precursor in acidic or basic conditions [31,32]. This method is very popular for silica preparation but is also employed for the development of some other metal oxide nanoparticles. The great advantage of this method is the mild conditions, which are fundamental for preparing inorganic nanoparticles bearing organic functional groups (organic–inorganic hybrid materials). By using the sol–gel process, some functional groups can also be introduced during the nanoparticle formation. An interesting example is the sol–gel reaction of zirconium alkoxide or titanium alkoxide in the presence of acrylic or methacrylic acid, giving rise to oxometal clusters bearing vinyl groups [33,34]. As will be discussed later, the functionalization of metal oxide nanoparticles already formed or during the nanoparticle formation will improve the nanoparticle dispersion, providing additional mechanical performance. Such materials are very interesting for the development of polymeric nanocomposites as they provide better interaction with the polymeric matrix. Moreover, the sol–gel process may also be used to prepare directly the nanoparticles in the presence of a polymer matrix, through the in situ hydrolysis/condensation, resulting in better nanoparticle dispersion. Some examples of metal oxide nanoparticle preparation with potential application on the development of rubber-based nanocomposites will be briefly described in this section.

9.2.1 Zinc oxide nanoparticle

ZnO is one of the most important metal oxides and presents interesting physicomechanical properties such as electroluminescence [35], cathodoluminescence [36], photoluminescence [37], piezoelectricity [38], and magnetic properties [39], which are important for applications in several fields as sensors [40,41], solar cells [42], and optoelectronic devices [43]. It is also one of the most important components in rubber formulations as will be considered in the following section.

ZnO nanoparticles have been prepared by different procedures, including the precipitation process from solution, sol–gel approach, hydrothermal methods, etc. Wang [44] published a review dealing with the preparation of ZnO nanostructures by the solid-vapor phase technique. Another nice review describing different routes for preparing ZnO nano- and microparticles was recently published by Kołodziejczak-Radzimska and Jesionowski [45]. For both precipitation and sol–gel processes, usually zinc acetate, zinc nitrate or zinc chloride are used as precursors. Some examples are summarized, as follows.

The precipitation process is considered one of the simplest methods for synthesizing ZnO nanoparticles in the laboratory and consists of the slow addition of the zinc salt precursor solution to a basic solution constituted by $(NH_4)_2CO_3$ [46–48], NH_4OH [49,50], and NaOH [51], among others. Some procedures also involve a calcination step to obtain the nanoparticles. However, other methodologies are described without the need of the calcination step. Wu et al. [51] describe a novel chemical route to prepare nano-ZnO particles from a $Zn(NO_3)_2$/NaOH system without performing the calcination step. According to the authors, nanorod morphology has been observed for systems prepared at 70°C. They proposed a reaction scheme with the formation of $Zn(OH)_2$ colloid as the intermediary, as illustrated:

$$Zn(NO_3)_2 \cdot 6H_2O + 2NaOH \rightarrow Zn(OH)_2 + 2NaNO_3 \cdot 6H_2O$$

$$Zn(OH)_2 + 2H_2O \rightarrow Zn^{2+} + 2OH^- + 2H_2O \rightarrow Zn(OH)_4^{2-} + 2H^+$$

$$Zn(OH)_4^{2-} \rightarrow ZnO + H_2O + 2OH^-$$

Lanje et al. [52] have used similar reactants [$Zn(NO_3)_2$ and NaOH] and starch as a stabilizing agent to reduce the particles agglomeration This procedure was performed under mild conditions and particles in spherical shape with an average size of 40 nm were obtained.

Sol−gel is also an easy process to obtain ZnO nanoparticles. Farhadi-Khouzani et al. [53] employed aqueous solution of $Zn(NO_3)_2 \cdot 6H_2O$ and citric acid in the presence of different amounts of ethylene glycol. After calcination of the white xerogel precursors, nanostructured ZnO species were obtained, whose morphology from nanoparticles to nanorods, strongly depends on the citric acid/ethylene glycol molar ratio.

9.2.2 Iron oxide nanoparticles

Iron oxide particles of nanometric size such as magnetite (Fe_3O_4) and/or maghemite (Fe_2O_3) are of great interest due to their magnetic, catalytic, and conducting properties resulting in important applications in the biological and biochemical fields, magnetic separation for industrial water treatment, and in the development of materials for electromagnetic interference shielding. Thus, many techniques for syntheses of these oxides have been reported in the literature, with the goal of controlling size, geometry, and size distribution of the nanoparticles [54]. The choice of the synthesis methodology will depend on the application, morphology, and size desired, and the chemical environment in which the particles will be employed. Among the chemical methods, coprecipitation and thermal decomposition are the two more studied methods to prepare iron oxide nanoparticles.

The thermal decomposition of iron complexes as precursors, such as ferric acetylacetonates, iron pentacarbonyl, and iron fatty acid salts, is performed in the presence of an organic solvent with a high boiling point and a surfactant such as oleic acid and hexadecylamine, and has been accepted as an effective method to synthesize iron oxide nanoparticles with small and controllable particle size [55−60]. The reducing agent is usually a diol as 1,2-hexanediol [61], or phenol [58] The disadvantage of this method is the use of the expensive reagents.

The coprecipitation technique is considered an easy, reproducible, and cheap route of synthesis and consists of preparing a mixture of ferrous and ferric salts in aqueous medium followed by a treatment with alkali solution (usually NaOH or $NH_3 \cdot H_2O$) until pH 8−14 to complete the precipitation [62−64]. As an alternative for wet chemical preparation methods, Herea et al. [65] have developed a procedure based on a salt-assisted solid-state chemical reaction, which involves the mechanical mixing of $FeCl_2$ and $FeCl_3$, and the treatment with NaOH as the precipitating agent. Another interesting variation of the coprecipitation method was developed by Zhu et al. [66]. The authors have performed the partial oxidation of $FeSO_4 \cdot 7H_2O$ in the presence of NaOH, $NH_3 \cdot H_2O$, or mixed alkali solution in the air without using nitrogen atmosphere and observed that the morphology of the iron oxide nanoparticles depended on the nature of the alkali used. In another approach, Qu et al. [67] have used ferric ions together with Na_2SO_3. Ferrous ions were formed in situ from ferric ions by partial reduction with Na_2SO_3. After mixing both reagents, the aqueous ammonia solution was added.

In order to provide a better hydrophobic surface and improve the dispersion of the iron oxide nanoparticles within an organic medium, as well as the compatibilization with polymer matrices, the iron nanoparticles are usually treated with oleic acid [62,68], as well as sodium oleate [66].

Magnetite may also be stabilized by forming a surface silica layer by the sol–gel process. Based on the same principle, another interesting process is the use of organosilanes (such as aminosilanes) for stabilizing the magnetic nanoparticles [69]. This process allows the functionalization with organic groups, which can interact with other materials, e.g., and provides better compatibility with polymers and elastomers.

9.2.3 Zirconium oxide nanoparticle

Zirconium dioxide (ZrO_2), also called zirconia, is an important transition-metal oxide, which finds applications in different fields as solid oxide for fuel cells, oxygen sensors, electrooptical materials, etc. [70,71], due to its interesting properties including high refractive index, excellent chemical resistance, thermal properties, ionic conductivity, and so on. There are different procedures for preparing nano-ZrO_2 particles. The most studied include hydrothermal and sol–gel processes.

The hydrothermal process consists of a chemical reaction of zirconium salts in an aqueous or organic–aqueous solution in the presence of an alkali or acid, under the action of temperature and pressure [72]. Using this process, Murase and Kato [73] have prepared zirconia whiskers from a mixture of zirconium tetrahydroxide and ZrO_2 particles, which were used as seed crystals. This mixture has been hydrothermally treated with sulfuric acid solution at 200°C in a closed reactor, giving rise to nano-ZrO_2 particles. The $Zr(OH)_4$ was prepared by reacting the $ZrClO_2$ precursor with ammonia in aqueous solution.

Zirconium alkoxides have been also employed as precursors for the synthesis of nano-ZrO_2 particles. Some reports describe the synthesis in the presence of *n*-hexanol, followed by a treatment with hydrochloric acid (HCl) solution [74,75]. The obtained solution was transferred to a Teflon crucible, placed in an autoclave and heated at 250°C for 7 h, to obtain ZrO_2 powder with a small particle size, in the range of 6–12 nm. The scheme for the hydrothermal synthesis is illustrated in Fig. 9.1. An adaptation of this technique was developed by Garnweitner et al. [76] who employed anhydrous benzyl alcohol as the solvent.

The sol–gel process has also been successfully employed for the preparation of nano-sized ZrO_2. An example consists of a treatment of $ZrOCl_2 \cdot 8H_2O$ with aqueous ammonia solution under stirring [77,78]. This step produces $Zr(OH)_4$ as a precipitate, which is dried and calcinated, giving rise to ZrO_2 particles with around 50 nm diameter and narrow size distribution. Because of the mild conditions involving the sol–gel process, hybrid organic–inorganic particles can be easily prepared by this technique. In fact, some reports in the literature describe some procedures for obtaining zirconium oxocluster bearing organic moiety by performing the hydrolysis/condensation of the zirconium tetraalkoxide (the precursor) in the

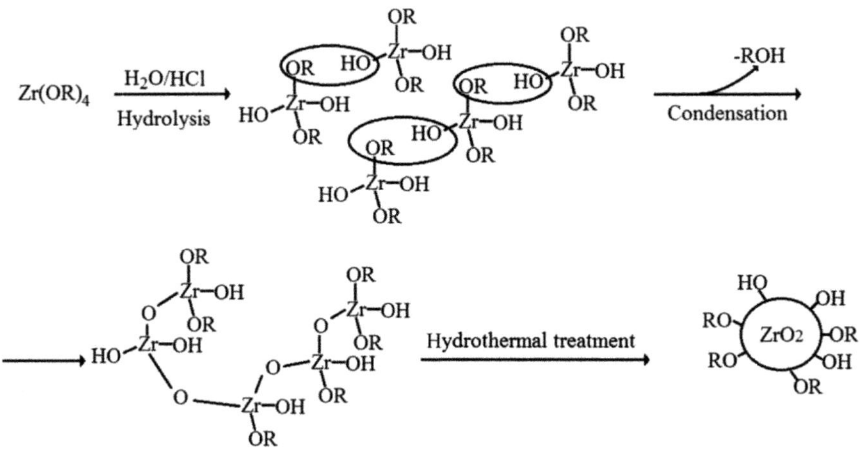

Figure 9.1 Scheme for the synthesis of ZrO_2 from zirconium alkoxide.

presence of carboxylic acid, which forms a bidentate carboxylate ligand linked to the zirconium [33,79]. By using methacrylic or acrylic acid, a polymerizable ZrO_2-based oxocluster, which acts as a building block system, is produced.

9.2.4 Titanium oxide nanoparticle

TiO_2 nanoparticles can be obtained in different crystalline forms as well as in an amorphous state. According to the literature, the highest photoactivity can be achieved for TiO_2 nanoparticles in the anatase crystalline form, combined with some amount of the rutile crystalline form [80,81]. The main reported procedures employ titanium inorganic salts or titanium alkoxide as precursors, which are hydrolyzed and condensed to form TiO_2 nanoparticles. Most of the works also employ ligands, surfactant, and peptization treatment to prevent the formation of agglomerates as well as to control the particle growth.

The sol–gel process is most often employed for this system. By using this method, titanium tetraisopropoxide was treated with aqueous HCl in toluene [82] or 1,4-dioxane [83] to form the TiO_2 nanoparticles. The authors have employed polystyrene–polyethylene oxide diblock copolymer as the template. Using a similar procedure, Kuang et al. [84] have used titanium tetrabutoxy dissolved in ethanol, in the presence of polyethylene glycol as surfactant. After stirring at room temperature and neutralized, the formed gel was calcinated at different temperatures. Arantes et al. [85] have also adapted the sol–gel process to obtain an anatase TiO_2 nanoparticle without the calcination process. They have also employed titanium tetraisopropoxide which was dissolved with diphenyl ether in the presence of oleic acid. After treating the solution with hydrogen peroxide and keeping the system at 120°C for 12 h, the anatase nanoparticle was obtained.

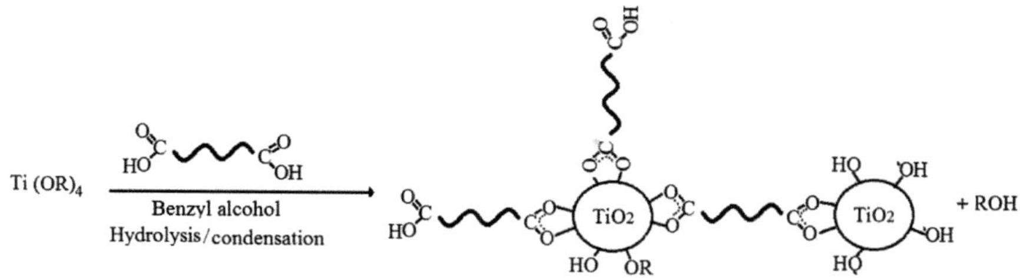

Figure 9.2 Scheme for the formation of the bidentate ligand between TiO_2 and carboxylated group.

The solvothermal approach is also an interesting strategy for obtaining TiO_2 nanoparticles with high crystallinity. In fact, Shan and Demopoulos [86] and Niederberger et al. [87] have obtaining anatase TiO_2 nanoparticles by controlling the hydrolysis of titanium tetrachloride in ethylene glycol or benzyl alcohol, respectively. Both systems were performed under pressure in an autoclave.

Spadavecchia et al. [88] reported a nice strategy based on the use of diacid-terminated polyethylene glycol in the sol–gel chemistry of titanium isopropoxide. They used benzyl alcohol as the solvent. The carboxylic group in the functionalized poly(ethylene glycol) is able to form a bidentade ligand with the formed TiO_2 nanoparticle, as illustrated in Fig. 9.2.

9.2.5 Aluminum oxide nanoparticle

Al_2O_3 or alumina constitutes a very important ceramic with several polymorphic species, with α-alumina being thermodynamically stable, which is usually formed by the calcination of aluminum hydroxide [89]. Different procedures are described in the literature to produce Al_2O_3 nanoparticles, including the well-known sol–gel and hydrothermal processes. In all these procedures, the calcination step is fundamental for obtaining pure Al_2O_3 with high crystallinity. A few examples are briefly discussed in this section.

Li et al. [90] have used an aqueous sol–gel process involving a mixture of $Al(NO_3)_3$ and citric acid. After treating the formed gel at 1000°C, a single α-phase of alumina was obtained. Hung-Low et al. [91] have employed $AlCl_3 \cdot 6H_2O$, which was treated with glycidol and dextran solution. After forming the gel, the material was calcinated at 800°C in order to increase the crystallinity of the formed Al_2O_3 nanoparticles. The presence of dextran contributes to the increase in surface area. Park et al. [92] have performed the reaction of aluminum isopropoxide with water, where the formed $Al(OH)_3$ was peptized by adding an aqueous solution of HCl and stirring for 24 h at 85°C. The process involves the transformation of Al[OCH $(CH_3)_2$] into $Al(OH)_3$ which is dehydrated to form the oxo-hydroxide aluminum intermediate (AlO(OH)), which constitutes the precursor for the sol–gel process, according to different authors [93]. The sol–gel reaction consisted of the transformation of (AlO(OH)) particle into an Al_2O_3 network via the dehydration process. To avoid or minimize the aggregation of the particles, bis-2-ethylhexyl sulfosuccinate sodium salt Na(AOT) was employed as surfactant. After calcination, nearly uniform α-Al_2O_3 nanoparticles were obtained.

$$2Al(OC_3H_7)_3 + 3H_2O \xrightarrow{Na(AOT)} \alpha\text{-}Al_2O_3 + 6C_3H_7OH$$

The sol–gel approach was also evaluated in systems constituted by $Al(NO_3)_3$ aqueous solution dispersed in cyclohexane as the oil phase and poly(oxyethylene) nonylphenyl ether as the nonionic surfactant. By adding aqueous ammonia to neutralize the system, $Al(OH)_3$ precursor was obtained. The material was calcinated to obtain Al_2O_3 [94].

The hydrothermal technique has been described as a very interesting approach for preparing Al_2O_3 nanoparticles. Qu et al. [95] have employed $Al(NO_3)_3 \cdot 9H_2O$ and sodium dodecyl sulfonate as anionic surfactant. After stirring for several minutes at 50°C, aqueous ammonia was added and the suspension was transferred into an autoclave and heated at 120°C for 90 h. The obtained solid was recovered and calcinated at 500–800°C.

Lee et al. [96] have used aluminum isopropoxide in the presence of oxalic acid as a chelating agent. The Al-oxalic complex was transferred to an autoclave and heated at 200°C for 24 h, to obtain a white powder corresponding to $Al(OH)_3$, which was calcinated at 1200°C.

9.2.6 Magnesium oxide nanoparticle

MgO nanoparticles have been prepared by different methods, including the sol–gel process [97–100] and the aqueous wet chemical approach [101,102]. The sol–gel process was performed by using magnesium alkoxide as the precursor, which was treated with ethanol and hydrolyzed with various aqueous inorganic or organic acids and bases [100,103]. The hydrolysis resulted on the production of Mg(OH)OR rather than $Mg(OH)_2$. The calcination step was important for obtaining MgO. Roy et al. [104] have used the sol–gel autocombustion procedure to prepare MgO nanoparticles. This technique involved the addition of citric acid in a solution of magnesium nitrate [$Mg(NO_3)_2 \cdot 6H_2O$]. A solution of ammonia was also added to adjust the pH value to about 4. After drying the solution the material was calcinated, thus obtaining nanoparticles with average size from 20 nm to 80 nm.

The wet chemical approach is a simple and cheap method for obtaining MgO nanoparticles. In this method, a magnesium nitrate solution is treated with ammonium hydroxide solution giving rise to a white precipitate, which after a calcination process, gives rise to the nano-sized MgO [101,102]. Also the decomposition of magnesium chelates and acetate has been described by Chhor et al. [105] as an economical method for preparing nanoscale MgO.

9.3 Preparation and processing of rubber nanocomposites containing different metal oxide nanoparticles

The development of rubber-based nanocomposites is difficult because of the high surface area of the nanoparticles and the incompatibility between particles and rubber matrices, which favor the agglomeration of the nanoparticles. To improve the particle dispersion and the interaction between particle and rubber, it is important to employ appropriate mixing conditions and modify the surface characteristics of the particle.

Melt mixing the filler inside the rubber matrix by physical–mechanical methods using two roll mixing mill or an internal chamber, known as a "top-down" approach, is very attractive from a technological point of view. ZnO, MgO, iron oxide, TiO_2, and some other metal oxide nanoparticles have been incorporated into rubber matrices using this technique. Sahoo et al. [106] have evaluated the dispersion degree of conventional rubber grade ZnO and nanoparticles in natural rubber (NR) vulcanizates using the laboratory two roll mill. As illustrated in Fig. 9.3 nanocomposites prepared with nano-ZnO (Fig. 9.3B) present a more uniform dispersion of the ZnO nanoparticles in comparison with that containing the conventional rubber grade ZnO (Fig. 9.3A).

Wang and Chen [107] have employed a double roller plasticator to compound the nano-ZnO into butadiene styrene rubber (SBR). They evaluated the effect of the ZnO particle size and the master batch technology, where 30 phr of nano-ZnO were first compounded with the ZnO particle size and the SBR. The transmission electron microscopy of the uncured systems containing different ZnO, shown in Fig. 9.4, reveal that the conventional ZnO is hardly dispersed in the rubber matrix (Fig. 9.4A), whereas the nano-ZnO presents better dispersion within the SBR. However, the best dispersion degree was achieved by using nano-ZnO prepared by the master batch technology in the mixing procedure (Fig. 9.4C), compared to

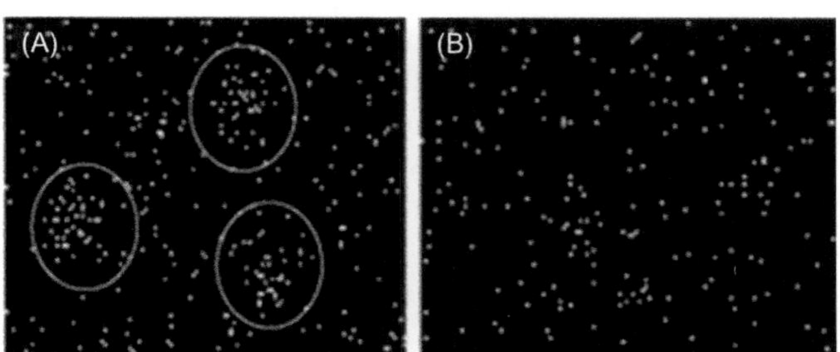

Figure 9.3 SEM/X-ray dot mapping of NR vulcanizate containing 5 phr of (A) conventional grade ZnO and (B) nano-ZnO.
Source: Sahoo S, Maiti M, Ganguly A, George JJ, Bhowmick AK. Effect of zinc oxide nanoparticles as cure activator on the properties of natural rubber and nitrile rubber. J Appl Polym Sci 2007b; 105: 2407–15. http://dx.doi.org/10.1002/app.26296

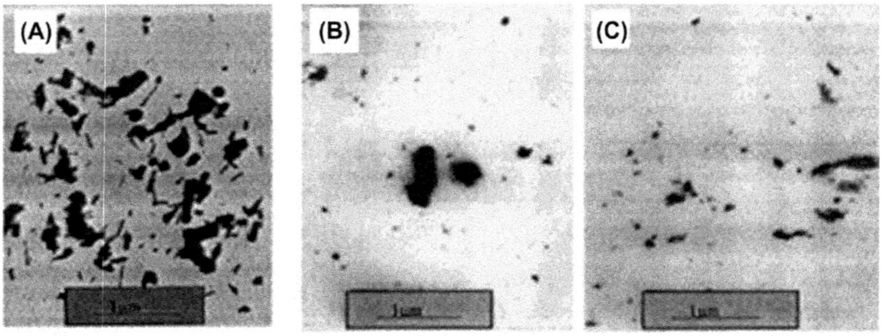

Figure 9.4 TEM micrographys of SBR–ZnO composites. (A) Conventional ZnO; (B) Nano-ZnO; (C) Nano-ZnO prepared by the master batch procedure.
Source: Wang J, Chen Y. Application of nano-Zinc oxide master batch in polybutadiene styrene rubber system. J Appl Polym Sci 2006; 101: 922–30. http://dx.doi.org/10.1002/app.23983.

the conventional way of mixing the nanoparticle, highlighting the importance of particle size and the methodology for dispersing such nanoparticles.

Although better nanoparticle dispersion is achieved by using the nanoparticles, a good dispersion is usually hard to achieve using the melt mixing procedure because of the high viscosity of the compound and the strong tendency to particle aggregation. In this sense, the use of organically functionalized silanes as coupling agents is considered one of the most popular approaches for introducing specific organic groups at the inorganic particles, which are able to interact with the organic polymeric matrix. This technique is usually employed to incorporate silica particles in a rubber matrix [29,108], but can also be efficient for improving the interaction between rubber matrix and other nanoparticles, such as some metal oxides [109]. This strategy was used by Wang et al. [110] to improve the dispersion of nano-ZnO in ethylene propylene diene monomer (EPDM) matrix. The authors compared the results of the composites prepared with untreated nano-ZnO with those involving the in situ modified nano-ZnO by adding bis-(3-three-ethoxy silylpropyl)-tetrasulfide as the silane coupling agent. They have used a high filler loading (180 phr) and observed the presence of many clusters in the untreated nano-ZnO filled composites, as illustrated in Fig. 9.5. The in situ modified nano-ZnO resulted in better dispersion with fewer agglomerates in the transmission electron microscopy (TEM) micrographs and the presence of some ZnO particles dispersed on the scale of less than 100 nm, as shown in Fig. 9.5B.

The colloidal route can be also considered a "top-down" approach and has been used by different researchers to develop homogeneous rubber-based nanocomposites. Arantes et al. prepared colloidal dispersions of TiO_2 [85] or ZrO_2 [111], in the presence of dodecyl benzenesulfonic acid as a surfactant, and added these dispersions to the commercial aqueous dispersion of SBR latex. The resulting mixture was homogenized at room temperature and poured in a Petri dish. After evaporating the water, self-sustained and highly homogeneous nanocomposite films were obtained.

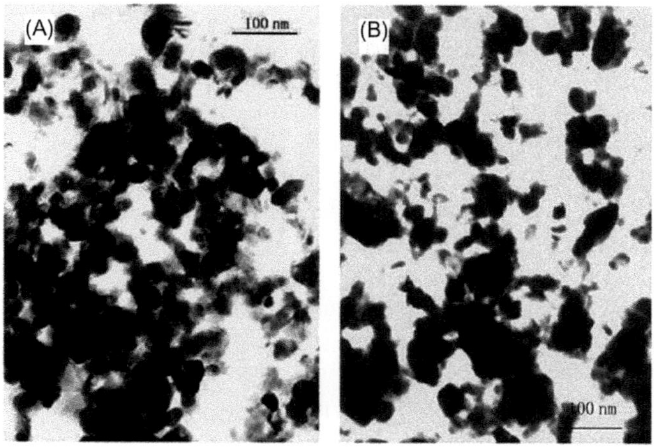

Figure 9.5 TEM micrographs of EPDM composites filled with 180 phr of untreated (A) and in situ modified (B) nano-ZnO particles.
Source: Wang Z, Lu Y, Liu J, Dang Z, Zhang L, Wang W. Preparation of nano-zinc oxide/ EPDM composites with both good thermal conductivity and mechanical properties. J Appl Polym Sci 2011b; 119: 1144–55. http://dx.doi.org/10.1002/app.32736.

Additionally, blending of rubber and metal oxides has been performed by dissolving the rubber matrix in an appropriate solvent and dispersing the nanoparticles in the same solvent. The combination of the solution and dispersion followed by casting on a substrate gives rise to films which can be used for different purposes. As an example, Kofod et al. [112] have developed mixtures constituted by hydrogenated styrene-*b*-butadiene block copolymer and TiO_2 for the development of elastomeric composites for dielectric actuators.

Another interesting strategy is based on the in situ generation of the inorganic particles by using the sol–gel process. This technique, known as the "bottom-up" process, has been reviewed by Messori [113] who described different approaches for preparing the rubber nanocomposites, most of them based on in situ generated silica from tetraethoxysilane (TEOS) as the precursor. Regarding metal oxide nanoparticles, some authors have used this technique to prepare rubber-based nanocomposites, mainly with titania or zirconia, because these metal oxides can be easily prepared by the sol–gel technique under mild conditions, without the need for heating treatment and calcination. Murugesan et al. [114] have developed poly (dimethylsiloxane) (PDMS) filled with zirconia (ZrO_2) and titania (TiO_2) using this technique. The authors have previously prepared the PDMS networks using TEOS as a cross-linking agent, generating reinforcing silica particle within the elastomer matrix. Then, in a second step, the PDMS films were swollen in a mixture containing zirconium *n*-propoxide, zirconium *n*-butoxide, titanium *n*-propoxide, or titanium *n*-butoxide and a small amount of di-butyl stannous diacetate or di-butyl stannous dilaurate as catalysts. The stannous compound facilitates the absorption of water from the atmosphere, thus promoting a controlled hydrolysis of the alkoxide precursors and initiating the sol–gel process. This procedure, involving the sol–gel with controlled hydrolysis gave rise to highly transparent nanocomposites with filler content as high as 22% in the case of TiO_2 and 28% for ZrO_2-based nanocomposites, suggesting very small particle size without significant aggregation.

Yang et al. [115] prepared a poly(ether-urethane) based network with titania particles as the cross-linking agent. This procedure involved the silane-terminated poly(ether-urethane) prepolymer which was reacted with titanium isopropoxide as the precursor for the preparation of titania nanoparticles by the in situ sol–gel process. The silane groups at the end of the polyurethane (PU) prepolymer can react with each other and with the titanium isopropoxide precursor via hydrolysis/condensation, thus forming the titania nanoparticle as the cross-linking point between the elastomeric PU. The sol–gel process was performed without the addition of water, because the moisture present in the atmosphere was enough to start the hydrolysis of the titanium alkoxide precursor.

In a different approach, Iketani et al. [116] have prepared TiO_2/PDMS hybrid films by performing the in situ hydrolysis/condensation of the titanium tetraisopropoxide in the presence of PDMS solubilized in tetrahydrofuran. The films containing up to 80% of TiO_2 were prepared by spin-coating the TiO_2/PDMS hybrid sols on poly(methyl methacrylate) (PMMA) substrates. By this technique, highly transparent films have been obtained even with an amount of TiO_2 as high as 80%.

9.4 Properties and applications of rubber nanocomposites filled with metal oxide nanoparticles

The incorporation of metal oxide nanoparticles into elastomers can significantly improve several properties, depending on the nature of the filler and its dispersion degree in the elastomer matrix. One of the important properties to be considered when incorporating metal oxide into elastomer matrix is the reinforcing effect. In this sense, metal oxide as nanoparticles provides superior reinforcing characteristics when compared with the micro-sized corresponding particles because of the higher surface area, which promotes better filler–matrix interactions. Besides reinforcement, other important properties can be achieved by incorporating such nanoparticles into the rubber matrices. The present section highlights these properties and the main applications of these nanocomposites as a function of the metal oxide nature.

9.4.1 The effect of nanometal oxide on the curing behavior of elastomer nanocomposites

Divalent metal oxides, such as ZnO, have been extensively used in the rubber industry as cure activators for the sulfur vulcanization of unsaturated rubbers [3,117] and as cross-linking agents for carboxylated elastomers [6,118] and CR [119].

As a cure activator, the divalent Zn^{2+} cation, generated from the action between ZnO and fatty acids, reacts with the accelerator producing a compound which is highly soluble in the rubber, accelerating the vulcanization process [120,121]. The cure reaction mechanism in the presence of ZnO is well described in the literature [106,122]. Due to the low specific area of the conventional ZnO, only a small part of ZnO is accessible to react with the accelerator. Large amounts of this compound remain unreacted in the rubber vulcanizates and are leached into the environment, which is considered toxic to aquatic organisms [123]. Therefore, the use of ZnO as nanoparticles is a very interesting strategy as it is possible to reduce the amount of this compound in rubber formulations. Moreover, the higher specific surface area of the ZnO nanoparticles favors the reaction with the accelerator to a larger extent, decreasing the amount of free ZnO able to be leached and thus polluting the aquatic environment. Some papers in the literature describe the efficiency of the ZnO nanoparticles as a cure activator in different rubber compounds. Sahoo et al. [106] have studied the effects of such nanoparticles (50 nm in size) as a cure activator of unfilled NR and nitrile rubber (NBR). The nanoparticles have been better dispersed inside the rubber matrices, providing higher maximum torque and higher difference in maximum and minimum torque values, indicating an increase of cross-link density. As summarized in Table 9.1. the rubber vulcanizates containing 3 phr of ZnO nanoparticles presented higher tensile strength and modulus than those prepared with conventional ZnO rubber grade, as a consequence of the higher cross-link density. This behavior confirms the ability of nano-ZnO to act as a reinforcing agent for rubber in lower amounts than the conventional ZnO particle.

The increase in cross-link density and improvement of mechanical properties by using very low amounts of nano-ZnO in comparison with conventional grade ZnO has been also reported for unsaturated rubber vulcanizates [120,124]. Mottaghi et al. [125] also studied the efficiency of nano-ZnO in increasing the cross-linking density of NR/BR and NR/SBR blends by using the swelling behavior technique.

In spite of the importance of nano-ZnO particles in improving the curing efficiency, cross-linking density, and reinforcement of rubber materials, the very high

Table 9.1 Effect of the ZnO nanoparticle on the main mechanical properties of NR and NBR vulcanizates [106]

ZnO		Tensile strength (MPa)		Modulus at 100% (MPa)	
Type	**Amount**	**NR**	**NBR**	**NR**	**NBR**
RGZnO	5	6.7	2.5	0.72	0.73
NZnO	3	7.6	4.3	0.83	1.0
NZnO	5	12.7	2.8	0.81	0.75

Formulation: stearic acid = 2.0 phr; 2,2,4-trimethyl-1,2-dihydroquinoline = 2.0 phr; *N*-cyclohexyl-benzothiazole-2-sulphenamide (CBS) = 0.8; tetramethyl thiuram disulfide (TMTD) = 0.2 phr; sulfur = 2.5 phr; RGZnO = ZnO rubber grade; NZnO = ZnO nanoparticle.

surface area of the nanoparticles favor their aggregation. Some strategies have been reported to minimize this problem. Thomas et al. [126] have developed new accelerators by modifying ZnO nanoparticles with *N*-benzylimine aminothioformamide and fatty acids. This modification reduces the agglomeration of ZnO nanoparticles and provides NR vulcanizates with improved curing and physicochemical properties. Wang and Chen [107] developed a master batch containing 30% of nano-ZnO and 70% of SBR and then the master batch was diluted to achieve formulations with low amounts of nano-ZnO (0.4–2.1 phr). They observed an increase of the scorch time which makes the processing safer. Also the master batch technology resulted in better filler dispersion and, consequently, better mechanical performance (Fig. 9.4).

MgO has been also employed as a curing activator for the sulfur-accelerated vulcanization of rubbers. It provides faster breakdown of the accelerator than ZnO, whereas higher cross-link density is usually achieved by using ZnO as the activator [127,128]. Moreover, MgO is less toxic than ZnO, which is very interesting from the environmental point of view. Several researchers have been studied the possibility of partially or totally substituting ZnO by MgO as an activator. In this context, MgO nanoparticles play an important role as the higher surface area of filler caused by the particle size in the nanometric scale should provide outstanding properties with a lower amount of filler. Such MgO nanoparticles have been used in rubber formulations for different purposes. Roy et al. [104] have studied the effect of the sol–gel synthesized MgO nanoparticles as a cure activator of sulfur-based vulcanized NR. Both scorch and optimum cure times were significantly lower using only 0.5 phr of nano-MgO as compared with the results obtained with 5 phr of conventional ZnO. Also the main tensile properties such as modulus, tensile strength, and elongation at break of NR cured in the presence of 1 phr of nano-MgO were superior to those achieved in systems prepared with conventional ZnO as the activator, indicating that nano-MgO accelerates the curing process and also acts as a reinforcing agent for sulfur-vulcanized NR. The thermal stability of NR compounded with 1 phr of nano-MgO was higher than that compounded with conventional ZnO. The outstanding performance of NR networks achieved by using nano-MgO particles was attributed to the higher affinity of MgO for the sulfur used as a cross-linking agent and the large surface area of the nanoparticle.

Guzman et al. [129] have prepared nanoparticles constituted by a mixture of ZnO and MgO ($Zn_{1-x}Mg_xO$) and studied their effect on the sulfur vulcanization accelerated by *N*-cyclohexylbenzothiazole-2-sulfenamide (CBS). By using squalene as a model molecule for NR, the authors observed that the mixed $Zn_{1-x}Mg_xO$ nanoparticles provoked faster consumption of the accelerator and sulfur than conventional ZnO. The breakdown process of the accelerator CBS was similar when using the mixed nanoparticle or conventional ZnO. However, the disappearance of CBS was much faster with the mixed nanoparticle. Regarding the cross-linking density, the mixed metal oxide nanoparticles also resulted in around 30% higher degree of cross-linking than conventional ZnO, with a higher proportion of disulphidic and lower proportion of monosulphidic cross-links. Summarizing, the mixed metal oxide nanoparticles are more active and may be a good option for reducing the total amount of zinc in the rubber compounds.

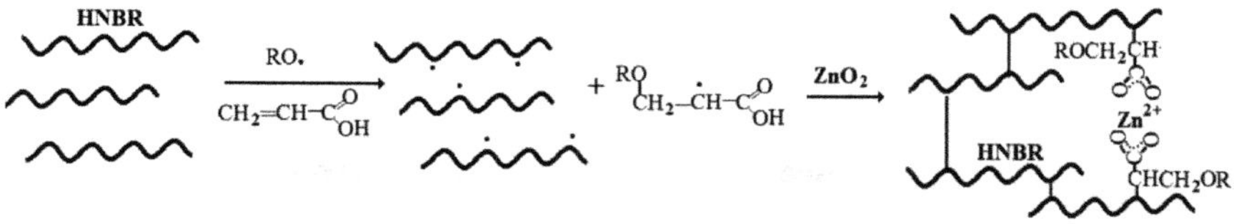

Figure 9.6 Scheme illustrating the reaction between acrylic acid and both rubber matrix and metal oxide.

Pingot and coworkers [130] have reported a comparative study involving different nanostructured metal oxides (ZnO, MgO, and CaO) as coagents in the peroxide curing process of hydrogenated butadiene-acrylonitrile rubber. They have employed acrylic acid or monoallyl maleate to form the salts. The highest cross-link density was observed for the peroxide/ZnO/acrylic acid curing system. The cross-link density analysis indicated that both ionic and covalent bonds are involved in the network. The unsaturated carboxylic acid acts as a bridge between the rubber matrix and the metal oxide nanoparticle as illustrated in Fig. 9.6.

Regarding the mechanical properties, the presence of the coagents resulted in a significant improvement of the tensile strength when compared with that observed for the peroxide cured system without coagents. Although the highest cross-link density was experienced with the combination of ZnO and AA as coagents, the highest tensile strength was observed for the system containing MgO in combination with monoallyl maleate. The improved mechanical properties were attributed not only to the increasing of the cross-linking density but also to the presence of the unsaturated carboxylic acid which acts as a compatibilizing agent between the inorganic nanoparticle and the rubber matrix.

MgO nanoparticles have been used by Kar and Bhowmick [131] as a cure activator for the CR rubber vulcanizates cured with ZnO. The mixture was prepared by dissolving the CR rubber in methyl ethyl ketone and dispersing the conventional ZnO and MgO (rubber grade or nano) particles into the rubber solution, following by drying at room temperature. Fig. 9.7 illustrates the tensile behavior of CR vulcanizate as a function of the MgO as activator. The presence of MgO accelerates the curing process of CR and promotes an improvement of tensile strength and modulus. However the best results are achieved by using the nano-MgO particles (Fig. 9.7C). According to TEM micrographs, the rubber grade MgO particles are not well distributed within the CR matrix. However, the MgO nanoparticles are uniformly distributed inside the rubber matrix with sizes in the range of 50–100 nm [131]. These results confirm the outstanding reinforcing effect on rubber materials by using nanoparticles.

As already discussed in the Introduction part of this chapter, ZnO can also act as a cross-linking agent for carboxylated rubber. The curing process is accepted to occur via the salt formation between the divalent metal oxide and the carboxylic groups, according to the scheme presented in Fig. 9.8. The resulting ionic cross-links form clusters within the rubber matrix and are also responsible for the enhanced tensile strength and elongation at break of the rubber vulcanizates [132].

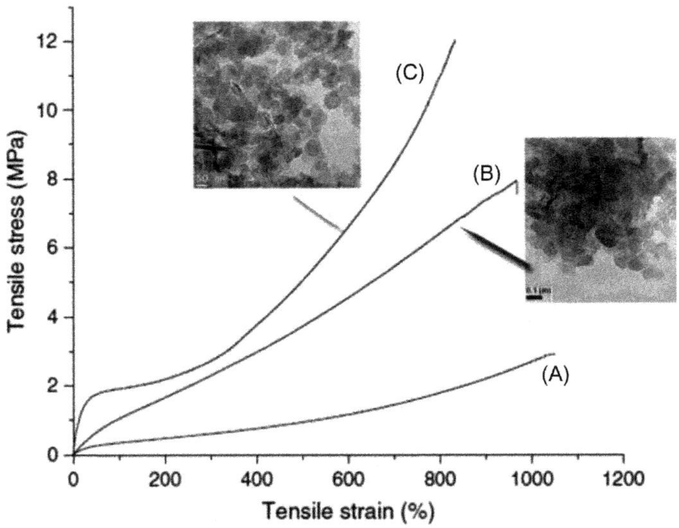

Figure 9.7 Tensile behavior and TEM micrographys of CR cured with 5% of rubber grade ZnO (A) without activator; (B) with 4% of rubber grade MgO; (C) with 4% of nano-MgO.
Source: Kar S, Bhowmick AK. Nanostructured magnesium oxide as cure activator for polychloroprene rubber. J Nanosci Nanotechnol 2009; 9: 3144–53. http://dx.doi.org/10.1166/jnn.2009.005.

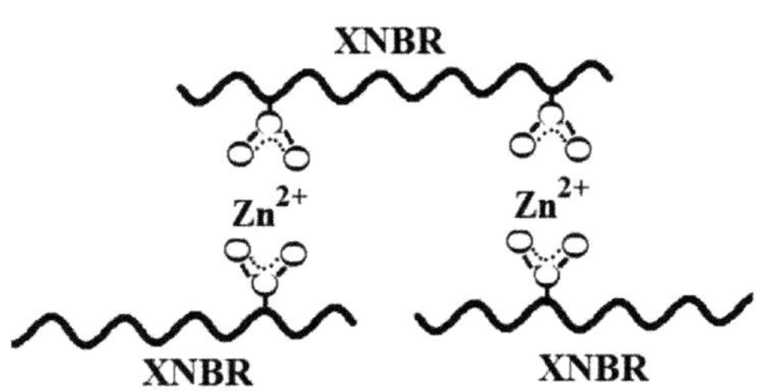

Figure 9.8 Scheme illustrating the probable interaction between carboxyl group of carboxylated nitrile rubber (XNBR) and Zn^{2+}.

This process is faster and usually results in higher cross-link density when the metal oxide particle size is small. Sahoo and Bhowmick [46] have studied the curing process of carboxylated NBR using nano- and conventional grade ZnO particles as the only cross-linking agent. A significant increase in cure rate index was observed by using nano-ZnO. Also, the cross-link density was higher, contributing to a higher tensile strength and modulus. This behavior was attributed to the larger number of interactions between ZnO nanoparticles and the –COOH groups present in rubber.

The effect of ZnO nanoparticles on the curing process and mechanical properties of carboxylated SBR and carboxylated NBR was also studied by Hamed and Hua [133]. Przybyszewska and Zaborski [134] have also used ZnOs with different morphologies (spheres, whiskers, snowflakes) as cross-linking agents for carboxylated nitrile elastomer, and observed that the three-dimensional snowflake-like ZnO nanoparticles resulted in higher cross-link density and tensile strength than the other ZnO nanoparticles at the same amount and suggested that the effective contact between the surface of the cross-linking agent and the rubber chain was responsible for this outstanding mechanical performance. The improved interaction in this system is promoted by the specific shape and complex structure of the ZnO aggregates which also improve the dispersion in the rubber matrix.

Besides the significant influence on the curing characteristics of rubber imparted by ZnO nanoparticles, they also act as reinforcing filler, thus improving the mechanical properties and contributing to an increase in thermal conductivity [110]. The modification of the surface of the ZnO nanoparticles with bis-(3-triethoxy silylpropyl)-tetrasulfide silane enhanced the interfacial interaction between rubber and filler, contributing to a better dispersion of the nano-ZnO particles. The intrinsic thermal conductive property of the ZnO nanoparticles and the better dispersion into the rubber matrix contribute to an efficient transport of the generated heat caused by the internal frictions of various networks from inside to outside the rubber compound, thus decreasing the heat build-up.

9.4.2 Properties and applications of aluminum oxide (Al_2O_3)-based rubber nanocomposites

Al_2O_3 has been used in rubber formulation to impart the desired reinforcement [135], and to improve electrical and dielectric properties [136], electric breakdown resistance [137], thermal conductivity, etc. Recently, the use of Al_2O_3 nanoparticles has been of increased interest because of the possibility of employing lower filler content to achieve a similar performance as that usually found with higher content of micro-Al_2O_3 particles.

Nano-Al_2O_3 particles have been compounded with silicone rubber to improve electric breakdown resistance. Outdoor insulation materials used in power transmission at high voltages are subjected to several stresses which cause degradation and aging, resulting in electric breakdown (tracking) of the material and loss of the insulation property. The problem of surface degradation can be minimized by incorporating proper inorganic fillers. Loganathan and Chandrasekar [136] have studied the effect of micro- (5–10 μm) and nano- Al_2O_3 (<80 nm) particles on the dielectric properties of silicone rubber composites aiming to develop materials with improved resistance to arcing, tracking, erosion, and mechanical stiffness to be applied in outdoor high voltage AC power transmission. According to Loganathan and Chandrasekar, silicone charged with nano-Al_2O_3 presented lower permittivity values, which was attributed to the presence of more interfaces causing an increasing in chain mobility restriction. Moreover, the presence of nano-Al_2O_3 resulted in an improvement of tracking (electrical breakdown) resistance, which is an important characteristic for insulating materials that are exposed to high voltages outdoors [136,137]. As illustrated in Fig. 9.9, the silicone rubber containing 5%

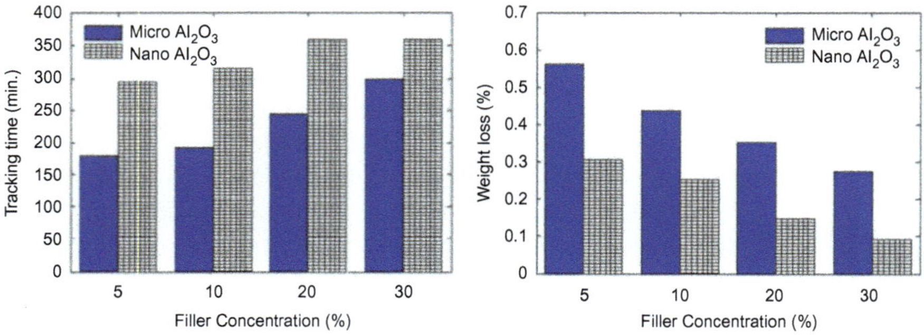

Figure 9.9 Tracking time of silicone rubber composite containing Al_2O_3 and weight loss of rubber after tracking test.
Source: Loganathan L, Chandrasekar S. Analysis of surface tracking of micro and nano size alumina filled silicone rubber for high voltage AC transmission. J Electr Eng Technol 2013; 8: 345–53. http://dx.doi.org/10.5370/JEET.2013.8.2.345.

of nano-Al_2O_3 particles presented similar tracking resistance as that containing 30% of micro-Al_2O_3 particles, highlighting the importance of the nanosized particles with high surface area in attaining this property. Also the nanocomposites presented improved erosion resistance, which was attributed to the better thermal stability of the nanosystems.

Al_2O_3 nanoparticles have been also used in rubber formulation to improve thermal conductivity. This characteristic is very important for decreasing the heat build-up that usually happens under dynamic conditions because of the internal frictions of the filler–filler, filler–rubber, and rubber–rubber networks [109]. In this context, Al_2O_3 nanoparticles, being thermal conductive filler, may be considered excellent filler for improving the thermal properties of rubber formulations. Zhou et al. [138] have studied the effect of Al_2O_3 particle sizes on thermal conductivity of silicone rubber end-functionalized with vinyl groups, which was cured with a peroxide-type curing agent at high temperature. They compared the efficiency of β-phase Al_2O_3 microparticles with α-phase Al_2O_3 nanoparticles (around 50 nm). The surface of the latter was modified with 3-methacryloyloxypropyltrimethoxysilane. For silicone rubber composites containing Al_2O_3 particles with size of 25 μm, the thermal conductivity increases slowly as the amount of filler increases because the particles are surrounded by the insulating rubber and a good response starts to appear at a Al_2O_3 concentration higher than 40 vol.%. Such high filler loading causes a decrease in the mechanical properties. On the other hand, the composites containing Al_2O_3 with particles at the nanometric size (around 50 nm) present higher thermal conductivity and also exhibit better mechanical performance. This behavior is justified by the very low size and higher surface area of the nanoparticles. Moreover, the functionalization of the Al_2O_3 nanoparticles with a methacryloyloxy propyl moiety imparts better interaction between filler and matrix because the surface of the inorganic particle containing vinyl groups can participate on the rubber network.

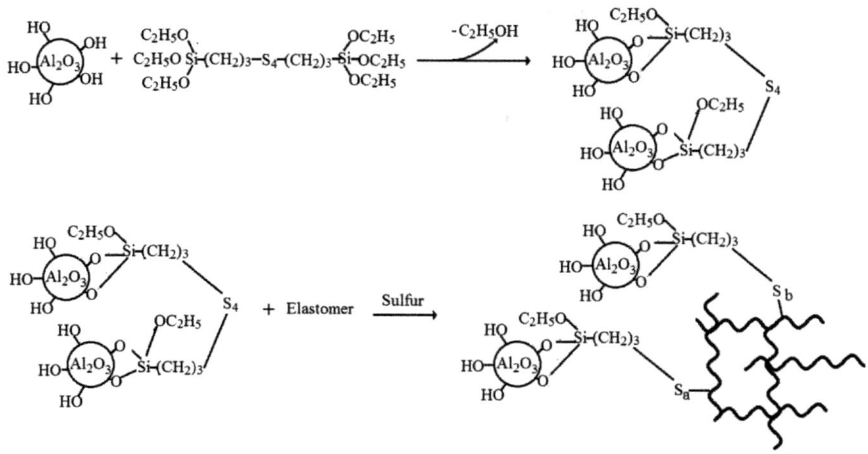

Figure 9.10 Scheme illustrating the coupling reaction of functionalized silane with rubber and Al_2O_3 nanoparticles.
Source: Adapted from Wang Z, Lu Y, Liu J, Dang Z, Zhang L, Wang W. Preparation of nanoalumina/EPDM composites with good performance in thermal conductivity and mechanical properties. Polym Adv Technol 2011c, 22, 2302–10.

Wang et al. [139] employed high loading of Al_2O_3 nanoparticles in ethylene propylene diene rubber and investigated the mechanical properties and thermal conductivity. They studied the influence of bis-(3-triethoxy silylpropyl)-tetrasulfide (Si69) as the coupling agent and also the surface modification of Al_2O_3 nanoparticles by stearic acid. The best particle dispersion without agglomerations was observed for systems containing silane as the coupling agent. This system also displayed better mechanical properties, probably because of the strong interfacial interaction between rubber and Al_2O_3 nanoparticles promoted by the functionalized silane, according to Fig. 9.10.

The thermal conductivity increased as the filler loading increased, but the surface modification of the particles exerts marginal effect on this property. According to the authors [139], the EPDM composites containing Al_2O_3 nanoparticles dispersed with the help of organosilane display a better combination of mechanical and thermal properties.

The effect of Al_2O_3 nanoparticles on the thermal, dielectric, and mechanical properties of silicone rubber filled with microsilicon nitride (Si_3N_4) particles was also investigated by Zha et al. [140]. Thermal conductivity of around 1.7 W/m K was achieved by using 60 vol% of micro-Si_3N_4 particles in silicone rubber formulations. However, a similar value was achieved by using a mixture of micro-Si_3N_4/nano-Al_2O_3 particles in a proportion corresponding to 26:4 vol.%. This behavior can be attributed to the ability of nano-sized Al_2O_3 particles to be located into the interstices of the micro-sized Si_3N_4 particles favoring the formation of thermal conductive pathways easily [140].

9.4.3 Properties and application of titanium oxide (TiO_2)-based rubber nanocomposites

TiO_2 or titania nanoparticles have been considered a promising candidate for developing rubber-based nanocomposites with interesting properties for special applications due to their inherent characteristics, which include antimicrobial and photocatalytic properties [141], UV absorbing ability [142], potential antifouling ability [143], relatively high electrical permittivity [18], chemical inertness, and nontoxicity.

The photocatalytic activity of TiO_2 is one of its important properties, which can be used to degrade rubber, to provide the antimicrobial activity of some membrane-based rubbers, or even to protect the organic substract from degradation. The photocatalysis of TiO_2 is based on the ejection of electrons (e^-) and the formation of positive holes, which react with O_2 and H_2O, respectively, to form active oxygen species, such as, O_2^- and HO^-, at the surface of the TiO_2 nanoparticles which can degrade the organic substances [16,17].

Arantes et al. [85] have prepared the TiO_2 nanoparticle in the pure anatase form with an average diameter of 7 nm by using the sol–gel process and employed a colloidal route to prepare styrene-butadiene rubber/TiO_2 nanocomposites. The films were exposed to UV light to observe the photocatalytic degradation by monitoring the appearance of carbonyl absorption by Fourier transform infrared (FTIR). The formation of these carbonyl groups was much faster in nanocomposites than in pure polymer, indicating the efficiency of small amounts of TiO_2 (around 0.2% in relation to the rubber matrix) in catalyzing the photodegradation of the rubber. Seentrakoon et al. [144] also prepared NR-based nanocomposites containing nano-TiO_2 and observed an improvement of the UV protection of the NR nanocomposites by the presence of titania nanoparticles. They also used the FTIR technique to evaluate the appearance of the carbonyl absorption and observed a decrease in the absorption intensity for the systems containing nano-TiO_2.

Lin et al. [17] blended two kinds of TiO_2 nanoparticles in NR or NBR using two roll milling and cured the nanocomposites with a sulfur-based curing system. The TiO_2 nanoparticles composed of 99% of anatase crystal were better dispersed in NR than those constituted by 70% of anatase and 30% rutile crystals. They analyzed the antimicrobial effect of the corresponding nanocomposites and concluded that all nanocomposites have good antimicrobial ability, but the best effect was found with nanocomposites constituted of NR containing TiO_2 nanoparticles made up of 70% anatase and 30% rutile crystals, which were not as well dispersed as those made up of pure anatase. According to the author, the rubber matrix involving the TiO_2 nanoparticles impedes the diffusion of electrons and holes to the surface, which find the adsorbed H_2O and O_2 to generate the active oxygen species.

Iketani et al. [116] developed highly transparent TiO_2/poly(dimethyl siloxane) hybrid films by using the in situ sol–gel technique, which were deposited on a PMMA substrate using the spin-coating technique. They observed a good photocatalytic activity for the decomposition of methylene blue and acetaldehyde whereas the photodegradation of the PMMA substrate was greatly suppressed.

Titania can be also used to develop materials with UV-blocking properties. In this sense, Yang et al. [115] prepared a poly(ether-urethane) based network with titania particles as the cross-linking agent by the in situ sol–gel process described previously and studied the UV and X-ray shielding efficiency. The corresponding hybrid films (thickness around 6 μm) containing up to 40% of TiO_2 exhibited excellent elastomeric properties and high transparence in the visible light range due to the nanometric dimension of the TiO_2 particles. However, they blocked most of the radiation along the entire UVB/UVA range (280–400 nm).

TiO_2 nanoparticles have been also compounded with silicone rubber in order to inhibit electrical tree propagation which causes degradation and premature failure of the insulating materials that are exposed to high voltages outdoors. Musa et al. [145] blended nano- TiO_2 with silicone rubber and analyzed the tree growth under an applied voltage and frequency of 10 kVrms AC and 50 Hz, respectively. They observed a significant increase in treeing breakdown time, which is very important for the development of high-voltage outdoor insulation materials.

Another important characteristic of titania nanoparticles is the high dielectric constant which can be important for the development actuators. Kofod et al. [112] blended TiO_2 nanoparticles with SEBS matrix by the solution cast method to develop films with a high dielectric constant for dielectric elastomer actuators [112]. Composites prepared with nano-TiO_2 (15 nm) presented higher permittivity than those constituted by micro-TiO_2 (300 nm). This effect was attributed to the much higher surface area of the nanoparticles.

TiO_2 nanoparticles have been also used as a reinforcing agent for NR compounds [146] and their blends with EPDM [147]. The NR/nano-TiO_2 composites presented higher tensile strength and modulus than those containing micro-TiO_2, which can be attributed to the larger particle surface area and homogeneous dispersion of the nanoparticle in the rubber matrix [144,146]. Regarding the NR/EPDM blends, the authors observed that the TiO_2 nanoparticles are preferentially located inside the NR phase.

9.4.4 Iron oxide (Fe_3O_4)-based rubber nanocomposites

Iron oxide nanoparticles are those with diameters between about 1 and 100 nanometers. The most popular form is magnetite (Fe_3O_4), which has found important applications due to its superparamagnetic characteristics. Magnetite nanoparticles have been incorporated into elastomer matrices aiming to develop new soft magnetic materials, which find applications in several fields of material science as actuactors, electromagnetic wave absorbers, magnetirheological elastomers, as well as in biomedical areas, including magnetic resonance imaging, drug delivery, hyperthermia, etc.

The literature reports several approaches for the magnetic elastomer preparation. The most common procedure involves the melt blending of the nanoparticles with the elastomer compound in a two roll mill [15,148–150]. To achieve better magnetic nanoparticle dispersion, other techniques involving solution casting, dispersion in rubber latex or the polymerization/curing of liquid elastomer in the presence of the magnetite NPs have been tried. Even employing specific mixing techniques, such as sonication or high-speed mixing, the homogeneous dispersion of magnetite NPs in a polymeric matrix is not an easy task due to the van der Waals forces and magnetic dipolar interactions, which force these particles to aggregate. In this sense, coating NP surfaces with a surfactant is considered essential for improving the particle–matrix interactions and avoiding segregation of the particle phase. A few examples involving these techniques is described in this section.

Kim et al. [151] have prepared magnetite NPs and treated their surface with oleic acid to impart a hydrophobic character to these NPs. These functionalized NPs were dispersed in silicone rubber dissolved in xylene using the sonication technique and they obtained an uniform dispersion of NP within the silicone matrix, with a particle average size of 10 nm.

Ashjari et al. [152] have also employed oleic acid-functionalized magnetite in thermoplastic elastomer PU matrix using the solution approach. The magnetite NPs were previously dispersed within tetrahydrofuran using ultrasonic irradiation of high intensity to produce a homogeneous dispersion. After the removal of the solvent, the obtained nanocomposite films presented homogeneous dispersion with a magnetite particle size of less than 100 nm. The presence of these magnetic NPs resulted in a decrease of the glass transition temperature (T_g) and a storage modulus below the T_g, suggested by dynamic-mechanical analysis. According to the authors, the magnetite particles interact with the PU matrix and should disrupt hydrogen bonding of the PU matrix, causing an increase in PU chain mobility [152]. Even with surface treatment, the adhesion between PU and magnetite was not significant, suggested by the decrease in tensile strength and elongation at break.

Abramchuk et al. [153] have prepared a dispersion of magnetite powder (particle size in the range of 0.2–0.3 μm) with a liquid silicone rubber, which is constituted by two components: silicone with vinyl groups and silicone oligomer with hydride groups. The polyaddition involving both components was performed in the presence of magnetite NP and catalyzed by a complex platinum catalyst. No morphological results have been presented. Nevertheless, the authors reported an increase of the shear modulus of the magnetic composite when an external magnetic field was applied.

Venkatanarasimhan and Raghavachari [154] developed a new procedure for epoxidized NR blended with magnetite in solution. They first dispersed the magnetite into NR dissolved into chloroform using sonication. Then, they performed the epoxidation of the NR phase in the presence of magnetite. The prepared magnetic epoxidized NR nanocomposites presented a high efficiency as an absorbent for petrol.

Ong et al. [155] produced iron oxide nanoparticles by the precipitation method, which is constituted by a mixture of Fe_3O_4 magnetite and γ-Fe_2O_3 maghemite, and dispersed within a NBR latex containing the curing ingredients. After casting on a petri dish and performing the curing process, a film with improved thermal stability and good magnetic properties was formed.

Another interesting work was recently reported by Tan and Bakar [156] and consisted of the synthesis of magnetite in situ in the presence of epoxidized NR. The average size of the magnetite dispersed in the epoxidized NR stayed in the range of 3–6 nm, which is significantly smaller than that observed for the bare Fe_3O_4, indicating that the epoxidized natural rubber matrix acts as a stabilizing agent to control the size and dispersion of the magnetite nanoparticles in the composite. In spite of the excellent dispersion no chemical interaction between particle and matrix was evidenced from FTIR and differential scanning calorimetry studies.

The incorporation of magnetite within elastomer matrices results in materials which present important responses under the action of an external magnetic field. These materials are considered intelligent because they can change their mechanical and rheological properties under the action of an external magnetic field. Under these conditions, the magnetic particles tend to align themselves in the direction of the magnetic field, forming chain-like microstructures which are responsible for a significant increase in the elastic modulus of the composite. These systems are considered more stable than those involving magneto-rheological fluids because the higher viscosity of the magnetic-active elastomers suppress the sedimentation of magnetic particles. Several reports deal with the use of magnetic nanoparticles on the development of such smart magnetic elastomeric materials. Abramchuk et al. [153] observed a significant increase in the shear modulus with the application of external magnetic field with relatively low intensity (below 0.3 T) for systems constituted by magnetite particles dispersed in silicone rubber, which was cured in the presence of the particles. Evans et al. [157] developed magnetic elastomers by blending magnetite nanoparticles within a copolymer constituted by amino-propylmethylsiloxane and dimethylsiloxane in solution. Measurements of modulus taken under compression indicated a significant increase under the action of a 300 mT magnetic field, suggesting the potential magnetic actuation applications. In another work reported by Zaborski and Maslowski [148], nano-sized iron oxide particles were dispersed within EPDM or NBR elastomeric matrices, together with the curing agent. Then, the blends were cured in two different ways: under the influence of magnetic field and without the magnetic field. The systems presented an increase of elastic modulus under the external field. However, those systems vulcanized in the presence of a magnetic field displayed better response under the action of the field. Also the composites containing nano-sized Fe_3O_4 resulted in higher elastic modulus.

Magnetite-based elastomer composites also have potential application as electro-magnetic wave absorbers and electromagnetic interference shielding materials due to their electrical and magnetic properties. When the electromagnetic radiation reaches a material, it may be transmitted, reflected, or absorbed by the material, as illustrated in Fig. 9.11. The overall electromagnetic interference shielding effectiveness (EMI SE) results from the combination of all these processes. The reflection is caused by the mobile charge carriers which interact with the electromagnetic field. The absorption is governed by the presence of electric or magnetic dipoles and the currents induced in the medium causing the heat of the material (ohmic losses). Al-Ghamdi et al. [15] have studied the influence of magnetite on the EMI SE in the microwave band 1 and 12 GHz, for nitrile butadiene rubber/magnetite nanocomposites. As illustrated in Fig. 9.12, an increase of shielding

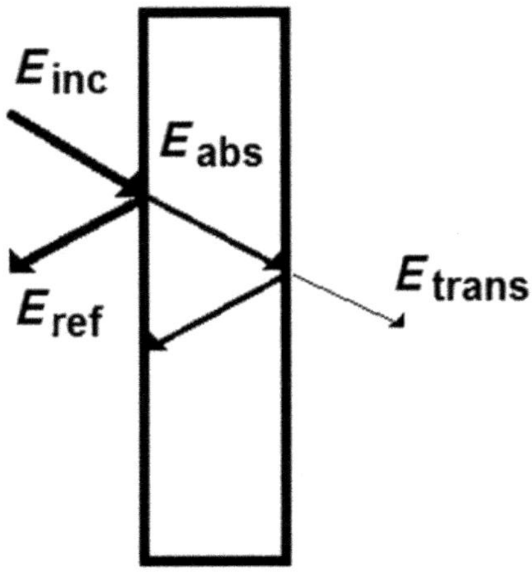

Figure 9.11 Scheme showing the phenomena which can occur when an electromagnetic wave reaches a material.

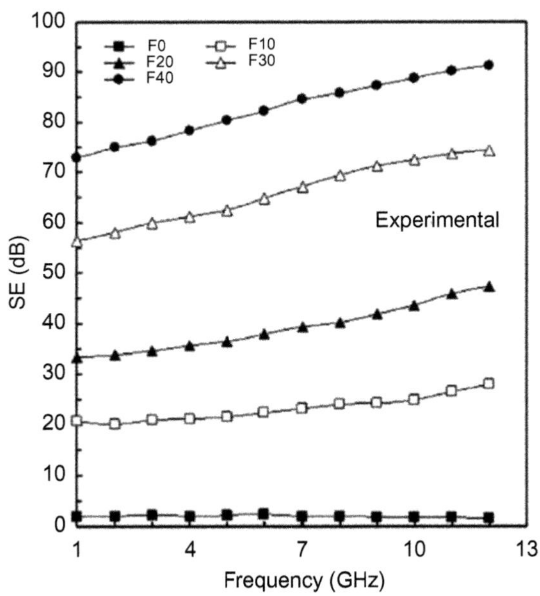

Figure 9.12 EMI SE for nitrile butadiene rubber/magnetite nanocomposites as a function of the magnetite content (0–40 phr of magnetite).
Source: Al-Ghamdi AA, Al-Hartomy OA, Al-Salamy F, Al-Ghamdi AA, El-Mossalamy EH, Abdel Daiem AM, et al. Novel electromagnetic interference shielding effectiveness in the microwave band of magnetic nitrile butadiene rubber/magnetite nanocomposites. J Appl Polym Sci 2012; 125: 2604–13. http://dx.doi.org/10.1002/app.36371.

effectiveness was observed as the amount of magnetite increased. At a magnetite concentration as high as 40 wt%, an EMI SE value of 92 dB was observed. According to the authors, this behavior is due to the high electrical conductivity and fast spin relaxation of superparamagnetic nanoparticles dispersed in the rubber matrix.

Another interesting application of magnetic elastomer nanocomposites is related to the development of NTCR (negative temperature coefficient of resistivity) thermistors, as the resistivity decreases with the temperature [149].

9.4.5 Zirconium oxide (ZrO_2)-based rubber nanocomposites

ZrO_2 presents excellent thermal and optical properties. Additionally, it is considered a semiconductor and tends to become more conductive with increasing temperatures. These characteristics make this metal oxide a promising candidate for rubber compounds with different applications as thermal protection systems, solid polymer electrolytes, etc. Different methodologies have been employed for the preparation of ZrO_2-based rubber nanocomposites, including melt blending and colloidal suspension mixing, and the in situ sol–gel process has been developed.

Yang et al. [158] have blended PDMS with nano-ZrO_2 in a two roll mill aiming to develop rubber compounds with improved ablative properties, which are very important for applications as thermal protection systems. In all formulations studied, SiO_2 and carbon fiber were also used as additional fillers. The presence of ZrO_2 resulted in a slight increase of tensile strength and an important improvement of thermal and ablation properties.

Arantes et al. [111] have mixed aqueous colloidal suspension of monoclinic ZrO_2 containing dodecylbenzene sulfonic acid with commercial SBR in the form of latex, and evaluated the effect of these nanoparticles on the photodegradation process of SBR. The particle size of the nano-ZrO_2 was in the range of 5–15 nm. However, no evaluation of the nanocomposite morphology was discussed.

The sol–gel process has been employed by some authors to generate the nano-ZrO_2 particles in the presence of rubber. Using this procedure, Katayama et al. [159] have used zirconium n-butoxide $(ZrOBu^n)_4$ as a zirconia precursor, which was mixed with ethyl acetoacetate, a chemical modifier of the metal alkoxide to prevent the precipitation of the oxide during the hydrolysis. The sol–gel was performed in the presence of PDMS dissolved in 2-ethoxyethanol. Using this procedure, a flexible and homogeneous PDMS-based inorganic/organic hybrid sheet with good mechanical properties was produced. Murugesan et al. [114] also employed the sol–gel process with controlled hydrolysis in systems containing zirconium tetra n-propoxide and n-butoxide within the PDMS matrix. They have obtained transparent films with an amount of ZrO_2 as high as 28%. Also the nanocomposites presented higher thermal stability and improved mechanical properties in terms of ultimate tensile strength.

9.5 Concluding remarks

The use of nanometal oxides to develop rubber-based nanocomposites is considered an interesting and important approach for achieving important properties with low amounts of filler compared to that of conventional metal oxides. This feature is due to the high surface area, and in several cases, the functionalization of the nanoparticle surface, which render better filler–matrix interactions, giving rise to a better particle dispersion. Different methodologies have been described to prepare such nanocomposites and include melt mixing, solution mixing, and in situ preparation of nanoparticle within the rubber matrix. In most cases discussed in this chapter involving different metal oxides, the particle size exerts significant influence on the main properties of the nanocomposites.

Acknowledgments

This work was sponsored by Conselho Nacional de Desenvolvimento Científico e Tecnológico – CNPq and Fundação de Amparo à Pesquisa do Estado do Rio de Janeiro – FAPERJ. The author would like to thank Dr. Jéssica A. Marins and Sara V. Oliveira for the contribution in the bibliography.

References

[1] Donnet JB, Custoderc E. Reinforcement of elastomers by particulate fillers. In: James EM, Burak E, Frederick RE, editors. Science and technology of rubber. 3rd ed. Builington: Academic Press; 2005.

[2] Kango S, Kalia S, Celli A, Njuguna J, Habibi Y, Kumar R. Surface modification of inorganic nanoparticles for development of organic-inorganic nanocomposites–a review. Prog Polym Sci 2013;38:1232–61.

[3] Dogdakin B, Shershnev VA. The interaction of tetramethylthiuram disulfide with rubber and with compounds containing a labile hydrogen atom. Rubber Chem Technol 1960;33:401–11.

[4] McGill WJ, Shelver SR. Effect of carboxylid acids on 2-bisbenzothiazole-2,2'-disulfide- and tetramethylthiuram disulfide-acellerated sulfur vulcanization. IV. Vulcanization of polyisoprene compounds with ZnO. J Appl Poly Sci 1999;72:1021–30.

[5] Ibarra L, Alzorriz M. Ionic elastomers based on carboxylated nitrile rubber and magnesium oxide. J Appl Polym Sci 2007;103:1894–9.

[6] Ibarra L, Marcos-Fernandez A, Alzorriz M. Mechanistic approach to the curing of carboxylated nitrile rubber (XNBR) by zinc peroxide/zinc oxide. Polymer 2002;43:1649–55.

[7] Chokanandsombat Y, Sirisinha C. MgO and ZnO as reinforcing fillers in cured polychloroprene rubber. J Appl Polym Sci 2013;128:2533–40.

[8] Datta S, De PP, De SK. Blends of ionomers. J Appl Polym Sci 1996;61:1839–46.

[9] Dutta S, De PP, De SK. Ionic elastomer blends of zinc salts of maleated EPDM rubber and carboxylated nitrile rubber. J Appl Polym Sci 1998;69:153–60.

[10] Antony P, Bandyopadhyay S, De SK. Synergism in properties of ionomeric polyblends based on zinc salts of carboxylated nitrile rubber and poly(ethylene-*co*-acrylic acid). Polymer 2000;4:787–93.
[11] Antony P, De SK. Ionomeric polyblends of zinc salts of maleated EPDM rubber and poly (ethylene-*co*-acrylic acid). I. Effect of blend ratio. J Appl Polym Sci 1999;71:1247–56.
[12] Janowska G, Kucharska-Jastrzabek A, Kasiczak A, Rzymski WM. Thermal properties and combustibility of cross-linked XNBR/CSM blends. J Therm Anal Calorim 2011;104:1107–15.
[13] Zhang Q, Tian M, Wu Y, Lin G, Zhang L. Effect of particle size on the properties of $Mg(OH)_2$–filled rubber composites. J Appl Polym Sci 2004;94:2341–6.
[14] Cherney EA. Silicone rubber dielectrics modified by inorganic fillers for outdoor high voltage insulation applications. IEEE Trans Dielectr Electr Insul 2005;12:1108–15.
[15] Al-Ghamdi AA, Al-Hartomy OA, Al-Salamy F, Al-Ghamdi AA, El-Mossalamy EH, Abdel Daiem AM, et al. Novel electromagnetic interference shielding effectiveness in the microwave band of magnetic nitrile butadiene rubber / magnetite nanocomposites. J Appl Polym Sci 2012;125:2604–13.
[16] Hagfeldt A, Graetzel M. Light-induced redox reactions in nanocrystalline systems. Chem Rev 1995;95:49–68.
[17] Lin G, Tian M, Lu Y, Zhang X, Zhang L. Morphology, antimicrobial and mechanical properties of nano-TiO_2/rubber composites prepared by direct blending. Polym J 2006;38:498–502.
[18] Carpi F, Rossi DD. Improvement of electromechnical actuating performance of a silicone dielectric elastomer by dispersion of titanium dioxide powder. IEEE Trans Dielectr Electr Insul 2005;12:835–43.
[19] Koo JH. Polymer nanocomposites: processing, characterization, and applications. New York: McGraw-Hill Companies Inc.; 2006.
[20] Mittal V, Kim JK, Pal K, editors. Recent advances in elastomeric nanocomposites. Berlin: Springer-Verlag; 2011.
[21] Bhowmick AK, Bhattacharya M, Mitra S, Kumar KD, Maji PK, Choudhury A, et al. Morphology-property relationship in rubber-based nanocomposites: some recent developments. Adv Polym Sci 2011;239:1–84.
[22] Maiti M, Bhattacharya M, Bhowmick AK. Elastomer nanocomposites. Rubber Chem Technol 2008;81:384–469.
[23] Sengupta R, Chakraborty S, Bandyopadhyay S, Dasgupta S, Muhopadhyay R. A short review on rubber/clay nanocomposites with emphasis on mechanical properties. Polym Eng Sci 2007;47:1956–74.
[24] Peddini SK, Bosnyak CP, Henderson NM, Ellison CJ, Paul DR. Nanocomposites from styrene-butadiene rubber (SBR) and multiwall carbon nanotubes(MWCNT) part 1: Morphology and rheology. Polymer 2014;55:258–70.
[25] Ponnamma D, Sung SH, Hong JS, Ahn KH, Varughese KT, Thomas S. Influence of non-covalent functionalization of carbon nanotubes on the rheological behavior of natural rubber latex. Eur Polym J 2014;53:147–59.
[26] Zhan Y, Lavorgna M, Buonocore G, Xia H. Enhancing electrical conductivity of rubber composites by constructing interconnected network of self-assembled graphene with latex mixing. J Mater Chem 2012;22:10464–8.
[27] Chua TP, Mariatti M, Azizan A, Rashid AA. Effects of surface-functionalized multi-walled carbon nanotubes on the properties of poly(dimethyl siloxane) nanocomposites. Compos Sci Technol 2010;70:671–7.

[28] Fletcher A, Gupta MC. Mechanical properties of elastomer nanocomposites for electromagnetic interference shielding applications. J Compos Mater 2014;48:1261–76.

[29] Pietrasik J, Gaca M, Zaborski M, Okrasa L, Boiteux G, Gain O. Studies of molecular dynamics of carboxylated acrylonitrile-butadiene rubber composites containing in situ synthesized sílica particles. Eur Polym J 2009;45:3317–25.

[30] Bandyopadhyay A, De Sarkar M, Bhowmick AK. Effect of reaction parameters on the structure and properties of acrylic rubber/silica hybrid nanocomposites prepared by sol–gel technique. J Appl Polym Sci 2005;95:1418–29.

[31] Sanchez C, Rozes L, Ribot F, Laberty-Robert C, Grosso D, Sassoye C, et al. "Chimie Doulce": a land of opportunities for the designed construction of functional inorganic and hybrid organic–inorganic nanomaterials. C R Chimie 2010;13:3–39.

[32] Debecker DP, Mutin PH. Non-hydrolytic sol–gel routes to heterogeneous catalysts. Chem Soc Rev 2012;41:3624–50.

[33] Trimmel G, Moraru B, Gross S, Di Noto V, Schubert U. Cross-linking of poly(methyl methacrylate) by oxozirconate and ozxotitanate clusters. Macromol Symp 2001; 175:357–66.

[34] Gross S. Oxocluster-reinforced organic–inorganic hybrid materials: effect of transition metal oxoclusters on structural and functional properties. J Mater Chem 2011; 21:15853–61.

[35] Alvi NH, Willander M, Nur O. The effect of the post-growth annealing on the electroluminescence properties of n-ZnO nanorods/p-GaN light emitting diodes. Superlattices Microstruct 2010;47:754–61.

[36] Du Y, Zeng F. Annealing effects on the cathodoluminescence properties of individual ZnO nanowire. Mater Lett 2011;65:2238–40.

[37] Xu N, Cui Y, Hui Z, Yu W, Sun J, Xu N, et al. Photoluminescence and low-threshold lasing of ZnO nanorods arrays. Opt Express 2012;20:14857–63.

[38] Lee Y-C, Lee T-Y, Jan J-H. Piezoelectric properties and microstructures of ZnO doped $Bi_{0.5}Na_{0.5}TiO_3$ ceramics. J Eur Ceram Soc 2011;31:3145–52.

[39] Wang XL, Lai KH, Rutuolo A. A comparative study on the ferromagnetic properties of undoped and Mn-doped ZnO. J Alloys Compd 2012;542:147–50.

[40] Shinde SD, Patil GE, Kajale DD, Gaikwad VB, Jain GH. Synthesis of ZnO nanorods by spray pyrolysis for H_2S gas sensor. J Alloys Compd 2012;528:109–14.

[41] Lim SK, Hwang S-H, Kim S, Park H. Preparation of ZnO nanorods by microemulsion synthesis and their application as a CO gas sensor. Sens Actuators B: Chem 2011; 160:94–8.

[42] Kim Y-T, Park J, Kim S, Park DW, Choi J. Fabrication of hierarchical ZnO nanostructures for dye-sensitized solar cells. Electrochim Acta 2010;78:417–21.

[43] Kiomrsipour N, Razavi RS. Characterization and optical property of ZnO nano-, submicro, and microrods synthesized by hydrothermal method on a large-scale. Superlattices Microstruct 2012;52:704–10.

[44] Wang ZL. Zinc oxide nanostructures: growth, properties and applications. J Phys: Condens Mater 2004;16:R829–58.

[45] Kołodziejczak-Radzimska A, Jesionowski T. Zinc oxide–from synthesis to application: a review. Materials 2014;7:2833–81.

[46] Sahoo S, Bhowmick AK. Influence of ZnO nanoparticles on the cure characteristics and mechanical properties of carboxylated nitrile rubber. J Appl Polym Sci 2007;106:3077–83.

[47] Chen CC, Liu P, Lu CH. Synthesis and characterization of nano-sized ZnO powders by direct precipitation method. Chem Eng J 2008;144:509–13.

[48] Hong R, Pan T, Qian J, Li H. Synthesis and surface modification of ZnO nanoparticles. Chem Eng J 2006;119:71–81.
[49] Dhage SR, Pasricha R, Ravi V. Synthesis of fine particles of ZnO at 100°C. Mater Lett 2005;59:779–81.
[50] Jia W, Dang S, Liu H, Zhang Z, Yu C, Liu X, et al. Evidence of the formation mechanism of ZnO in aqueous solution. Mater Lett 2012;82:99–101.
[51] Wu C, Qiao X, Chen J, Wang H, Tan F, Li S. A novel chemical route to prepare ZnO nanoparticles. Mater Lett 2006;60:1828–32.
[52] Lanje AS, Sharma SJ, Ningthoujam RS, Ahn JS, Pode RB. Low temperature dielectric studies of zinc oxide (ZnO) nanoparticles prepared by precipitation method. Adv Powder Technol 2013;24:331–5.
[53] Farhadi-Khouzani M, Fereshteh Z, Loghman-Estarki MR, Razavi RS. Different morphologies of ZnO nanostructures via polymeric complex sol–gel method: synthesis and characterization. J Sol–Gel Sci Technol 2012;64:193–9.
[54] Duran N. Nanotecnologia: introdução, preparação e caracterização de nanomateriais e exemplos de aplicação / Nelson Duran; Luiz Henrique Mattoso; Paulo Cezar de Morais–São Paulo: Artliber Editora, 2006.
[55] Park J, Lee E, Hwang N-M, Kang M, Kim SC, Hwang Y, et al. One-nanometer scale size-controlled synthesis of monodisperse magnetic iron oxide nanoparticles. Angew Chem Int Ed 2005;44:2872–7.
[56] Singamaneni S, Bliznyuk VN, Binek CV, Tsymbal EY. Magnetic nanoparticles: recent advances in synthesis, self-assembly and applications. J Mater Chem 2011;21:16819–45.
[57] Sun S, Zeng H. Size-controlled synthesis of magnetic nanoparticles. J Am Chem Soc 2002;124:8204–5.
[58] Wang Y, Zhu Z, Xu F, Wei X. One-pot reaction to synthesize superparamagnetic iron oxide nanoparticles by adding phenol as reducing agent and stabilizer. J Nanopart Res 2012;14:755.
[59] Granitzer P, Rumpf K, Venkatesan M, Cabrera L, Roca AG, Morales MP, et al. Magnetic behavior of a magnetite/silicon nanocomposite. J Nanopart Res 2011; 13:5685–90.
[60] Li D, Jiang D, Chen M, Xie J, Wu Y, Dang S, et al. An easy fabrication of monodisperse oleic acid-coated Fe_3O_4. Mater Lett 2010;64:2462–4.
[61] Sun S, Zeng H, Robinson DB, Raoux S, Rice PM, Wang SX, et al. Monodisperse MFe_2O_4(M = Fe, Co, Mn) nanoparticles. J Amer Chem Soc 2004;126:273–9.
[62] Petcharoen K, Sirivat A. Synthesis and characterization of magnetic nanoparticles via the chemical co-precipitation method. Mater Sci Eng B 2012;177:421–7.
[63] Babes L, Denizot B, Tanguy G, Le Jeune JJ, Jallet P. Synthesis of iron nanoparticles used as MRI contrast agent: a parametric study. J Colloid Interface Sci 1999; 212:474–82.
[64] Kim DK, Mikhaylova M, Zhang Y, Muhammed M. Protective coating of superparamagnetic iron oxide nanoparticles. Chem Mater 2003;15:1617–27.
[65] Herea DD, Chiriac H, Lupu N. Preparation and characterization of magnetic nanoparticles with controlled magnetization. J Nanopart Res 2011;13:4357–69.
[66] Zhu L, Li C, Wang J, Zhang H, Zhang J, Shen Y, et al. A simple method to synthesize modified Fe_3O_4 for the removal of organic pollutants on water surface. Appl Surface Sci 2012;258:6326–30.
[67] Qu S, Yang H, Ren D, Kan S, Zou G, Li D, et al. Magnetite nanoparticles prepared by precipitation from partially reduced ferric chloride aqueous solutions. J Colloid Interface Sci 1999;215:190–2.

[68] Zhang H, Zhang Q, Zhang B, Xu H, Fan X. Preparation of inner asymmetric composite microspheres. Polym Int 2012;61:990–3.
[69] Laurent S, Forge D, Port M, Roch A, Robic C, Vander Elst L, et al. Magnetic iron oxide nanoparticles. Synthesis, stabilization, vectorization, physicochemical characterizations and biological applications. Chem Rev 2008;108:2064–110.
[70] Kickelbick G. Concepts for the incorporation of inorganic building blocks into organic polymers on a nanoscale. Prog Polym Sci 2003;28:83–114.
[71] Kumari L, Du G-H, Li W-Z, Vennila RS, Saxena SK, Wang D-Z. Synthesis, microstructure and optical characterization of zirconium oxide nanostructures. Ceram Int 2009;35:2401–8.
[72] Ataie A, Piramoon MR, Harris IR, Ponton CB. Effect of hydrothermal synthesis environment on the particle morphology, chemistry and magnetic properties of barium hexaferrite. J Mater Sci 1995;30:5600–6.
[73] Murase Y, Kato E. Preparation of zirconia whiskers from zirconium hydroxide in sulfuric acid solutions under hydrothermal conditions at 200°C. J Am Ceram Soc 2001;84:2705–6.
[74] Asilturk M, Burunkaya E, Sayilkan F, Kiraz N, Arpaç E. Structural and optical properties of thin films prepared from surface modified ZrO_2. J Non-Cryst Solids 2011;357:206–10.
[75] Sayilkan F, Asilturk M, Burunkaya E, Arpaç E. Hydrothermal synthesis and characterization of nanocrystalline ZrO_2 and surface modification with 2-acetoacetoxyethyl methacrylate. J Sol–Gel Sci Technol 2009;51:182–9.
[76] Garnweitner G, Goldenberg LM, Sakhno OV, Antonietti M, Niederberger M, Stumpe J. Large-scale synthesis of organophilic zirconia nanoparticles and their application in organic–inorganic nanocomposites for efficient volume holography. Small 2007; 3:1626–32.
[77] Wang X, Wu L, Li J. Influence of nanozirconia on the thermal stability of poly(methyl methacrylate) prepared by in situ bulk polymerization. J Appl Polym Sci 2010; 117:163–70.
[78] Caracoche MC, Rivas PC, Cervera MM, Caruso R, Benavidez E, Sanctis O, et al. Zirconium oxide structure prepared by the sol–gel route: I, the role of the alcoholic solvent. J Am Ceram Soc 2000;83:377–84.
[79] Schubert U, Arpac E, Glaubit W, Helmerich A, Chau C. Primary hydrolysis products of methacrylate-modified titanium and zirconium alkoxides. Chem Mater 1992;4:291–5.
[80] Song XM, Wu JM, Yan M. Photocatalytic degradation of selected dyes by titania thin films with various nanostructure. Thin Solid Films 2009;517:4341–7.
[81] Kolenko YV, Churagulov BR, Kunst M, Mazerolles L, Clbeau-Justin C. Photocatalytic properties of titania powder prepared by hydrothermal method. Appl Catal B: Environ 2004;54:51–8.
[82] Gutierrez J, Tercjak A, Mondragon I. Conductive behavior of high TiO_2 nanoparticle content of inorganic/organic nanostructured composites. J Am Chem Soc 2010; 132:873–8.
[83] Cheng YJ, Gutmann JS. Morphology phase diagram of ultrathin anatase TiO_2 films templated by a single PS-*b*-PEO block copolymer. J Am Chem Soc 2006;28:4658.
[84] Kuang X, Gao Q, Zhu H. Effect of calcination temperature of TiO_2 on the crystallinity and the permittivity of PVDF-TrFE/TiO_2 composites. J Appl Polym Sci 2013; 129:296–300.
[85] Arantes TM, Leite ER, Longo E, Camargo ER. Nanocomposites of styrene-butadiene rubber and synthetic anatase obtained by a colloidal route and their photooxidation. J Appl Polym Sci 2009;113:1898–904.

[86] Shan GB, Demopoulos GP. The synthesis of aqueous-dispersible anatase TiO_2 nanoplatelets. Nanotechnology 2010;21 025604 (9pp).
[87] Niederberger M, Barti MH, Stucky GD. Benzyl alcohol and titanium tetrachloride. A versatile reaction system for the non-aqueous and low-temperature preparation of crystalline and luminescent titania nanoparticles. Chem Mater 2002;14:4364.
[88] Spadavecchia J, Boujday S, Landoulsi J, Pradier CM. nPEG-TiO_2 nanoparticles: a facile route to elaborate nanostructured surfaces for biological applications. Appl Mater Interfaces 2011;3:2637−42.
[89] Xie ZP, Lu JW, Huang Y, Cheng YB. Influence of α-alumina seed on the morphology of grain growth in alumina ceramics from Bayer aluminum hydroxide. Mater Lett 2003;57:2501−8.
[90] Li J, Pan Y, Xiang C, Ge Q, Guo J. Low temperature synthesis of ultrafine α-Al_2O_3 powder by a simple aqueous sol−gel process. Ceram Int 2006;32:587−91.
[91] Hung-Low F, Peterson GR, Davis M, Hope-Weeks LJ. Rapid preparation of high surface area iron oxide and alumina nanoclusters through a soft templating approach of sol−gel precursor. New J Chem 2013;37:245−9.
[92] Park YK, Tadd EH, Zubris M, Tannenbaum R. Size-controlled synthesis of alumina nanoparticle from aluminum alkoxide. Mater Res Bull 2005;40:1506−12.
[93] Bokhimi X, Toledo-Antoni JA, Guzman-Castillo ML, Mar-Mar B, Hernandez-Beltran F, Navarrete J. Dependence of boehmite thermal evolution on its atom bond lengths and crystalline size. J Solid State Chem 2001;161:319−26.
[94] Chandradass J, Bae DS. Synthesis and characterization of alumina nanoparticles by IGEPAL CO-520 stabilized reverse micelle and sol−gel processing. Mater Manuf Processes 2008;23:494−8.
[95] Qu L, He C, Yang Y, He Y, Liu Z. Hydrothermal synthesis of alumina nanotubes templated by anionic surfactant. Mater Lett 2005;59:4034−7.
[96] Lee JS, Kim HS, Park NK, Lee TJ, Kang M. Low temperature synthesis of α-alumina from aluminum hydroxide hydrothermally synthesized using $[Al(C_2O_4)_X(OH)_Y]$ complexes. Chem Eng J 2013;230:351−60.
[97] Utamapanya S, Klabunde KJ, Schlup JR. Nanoscale metal oxide particles/clusters as chemical reagents. Synthesis and properties of ultrahigh surface area magnesium hydroxide and magnesium oxide. Chem Mater 1991;3:175−81.
[98] Chadwick AV, Poplett IJF, Maitland DTS, Smith ME. Oxygen speciation in nanophase MgO from solid state ^{17}O NMR. Chem Mater 1998;10:864−70.
[99] Jiu J, Kurumada K-I, Tanigaki M, Adachi M, Yoshikawa S. Preparation of nanoporous MgO using gel as structure-direct template. Mater Lett 2003;58:44−7.
[100] Lopez T, Gomez R, Navarrete J, Lopez-Salina E. Evidence for lewis and bronsted acid sites on MgO obtained by sol−gel. J Sol−Gel Sci Technol 1998;13:1043−7.
[101] Bhargava A, Alarco JA, Mackinnon IDR, Page D, Ilyushechkin A. Synthesis and characterization of nanoscale magnesium oxide powders and their application in thick films of $Bi_2Sr_2CaCu_2O_8$. Mater Lett 1998;34:133−42.
[102] Khairallah F, Glisenti A. Synthesis, characterization and reactivity study of nanoscale magnesium oxide. J Mol Catal A: Chem 2007;274:137−47.
[103] Bokhimi X, Morales A, Lopez T, Gomez R. Crystalline structure of MgO prepared by the sol−gel technique with different hydrolysis catalysts. J Solid State Chem 1995;115:411−15.
[104] Roy K, Alam MN, Mandal SK, Debnath SC. Preparation of zinc oxide—free natural rubber nanocomposites using nanostructured magnesium oxide as cure activator. J Appl Polym Sci 2015;132:42705.

[105] Chhor K, Bocquet JF, Pommier C. Synthesis of submicron magnesium oxide powders. Mater Chem Phys 1995;40:63–8.
[106] Sahoo S, Maiti M, Ganguly A, George JJ, Bhowmick AK. Effect of zinc oxide nanoparticles as cure activator on the properties of natural rubber and nitrile rubber. J Appl Polym Sci 2007;105:2407–15.
[107] Wang J, Chen Y. Application of nano-Zinc oxide master batch in polybutadiene styrene rubber system. J Appl Polym Sci 2006;101:922–30.
[108] Fukuda T, Fujii S, Nakamura Y, Sasaki M. Mechanical properties of sîlica particle-filled styrene-butadiene rubber composites containing polysulfide-type silane coupling agents: influence of loading method of silane. J Appl Polym Sci 2013;130:322–9.
[109] Wang Q, Gao W, Zhang L. Research of styrene-butadiene rubber/silicon-aluminum oxide nanotube binary nanocomposites. J Appl Polym Sci 2011;120:3196–203.
[110] Wang Z, Lu Y, Liu J, Dang Z, Zhang L, Wang W. Preparation of nano-zinc oxide/EPDM composites with both good thermal conductivity and mechanical properties. J Appl Polym Sci 2011;119:1144–55.
[111] Arantes TM, Sala RL, Leite ER, Longo E, Camargo ER. Comparison of the nanoparticles performance in the photocatalytic degradation of a styrene-butadiene rubber nanocomposites. J Appl Polym Sci 2013;128:2368–74.
[112] Kofod G, Stoyanov H, Kollosche M, Risse S, Ragusch H, Rychkov D, et al. Nano-scale materials science for soft dielectrics: composites for dielectric elastomer actuators. 2010 IEEE-International Conference on Solid Dielectrics, Potsdam, Germany, 2010, July 4–9, 1–4.
[113] Messori M. In situ synthesis of rubber nanocomposites. In: Mittal V, Kim JK, Pal K, editors. Recent advances in elastomeric nanocomposites. Berlin: Springer; 2011. p. 57–85.
[114] Murugesan S, Sur GS, Mark JE, et al. In situ catalyst generation and controlled hydrolysis in the sol–gel precipitation of zirconia and titania particles in poly (dimethylsiloxane). J Inorg Organometallics Polym 2004;14:239–52.
[115] Yang C, Tang YH, Lam WM, Lu WW, Gao P, Zhao CB, et al. Moisture-cured elastomeric transparent UV and X-ray shielding organic-inorganic hybrids. J Mater Sci 2010;45:3588–94.
[116] Iketani K, Sun RD, Toki M, Hirota K, Yamaguchi O. Sol–gel derived TiO_2/poly (dimethylsiloxane) hybrid films and their photocatalytic activities. J Phys Chem Solids 2003;64:507–13.
[117] Beniska J, Dogadkin B. The influence of activators on the vulcanization process. I. the influence of zinc oxide on the rate of combination of sulfur with rubber. Rubber Chem Technol 1959;32:774–9.
[118] Mandal UK, Tripathy DK, De SK. Dynamic mechanical spectroscopic studies on plasticization of an ionic elastomer based on carboxylated nitrile rubber by ammonia. Polymer 1996;37:5739–42.
[119] Miyamata Y, Atsumi M. Zinc oxide cross-linking reaction of polychloroprene rubber. Rubber Chem Technol 1989;62:1–12.
[120] Roy K, Alam MN, Mandal SK, Debnath SC. Sol–gel derived nano zinc oxide for the reduction of zinc oxide level in natural rubber compounds. J Sol–Gel Sci Technol 2014;70:378–84.
[121] Debnath SC, Basu DK. Studies on the effect of thiuram disulfide on NR vulcanization accelerated by thiazole-based accelerator systems. J Appl Polym Sci 1996;60:845–55.
[122] Coran AY. In: Erman B, Mark JE, Roland CM, editors. The science and technology of rubber. 4th ed. Amsterdam: Elsevier; 2013. p. 355.
[123] Fosmire GJ. Zinc toxicity. Am J Clin Nutr 1990;51:225–7.

[124] Zimmer RJ, Boes CEF. Rubber composition containing nanoscaled zinc oxide particles. United States Patent US, 6,972,307 B2 in December 6, 2005, Assigned by The Goodyear Tire & Rubber company.

[125] Mottaghi M, Khorasani SN, Esfahany MN, Farzadfar A, Talakesh MM. Comparison of the effect of nano ZnO and conventional grade ZnO on the cross-linking densities of NR/BR and NR/SBR blends. J Elastomers Plast 2012;44:443–51.

[126] Thomas SP, Mathew EJ, Marykutty CV. Synthesis and effect of surface modified nano ZnO in natural rubber vulcanization. J Appl Polym Sci 2012;124:3099–107.

[127] Garreta E, Agulló N, Borrós S. The role of activator during the vulcanization of natural rubber using sulfonamide accelerator type. Kautsch Gummi Kunstst 2002;55:82–5.

[128] Heideman G, Noordermeer JWM, Datta RN, van Baarle B. Effect of metal oxides as activator for sulphur vulcanization in various rubbers. Kautsch Gummi Kunstst 2005;58:30–42.

[129] Guzman M, Reyes G, Agulló N, Borrós S. Synthesis of Zn/Mg oxide nanoparticles and its influence on sulfur vulcanization. J Appl Polym Sci 2011;119:2048–57.

[130] Pingot M, Pingot T, Zaborski M. Nanostructured metal oxide and unsaturated acid as co-agent in peroxide cross-linking of hydrogenated butadiene-acrylonitrile rubber. Prog Colloid Polym Sci 2011;138:147–50.

[131] Kar S, Bhowmick AK. Nanostructured magnesium oxide as cure activator for polychloroprene rubber. J Nanosci Nanotechnol 2009;9:3144–53.

[132] MacKnight WJ, Lundberg RD. Elastomeric ionomers. Rubber Chem Technol 1984;57:652–63.

[133] Hamed GR, Hua K-C. Effect of ZnO particle size on the curing of carboxylated NBR and carboxylated SBR. Rubber Chem Technol 2004;77:214–26.

[134] Przybyszewska M, Zaborski M. The effect of zinc oxide nanoparticle morphology on activity in cross-linking of carboxylated nitrile elastomer. Express Polym Lett 2009;3:542–52.

[135] El-Sabbagh. Preparation and characterization of high performance rubber vulcanizates loaded with modified aluminum oxide. Pigm Resin Technol 2006;35:119.

[136] Loganathan L, Chandrasekar S. Effects of nano and micro Al_2O_3 filler on dielectric properties of silicone composites. J Mater Sci Eng 2010;4:51–6.

[137] Loganathan L, Chandrasekar S. Analysis of surface tracking of micro and nano size alumina filled silicone rubber for high voltage AC transmission. J Electr Eng Technol 2013;8:345–53.

[138] Zhou W, Qi S, Tu C, Zhao H, Wang C, Kou J. Effect of the particle size of Al_2O_3 on the properties of filled heat-conductive silicone rubber. J Appl Polym Sci 2007;104:1312–18.

[139] Wang Z, Lu Y, Liu J, Dang Z, Zhang L, Wang W. Preparation of nanoalumina/EPDM composites with good performance in thermal conductivity and mechanical properties. Polym Adv Technol 2011;22:2302–10.

[140] Zha J, Dang Z, Li W, Zhu Y, Chen G. Effect of micro-Si_3N_4-nano-Al_2O_3 co-filled particles on thermal conductivity, dielectric and mechanical properties of silicone rubber composites. IEEE Trans Dielectr Electr Insul 2014;21:1989–96.

[141] Minabe T, Tryk DA, Sawunyama P, Kikuchi Y, Hashimoto K, Fujishima A. TiO_2-mediated photodegradation of liquid and solid organic compounds. J Photochem Photobiol, A 2000;137:53–62.

[142] Nasu A, Otsubo Y. Rheology and UV-protecting properties of complex suspension of titanium dioxides and zinc oxides. J Colloid Interface Sci 2007;310:617–23.

[143] Razmjou A, Resosudarmo A, Holmes RL, Li H, Mansouri J, Chen V. The effect of modified TiO_2 nanoparticles on the polyethersulfone ultrafiltration hollow fiber membranes. Desalination 2012;287:271–80.
[144] Seentrakoon B, Junhasavasdikul B, Chavasiri W. Enhanced UV-protection and antibacterial properties of natural rubber/rutile-TiO_2 nanocomposites. Polym Degrad Stab 2013;98:566–78.
[145] Musa M, Arief YZ, Abdul-Malek Z, Ahmad MH, Abd Jamil AA. Influence of nano-titanium dioxide (TiO_2) on electrical tree characteristics in silicone rubber based nanocomposites. IEEE Annual report Conference on Electrical Insulation and dielectric Phenomena 2013; 498–501: 20.
[146] Hayeemasae N, Rathnayake WGIU, Ismail H. Nano-sized TiO_2- reinforced natural rubber composites prepared by latex compounding method. J Vinyl Addit Technol 2015. Available from: http://dx.doi.org/10.1002/vnl.21497.
[147] Motaung TE, Luyt AS, Thomas S. Morphology and properties of NR/EPDM rubber blends filled with small amounts of titania nanoparticles. Polym Compos 2011;32:1289–96.
[148] Zaborski M, Maslowski M. Magnetorheological elastomer composites. Prog Colloid Polym Sci 2011;138:21–6.
[149] Al-Juaid SS, El-Mossalamy EH, Arafa HM, Al-Ghamdi AA, Abdel Daiem AM, El-Tantawy F. Novel functional nitrile butadiene rubber/ magnetite nano composites for NTCR thermistors application. J Appl Polym Sci 2011;121:3604–12.
[150] Kong I, Ahmad SH, Abdullah MH, Hui D, Yusoff AN, Puryanti D. Magnetic and microwave absorbing properties of magnetite-thermoplastic natural rubber nanocomposites. J Magn Magn Mater 2010;322:3401–9.
[151] Kim CY, Xu L, Lee EH, Choa YH. Magnetic silicone composites with uniform nanoparticle dispersion as a biomedical stent coating for hyperthermia. Adv Polym Technol 2013;32:E714–23.
[152] Ashjari M, Mahdavian AR, Ebrahimi NG, Mosleh Y. Efficient dispersion of magnetite nanoparticles in the polyurethane matrix through solution mixing and investigation of the nanocomposite properties. J Inorg Organometallics Polym 2010;20:213–19.
[153] Abramchuk SS, Grishin DA, Kramarenko EY, Stepanov GV, Khokhlov AR. Effect of a homogeneous magnetic field on the mechanical behavior of soft magnetic elastomers under compression. Polym Sci, Ser A 2006;48:138–45.
[154] Venkatanarasimhan S, Raghavachari D. Epoxidized natural rubber–magnetite nanocomposites for oil spill recovery. J Mater Chem A 2013;1:868–76.
[155] Ong HT, Julkapli NM, Hamid SBA, Boondamnoen O, Tai MF. Effect of magnetic and thermal properties of iron oxide nanoparticles (IONS) in nitrile butadiene rubber (NBR) latex. J Magn Magn Mater 2015;395:173–9.
[156] Tan WL, Bakar MA. Synthesis, characterization and impedance spectroscopy study of magnetite/epoxidized natural rubber nanocomposites. J Alloys Compd 2013; 561:40–7.
[157] Evans BA, Fiser BL, Prins WJ, Rapp DJ, Shields AR, Glass DR, et al. A highly tunable silicone-based magnetic elastomer with nanoscale homogeneity. J Magn Magn Mater 2012;324:501–7.
[158] Yang D, Zhang W, Jiang B, Guo Y. Silicone rubber ablative composites improved with zirconium carbide or zirconia. Composites, Part A 2013;44:70–7.
[159] Katayama S, Kubo Y, Yamada N. Characterization and mechanical properties of flexible dimetylsiloxane-based inorganic/organic hybrid sheets. J Am Ceram Soc 2002;85:1157–63.

Rubber blend nanocomposites

10

K. Pal, V. Panwar and J. Bahadur
Indian Institute of Technology, Roorkee, Uttarakhand, India

10.1 Introduction to elastomers

Elastomers are viscoelastic polymeric materials having very weak intermolecular forces, which are responsible for their poorer elastic modulus and higher failure strain than other polymers, metals, and ceramics, i.e., they show significant perpetual deformation even after being elongated continually twice from their inventive shape [1]. These are also well known as rubber or elastic polymers and as a subset of amorphous polymers which exist above their glass transition temperature and allow considerable movement of small chain segments. Elastomers may be used in diverse applications depending on the required characteristics, such as hardness, energy absorption, etc. These materials can be prepared via cross-linking of polymer chains through chemical bonds. Elastomers have been considered as most versatile engineering materials which can efficiently recover from large deformations. They exhibit a unique combination of resilience and elasticity when subjected to tensile, compression, or twisting. In addition, most elastomers possess low permeability to air, water, or gases; good insulative (electrical and thermal) properties; good mechanical properties; and good bonding ability with metals, fibers, and plastics, due to the presence of long chains of carbon, hydrogen, and oxygen atoms which facilitate high degree of cross-linking among neighboring chains.

10.1.1 Basic properties of elastomers

1. Unable to melt when passed through gaseous phase
2. Offer swelling in the presence of good solvent
3. Insoluble in nature
4. Elastomeric blends facilitate flexibility
5. Thermosetting elastomers show smaller creep resistance in comparison to thermoplastics

10.1.2 Types of elastomers

Elastomers can be mainly classified as thermosets and thermoplastics. Thermosetting elastomers are highly cross-linked polymers having excellent mechanical properties, such as tensile strength, dimension stability, good thermal resistance, and easy processing. Cross-linking of polymer chains plays a very crucial role in affecting the morphology, miscibility, and composition of thermosetting polymers. Some common thermosetting elastomers include natural rubber (NR), silicone, styrene butadiene,

Progress in Rubber Nanocomposites. DOI: http://dx.doi.org/10.1016/B978-0-08-100409-8.00010-3

chloroprene, fluorocarbon, ethylene-propylene, polyurethane, etc. Still their low toughness, resistance to temperature and crack initiation restricts their extensive applicability [2]. On the other hand, thermoplastics exhibit a combination of thermoplastic and elastomeric properties. Some typical thermoplastic elastomers include thermoplastic urethane, styrenic block copolymers, copolyether ester elastomers, polyester amide, etc. These elastomeric materials facilitate different noble electrical properties, such as excellent water, ozone, weather, heat, and steam resistance and have smaller temperature flexibility. These are having wide application in automobile industries. The main advantages of thermoplastic elastomers over rubber are easy processing and the ability to change properties due to recycling and reuse. These elastomers have a strong ability to retain their shape after being deformed, and have a high melting temperature and zero-creep phenomenon.

10.1.3 Advantages, applications, and limitations of some of the most commonly used elastomers

The most commonly used elastomer, which is popularly known as NR, is preferred due to its high abrasion and resilience, good compression set resistance, and better surface friction properties but its poor resistance to petroleum oils, ozone, and sunlight limits its practical applicability. However, they were found to be highly appropriate for automotive diaphragms, beverage seals, etc. Moreover, silicone exhibits higher temperature properties and compression set resistance for production of goods but still faces poor resistance against ozone and ultraviolet radiation. These were also noticed to be competitive with NR for similar area of applications, i.e., automotive diaphragms, food and beverage seals, etc. Polyurethane is another thermoplastic elastomer which exhibits higher tensile strength, low barrier for water, and resistance for petroleum oils, weather, and good abrasion resistance. With an increase in temperature it softens and consequently drops strength and fluid resistance. These elastomers are commonly used as seals for hydraulic pressure and stressed components in wear prone zones. Styrene butadiene rubber (SBR) is a synthetic rubber having excellent resistance for water and brake fluids; and low resistance to weather and petroleum oils. SBR has its main applications for plumbing, diaphragms, etc. Another type of rubber elastomer is ethylene-acrylic which facilitates better resistance to water and alkalis, good vibration damping, heat aging properties, and excellent dynamic properties for a large range of temperature. Ethylene-acrylic has been commonly used in automotive sealing, transmission, etc. Polyacrylate rubber also comes in the category of synthetic rubber which exhibits excellent resistance for petroleum fuel and oils, ozone heating, and flex cracking but faces low resistance to compression set and water. The main applications of polyacrylate lie in automotive transmissions. Neoprene (CR), also known as polychloroprene, offers excellent resistance to ozone, sunlight, and ammonia; and exhibits low resistance to water. It has been used in seals, wires, and refrigeration industries. Fluorocarbon-based elastomers offer excellent mechanical properties, low resistance to hot water, as well as good resistance to chemical, compression set, and heat. These elastomers are found to be effective in aircraft engine seals and automotive fuel handling. Synthetic nitrile rubbers offer lower resistance to weather and exhibit good resistance for compression set, tear, and abrasion. These are commonly used in automotive and aircraft components.

Although rubber and elastomers are often used interchangeably, more specifically, an elastomer is termed as any polymer having the elastic properties of rubber. On the other hand, the term rubber is more appropriate for vulcanizates.

10.2 Rubber blends and composites

Various rubber/rubber or rubber/polymer blends as well as rubber-based composites reinforced with different fillers, i.e., clay, TiO_2, silica, nanofibers, nanotubes, nanographite, etc., offers significant improvement in morphological, electrical, thermal, and mechanical properties over pristine rubber. These rubber-based blends and composites facilitate wide applicability of rubber in industrial uses. The phenomenon of the blending process includes incorporation of two or more types of components and is considered as a useful method for preparing the materials with better properties comparing to those of individual constituents. The chemical configuration and composition of each constituent in the blend has its own advantages for precise applications. Blending of rubber provides novel materials with a wide variety of applications because it combines the desirable properties of the blend constituents and offers lower cost than virgin polymers or composites having equivalent mechanical properties [3–5]. A broad classification of polymer blends is illustrated in Fig. 10.1A of which this chapter will be focused on rubber/rubber and rubber/plastic blends.

The morphology of rubber/rubber blends has been generally considered as a three-phase composition in which there exists an interface between the regions of different constituents. On this basis, they are also considered to be heterogeneous blends. However, the quantitative characterization of interfaces in these blends was found to be impossible due to the similar structure of two or more components. On the other hand, rubber/plastic blends emerged as thermoplastic elastomers having the attractive properties of rubber and the possibility for processing as thermoplastics. In addition, they were not required to be vulcanized for application in different components. This offered quite attractive advantages in the era of rubber blends.

10.2.1 Techniques for preparing rubber blends

Some common techniques which were found appropriate for the blending of elastomers are illustrated in Fig. 10.1B. Dynamic vulcanization (DV) is the process of cross-linking one constituent with another polymer or rubber in its molten state. This process offers good dispersion and cross-linking of two constituents. Peroxide-induced DV is a two-step processing technique by which the sample is prepared, i.e., polypropylene (PP)/ethylene-propylene-diene monomer (EPDM)/ZDMA blends

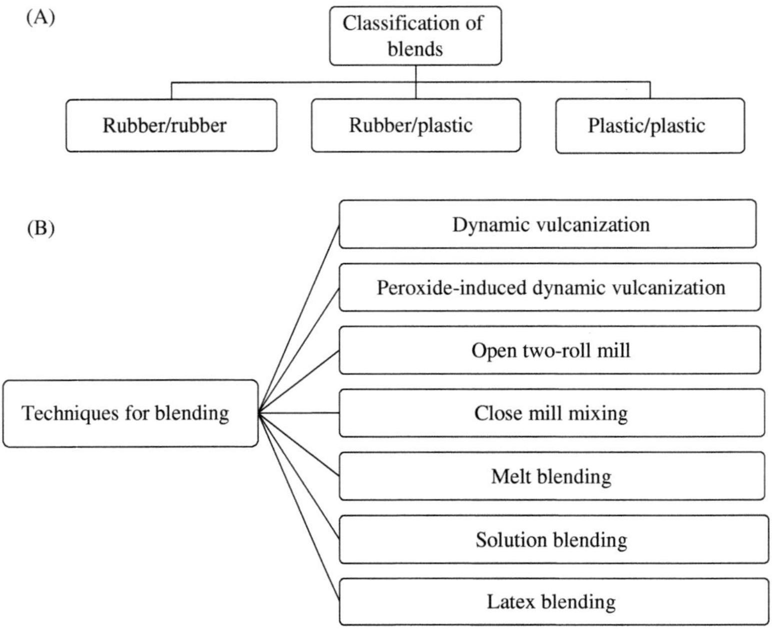

Figure 10.1 (A) Different types of blends; (B) Various techniques used for preparing rubber blends.

have been prepared using this technique. The open two roll mills have been used for the preparation of these composite samples. However, the main disadvantage of this technique is incorporation of dust particles during the synthesis of blends in an open environment.

In the open two roll mill blending process mixing is carried out between two contrary rotors. In this process the rubber component encounters significant displacement and pressure due to the variation in tangential speed and nip constriction which increases the displacement energy to break the bonding between rubber molecules. This enables uniform mixing of carbon and other chemicals with rubber. It can be open or close mill mixing depending on the roll mills are completely open or closed to air and dust. Although, protection against dust is the serious disadvantage in open mill mixing, this does provide access making it easy to clean devices after mixing. It has been noticed that the effectiveness of mixing in these techniques may drastically vary, depending on the operator's skill as well as the quality of components.

Melt mixing is a comparatively simple and cost-effective approach for the preparation of different rubber-based blends using an internal mixture. Some of the main advantages of this technique are the relatively low contamination of samples and automatic removal of water or solvent. However, its limitations are the high cost of equipment and less homogeneity via this method.

Moreover, collection of blended components and cleaning of the mixture to reuse it to blend new compositions are highly tedious and imprecise. Solution mixing is another similar approach which was noticed to be a very useful technique for preparing extremely viscous blends. During mixing it has been observed that the morphology of blend strongly depends on the nature of solvent, reaction temperature, and concentration. This is practically considered as a casting technique and is highly appropriate when the blending is to be carried out through very small quantities of constituents. The main limitation of this technique is the availability of a suitable common solvent for all the components to be blended. Latex blending is one of the techniques which has been seldom used for the preparation of most common polymer blends or composites. However, it has been found to be promising for emulsion polymerization of rubber toughened plastic blends. The absence of water contents and processing of micro-sized constituents in this technique, assist in the excellent distribution of different phases.

10.2.2 Some different types of trendy rubber blends

Blending of rubber with rubbers or polymers was intended to produce novel materials with desired properties and a wide range of applications. A lot of work had been reported by various researchers for different types of possible blends, their optimum properties, and suitable applications. Nowadays, some of those are extensively used in automobile and household appliances. Xu et al. [6] have prepared highly toughened poly-(vinylidene fluoride)/nitrile butadiene rubber (PVDF/NBR) blends under different composition via peroxide-induced DV. They observed that the integration of NBR increased impact strength and elongation at break but conversely it was also responsible for a slight reduction in tensile strength and flexural strength of the blend. The Izod-impact strength of 28 wt% NBR-PVDF blend was found to be 77.5 KJ/m^2, indicating a 6–7 times increase in comparison to pure PVDF.

Pongtanayut et al. [7] prepared blends of poly-(lactic acid) (PLA) with NR and epoxidized natural rubber (ENR), via the melt blending technique. It has been well known that PLA is biodegradable, due to which its applications are restricted in the fields requiring high brittleness and it shows poor crystallization performance. However, researchers overcame these demerits of PLA through blending with rubber. The integration of NR has shown better results for improvement in crystallization ability of PLA, in comparison to ENR/PLA blends, and it has been observed that its thermal stability was reduced with both types of rubbers. However, the ductility and toughness of the NR/PLA blend were significantly improved. Bitinis et al. [8] had also investigated NR/PLA blend prepared via melt blending technique and noticed that the ductility of PLA was significantly improved. Manshaie et al. [9] reported that NR/SBR blend prepared via two roll mills and cured through irradiation showed an increase in tensile strength, hardness, thermal stability, and abrasion resistance with electron beam irradiation but simultaneously its elongation at break, compression set and resilience were found to decrease. Table 10.1 shows some typical types of rubber blends and the corresponding techniques used for preparation of those blends.

Table 10.1 Various types of rubber based blends prepared via different techniques

Composition of blends	Techniques used	References
PVDF/silicone rubber/fluro-rubber	Peroxide dynamic vulcanization	[10]
PVDF/silicone rubber	Dynamic vulcanization	[11]
NR/SBR	Mechanical mixing, solution blending	[12]
Poly (ethylene-naphthalate)/glass fibers/NBR	Melt mixing	[13]
Ternary polystyrene/SBR/polyethylene	Melt mixing	[14]
Ethylene-propylene block copolymer/ EPDM rubber	Melt mixing	[15]
Poly (vinyl chloride)/epoxide natural rubber	Melt blending	[16]
Isotactic polypropylene/acrylonitrile-co-butadiene rubber	Melt mixing	[17]
SBR/EPDM	Melt mixing	[18]
Poly(vinyl chloride)/epoxidized natural rubber	Melt blending	[19]
Poly(butylene terephthalate)/epoxide-containing rubber	Melt blending	[20]
Maleated natural rubber/NR	Two roll mill	[21]

10.2.3 Types of fillers used for preparing rubber based composites

As discussed earlier, elastomeric materials have been widely used in different areas of applications due to their excellent properties, i.e., light weight, design flexibility, and processability. However, their low strength and elastic modulus as compared to metals and ceramics limits their practical applicability. It had been noticed that the addition of fillers/nanofillers changes one or more properties of elastomeric composites, i.e., magnetic, optical, electrical, thermal, mechanical, rheological, surface characteristics, and durability, directing their physical suitability as per the required dynamics. Studies revealed that these properties strongly depend on the dispersion behavior of filler and chemical interaction between the two or more constituents. It has also been observed that the physical and mechanical properties of composites were extremely dependent on the nature of fillers. The distinguished fillers may be characterized on the basis of different parameters such as color, source, and activity of filler [22]. Fig. 10.2 represents typical classification of fillers based on their behavioral aspects, though some of these fillers may simultaneously exist in two or more categories.

Studies covered various particulates, fibers, whiskers, etc. as reinforcements in rubber matrices. These studies were intended to modify the physical properties of original matrices and to optimize corresponding processing–performance

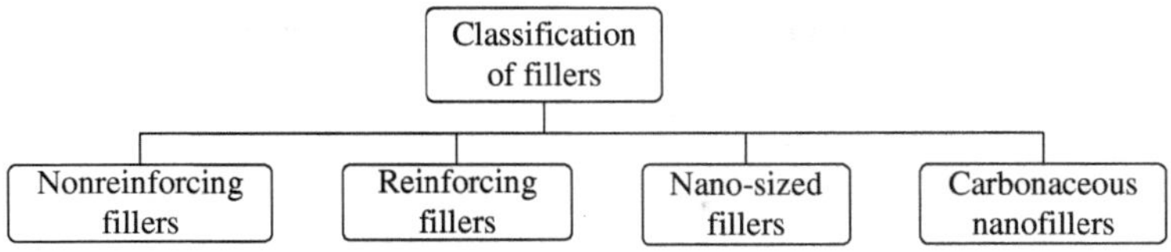

Figure 10.2 Types of fillers for rubber-based composites.

correlations. The particulate fillers have been classified into nonreinforcing and reinforcing fillers. Generally, nonreinforcing fillers tend to increase the viscosity of the composites which results in reducing the mechanical properties of the vulcanizates. However, in some other cases, nonreinforcing fillers, such as calcium silicate, chalk powder, titanium dioxide, china clay, calcium carbonate, barium sulfate, and zinc [11–15,23–27], were noticed to play vital roles in improving the mechanical and thermal properties under low overall cost of synthesis [28]. China clay and calcium carbonate had been intensively studied and found effective for many applications. On the basis of their particle size (d), they may be characterized as soft and hard clay ($d < 2\ \mu m$ and $d > 2\ \mu m$, respectively). Particularly, it had been noticed that clay helps to improve the hardness and abrasion resistance of rubber composites. The introduction of reinforcing fillers was also found to improve abrasion resistance and hardness of composites but additionally they assist to advance tear resistance and stiffness of the vulcanized composites [29]. Carbon black (CB) and silica are the most widely used reinforcing fillers in the rubber industry. As reinforcing fillers, these were found to facilitate significant improvement in typical composite properties, i.e., tensile strength, modulus, thermal stability, damping characteristics, etc. Currently, nano-sized fillers, such as carbon nanotubes (CNTs), exfoliated graphite, carbon nanofibers (CNFs), sodium montmorillonite, organoclay, etc., are the traditional fillers. Some researchers have shown that the addition of reinforcing elements like CB and nano-sized soft clay resulted in an enormous improvement in desired properties of rubber-based composites [30–31]. Most renowned carbonaceous fillers include CB, carbon fiber/nanofibers, graphite, fullerenes, CNT, layered nanodimensional fillers, and exfoliated graphite. Among these fillers nanotubes have gained the most of attention. The discovery of CNTs in 1991 in by Iijima [32] brought about a new revolution in the field of research. Fundamental and application oriented research on these tubular structures determined and explored the unique chemical structure, morphology, and various distinctive properties of these tubes. Owing to the unique mechanical, chemical, optical, and tuneable electronic properties, they have been proposed to be attractive nanofillers for preparing nanocomposites. They are composed of sp^2 bonded hexagonal arrangements of carbon atoms in the form of concentric tubes (several graphitic shells) of about 1 nm diameter and individual tubes are separated by a distance of 0.34 nm, which facilitates the excellent aspect ratio of these nanotubes. In addition, these were considered as the lightest and stiffest materials so far discovered. The excellent properties of CNTs offered themselves as convention nanofillers for elastomeric composites. Arc discharge, laser ablation, and chemical vapor deposition were the three most common methods for production of nanotubes [33]. Among these methods, the chemical vapor deposition was noticed to be a qualitative and cost-effective approach due to the one step synthesis technique and high yield of CNTs. Furthermore, nanofibers, also known as carbon filament, were also found to be promising as reinforcing elements in rubbers [34]. These were mainly synthesized using carbonaceous gases or through interaction between methane and metal surfaces for the growth of graphitic carbon [34,35]. Since the 1980s many researchers have worked to ensure the physical adaptability of CNFs/filaments as reinforcement in different rubber matrices and noticed the promising results for the use of such composites in a wide area of applications. Some common CNFs, such as Polyacrylonitrile (PAN), rayon, and pitch, were differentiated on the basis of their structural aspects. The carbon filament PAN contains long main units parallel to the fiber axis. These units are bonded together to form stretched networks of monolayer sp^2 carbon where the carbon monolayers seem to be 6 nm to several hundred nanometers wide nanoribbons [36,37]. The crystalline structure of these ribbons appears turbostratic and these ribbons were assembled in parallel to form microfibrils.

Besides these carbon-based fillers, layer dimensional fillers like mica and talc were also investigated as reinforcing elements in elastomeric materials to enhance the properties of base matrices. Due to their relative low cost, easily availability, and excellent chemical, heat, and electrical resistance properties these fillers are found to be efficient to be used as reinforcement in elastomers. The layered structure of these fillers facilitates excellent shock-absorbing and vibration-damping properties. In addition, the anisotropic nature of mica made it suitable for barrier applications. Yoshikawa et al. [38] investigated nitrile rubber composites and studied swelling, permeability, and low temperature properties of untreated mica filler reinforced composites. Debnath et al. [39] have synthesized mica-styrene butadiene rubber composites and investigated the network structure, matrix–filler interaction, and physical properties of these composites. Ahead of the successful carbon based nanofillers, alumina silicate has also been found another candidate in this category which shows tremendous improvement in mechanical and physical properties of layered nanofiller reinforced composites. The addition of nanophase provides better dimensional stability and improvements in tensile and flexural strength but still diminishes the thermal stability, electrical conductivity, and surface appearance. Another prominent type of layered dimensional filler is nanoclay. The analysis of the structural property relations of nanoclay–rubber nanocomposites assists in designing materials with desired properties for different area of applications. It has been noticed that the improvement in properties of composite depend on three main factors, i.e., quality of layered inorganic fillers, dispersion of clay in matrix, and bonding between matrix and filler. Montmorillonite (MMT), hectorite, and saponite are some commonly used natural silicate layered nanofillers [40–42]. The cation exchange reaction replaces Na, K, and Ca ions with alkyl ammonium long-chain ions to make MMT well-suited with an organic matrix. This reaction basically changes the nature of clay from hydrophilic to hydrophobic which is responsible for the improved compatibility of MMT for organic matrices. Clay nanoplatelets having length 0.5 to 1 μm and thickness 1 nm, offer a high aspect ratio of the nanofillers. However, the diameter of silicate sheets is about 20 to 200 nm and the thickness is in the order of 1 nm. The spacing between silicate sheets is less than the radius of gyration of typical rubber matrices which offers a large entropic barrier for penetration of matrix material into the gap and thereby proper dispersion of clay within the matrix. The proper dispersion of these high aspect ratio platelets helps to form a crystalline structure of the composite material.

Nowadays researchers are much more focused towards graphite, which is mainly interesting due to the number of graphene sheets stacked together and because it has a very high aspect ratio within the range of 1000 to 1500 [43]. The basic structure of graphene provides outstanding thermal conductivity, electrical, and mechanical properties, which help it as a reinforcing filler in elastomers. Although direct synthesis of single- or few-layered graphene has been found very difficult, costly, and highly nonuniform, chemical synthesis of graphene oxide (GO) and its subsequent reduction has been found to be an alternative successful approach for achieving high quality graphene derivative. Among various methods available for synthesis of GO, modified Hummers method [44], improved Hummers method [45], and HNO_3 introduced improved Hummers method [46] have been found to be really competent. GO has a hexagonal layered arrangement of carbon atoms consisting of different oxygen functionalities, such as –COOH, –OH, etc., on the basal plane and edges of GO. The outstanding properties of graphene enable it to provide opportunities in a wide area of applications, such as energy conversion and storage, sensor and nanoelectronics based semiconductor devices, structural components, etc.

Among the wide variety of available fillers/nanofillers it seems to be very difficult to extract the specific one for a particular application. While the available literature really helps a lot in this regard, before finding out the appropriate reinforcing filler for a matrix one must find out the suitable particle size, structure, and chemical nature of the filler for the corresponding matrix. These are considered as the three main characteristics of fillers which play vital roles in rubber/polymer matrix composites.

10.2.4 Typical properties of rubber based blends and composites

We have already discussed the advantages of and the need for blending rubber with different rubber or plastics, as well as the addition of various fillers to the rubber matrices in the previous sections. Here, we will discuss some typical properties of rubber blends and rubber matrix composites to present a platform for understanding the effect of the addition of these materials to the base matrix.

Sreeja et al. [47] reported cure characteristics and some mechanical properties of SBR/reclaimed rubber blends. They observed improvement in tensile strength from 1.9 N/mm^2 to 5.07 N/mm^2 with the addition of 80 parts of reclaimed rubber. The poor tensile strength of SBR was believed to be due to its noncrystallizing nature. The tear strength, compression set, and elongation at break of this binary blend were also increased with the increase in reclaimed content. However, at the same time the introduction of reclaimed rubber showed a decrease in resilience. This was thought to be due to some change in energy dissipation through loss at interfaces, friction between intermolecular chains, or breakdown of the basic structure of filler. Other properties like tear strength, elongation at break, compression set, and abrasion loss were observed to improve with the blending of SBR with reclaimed rubber. da Costa et al. [48] investigated the mechanical properties of PP/EPDM rubber: GTR (ground tire rubber, also known as scrap tire rubber) ternary blend with the help of response surface methodology. They predicted the optimal composition of PP, EPDM, and GTR through a minimum number of experiments. Results have shown that the mechanical properties of this ternary blend deteriorate with the increase of GTR particles, due to the poor adhesion between PP and GTR matrices. Here, EPDM was found to act as a compatibilizer which improves the interface between PP and GTR. The ternary blend showed highest impact strength when the content of EPDM and GTR were maintained at 25%, i.e., PP/EPDM:GTR 50/50 (1:1). Lima et al. [49] have also studied the morphology and mechanical properties of the PP/EPDM/GTR ternary blend. This work confirmed the complimentary effect of GTR in the PP/EPDM binary blend. The results of the PP/EPDM blend showed a decrease in hardness, tensile strength, and Young's modulus with increasing content of EPDM. EPDM encapsulated GTR showed better hardness and toughness of the ternary blend. This study has also shown that the size of the GTR particles and their content are the key factors to be considered for successful encapsulation. EPDM content should be increased for larger sized GTR particles.

Mostafa and his coworkers [50] studied the swelling properties and compression set behavior of CB reinforced SBR and NBR rubber composites. They observed the effect of varying time, temperature, and filler loading on these properties. The addition of filler particles resulted in a decrease in swelling percentage and an increase in compression set values. They also observed that the properties of CB/NBR composites were better than CB/SBR composites due to better cross-linking density of the NBR-based composites. This shows the effect of cross-linking density in the case of rubber composites. Devi et al. [51] studied the change in morphology and mechanical properties of clay/SBR nanocomposites due to the addition of organoclay as a reinforcing element. In this case, clay nanoparticles (6–7 nm) showed a significant effect in the reduction in optimum cure time; an improvement in rigidity, ductility, tensile strength, and tear strength; and a high resistance to abrasion. The samples prepared with 1, 3, and 5 phr Cloisite 20 A showed an increase in tensile strength by 61%, 110%, and 164%, respectively. In an out-of-trend manner, Li et al. [52] studied the effect of particle size in rice bran carbon (RBC)/nitrile rubber (NBR) composites. It was noticed that the mechanical properties of the composite deteriorate with the increase in particle size of RBC. They found the upper range for optimum content of RBC in NBR to be 60 phr, above which the dispersion of particles in the matrix as well as the interfacial bonding became poorer. The incorporation of the optimum concentration of suitable sized RBC particles in the NBR matrix resulted in a considerable improvement in mechanical and antiskid properties.

Mao et al. [53] reported the effect of GO as a filler in rubber composites and observed that the addition of 2 vol% GO in SBR improved tensile and tear strength of the composite by as much as 700% and 208%, respectively, as can be seen in Fig. 10.3A. Fig. 10.3B shows the variation in modulus at 100% and 300% elongation at different filler concentration. From the tensile test results it has been noticed that the addition of GO has a significant effect on the improvement in the tensile

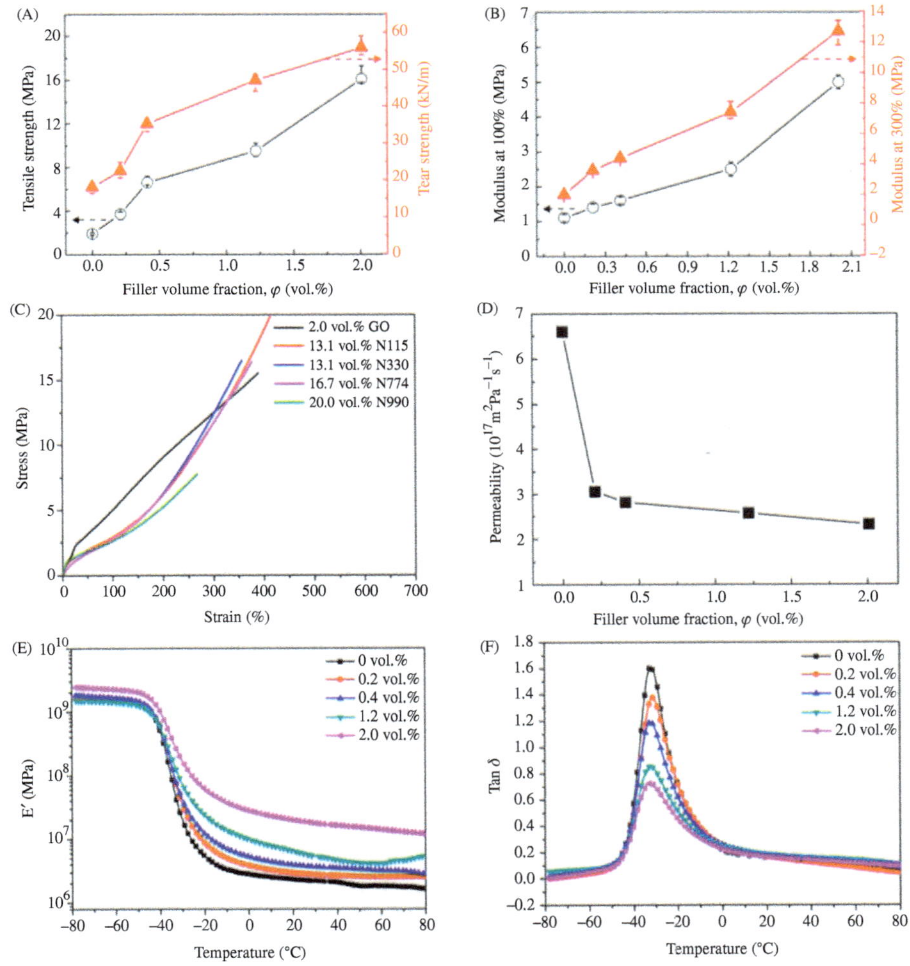

Figure 10.3 Properties of GO/SBR composite at different GO volume fractions: (A) Tensile strength and tear strength; (B) Modulus at 100% and 300% elongation of GO/SBR composite at different GO volume fractions. Symbols denote experimental values, and experimental errors were derived from measurements acquired from different regions of the same sample; (C) Stress–strain curves of SBR composites containing various fillers at different filler volume fractions; (D) Gas permeability of GO/SBR composite as a function of GO loading; (E) Storage modulus versus temperature for neat SBR and SBR/GO composite at different GO contents; (F) Tan delta versus temperature for neat SBR and SBR/GO composite at different GO contents. *Source*: Reproduced from Mao Y, Wen S, Chen Y, Zhang F, Panine P, Chan TW, et al. High performance graphene oxide based rubber composites. Sci Rep 2013;3:2508.

properties of composites. Although the percolation threshold of GO reinforced composites was noticed at around 0.5 vol%, the tensile properties were found to be improving with increasing concentration of filler. Viscoelastic properties of composites were also noticed to be better with the addition of GO, as shown in Fig. 10.3E,F. With an increase in the GO contents, storage modulus increases and peak of tan δ decreases, which indicates better load bearing capability, reduced heat buildup, and enhanced damping of GO/SBR nanocomposites. In a similar work, Song et al. [54] reported modified graphite filled NR composites and observed that fine dispersion of filler into the composite matrix offers an improvement in modulus, thermal conductivity, tensile strength, and tear strength at 100% and 300% elongation. Wang et al. [55] has reported the effect of expanded graphite (EG) on the properties of EG/NBR nanocomposites. These nanocomposites have shown superior tensile and dynamic properties over pristine NBR. They noticed that the nano-sized dispersed graphite exhibited better adhesion between graphite–rubber interfaces, excellent filler volume fraction, and a good reinforcement to rubber. Moreover, dispersion of nano-sized particles offers lower coefficient of friction and higher wear resistance than microfillers through forming uniform, stable, and continuous lubricant films under heavy loads and high sliding velocity.

Yan et al. [56] prepared reduced graphene oxide (rGO) filled NR composites through a combination of the latex cocoagulation approach with static hot pressing and the two roll mixing process. Results have shown that the segregated rGO network in NR composite provided considerable improvement in tensile strength and a reduction in strain at failure. Permeability analysis, as shown in Fig. 10.4, revealed that the segregated rGO network in NR provides significant advantages as a barrier against oxygen and water vapor permeability in comparison to nonsegregated rGO filler composites. This study claimed that the morphology of the filler plays a crucial role in the determination of bulk properties in rubber-based composites and must be considered as an important parameter for tailoring the properties in multifunctional materials.

Kim et al. [57] studied the effect of ionic liquid (IL)-based single-walled carbon nanotubes (SWNTs) and nitric acid treatment on sheet resistance (R_s) and conductivity of silicon rubber (SR) matrix nanocomposites. They noticed that the R_s significantly reduces with IL-based SWNTs and further with nitric acid doping. The addition of 4 wt% filler showed R_s and conductivity of IL-based SWNTs/SR composites to be 412 $\Omega\ sq^{-1}$ and 0.34 $S\ cm^{-1}$, respectively. While R_s of nitric acid treated composites reduced to 50 $\Omega\ sq^{-1}$, this must be due to the increased number of charge carriers after doping. In order to verify the applicability of these composites as stretchable and conductive composites, they studied the ability of these composites to absorb high strains and the corresponding degradation in their electronic properties. The effect of stretching on R_s and conductivity of IL-based and acid treated composites can be clearly observed with the help of Fig. 10.5. It has been reported and also can be observed through the corresponding figure that these SWNTs/SR composites were conductive even after the 20th cycle of their stretching. This proves the effectiveness of SWNTs for providing a highly conductive network in rubber-based composites.

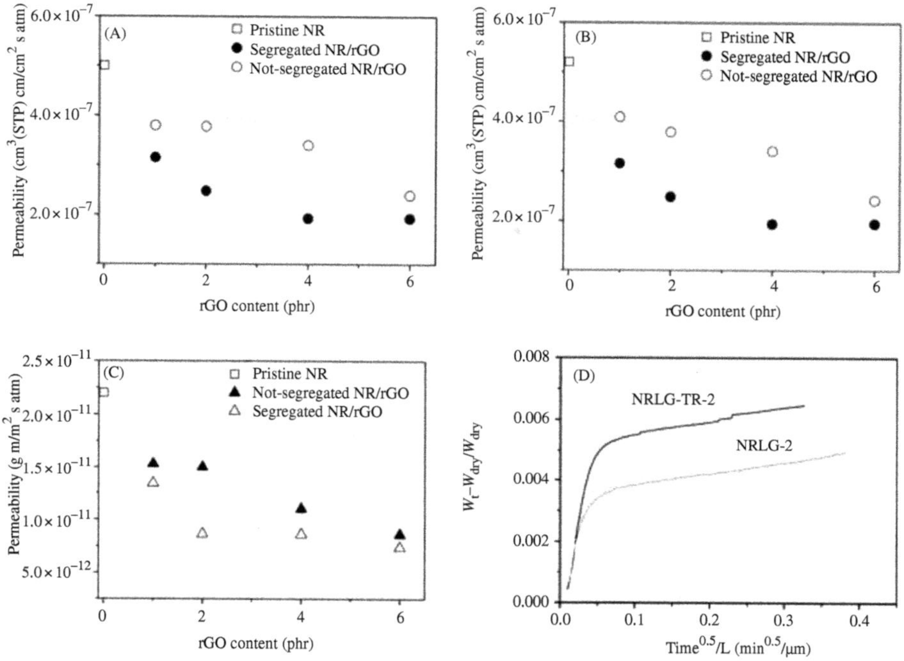

Figure 10.4 Gas permeability of rGO/NR film as a function of rGO content: oxygen permeability measured at 38°C: (A) 0% RH, (B) 50% RH, and (C) water vapor permeability of rGO/NR film measured at 38°C. (D) Water vapor sorption isotherms for rGO/NR composites with both the segregated (NRLG-2) and not-segregated morphologies (NRLG-TR-2).
Source: Reproduced with permission from Yan N, Buonocore G, Lavorgna M, Kaciulis S, Balijepalli SK, Zhan Y, et al. The role of reduced graphene oxide on chemical, mechanical and barrier properties of natural rubber composites. Compos Sci Technol 2014;102:74–81.

Geng et al. [58] also reported the impedance characteristics and electrical properties of MWNTs/SR composites prepared via the solvent evaporation method. They observed the percolation threshold p_c of nanotubes in SR matrix lies between 2 wt% to 2.9 wt%. This can also be observed from Fig. 10.6. Fig. 10.6 shows that the samples with low concentrations of MWNTs, i.e., 0.5 wt%, 1.0 wt%, and 2.0 wt%, offer high impedance, whereas samples with high concentrations, i.e., 2.9 wt%, 4.7 wt%, and 6.5 wt%, show relatively low impedance values. They affirmed the validity of their modeling approach by comparing the results with experimentations. Zhou et al. [59] reported detailed mechanical properties of SBR matrix composites reinforced with CB and CNTs. They observed that SBR filled with 20 phr CNTs provide tensile and tear strength of 5.0 MPa and 26 kN/m, respectively. These mechanical properties are about three times higher as compared to pristine SBR. In addition, rheological properties showed higher rigidity of CNT filled composites. Although CB filled composites also showed an improvement in mechanical properties, those increments were far less than those observed through CNT/SBR nanocomposites. This proved the effectiveness of CNT over CB in rubber matrix composites.

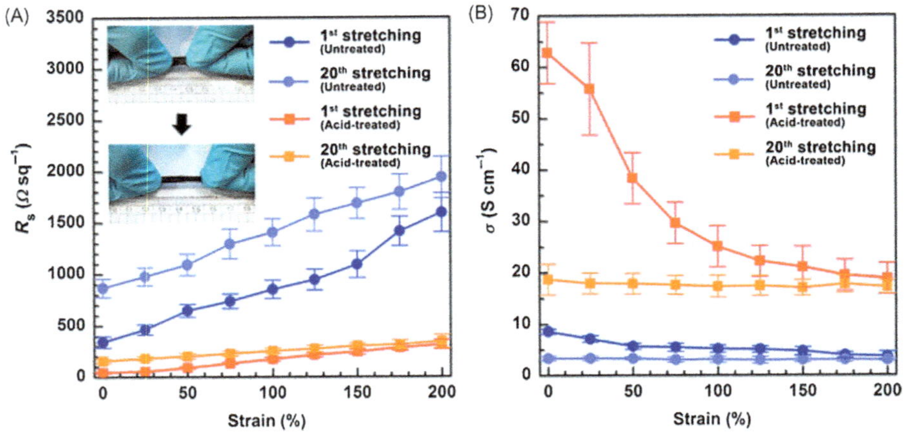

Figure 10.5 (A) Sheet resistance of the composites as a function of tensile strain. (B) Conductivity of the composites as a function of tensile strain when each stretching cycle is applied.
Source: Reproduced with permission from Kim TA, Kim HS, Lee SS, Park M. Single-walled carbon nanotube/silicone rubber composites for compliant electrodes. Carbon N Y 2012;50: 444–9.

10.2.5 Advantages of blends and composites over pristine rubber

- Improvement in physical and chemical properties
- Superior aging resistance of the former without losing mechanical properties
- Even better mechanical performance can be achieved by using a functionalized copolymer
- Depending on blending ratio, compatibility, and concentration of reinforcing fillers, the properties of blends or composites can be tailored as per application requirements
- Blending or reinforcing provides a lightweight material having cost-effective advancements in properties

10.2.6 Formulations relating to blends and composites

1. *Rubber–Polymer blends*

   Blending of two similar/dissimilar matrix components mainly depends on the uniformity in distribution of the minor constituent. In the case of rubber blends, the overall morphology of the blend may vary from dispersed rubber phase to continuous rubber phase, depending on the contribution of rubber in the blend. Extremely high content of rubber in the blend results in a continuous rubber phase, whereas low content of rubber shows a dispersed rubber phase. The existence of nearly equal amounts of rubber and a polymer shows a cocontinuous phase. The phase continuity of blends follows the Eq. 10.1 [60].

$$\frac{\sigma_a}{\sigma_b} = \frac{\eta_a}{\eta_b} \tag{10.1}$$

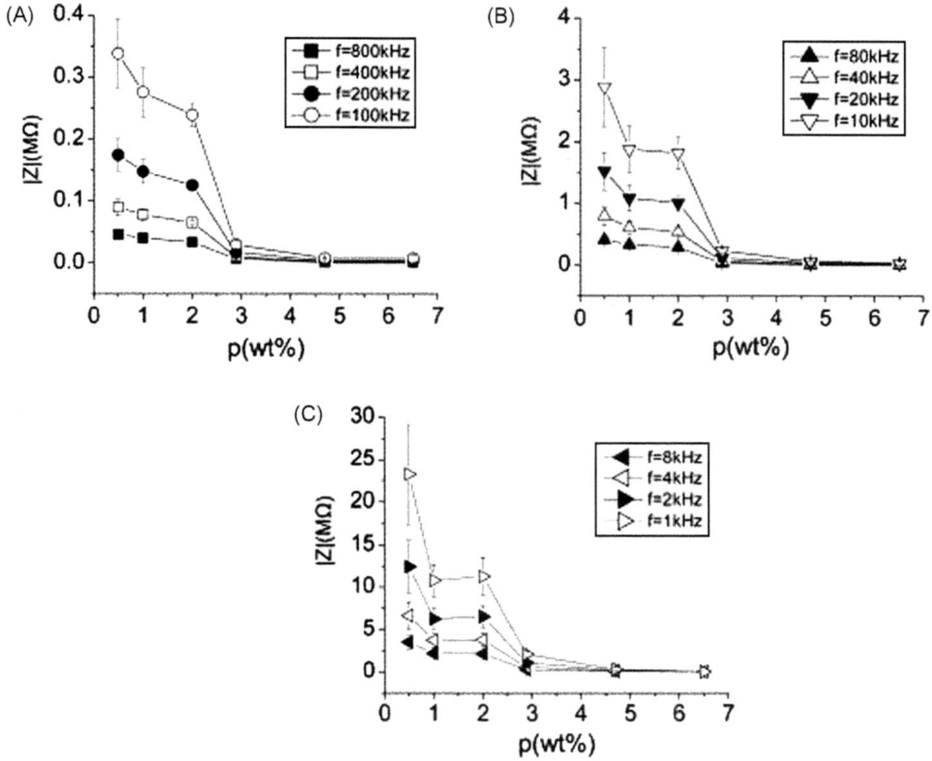

Figure 10.6 Magnitude of impedance versus MWNTs concentration in the frequency range of (A) 800 k–100 kHz; (B) 80 k–10 kHz; and (C) 8 k–1 kHz.
Source: Reproduced with permission from Geng S, Wang P, Ding T. Impedance characteristics and electrical modeling of multi-walled carbon nanotube/silicone rubber composites. Compos Sci Technol 2011;72:36–40.

where,

σ = shear stress

η = viscosity

A and B are two similar/dissimilar matrices.

Typically, in the case of blending of a polymer and rubber, the mechanical properties (i.e., tensile strength and ductility) depend upon the relative proportions of the two constituents as well as the composition of rubber. A higher proportion of rubber gives a blend with high ductility but reduces the tensile strength and vice versa. Considering the composition of rubber, a high level of ethylene increases the crystallinity of the blend and consequently the use of crystalline rubber improves the tensile strength as well as the ductility of blends, although less improvement in tensile strength has been observed. These factors have arisen out of the need for some fillers in rubber blends that may result in a significant improvement in tensile strength of the rubber blend composites.

2. *Matrix–Filler composites*

The "Rule of mixtures" has been used to consider the volume fractions of fillers (V_f) and matrix (V_m), theoretical density of composite (ρ), and total cost per unit weight of the composite (C). The expressions for these parameters are shown in Eqs. (10.2), (10.3), (10.5), and (10.6), respectively

$$V_f = v_f/(v_f + v_m) \tag{10.2}$$

$$V_m = v_m/(v_f + v_m) \tag{10.3}$$

$$V_f + V_m = 1 \tag{10.4}$$

$$\rho = V_f\rho_f + (1 - V_f)\rho_m \tag{10.5}$$

$$C = V_f\rho_f/\rho C_f + (1 - V_f)\rho_m/\rho C_m + C_i \tag{10.6}$$

where,
C_i = Cost of compounding per unit weight of the composite.

Sometimes equations for the "Rule of mixtures" have also been used to describe mechanical and thermal properties of blends and composites.

10.2.7 Role of nanoparticles in compatibility and miscibility of blends

Compatibilization, structuration, and selective localization of nanofillers are the three main considerable features of filler reinforced elastomeric blends. Compatibility and miscibility are two close terms which are often confused while discussing the blending of elastomers. Technically, two or more elastomers are compatible if their combination provides desired properties, whether the same are completely miscible or not. Some of the major factors which are responsible for influencing the compatibilizing mechanism are kinetic and thermodynamic effects, localization, and transfer of particles. It is mainly a difference in polarity or high interfacial tension that causes incompatibility of elastomeric blends. Compatibilization is essentially an effort to modify interfacial properties of immiscible blends which reduces the coefficient of interfacial tension and provides stable morphology of the blended composite. There exist a lot of research articles and reviews in which researchers have explored the need and phenomenon of compatibilization in polymer blends. Reactive compatibilization, addition of multifunctional copolymers, and addition of block or graft copolymers have been proved to be some of the most promising techniques for this purpose. Beyond these techniques, the addition of fillers to serve as compatibilizers has been noticed to be a new approach which has attracted lot of interest. It is a broad topic to discuss different compatibilizers in detail. So, here we will focus only on the compatibilization ability of nanoparticles in rubber blends. The addition of certain nanoparticles as additional components was found to bring positive changes and therefore these are also termed as compatibilizers. The introduction of these nanoparticles to elastomer blends increases the viscosity of matrix and reduces interfacial tension, thereby improving the compatibility of blends. Among the various clay based nanoparticles which are found to be quite effective compatibilizers in rubber/rubber and rubber/plastic blends, organically modified nanoclay has been noticed to be very effective in many immiscible blends. The basic phenomenon behind the effectiveness of clay is the intercalation and exfoliation of these particles after being added to the blend which assists in the appreciable material properties of elastomer blend nanocomposites. This is the case not only in blends, similar problems have been noticed with certain CB and CNTs reinforced polymer/elastomer composites and studies have shown that the addition of different types of nanoclay really improved the dispersion behavior of filler in these matrices.

Selective localization of nanofillers plays a crucial role in tailoring the morphology of composites in such a way that offers the desired mechanical, thermal, and rheological properties. The kinetic effects for dispersion of nanofillers can be studied through simple measurements of electrical conductance but the localization of particles in binary and ternary blends is based on the wetting concept. However, to understand the quantitative localization of fillers in a blend the Z-model has been proposed. The comparative study of these two phenomena provides a clear insight into the transfer of filler within the matrix during the mixing process. Le et al. [61] have explained both these concepts in a very clear and descriptive way. In short, the wetting concept provides the filler localization in binary and ternary blends and the Z-model is based on the thermodynamic criterion using a master curve prepared through surface tension values of the blended components to represent localization of filler in the blend. Blending components having a high filler wetting speed give a good localization of filler and its interaction with the corresponding blending components via relocalization of fillers within the blend. This relocalization of filler continues to approach the thermodynamic state predicted by the Z-model and results in a change in surface tension. This effect of the addition of fillers or coupling agents can be observed by comparing the surface tension with that forecasted by the master curve.

10.3 Importance of rubber blend nanocomposites

Blending of two or more rubbers or rubbers with polymers has been a novel technique to obtain the desired properties from blends at low cost. NR has been most commonly used to toughen the polymer matrices, thus producing blends with high ductility. The two main factors for the remarkable changes in mechanical, thermal, and barrier properties of polymers are the nature of the dispersion of the filler, which may be via exfoliation or intercalation, and the compatibility of fillers with matrix. Rubber blends have been proved to yield high improvements in ductility. It has been noticed that the reduction in coalescence rate between particles in the melt mixing technique tends to decrease the size of rubber particles which is a key issue towards achieving high toughness [62,63]. Optimal toughening of blends depends on various factors, such as type of rubber, molecular weight, and polyamide type [64,65]. These factors decide the upper and lower limits for rubber particle size.

It has been shown here that super-tough polyamide was obtained when the rubber particle size rages from 0.1 to 1 mm, depending on the molecular weight of another constituent (Nylon 6). The properties of different types of NR and their blends with polyurethane, reinforced with layered silicates, have been studied recently [66–68]. Layered silicates consist of platelet type structure which assists for significant reduction in permeant diffusion and therefore enhances the barrier properties of blended composites. Moreover, barrier properties have been noticed comparatively better under dispersion via exfoliation. This is due to the high aspect ratio of clay platelets in exfoliated systems. Polymer composites reinforced with micro-fibers shown good tensile strength and stiffness but these fillers reduces ductility as well as impact resistance. These problems have been resolved upto considerable extent through rubber toughening in polymers to improve the overall performance of the rubber blend composites.

Angellier et al. [69] prepared NR-starch nanocomposites where waxy maize starch nanocrystals have been used as reinforcement and their effect on swelling behavior of NR has been examined. It has been observed that the swelling in NR significantly reduces after adding reinforcing filler, which has been credited to the three-dimensional network of starch nanocrystals. Also, the rate of absorption has been found to be inversely dependent on starch content. It has already been proved many times that the properties of composites depend upon a critical concentration of fillers known as the percolation threshold. Especially in the case of polymer composites, the need for achieving specific mechanical or thermal properties increases the percolation threshold of fillers. The fillers for reinforcing rubbers may be in the form of particulates, inorganic compounds, or fibrous materials. Some of these may be micro- or nanofillers. Compared to others, the use of nanofillers, i.e., clay, CNTs, and graphene, has been proven to be much more effective as reinforcements in polymer composites but their high cost and their difficult processing limits their industrial applications. Reinforcing such nanofillers in blends of polymers and rubbers assists in the reduction in the amount of required nanofillers [70–72]. Multiwall carbon nanotubes (MWCNTs) have been found to be very effective for tuning the properties of polymer blends [73–76]. The use of compatibilizer has been found to be very effective in the case of blends. Nanoclay [77–79] and GO sheets [80] have received a great deal of attention as compatibilizing agents in polymer blends. Even though organically modified clay (layered silicates) has been very effective in this regard, the major challenge remains for separation of individual layers of clay within the matrix. Fenouillot et al. [81] observed that the state of dispersion and location of nanofillers in polymer blends affect the dynamics and coalescence of blend components.

10.3.1 Various types of rubber blend nanocomposites

As discussed in the previous section, the blending of rubbers assists in achieving optimum physical, thermal, and rheological properties using reduced amounts of costly nanofillers and therefore reducing the costs with simplified processability. Clay has been extensively used in pristine as well as organically modified forms in rubber and polymer composites. Researchers have reported exciting advancements in the properties of rubber nanocomposites through reinforcing reactive and nonreactive nanoclay into a variety of rubber matrices, i.e., NR, styrene-butadiene, ethylene-propylene, polybutadiene, synthetic polyisoprene, and nitrile rubbers, etc. [82,83]. Various techniques, such as melt intercalation, in situ polymerization, latex compounding, and solution blending, have been reported to prepare rubber based nanocomposites. Thermoplastic nanocomposites based on polymer blends seem to be a new approach in the nanocomposites studies. Chow et al. [84] prepared PA6/PP blends reinforced with 4 phr organoclay (organophilic modified montmorillonite) and used maleic anhydride grafted ethylene-propylene rubber (EPRgMA) as a compatibilizer. PA6 has been chosen due to its good mechanical strength and PP has a high moisture resistance and supports easy processability. The decrease in melt flow index, increase in strength and ductility, improvement in rheological properties, and fine dispersion of PP has been observed due to the addition of a compatibilizer. In a similar work, maleated polypropylene (PPgMA) as a compatibilizer in PA6/PP blend resulted in a considerable improvement in the tensile strength and stiffness of composites [85]. Voulgaris et al. [86] have also shown the effect of organoclay as an emulsifier in a polystyrene/poly(ethyl methacrylate) (PS/PEMA) blend. Gelfer et al. [87] have prepared composites using organoclay reinforced to a polystyrene/poly(methyl methacrylate) (PS/PMMA) blend. A huge reduction in domain size of PS/PMMA blends has been noticed due to partial compatibilization by surfactants and an increase in viscosity. Moreover, in the case of a high-density polyethylene/N6 blend, the addition of clay (5 wt%) reduced the dispersed nylon domains and their shape changed from spherical to laminar [88]. Although, significant changes in kinetics and morphological development in phase behavior of organoclay/PS/poly(vinyl methyl ether) has been observed by Yurekli et al. [89,90], the low critical solution transition observed in PS/poly(vinyl methyl ether) blend shows negligible change after the addition of organoclay (4 wt%). Significant improvement in modulus at low filler loadings has also been noticed in the case of N6 based nanocomposites [91,92] but with a high content of nano-sized particles there has been a decrease in impact strength and a sharp increase in the ductile–brittle transition temperature [91].

Another factor that must be considered for the blending of polymers and rubbers is the rubber toughening process. The applicability of most polymer nanocomposites has been restricted by their low toughness. This factor invites the high toughening property of rubber to be merged here for its end-use applications. Bitinis et al. [93] carried out the toughening phenomenon in PLA bionanocomposites through blending it with NR. They have prepared PLA/NR/MMT nanocomposites through the melt blending technique to study their physicochemical properties. In 2006, Frounchi et al. [94] prepared nanocomposites through reinforcing montmorillonite-based organoclay to a PP-EPDM blend (50:50) via the solvent blending method. They were the first to investigate the barrier properties of nanoclay reinforced PP/EPDM blend and observed an improvement in barrier properties against oxygen and carbon dioxide. However, the crystallinity of the blend decreased after the addition of organoclay platelets. Razak et al. [95] investigated the effect of graphene nanoplatelets (GNPs) on the mechanical properties and fracture morphology of NR-EPDM rubber blends. About a two-fold improvement in tensile strength along with significant improvement in ductility has been noticed. In 2015, Razak et al. [96] also studied the effect of surface modified GNPs in a NR-EPDM rubber blend. At a comparatively lower GNPs loading ($\leq$1 wt%) the tensile strength and percentage elongation have been noticed to improve by 64.55% and 14.2%, respectively, compared to unfilled NR-EPDM blend. The morphological and thermal behavior of epoxy/clay/carboxyl terminated poly(butadiene-co-acrylonitrile) liquid rubber (CTBN) nanocomposites were investigated by Vijayan et al. in 2013 [97]. Mahmood et al. [98] reported MWCNTs reinforced thermoplastic polyurethane-urea (TPUU)/carboxylated acrylonitrile butadiene rubber (XNBR) blend nanocomposites. Significant improvement in tensile modulus as well as tensile strength has been observed due to the addition of CNTs. It has been observed that the behavior of thermal degradation is mainly dependent on dispersion rather than the distribution of clay platelets in the blend. Nair et al. [99] observed the effect of blending PP and NR to reduce the percolation threshold of CNTs in blend. They have observed varying percolation mechanisms in each blend composition. Nanoreinforcement in rubber–polymer blend nanocomposites has been noticed as an emerging and exciting trend to reduce the percolation threshold of nanofillers.

10.3.2 Properties of rubber blend nanocomposites over conventional rubber based blends and composites

Various types of modified clays, CNT, graphene based nanofillers, etc. have been found to be highly effective in rubber blends. Polymer based nanocomposites have been a great area of interest in research for a long time. However, they are still not fully able to meet the mechanical properties of metals. With the use of nanofillers they came very near but the costs were incompatible. It was briefly discussed earlier that blending of rubber with polymers improves the ductility of blends which assists in reducing the amount of nanofillers required. Therefore, this combination may be very useful for structural and automotive applications. As not much research has been available until now on rubber–polymer blend nanocomposites, it provides a fine space to explore this area. In this regard, this section will focus on the main achievements in properties obtained due to the addition of nanofillers in the rubber-based blends that have been discussed in the previous section.

The rheological properties of PA6/PP/organoclay nanocomposite [84] in the presence of EPRgMA compatibilizer are shown in Fig. 10.7. It has been clearly observed that the storage and loss modulus both increase with the addition of organoclay, resembling the enhanced interaction between silicate layers and the polymer. The intercalation of organoclay has been indicated through a shoulder peak in X-ray diffraction (XRD). Furthermore, EPRgMA compatibilizer shows effectiveness at low frequencies towards improvement in both the modulus values. Mechanical properties show that organoclay improves the stiffness but reduces the

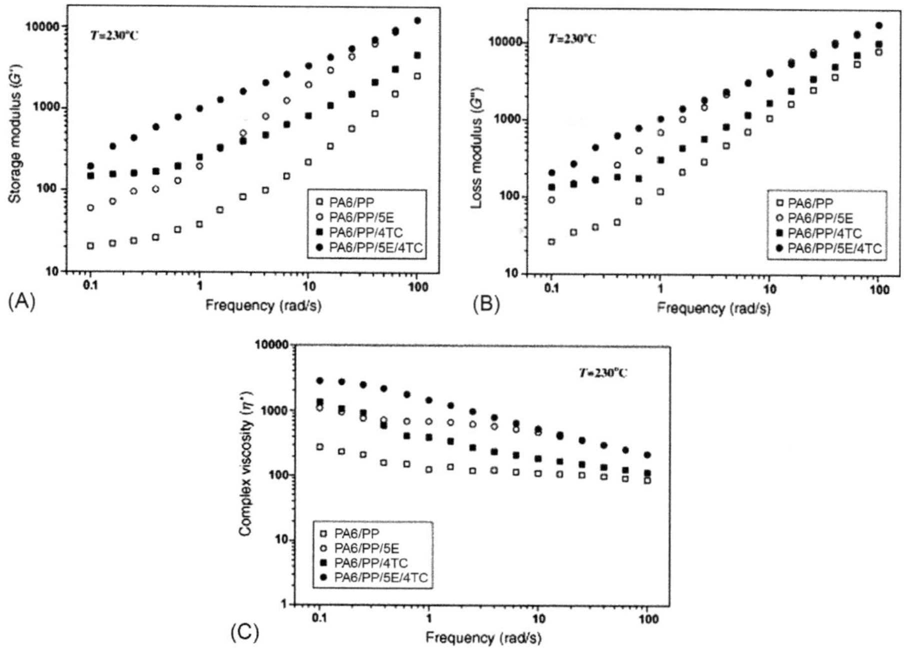

Figure 10.7 (A) Storage modulus (G') versus frequency plot for the PA6/PP blend and PA6/PP/organoclay nanocomposite; (B) Loss modulus (G'') versus frequency plot for the PA6/PP blend and PA6/PP/organoclay nanocomposite; (C) Complex viscosity (η^*) versus frequency plot for the PA6/PP blend and PA6/PP/organoclay nanocomposite.
Source: Reproduced with permission from Chow WS, Bakar AA, Mohd-Ishak ZA, Karger-Kocsis J, Ishiaku US. Effect of maleic anhydride-grafted ethylene–propylene rubber on the mechanical, rheological and morphological properties of organoclay reinforced polyamide 6/polypropylene nanocomposites. Eur Polym J 2005;41:687–96.

ductility of composites. The addition of compatibilizer helps to overcome that problem by an improvement in strength as well as ductility. Khatua et al. [100] studied the barrier effect of organoclay on N6/EPR blends through morphological analysis. They have also noticed that the dispersed domain size reduces due to the effect of organoclay in incompatible polymer blends. Such a phenomenon has already been cited earlier with other blends. However, in this work researchers also highlighted that this reduction in dispersed domain size does not ensure compatibility between the two different polymers because organoclay platelets have not been observed near the interfacial sites.

Variation of clay in epoxy/CTBN blends shows the effect of clay on various properties [97]. Thermal analysis has shown that the degradation of these nanocomposites depends on the dispersion of clay rather than its distribution in the blend. Fracture toughness of composites with and without CTBN is shown in Fig. 10.8.

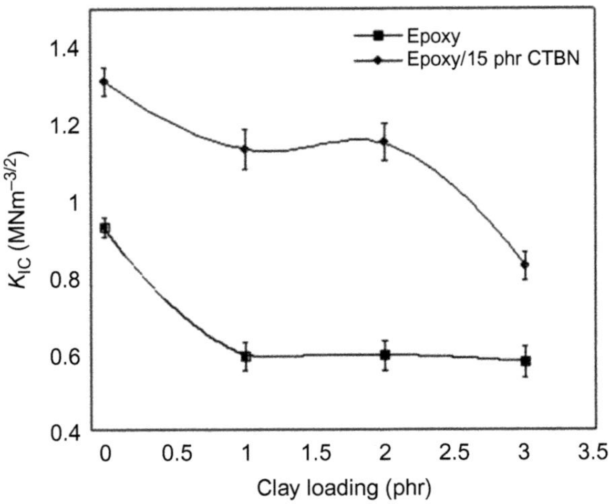

Figure 10.8 Variation of fracture toughness (KIC) with clay loadings in epoxy/clay and epoxy/clay/15 phr CTBN nanocomposites.
Source: Reproduced with permission from Vijayan PP, Puglia D, Maria HJ, Kenny JM, Thomas S. Clay nanostructure and its localisation in epoxy/liquid rubber blend. RSC Adv 2013;3:24634–43.

It has been reported that the fracture toughness increases by 41.13% with 15 phr CTBN. Recently, Mahmood et al. observed the effect of varying fraction of CNTs on mechanical, rheological, morphological, and electrical properties of the XNBR/TPUU blend [98]. Significant improvement in elastic modulus and strain at break shows the effectiveness of CNTs in the blend. Moreover, direct current (DC) electrical percolation behavior, shown in Fig. 10.9, reveals that even a small amount of CNTs gives considerable rise to DC electrical conductivity. This clearly revealed the effectiveness of reinforcing nanofillers in rubber blends.

Stephen et al. [101] studied solvent resistant properties of layered silicates (sodium bentonite and sodium fluorohectorite) filled NR/XSBR (carboxylated styrene butadiene rubber) blends through the equilibrium swelling method using benzene, toluene, and *p*-xylene as solvents. They observed an intercalated structure in the case of sodium bentonite and an exfoliated structure in the case of sodium fluorohectorite with the help of XRD analysis. Fig. 10.10 depicts the increase in *d*-spacing indicating the good intercalation of polymer chains within layered silicates. The reduction in swelling rate has been observed in nanocomposites prepared with NR, XSBR, and their blends. This may be due to the tortuous path and low transport area in membrane. Also, the intercalation of rubber into nanostructured silicate layers increases the solvent resistance of rubber/polymer blended nanocomposites. This study observed the dependence of solvent uptake of polymer on their permeant molecule size and found that benzene is suitable for very small and *p*-xylene is most suitable for very large sized permeant molecules.

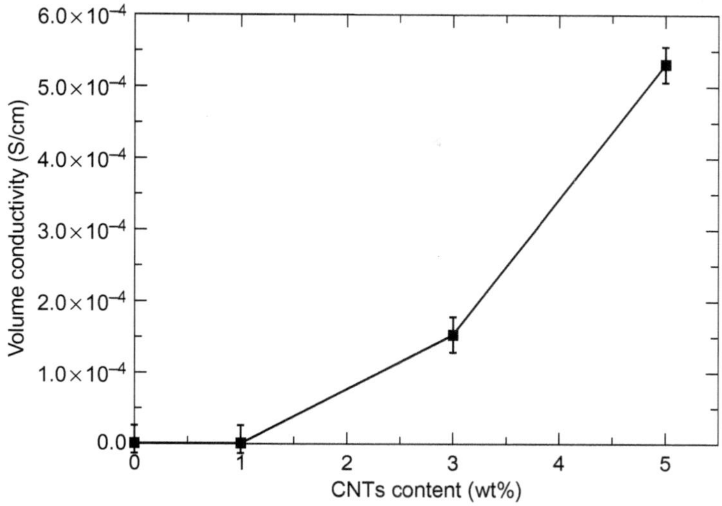

Figure 10.9 Dependence of DC electrical conductivity on the CNTs content of TPUU: XNBR blend nanocomposite.
Source: Reproduced with permission from Mahmood N, Khan AU, Stockelhuber K, Das A, Jehnichen D, Heinrich G. Carbon nanotubes-filled thermoplastic polyurethane-urea and carboxylated acrylonitrile butadiene rubber blend nanocomposites. J Appl Polym Sci 2014;131(11):40341.

10.3.3 Advanced applications of rubber blend nanocomposites

- Structural and home decor applications such as bridges, wall mountings, lightweight hangings, etc.
- Heating and sensing applications.
- Electronic equipment, pressure sensitive switches, important strategic materials, such as EMI shielding, floor heating elements, etc.

10.4 Summary and future scope

Rubber-based composites have been considered suitable for several essential applications for many decades due to the ductile behavior of rubber, while the lack of other mechanical properties, i.e., strength, fracture toughness, etc., have limited their applicability. The reinforcing of rubber with different fillers and blending of rubber with polymers have been found to overcome certain limitations. In the past few years, research has advanced towards the addition of popular nanofillers like CNTs and graphene to different blends of rubber–polymer and rubber–rubber. These materials are well known as rubber blend nanocomposites and have proved to be highly suitable for barrier, structural, and electronic applications. Although, initial steps along this path have been carried-out, the optimization of various

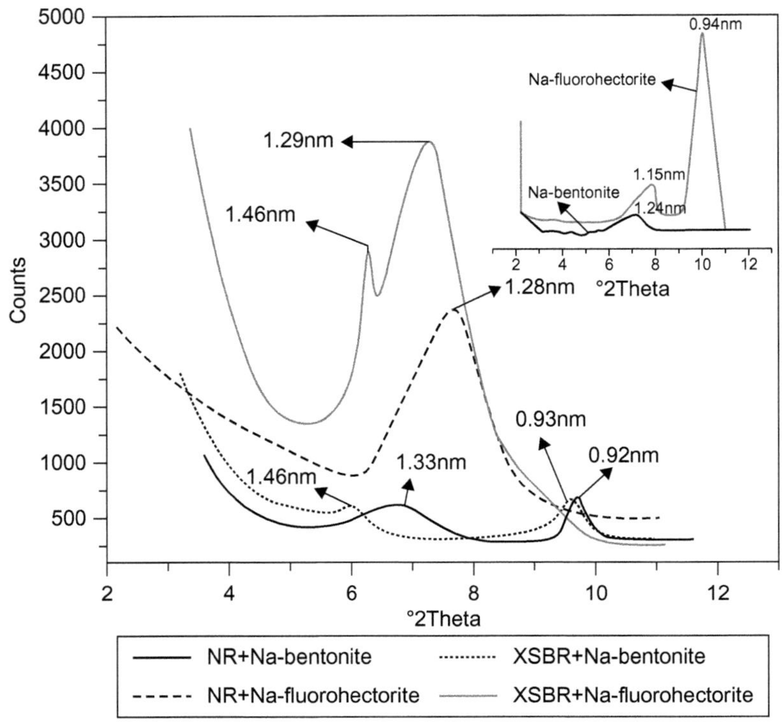

Figure 10.10 X-ray diffraction patterns of layered silicates and latex nanocomposites (2.5 phr).
Source: Reproduced with permission from Stephen R, Varghese S, Joseph K, Oommen Z, Thomas S. Diffusion and transport through nanocomposites of natural rubber (NR), carboxylated styrene butadiene rubber (XSBR) and their blends. J Membr Sci 2006;282:162–70.

mechanical, thermal, rheological, and electronic properties for its end-use applications is still lacking.

References

[1] Drobny JG. Handbook of thermoplastic elastomer. Norwich (NY): William Andrew Publishing; 2007. ISBN 978-0-8155-1549-4.
[2] Rubber-modified thermoset resins. In: Riew CK, Gillham JK, editors. Advances in chemistry, vol. 208. Washington DC: ACS; 1984.
[3] El-Sabbagh SH. Compatibility study of natural rubber and ethylenepropylene-diene rubber blends. Polym Test 2003;22(1):93–100.
[4] Chang YW, Shin YS, Chun H, Nah C. Effects of trans-polyoctylene rubber (TOR) on the properties of NR/EPDM blends. J Appl Polym Sci 1999;73(5):749–56.
[5] Sae-oui P, Sirisinha C, Thepsuwan U, Thapthong P. Influence of accelerator type on properties of NR/EPDM blends. Polym Test 2007;26(8):1062–7.

[6] Xu C, Wang Y, Chen Y. Highly toughened poly(vinylidene fluoride)/nitrile butadiene rubber blends prepared via peroxide-induced dynamic vulcanization. Polym Test 2014;33:179–86.
[7] Pongtanayut K, Thongpin C, Santawitee O. The effect of rubber on morphology, thermal properties and mechanical properties of PLA/NR and PLA/ENR blends. Energy Procedia 2013;34:888–97.
[8] Bitinis N, Verdejo R, Cassagnau P, Lopez-Manchado MA. Structure and properties of polylactide/natural rubber blends. Mater Chem Phys 2011;129(3):823–31.
[9] Manshaie R, Khorasani SN, Veshare SJ, Abadchi MR. Effect of electron beam irradiation on the properties of natural rubber (NR)/styrene–butadiene rubber (SBR) blend. Radiat Phys Chem 2011;80(1):100–6.
[10] Wang Y, Jiang X, Xu C, Chen Z, Chen Y. Effects of partial replacement of silicone rubber with flurorubber on properties of dynamically cured poly(vinylidene fluoride)/silicone rubber/flurorubber ternary blends. Polym Test 2013;32(8):1392–9.
[11] Wang Y, Fang L, Xu C, Chen Z, Chen Y. Preparation and properties of dynamically cured poly(vinylidene fluoride)/silicone rubber blends. Polym Test 2013;32(6):1072–8.
[12] Mansilla MA, Marzocca AJ. About the cure kinetics in natural rubber/styrene Butadiene rubber blends at 433K. Physica B 2012;407(16):3271–3.
[13] Xu C, Cao L, Chen Y. Ternary blends based on poly (ethylene-naphthalate)/glass fibers/nitrile rubber: preparation, properties and effect of dynamic vulcanization. Polym Test 2013;32(8):1529–37.
[14] Luzinov I, Xi K, Pagnoulle C, Huynh-Ba G, Jerome R. Composition effect on the core–shell morphology and styrene/styrene–butadiene rubber/polyethylene blends. Polymer (Guildf) 1999;40(10):2511–20.
[15] Maity AK, Xavier SF. Rheological properties of ethylene-propylene block copolymer and EPDM rubber blends using a torque rheometer. Eur Polym J 1999;35(1):173–81.
[16] Ratnam CT, Nasir M, Baharin A, Zaman K. Electron-beam irradiation poly (vinyl chloride)/epoxidized natural rubber blend in the presence of Irganox 1010. Polym Degrad Stabil 2001;72(1):147–55.
[17] Georgea S, Varughese KT, Thomas S. Molecular transport of aromatic solvents in isotactic polypropylene/acrylonitrile-co-butadiene rubber blends. Polymer (Guildf) 2000;41(2):579–94.
[18] Naseri ASZ, Jalali-Arani A. A comparison between the effects of gamma radiation and sulfur cure system on the microstructure and crosslink network of (styrene butadiene rubber/ethylene propylene diene monomer) blends in presence of nanoclay. Radiat Phys Chem 2015;115:68–74.
[19] Ratnam CT, Kamaruddin S, Sivachalam Y, Talib M, Yahya N. Radiation crosslinking of rubber phase in poly(vinyl chloride)/epoxidized natural rubber blend: effect on mechanical properties. Polym Test 2006;25(4):475–80.
[20] Martin P, Maquet C, Legras R, Bailly C, Leemans L, van Gurp M, et al. Particle-in-particle morphology in reactively compatibilized poly(butylene terephthalate)/epoxide-containing rubber blends. Polymer (Guildf) 2004;45(10):3277–84.
[21] Nakason C, Kaesaman A, Samoh Z, Homsin S, Kiatkamjornwong S. Rheological properties of maleated natural rubber and natural rubber blends. Polym Test 2002;21(4):449–55.
[22] Kraus G. Reinforcement in elastomers. New York, NY: Wiley-Interscience; 1965.
[23] Spurr RT. Fillers in friction materials. Wear 1972;22(3):367–72.
[24] Debnath S. Mica as reinforcing filler in elastomeric composites. Ph.D. thesis. Kharagpur: Rubber Technology Center, Indian Institute of Technology; 2000.

[25] Bajaj P, Babu GN, Khanna DN, Varshney SK. Room temperature-vulcanized silicone elastomer: effect of curing conditions and the nature of filler on mechanical and thermal properties. J Appl Poly Sci 1979;23(12):3505–14.
[26] Kwei TK, Kunins CA. Polymer-filler interaction: vapor sorption studies. J Appl Poly Sci 1964;8(3):1483–90.
[27] Krysztafkiewicz A. Modified calcium silicates as active rubber fillers. J Mat Sci 1987;22(2):478–82.
[28] Morton-Jones DH. Polymer processing. London: Chapman and Hall; 1989.
[29] Fowkes FM. Attractive forces at interfaces. Ind Eng Chem 1964;56(12):40–52.
[30] Studebaker ML. A useful new system for categorizing rubber grade carbon blacks. Technical Bulletin-Phillips Petroleum Co.; 1971.
[31] Wagner MP. Non-black reinforcement fillers for rubber. Rubber World 1971;164(5):46.
[32] Iijima S. Helical microtubules of graphitic carbon. Nature 1991;354:56–8.
[33] Khare R, Bose S. Carbon nanotube based composites-a review. J Miner Mater Charact Eng 2005;4:31–46.
[34] Hughes TV, Chambers CR. Manufacture of carbon filaments; 1889. U.S. Patent: No. 405480.
[35] Robertson SD. Graphite formation from low temperature pyrolysis of methane over some transition metal surfaces. Nature 1969;221:1044–6.
[36] Schouten FC, Kaleveld EW, Bootsma GA. AES-LEED-ellipsometry study of the kinetics of the interaction of methane with Ni(110). Surf Sci 1977;63:460–74.
[37] Zaikovskii VI, Chesnokov VV, Buyanov RA. High-resolution electron microscopic study of the structure of filamentary carbon on iron and nickel catalysts. Appl Catal 1988;38:41–52.
[38] Yoshikawa M, Niwa H, Fukuura Y, Naito K. Low permeable nitrile rubber hose; 1995. U.S. Patent: No. 5476121.
[39] Debnath S, De SK, Khastgir D. Effect of silane coupling agent on vulcanization, network structure, polymer-filler interaction, physical properties and failure mode of mica-filled styrene-butadiene rubber. J Mat Sci 1987;22:4453–9.
[40] Brown JM, Curliss D, Vaia RA. Thermoset-layered silicate nanocomposites. Quaternary ammonium montmorillonite with primary diamine cured epoxies. Chem Mater 2000;12:3376–84.
[41] Zanetti M, Camino G, Canavese D, Morgan AB, Lamelas FJ, Wilkie CA. Fire retardant halogen-antimony-clay synergism in polypropylene layered silicate nanocomposites. Chem Mater 2002;14:189–93.
[42] Chang JH, Jang TG, Ihn KJ, Lee WK, Sur GS. Poly(vinyl alcohol) nanocomposites with different clays: pristine clays and organoclays. J Appl Poly Sci 2003;90:3208–14.
[43] Sridhar V, Chaudhary RNP, Tripathy DK. Effect of fillers on the relaxation behavior of chlorobutyl vulcanizates. J Appl Polym Sci 2006;100:3161–73.
[44] Hummers Jr. WS, Offeman RE. Preparation of graphitic oxide. J Am Chem Soc 1958;80. p. 1339–1339.
[45] Marcano DC, Kosynkin DV, Berlin JM, Sinitskii A, Sun Z, Slesarev A, et al. Improved synthesis of graphene oxide. ACS Nano 2010;4:4806–14.
[46] Panwar V, Chattree A, Pal K. A new facile route for synthesizing of graphene oxide using mixture of sulfuric-nitric-phosphoric acids as intercalating agent. Physica E 2015;73:235–41.
[47] Sreeja TD, Kutty SKN. Styrene butadiene rubber/reclaimed rubber blends. Int J Polym Mater 2003;52:599–609.

[48] da Costa HM, Ramos VD, da Silva WS, Sirqueira AS. Analysis and optimization of polypropylene (PP)/ethylene-propylene-diene monomer (EPDM)/scrap rubber tire (SRT) mixtures using RSM methodology. Polym Test 2010;29:572–8.
[49] Lima P, Oliveira J, Costa V. Partial replacement of EPDM by GTR in thermoplastic elastomers based on PP/EPDM: effects on morphology and mechanical properties. J Appl Polym Sci 2014;131(8):40160.
[50] Mostafa A, Abouel-Kasem A, Bayoumi MR, El-Sebaie MG. Effect of carbon black loading on the swelling and compression set behavior of SBR and NBR rubber compounds. Mater Des 2009;30:1561–8.
[51] Devi KSU, Ponnamma D, Causin V, Mari HJ, Thomas S. Enhanced morphology and mechanical characteristics of clay/styrene butadiene rubber nanocomposites. Appl Clay Sci 2015;114:568–76.
[52] Li M-C, Zhang Y, Cho UR. Mechanical, thermal and friction properties of rice bran carbon/nitrile rubber composites: influence of particle size and loading. Mater Des 2014;63:565–74.
[53] Mao Y, Wen S, Chen Y, Zhang F, Panine P, Chan TW, et al. High performance graphene oxide based rubber composites. Sci Rep 2013;3:2508.
[54] Song J, Ma L, He Y, Yan H, Wu Z, Li W. Modified graphite filled natural rubber composites with good thermal conductivity. Chin J Chem Eng 2015;23:853–9.
[55] Wang L, Zhang L, Tian M. Effect of expanded graphite (EG) dispersion on the mechanical and tribological properties of nitrile rubber/EG composites. Wear 2012;276–277:85–93.
[56] Yan N, Buonocore G, Lavorgna M, Kaciulis S, Balijepalli SK, Zhan Y, et al. The role of reduced graphene oxide on chemical, mechanical and barrier properties of natural rubber composites. Compos Sci Technol 2014;102:74–81.
[57] Kim TA, Kim HS, Lee SS, Park M. Single-walled carbon nanotube/silicone rubber composites for compliant electrodes. Carbon N Y 2012;50:444–9.
[58] Geng S, Wang P, Ding T. Impedance characteristics and electrical modelling of multi-walled carbon nanotube/silicone rubber composites. Compos Sci Technol 2011;72:36–40.
[59] Zhou XW, Zhu YF, Liang J. Preparation and properties of powder styrene-butadiene rubber composites filled with carbon black and carbon nanotubes. Mater Res Bull 2007;42:456–64.
[60] Patel RP, Shin J. Compounding and processing of plastic/rubber blends. In: Isayev AI, editor. Encyclopedia of polymer blends. vol. 2: processing. Akron, USA; 2011.
[61] Le HH, Ilisch S, Heinrich G, Radusch H-J. Filler migration in natural rubber blends during the mixing process. In: Thomas S, Chan CH, Pothen LA, Rajisha KR, Maria HJ, editors. Natural rubber materials, vol. 1. Cambridge, UK: Blends and IPNs; 2014.
[62] Majumdar B, Keskkula H, Paul DR. Morphology development in toughened aliphatic polyamides. Polymer (Guildf) 1994;35:1386–98.
[63] Sundararaj U, Macosko CW. Drop breakup and coalescence in polymer blends: the effects of concentration and compatibilization. Macromolecules 1995;28:2647–57.
[64] Oshinski AJ, Keskkula H, Paul DR. Rubber toughening of polyamides with functionalized block copolymers: 1. Nylon-6. Polymer (Guildf) 1992;33:268–83.
[65] Oshinski AJ, Keskkula H, Paul DR. The role of matrix molecular weight in rubber toughened nylon 6 blends: 2. Room temperature Izod impact toughness. Polymer (Guildf) 1996;37:4909–18.

[66] Varghese S, Karger-Kocsis J, Gatos KG. Melt compounded epoxidized natural rubber/layered silicate nanocomposites: structure-properties relationships. Polymer (Guildf) 2003;44:3977−83.
[67] Varghese S, Karger-Kocsis J. Natural rubber-based nanocomposites by latex compounding with layered silicates. Polymer (Guildf) 2003;44:4921−7.
[68] Varghese S, Gatos KG, Apostolov AA, Karger-Kocsis J. Morphology and mechanical properties of layered silicate reinforced natural and polyurethane rubber blends by latex compounding. J Appl Polym Sci 2004;92:543−51.
[69] Angellier H, Molina-Boisseau S, Lebrun L, Dufresne A. Processing and structural properties of waxy maize starch nanocrystals reinforced natural rubber. Macromolecules 2005;38:3783−92.
[70] Huang JC. Carbon black filled conducting polymers and polymer blends. Adv Polym Technol 2002;21:299−313.
[71] Gubbels F, Jbrome R, Teyssie Ph, Vanlathem E, Deltour R, Calderone A, et al. Selective localization of carbon black in immiscible polymer blends: a useful tool to design electrical conductive composites. Macromolecules 1994;27:1972−4.
[72] Rohini R, Bose S. Electromagnetic interference shielding materials derived from gelation of multiwall carbon nanotubes in polystyrene/poly(methyl methacrylate) blends. ACS Appl Mater Interfaces 2014;6:11302−10.
[73] Tao F, Auhl D, Baudouin AC, Stadler FJ, Bailly C. Influence of multiwall carbon nanotubes trapped at the interface of an immiscible polymer blend on interfacial tension. Macromol Chem Phys 2013;214:350−60.
[74] Zonder L, Ophir A, Kenig S, McCarthy S. The effect of carbon nanotubes on the rheology and electrical resistivity of polyamide 12/high density polyethylene blends. Polymer (Guildf) 2011;52:5085−91.
[75] Bose S, Bhattacharyya AR, Kulkarni AR, Pötschke P. Electrical, rheological and morphological studies in co-continuous blends of polyamide 6 and acrylonitrile-butadiene-styrene with multiwall carbon nanotubes prepared by melt blending. Compos Sci Technol 2009;69:365−72.
[76] Monemian S, Jafari SH, Khonakdar HA, Goodarzi V, Reuter U, Potschke P. MWNT-filled PC/ABS blends: correlation of morphology with rheological and electrical response. J Appl Polym Sci 2013;130:739−48.
[77] Si M, Araki T, Ade H, Kilcoyne ALD, Fisher R, Sokolov JC, et al. Compatibilizing bulk polymer blends by using organoclays. Macromolecules 2006;39(14):4793−801.
[78] Ray SS, Pouliot S, Bousmina M, Utracki LA. Role of organically modified layered silicate as an active interfacial modifier in immiscible polystyrene/polypropylene blends. Polymer (Guildf) 2004;45:8403−13.
[79] Hong JS, Namkung H, Ahn KH, Lee SJ, Kim C. The role of organically modified layered silicate in the breakup and coalescence of droplets in PBT/PE blends. Polymer (Guildf) 2006;47:3967−75.
[80] Cao Y, Zhang J, Feng J, Wu P. Compatibilization of immiscible polymer blends using graphene oxide sheets. ACS Nano 2011;5:5920−7.
[81] Fenouillot F, Cassagnau P, Majeste JC. Uneven distribution of nanoparticles in immiscible fluids: morphology development in polymer blends. Polymer (Guildf) 2009;50:1333−50.
[82] Mohammad A, Simon GP. Rubber-clay nanocomposites. In: Mai Y, Yu Z, editors. Polymer nanocomposites. Abington Hall, Abington, Cambridge, England: Woodhead Publishing Limited; 2006.

[83] Utracki LA. Clay-containing polymeric nanocomposites, vol. 2. Shawbury, Shrewsbury, Shropshire, SY4 4NR, UK: Rapra Technology Limited; 2004.
[84] Chow WS, Bakar AA, Mohd-Ishak ZA, Karger-Kocsis J, Ishiaku US. Effect of maleic anhydride-grafted ethylene–propylene rubber on the mechanical, rheological and morphological properties of organoclay reinforced polyamide 6/polypropylene nanocomposites. Eur Polym J 2005;41:687–96.
[85] Chow WS, Mohd-Ishak ZA, Karger-Kocsis J, Apostolov AA, Ishiaku US. Compatibilizing effect of maleated polypropylene on the mechanical properties and morphology of injection molded polyamide 6/polypropylene/organoclay nanocomposites. Polymer (Guildf) 2003;44:7427–40.
[86] Voulgaris D, Petridis D. Emulsifying effect of dimethyldioctadecylammonium-hectorite in polystyrene/poly(ethyl methacrylate) blends. Polymer (Guildf) 2002;43:2213–18.
[87] Gelfer MY, Song HH, Liu L, Hsiao BS, Chu B, Rafailovich M, et al. Effects of organoclays on morphology and thermal and rheological properties of polystyrene and poly(methyl methacrylate) blends. J Polym Sci Part B Polym Phys 2003;41(1):44–54.
[88] Mehrabzadeh M, Kamal MR. Polymer-clay nanocomposites based on blends of polyamide-6 and polyethylene. Can J Chem Eng 2002;80:1083–92.
[89] Yurekli K, Karim A, Amis EJ, Krishnamoorti R. Influence of layered silicates on the phase-separated morphology of PS-PVME blends. Macromolecules 2003;36: 7256–67.
[90] Yurekli K, Karim A, Amis EJ, Krishnamoorti R. Phase behavior of PS-PVME nanocomposites. Macromolecules 2004;37:507–15.
[91] Cho JW, Paul DR. Nylon 6 nanocomposites by melt compounding. Polymer (Guildf) 2001;42:1083–94.
[92] Fornes TD, Yoon PJ, Hunter DL, Keskkula H, Paul DR. Effect of organoclay structure on Nylon-6 nanocomposite morphology and properties. Polymer (Guildf) 2002;43:5915–33.
[93] Bitinis N, Verdejo R, Maya EM, Espuche E, Cassagnau P, Lopez-Manchado MA. Physicochemical properties of organoclay filled polylactic acid/natural rubber blend bio-nanocomposites. Compos Sci Technol 2012;72:305–13.
[94] Frounchi M, Dadbin S, Salehpour Z, Noferesti M. Gas barrier properties of PP/EPDM blend nanocomposites. J Membr Sci 2006;282:142–8.
[95] Razak JA, Ahmad SH, Ratnam CT, Mahamood MA, Yaakub J, Mohamad N. Graphene nanoplatelets-filled NR/EPDM rubber blend: effects of GNPs loading on blend processability, mechanical properties and fracture morphology. Polym Res J 2015;9:1–14.
[96] Razak JA, Ahmad SH, Ratnam CT, Mahamood MA, Lau KT, Ab Maulod HE, et al. Non-covalent polymeric wrapping of IGEPAL C0890 for graphene nanoplatelets (GNPs-C0890) Filled NR/EPDM rubber blend nanocomposites. Appl Mech Mater 2015;761:385–90.
[97] Vijayan PP, Puglia D, Maria HJ, Kenny JM, Thomas S. Clay nanostructure and its localisation in epoxy/liquid rubber blend. RSC Adv 2013;3:24634–43.
[98] Mahmood N, Khan AU, Stockelhuber K, Das A, Jehnichen D, Heinrich G. Carbon nanotubes-filled thermoplastic polyurethane-urea and carboxylated acrylonitrile butadiene rubber blend nanocomposites. J Appl Polym Sci 2014;131(11):40341.
[99] Nair ST, Vijayan PP, Xavier P, Bose S, George SC, Thomas S. Selective localization of multi walled carbon nanotubes in polypropylene/natural rubber blends to reduce the percolation threshold. Compos Sci Technol 2015;116:9–17.

[100] Khatua BB, Lee DJ, Kim HY, Kim JK. Effect of organoclay platelets on morphology of Nylon-6 and Poly(ethylene-ran-propylene) rubber blends. Macromolecules 2004;37:2454–9.

[101] Stephen R, Varghese S, Joseph K, Oommen Z, Thomas S. Diffusion and transport through nanocomposites of natural rubber (NR), carboxylated styrene butadiene rubber (XSBR) and their blends. J Membr Sci 2006;282:162–70.

Hybrid filler systems in rubber nanocomposites

11

M. Galimberti[1], *S. Agnelli*[2] *and V. Cipolletti*[3]
[1]Politecnico di Milano, Milano, Italy, [2]University of Brescia, Brescia, Italy, [3]Pirelli Tyre, Milano, Italy

11.1 Introduction

11.1.1 Objective and structure of the chapter

Reinforcing fillers used at the commercial level in rubber compounds, such as carbon black (CB) and silica, are nanostructured: they are made by aggregates of spherical primary particles that can not be separated by thermomechanical mixing [1]. On the contrary, nanofillers [2,3] have primary particles with at least one dimension below 100 nanometers [4], that can be individually dispersed in the rubber matrix. In recent decades, nanofillers are one the most important novelties in the rubber field. The three main families of nanofillers are the subject of large research efforts in academic and industrial laboratories and have already found commercial applications: clays (C) and organically modified clays (organoclays, OC), [2,3,5–10], carbon nanotubes (CNTs), [2,3,11] and graphene (G) or graphitic nanofillers made by a few layers of graphene (named as graphite nanoplatelets (GNPs), nanosheets, or nanoflakes) [3,12–16].

It is nowadays clear that nanofillers can hardly replace the nanostructured ones in large-scale applications. However, the combination of both families of fillers can lead to remarkable properties.

This chapter is dedicated to rubber composites based on the combination of nano- and nano-structured fillers. It makes reference to what has been published in the scientific literature but it is not a directory of what is available. In fact, it is mainly focused on systems that have been remarkably investigated in the literature and that have already found some commercial applications. They are those based on CB and either OC or CNT or nanographite. Systems based on silica and CNT are also considered. For other families of combinations, in particular, combinations of nanofillers, that have been studied only to a minor extent, reference is given to the specific literature.

In particular, the authors' intention was to analyze synergistic effects of hybrid fillers, with the aim to model them. Results reported in the literature are based on a large variety of fillers, polymer matrices, as well as experimental conditions. This does not allow to attempt an overall rationalization. However, at the end of the chapter, a model proposed by the authors for the rationalization of the mechanical reinforcement is discussed.

Progress in Rubber Nanocomposites. DOI: http://dx.doi.org/10.1016/B978-0-08-100409-8.00011-5

11.1.2 Hybrid filler systems?

Words are important in the scientific field. They have semantic implications and, most of all, they should be used correctly. How to define filler systems made by the combination of nano- and nanostructured fillers? In most papers, authors use the expression "hybrid filler system." Is this expression correct? Or should it be more appropriate to use dual filler or three phase composite? The term "hybrid" commonly means something composed of mixed parts and, to the authors' opinion, the definition "hybrid filler system" gives an added value to the concept of a three phase composite. In fact, the term "hybrid" underlines that the two different fillers are intentionally added together to the matrix to obtain specific properties and, if possible, synergistic effects between the two fillers.

The IUPAC definition of "hybrid material" [17] is a "material composed of an intimate mixture of inorganic components, organic components, or both types of component." An important note states that "the components usually interpenetrate on scale of less than 1 μm." In the light of this definition, systems based on nano- and nanostructured fillers can be defined as "hybrid filler systems."

Different definitions of hybrid material are given and revised in literature [18]: some of them relate to the submicrometer size of the mixing, some others are more focused on the creation of new chemical bonds or new functions. A new systematic classification of hybrid materials is proposed [18], which includes three categories, depending on the purpose of the mixing of materials:

1. "structurally-hybridized materials," when the purpose of the combination or mixture of materials is hybridization of the macroscopic structure;
2. "materials hybridized in chemical-bond," when the purpose of mixing is the creation of new chemical bonds between different materials in order to obtain superior performance and/or new functions in the hybrid materials;
3. "functionally-hybridized materials," when the aim is the creation of new functions or super functions.

Nanocomposites are therefore included at least in the first definition, if not in the second one, when chemical bonds at the interface between the component materials play a key role in material development.

A specific definition has not been established, yet, for rubber composites filled with at least two different fillers or nanofillers. As a consequence, different expressions are reported in the literature works reviewed in this chapter. The authors of this chapter propose to refer to these materials as "hybrid filler systems," but, the terms used by the literature will be uncritically reported, trying to clarify the author's choice of terminology.

11.1.3 Synergistic effects of different fillers on the properties of rubber composites

The challenging goal of the research activity on hybrid filler systems is to develop synergistic effects between two different fillers, achieving substantial improvement of the material properties. In spite of such a stimulating goal, only a few works in the literature explicitly point out synergistic effects. Moreover, such effects are addressed only qualitatively and are not quantified.

According to the common definition, e.g., reported in the Oxford dictionary, a synergistic effect arises from the interaction or cooperation of two or more substances and leads to produce a combined effect greater than the sum of the substances separate effects. This definition holds also in the case of polymer composites: in this case, fillers are the substances. To realize the occurrence of a synergistic effect, it is crucial to define to what extent it should be greater the effect arising from the cooperation of different substances. In the literature, synergistic effects are defined in different ways. Most of them refer to the higher properties obtained in hybrid filled systems in comparison to systems containing only the nanostructured filler at the same overall filler amount [19–22]. Other authors compare hybrid and single-filler systems with different filler amounts but having the same level of a specified property, such as hardness [23] or tensile properties [24]. Synergistic effects are also claimed when a nanofiller is added to a composite containing a given amount of nanostructured filler. In this chapter, the synergistic effects explicitly reported in the literature works will be quoted, along with the criterion used to establish them.

An attempt to quantify synergistic effects (cautiously defined "interactive effects") on the initial modulus of rubber composites has been reported only by Agnelli et al. [25]. Elaboration of such an attempt is discussed below in the chapter.

11.1.4 Self-assembly of nanofillers in rubber matrix

In the prior art, aggregation and self-assembly have been first associated with micellar and colloidal systems and then also with molecules systems, as a consequence of the advent and success of supramolecular chemistry. Synonyms are adopted to describe self assembly: self-organization, molecular self-assembly, dynamic self-assembly. To rationalize the terminology, IUPAC came out with "Terminology for aggregation and self-assembly in polymer science" (IUPAC Recommendations 2013). In [26], it is reported that self assembly is the spontaneous and reversible organization of molecular entities by noncovalent interactions, where typical noncovalent interactions are van der Waals interactions, $\pi-\pi$ interactions, electrostatic interactions, and hydrogen bonds. Note 2 appears particularly interesting in the light of what reported in this chapter, as it reports that self-assembly is a process in which a system of preexisting components, under specific conditions, adopts a more organized structure through interactions between the components themselves. Indeed, the organization of fillers in a rubber matrix plays a dominant role for the final properties of the composite materials.

In the light of what is defined by IUPAC and reported in the literature, it can be said that the self-assembly of fillers indicates the driving mechanism for the aggregation of fillers in a solvent that steers its spatial arrangement. To achieve a controlled filler dispersion in polymer matrix composites, different self-assembly mechanisms can be exploited, which are driven by different forces: shear-flow forces [27] in blending methods, and physical interactions among particles, such as electrostatic attraction, hydrophobic interactions, and H bonding [28]. In rubber matrices, self-assembly methods were used to achieve uniform dispersion in NR latex of CNT [28] and of chemically reduced graphene [29], also achieving some degree of orientation.

Self-assembly of hybrid fillers is not explicitly addressed in the literature on rubber composites. However, as it will be pointed out below in the text, interaction between nano- and nanostructured fillers is clearly documented. The presence of the second filler might be thus a driving force for self-assembly. Hybrid nanofillers can exert synergistic effects that lower the percolation threshold [30]. The lower percolation threshold of nanofillers was clearly documented in the presence of CB, in the case of organoclay, CNTs, and nano-sized graphite [31], even though the phenomenon of self-assembly was not explicitly mentioned. In [32] the interaction between alumina and clay in cross-linked polyethylene gave rise to a unique filler architecture. In polyacrylonitrile nanofibers [33], self-assembly of montmorillonite/graphene oxide (GO) hybrid nanofillers led to improved properties due to the mutual-assisted dispersion of the two fillers in solvent. In a polyamide-6 [34], percolation threshold substantially shifted to lower CB contents in the presence of organoclay, since organoclay acted like a dispersion control agent to induce self-assembly of CB network. As discussed below in the text, OC and CB were reported to create "nanounits."

11.2 Nanocomposites based on organoclays and carbon black

A good number of papers are available in the scientific literature on rubber nanocomposites based on OC and CB. They are collected in Table 11.1, that shows the type of rubber, CB and OC, the blending technology, and the main aspects investigated in the work.

11.2.1 Structure of OC–CB nanocomposites

In the literature, the structure of nanocomposites based on CB and OC has been investigated by means of transmission electron microscopy (TEM), scanning electron microscopy (SEM), and X-ray diffraction (XRD).

11.2.1.1 OC organization in the rubber matrix

A clay mineral, pristine or modified with cations different from the naturally occurring ones, has a multiscale organization in a polymer matrix. Clay minerals are layered: they can be present in the rubber matrix as individual layers or in stacks. This is the lowest level of clay organization. Distribution and dispersion of clay aggregates are at the upper level of clay organization. Moreover, clays are made by platelets with high aspect ratio that can be oriented in the rubber matrix, as a consequence of different process conditions.

Table 11.1 **Papers on OC–CB hybrid fillers: type of rubber, CB and OC, blending technology and main aspects investigated in the work**

References	Rubber	CB type	OC type	Blending method	Analyses
[35]	E-SBR (S = 23.75%) Buna VSL 5025-2	N234	Cloisite C20A: 2HT	Emulsion blending for masterbatch. Melt blending	TEM, XRD, cure properties, dynamic-mechanical analysis (strain sweep and temperature sweep in tensile mode), tensile test, tear test, Shore A hardness
[36]	IR	N326	Dellite 67G: 2HT	Melt blending	XRD, TEM, DSC, dynamic-mechanical analysis (shear sandwich mode, strain sweep, recovery test, master curve on uncrosslinked material)
[37]	BIMS	N660	Cloisite C20A: 2HT	Melt blending	TEM, SAXS, permeability
[38]	CR	N550	Cloisite 15A: 2HT	Melt blending	TEM, XRD, cure characteristics (by MDR), tensile test, tear test, electrical properties
[39]	ENR25	N330	Cloisite 30B: T $(CH_2CH_2OH)_2$	Melt blending (two roll mill)	SAXS, High resolution TEM, swelling, tensile test, dynamic-mechanical analysis (temperature sweep in tensile mode)
[22]	IR	N326	Mt/2HT	Melt blending	TEM, mooney viscosity, DSC, bound rubber, density, cure characteristics (by MDR), dynamic-mechanical analysis (strain sweep in shear mode)
[40]	ESBR 1502 (S = 23) XSBR	N774	Cloisite 15A: 2HT Cloisite 20 A: 2HT Cloisite 30B: T $(CH_2CH_2OH)_2$	Solution blending XSBR for masterbatch. Melt blending with E-SBR	Cure characteristics (by MDR), WAXD High resolution TEM, SEM, TGA, dynamic-mechanical analysis (temperature sweep in tensile mode), tensile test, tear test

(Continued)

Table 11.1 **(Continued)**

References	Rubber	CB type	OC type	Blending method	Analyses
[41]	ESBR 1502 (S = 23)	N330	Cloisite 20A: 2HT	Melt blending	SAXS, FTIR, high resolution TEM, tensile test, tear test, abrasion test, heat build up, dynamic-mechanical analysis (temperature sweep in tensile mode, frequency and strain sweep in shear mode)
[42]	NR	N234	Clay Rockwood + 2HT or ODA	Melt blending	XRD, mooney viscosity, cure characteristics and kinetics, dynamic-mechanical analysis (strain sweep in shear mode), tensile test, hardness
[43]	X-NBR	N330	Nanomer I30E: ODA	Melt blending (two roll mill)	WAXD, TEM, curing characteristics, tensile test
[44]	IR	N326	Dellite 67 G: 2HT	Melt blending	Curing characteristics (by ODR), TEM, WAXD, density, tensile tests, dynamic-mechanical analysis (strain sweep in shear mode)
[45]	EPDM: E/P = 6/4, ENB = 9.4 XSBR	N774	Cloisite 15A: 2HT Cloisite 30B: T $(CH_2CH_2OH)_2$	Solution blending XSBR + Clay. Melt blending with EPDM	Curing characteristics, XRD, high resolution TEM, SEM, dynamic-mechanical analysis (temperature sweep in tension mode), tensile test, TGA
[46]	ENR	N550, N330, N220, N110	Nanomer I30E ODA	Melt blending	SAXS, high resolution TEM, dynamic contact angle, dielectric measurements, AC resistance, volume resistivity, tensile test Dynamic-mechanical analysis (temperature and strain sweep in tension mode)

[47]	E-SBR 1502	N330	Cloisite 15A: 2HT	Melt blending (two roll mill)	XRD, TEM, tensile test, dynamic-mechanical analysis, heat build up, abrasion test
[48]	NR	N330, N754	Closite 15A: 2HT Cloisite 20 A: 2HT	Melt blending (two roll mill)	TEM, SEM, tensile test, tear test, fatigue test
[24]	NR (SMR)	N330	Mt/Distearyl	Melt blending	XRD, TEM, cure characteristics, tensile test, Shore A hardness, dynamic-mechanical analysis (temperature sweep in tension mode)
[49]	IR, NR, BR, E-SBR 1721, S-SBR SOL R72612, Buna 5025-0	N326	Dellite HPS, Dellite 67G: 2HT	Melt blending	XRD, TEM, Mooney viscosity, cure characteristics (by MDR), tensile test, hardness, dynamic-mechanical analysis (temperature sweep in tensile mode)
[50]	E-SBR-1502	N330	Nanomer I30E: ODA	Melt blending (two roll mill)	XRD, TEM, tensile test, dynamic-mechanical analysis (temperature sweep in tensile mode)
[15]	Chlorobutyl, CIIR	N330	Cloisite 15 A: 2HT	Solution blending Rubber + clay Melt blending Masterbatch + CB	curing characteristics (by ODR), tensile test, tear test, Shore A hardness, dynamic-mechanical analysis (temperature sweep in tensile mode), positron annihilation, bound rubber
[51]	SIR20 E-SBR 1712	N339	Mt/ tetraalkylammonium	Solution blending Rubber + clay Melt blending Masterbatch + CB	FT-IR, curing characteristics, tensile test, tear test, abrasion test, heat build up, De Mattia crack initiation test
[52]	poly(isobutylene-co paramethylstyrene) styrene-butadiene rubber	N330, N550, N660	ODA	Solution blending	XRD, tensile test, dynamic-mechanical analysis (temperature sweep in tensile mode)

11.2.1.2 Lowest level of clay organization: stacking or exfoliation

As it will be shown below, in most of the TEM micrographs available in the papers of Table 11.1, both individual lamellae and tactoids made by few layers are visible. Comments are proposed on what is intercalated in the space available between two opposite layers: only the ammonium cation or also the polymer chains? In the scientific literature, many reports are available on this point [5,6,53] for rubber clay nanocomposites in the absence of CB. In order to rationalize the clay organization in the rubber matrix in the presence of CB, it is worthwhile to take this literature into consideration and, in particular, it is useful to summarize the crystalline organization of pristine OC, prior to the mixing with the rubber. In fact, as it will be discussed below, the presence of CB does not substantially alter the situation.

Fig. 11.1A shows the XRD patterns of Na-Mt: (a) pristine; (b) modified with T (2-hydroxyethyl); and (c) 2HT. Fig. 11.1B shows the structure of the organophilic cations used in the papers of Table 11.1 for clay modification.

In the XRD pattern, (00ℓ) indexes refer to a crystalline order in the direction perpendicular to the structural layers: (001) reflection is due to regularly stacked layers whereas higher order reflections, e.g., (002) and (003), originate from the regular arrangement of intercalants. An expansion of d interlayer spacing corresponds to a shift of (001) reflection towards lower 2θ angle values. In Fig. 11.1A, it is evident that the intercalation of organophilic cations leads to expansion of interlayer distance and to the appearance of higher order reflections in the XRD pattern, when the ammonium cation has long chain substituents such as 2 talloyl groups.

Meaningful data on the organization in the interlayer space have been recently proposed [54]. Reaction of Mt with 2HTCl was performed either at the solid state in the absence of any other chemical substance, or in the presence of solvents, or in

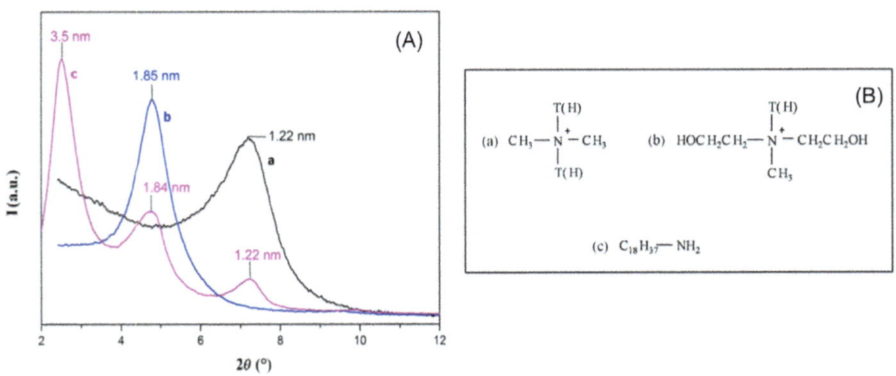

Figure 11.1 (A) XRD patterns of Na-Mt: pristine (Cloesite®) (a), modified with T(2-hydroxyethyl) (Cloesite® 30B) (b), 2HT (Dellite® 67 G) (c). (B) organophilic clay compensating cations: 2HT (a), $T(CH_2CH_2OH)_2$ (b), ODA (c).
Source: Reprinted with permission from Galimberti M. Rubber clay nanocomposites: science, technology, applications, 1st ed., John Wiley and Sons; 2011.

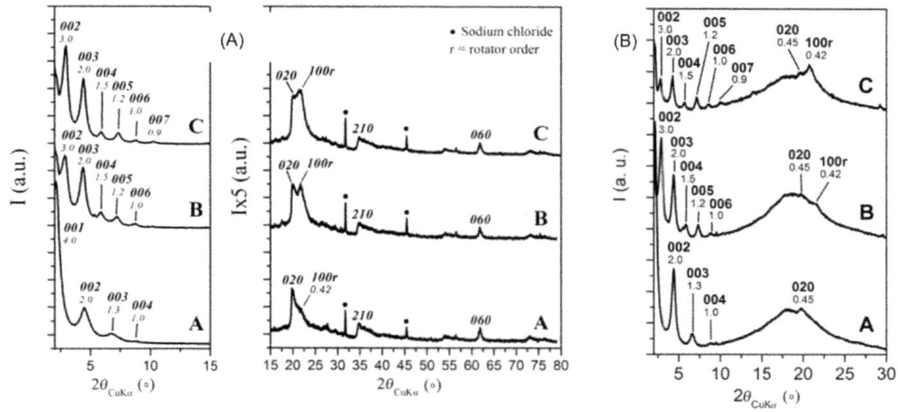

Figure 11.2 XRD patterns of: (A) (a) Mt/2HT, (b) Mt/2HT/SA, (c) Mt/2HT/SAES, prepared from neat synthesis. Asterisks indicate reflections due to NaCl and symbol *r* indicate reflections relative to the rotator order; (B) OC samples prepared in IR as the reaction medium: (a) Mt/2HT, (b) Mt/2HT/SA, (c) Mt/2HT/SAES.
Source: Reprinted with permission from Cipolletti V, Galimberti M, Mauro M, Guerra G. Organoclays with hexagonal rotator order for the paraffinic chains of the compensating cation. Implications on the structure of clay polymer nanocomposites. Appl Clay Sci 2014;87:179–88.

poly(1,4-cis-isoprene) as the reaction medium, optionally adding either stearic acid (SA) or 2-stearamidoethyl stearate (SAES). In Fig. 11.2A, the XRD patterns of OC samples prepared through neat synthesis, in the absence of solvents, reveal two interlayer distances: 3.6–4.0 nm for Mt/2HT and about 6 nm for Mt/2HT/SA and Mt/2HT/SAES. The 3.6–4 nm value was explained with the bilayer placement of the tilted paraffinic chains of the ammonium cation [55] and the 6 nm value with the perpendicular placement of the paraffinic chains, associated with the presence in the interlayer space of either the acid or the amide. For the first time, it was documented the occurrence of hexagonal rotator order in the packing of the long hydrocarbon tails of the 2HT compensating cation. SA and, in particular, SAES were found to enhance this type of order. Fig. 11.2B shows XRD patterns of OC samples prepared in IR as the reaction medium: (A) Mt/2HT, (B) Mt/2HT/SA, (C) Mt/2HT/SAES. Exactly the same reflections were obtained for crystalline OC and the rotator order was detected.

These findings demonstrate that low molecular mass chemicals are able to get in the interlayer space, together with the compensating cation, leading to a further expansion of the interlayer distance. Moreover, they demonstrate that the same OC crystalline structures are formed in the absence and in the presence of polymer chains, thus suggesting that polymer intercalation is not such a likely phenomenon as it is widely considered in the literature. The maximum value reported in the literature, for the interlayer distance of crystalline OC, is 6 nm. XRD analysis on composites of Table 11.1 was performed on polymer composites containing low molecular mass chemicals and d_{001} spacing is, in all papers, well below 6 nm.

Hence, it is hard to state who is to blame for the expansion of the interlayer distance. However, the objective of this paragraph is to assess if CB plays a role in modifying the OC organization in the polymer matrix.

Some papers indicate a minor influence, if any, by CB on the clay organization in the interlayer space. Reflections due to OC(2HT), in XRD patterns of composites based on IR and containing CB, were found at the same 2θ values [36,44] as in XRD patterns of pristine OC. In NR as the matrix, the expansion of the OC interlayer distance was substantially a function of the low molecular mass guests in the interlayer space; it was larger for 2HT than for ODA [42], whereas OC(2HT) exfoliation was promoted by preparing an "expanded clay," by intercalating a long chain fatty acid together with the ammonium cation [24]. Expansion of interlayer distance was found by preparing OC(2HT)/E-SBR blend through latex blending [35], and the expansion of the interlayer distance was absent or very minor after melt blending with S-SBR/BR and CB. The XRD patterns of nanocomposites (OC with 2HT) based on BIMS and CB had general features that were similar to those of nanocomposites without CB [37]. In XNBR matrix, the key role for the OC(ODA) organization was played by a low molar mass liquid rubber, carboxyl-terminated copolymer of butadiene and acrylonitrile, rather than by CB [43]; the intercalation of the liquid rubber led to further expansion and exfoliation of OC.

Only a minor number of reports suggest an effect of CB for clay intercalation. In the case of silicone butadiene rubber (SBR) based nanocomposites (OC with ODA), further expansion of the interlayer distance was observed in the presence of CB and was attributed to the enhancement of polymer intercalation [50]. For composites based on CR, addition of CB was reported to favor OC(2HT) dispersion and intercalation [38]. In the case of composites based on ESBR and OC(2HT), the addition of calcium stearate led to a further expansion of the interlayer distance, that was attributed to the specific ionic interaction existing between anionic surface of layered silicates and Ca^{2+} and interpreted with a more pronounced intercalation [41].

The effect of CB on OC exfoliation seems to be different, in different papers. The addition of CB to composites based on BIMS hindered the OC(2HT) exfoliation process (unfunctionalized BIMS) or at least did not enhance the exfoliated structure (functionalized BIMS) [37].

Composites based on either E-SBR [40] or EPDM [45] and CB did not reveal reflections due to OC. However, OC was first dispersed in a masterbatch with carboxylated styrene butadiene rubber (XSBR) via solution blending. In ENR25 the exfoliation of OC(ODA, T2HE) was found to depend on the compensating cation. It was reported that the interactions between CB and OC(ODA) were more intimate than that of CB and OC(2THE) and such larger interactions led to larger exfoliation [39].

11.2.1.3 Upper level of clay organization: distribution and dispersion of OC

Both CB and OC achieve better dispersion in composites based on the binary CB–OC system. In fact, CB was reported to favor OC (ODA) dispersion, through the formation of the so-called "nanounit" [50]. Prevailingly delaminated OC(2HT) was observed to promote a remarkable improvement of CB dispersion [22,44].

OC distribution and dispersion in rubber matrices with CB are affected by the nature of the polymer matrix, by the blending technology, by the degree of OC exfoliation, and by the addition of promoters of dispersion. The overall OC(2HT) distribution was reported to be homogenous in rubber composites based on an apolar matrix such as the isoprene rubber [36,44]. Dispersion of OC, with larger exfoliation, was improved by using a polar Mt modifier, such as T2HE instead of 2HT, for the preparation of masterbatches with a polar rubber such as XSBR [40]. When a masterbatch of OC in E-SBR was prepared via latex blending, the successive melt blending in BR led to a good OC dispersion [35].

OC exfoliation was reported to favor the homogenous clay distribution. In [39], composites were compared, based on HNR-25, CB N330, and either OC(ODA) or OC(T2HE). A larger degree of exfoliation was achieved with ODA as the modifier and clay dispersion was observed to be more uniform.

Calcium stearate has been reported to act as a dispersion promoter for OC(2HT) in SBR matrix [41].

11.2.1.4 Orientation

Orientation of OC(2HT) in nanocomposites based on BIMS was studied as a function of the type of flow [37]. In the extrusion flow, the extensional component aligned the organoclay particles along the flow direction and the squeeze flow of the melt compression process aligned the sheet-like organoclay particles along the planar direction. In the case of unfunctionalized BIMS, in spite of the tendency of clays to wrap around CB particles, most clay tactoids maintained large axial orientation and they did not curve notably. When BIMS was functionalized, either with triethylamine or with dimethyl benzoic acid, the presence of CB decreased the organoclay orientation parallel to the plane direction of the polymer film.

In IR based composites processed through a two roll mill and pressed for the vulcanization step [22], isolated silicate layers or stacks (OC with 2HT) appeared in TEM micrographs oriented along a preferential direction.

In HNBR, with either 5 or 15 phr of CB [56] and OC(2HT) either in place of the 5 CB phr or added to the 15 CB phr, the mechanical properties of nanocomposites could be explained by the formation of an anisotropic, oriented filler–filler network.

11.2.1.5 Fractured surfaces

Tensile fractured surface of hybrid nanocomposites showed highly rougher surface morphology compared to the composites based on rubber, either neat or containing OC. This was attributed to the dispersion of CB and OC(ODA, T2HE) platelets in the rubber matrix, that alter the crack path depending upon their orientation [45].

11.2.2 Affinity of OC for CB

TEM micrographs demonstrate that OC have affinity for CB; OC individual layers and tactoids lie close to CB particles. It is reported that OC and CB form hybrid filler networks. These findings are independent of the type of polymer matrix and clay modifier.

Fig. 11.3a,B show TEM micrographs of nanocomposites based on IR [44] and BIMS [37] as the polymer matrix and OC with 2HT as the compensating cation.

In the micrographs of Fig. 11.3, individual lamellae and tactoids can be detected, for both nanocomposites. As reported in [37], the majority are observed around CB clusters, with only a small fraction laying in the polymer matrix.

In NR matrix [48] physical contact network between OC(2HT) and CB was observed, with clay layers bridging CB aggregates, and the formation of a hybrid network was also proposed. Favorable interaction between CB and OC(2HT) has been observed also in CR [38].

OC(ODA) and CB were reported to form a dual structure, named a "nanounit," in E-SBR 1502 [50] and in ENR [39,46]. As mentioned by the authors of these papers, the OC–CB "nanounit" was first found in polyamide-6/OC(2T2HE)/CB ternary hybrids [34]. The nanounit was described as composed of stacked clay platelets that deform to wrap partially around one or two primary CB aggregates. This elementary nanounit structure induced self-assembly of a CB network within polyamide 6 matrices. Such a structure was prevalent throughout the polymer matrix and remained robust under a wide range of thermal deformation histories.

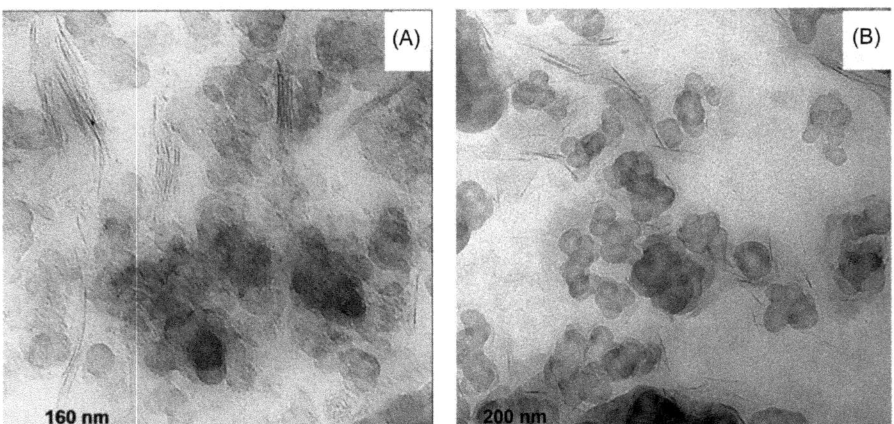

Figure 11.3 TEM micrographs of OC–CB composites: (A) IR as the polymer, N326 as CB, 7.5 as OC phr, 2HT as the cation; (B) BIMS as the polymer, N660 as CB, 2HT as the cation. *Source*: For (A) Reprinted with permission from Galimberti M, Coombs M, Cipolletti V, Riccio P, Ricco T, Pandini S, et al. Enhancement of mechanical reinforcement due to hybrid filler networking promoted by an organoclay in hydrocarbon based nanocomposites. Appl Clay Sci 2012;65–66:57–66. For (B) Reprinted with permission from Nawani P, Burger C, Rong L, Hsiao BS, Tsou AH. Structure and permeability relationships in polymer nanocomposites containing carbon black and organoclay. Polymer (Guildf) 2015;64:19–28.

The effect of OC organization on the formation of the hybrid filler structure was investigated [36]. OC(2HT) was delaminated through ball milling and composites based on IR and OC, either pristine or delaminated, were prepared. As revealed by TEM micrographs, the larger volume occupied by pristine OC stacks helped to form bridges between CB clusters, whereas individual clay layers obtained via ball milling simply adhered to CB particles.

11.2.2.1 The origin of the affinity of OC for CB

In the case of nanocomposites in nylon 6 matrix [34], the attraction of CB toward the organoclay platelet was explained with a combination of weaker (London dispersion forces) and stronger (hydrogen bonding) interactions between the amide group of nylon 6 chains, negatively charged pristine clay and functional groups attached to the organic modifier. It was written that competitive adsorption on CB rough surfaces occurs between the end tail sides of nylon 6 chains and organoclay. Thanks to the improved CB dispersion, the electrical CB percolation threshold was substantially shifted to lower volume fractions.

The dispersion of CB was found to be dramatically improved in IR as the matrix in the presence of only the ammonium salt 2HTCl. The favorable interaction between the organophilic cation and CB was explained with the cation-pi interaction, well known in organic chemistry, that was thus proposed at the origin of OC–CB affinity [22]. Such favorable interaction is well documented in the scientific literature.

Even distribution of OC and CB should be expected also on the basis of the Zeta potential (ZP) values of nanoparticles [57]. ZP is the measure of the repulsion between particles that are similarly charged. In rubber composites, ZPs of nanofillers, at typical pH levels, are negative, whereas ZPs of CBs are positive [58–62]. ZP can be applied also to highly viscous nonaqueous fluids, such as polymer melts. In the case of CB hybrid systems with clays, studies are reported on clay bonhomie [63,64].

11.2.3 Curing

Curing of nanocomposites based on OC and CB of Table 11.1 was performed with sulfur based systems.

It is widely acknowledged that CB promotes acceleration of curing reactions of diene rubbers, promoted by sulfur [65,66]. Explanations are based on the increased thermal conductivity of the compound as well as on the supposed basic pH of CB.

OC also gives rise to faster curing [2,67–69]. The formation of amines from the ammonium cations and the increased mobility of sulfur accelerating anionic species are considered to be at the origin of this phenomenon [70].

Results available in the papers of Table 11.1 demonstrate that the addition of OC, to rubber composites containing CB, brings faster curing, with reductions of scorch and optimum times of vulcanization. This is reported for many different systems, such as NR-OC(2HT, ODA) [42] NR-OC(2HT) [24], IR-OC(2HT) [44], SBR-OC(2HT) [41], XNBr-OC(ODA) [43], NR/BR blend (OC with quaternary ammonium) [51], and EPDM-OC(2HT, T2HE) [45].

Results are also available that show a synergistic effect of OC and CB: in CR matrix [38], lower scorch time was observed for the composite with the hybrid filler system with respect to those based on only either CB or OC(2HT). A prevulcanization inhibitor, such as a thiophtalimide, was proved to be efficient for controlling the reduction of scorch time [44].

The ammonium cation acts thus as an accelerator, but the curing reaction relies on the traditional ingredients, such as, e.g., divalent Zn cations. In fact, the addition of a small amount of calcium stearate to composites containing ZnO [41] led to a further increase of curing rate, thanks to the formation of soluble Zn soap, whereas further addition deactivated the reaction, capturing zinc.

The value of the maximum modulus depends on the amount of OC. M_H increases for low OC values and then decreases, for OC(2HT, ODA) [42], OC (2HT) [24], (OC with quaternary ammonium) [51] or reaches a plateau value for OC(2HT) [44]. The decrease of M_H value after a given amount of OC is explained with the plasticizing effect of the long alkyl chain substituents of the ammonium cations. Plasticization can be also due to other OC modifiers, such as the carboxyl-terminated copolymer of butadiene and acrylonitrile used to compatibilize OC (ODA) with carboxylated acrylonitrile–butadiene rubber matrix [43] or the SA used to prepare the so-called expanded OC(2HT) [24]. Larger values of M_H were also ascribed to different dispersions of OC in the matrix: e.g., in EPDM [45], OC with T2HE was reported to be better dispersed than OC with 2HT.

Swelling measurements, performed on ENR based nanocomposites [39] revealed that OC(T2HE) helped the swelling, however, in the presence of poor OC dispersion.

11.2.4 Mechanical reinforcement as a function of fillers' features

Reinforcement of a rubber is obtained when a filler(s) with a much higher modulus is added to the soft matrix. According to theories developed for *98 InTech* CB as the filler, the modulus of a filled rubber compound can be seen as the sum of different contributions, as it is shown in Fig. 11.4.

Contributions that do not depend on the strain amplitude are due to the polymer network, hydrodynamic effects, and immobilization of rubber on filler particles. The contribution that depends on the strain amplitude is the so-called filler network: filler particles are joined together either directly or through polymer layers. To describe mechanical reinforcement, micromechanical models have been developed [1,71,72] based on the replacement of part of the soft matrix with filler particles, that take into account the nonlinear dependence of the modulus on the filler concentration as the result of interactions of stress fields of each filler particle. The Smallwood–Guth–Gold equation (Eq. (11.1)) [73] expresses the enhancement of the matrix elastic modulus as a function of volume fraction of the added filler:

$$G_c G_c / G_m = 1 + 2.5\Phi + 14.1\Phi^2 \tag{11.1}$$

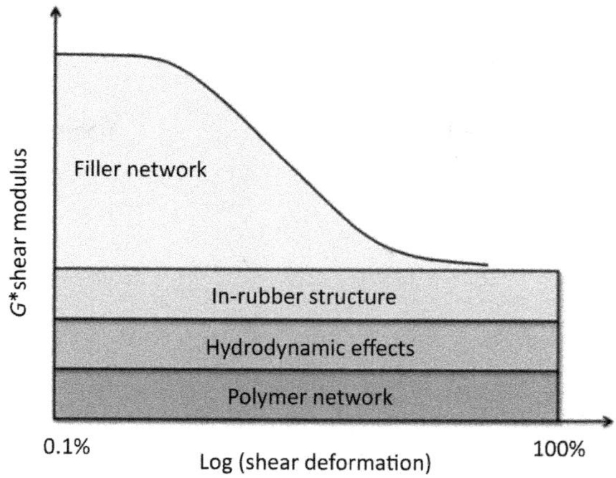

Figure 11.4 Modulus of a rubber compound filled with a particulate filler as the sum of different contributions.
Source: From Galimberti M. Rubber clay nanocomposites: science, technology, applications. 1st ed. John Wiley and Sons; 2011.

where G_c and G_m are the elastic moduli of the composite and of the neat elastomer, respectively, Φ is the filler volume fraction and the quadratic term accounts for the mutual disturbance of filler particles. For particle aggregates and for nonspherical particles, another equation was proposed by Guth [74] (Eq. (11.2)):

$$G_c/G_m = 1 + 0.67f\Phi + 1.62f^2\Phi^2 \quad (11.2)$$

where f is the shape factor, given by the length to width ratio of aggregate or particle. To take into account the occlusion of rubber in the filler voids, Φ is replaced by Φc, calculated through measurements of oil absorption by the filler.

Modulus at minimum strain essentially depends on the polymer–filler interfacial area, calculated as $A_i \times \rho \times \phi$, where A_i is the filler surface area and ρ is the filler density. Modulus at large strain mainly depends on Φc. Compounds based on fillers with large surface area have high modulus at minimum strain and remarkable reduction of modulus as the strain amplitude increases, a phenomenon known as the Payne effect [71]. Moreover, their percolation threshold in the polymer matrix is lower than for fillers with larger surface area.

As mentioned in the Introduction, most of the reinforcing fillers for rubbers used are nanostructured: CB and silica. Their aggregates have dimensions of hundreds of nanometers and can not be split into individual particles. On the contrary, the so-called nanofillers, such as clays, CNTs, graphene, and nanographites, have individual particles, with at least one dimension below one hundred nanometers, that can be individually dispersed in the rubber matrix and are characterized by a high or even very high aspect ratio. Nanofillers are thus expected to have large/very large surface area and they should promote high modulus at minimum strain. The

Table 11.2 Percolation threshold (Φ_p) of OC, CNT and nanographite in rubber matrix[a]

Nanofiller	Matrix	Φ_p (phr)	References
OC	Neat SBR	12[b]	[76]
	Neat NR	8	[77]
	Neat IR	6	[44]
	IR + 60 phr CB	3	[44]
CNT	Neat IR	7	[78]
	Neat IR	9	[25]
	IR + 60 phr CB	3	[78]
nanoG	Neat IR	21	[16]
	Neat IR	17	[25]
	IR + 60 phr CB	8	[16]

[a]The percolation threshold was calculated through the Huber-Vilgis method. Round figures are indicated in the Table.
[b]Estimated value.

mentioned nanofillers have indeed high surface area but do not have the so-called structure, that means they are not able to occlude rubber chains. Hence they are not able to transform the soft rubber into a rigid undeformable phase. In the light of these findings, nanofillers with high surface area promote a pronounced nonlinearity of the viscoelastic modulus. Nanofillers are also characterized by the low concentration in the rubber matrix required to achieve percolation. Table 11.2 reports values of percolation thresholds for nanofillers and for CB in neat rubber matrices. Such thresholds were estimated by applying to the initial modulus values a model proposed by Huber and Vilgis [75] for rubbers filled with spherical nanostructured CB. According to this model, the excess of initial modulus $(E-E_0)/E_0$ scales linearly with the filler content below the percolation threshold, whereas above this limit it increases following a power law with exponent 4. The different exponents calculated for nanofillers can be attributed to their deviation from sphericity.

Bearing in mind the overall picture, a peculiar feature of OC has to be carefully taken into account. As reported in section 11.2.3, OC are effective accelerators. The study of the mechanical reinforcement brought about by OC should always be performed bearing in mind that OC, thanks to the ammonium cations, promotes remarkable variation in the structure of the cross-link network, essentially favoring the formation of shorter sulfur bridges and higher cross-linking density. Such a variation of the crosslink network is responsible, per se, for the enhancement of the modulus of the rubber nanocomposites.

11.2.5 Dynamic-mechanical properties

In most papers, the filler networking phenomenon was investigated by applying sinusoidal stresses to nanocomposites, typically in the shear mode, increasing the strain amplitude (strain sweep test). Results reported in many papers, referring to different

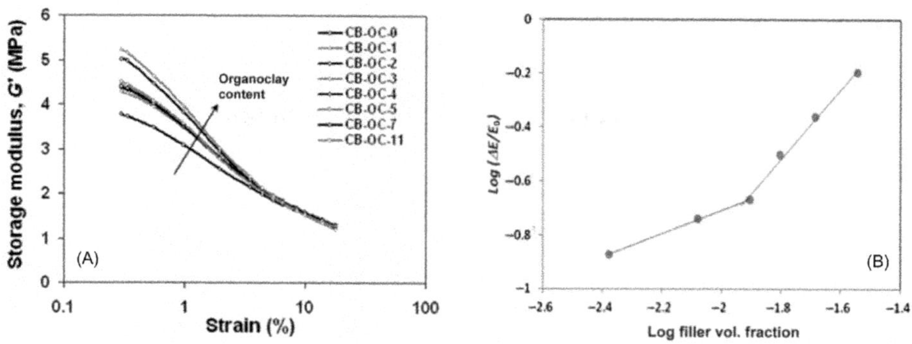

Figure 11.5 (A) Dynamic storage modulus G' in shear mode at 1 Hz versus strain amplitude for CB (60 phr)-OC hybrid composites. (B) Double logarithmic plot of the excess modulus, with respect to the matrix with 60 phr CB, as a function of the OC volume fraction (Huber-Vilgis plot).
Source: For (A) Reprinted with permission from Galimberti M, Coombs M, Cipolletti V, Riccio P, Ricco T, Pandini S, et al. Enhancement of mechanical reinforcement due to hybrid filler networking promoted by an organoclay in hydrocarbon based nanocomposites. Appl Clay Sci 2012;65–66:57–66.

rubbers, consistently agree on the remarkable increase of the Payne effect when increasing amounts of OC were added to rubber composites in the presence of CB. Data are available for OC(2HT) in SBR [41,47], in NR [24], in IR [44], for OC (ODA) in ENR [46], and for masterbatches of OC(2HT) in E-SBR, melt blended with CB and SSBR/BR [35]. In the case of nanocomposites based on NR and 50 phr of CB [42], the replacement of CB with OC(2HT, ODA) (in the same amount by mass) apparently led to the decrease of the Payne effect. However, taking into consideration the higher density of OC, also this result seems to confirm the overall picture.

It is definitely very important to carefully consider the effect of the amount of both OC and CB. In fact, in the case of the hybrid OC(2HT)/CB filler system in SBR [41], the replacement of 3 out of 45 phr of CB with 3 phr of OC led to a reduction of the Payne effect. In Fig. 11.5A, that refers to IR based composite containing 60 phr of CB, it is evident that a remarkable Payne effect occurs after a sort of threshold for the OC(2HT) content [44].

The existence of such a threshold could mean the existence of two regimes of mechanical reinforcement. An investigation was performed [44] by applying the Huber-Vilgis model to dynamic storage G' moduli at minimum strain and to initial E moduli from stress–strain curves. The double logarithmic plot of the excess of E modulus versus OC content is shown in Fig. 11.5B. Two different straight lines are in the graph of Fig. 11.5B, indicating the discontinuity in the regime of mechanical reinforcement, as if at a given OC threshold value (3 phr in the figure, lower than in the neat rubber matrix, 6 phr), OC and CB were able to form a hybrid filler system. The continuous hybrid system was visualized by TEM micrographs.

The effect of OC organization on the Payne effect of a composite based on IR and CB was studied [36]. OC(2HT) was used in its pristine state, with stacked layers, or delaminated via ball milling. It was demonstrated that the delaminated OC brought about a much lower Payne effect, than the stacked OC. As mentioned in section 11.2.2, this result was attributed to the large volume occupied by OC stacks, that made the formation of hybrid filler networks easier. As mentioned in the same section, the affinity of OC for CB was attributed to the cation-pi interaction between the compensation ammonium and the aromatic rings of CB [22] and was reported to cause the synergistic effect by the two fillers on the mechanical reinforcement. A very low amount of the ammonium salt was needed to cause a remarkable enhancement of the composite initial modulus and of the filler networking process. In fact, maximum enhancement was estimated to occur with an ammonium salt layer, with the hydrocarbon chains lying flat on CB surface.

The storage modulus of nanocomposites was investigated from below to above the glass transition temperature (T_g).

Most results indicate that the modulus increases, below the glass transition, by adding OC to the CB based composite and such an increase depends on the type of substituents. OC(T2HE) and OC(2HT), added in 3 phr amount to SBR-based composite, to replace 5 out of 40 phr of CB, brought a clear increase of storage modulus, to different extents [40]. The same findings were obtained by adding the same OCs to an EPDM based composite, to replace 3 out 40 phr of CB [45]. In both composites, the largest increase was observed by using OC(T2HE). More subtle differences appear for composites based on XNBR, with larger values for the OC(ODA) compatibilized thanks to a low molecular-weight liquid rubber, carboxyl-terminated copolymer of butadiene and acrylonitrile [43]. An increase of modulus was observed also in NR by adding to a composite containing either 30 or 50 phr of CB an OC(2HT) "expanded" with a different amount of SA [24]. It was reported that the "expanded OC" was easily exfoliated and the increase of modulus was supposed to be the consequence of the large interaction with the composite matrix of individual lamellae. Moreover, by the use of hybrid fillers, synergistic effects were obtained on mechanical properties: with the proper filler content (10 phr SA, 30 phr CB, and 5 phr OC), CB content and rolling resistance could be reduced at the same time without a decrease of other mechanical properties with respect to 50 phr CB filled NR. It was also reported that the addition to a composite based on ENR and 20 phr of CB of 15 phr of OC(ODA) led to an increase of T_g, whereas 15 phr of OC(T2HE) led to modulus reduction [39]. As expected, the addition of 25 phr of CB to a composite based on ESBR and containing 6 phr of OC (2HT) led to a clear increase of the modulus [47].

A similar trend was observed above glass transition temperature for the composite based on ENR and OC(ODA, T2HE) [39]. Increasing amounts of OC(ODA) led to higher values of E' above glass transition temperature, in XNBR Matrix, compatibilizing OC with a low molecular-weight carboxyl-terminated copolymer of butadiene and acrylonitrile [43]. Differences between OCs (2HT, T2HE) were subtle in SBR [40] and EPDM matrices [45].

In HNBR, with either 5 or 15 phr of CB [56] and OC(ODA) either in place of the 5 CB phr or added to the 15 CB phr, the larger values of E' and the lower Tan δ values were interpreted with the strong filler–matrix interaction.

The main drawback of OC appears to be the strong reduction of the storage modulus above room temperature. The replacement of 10 out of 60 phr of CB with the same volume amount of OC(2HT) in composites based on either IR or NR led to remarkably larger E' reduction (ΔE), passing from 23°C to 70°C: from 3.3 to 4.4 in the case of NR, from 2.7 to 3.3 in the case of IR. The replacement of 5 and 10 phr out of 60 phr of CB in SBR-based compound with the same amount of OC increased $\Delta E(23°-70°)$ from 7.8 to 8.6 to 10.2 [49]. It is well documented in the literature that crystalline OC show endothermic phenomena above room temperature [79,80–82].

In a composite based on NR and OC(2HT), a secondary transition evidenced in the modulus versus temperature plots of the nanocomposites above the percolation threshold of OC was observed and was attributed to a reduction in the efficiency of load transfer from matrix to clay due to the "melting" of onium chains at the polymer–filler interface [77].

The effect of OC on the storage modulus of rubber composites based on CB appears clear: enhancement of storage modulus and enhancement of modulus reduction as the temperature increases. Differences and discrepancies among various results could be attributed to the different organization of OC in the rubber matrix.

The effect of OC(2HT) organization on storage modulus in the plateau region was studied [36] performing multifrequency dynamic-mechanical analysis on masterbatches containing only PI, CB, OC, and 3-octanoylthio-1-propyltriethoxysilane as the silane, obtaining a master curve representation of G' as a function of frequency. Two types of OC were used: either pristine, with some layers in crystalline stacks, or delaminated (D-OC), without any (00ℓ) reflection in the XRD pattern. It was shown that the master curves of masterbatches with only CB as the filler and with CB/D-OC hybrid filler system were almost overlapped, with the system incorporating only CB revealing a shorter and steeper plateau, with higher modulus values on the high frequencies region. The masterbatch containing pristine OC stacks was clearly different: the rubbery plateau had a major extension and a lower slope, with expansion of the low frequency end of the master curve.

OC organization, in stacks or in individual lamellae, was shown to have a profound effect also on the variation of the storage modulus with the temperature. OC, as stacks or delaminated, were used in IR based composites in the presence of CB [83,84]: $\Delta E'$ of the composites, between low (0, 10, 23°C) and high (70, 100°C) temperatures was appreciably lower when D-OC was used in place of pristine stacked OC.

It could be thus commented that stacks of OC promote larger modulus enhancement and they seem to be able to give more effective filler networking via the immobilization of polymer chains. A key role is thus played by the extent of OC exfoliation, which depends on the type of low molecular mass chemicals that interact with the clay layers and on the processing conditions.

Tan delta, that means the ratio between loss and storage modulus is investigated in many papers. It is reported that variation in shape and strength of the tan δ peaks might suggest that T_g of the samples are affected by the extent of polymer–filler interaction. Actually, this should be taken much with caution. As demonstrated in

[85], tan δ in the glass-to-rubber softening region is influenced not only by the local segmental dynamics, reflected in the magnitude of the loss modulus toward lower T, but also by filler induced changes of both G' and G''. In other words, the polymer—filler interaction should be better represented by G". However, a clear effect of OC on tan δ was shown. For example, composites based on OC(ODA, T2HE) and CB that revealed the largest enhancement of E' (see above) [39], showed also the maximum broadening of tan δ peak, interpreted with the highest effectiveness of OC and CB for the enhancement of volume of polymer—OC interface. Reduction in the height of tan δ peak was observed in the case of composites based on SBR [40] and on EPDM [45] (OC with 2HT or T2HE) and was interpreted with physical and chemical adsorption of the E-SBR molecules on the filler surface. In the case of IR, NR, and E-SBR based composites with 60 phr of CB, the replacement of 10 phr of CB with the same volume fraction of OC(2HT) led to very similar tan δ at low temperatures (10 and 23°C) and higher tan δ at 70°C [49].

The effect of OC(2HT) on the dynamic-mechanical properties of the nanocomposites was investigated [35] in the light of the application as a tire tread compound. Different amounts of OC, from 5 to 20 phr, were added to a composite based on E-SBR, S-SBR and BR and 30 phr of CB. The remarkable increase of tan δ at −10 and 10°C allowed to envisage better performances on ice and wet road and the higher E' at 30°C was assumed as indication of good dry handling. However, the much higher values of E' at −20°C and of tan δ at 60°C suggested poor winter traction and worse rolling resistance.

To reduce the heat build up of composites based on SBR, 25 phr of CB and 6 phr of OC(2HT) were used in place of 40 phr of CB, achieving a good set of mechanical properties [47]. Delamination of OC(2HT) was reported to be an effective way for reducing the hysteresis of nanocomposites based on NR [83,84]. Larger exfoliation of OC(2HT) in SBR, thanks to calcium stearate as a dispersion promoter [41] led to lower heat build up.

11.2.6 Tensile properties

It is worth considering at first, nanocomposites based on traditional, large amounts of CB, at least 60 phr.

In E-SBR based composites [49] with 60 phr of CB, the replacement of CB with the same volume fraction of OC(2HT) (5 and 10 phr) led to the reduction of moduli at 100% and 300% elongation, to larger elongation and energy at break and to larger tensile strength, with more improvement of ultimate properties with 10 phr of OC. In IR and NR-based compounds with 60 phr of CB, the replacement of CB with OC(2HT) (10 phr, same volume fraction) led to very similar results but for the increase of moduli at short deformation.

OC(quaternary ammonium) addition (5, 10, 15 phr) to a compound containing comparable amounts of NR and SBR and 75 phr of CB [51] did not cause appreciable changes in the tensile strength, but led to a gradual reduction in the elongation at break, with 15% reduction for 10 phr of OC, interpreted as the compound stiffening.

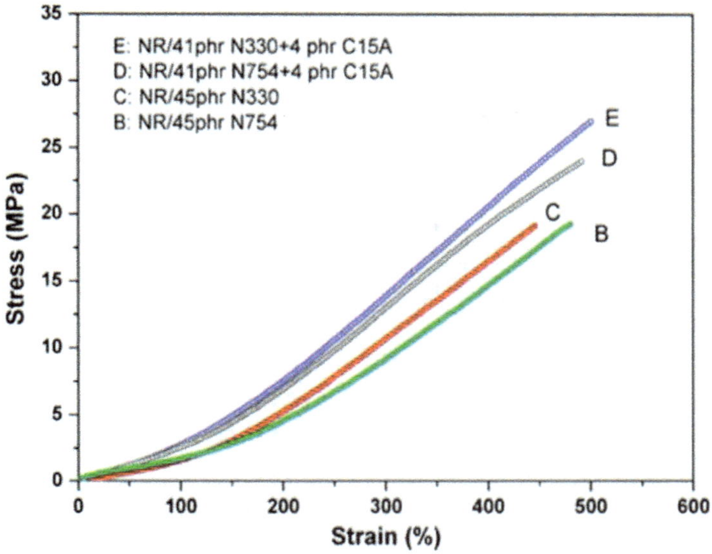

Figure 11.6 Stress−strain curves for composites based on either only CB or OC/CB filler systems.
Source: Reprinted with permission from Liu Y, Li L, Wang Q, Zhang X. Fracture properties of natural rubber filled with hybrid carbon black/nanoclay. J Polym Res 2011;18:859−67.

In the presence of a lower CB amount (45 phr), in a NR-based compound, the replacement of a minor amount of CB with OC(2HT) led to higher stresses at all elongations and better ultimate properties, as it is shown in Fig. 11.6 [48].

Again in NR matrix, modulus at 100% and tensile strength increased when 5 phr OC(2HT) were added to composites containing 20, 30, 40, and 50 phr of CB, in the presence of a slight reduction of elongation at break [24].

The replacement of CB, from 5 to 15 phr out of 50, with the same mass amount of OC (2HT, ODA) was reported to lead to a decrease of 300% elongation and tensile strength, for the highest OC contents [42]. The reinforcing effect, lower than expected, was attributed to insufficient filler dispersion.

In SBR-based composite, the replacement of 5 out 45 phr of CB with 3 phr of different OC ("HT, T2HE) led to maximum improvements of about 50% for the tensile strength and 34% for the modulus at 300%, in comparison with composites with only CB [40].

In ESBR based composites, higher EB and lower tensile strength were observed by replacing from 3 to 10 phr out of 45 phr of CB with the same mass amount of OC(2HT) [41]. It was speculated that the better clay dispersion promoted by calcium stearate helped the slippage of polymer chains on the filler surface.

The mechanical properties of EPDM filled with the OC(2HT, T2HE)-CB system were better than the properties of the EPDM−CB compound [45]. A composite containing 6 phr of OC and 30 phr of CB showed higher moduli at 50% and 300% elongation, and better tensile strength and elongation at break than a composite containing 40 phr of CB. Better results were obtained with the OC(T2HE), which achieved the best dispersion.

Both in BIMS and in ESBR, stresses at a given strain, moduli at 50% and 100%, tensile strength, and elongation at break at a particular strain were observed to increase by adding 4 phr of OC(ODA) to composites containing either 10 or 20 phr of CB [52].

OC(ODA) was added to X-NBR based composites containing 20 phr of CB, in the range from 2.5 to 10 phr [43]. The tensile strength and the elongation at break showed a maximum at 2.5 phr as added OC. The increase of tensile stresses was not necessarily associated with a reduction of elongation at break.

Larger elongation at break and tensile strength were observed in ENR by combining 20 phr of CB and 15 phr of OC(ODA, T2HE) [39].

In NR, the combination of 20% mass of CB and OC(2HT) from 2−15% mass gave synergistic effects on tensile properties [86]. A synchrotron WAXD study proved the occurrance of strain-induced crystallization [87].

In HNBR, when 5 phr of OC(2HT) are used instead of 5 phr of CB, moduli at 50% and 100% elongation are higher but ultimate properties are, to a minor extent, worse (OC-CB-2006-Herman). Correspondingly, Shore A increases from 66 to 72. When 5 and 15 phr of OC are added to a composite containing 15 phr of CB, moduli increase at all the elongations, but elongation at break decreases.

In NR, various type of CB (in minor amounts: from 15 to 22.5 phr) were combined with a calcined clay (54 phr), not modified with an organophilic cation [88]. Hardness, 100% modulus and elongation at break were not affected by the type of CB, whereas tensile strength appreciably increased by decreasing the size of CB individual particle (N330 vs N774).

11.2.7 Tear resistance

Tear resistance was studied in composites based on a comparable amount of NR and SBR and 75 phr of CB [51]. It was unchanged up to 5 phr loading of OC(quaternary ammonium) and then became slightly lower than the reference value, with a reduction that was negligible at 10 phr and appreciable at 15 phr. It was reported that such a reduction was lower than expected, in consideration of the high stiffening effect of OC, as stiffer compounds in general show worse tear resistance. Also De Mattia crack initiation tests did not show any particular differences in the absence of in the presence of OC.

On the contrary, in the case of nanocomposites based on NR [48] and 45 phr of CB, better crack resistance was found, in the higher tearing energy region, for specimens with 4 phr of OC(2HT) in place of 4 phr of CB. It was hypothesized that either more energy dissipation occurs in the presence of the hybrid filler network or that clays promote enhanced crystallization under strains [87].

11.2.8 Abrasion resistance

The addition of OC(quaternary ammonium) (from 5 to 15 phr) to an NR-based composite containing 75 phr of CB did not lead to significant differences of abrasion resistance at the DIN (Deutsches Institut für Normung) abrader [51].

In ESBR based nanocomposites, a slight improvement of abrasion resistance was obtained by replacing 3 out of 45 phr of CB with 3 phr of OC(2HT) [41] and adopting the combination of 25 phr of CB and 6 phr of OC(2HT) in place of 40 phr of CB [47].

11.2.9 Impermeability

It is widely acknowledged that clays can promote larger impermeability of polymer composites [7,10], as permeating molecules have to go through a tortuous path. Permeability is particularly reduced if clays are largely exfoliated and are oriented perpendicular to the molecule's flow. Orientation appears a critical aspect in rubber products, such as e.g., tire compounds, that experience important dynamic-mechanical stresses.

However, the increase of impermeability of rubber products such as tire liner has great importance and large research efforts have been spent in this direction, with many patents filed on composites based on OC and CB [10]. These documents are mainly focused on the type of OC and processing conditions, but there is no indication if OC and CB have a synergistic effect on impermeability.

In the scientific literature, a recent paper on brominated poly(isobutylene-co-p-methylstyrene) (BIMS) [37] and OC(2HT) revealed that nanocomposites having both CB and organoclay had the lowest permeability, due to a synergy between the two fillers. Rubber films obtained through squeeze flow in melt compression decreased the permeability of the film, thanks to layer orientation. The key role played by exfoliation was proved. BIMS was functionalized with either triethylamine or dimethyl benzoic acid: the former chemical gave larger exfoliation and, as a consequence, lower permeability.

In NR, impermeability was observed to increase when a calcined clay (54 phr) was mixed with CB with a larger size of the individual particle: N774 was better than N550 and N330 [88].

11.2.10 Thermal stability

Ammonium cations could undergo thermal degradation in the temperature ranges experienced by rubber compounds [70,89].

It is known that alkyl ammonium salts degrade for temperatures in temperature range from 150 to 200°C [70,89], whereas imidazolinium salts are stable up to about 450–550°C. The thermal stability of nanocomposites based on EPDM was studied [45] and it was reported that compounds loaded with the hybrid OC-CB filler system had improved thermal stability. This was attributed to the heat shielding effect of the hybrid system. In particular, better stability was observed for composites with the same OC(T2HE), that gave the largest enhancement of mechanical properties.

11.2.11 Electrical properties

CR based nanocomposites loaded with CB N550 and OC(2HT) were investigated [38], determining water absorption and volume electrical resistivity, in view of the underwater application of products based on this type of rubber. The electrical resistivity requirements of rubber vulcanizates were maintained after the filler's addition.

Electrical permittivity was studied for composites based on ENR and containing CB and OC(ODA) [46]. The addition of OC led to sudden drop of relative permittivity at lower (100–100,000 Hz) and higher (100,000 Hz and 500,000 Hz) frequency ranges. However, it was surprisingly found that a composite with 20 phr of CB and 15 phr of OC(ODA) had the minimum drop of relative permittivity in both frequency ranges. This was explained with the electron tunneling through well connected network established by the so-called "nanounit". The tunneling effect was invoked to explain the electrical conductivity in nylon 6 matrix [34].

11.3 Nanocomposites based on CNT and carbon black

11.3.1 Introduction

Papers available in the scientific literature, considered for this summary, are listed in Table 11.3. Beyond the CB and CNT acronyms there is a good number of different materials, that, often, are not completely characterized (CNT in particular) in the literature. Information is lacking mainly for surface area and, in particular, activity, features that are indeed relevant for the composites' properties. It is thus interesting to analyze if it is possible to identify general behaviors, in spite, in some cases, of such a lack of information.

11.3.2 Structure

In this section, the crystalline organization of CNT, interaction of CNT with the rubber matrix, and structure of the nanocomposites are discussed, highlighting the role of CB. Studies available in the scientific literature were performed with XRD, Raman spectroscopy, TEM, SEM, and atomic force microscopy (AFM).

11.3.2.1 Crystallinity of carbon allotropes

Carbon allotropes made by sp^2 carbon atoms, such as CB, CNT, and graphite, have a common feature: they are made by graphene sheets, as it is shown in Fig. 11.7, either stacked (CB, graphite) or wrapped. One or more graphene layers are wrapped in single walled carbon nanotube (SWCNT) or multiwall carbon nanotubes (MWCNT), respectively.

In the case of CB and graphite, stacks are characterized by different number of layers, crystallinity inside the layers and shape anisotropy, that means the ratio between the crystallites dimensions in directions parallel and orthogonal to the layers [100]. In Fig. 11.8A are shown the 002 reflections of XRD patterns of CB, CNT and

Table 11.3 **Papers on CNT-CB hybrid fillers: type of rubber, CB and CNT, blending technology and main aspects investigated in the work**

References	Rubber	CB type	CNT type	Blending method	Analyses
[90]	NR	N330	MWCNTs (Nanocyl™ NC7000)	Melt blending by a two roll mill after solvent dispersion of CNTs	Field emission SEM, helium ion microscopy, Shore A hardness, abrasion resistance, tensile test, “edge-cut” tensile strength
[91]	SBR	CB (specific surface area 110 m^2/g)	MWCNTs	MWCNT/SBR masterbatch prepared by a coagulation process followed by melt mixing in a high shear twin screw extruder. The masterbatch was then diluted and other ingredients added by melt compounding in a Haake batch mixer	Constrained tear test, tensile test, hysteresis measurements, equilibrium swelling measurements
[23]	NR	N234; N330	Highly one-dimensional aligned multi-walled carbon nanotube bundles (CNTBs) (FloTube TM 7000)	Mechanical blending by a two roll mill	TEM, tensile test, tear test, hardness, dynamic-mechanical analysis (strain sweep in shear mode, temperature sweep in tensile mode), fracture resistance, fatigue resistance
[31]	IR, NR	N326	MWCNTs (Baytubes C150 P)	Melt blending by an internal mixer	WAXD, TEM, tensile test, dynamic-mechanical analysis (shear mode)
[92]	NR	N330; Printex XE2-B	MWCNTs (Nanocyl-7000)	Melt blending by two roll mill	Mooney viscosity, rheometric curve, Shore A hardness, tensile test, tear test, rebound resilience, compression set, dynamic-mechanical analysis (in tension mode, strain sweep), volume

(Continued)

Table 11.3 (Continued)

References	Rubber	CB type	CNT type	Blending method	Analyses
					resistivity, thermal conductivity, SEM
[25]	IR	N326	MWCNTs (Baytubes C150 P)	Melt blending by an internal mixer	TEM, WAXD, dynamic-mechanical analysis (in shear mode, strain sweep)
[93]	IR	N326	MWCNTs (Baytubes C150 P)	melt blending by an internal mixer	WAXD, TEM, tensile test
[19]	silicone	Printex XE2	MWCNTs	Mixing by a planetary ball mill	Rheological temperature sweep measurements, electrical resistance, tensile test, fatigue tests (for electrical resistance), pressure sensitivity tests, TEM
[78]	IR	N326	MWCNTs (Baytubes C150 P)	Melt blending by an internal mixer	Mooney viscosity, rheological temperature sweep measurements, density, TEM, tensile test, dynamic-mechanical analysis (in shear mode, strain sweep), electrical conductivity
[94]	NBR	Conductex SC ultra	MWCNTs (CM-100), acid and plasma treated	Solution blending	Tensile test, DSC, TGA
[21]	NR	N330	Hydroxyl MWCNTs	CNTs/NR masterbatch by ultrasonic assisted latex mixing process + melt blending by an open two roll mill	Tensile test, tear test, compression test, hysteresis tests, dynamic-mechanical analysis (in tensile mode, temperature sweep), SEM, swelling test for cross-link measurement, dynamic rheological tests (in shear mode, frequency sweep)

[95]	NR	N330	MWCNTs (by Sun Nanotech)	Melt blending by two roll mill	Cure characteristics, swelling, SEM, tensile test, fatigue test, TGA
[96]	S-SBR/ BR, NR, EPDM	N550	MWCNTs (Nanocyl-7000)	Dry mixing or mixing after solvent dispersion of CNTs, by internal mixer	Reflected light microscopy, TEM, tensile test, dynamic-mechanical tests (in shear mode, temperature sweep), dielectric measurements, fatigue tests, thermal diffusivity
[20]	NR	N330	Hydroxyl MWCNTs functionalized with coupling agent Si-69	Melt blending by two roll mill	Cure characteristics, swelling, SEM, tensile test, tear and compression tests, dynamic-mechanical analysis (in tensile mode, temperature sweep)
[97]	NR	N375	MWCNTs (Nanocyl-7000)	Melt blending by an internal mixer	Mooney viscosity, curve kinetics, tensile test, hardness, heat buildup, dynamic-mechanical analysis (in tensile mode, temperature sweep)
[98]	SBR	N330	MWCNTs (by Nanocyl)	Solution blending	TEM, electrical volume resistivity
[99]	SBR (25% styrene)	N330; Ensaco 250 G	MWCNTs (by Arkema)	Solution blending	TEM, AFM, tensile test, Raman spectroscopy, TGA, electrical resistivity

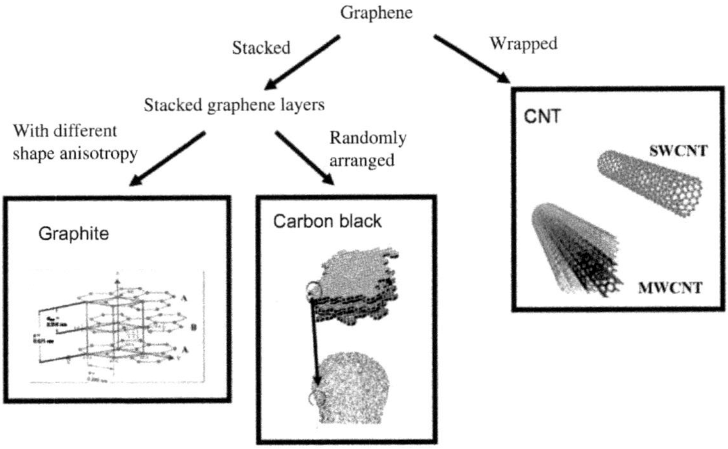

Figure 11.7 Layered structure as the common feature of carbon allotropes with sp^2 carbon atoms.

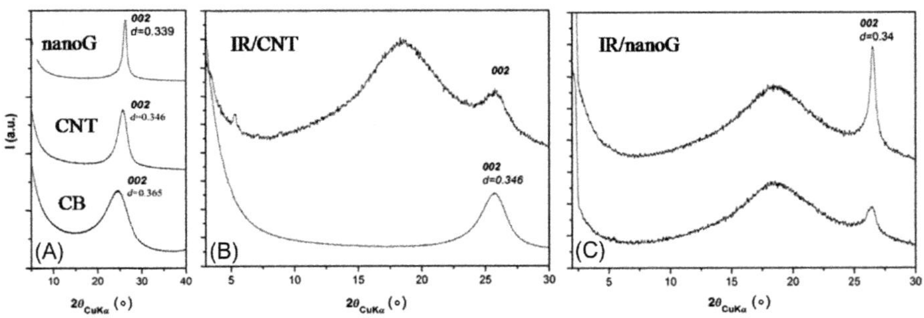

Figure 11.8 XRD patterns of carbon allotropes (A) and of IR based nanocomposites with CNT (B) and nanoG (C), uncrosslinked (lower curves) and crosslinked (upper curves). *Source*: From Galimberti M, Cipolletti V, Musto S, Cioppa S, Peli G, Mauro M, et al. Recent advancements in rubber nanocomposites. Rubber Chem Technol 2014;87(3):417–42.

nanoG [100]. As reported in [100], through the peak analysis it is possible to estimate the number of stacked layers, i.e., about 5, 10, and 35 for CB, CNT, and nanoG, respectively, reported in the figure. In the case of CNT, this number is an indication of the average number of wrapped graphene layers. In Fig. 11.8B and Fig. 11.8C, are the reported XRD patterns of nanocomposites based on IR as the rubber and either CNT or nanoG, respectively. It was calculated [31] that a higher number of wrapped layers are in the cross-linked nanocomposites rather than in the uncross-linked ones. This finding, which is quite surprising, suggests that the energy given to the compound in the vulcanization step brings a higher degree of order in CNT. The same result was obtained by taking XRD analysis on nanocomposites containing CNT and CB [78].

11.3.2.2 Interaction of CNT with the rubber matrix

Raman spectroscopy has been adopted to investigate the interaction between the polymer matrix and CNT. Nanocomposites have been prepared with CNT and SBR as the rubber [99]. In the reported Raman spectra, the G peak, characteristic of bulk crystalline graphite (graphene), shifted to higher frequencies, when CB particles or nanotubes were incorporated into the elastomeric matrix. In the case of MWNTs, the shift to higher frequencies was explained by a disentanglement of the CNTs and a subsequent dispersion in the polymer as a consequence of polymer penetration into the bundles during solution mixing. Addition of CB led to further upshifting of the G band and it was reported that better dispersion of CNT was obtained thanks to the dual phase.

11.3.2.3 Distribution and dispersion of carbon allotropes

In a nutshell, reports available in the literature clearly show the difficulty of distributing and dispersing CNT. However, they also show that CB aids CNT deagglomeration and that, upon achieving uniform distribution and dispersion of CNT, a hybrid filler network is formed.

Very poor dispersion of MWCNT was observed in SEM fracture images of nanocomposites based on NR, prepared via melt blending and containing CB (N330, total CB + CNT content equal to 25 phr); CNT (3 phr) were agglomerated into bundles. However, bundles were not observed by modifying CNT with sylane (3-triethoxysilylpropyl)-tetrasulfide (TESPT) [20].

SEM analysis of nanocomposites based on NR, prepared via melt blending and containing CB and MWCNT (N330, total amount of CB + CNT = 30 phr) revealed that acceptable fillers dispersion was obtained only for CNT content equal to 0.5, whereas it worsened by increasing the CNT amount [95].

For nanocomposites based on NR and CB (N330, 40 phr), prepared via melt blending, SEM analysis revealed that MWCNT (up to 4 phr) were not well distributed throughout the rubber matrix but were gathered in small areas [92]. CNT caused however considerable increase of CB dispersion.

The effect of mixing technology was shown for nanocomposites based on NR and containing CB (N330, up to 25 phr) and MWCNT (up to 7 phr) [21]. When only a two roll mill was used, SEM pictures revealed some agglomerates of CNT. Still some agglomerates were observed by dispersing CNT in latex. Uniform dispersion was achieved by using ultrasonically assisted dispersion of CNT in latex.

FE-SEM of composites based on NR prepared via melt blending after solvent dispersion of CNTs revealed MWCNT agglomerates, in the absence of CB, for CNT loading more than 2 phr, with the size of agglomerates increasing with CNT amount [90]. However, in the presence of CB (N330, 30 phr), large agglomerates of CB and CNT were not observed and this was attributed to the higher mixing shear.

As mentioned above, a good dispersion of CNT was obtained thanks to the mixing with CB and a hybrid filler network was formed. The simultaneous incorporation of CB (N330, 5 phr) and MWCNT in SBR, via solution blending, substantially reduced the size of CNT bundles observed by TEM analysis in

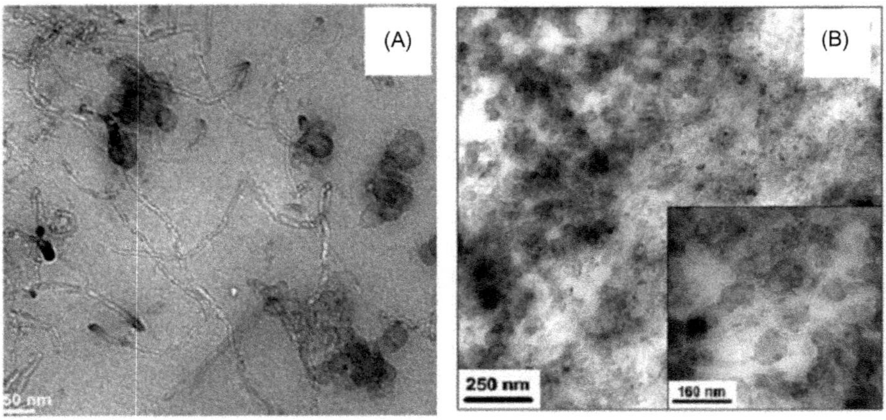

Figure 11.9 TEM images of composites containing a hybrid filler system: (A) SBR with 5 phr CB + 5 phr MWCNT ; (B) NR filled with 5.7 phr CNT and 60 phr CB.
Source: For (A) Reprinted with permission from Bokobza L, Rahmani M, Belin C, Bruneel JL, El Bounia NE. Blends of carbon blacks and multiwall carbon nanotubes as reinforcing fillers for hydrocarbon rubbers. J Polym Sci Part B Polym Phys 2008;46:1939–51. For (B) Reprinted with permission from Galimberti M, Coombs M, Riccio P, Riccò T, Passera S, Pandini S, et al. The role of CNTs in promoting hybrid filler networking and synergism with carbon black in the mechanical behavior of filled polyisoprene. Macromol Mater Eng 2013;298:241–51.

nanocomposites with only CNT. As it is shown in Fig. 11.9A, individual CNT are attached at the surface of CB aggregates [98,99]. The dual filling CB/CNT system exhibited in AFM images [99] a less rougher surface, suggesting a more homogeneous distribution of both fillers.

CNT/CB filler networks were observed in TEM micrographs and for the first time defined as hybrid filler networks [78] in nanocomposites prepared via melt blending, based on IR and CB (N326, 60 phr) and containing MWCNT in a range from 0 to 15 phr. A threshold for the CNT content was identified for the formation of the hybrid filler networks. As it is possible to see in Fig. 11.9B, above such threshold CNT webs and isolated tubes are within and inside CB aggregates and CB particles can be observed inside CNT webs, suggesting a strong CB/CNT interaction. As reported in the inset of Fig. 11.9B, CNT are also wrapped around CB aggregates and are bridges among CB agglomerates [31,93]. A continuous hybrid CNT-CB network was clearly visible in IR based nanocomposites containing the same amount of CB (N326 and MWCNT (15 phr). It was reported that the high aspect ratio of CNT tubes and their ability to wrap around CB aggregates appear to play a key role in creating the hybrid network [25].

TEM analysis of nanocomposites prepared via mixing by a planetary ball mill, based on silicone rubber, containing CB (Printex, 2.5 phr) and MWCNT (1 phr), revealed that CB particles linked the gaps between the unconnected CNT, resulting in the formation of conducting networks [19].

A rather homogeneous distribution of MWCNT (up to 5 phr) was obtained in NR-based nanocomposites having a constant hardness, prepared via melt blending, in the presence of CB (CB N234, up to 25 phr) [23]. TEM images showed that the CB aggregates were bridged by highly one-dimensional aligned CNT to form a hybrid filler network.

11.3.3 Rheology

The addition of CNT to composites containing CB leads to an increase of viscosity. The data reported in this section have arisen from composites based on NR. For composites containing 40 phr of CB (N330) [92], the increase of viscosity due to MWCNT addition was larger than that due to CB addition for amounts of added filler larger than 2 phr; for 4 phr filler addition, Mooney viscosity (ML $(1+4)100°C$) was about 60 in the case of CNT and about 50 in the case of CB. This was attributed to the high aspect ratio of CNT, that favors the entanglement and, in turns, flow restriction. The role of shape factor of CNT in enhancing the viscosity of CB/CNT composites was highlighted also in [97] and it was reported that the value of shape factor was not constant, passing from about 53 to about 34 for CNT content from 2.5 to 15 phr. The shape factor was evaluated by fitting the Einstein-Guth equation [74] to experimental results of Mooney viscosity.

The effect of shear rate on complex viscosity was studied [21] and results were correlated with the occurrence of the percolation network. For composites based only on CNT, it was seen that the complex viscosity increased with the loading level of CNT and the shear viscosity decreased with an increase in the shear rate. In the region of low frequencies, a pseudoplastic shear thinning behavior appeared (Newtonian plateau disappeared) by increasing the CNT content and this was attributed to the formation of a percolated CNT network. For CB/MWCNT composites (CB N330, 25 phr as total amount of filler) high shear thinning behavior was clearly observed.

11.3.4 Curing

In the literature, information is available on cross-linking of rubber nanocomposites, containing CB and CNT, promoted by either peroxide or sulfur based systems. As mentioned previously, the surface characteristics of carbon allotropes, such as surface area and activity, play a fundamental role, in particular in vulcanization reactions. They are not always reported in the papers on diene rubbers vulcanization. The summary reported in the following sections is thus based on the available data.

11.3.4.1 Crosslinking with peroxide

Nanocomposites containing CB and CNT in 1:1 as the mass ratio (and as volume ratio as well, as they had the same density) from 2.5 to 30 as the total filler content were cross-linked with dicumyl peroxide [31], without observing any appreciable difference, either among the different composites with the hybrid filler system or with the composites with only one carbon allotrope. Characteristics of curing could be attributed to peroxide, a low induction time and absence of reversion.

11.3.4.2 Crosslinking with sulfur based system

As already mentioned in this Chapter, it is widely acknowledged that CB promotes faster sulfur based vulcanization reactions [65,66]. This is attributed to the enhanced heat transfer in the rubber matrix, thanks to CB. Moreover, it is reported that both the degree of cross-linking and cure rate increase with increasing surface area and sulfur content, whereas the optimum cure time and scorch time decrease [101]. In the case of CNT, SWCNT were shown to reduce the activation energy and to accelerate NR vulcanization [102]. The scorch time of NR vulcanization decreased with the addition of acid treated CNTs [103], without substantial effect on the optimum cure time. On the contrary, carboxylic acid groups and quinone oxygen atoms were reported [104,105] to delay NR vulcanization, as a consequence of the absorption of basic accelerator species.

Acceleration of curing was found by increasing the amount of MWCNT (from 2.5 to 15 phr) in NR-based composites containing CB (N375, from 50 to 65 phr) [97]. A consistent decrease of t_{50} and scorch time (scorch safety) measured at 127°C was observed. Such findings were commented to be unexpected and were justified with the residual amount of catalyst used for CNT synthesis.

In the case of NR/CB composites, containing CB and MWCNT (N330, total amount of CB + CNT = 30 phr) [95], the increase of CNT relative amount (CB/CNT ratio from 29.5/0.5 to 25/5) led to a consistent reduction of scorch time and curing time, as well as to the increase of the maximum M_H modulus.

Vulcanization was studied for IR/MWCNT nanocomposites containing from 1.1 to 13.3. phr of CNT, in the absence and in the presence of CB (N326, 60 phr) and the increase of the cross-linking reaction rate was observed in the presence of CB [78].

For nanocomposites based on NR and CB (N330, 40 phr) [92], scorch and curing times, minimum and maximum torque, and the torque difference were found to increase with increasing CNT content (up to 4 phr), though the dispersion of the nanotubes is not good.

In the case of composites based on NR/CB (N330, 30 phr) [90], scorch and cure times slightly decreased with increasing MWCNT content, while their torque difference (M_H-M_L), increased. The dilution effect and thermal conductivity enhancement of MWCNT were commented to be responsible for the obtained results.

Instead, in the case of NR/CB based composites (N330, total CB + CNT amount equal to 25 phr) when CNT were modified with TESPT as the silane coupling agent, the vulcanization and the scorch times increased with increasing content of CNT. These results were explained with the absorption of curing chemicals on the CNT surface [20].

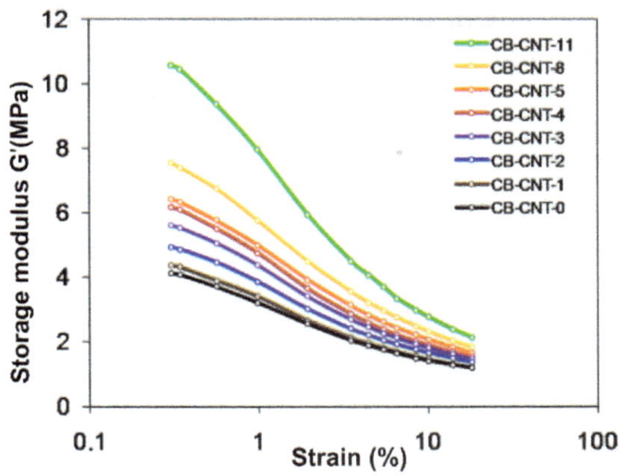

Figure 11.10 Strain dependence of storage modulus G' for CB/CNT hybrid composites. *Source*: Reprinted with permission from Galimberti M, Coombs M, Riccio P, Riccò T, Passera S, Pandini S, et al. The role of CNTs in promoting hybrid filler networking and synergism with carbon black in the mechanical behavior of filled polyisoprene. Macromol Mater Eng 2013;298:241–51.

11.3.5 Dynamic-mechanical properties

Dynamic-mechanical properties of cross-linked composites based on CNT and CB were studied by applying sinusoidal stresses both in the shear and tensile mode and cross-linking the composites both with peroxide and sulfur based systems.

From strain sweep tests, it was seen that when CNT were added to neat polymer matrix (IR), they gave rise to remarkable filler networking [31], in the presence of a cross-linking network made by short bridges such as the one formed by a peroxide. Filler networking drastically increased above the percolation threshold but was detectable also below such a threshold. In the presence of CB (N326, 60 phr) [31,78], in IR based composites cross-linked with sulfur, CNT increased the filler networking, as it is shown in Fig. 11.10, in particular above a critical concentration threshold, as in the neat IR matrix.

Interestingly, it was found that CNT increased the low amplitude storage modulus of the matrix, in the presence or in the absence of CB, by a factor depending only on its content. This aspect will be discussed more in detail, to attempt a rationalization of literature data, below in the text.

Composites based on NR and containing 25 phr as total level of carbon allotropes (CB N234, MWCNT) revealed a clear increase of filler networking when CB was replaced by the same mass amount of CNT, from a level of at least 3 phr [23]. High hysteresis was observed by replacing CB with 5 phr of CNT. To NR-based composites with 40 phr of CB (N 330), CNT was added (from 2 to 4 phr) and promoted much more filler networking than a conductive CB [92].

CNT/CB/NR nanocomposites were prepared by latex blending of CNT (containing 3% of hydroxyl groups) and NR and successive melt blending of this masterbatch with NR and CB, keeping the total amount of filler at 25 phr with increasing amount of CNT, from 1 to 7 phr [20,21]. An increase of storage modulus, in particular above −40°C, and increase of tan δ peak were observed, and reported as synergistic effects. MWCNT were shown to promote a remarkable increase of heat build up [97] when they were added to NR/CB (N375, 50 phr) composites, from 2.5 to 15 phr. It was hypothesized that the high internal friction was due to the poor interaction between the nanotubes and the polymer matrix.

11.3.6 Tensile properties

The hybrid CNT-CB filler systems are able to promote substantial improvements of modulus, tensile strength, and strain at break, with respect to composites filled with a single carbon allotrope. Data available in the literature show that the amount and the extent of dispersion of CNT are crucial to achieve improvements rather than a worsening of the mechanical properties.

SBR-based composites were prepared with CB (N330), MWCNT (5 phr), or a mixture of both fillers (CB 10 + CNT 5) [99]. With the latter composite substantial improvement of mechanical properties was obtained. Also by keeping constant the total amount of filler (10 phr), better mechanical properties, in particular ultimate properties, were obtained in the presence of the dual filler system. Values of stresses were plotted using the semiempirical Mooney-Rivlin equation (Eq. (11.3)) [106].

$$\sigma * = 2C_1 + 2C_2\alpha^{-1} \tag{11.3}$$

in which C_1 and C_2 are constants independent of α, and σ^* is the reduced stress, i.e.,

$$\sigma * = \sigma/(\alpha - \alpha^{-2}) \tag{11.4}$$

The plot shown in Fig. 11.11 reveals an upturn in the modulus at a lower extension ratio and a large increase in the reduced stress, that were correlated with the increase in the cross-linking density created by polymer–filler interactions.

On the basis of the curves in Fig. 11.11, it was commented that the presence of CB leads to a better interaction between CNT and polymer chains. It was also investigated as the Mullins effect, a hysteretic mechanism occurring during deformation of the material, that corresponds to a decrease in the number of elastically effective networks chains [107].

Stress softening was more pronounced in the sample containing the dual filler system, confirming improved interactions with polymer chains, when CB and CNT were dispersed together in the rubber matrix.

In composites based on NR and CB (N330, 30 phr) [90], the addition of CNT (from 2 to 8 phr) led to an increase of hardness and of 100% modulus. The tensile strength was not modified by the addition of CNT to CB filled NR, instead it was observed to decrease when CNT were added to the neat polymer matrix. These results were attributed to the increase of cross-linking density and to the better dispersion of CNT in the presence of CB.

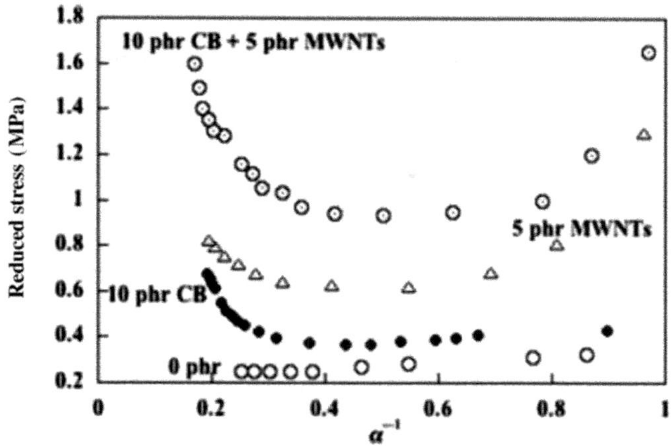

Figure 11.11 Mooney-Rivlin plots for pure SBR and for SBR composites.
Source: Reprinted with permission from Bokobza L, Rahmani M, Belin C, Bruneel JL, El Bounia NE. Blends of carbon blacks and multiwall carbon nanotubes as reinforcing fillers for hydrocarbon rubbers. J Polym Sci Part B Polym Phys 2008:46;1939–51.

Increases of tensile stress at all elongations and of tensile strength and the occurring of failure at higher stresses were observed by adding 3 phr of MWCNT to 40 phr of CB in SBR-based composites [91]. CNT were commented to act as crack bridging elements that inhibited tear propagation. This was attributed to strong surface interaction between CNT and SBR.

Nanocomposites based on NR with a fixed hardness level, obtained by partially replacing 25 phr of CB with CNT, revealed synergetic reinforcing effects: substantial increases of stresses at 100 and 300%, tensile strength, and tear strength were observed when CNT were used in place of CB (1 phr CNT for 3 phr of CB), up to 5 phr of CNT [23]. Strain-induced crystallization was also enhanced by the addition of CNT, that was hypothesized to lead to greater strain amplification near the crack tip.

Clear enhancement of stresses at all elongations was observed by adding MWCNT (from 1 to 9 phr) to IR based composites containing 60 phr of CB (N326) [31,78]. For comparison, composites with the same amount of MWCNT but without CB were prepared. These data allowed to estimate the initial modulus as the slope at the origin of stress–strain curves and to evaluate the modulus enhancement due to CNT addition with respect to the composite free of CNT (with or without CB). Such enhancement was found to depend only on the amount of CNT, irrespective of CB addition. This point is discussed below in the text, in section 11.6.

Increases of hardness, 100% modulus, and tensile strength were observed for composites based on NR and containing 25 phr as total level of carbon allotropes (CB N234 and MWCNT), when CB was replaced by the same mass amount of CNT (up to 4 phr) [92]. The higher cross-linking density and the entanglement of polymer chains with CNT were commented to be at the origin of these results.

Increases in tensile strength and strain to failure, with respect to the neat rubber matrix, were obtained by adding CB and CNT to silicone rubber [19]. Better results were achieved with 2.5 CB and 1 CNT with respect to 3 CB or 4 CNT.

The tensile strength and modulus enhancement of nanocomposites based on NBR, conductive CB (20 or 40 phr), and MWCNT treated with acid treatment or plasma (up to 9 phr) [94] gradually increased with increasing CNT content, when compared with the matrix and with the rubbers filled with the same amount of CB. In particular, the tensile strength and modulus of the composite with a CNT content of 9 phr increased up to 31% and 91%, respectively, with respect to the matrix.

In CNT/CB/NR nanocomposites, very uniform dispersion of CNT (containing 3% of hydroxyl groups) was obtained by latex blending [21] and this led to synergistic reinforcing effects. The nanocomposite (prepared by melt blending) containing 20 phr of CB and 5 phr of CNT reached the best mechanical properties: stress at 300%, tensile strength, and elongation at break. Larger amounts of CNT led to worse properties. CNT was described to bring ideal toughening and reinforcing effects, thanks to its interfacial interaction with the NR matrix.

In the case of nanocomposites based on NR, prepared via melt blending and containing MWCNT and CB (N330, total amount of CB + CNT = 30 phr) [95], the enhancement of tensile strength and elongation at break was obtained only for a low content of CNT (0.5 phr) and the decrease for larger CNT amounts was attributed to the filler–filler interaction, which was preferential with respect to the filler–polymer interaction, that favored the formation of filler agglomerates.

In EPDM as the matrix, the addition of 3 phr of MWCNT, assisted by ethanol, to 40 phr of CB (N550) led to a clear enhancement of stresses at any elongation, whereas the addition in the absence of ethanol caused a dramatic worsening of reinforcement and elongation at break [96]. This was correlated with the formation of agglomerates in the latter case.

MWCNT were added to NR/CB (N375, 50 phr) composites, from 2.5 to 15 phr [97]. It was hypothesized that the high internal friction was due to the poor interaction between the nanotubes and the polymer matrix and caused a dramatic increase of hardness and moduli and a clear reduction of elongation at break. The modulus was nearly doubled at 10 and 15 phr as the CNT content. Such a reinforcing effect was explained with the hydrodynamic effect and the Einstein-Guth equation [108] was used to calculate the shape factor of CNT, which achieved a value of about 50, one order of magnitude larger than that typical of CB.

Fig. 11.12 shows the comparison of stress–strain curves for composites based on 60 phr of CB (N326) and various added fillers, cross-linked with peroxide [31]. It appears that CNT leads to larger stress values at all strains. It is interesting to observe that an organoclay (with 2HT as the compensating cation) leads to lower reinforcement even with respect to the neat polymer matrix.

These findings indicate that CNT per se promote mechanical reinforcement, whereas OC appear to need the interaction of the compensating cation with a sulfur-based cross-linking system.

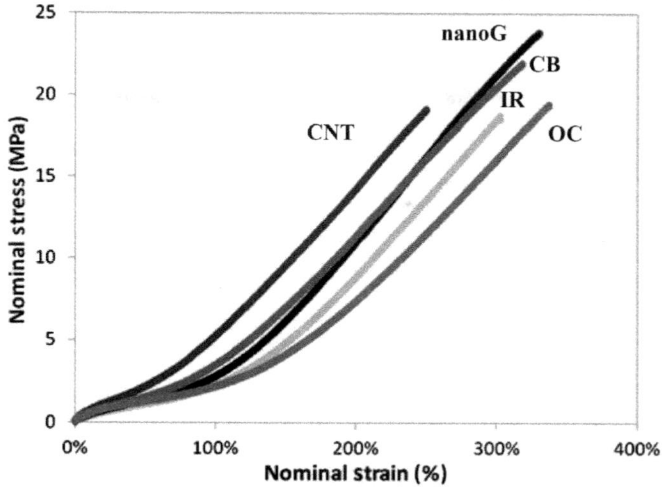

Figure 11.12 Nominal stress versus nominal strain curves for peroxide cross-linked IR based composites. The matrix is: IR + 60 phr of CB. Curves refer to composites without any added filler (IR) or containing 60 phr of CB and 12 phr of: either OC, or CNT or CB.
Source: Reprinted with permission from Galimberti M, Cipolletti V, Musto S, Cioppa S, Peli G, Mauro M, et al. Recent advancements in rubber nanocomposites. Rubber Chem Technol 2014;87(3):417–42.

11.3.7 Fracture resistance

The fracture resistance of composites based on CNT and CB was studied both under quasistatic and under cyclic (fatigue) loading. The addition of both CNT and CB was found to increase quasistatic fracture resistance, while an increase of fatigue resistance was observed only for small amounts of CNT.

The addition of 40 phr CB to a SBR matrix causes an increase in tear energy (evaluated from constrained tear test) of about 60% with respect to the near matrix, while adding both 40 phr CB and 3 phr MWCNT causes an increase of 280% [91]. The higher energy was achieved by an increase of both stress and elongation at break. This substantial improvement was attributed to a crack bridging effect exerted by CNTs during fracture propagation, to strong surface interactions between MWCNT and SBR matrix, and to the uncoiling of curved CNT during the stretching process. Hybrid filler NR-based composites were prepared by partially replacing 25 phr of CB (N234) with increasing amounts of highly one-dimensional aligned CNTs bundles (CNTBs) up to 5 phr, by keeping the composite hardness fixed [23], in order to evidence synergistic effects by the two fillers. J-integral testing on single edge notched in tension specimens showed an enhanced fracture resistance both at initiation and propagation by CNTB addition. In the same work, fatigue resistance was found to increase up to 3 phr CNTB, while higher contents of CNTB (5 phr) led to higher hysteresis, which accelerated the rupture of material under dynamic conditions, and thus decreased fatigue crack growth resistance.

MWCNT were added up to 5 phr to a CB (N330)/NR matrix [95]. The total amount of hybrid filler was fixed to 25 phr. Fatigue life of hybrid nanocomposites increased for the composite with 25.5 and 0.5 phr of CB and CNT, respectively. At higher amounts of CNT, it was hypothesized that agglomeration of CNT could be responsible for the reduced fatigue life.

11.3.8 Abrasion resistance

The abrasion resistance was found to increase by increasing MWCNT content (from 2 to 8 phr) for NR-based composites containing 30 phr of CB (N330), with respect to nanocomposites containing only CNT [90]. This was attributed to the increase in hardness and modulus and to the better dispersion of CNT, due to the presence of CB.

11.3.9 Electrical properties

The addition of MWCNT, from 2 to 10 phr, to SBR-based composites containing 10 phr of CB (N330 Cabot, or Timcal, more conductive) [99] reduced the resistivity and the percolation threshold for both CB, at a smaller CNT concentration for the more conductive CB. The interconnected filler network was formed for CNT content between 2 and 3 phr and was also formed by feeding to SBR 5 phr CNT and 5 phr CB N330 [98]. It was commented that the incorporation of CNT promoted the formation of connected structures by bridging the uncontacted CB particles, as confirmed by analysis of TEM images (see Fig. 11.9 above). A gradual increase in resistivity was obtained upon applying increased deformations and this was explained by the breakdown of the conductive networks and the orientation of CB aggregates and CNT. The filler network was not reformed upon releasing the stress, as revealed by the high resistivity values. However, new stretching led to the formation of new conductive pathways with decreases in resistivity. Much higher conductivity is obtained by adding CNT to CB rather than by increasing the CB content. This appears from results reported in Ref. [98] and in Ref. [78] for IR based composites. Fig. 11.13A,B are taken from such refs.

In Fig. 11.13B, the DC conductivity monotonously increases with CNT content and the curve does not show any discontinuity, as instead was observed for the excess of initial modulus, as mentioned above discussing the dynamic-mechanical properties.

Conductive silicone rubber was obtained by ball milling, adding 2.5 phr of CB and 1 phr of MWCNT [19]. Due to the synergistic effects, the electrical resistance was lower when hybrid fillers were used than when a single filler was used with a similar overall concentration. This was attributed to the fact that CB particles effectively link the gaps between the unconnected CNT, resulting in the formation of conducting networks. A high sensitivity of electrical resistivity

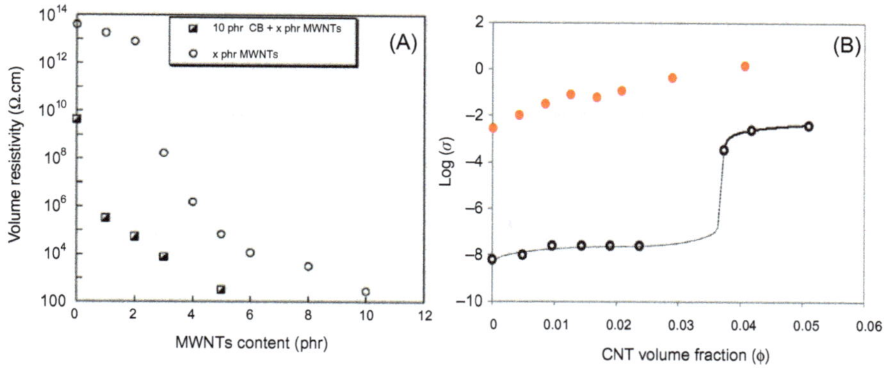

Figure 11.13 (A) Volume resistivity versus filler loading for SBR composites with 10 phr CB and CNT. (B) DC electrical conductivity versus CNT content for composites containing CNT. Composites without CB are indicated by ○ symbol in both figures.
Source: For (A) Reprinted with permission from Bokobza L. Mechanical, electrical and spectroscopic investigations of carbon nanotube-reinforced elastomers. Vib Spectrosc 2009;51:52–9. For (B) Reprinted with permission from Galimberti M, Coombs M, Riccio P, Riccò T, Passera S, Pandini S, et al. The role of CNTs in promoting hybrid filler networking and synergism with carbon black in the mechanical behavior of filled polyisoprene. Macromol Mater Eng 2013;298:241–51.

was observed to tensile and compressive stresses. However, long-term measurements revealed that excellent electrical properties were maintained after different strain histories.

Electrical percolation was obtained in NR by adding 40 phr of CB (N330). The further addition of CNT (up to 4 phr) led to a gradual reduction of the volume resistivity [92]. Also, in these nanocomposites CNT gave lower volume resistivity than CB.

11.3.10 Thermal properties

The thermal stability of nanocomposites containing CB and CNT was studied. In SBR-based composites, CB (N330, 10 phr), MWNT (5 phr), or a mixture of both fillers (10 phr CB + 5 phr CNT) appeared to have very little effect on the thermal stability of the composites [99]. A very minor effect was observed in the case of nanocomposites based on NR and containing CB (N330) and MWCNT (0.5 or 5 phr) (total amount of CB + CNT = 30 phr) [95]. A slight improvement of thermal stability was reported for composites based on NBR (35 phr), conductive CB (20, 40 phr), and MWCNT treated with acid treatment or plasma (up to 9 phr) [94]. The degradation temperature of matrix was 455°C, whereas that of the sample with 40 phr CB and 9 phr CNT was 464°C.

An increase of thermal conductivity was measured for composites based on NR and CB (N330, 40 phr) with MWCNT up to 4 phr [92].

11.4 Nanocomposites based on nanographite and carbon black

11.4.1 Introduction

Papers available in the scientific literature, considered for this summary, are listed in Table 11.4. As mentioned in the introduction in Section 11.1, graphitic nanofillers can be made by few layers of graphene and are named as nanographite, GNPs, nanosheets, or nanoflakes. Careful attention has thus to be made to the specific grade of nano-sized graphite used for the preparation of the nanocomposite.

11.4.2 Structure

Dispersion of graphitic nanofillers in the rubber matrix was remarkably increased either by blending the nanofiller with rubber latex or by using a solvent. NBR/expanded graphite (EG)/CB micro- and nanocomposites were prepared: the former by simply melt blending and the latter by latex blending EG with NBR latex, then adding CB via melt blending [113]. SEM micrographs of worn surfaces revealed, in the case of microcomposites, large graphite particles and, in the case of nanocomposites, smooth and flat surfaces because of the good graphite–rubber interfacial adhesion. The nano-size of graphite particles was crucial to avoid the formation of lubricant graphite films that reduced in microcomposites the direct contact between the matrix and the polymer. EG and modified EG (MEG) were blended (3 and 6 phr) with E-SBR (styrene = 25%) and CB (N234, 30 and 40 phr) [112]. EG was obtained from natural graphite by thermal expansion and was further oxidized with H_2SO_4/HNO_3 to prepare MEG, that contains larger numbers of polar groups on the outer surface. Delaminated EG/MEG flakes were largely prevailing in TEM micrographs when solution blending was adopted for SBR and EG, whereas agglomerates were observed in the SBR matrix when only melt blending was used.

The dispersion of graphitic layers was also much improved in the presence of CB. Hybrid nanocomposites were developed by dispersing CB (Printex XE2) and GNPs at 4–6 and 12–14 wt%, respectively, into rubbery epoxy resin [114]. CB particles improved the dispersion of GNP in the hybrid composite, as revealed by SEM analysis. In composites based on NR and CB (N330, 40 phr), a minor part of CB was replaced by the same amount of (GO) or reduced graphene oxide (RGO), from 0.5 to 5 phr [110]. TEM micrographs revealed that graphitic nanosheets were dispersed among the CB particles, as shown below in Fig. 11.14A, and that CB aggregates became smaller in the presence of nanosheets. A more developed filler network was observed in composites with rGO. However, both GO and rGO tended to aggregate at a high concentration.

Composites based on E-SBR (25% styrene), CB (10 phr) and RGO (from 1 to 3 phr), formed by in situ reduction of GO by a para-phenylene-diamine in CB/GO water suspension, were prepared by two roll milling [111]. Homogeneous

Table 11.4 Papers on graphitic nanofillers and CB hybrid fillers: type of rubber, CB, and graphitic nanofiller, blending technology, and main aspects investigated in the work

References	Rubber	CB type	Graphitic nanofiller type	Blending method	Analyses
[109]	Liquid silicone rubber	High structure CB	Graphene nanoplatelets (GNPs)	Liquid mixing method	TEM, electrical resistivity
[110]	NR	N330	Graphene oxide (GO) or reduced graphene oxide (RGO)	Solution blending	Cure characteristics, dynamic-mechanical analysis on uncured compounds (strain sweep in shear mode), TEM, tensile test, dynamic-mechanical analysis (in tensile mode, temperature sweep), Shore A hardness, swelling test, flexing resistance test (De Mattia), dynamic compression heat build-up test
[111]	Emulsion SBR		Reduced graphene oxide (RGO)	Melt blending (two roll mill)	AFM, SEM, XRD, tensile test, volume resistance test, dynamic-mechanical analysis (in tensile mode, temperature sweep), TGA
[112]	Emulsion SBR, and Oil extended carboxylated styrene butadiene	N234	Expanded graphite (EG) and modified EG (MEG)	Solution mixing or melt blending of EG and MEG + XSBR. Melt blending (two roll mill) of filled XSBR + SBR	Cure characteristics, XRD, High Resolution TEM, dynamic-mechanical analysis (in tensile mode, temperature sweep), tensile test, tear test, abrasion test,

(Continued)

Table 11.4 (Continued)

References	Rubber	CB type	Graphitic nanofiller type	Blending method	Analyses
	rubber (XSBR)				Shore A hardness, SEM, TGA, FT-IR
[16]	IR	N326	Nanographite (nanoG)	Melt blending	Density, TEM, XRD, tensile test, dynamic-mechanical analysis (strain sweep in shear mode)
[31]	IR	N326	Nanographite (nanoG)	Melt blending	XRD, tensile test, dynamic-mechanical analysis (strain sweep in shear mode)
[113]	NBR	N330	Expanded graphite (EG)	Melt blending (two roll mill) or latex compounding	Cure characteristics, tensile test, dynamic-mechanical analysis (in tensile mode, temperature sweep), Shore A hardness, wear test, SEM
[114]	Rubbery epoxy resin	Printex XE2 (with or without silane functionalization)	Graphite nanoparticles (GNPs) (with or without silane functionalization)	Conventional mechanical mixing (MM) or dual asymmetric centrifuge mixing by speed mixer (SM)	Viscosity test, FE-SEM, thermal conductivity, electrical conductivity, compression test

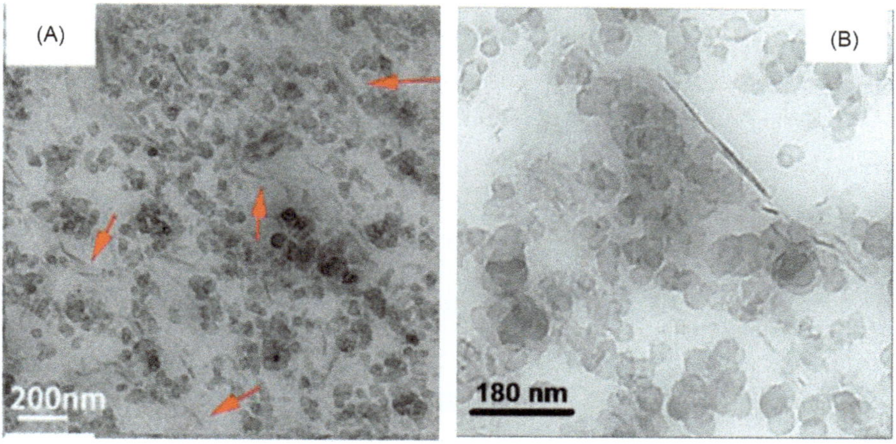

Figure 11.14 TEM micrographs of nanocomposites with: (A) 10 phr CB and 1 phr RGO; (B) 60 phr CB and 12 phr of nanoG.
Source: For (A) Reprinted with permission from Yang G, Liao Z, Yang Z, Tang Z, Guo B, Effects of substitution for carbon black with graphene oxide or graphene on the morphology and performance of natural rubber/carbon black composites. J Appl Polym Sci 2015;132 (15):41832. For (B) Reprinted with permission from Galimberti M, Kumar V, Coombs M, Cipolletti V, Agnelli S, Pandini S, et al. Filler networking of a nanographite with a high shape anisotropy and synergism with carbon black in poli(1,4-cis-isoprene) based nanocomposites. Rubber Chem Technol 2014;87(2):197–218.

dispersion of graphene sheets in SBR matrix was observed thanks to the hybrid filler system: small CB agglomerates were adsorbed on graphene surfaces. CB acted as a barrier preventing the RGO sheets from restacking.

Nanocomposites based on IR, CB (N326, 60 phr) and a nano-sized graphite with very high surface area (300 m^2/g) and high shape anisotropy, defined as the ratio between the crystallite dimensions in directions orthogonal and parallel to structural layers, were prepared via melt blending [16,31]. Even nanoG distribution with nanoG aggregates made by few-layers nanoG can be observed in Fig. 11.14B. NanoG layers are preferentially close to CB agglomerates.

11.4.2.1 Crystalline organization of G in G/CB nanocomposites

Results reported in the scientific literature demonstrate that the crystalline organization of graphitic fillers is not affected by blending with rubber matrices, both in the absence and in the presence of CB.

EG and MEG revealed larger interlayer distances between the graphene sheets with respect to pristine graphite [112]. The stacking order was retained in SBR-based nanocomposites.

When RGO (from 1 to 3 phr) was blended with CB (10 phr) and E-SBR (25% styrene) [111], composites with 2 and 3 phr content of RGO showed in XRD pattern (001) peak at the same 2θ value as in the pattern of pristine GO.

As mentioned above, when nano-sized graphite was blended with IR and CB (N326, 60 phr), the interlayer distance in graphite stacks was not modified [16,31]. The number of layers stacked in the crystalline domain increases upon applying pressure (during vulcanization) on the nanocomposite.

11.4.3 Rheology

The addition of GNP to rubbery epoxy resin, both in the absence and in the presence of CB, brought about a remarkable increase of viscosity, that was clearly dependent on the shear rate [114]. Reduction of viscosity was obtained by modification of GNP with 3-aminopropyltriethoxysilane.

Minimum torque (ML) in rheometric curves was increased by the addition of EG or MEG (3 and 6 phr) to E-SBR (styrene = 25%)/CB (N234, 30 and 40 phr) composites [112].

11.4.4 Curing

Graphitic nanofillers affect the course of sulfur-based vulcanization like the other carbon allotropes: they promote a faster vulcanization reaction. However, a key role is played by the chemical nature of their surface: the presence of polar groups can reverse this behavior.

Both scorch and optimum time of vulcanization were clearly reduced by adding to IR (in the absence of CB) increasing amount of nanographite [16]. An increase, even though less evident, was observed also in the presence of CB.

EG and MEG (3 and 6 phr) promoted reduction of scorch and optimum cure time when added to neat E-SBR (styrene = 25%) or to the E-SBR blend with CB (N234, 30 and 40 phr) [112].

On the contrary, a retarding effect on curing was observed by replacing CB (N330, 40 phr) with the same amount of GO (from 0.5 to 5 phr) [110]. This was attributed to the interaction of oxygenated polar groups on the GO surface with vulcanizers. Replacement of CB with RGO did not have any effect on scorch time but delayed the optimum time of vulcanization. This was explained with the limited diffusion of vulcanizers. At 3 phr of graphitic nanofiller, a large cross-linking density was obtained with RGO.

11.4.5 Dynamic-mechanical properties

Graphite nanofillers promote enhancement of storage modulus in the glassy region, the shift of tan δ to higher temperatures, and the increase of filler networking. A key role appears to be played by the chemical nature of the graphite surface.

The storage modulus in the glassy region of rubber nanocomposites based on SBR (E-SBR, XSBR) and CB (N234, 30 and 40 phr) was drastically enhanced by the addition of EG and MEG [112]: from about 30% (SBR/EG) to about 150% (XSBR/MEG). A substantial shift in the tan δ_{max} peak towards higher temperature was observed for EG and MEG loaded nanocomposites.

In the case of SBR (25% styrene)/CB composites, T_g shifted only slightly to higher temperature on increasing the CB amount from 10 phr to 13 phr and was found almost at the same temperature level when 1 phr of RGO was in place of 3 phr of CB. By increasing the RGO loading, T_g shifted to higher temperatures [111].

In composites based on NR and CB (N330, 40 phr), the replacement of a minor part of CB with GO (from 0.5 to 5 phr) [110] led to a remarkable increase of filler networking, as revealed by strain sweep tests performed in the shear mode. Lower nonlinearity of the storage modulus was observed by using RGO. A significant increase in flex cracking resistance was obtained with RGO, that gave much lower heat build up than GO.

11.4.6 Tensile properties

Graphite nanofillers lead to improvement of stresses at every strain when they are added to rubber composites in addition or in minor replacement of CB. The presence of polar groups on the surface can favor aggregation and, in turn, worsen mechanical properties.

In composites based on IR and CB (N326, 60 phr), the addition of nano-sized graphite (from 1 to 12 phr) [16] caused the increase of stresses at all the strains, with a progressive reduction of elongation at break.

In composites based on NR and CB (N330, 40 phr), improvement of both stress at 300% strain and tensile strength were obtained by replacing CB with a little amount of either GO or RGO (up to 2 phr) [110]. In the presence of larger amounts (up to 5 phr), improvement was observed only for stress at 300% by using RGO, whereas GO led to worsening of both static properties, as a consequence of the aggregation of GO or RGO sheets. In Mooney-Rivlin plots, the upturn of the curves, a consequence of the limitation of chain movements by neighboring fillers, occurs at smaller strains for GO content of 1 phr and RGO content of 3 phr. The upturning point increases firstly at lower CB substitution contents and then decreases gradually.

The tensile strength of rubber nanocomposites based on SBR (E-SBR, XSBR) and CB (N234, 30 and 40 phr) was appreciably increased by the addition of EG and MEG [112], from about 10% (SBR/EG) to about 150% (XSBR/MEG), whereas the increase of modulus at 300% was from about 15% to about 30% [112]. In line with these findings, hardness values were found to increase. MEG was commented to act as a better reinforcing filler (with and without CB), as a consequence of the larger interlayer spacing and the presence of polar groups on its surface. 3 phr was reported to be the optimum amount of the graphitic filler.

Tensile properties of SBR (25% styrene)/CB-RGO composites (see above for the in situ reduction of RGO) were much better compared to the SBR/CB blends [111]. For example, the moduli at 200% elongation of the SBR/CB-RG blends were higher than those of the SBR/CB blends, without reduction of elongation. The elongation at break increased from 260% to 300% when 1 phr of RGO was added to SBR/CB with 10 phr CB and the tensile strength increased by about 40% after addition of 1 phr RGO.

Compression stress–strain curve testing of rubbery epoxy resin revealed that the compressive modulus of the hybrid composites with 4 wt% CB (Printex XE2) and 14 wt% of GNP had a two-fold increase with respect to the composite with only 4 wt% CB [114]. This was attributed to the close interactions between GNPs and CB particles and to the inherent high modulus of GNPs.

11.4.7 Abrasion resistance

EG/MEG loaded SBR composites, in the presence and absence of CB, revealed lower mass loss (DIN abrasion) than composites with only CB [112].

11.4.8 Thermal and electrical properties

Thermal stability of SBR (25% styrene)/CB composites [111] was not appreciably increased by feeding RGO. With respect to composites containing RGO, the elimination of oxygenated groups from the graphitic surface was crucial.

The thermal conductivity of rubbery epoxy resin was four times improved by adding 4 wt% CB (Printex XE2) and 14 wt% of GNP [114]. Functionalization of GNP with sylane, which was reported above to reduce the compound viscosity, decreased the composite's thermal conductivity by about 15% and made the hybrid composite highly electrically insulating.

In the case of SBR (25% styrene)/CB composites, the volume resistivity was more or less the same in the presence of either 10 or 13 phr of CB (about 10^{13} Ω cm) [111]. The addition of 1 phr of RGO led to lower resistivity (about 10^{12} Ω cm), without substantial differences for a further addition of GO.

In liquid silicone rubber [109], a lower percolation threshold was found when both GNPs and CB were added to the matrix, rather than using only CB. This result was attributed to a synergetic effect of mixed GNPs and CB, indicating that the addition of GNPs increased the electrical conductivity of CB filled silicone rubber.

11.5 Nanocomposites based on CNT and silica

11.5.1 Introduction

Only a few works are available in the literature on silica–CNT hybrid filler rubber composites. They are shown in Table 11.5. Although the limited number of works does not provide sufficient data for the generalization of results, the properties obtained with hybrid fillers are briefly summarized in the following paragraphs.

11.5.2 Structure

A good dispersion of CNTs was achieved by melt blending technology for systems with a high amount of silica. Up to 3 phr of CNTs were successfully dispersed

Table 11.5 Papers on nanocomposites based on silica and CNT hybrid fillers: type of rubber, silica, CNT, blending technology, and main aspects investigated in the work

References	Rubber	Silica type	CNT type	Blending method	Analyses
[115]	SBR	Silica (Oriental Silicas Corporation, Tokusil 255) + coupling agent TESPT	Three kinds of MWCNTs with different length (M85 by Carbon Nanomaterial Technology Co., Ltd.; CM-150 and CM-250 by Hanwha Nanotech Co.)	Melt blending	Mooney viscosity, cure characteristics, tensile test, dynamic-mechanical analysis (in tensile mode, temperature and strain sweep), SEM, swelling test for cross-link measurement, heat build up, abrasion resistance, electrical impedance
[116]	NR	Highly dispersible nanoscale silica (Ultrasil 7000 GR, Evonik Degussa), silanized by TESPT	MWCNTs (Nanocyl-7000)	Melt blending	Light microscopy, TEM, dielectric analyses, tensile test, dynamic-mechanical analysis (in torsion mode, temperature and strain sweep), fatigue tests
[96]	BR/SBR, NR, EPDM	Silica Ultrasil GR 7000, Evonik + organo-silane Si 69	MWCNTs (Nanocyl-7000)	Dry mixing or mixing after solvent dispersion of CNTs, by internal mixer	Reflected light microscopy, TEM, tensile test, dynamic-mechanical tests (in shear mode, temperature sweep), dielectric measurements, fatigue tests, thermal diffusivity

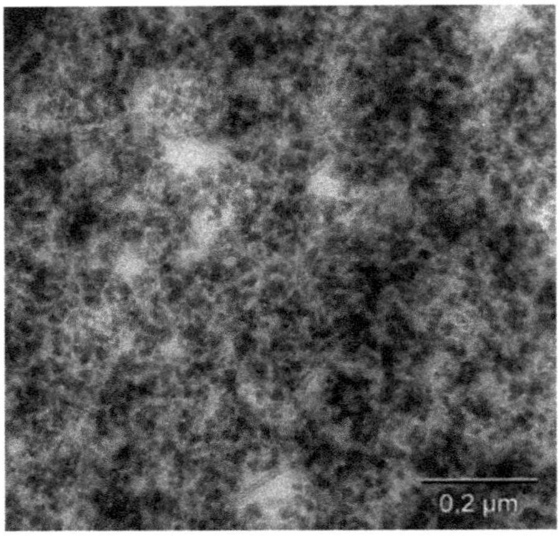

Figure 11.15 TEM image of NR filled with 6 phr CNT and 84 phr silica.
Source: Reprinted with permission from Fritzsche J, Lorenz H, Klueppel M. CNT based elastomer-hybrid-nanocomposites with promising mechanical and electrical properties. Macromol Mater Eng 2009;294:551–60.

within SBR with 80 phr silica in Ref. [115], but bundles of CNTs were observed by SEM analysis for higher amounts. In [116] no agglomeration of nanotubes was observed by TEM analysis, up to 10 phr of CNTs added to OSC/NR composites, for a total filler amount of 90 phr (Fig. 11.15).

The interaction between SBR matrix and CNTs was investigated by swelling measurements [115], but, since the formation of any chemical bonds was excluded, the reduced degree of swelling was interpreted as an increase of apparent cross-linking density.

11.5.3 Rheology and curing

It was reported that the addition of MWCNTs by melt blending reduced the vulcanization time. The cure time evaluated by t_{90} with a vulcameter was reduced by 3 phr MWCNTs added to a SBR/BR composite by dry mixing, and this result was attributed to the huge specific surface area of the CNT [96]. When up to 7 phr were added in a SBR matrix filled with 80 phr silica [115] CNTs definitely increased the cure rates due to enhanced thermal conduction, and considerably increased the Mooney viscosity due to the hydrodynamic reinforcement. The high aspect ratio and large interfacial area of CNTs were commented to be the reason for a restriction in the matrix fluidity and therefore for a viscosity increase.

11.5.4 Dynamic-mechanical properties

Dynamic-mechanical measurements were performed on hybrid filler silica–CNT systems with a high amount of silica. Dynamic moduli were obtained both in temperature sweep and in strain sweep, to study the Payne effect. Generally, CNT addition reduced the loss factor peak at the glass transition temperature and increased the Payne effect.

Dynamic-mechanical tests were performed in tensile mode to evaluate dynamic properties related to tire performance in SBR composites filled with 80 phr silica and up to 7 phr MWCNTs [115], obtained by melt blending. Three different kinds of CNT with different lengths were used. The Payne effect was found to be increased by increasing CNT content, but only when using longer CNTs (30 and 100 μm long), while shorter CNTs (5 μm long) did not cause any remarkable effect (Fig. 11.16). Adverse effects of longer CNT addition above 4 phr were recorded also for other dynamic properties: frictional hysteresis increased, loss factor at 0°C (related to wet traction) decreased, and loss factor at 70°C (related to rolling resistance) increased. On the contrary, the addition of short CNTs with low loading enhanced the wet traction of the silica/SBR composite. The influence of longer CNTs was attributed to a scarce dispersion of CNTs, which resulted in the segregation of silica particles and aggregation of CNTs. It was commented that both the amounts of the fillers and the length of CNTs should be considered as parameters influencing the mechanical properties of SBR compounds.

Dynamic-mechanical properties were measured also for NR-based composites with a fixed total amount of hybrid filler (silica + MWCNTs up to 10 phr), obtained by melt blending [116]. Dynamic moduli measured during temperature sweeps

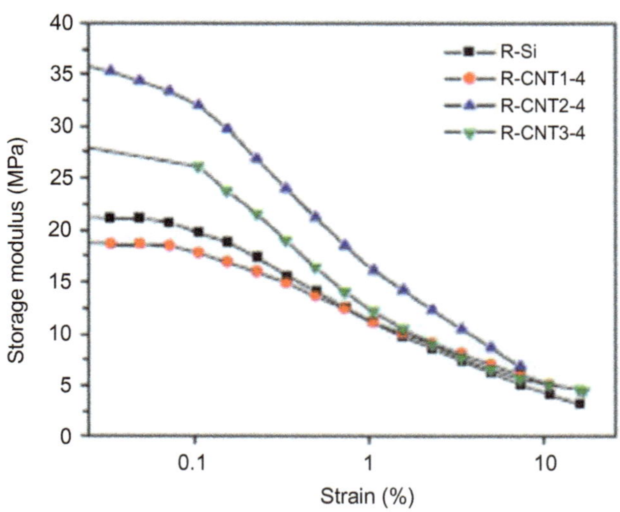

Figure 11.16 The Payne effect of SBR composites filled with 80 phr silica and up to 7 phr MWCNTs of different lengths (CNT1: 5 μm, CNT2: 30 μm; CNT3: 100 μm).
Source: Reprinted with permission form Park SM, Lim YW, Kim CH, Kim DJ, Moon WJ, Kim JH, et al. Effect of carbon nanotubes with different lengths on mechanical and electrical properties of silica-filled styrene butadiene rubber compounds. J Ind Eng Chem 2013;19:712–19.

revealed that the increase of CNT content reduced the peak of loss factor at the glass transition temperature: this was commented to be an effect of the polymer–filler interface, increasing with the nanofiller amount. The Payne effect was increased by the addition of CNTs, particularly at the highest CNT content (10 phr) and this effect was attributed to the formation of bound or trapped rubber with CNT.

11.5.5 Tensile properties

Papers show common effects of CNT addition on tensile properties, independently of the matrix considered: CNT increase stresses in the tensile stress–strain curve, while elongation at break is negatively affected after a certain CNT loading, depending on the composite considered.

A clear increase of 300% modulus was obtained by the addition to a SBR matrix filled with 80 phr silica with three kinds of MWCNTs (up to 7 phr) of different lengths (5, 30, and 100 μm) [115]. This effect was explained by the hydrodynamic effect. Tensile strength was not clearly affected by the CNTs presence, while elongation at break was reduced when increasing the nanofiller content, already at 1 phr CNT.

The stress values of stress–strain curves were increased by MWCNTs content also in composites based on NR and filled with a total amount of 90 phr of silica + MWCNTs (up to 10 phr) [116]. The elongation at break did not decrease by the addition of CNTs up to 6 phr. At higher CNT content the detrimental effect on elongation at break was attributed to the presence of agglomerates which behave as stress concentrators (Fig. 11.17).

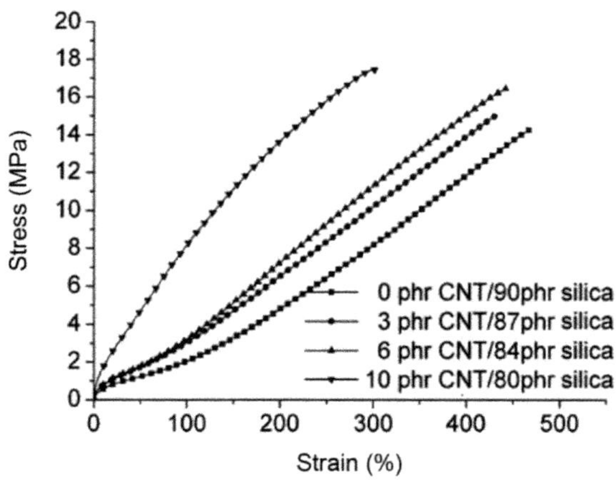

Figure 11.17 Quasistatic stress–strain behavior of composites with varying amounts of silica and CNT.
Source: Reprinted with permission from Fritzsche J, Lorenz H, Klueppel M. CNT based elastomer-hybrid-nanocomposites with promising mechanical and electrical properties. Macromol Mater Eng 2009;294:551–60.

Similar results were obtained by adding 3 phr MWCNTs either to a silica filled SBR/BR or to a silica filled NR (silica content: 40 or 60 phr) [96]. For both matrices, mechanical stresses were increased, while failure strains decreased by CNT addition. In the case of SBR/BR-based composites, an insufficient filler–polymer interaction was invoked.

11.5.6 Fracture resistance

Fatigue resistance was investigated by dynamic fracture mechanical investigations, performed on notched strip samples in tension mode.

The partial replacement of 90 phr silica with MWCNTs (up to 10 phr) in NR matrix led to the increase of the crack propagation rate [116] with respect to the silica filled NR. This effect was larger at higher tearing energies, corresponding to higher elongations, in agreement with a decrease in elongation at break by CNT addition.

Qualitatively similar results were obtained from the comparison of fatigue resistance of NR + 3 phr MWCNTs with or without 40 phr silica in Ref. [96].

11.5.7 Abrasion

Abrasion properties of hybrid filled CNT–silica composites were presented only in Ref. [115]. An SBR matrix was added with 80 phr silica and 1 to 10 phr MWCNTs of three different lengths. Abrasion resistance was measured by a Lambourn abrasion tester (Ueshima) according to ISO 23794. The addition of MWCNTs led to a decrease of abrasion resistance and an increase of heat generation at repeated dynamic impacts (measured by a Goodrich flexometer). This result was related to the decrease of elongation at break and the increase of the irreversible deformation by the CNT addition.

11.5.8 Electrical properties

The effect on electrical conductivity of the addition of 1 to 7 phr of three different MWCNT with different length to SBR filled with 80 phr was studied in Ref. [115]. The higher the MWCNT length (5, 30, and 100 μm) then the lower the amount of CNTs necessary to achieve the electrical percolation threshold (4, 2, and 2 phr, respectively). This effect was attributed to the higher effectiveness of longer lengths, which allow to achieve more easy connection points between them.

The conductivity was also significantly increased with the amount of CNT in NR filled with 90 phr of silica + MWCNT (1 to 10 phr) [116]. The percolation threshold was found below 0.6 vol% MWCNTs (equal to about 3 phr) and this indicated a quite good dispersion of the nanofiller. By the measurement of dielectric properties as a function of frequency, and by an interpretation of results based on the percolation theory, further insights into the qualitative properties of conduction mechanism in CNT/polymer composites are given. The conduction mechanism found was the same proposed for CB filled composites. The characteristic

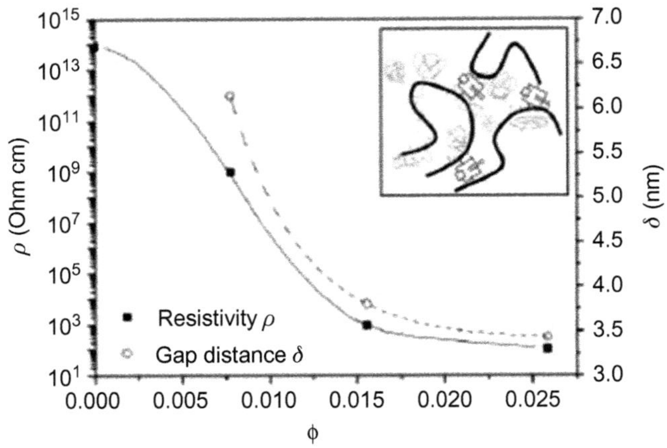

Figure 11.18 DC resistivity and distance between CNT as a function of CNT volume fraction.
Source: Reprinted with permission from Fritzsche J, Lorenz H, Klueppel M. CNT based elastomer-hybrid-nanocomposites with promising mechanical and electrical properties. Macromol Mater Eng 2009;294:551–60.

frequency of the relaxation transition observed at high frequencies can be related to the distance between adjacent filler particles. Fig. 11.18 shows that the gap distance between nanotubes exponentially decreases with the increase in CNT content and does reach a plateau value. Moreover, the decrease in gap distance is related to a decrease in resistivity. This indicates that adjacent tubes are not in direct contact, but are separated by thin polymer layers which hinder the charge transport.

The presence of silica in CNT filled SRB/BR or NR seems to increase electrical conductivity: for SBR/BR, the addition with 3 phr MWCNTs gave an electrical conductivity of 0.10 S/m, which was increased to 0.15 S/m by further addition of 40 phr silica [96]. Silica was commented to help dispersion of the CNT and accordingly, to lower the percolation threshold. A higher amount of silica (60 phr) led to lower conductivity, probably due to lower CNT volume fraction. A similar effect of the presence of both silica and CNTs was found in NR matrix, but the conductivity values were significantly lower than for SBR/BR matrix, possibly due to the different CNT blending technique ("drying mixing" for NR, mixing with ethanol for SBR/BR).

11.6 Rationalization of the mechanical reinforcement

As shown in previous paragraphs, a good number of reports are available on rubber composites with hybrid filler systems. However, they are characterized by a large variety of preparation method, type of ingredients, filler amount, degree of filler dispersion, and properties explored. Therefore, it appears difficult to attempt a rationalization. However, limiting the attention only to mechanical reinforcement, a contribution to rationalization was given by the authors of this chapter. By analyzing the increase of the initial modulus, the synergistic effect between different fillers was pointed out and a key for the interpretation was proposed.

Mechanical reinforcement effects promoted by a single filler in a rubber matrix have been widely explored in the literature. Brief comments were proposed in section 11.2.4. Rigid particles embedded in a soft, elastomeric matrix increase the material stiffness. Literature works analyzed the dependence of stiffness on filler content, and Huber and Vilgis [75] identified the presence of at least two regimes: by plotting the modulus increase versus the filler volume fraction on a double logarithmic plot, at low filler content the modulus increases by a linear law, at high filler content the modulus increases by a steeper linear law. The filler content at which the change of regime occurs is related to the mechanical percolation threshold (Φ_c). Above the percolation threshold, the distance among filler particles decreases and a sort of continuous network occurs, which allows the filler to be more effective toward the reinforcement.

The authors have measured the percolation threshold by the Huber-Vilgis plot for single nanofiller systems. Surprisingly, also when the nanofillers were added with 60 phr of CB N326 to the same matrix a discontinuity of the modulus–filler content plot could be observed, and a pseudopercolation threshold measured. The pseudothreshold value was lower than that found for single-filler systems, as if the nanofiller was able to establish, with CB, a continuous hybrid network, which was shown by TEM analyses. As it was shown in Table 11.2, the percolation threshold of CNT in IR was reduced from 7 to 3 phr by the addition of CB [78], that of OC in IR from 6 to 3 phr [44], and that of nanoG in IR from 21 to 18 phr [16]. This result pointed out that CB favored the dispersion of the nanofiller and the formation of a continuous hybrid filler network, which is more efficient that the one based on a single filler.

With the aim to study the strength of interaction of nano- and nanostructured fillers, in a first approach the authors analyzed composites prepared by adding various amounts of nanofillers (either OC, CNT, or nanoG) to a neat IR matrix or to composites based on IR and 60 phr of CB [16,44,78]. The strength of interaction between nanofiller and CB particles in the hybrid filler systems was quantified by comparing the measured moduli of the hybrid systems with those computed by a simple additive rule, indicated as the "additive model" [117]. Such a rule was first proposed to study the influence of the mutual interaction of two different fillers on the mechanical reinforcement of an elastomeric matrix.

The additive model provides moduli values calculated under the hypothesis that no interaction occurs between two different fillers, therefore the measurement of experimental values higher than the calculated ones can indicate an interaction between the two fillers. The predicted moduli, $E(\Phi_{CB},\Phi_{nanofiller})$ for CB and a nanofiller, are obtained as the sum of different contributions: the matrix modulus and the modulus enhancements independently originated by each single filler with respect to the neat matrix, for a given filler content. $E(\Phi_{CB},\Phi_{nanofiller})$ is calculated by the Eq. (11.5) (also referred to as "additive model"):

$$E(\Phi_{CB},\Phi_{nanofiller}), = E(\Phi_{CB}) + E(\Phi_{nanonanofiller}) - E(IR) \quad (11.5)$$

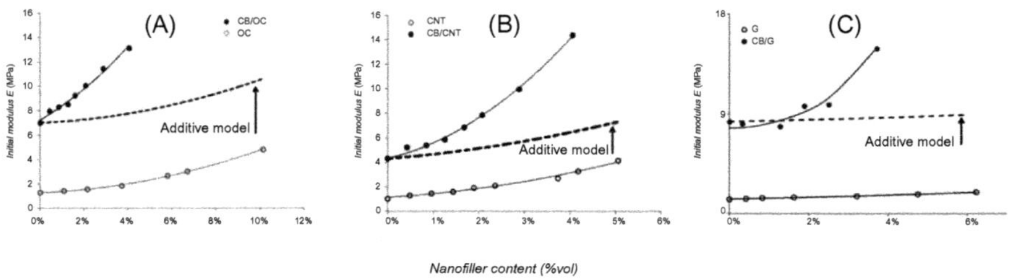

Figure 11.19 *E* initial modulus values for nanocomposites containing: (A) OC; (B) CNT; (C) nanoG. Experimental values: ●, ○. Values calculated according to the "additive model" (Eq. (11.5)): —. Experimental values: ○ refer to nanocomposites with nanofiller in the neat polymer matrix, ● refer to nanocomposites with nanofiller in the polymer matrix containing 60 phr of CB.
Source: Reprinted with permission from Galimberti M, Cipolletti V, Musto S, Cioppa S, Peli G, Mauro M, et al. Recent advancements in rubber nanocomposites. Rubber Chem Technol 2014;87(3):417−42.

where *E*(IR) is the measured elastic modulus of the neat IR matrix, $E(\Phi_{CB})$ and $E(\Phi_{nanonanofiller})$ are the moduli of IR composites containing only one filler (CB or nanofiller, respectively). The subscripts refer to the filler content considered, expressed as volume fraction, Φ. Fig. 11.19 shows both the measured and the calculated moduli values for the investigated systems.

For all of the three nanofillers (OC, CNT, and nanoG), interactive effects are shown: the combination of CB and the nanofiller leads to initial modulus values much higher than those calculated through the simple addition of the two initial moduli of the composites containing only CB (60 phr) and only the nanofiller.

However, the systems analyzed do not allow to clarify whether the nanofiller promotes a real synergistic effect, or rather the modulus enhancement is merely due to an increase of the overall filler concentration. Therefore, a further, more systematic work was performed [25] on IR based systems filled with only carbonaceous fillers. Single-filler systems, obtained by adding only CB, CNT, or nanoG at increasing contents were prepared, and, for comparison, also hybrid filler systems were prepared having the same total contents, but made of equal parts of CB and CNT or of CB and nanoG.

The dynamic shear moduli at small deformation amplitude ($G'\gamma_{min}$) were measured and compared to the moduli calculated with the additive rule. The difference between the experimental and the predicted values was labeled "interactive term," $\Delta G'_{int}$. Such a term, indicating the strength of interaction between two fillers with each other, is dependent on the type of matrix and fillers, and on the total and relative filler amount. In order to have reference values for the interactive terms, they were calculated also for single-filler systems, considering the total filler amount (Φ) as made by two equal parts of the same filler (Φ/2) rather than two equal parts of different fillers. Therefore, autointeractive terms indicate the modulus enhancement originated by the interaction between the two parts of the same filler, at the same

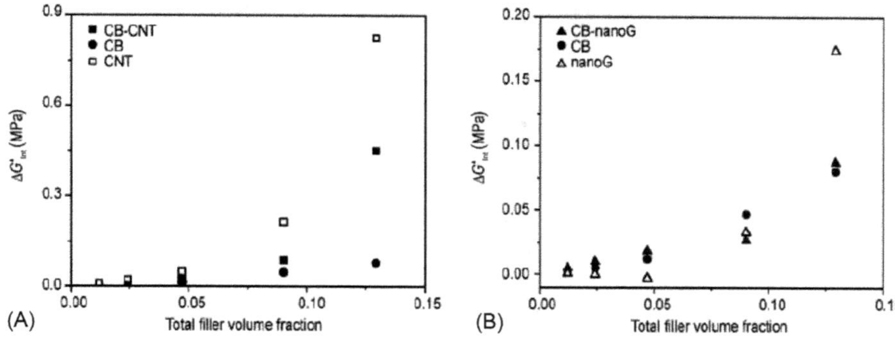

Figure 11.20 Interactive and autointeractive terms as a function of the total filler volume fraction for the investigated systems: (A) CB and/or CNT filled systems; (B) CB and/or nanoG filled systems.
Source: Reprinted with permission from Agnelli S, Cipolletti V, Musto S, Coombs M, Conzatti L, Pandini S, et al. Interactive effects between carbon allotrope fillers on the mechanical reinforcement of polyisoprene based nanocomposites. Express Polym Lett 2014;8 (6):436–49.

total and relative filler volume fractions realized in the hybrid composites. A comparison of interactive and autointeractive terms should evidence the contribution to the interaction term due to the peculiarities of the two different fillers.

Fig. 11.20 shows interactive and autointeractive terms obtained for CB–CNT and CB–nanoG systems as a function of the total filler volume fraction.

CNT promote the highest autointeractive terms, and are responsible for the reinforcement of the hybrid filler system as well: in fact, the interactive terms of the hybrid filler composites with CNT are between those of the single-filler composites, while CB exerts a moderate contribution. On the other hand, nanoG shows a significant autointeractive term only at the highest filler content (33 phr), and it is not able to promote significant interactions with CB in the hybrid filler system, at the contents investigated.

The superior interaction ability of CNT, both with itself and with other particles, was interpreted by the authors with the highest specific surface area of CNT available to the contact with the elastomeric matrix. In fact, by plotting the moduli of single and hybrid filler systems as a function of specific surface area, a common correlation between the reinforcing ability of a nanofiller CNT and a nanostructured filler CB was found. Fig. 11.21 shows such a moduli as a function of the specific surface area, evaluated through Eq. (11.6):

$$\text{Specific surface area (for hybrid fillers)} = A_{\text{CB}} \times \rho_{\text{CB}} \times \Phi_{\text{CB}} + A_{\text{nanofiller}} \times \rho_{\text{nanofiller}} \times \Phi_{\text{nanofiller}} \tag{11.6}$$

where A is assumed to be equal to BET surface area (area per unit of filler weight), ρ is the filler density, and Φ is the filler volume fraction.

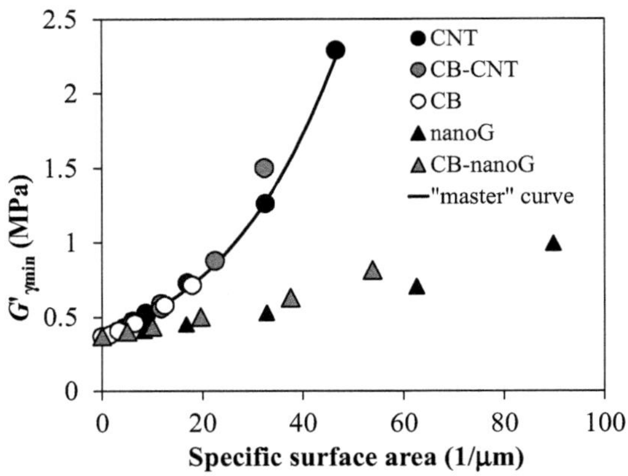

Figure 11.21 Shear modulus as a function of specific surface area for single and hybrid filler systems.
Source: Reprinted with permission from Galimberti M, Cipolletti V, Musto S, Cioppa S, Peli G, Mauro M, et al. Recent advancements in rubber nanocomposites. Rubber Chem Technol 2014;87(3):417–42.

Fig. 11.21 shows also that data of systems filled with nanoG do not lie on the "master" curve common to CNT and CB. The authors attributed this discrepancy to the layered structure of nanoG: the high surface area of filler, measured by gas adsorption method, is not fully accessible to the polymer matrix, lowering its reinforcing efficiency.

Finally, a further analysis was performed by the authors on the systems investigated in [16,25,31,44,78]. The modulus reinforcement promoted by nanofillers was evaluated by a *k* factor, defined as the ratio of the composite modulus over the matrix modulus. For single-filler composites, the neat polymer was considered as the matrix, for hybrid filler systems, the polymer added with CB was considered as the matrix. Interestingly, by plotting *k* factors as a function of nanofiller volume fraction, the data for single and hybrid filler composites filled with the same nanofiller are overlapping, as shown in Fig. 11.22.

Although it was not possible to provide a physical motivation for this finding, it is clear that nanofillers enhance the "matrix" modulus by a multiplication factor, *k*, that depends on the nanofiller type and content, independently of the "matrix" composition, which can be both a neat elastomer and the same elastomer filled with CB.

Such a result brings about also convenient consequences. In fact, provided that the *k* factor is the same for single and hybrid filler systems, i.e.,

$$k = G'_{\mathrm{nanofiller}}(\Phi_{\mathrm{nanofiller}})/G'_{\mathrm{m}} = G'_{\mathrm{c}}(\Phi_{\mathrm{nanofiller}}, \Phi_{\mathrm{CB}})/G'_{\mathrm{CB}}(\Phi_{\mathrm{CB}}) \tag{11.7}$$

then the modulus of a hybrid filler composite can be calculated as a product of the moduli of the single-filler composites over the matrix modulus. In fact, by rearranging Eq. (11.7), the following relationship is found:

$$G'_{\mathrm{c}}(\Phi_{\mathrm{nanofiller}}, \Phi_{\mathrm{CB}}) = G'_{\mathrm{nanofiller}}(\Phi_{\mathrm{nanofiller}}) * G'_{\mathrm{CB}}(\Phi_{\mathrm{CB}})/G'_{\mathrm{m}} \tag{11.8}$$

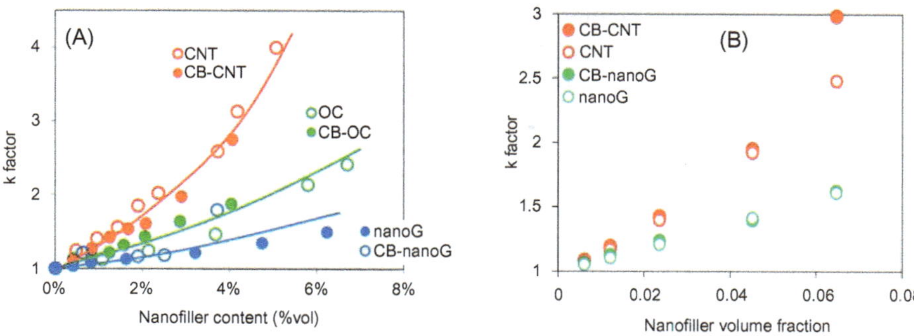

Figure 11.22 *k* factors for nanocomposites containing either OC, CNT, or nanoG as the nanofillers. In both (a and b): in the neat polymer matrix (○). In (A): in the polymer matrix containing 60 phr of CB (●). In (B): in the polymer matrix containing the same amount of CB as the nanofiller (●).

Although limited to the systems investigated by the authors of the chapter, this finding can suggest a general rule for hybrid filler systems, which could help the prediction of the moduli for the hybrid filler composites.

11.7 Hybrid systems made by different nanofillers

Papers are increasingly appearing in the literature on hybrid filler systems based on nanofillers: OC and graphite nanoparticles were combined with CNT. They do not yet allow either a rationalization or even a discussion. However, they already show interesting indications.

As discussed above in this chapter, a relevant issue is the CNT debundling and dispersion in the polymer matrix. The combination of CNTs and lamellar nanoparticles, either graphitic [118] or clays [119], was found to favor CNT dispersion.

A great opportunity is given by the synergistic effect brought about by different types of fillers. Mixed filler networks and consequently synergistic effects on both electrical and mechanical properties were observed with the combination of two graphitic nanofillers, such as CNTs and graphene nanoparticles [120,121]. Synergistic effects were found also from the combination of CNTs and clay on thermal degradation behavior [122] and on mechanical properties [123].

11.8 Conclusions and perspectives

A few words have to be used to summarize the main results reported in previous paragraphs.

Rubber composites based on hybrid filler systems made by nanofillers such as OC, CNT, graphene, and nanographite and nanostructured fillers, such as CB and silica, do show remarkable improvements in properties compared to composites based on only one family of fillers, i.e., improvements in mechanical reinforcement and electrical and thermal conductivity.

Nano- and nanostructured fillers show synergistic effects, e.g., for the improvement of mechanical properties.

The behavior of nano- and nanostructured fillers can be rationalized upon identifying suitable filler features. For example, a common master curve can be drawn for the improvement of the composite initial modulus as a function of the filler–polymer interfacial area.

List of abbreviations

6PPD	N-(1,3-dimethylbutyl)-N′-phenyl-p-phenylene diamine
AFM	atomic force microscopy
CB	carbon black
CBS	N-cyclohexyl-2-benzothiazole sulfenamide
CDDC	copper dimethyl dithio carbamate
CNTs	carbon nanotubes
CNTBs	carbon nanotube bundles
DCUP	dicumyl peroxide
DSC	differential scanning calorimetry
EPDM	ethylene-propylene diene monomer
FE-SEM	filed emission scanning electron microscopy
IPPD	N-isopropyl-N′-phenyl-p-phenylene diamine
IR	isoprene rubber
MBT	2-Mercaptobenzothiazole
MBTS	2.2′-Dithio-dibenzothiazole
MWCNTs	multiwalled carbon nanotubes
NBR	nitrile-butadiene rubber
NR	natural rubber
PBN	phenyl-β-naphthylamine
phr	per hundred resin
S	sulfur
SA	stearic acid
SBR	styrene-butadiene rubber
sDPA	substituted diphenyl amine
SEM	scanning electron microscopy
SWCNTs	single-walled carbon nanotubes
TAIC	triallyl isocianurate
TBBS	N-tert-butyl-2-benzothiazole sulfenamide
TEM	transmission electron microscopy
TGA	thermogravimetric analysis
TMTD	Bis(dimethylthiocarbamyl) disulfide
TOTM	trioctyl trimellitate 77, 78
WAXD	wide-angle X-ray diffraction

XRD X-ray diffraction
ZDC zinc dithio carbamate
ZDDC zinc dimethyl dithio carbamate
ZDEC zinc diethyl carbamate
ZMMBI zinc methyl mercaptobenzimidazole

References

[1] Donnet JB, Custodero E. Reinforcement of elastomers by particulate fillers. In: Mark JE, Erman B, Eirich FR, editors. The science and technology of rubber. 3rd ed. San Diego: Academic Press; 2005. p. 367–400.

[2] Maiti M, Bhattacharya M, Bhowmick AK. Elastomer nanocomposites. Rubber Chem Technol 2008;81:384–469.

[3] Galimberti M, Cipolletti V, Kumar V. Nanofiller in natural rubber. In: Thomas S, Chan CH, Pothan L, Joy J, Maria H, editors. *Natural rubber materials: volume 2: composites and nanoco*mposites. Royal Society of Chemistry; 2013. p. 34–72.

[4] Vocabulary – Nanoparticles, 2005, PAS 71-2005 BSI.

[5] Ray SS, Okamoto M. Polymer/layered silicate nanocomposites: a review from preparation to processing. Prog Polym Sci 2003;28:1539–641.

[6] Chen B, Evans JRG, Greenwell HC, Boulet P, Coveney PV, Bowden AA, et al. A critical appraisal of polymer–clay nanocomposites. Chem Soc Rev 2008;37:568–94.

[7] Paul DR, Robeson LM. Polymer nanotechnology: nanocomposites. Polymer (Guildf) 2008;49:3187–204.

[8] Galimberti M. Rubber clay nanocomposites: science, technology, applications. 1st ed. John Wiley and Sons; 2011.

[9] Galimberti M. Rubber clay nanocomposites in advanced elastomers - technology, properties and applications. In: Boczkowska A, editor; 2012. p. 91–20. Available from: http://dx.doi.org/0.5772/51410.

[10] Galimberti M, Cipolletti V, Coombs M. Application of clay polymer nanocomposites. In: Bergaya F, Lagaly G, editors. Handbook of clay science. 2nd ed. Amsterdam: Elsevier; 2013 (Part B, Chapter 4.4 ISBN: 9780080993645).

[11] Bokobza L. Multiwall carbon nanotube elastomeric composites: a review. Polymer (Guildf) 2007;48(17):4907–20.

[12] Al-Solamya FR, Al-Ghamdib AA, Mahmoud WE. Piezoresistive behavior of graphite nanoplatelets based rubber nanocomposites. Polym Adv Technol 2012;23(3):478–82.

[13] Bhowmick AK, Bhattacharya M, Mitra S. Exfoliation of nanolayer assemblies for improved natural rubber properties: methods and theory. J Elastom Plast 2010;42(6):517–37.

[14] Bhattacharya M, Maiti M, Bhowmick AK. Tailoring properties of styrene butadiene rubber nanocomposite by various nanofillers and their dispersion. Polym Eng Sci 2009; 49(1):81–98.

[15] Sridhar V, Xu D, Pham TT, Mahapatra SP, Kim JK. Dielectric and dynamic mechanical relaxation behavior of exfoliated nano graphite reinforced flouroelastomer composites. Polym Compos 2009;30(3):334–42.

[16] Galimberti M, Kumar V, Coombs M, Cipolletti V, Agnelli S, Pandini S, et al. Filler networking of a nanographite with a high shape anisotropy and synergism with carbon black in poli(1,4-cis-isoprene) based nanocomposites. Rubber Chem Technol 2014;87 (2):197–218.

[17] McNaught D, Wilkinson A. *IUPAC. compendium of chemical terminology*, (the "Gold Book"). 2nd ed. Oxford: Scientific Publications; 1997XML on-line corrected version: http://goldbook.iupac.org (2006-) created by M. Nic, J. Jirat, B. Kosata; updates compiled by A. Jenkins. ISBN 0-9678550-9-8. Available from: http://dx.doi.org/10.1351/goldbook.

[18] Nanko M. Definitions and categories of hybrid materials. Adv Technol Mater Mater Process 2009;11(1):1–8.

[19] Witt N, Tang Y, Ye L, Fang L. Silicone rubber nanocomposites containing a small amount of hybrid fillers with enhanced electrical sensitivity. Mater Des 2013;45:548–54.

[20] Yan N, Wu JK, Zhan YH, Xia HS. Carbon nanotubes/carbon black synergistic reinforced natural rubber composites. Plast Rubber Compos 2009;38(7):290–6.

[21] Zhan YH, Liu GQ, Xia HS, Yan N. Natural rubber/carbon black/carbon nanotubes composites prepared through ultrasonic assisted latex mixing process. Plast Rubber Compos 2011;40(1):32–9.

[22] Galimberti M, Coombs M, Cipolletti V, Giannini L, Conzatti L. The origin of synergism between an organoclay and carbon black. Appl Clay Sci 2013;83–84:449–56.

[23] Dong B, Liu C, Lu Y, Wu Y. Synergistic effects of carbon nanotubes and carbon black on the fracture and fatigue resistance of natural rubber composites. J Appl Polym Sci 2015;132(25):42075.

[24] Das A, Stoeckelhuber KW, Rooj S, Wang DY, Heinrich G. Synergistic effects of expanded nanoclay and carbon black on natural rubber compound. Kautschuck Gummi Kunstoffe 2010;63(7–8):296–302.

[25] Agnelli S, Cipolletti V, Musto S, Coombs M, Conzatti L, Pandini S, et al. Interactive effects between carbon allotrope fillers on the mechanical reinforcement of polyisoprene based nanocomposites. Express Polym Lett 2014;8(6):436–49.

[26] Jones RG, Ober CK, Hodge P, Kratochvíl P, Moad G, Vert M. Terminology for aggregation and self-assembly in polymer science (IUPAC Recommendations 2013). Pure Appl Chem 2013;85(2):463–92.

[27] Huang JR, Zhu YT, Jiang W, Yin JH, Tang QX, Yang XD. Parallel carbon nanotube stripes in polymer thin film with remarkable conductive anisotropy. ACS Appl Mater Interfaces 2014;6(3):1754–8.

[28] Peng Z, Feng C, Luo Y. Natural rubber/multiwalled carbon nanotube composites developed with a combined self-assembly and latex compounding technique. J Appl Polym Sci 2012;125(5):3920–8.

[29] Yang H, Liu P, Zhang T, Duan Y, Zhang J. Fabrication of natural rubber nanocomposites with high graphene contents via vacuum-assisted self-assembly. RCS Adv 2014;4(53):27687–90.

[30] Zhao Y, Milano G, Cong Y, Yu N, He Y, Cong Y, et al. Self-assembled morphologies and percolation probability of mixed carbon fillers in the diblock copolymer template: hybrid particle-field molecular dynamics simulation. J Phys Chem C 2015;119(44):25009–22.

[31] Galimberti M, Cipolletti V, Musto S, Cioppa S, Peli G, Mauro M, et al. Recent advancements in rubber nanocomposites. Rubber Chem Technol 2014;87(3):417–42.

[32] Jose JP, Thomas S. XLPE based Al_2O_3–clay binary and ternary hybrid nanocomposites: self-assembly of nanoscale hybrid fillers, polymer chain confinement and transport characteristics. Phys Chem Chem Phys 2014;16(37):20190–201.

[33] Wang Q, Li G, Zhang J, Huang F, Lu K, Wei Q. PAN nanofibers reinforced with MMT/GO. J Nanomater 2014; no. 298021.

[34] Konishi Y, Cakmak M. Nanoparticle induced network self-assembly in polymer–carbon black composites. Polymer (Guildf) 2006;47(15):5371–91.
[35] Bao Z, Flanigan C, Beyer L, Tao J. Processing optimization of latex-compounded montmorillonite/styrene-butadiene rubber-polybutadiene rubber. J Appl Polym Sci 2015;132 (8):41521/1–41521/8.
[36] Galimberti M, Coombs M, Pandini S, Riccò T, Cipolletti V, Conzatti L, et al. Delamination of organically modified montmorillonite for reducing the filler networking with carbon black in poly(1,4-cis-isoprene) based nanocomposites. Appl Clay Sci 2015;104:8–17.
[37] Nawani P, Burger C, Rong L, Hsiao BS, Tsou AH. Structure and permeability relationships in polymer nanocomposites containing carbon black and organoclay. Polymer (Guildf) 2015;64:19–28.
[38] Annadurai P, Mukundan T, Joseph R. Influence of carbon black in polychloroprene organoclay nanocomposite with improved mechanical, electrical and morphology characteristics. Plast Rubber Compos 2013;42(9):379–84.
[39] Chattopadhyay PK, Praveen S, Das NC, Chattopadhyay S. Contribution of organomodified clay on hybrid microstructures and properties of epoxidized natural rubber-based nanocomposites. Polym Eng Sci 2013;53(5):923–30.
[40] Malas A, Das CK. Selective dispersion of different organoclays in styrene butadiene rubber in the presence of a compatibilizer. Mater Des 2013;49:857–65.
[41] Mohanty TR, Bhandari V, Chandra AK, Chattopadhyay PK, Chattopadhyay S. Role of calcium stearate as a dispersion promoter for new generation carbon black-organoclay based rubber nanocomposites for tyre application. Polym Compos 2013;34(2):214–24.
[42] Sapkota J, Poikelispa M, Das A, Dierkes W, Vuorinen J. Influence of nanoclay-carbon black hybrid fillers on cure and properties of natural rubber compounds. Polym Eng Sci 2013;53(3):615–22.
[43] Sreenivasan P, Ratna D, Albert P, Somashekaran J, Raut R, Chakraborty BC. A new liquid rubber-assisted dispersion of organoclay in carbon black filled carboxylated acrylonitrile–butadiene rubber matrix. J Appl Polym Sci 2013;128(4):2414–23.
[44] Galimberti M, Coombs M, Cipolletti V, Riccio P, Ricco T, Pandini S, et al. Enhancement of mechanical reinforcement due to hybrid filler networking promoted by an organoclay in hydrocarbon based nanocomposites. Appl Clay Sci 2012;65–66: 57–66.
[45] Malas A, Das CK. Carbon black–clay hybrid nanocomposites based upon EPDM elastomer. J Mater Sci 2012;47:2016–24.
[46] Chattopadhyay PK, Das NC, Chattopadhyay S. Influence of interfacial roughness and the hybrid filler microstructures on the properties of ternary elastomeric composites. Compos Part A Appl Sci Manuf 2011;42(8):1049–59.
[47] Gopi JA, Patel SK, Chandra AK, Tripathy DK. SBR-clay-carbon black hybrid nanocomposites for tire tread application. J Polym Res 2011;18:1625–34.
[48] Liu Y, Li L, Wang Q, Zhang X. Fracture properties of natural rubber filled with hybrid carbon black/nanoclay. J Polym Res 2011;18:859–67.
[49] Galimberti M, Senatore S, Lostritto A, Giannini L, Conzatti L, Costa G, et al. Reinforcement of diene elastomers by organically modified layered silicates. e-Polymers 2009;1–16 no. 057.
[50] Praveen S, Chattopadhyay PK, Albert P, Dalvi VG, Chakraborty BC, Chattopadhyay S. Synergistic effect of carbon black and nanoclay fillers in styrene butadiene rubber matrix. Development of dual structure. Compos Part A 2009;40:309–16.

[51] Cataldo F. Preparation and properties of nanostructured rubber composites with montmorillonite. Macromol Symp 2007;247:67–77.
[52] Maiti M, Sadhu S, Bhowmick AK. Effect of carbon black on properties of rubber nanocomposites. J Appl Polym Sci 2005;96:443–51.
[53] Galimberti M, Cipolletti V, Giudice S. Morfology of rubber-clay nanocomposites. In: Galimberti M, editor. Rubber clay nanocomposites – science, technology and applications. New York, NY: Wiley and Sons; 2011. p. 181–240.
[54] Cipolletti V, Galimberti M, Mauro M, Guerra G. Organoclays with hexagonal rotator order for the paraffinic chains of the compensating cation. Implications on the structure of clay polymer nanocomposites. Appl Clay Sci 2014;87:179–88.
[55] Lagaly G, Ogawa M, Dekany I. Clay mineral–organic interactions. In: Bergaya F, Lagaly G, editors. *Handbook of* clay *science, Second Edition.* Amsterdam: Elsevier; 2013. p. 435–505.
[56] Herrmann W, Uhl C, Heinrich G, Jehnichen D. Analysis of HNBR-montmorillonite nanocomposites. Morphology, orientation and macroscopic properties. Polym Bull 2006;57:395–405.
[57] Bhattacharya M, Bhowmick AK. Synergy in carbon black-filled natural rubber nanocomposites. Part I: mechanical, dynamic mechanical properties, and morphology. J Mater Sci 2010;45:6126–38.
[58] Kvande I, Øye G, Hammer N, Rønning M, Raaen S, Holmen A, et al. Deposition of Au colloids on plasmachemically modified carbon nanofibers. Carbon N Y 2008;46(5):759–65.
[59] Sabah E, Mart UC, Nar MC, Elik MS. Zeta potentials of sepiolite suspensions in concentrated monovalent electrolites. Sep Sci Technol 2007;42(10):2275–88.
[60] Gates WP. Crystalline swelling of organo-modified clays in ethanol-water solutions. Appl Clay Sci 2004;27:1–12.
[61] Lin JJ, Chu CC, Chiang ML, Tsai WC. First isolation of individual silicate platelets from clay exfoliation and their unique self-assembly into fibrous arrays. J Phys Chem B 2006;110(37):18115–20.
[62] Xu R, Wu C, Xu H. Particle size and zeta potential of carbon black in liquid media. Carbon N Y 2007;45(15):2806–9.
[63] Konishi Y, Cakmak M. Structural hierarchy developed in injection molding of nylon 6/clay/carbon black nanocomposites. Polymer (Guildf) 2005;46(13):4811–26.
[64] Feller JF, Bruzaud S, Grohens Y. Influence of clay nanofiller on electrical and rheological properties of conductive polymer composite. Mater Lett 2004;58(5):739–45.
[65] Thomas S, Stephen R. Rubber nanocomposites: preparation, properties and applications. John Wiley & Sons; 2010.
[66] Semsarzadeh MA, Bakhshandeh GR, Ghasemzadeh-Barvarz M. Effect of carbon black on rate constant and activation energy of vulcanization in EPDM/BR and EPDM/NR blends. Iran Polym J 2005;14(6):573–8.
[67] Avalos F, Ortiz JC, Zitzumbo R, López-Manchado MA, Verdejo R, Arroyo M. Effect of montmorillonite intercalant structure on the cure parameters of natural rubber original research article. Eur Polym J 2008;44(10):3108–15.
[68] Sengupta R, Chakraborty S, Bandyopadhyay S, Dasgupta S, Mukhopadhyay R, Auddy K, et al. A short review on rubber/clay nanocomposites with emphasis on mechanical properties. Polym Eng Sci 2007;47(11):1956–74.
[69] Verdejo R, Hernandez M, Bitinis N, Kenny JM, Lopez-Manchado MA. Vulcanization characteristics and curing kinitects of rubber-organoclay nanocomposites. In: Galimberti M, editor. Rubber clay nanocomposites – science, technology and applications. New York, NY: Wiley and Sons; 2011. p. 275–303.

[70] Giannini L, Citterio A, Galimberti M, Cozzi D. Chemistry of rubber-organoclay nanocomposites. In: Galimberti M, editor. Rubber clay nanocomposites – science, technology and applications. New York, NY: Wiley and Sons; 2011. p. 127–44.
[71] Payne AR, Whittaker RE. Low strain dynamic properties of filled rubbers. Rubber Chem Technol 1971;44(2):440–78.
[72] Medalia AI. Effect of carbon black on dynamic properties of rubber vulcanizates. Rubber Chem Technol 1978;51(3):437–523.
[73] Guth E, Gold O. On the hydrodynamical theory of the viscosity of suspensions. Phys Rev 1938;53:322.
[74] Guth E. Theory of filler reinforcement. Rubber Chem Technol 1945;18(3):595–604.
[75] Huber G, Vilgis TA. Universal properties of filled rubbers: mechanisms for reinforcement on different length scales. Kautschuk Gummi Kunststoffe 1999;52(2):102–7.
[76] Schön F, Gronski W. Filler networking of silica and organoclay in rubber composites: reinforcement and dynamic-mechanical properties. Kautschuk Gummi Kunststoffe 2003;56:166–71.
[77] Ramorino G, Bignotti F, Pandini S, Riccò T. Mechanical reinforcement in natural rubber/organoclay nanocomposites. Compos Sci Technol 2009;69:1206–11.
[78] Galimberti M, Coombs M, Riccio P, Riccò T, Passera S, Pandini S, et al. The role of CNTs in promoting hybrid filler networking and synergism with carbon black in the mechanical behavior of filled polyisoprene. Macromol Mater Eng 2013;298:241–51.
[79] Gelfer M, Burger C, Fadeev A, Sics I, Chu B, Hsiao BS, et al. Thermally induced phase transitions and morphological changes in organoclays. Langmuir 2004;20(9):3746–58.
[80] Osman MA, Ploetze M, Skrabal P. Structure and properties of alkylammonium monolayers self-assembled on montmorillonite platelets. J Phys Chem B 2004;108(8):2580–8.
[81] Vaia RA, Teukolky RK, Giannelis EP. Interlayer structure and molecular environment of alkylammonium layered silicates. Chem Mater 1994;6(7):1017–22.
[82] Klatte SJ, Beck TL. Molecular dynamics of tethered alkanes: temperature-dependent behavior in a high-density chromatographic system. J Phys Chem 1993;97(21):5727–34.
[83] Galimberti M, Guerra G, Lostritto A. Tire and crosslinkable elastomeric composition' European Patent application 2027199 A1 (WO 2007/144012) to Pirelli Tyre S.p.A; 2007.
[84] Galimberti M, Guerra G, Lostritto A, Giannini L. Tire and crosslinkable elastomeric composition European Patent application 2061833 A1 (WO 2008/009304 A1) to Pirelli Tire S.p.A; 2008.
[85] Robertson CG, Lin CJ, Rackaitis M, Roland CM. Influence of particle size and polymer– filler coupling on viscoelastic glass transition of particle-reinforced polymers. Macromolecules 2008;41(7):2727–31.
[86] Qu L, Huang G, Zhang P, Nie Y, Weng G, Wu J. Synergistic reinforcement of nanoclay and carbon black in natural rubber. Polym Int 2010;59(10):1397–402.
[87] Carretero-González J, Retsos H, Verdejo R, Toki S, Hsiao BS, Giannelis EP, et al. Effect of nanoclay on natural rubber microstructure. Macromolecules 2008;41(18):6763–72.
[88] Rattanasom N, Prasertsri S. Relationship among mechanical properties, heat ageing resistance, cut growth behavior and morphology in natural rubber: partial replacement of clay with various types of carbon black at similar hardness level. Polym Test 2009;28(3):270–6.

[89] Galimberti M, Martino M, Guenzi M, Leonardi G, Citterio A. Thermal stability of ammonium salts as compatibilizers in polymer/layered silicate nanocomposites. e-Polymers 2009;1–14 056.

[90] Yunkong P, Kueseng P, Wirasate S, Huynh C, Rattanasom N. Cut growth and abrasion behavior, and morphology of natural rubber filled with MWCNT and MWCNT/carbon black. Polym Test 2015;41:172–83.

[91] Peddini SK, Bosnyak CP, Henderson NM, Ellison CJ, Paul DR. Nanocomposites from styrene -butadiene rubber (SBR) and multiwall carbon nanotubes (MWCNT) part 2: mechanical properties. Polymer (Guildf) 2015;56:443–51.

[92] Thaptong P, Sirisinha C, Thepsuwan U, Sae-Oui P. Properties of natural rubber reinforced by carbon black-based hybrid fillers. Polym Plast Technol Eng 2014;53(8): 818–23.

[93] Galimberti M, Coombs M, Cipolletti V, Riccò T, Agnelli S, Pandini S. The role of nanofillers in promoting hybrid filler networking and synergism with carbon black in a hydrocarbon rubber. Kautschuk Gummi Kunstoffe 2013;7–8:31–6.

[94] Ryu S-R, Sung J-W, Lee D-J. Strain-induced crystallization and mechanical properties of NBR composites with carbon nanotube and carbon black. Rubber Chem Technol 2012;85:207–18.

[95] Ismail H, Ramly AF, Othman N. The effect of carbon black/multiwall carbon nanotube hybrid fillers on the properties of natural rubber nanocomposites. Polym Plast Technol Eng 2011;50(7):660–6.

[96] Lorenz H, Fritzsche J, Das A, Stöeckelhuber KW. Advanced elastomer nanocomposites based on CNT-hybrid filler systems. Compos Sci Technol 2009;69: 2135–43.

[97] Cataldo F, Ursini O, Angelini G. MWCNTs elastomer nanocomposite, Part 1: the addition of MWCNTs to a natural rubber-based carbon black-filled rubber compound. Fuller Nanotub Carbon Nanostruct 2009;17(1):38–54.

[98] Bokobza L. Mechanical, electrical and spectroscopic investigations of carbon nanotube-reinforced elastomers. Vib Spectrosc 2009;51:52–9.

[99] Bokobza L, Rahmani M, Belin C, Bruneel JL, El Bounia NE. Blends of carbon blacks and multiwall carbon nanotubes as reinforcing fillers for hydrocarbon rubbers. J Polym Sci Part B Polym Phys 2008;46:1939–51.

[100] Mauro M, Cipolletti V, Galimberti M, Longo P, Guerra G. Chemically reduced graphite oxide with improved shape anisotropy. J Phys Chem C 2012;116:24809–13.

[101] Li ZH, Zhang J, Chen SJ. Effects of carbon blacks with various structures on vulcanization and reinforcement of filled ethylene-propylene-diene rubber. Express Polym Lett 2008;2(10):695–704.

[102] Lopez-Manchado MA, Biagiotti J, Valentini L, Kenny JM. Dynamic mechanical and Raman spectroscopy studies on interaction between single-walled carbon nanotubes and natural rubber. J Appl Polym Sci 2004;92:3394–400.

[103] Sui G, Zhong W, Yang X, Zhao S. Processing and material characteristics of a carbon-nanotube-reinforced natural rubber. Macromol Mater Eng 2007;292(9): 1020–6.

[104] Shanmugharaj A, Bae J, Lee K, Noh W, Lee S, Ryu S. Physical and chemical characteristics of multiwalled carbon nanotubes functionalized with aminosilane and its influence on the properties of natural rubber composites. Compos Sci Technol 2007;67(9): 1813–22.

[105] Falco AD, Marzocca AJ, Corcuera MA, Eceiza A, Mondragon I, Rubiolo GH, et al. Accelerator adsorption onto carbon nanotubes surface affects the vulcanization process of styrene–butadiene rubber composites. J Appl Polym Sci 2009;11(5):2851–7.
[106] Treolar LRG. The physics of rubber elasticity. 3rd ed. Oxford: Clarendon Press; 1975.
[107] Mullins L. Softening of rubber by deformation. Rubber Chem Technol 1969;42(1): 339–62.
[108] Boonstra BB. Reinforcing fillers. In: Blow CM, Hepburn C, editors. *Rubber technology and man*ufacture. London: Butterworths; 1987.
[109] Liu P, Huang Y, Liu CX, Wang Y, Guo XH, Zhang YG, et al. Enhanced electrical conductivity and mechanical stability of flexible pressure-sensitive GNPs/CB/SR composites: synergistic effects of GNPs and CB. J Mater Res Soc 2015;30(22):3394–402.
[110] Yang G, Liao Z, Yang Z, Tang Z, Guo B. Effects of substitution for carbon black with graphene oxide or graphene on the morphology and performance of natural rubber/carbon black composites. J Appl Polym Sci 2015;132(15):41832.
[111] Zhang H, Wang C, Zhang Y. Preparation and properties of styrene-butadiene rubber nanocomposites blended with carbon black-graphene hybrid filler. J Appl Polym Sci 2015;132(3):41309.
[112] Malas A, Pal P, Giri S, Mandal A, Das KC. Synthesis and characterizations of modified expanded graphite/emulsion styrene butadiene rubber nanocomposites: mechanical, dynamic mechanical and morphological properties. Compos Part B 2014;58: 267–74.
[113] Wang LL, Zhang LQ, Tian M. Mechanical and tribological properties of acrylonitrile–butadiene rubber filled with graphite and carbon black. Mater Des 2012;39:450–7.
[114] Raza MA, Westwood A, Stirling C. Carbon black/graphite nanoplatelet/rubbery epoxy hybrid composites for thermal interface applications. J Mater Sci 2012;47:1059–70.
[115] Park SM, Lim YW, Kim CH, Kim DJ, Moon WJ, Kim JH, et al. Effect of carbon nanotubes with different lengths on mechanical and electrical properties of silica-filled styrene butadiene rubber compounds. J Ind Eng Chem 2013;19:712–19.
[116] Fritzsche J, Lorenz H, Klueppel M. CNT based elastomer-hybrid-nanocomposites with promising mechanical and electrical properties. Macromol Mater Eng 2009;294:551–60.
[117] Sternstein SS, Ramorino G, Jang B, Zhu A-J. Reinforcement and nonlinear viscoelasticity of polymer melts containing mixtures of nanofillers. Rubber Chem Technol 2005;78:258–70.
[118] Hu H, Zhao L, Liu J, Liu Y, Cheng J, Luo J, et al. Enhanced dispersion of carbon nanotube in silicone rubber assisted by graphene. Polymer (Guildf) 2012;53:3378–85.
[119] Rooj S, Das A, Stöckelhuber KW, Wießner S, Fischer D, Reuter U, et al. Expanded organoclay' assisted dispersion and simultaneous structural alterations of multiwall carbon nanotube (MWCNT) clusters in natural rubber. Compos Sci Technol 2015;107:36–43.
[120] Das A, Kasaliwal GR, Jurk R, Boldt R, Fischer D, Stöckelhuber KW, et al. Rubber composites based on graphene nanoplatelets, expanded graphite, carbon nanotubes and their combination: a comparative study. Compos Sci Technol 2012;72:1961–7.
[121] Ponnamma D, Sadasivuni KK, Strankowski M, Guob Q, Thomas S. Synergistic effect of multi walled carbon nanotubes and reduced graphene oxides in natural rubber for sensing application. Soft Matter 2013;9:10343–53.
[122] Chen S, Yu H, Ren W, Zhan Y. Thermal degradation behavior of hydrogenated nitrile-butadiene rubber (HNBR)/clay nanocomposite and HNBR/clay/carbon nanotubes nanocomposites. Thermochimica Acta 2009;491:103–8.

[123] Ivanoska-Dacikj A, Bogoeva-Gaceva G, Rooj S, Wießner S, Heinrich G. Fine tuning of the dynamic mechanical properties of natural rubber/carbon nanotube nanocomposites by organically modified montmorillonite: a first step in obtaining high-performance damping material suitable for seismic application. Appl Clay Sci 2015; 118:99–106.

[124] Osman MA, Ernst M, Meier BH, Suter UW. Structure and molecular dynamics of alkane monolayers self-assembled on mica platelets. J Phys Chem B 2002;106(3): 653–62.

Manufacturing and Structure of Rubber Nanocomposites

12

A. Kato[1], A. Tohsan[2], S. Kohjiya[3], T. Phakkeeree[4], P. Phinyocheep[5] and Y. Ikeda[4]

[1]NISSAN ARC, LTD, Yokosuka, Kanagawa, Japan, [2]King Mongkut's University of Technology North Bangkok, Bangsue, Bangkok, Thailand, [3]Kyoto University, Umezu, Ukyou-ku, Kyoto, Japan, [4]Kyoto Institute of Technology, Matsugasaki, Kyoto, Japan, [5]Mahidol University, Ratchthewee, Bangkok, Thailand

Nano has recently been a fashionable word, and nanotechnology is expected to be one of the key technologies of the 21st century [1,2]. However, the initial claim of molecular engineering in general [3,4] has faced critical comments [5,6]. In spite of the fact that lots of critical propositions associated with nanotechnology have been reported, however, so far the popularity of rubbery nanocomposite has escaped from any damage. The reason may be simple and clear: consider the historical fact that rubbery composites of natural rubber (NR) and a nanoparticle, e.g., carbon black (CB), have been establishing their presence among industrial soft materials since early in the 1st quarter of the 20th century. Note that Ford Model-T was marketed in 1908, and this series of Ford cars, which were equipped with the pneumatic tires manufactured from NR and CB, continued to be the most popular automobile until 1925 [7,8]. In other words, NR/CB nanocomposites were on the market well before that of the fiber-reinforced plastics (FRPs), and long before the time when the word "nanotechnology" was coined and came onstage.

Therefore, the NR/CB composites are reasonably proved to be the so-called "precursor of composites," or historically the first commercial product of nanotechnology. The history of the present human beings (*Homo sapiens*) is periodically divided in accordance with what kind of materials they have been mainly using: The Stone Age, the Bronze Age, and the Iron Age, which may possibly be followed by the Composite Age. Consequently, the NR composite might be destined to be the pioneer of the next historical period of us, in terms of materials.

Composites, where the dispersant of which are of nanometer size (e.g., CB in the abovementioned automobile tires) are called nanocomposites. They have, as mentioned, a long history, and yet they are reasonably expected to be one of the main players in the 21st century, due to the strong demand for automobile and aircraft tires in the Transportation Age [7,8]. Here, it is notable for the rubbery composites that the nanofiller usually brings about much reinforcement into rubber [9], while compounding of micron-meter sized fillers into rubber does not result in much reinforcement. The latter are usually called nonreinforcing fillers among rubber technologists. In regard to the significance of nanofillers, we can also mention a recent nonrubbery example among FRPs: FRP using carbon nanofiber as dispersant

Progress in Rubber Nanocomposites. DOI: http://dx.doi.org/10.1016/B978-0-08-100409-8.00012-7

is currently under active development for use as automobile and airplane bodies where traditionally metals have been used [10].

Taking these situations into account, we are focusing on the manufacturing and structural elucidation of the soft nanocomposites in this chapter. Manufacturing or the preparation method of rubbery nanocomposites has so far been dependent on the mechanical mixing method. As will be discussed in Section 12.1, we toned to achieve a certain morphology of nanofiller in the rubbery matrix in order to enjoy a good reinforcement. On the base of this consideration, we have tried to prepare rubber/nanofiller composites not by the mechanical mixing, but by a "soft" process.

In this trial, we have used two dispersants: one is particulate silica generated in situ [11–19]. Unlike CB which is a petrochemical product, silica is not of petroleum origin, thus being preferred from the sustainable development (SD) point of view. The rubber tires made of silica instead of CB are to become more and more popular in the near future. The other is lignin, a natural polymer and one of the main components of wood [20,21]. The lignin used in this study is waste from paper-pulping industries, and much effort has been paid to the effective utilization of the lignin waste [21–23]. As a promising reuse of the lignin waste, trials of using it as a filler component in rubbery composites have been carried out extensively [24–29]. When NR is used as the matrix, a lignin/NR biocomposite which is again a favorable material from the SD standpoint since the product is carbon neutral.

We have recently reported on the new soft process for manufacturing the lignin/NR biocomposites from NR latex: Lignin was dissolved in an alkaline water, and mixing of the lignin *aq.* solution with NR latex gave a liquid mixture of lignin and NR. The evaporation afforded a soft composite film, which was prepared without any solid-state mechanical mixing at all [30]. This liquid-state soft process for NR composites using NR latex was also found to be applicable to in situ silica/NR composite [31], which may give rise to a highly efficient manufacturing process for polymer composites in general when we start the preparation from rubber latex, suspension or emulsion.

The other topic of this chapter is a modern morphologic study: nanofiller/NR composites are very traditional, but the dispersion of the nanofillers in rubbery matrix is still under active discussion among rubber technologists all over the world due to the utmost importance of rubber reinforcement by nanofillers. A somewhat problematic situation associated with this issue is commented in Section 12.1. This comment is the starting line of the following study described in the second half of this chapter. We have successfully applied a three-dimensional (3D) structural elucidation by transmission electron microscopy (TEM) [32–34] to the dispersed particulate silica and CB in the rubbery matrix [11,35], and the reinforcement effect of CB is preliminary discussed on the basis of the 3D-TEM results [36,37](Electron tomography is also used interchangeably, instead of 3D-TEM.). These results are extensively discussed in the present chapter, especially paying attention to the pseudo network structure of the nanofillers in the 3D rubbery matrix as well as the traditional idea of bound rubber which has been studied as one of the factors for rubber reinforcement.

12.1 A comment on rubber reinforcement by the nanofiller

In this section, we first discuss the unique behavior of nanofiller in the rubbery matrix in general that would be closely related to the reinforcement effect in rubbery soft nanocomposites. The expected function of rubber composites is quite multifarious: Taking pneumatic tires of automobiles and aircrafts, e.g., the tire rubber has to display not only mechanical strength but also appropriate moduli up to a high elongation, resistance against cracking, grip (holding) of road surface via an appropriate friction and with a moderate frictional wear, skid (moving without rotation of the tire) resistance especially on wet surface, good heat resistance and low heat generation when the automobile is moving particularly at high speed, to name several. To simply express the complex and relentless requirements for the tire rubber, the one word may be chosen, "tenacity." In order to meet this requirement, NR itself might have failed to be the best choice for the automobile and aircraft tires without any nanofillers for pneumatic tires, as described in the introduction and in Chapter 12 in Ref. [7]. Fortunately, CB was found to be well-suited to NR, early in the 20th century, and the use of CB has continued and will be the case for a while.

It has been well recognized by rubber technologists that for better rubber reinforcement the diameter of fillers (usually assumed to be spherical) should be as small as possible. However, it is also known that nanometer (nm) size fillers, such as CBs of high abrasion furnace (HAF) and intermediate super abrasion furnace (ISAF) grades, are subjected to cluster formation leading to so-called "structure" in the rubbery matrix. These CBs are reinforcing fillers for rubber and they are designated as CB in this chapter. Note this structure is not a scientific term in chemistry and physics, and traditionally is used among rubber technologists for "clustering." The "structure" was known very early, and it has popularly been believed that to disperse each of the nanoparticles homogeneously in the rubber should be the most favorable dispersion for rubber reinforcement. In a sense, a counter comment on this more or less traditional belief is given here.

A schematic representation of nanofiller morphology is shown in Fig. 12.1 [35,36]. In Fig. 12.1A, one particulate nanofiller, e.g., CB, is displayed, the surface of which is encircled by a rubbery layer. Rubber in the layer has been well recognized among rubber chemists and technologists as "bound rubber" [9]. Actually, CB particles are associated during their industrial manufacturing process due to so-called filler-to-filler interaction (quantum mechanical dispersion force, i.e., the van der Waals force is considered here), and the resulting CB aggregate is encircled by the bound rubber during the mechanical mixing into rubbery matrix as shown in Fig. 12.1B. Aggregate formation is due to the filler's huge surface area based on the nm size. Hence, the interaction between the fillers is reasonably assumed to be larger than the filler-to-rubber interaction. At the same time, relatively higher filler-to-rubber interaction compared with nonreinforcing fillers such as calcium carbonate is an important characteristic of rubber-reinforcing nanofillers, and gives rise to the bound rubber formation. The aggregates of calcium carbonate, if any, are

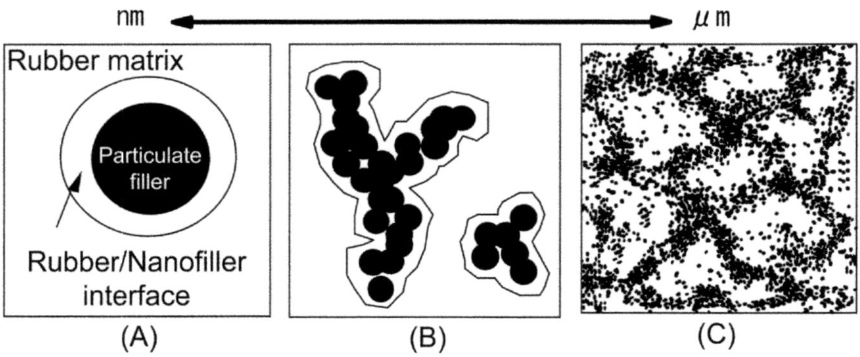

Figure 12.1 Schematic representation of nanofiller morphology in the rubbery matrix. (A) Particulate filler size from ca. 3 to 100 nm. (B) Aggregate size from ca. 20 to 1000 nm. (C) Agglomerate size from ca. 1000 nm to 1000 μm.
Source: From Fig. 1 in Kohjiya S, Kato A, Suda T, Shimanuki J, Ikeda Y. Visualisation of carbon black networks in rubbery matrix by skeletonisation of 3D-TEM image. Polymer 2006;47:3298, with permission.

broken during the mechanical mixing, while the CB aggregates do not segregate by the mechanical stress.

In the case of reinforcing nanofiller like CB and particulate silica, the aggregates are further subjected to clustering, which give rise to an agglomerate. The agglomerate, i.e., the cluster of aggregates, may often be of a pseudo network structure as suggested in Fig. 12.1C. Note that both aggregate and agglomerate are a so-called van der Waals cluster, and even the agglomerates are encircled by the bound rubber layers in the rubbery matrix. The bound rubber is also called "immobilized rubber," because it is very difficult to remove from the filler surface even by extraction with a good solvent of rubber, together with less active thermal micro-Brownian motion in the bound rubber layer than that of rubbery matrix. Some rubber technologists use the words "carbon gel" or "filler gel" for bound rubber. However, we do not use them, because we are suggesting a network structure (sometimes confused with "gel") of nanofiller in the rubbery matrix.

The morphology shown in Fig. 12.1 is well compatible with the recent experimental results reported by Koga et al. using the most sophisticated scattering techniques [38]. One legitimacy of the more or less network-like structures for the agglomerates of nanofiller has been given by the failure of the Guth–Gold equation [39,40] in explaining the reinforcing effect of nanofillers. The equation was originally derived for the presence of isolated sphere particles in liquid (Fig. 12.1A), but it has managed to explain only the effect of micron-sized fillers like calcium carbonate [41]. The equation was improved by introducing a shape factor (e.g., the aspect ratio) to accommodate the effect of nonsphere fillers or the aggregates [40]. However, it again failed to explain the reinforcement effect of nanofillers, which suggests the morphology shown in Fig. 12.1B is not valid to explain the typical nanofiller's effect. As already mentioned, it is also well known that the aggregates of nonreinforcing fillers like calcium carbonate are broken down during mechanical mixing with rubber. The filler-to-filler interaction in calcium carbonate is not strong enough to maintain its aggregated morphology against the stress due to mechanical mixing. Thus, the much high filler-to-filler interaction in nano-size sphere fillers is basically due to their large surface area. The lower rubber reinforcement of calcium carbonate than CB has so far been simply explained by the difference in their particle sizes: the larger the size, the lower the reinforcement.

Unfortunately, this belief has given rise to a more or less misunderstood hypothesis: We might enjoy better rubber reinforcement, once we do succeed in mixing CB particles not in the aggregated state but in the completely isolated state (as shown in Fig. 12.1A) against the strong filler-to-filler interaction. In other words, simply "The better the dispersion, the greater the reinforcement" has been a popular belief among some rubber technologists.

This hypothesis is wrong, is one of the conclusions of this chapter. Not the highest dispersion, i.e., complete isolation of each particle, but achieving a certain dispersion that may be favorable to form a network-like structure of nanofiller as shown in Fig. 12.1C is the best dispersion, in terms of rubber reinforcement. (Precisely speaking, we could not exclude the more or less organized other type of structure as an alternative to the networks.) Accordingly, it may be of much use to try mixing procedures other than the conventional mechanical methods for nanofiller mixing into rubbery matrix. Thus, structural determination of agglomerate of nanofiller in the 3D space may much help us to understand the best morphology of nanofiller for rubber reinforcement. These subjects are to be under consideration in this chapter.

We have so far discussed rubber reinforcement by nanofillers. Here, some complemental considerations are to be given to bound rubber, which is one of the factors of rubber reinforcement. Bound rubber has been known for a long time [9], and has been a central focus since then in rubber science and technology, in parallel with the so-called "structure" of CB for the rubber reinforcement. Kraus seemed to show a little reserved attitude to the significance of the bound rubber idea on reinforcement, though he did give excellent study on it in Chapter 4 of Ref. [9], and our comment described above seems to be on the "structure" side, and not much on the "bound rubber" side.

Among so many studies on bound rubber, one of the most concerns to our morphological interests is the thickness of bound rubber layer, and the difference between CB and particulate silica in terms of rubber reinforcement. At present, thicknesses between a few nm and several nm are considered to be a reasonable value, which will be mentioned in the later sections. On bound rubber, the majority of research has been on CB, and relatively few are devoted to silica. In the case of CB, the affinity between filler and rubber is explained by physical and chemical adsorption of the rubber molecules onto the surface of CB [9]. Macromolecularity (being of very high molecular weight) of rubber molecular chains helps the adsorption: Even when a segment is adsorbed on the surface, it may be detached by the thermal segmental motion of the chain, but when not one but two or more segments of the same rubber molecule are adsorbed, the molecule would be more or less stabilized on the surface to form a layer, which ultimately gives rise to "bound rubber." From this picture, it is reasonably assumed that the higher the molecular weight of the rubber molecule, the stronger is the adsorption, and physical adsorption by chance gradually leading to chemisorption.

The inferiority of silica to CB in terms of rubber reinforcement has been supposed to be due to its lower tendency for bound rubber formation. Silica carries silanol groups on its surface to result in much filler-to-filler interaction due to hydrogen bonding. Also, some of the reagents in sulfur vulcanization systems are adsorbed onto its surface to result in a retardation of vulcanization. To overcome this, some reagents such as ethylene glycol have been added in the sulfur cure formulations of silica compounds. This drawback of silica in physical properties and in vulcanization has been overcome by various means. Surface modification of the silica to be more hydrophobic has been done to modify the physical performances of silica as well as curing ones. Recently, the comparison of hydrophobic silica and hydrophilic silica has been reported [42]. Also, to increase the filler-to-rubber interaction by using a silane coupling agent is proved to be effective, though the cost of the reagent is more or less additional from the industrial view point. "Si-69" or bis(triethoxysilylpropyl) tetrasulfide is one of the coupling agents, and it has given rise to the increase of bound rubber by involving silica filler in sulfur vulcanization reaction [15,16]. Thus, silica may be an example to become equivalent to CB in rubber-reinforcing ability by using a coupling agent, without much damage to its tendency to aggregate into network-like structure. Also, it has to be added that in situ generation of silica by the sol–gel reaction (see the later sections) is the other factor to have promoted the utilization of silica in rubber science and technology [11–19,31].

The two factors, i.e., bound rubber and network-like agglomerate formations are important for rubber reinforcement. CB and particulate silica by some suitable contrivance, make an excellent reinforcer due to their ability to lead to these structures and morphology. From the above discussions, we may assume that the network-like agglomerate has something to do with bound rubber. Theoretical reasoning of the relationship [43,44] is to be developed further, but our experimental observation which is described in the later sections strongly suggests that bound rubber formation is also important for network formation of nanofillers in the rubbery matrix. In the network structure of CB, the surface-to-surface distance between two CB aggregates is estimated to be 3 nm (of bound rubber layer), and this separation allows electron hopping (possibly by quantum tunneling effect) to afford electroconductivity of the rubbery matrix with a CB loading of a higher amount than the threshold value of percolation. In the CB networks, the connection between the CB aggregates is by intermediating bound rubber layer of 3 nm thickness, as well as the direct contact of the CB particles inside the aggregates. Without the presence of the bound rubber, the agglomerate from nanofillers might have been of simply an irregular, randomized shape, not a network-like one, which is described in the later sections of this chapter.

At the same time, it has to be highly emphasized that the intermediating rubber layer may give some tenacity, flexibility, and deformability to the filler network-like structure in the rubbery matrix. This tenacity is of superb importance among the characters of the network-like structure formed by the nanofillers in the rubbery matrix: Rigid networks, such as those of thermoset resins, would not work for reinforcement. They simply damage the rubber elasticity.

Being a nanofiller is surely a precondition, but the traditional belief, i.e., a nanofiller at its isolated state (i.e., without network-like clustering), could be the most favorable reinforcing state, does not seem to be the case as the discussion here suggested.

12.2 Features of preparing the rubber composite by the soft processing method

The premise of excellent reinforcement of rubber composites, where very fine filler particles, usually nanofillers, are organized into a network-like morphology in the rubbery matrix, in accord with the discussions developed in Section 12.1. Thus, soft processing would be regarded as one of the possible techniques to achieve a good dispersion state of nanofiller. The words "soft processing method" are used for the procedures carried out through a liquid state (not involving a solid state), where rubber latex can conveniently be used as a source material.

NR latex is, therefore, focused on as the liquid state source of biomass NR, and is preferred in the soft processing from the SD point of view. Nowadays, several kinds of NR latex are available, e.g., high ammonia latex (HA), low ammonia latex, deproteinized latex, and compounded prevulcanized latex, etc. These concentrated NR latexes are marketed for producing various specific products. For example, deproteinized latex is generally used to manufacture medical gloves for surgical operations in order to minimize the allergic response to the patients and the medical staffs [45]. The protein presenting in the NR latex is supposed to be an allergen. HA is the most widely used latex due to its higher mechanical stability than the others, and is here employed in the study of the soft-processing technique. Additionally, the high ammonia content is found to promote an interaction between in situ silica and rubber particles in the NR latex, which is explained later in this chapter.

The manufacturing of soft composites by the soft process is generally started by adding liquid state precursor of filler or filler solution onto the latex. Before the addition, it is necessary to check the colloidal stability of the mixture to avoid coagulation. Then, the effect of the mixing of suitable cross-linking reagents is checked, too. Sometimes modification of the recipe for vulcanization may be needed in order to maintain the colloidal stability during the processing. After all reagents are mixed into the latex, the mixture of the liquid state is subjected to casting onto a clean flat plate maintained horizontally in order to form a film. Although a dipping technique into the latex mixture may be of use for a tube or some other shaped films, it may be recommended only when such a specific shape is necessary in the industrial manufacturing.

In situ silica/NR composite manufactured by the soft processing method did show a network-like or biphasic structured morphology, which is shown in Fig. 12.2 as observed by scanning probe microscopy (SPM) [46]. Two samples having different silica contents (10 phr and 17 phr) generated in NR matrix are Si10 and Si17, respectively. In the phase images, bright and dark phases are attributed to hard and soft phases which are silica and rubber phases, respectively. It can be seen that around rubber particles, particulate silica was generated to form bilayer structure composites of NR matrix and silica layers. This specific silica phase may be regarded as a good model of filler network, which is the origin of the reinforcement effect by this filler (See Section 12.1).

In Si10, it was observed that the in situ silica primary particles appeared to be mutually fused to form their higher ordered structures, i.e., the aggregates, and/or the

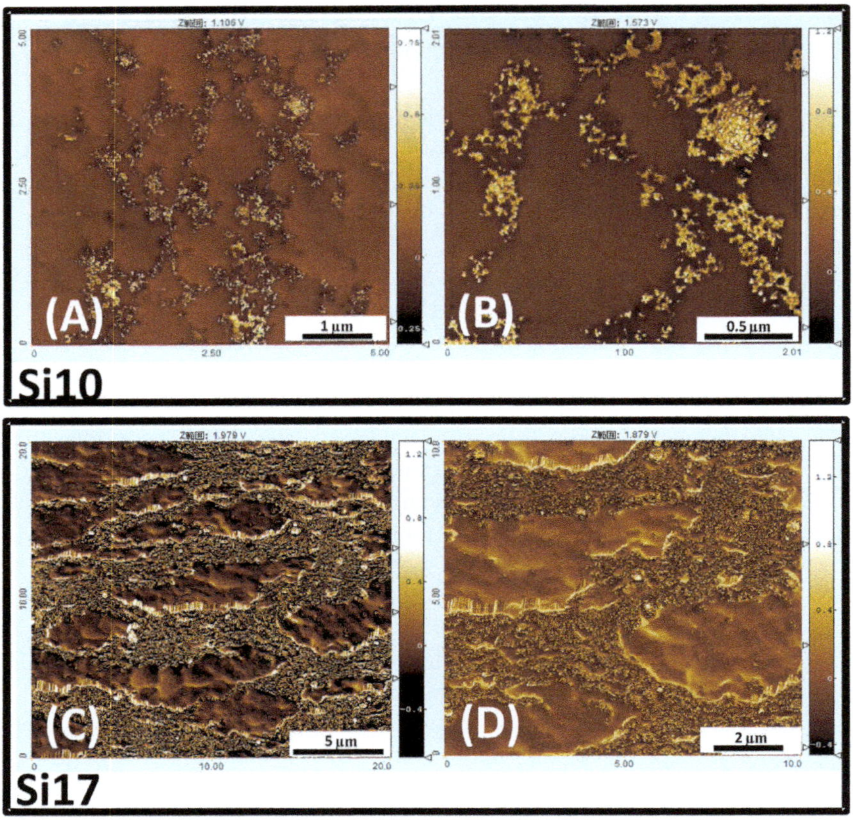

Figure 12.2 SPM photographs of in situ silica filled peroxide cross-linked nanocomposites. Upper and lower images show Samples Si10 and Si17, respectively. (A), (B), (C), and (D) were observed at the magnifications of 25 μm^2, 4 μm^2, 400 μm^2, and 100 μm^2, respectively. *Source*: From Fig. 3 in Tohsan A, Kishi R, Ikeda Y. A model filler network in nanocomposites prepared by in situ silica filling and peroxide cross-linking in natural rubber latex. Colloid Polym Sci 2015;293:2083, with permission.

agglomerates. They seemed to be further clustered into a network in the rubber matrix. Additionally, it should be pointed out that the rubber phases, which were covered by in situ silica network, were simply rubber, as we can see no silica particles inside the network. In Si17, the size of the filler network seems to be larger than that of Si10. When comparing Si10 and Si17, some different characteristics of filler network are reasonably detected: The size of the filler network in Si17 was larger and clearer than in Si10. Careful check of SPM images suggests a little hard interface of rubber particles inside the filler networks. From the size of the NR particle (about 1 μm), therefore, the rubber phases seem to consist of a cluster of a few rubber particles having originally existed in the latex. The clustering of rubber particles may be due to the destabilization of rubber particles during the sol–gel reaction. Further details of manufacturing of in situ silica composites will be explained later.

The soft processing technique is found to be very useful for preparing NR nanocomposites of a network-like state of the contained nanofiller. This technique can be of importance not only for academic study but also for industrial manufacturing of soft nanocomposites.

12.3 Manufacturing by soft process and conventional mixing method

12.3.1 Preparation of in situ silica/rubber composites

12.3.1.1 Sulfur cross-linked composites

In the last decades, several methods for preparing sulfur cross-linked in situ particulate silica/rubber composites were proposed [16,19]. Among them, the liquid state mixing technique, which brings about a unique morphological structure and mechanical properties, is introduced in this section. The liquid-state mixing of NR latex with tetraethoxysilane (TEOS) for the sol–gel reaction was conducted using *n*-butylamine as a catalyst at room temperature (r.t.), followed by the mixing with the sulfur cross-linking reagents. The films were formed *via* casting the liquid mixture. The obtained film was subjected to cross-linking reaction at 70°C for 4 h, and dried at r.t. under a reduced pressure. Recipes and silica contents of the three samples are listed in Table 12.1.

Morphological features of the composites were observed by TEM as shown in Fig. 12.3. By increasing the silica contents, more aggregates and/or agglomerates of

Table 12.1 Compositions of cross-linked and uncross-linked composite films in phr[a]

Sample code	Content of in situ silica	20% w/v K-oleate[b]	10% w/v KOH[c]	50% w/v Sulfur	50% w/v ZDEC[d]	50% w/v ZnO[e]
NR20	–	–	–	–	–	–
NR20-Si10	10	–	–	–	–	–
NR20-Si19	19	–	–	–	–	–
NR20-Si35	35	–	–	–	–	–
NR-S	–	0.4	0.8	1.5	1.8	1.0
NR20-Si10-S	10	0.4	0.8	1.5	1.8	1.0
NR20-Si19-S	19	0.4	0.8	1.5	1.8	1.0
NR20-Si35-S	35	0.4	0.8	1.5	1.8	1.0

[a]Parts per one hundred rubber by weight.
[b]Potassium oleate.
[c]Potassium hydroxide.
[d]Zinc diethyl dithiocarbamate.
[e]Zinc oxide.
Source: From Table 1 in Tohsan A, Phinyocheep P, Kittipoom S, Pattanasiriwisawa W, Ikeda Y. Novel biphasic structured composite prepared by in situ silica filling in natural rubber latex. Polym Adv Technol 2012;23: 1335, with permission.

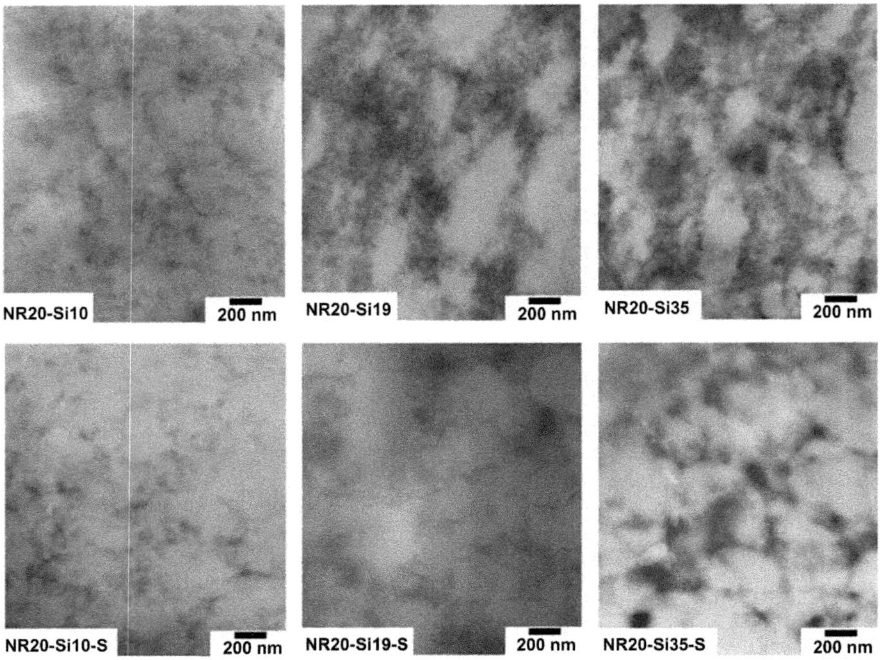

Figure 12.3 TEM photographs of the uncross-linked and cross-linked in situ silica filled composites. "S" in the end of sample code means the vulcanizate.
Source: From Fig. 2 in Tohsan A, Phinyocheep P, Kittipoom S, Pattanasiriwisawa W, Ikeda Y. Novel biphasic structured composite prepared by in situ silica filling in natural rubber latex. Polym Adv Technol 2012;23:1335, with permission.

the silica were observed. It is worth noting that the formation of in situ silica network was observed in all samples. Even in the case of 10 phr in situ silica filled NR latex, the filler network was also detected. In general, the formation of filler network is observed only when a high content of filler was used. Therefore, this preparation technique was found to be very effective to produce the silica filler network structure even when the content of filler was low.

The TEM observation did not clearly give us a definite size of in situ silica. But, the generated silica was found to be dispersed around the deformed spherical-shape rubber phases, and little silica was observed inside the rubber phases (the bright phases in the TEM photographs) in all samples. In addition, the size of rubber phases was found to be of a few hundred nanometers in size. These results suggest that the sol–gel reaction occurred outside the rubber particles in NR latex. In other words, the locally dispersed (around the rubber particle surfaces) silica filling was achieved by using NR latex, which is a new morphology by the soft filling method.

In this study, the sulfur cross-linking reaction was carried out after the preparation of in situ silica filled solid film by evaporation of the solvent followed by the cross-linking reaction with sulfur. Thus, it would be interesting to clarify how the cross-linking reaction progresses during the film-forming process. At first, X-ray

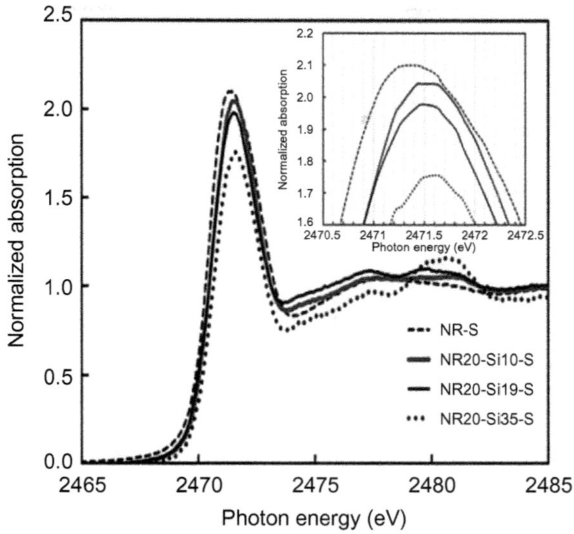

Figure 12.4 Sulfur K-shell XANES spectra of sulfur cross-linked composites and sulfur cross-linked NR.
Source: From Fig. 3 in Tohsan A, Phinyocheep P, Kittipoom S, Pattanasiriwisawa W, Ikeda Y. Novel biphasic structured composite prepared by in situ silica filling in natural rubber latex. Polym Adv Technol 2012;23:1335, with permission.

absorption near-edge structure (XANES) spectroscopy was carried out in order to investigate a characteristic of sulfidic linkage formed in the composite. The XANES results were identified that polysulfidic linkages ($C{-}S_x{-}C$, $x \geq 2$) are the main contribution in the in situ silica/NR latex composite [31]. In addition, we also found that the portion of shorter polysulfidic linkage seemed to be increased (that of longer ones decreased) by increasing the contents of in situ silica filling in NR latex as shown in Fig. 12.4. The change of sulfidic linkage may be ascribable to the absorption of sulfur on the silica particles: "The higher the silica content, the higher the absorption of sulfur on silica" has resulted in the decrease of sulfur in the cross-linking reaction, followed by the decrease of sulfur in the polysulfidic linkages in the cross-linked composites.

According to the evidence obtained from TEM photographs and XANES analysis, mechanism of the film-forming process is proposed in this study as follows: Fig. 12.5 shows a speculated mechanism of biphasic-structured composite formation of silica generated in situ in NR latex which occurred after the sol–gel reaction. During the film formation, aggregation of in situ silica occurred as detected by TEM in Fig. 12.3. The formation of this morphological feature is supported by the reported phenomenon of gelation of silica particles [47,48]. In the sol–gel reaction, amines or ammonia of a very low concentration (10^{-6}–10^{-3} M) are reported to be a good catalyst for gelling silica colloids effectively. A strong base may also deprotonate the silanol groups to leave a negatively charged silica surface and quaternary ammonium cations in solution. The quaternary ammonium ions then behave like a "bridge," electrostatically linking the negatively charged vicinal sites on the silica

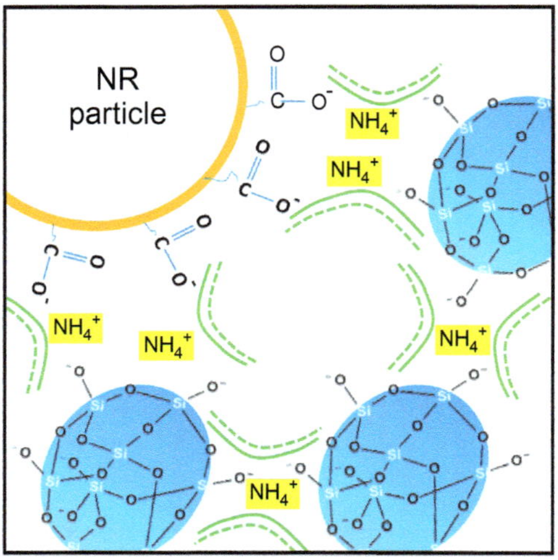

Figure 12.5 Speculated formation mechanism of biphasic structured composite of silica generated in situ in NR latex.
Source: From Fig. 2.4 in Tohsan A. Characterization of biphasic structured natural rubber/in situ silica composites (PhD Thesis at Kyoto Institute of Technology), 2013, with permission.

surface. In essence, negatively charged silica particles become connected by the bridge, thus aggregations of silica particles were formed.

In our study, *n*-butylamine (4.98×10^{-4} M, soluble in rubber phase) was used as a catalyst for the sol–gel reaction. Therefore, the gelling phenomenon is expected in our samples. In addition, NR latex preserved by NH_4OH was used. The ammonia-preserved NR particles are covered with negatively charged fatty acids. Not only in situ silica particles but also NR particles seemed to be aggregated during drying. Therefore, a Coulomb interaction between silica and rubber particles results in a specific morphology in the NR composites as schematically displayed in Fig. 12.5 [49]. When NR particles were dominantly aggregated, the elliptical shaped rubber phases were formed as detected in the TEM photographs.

The rubber particles got closer during drying, and consequently the densely-packed in situ silica became dispersed around the rubber particles as in Fig. 12.2. It was however noted that the above consideration is valid mostly in liquid state, or just before the film is completely dried. Once the film is dried, there appears to be hydrogen-bonding between carboxyl or amine groups and silanol groups of silica in the composites. The speculated hydrogen-bonding formation is schematically displayed in Fig. 12.6 [49].

12.3.1.2 Peroxide cross-linked composites

Preparation conditions of the peroxide cross-linked composites are shown in Table 12.2. Commercial grade high ammonia NR latex was diluted to 20% dry rubber content (DRC). It was then mixed with tetraethylenepentamine (TEPA)

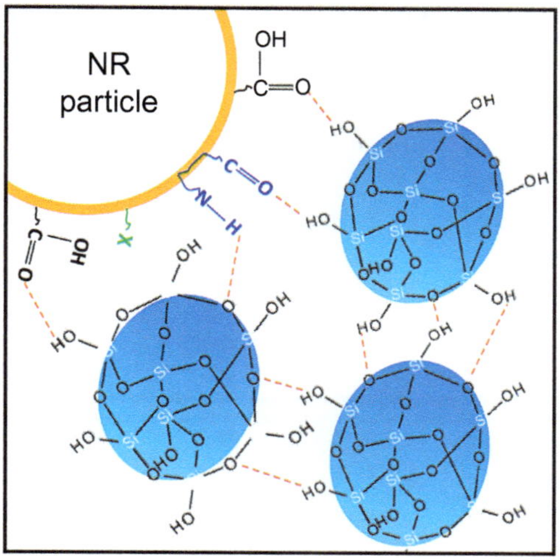

Figure 12.6 Speculated formation mechanism of hydrogen-bonding in the composites of silica generated in situ.
Source: From Fig. 2.6 in Tohsan A. Characterization of biphasic structured natural rubber/in situ silica composites (PhD Thesis at Kyoto Institute of Technology), 2013, with permission.

Table 12.2 Recipe for the preparation of peroxide cross-linked nanocomposites in phr[a]

Sample	PLE[b]	*t*-BHP[c]	TEPA[d]	TEOS[e]	TEVS[f]	In situ SiO_2
Si0	0.15	1	1	0	0	0
Si10	0.15	1	1	35	3.46	10
Si17	0.15	1	1	104	10.4	17

[a]Parts per one hundred rubber by weight.
[b]Polyoxyethylene lauryl ether.
[c]*t*-Butyl hydroperoxide.
[d]Tetraethylenepentamine.
[e]Tetraethoxysilane.
[f]Triethoxyvinylsilane.
Source: From Table 2 in Tohsan A, Kishi R, Ikeda Y. A model filler network in nanocomposites prepared by in situ silica filling and peroxide cross-linking in natural rubber latex. Colloid Polym Sci 2015;293:2083, with permission.

and *t*-butyl hydroperoxide (*t*-BHP), which was dispersed in polyoxyethylene lauryl ether (PLE) at r.t. to obtain a uniformly emulsified latex mixture. Thereafter, TEOS was added to the mixture and kept in an oven at 40°C for 5 h to progress the sol–gel reaction. After the reaction, a silane coupling agent, triethoxyvinylsilane (TEVS) was added at r.t. and the mixture was stirred at r.t. for 0.5 h. Peroxide cross-linked in situ silica filled NR nanocomposite film was formed by pouring the mixture onto a Teflon mold, followed by heating at 80°C for 0.5 h and drying at r.t. Subsequently, it was dried at 80°C under a reduced pressure to the constant weight [46].

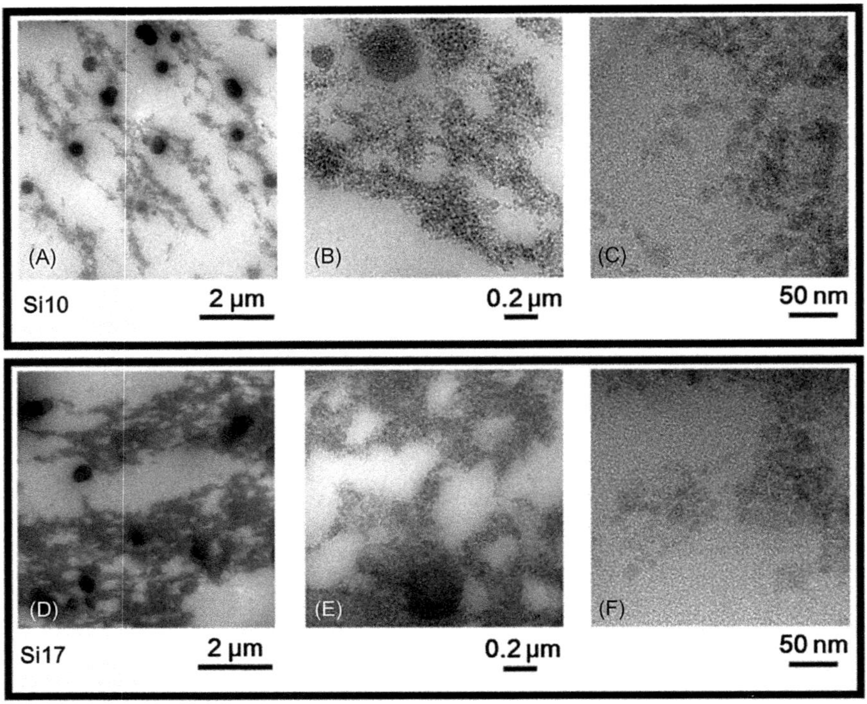

Figure 12.7 TEM photographs of in situ silica filled peroxide cross-linked nanocomposites. *Source*: From Fig. 2 in Tohsan A, Kishi R, Ikeda Y. A model filler network in nanocomposites prepared by in situ silica filling and peroxide cross-linking in natural rubber latex. Colloid Polym Sci 2015;293:2083, with permission.

TEM analysis was carried out for the composites as shown in Fig. 12.7. The dark parts in the micrographs show in situ silica. It can be seen that a filler network structure was recognized in both nanocomposites. A filler network of silica was detected more clearly in Sample Si17, as was similarly observed in our previous work [31]. It should be pointed out that the formation of filler network in the rubbery matrix is due to the processing technique by which NR particles were utilized as a template in the present study. In the high magnification images, i.e., Fig. 12.7(C) and (F), individual in situ silica particles may be seen, and the particle size was about 20 nm for both composites. Moreover, spherical aggregates with an average size of 500 nm were observed, too. The generation of aggregates was assisted by the presence of the surfactant, which was added when mixing the peroxide before the sol–gel reaction. Namely, micelles may have been formed at first from the surfactant in the NR latex, and then TEOS and the catalyst may have penetrated into the micelles before the sol–gel reaction of TEOS started. This mechanism is an opposite case to the phenomenon reported in our previous papers [50,51] and Miloskovska's paper [52], where the inverse micelles were supposed to be formed in TEOS.

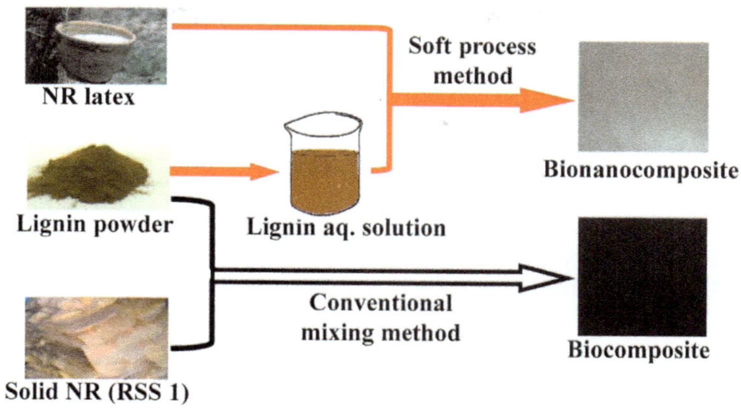

Figure 12.8 Schematic diagram for preparations of lignin/natural rubber biocomposites.

12.3.2 Preparation of lignin/natural rubber biocomposite by the soft process from latex

Lignin/natural rubber (lignin/NR) biocomposite was prepared by the soft process from the NR latex. By this soft processing, a vulcanizate was obtained from lignin and NR latex as well as by a milling method as shown in Fig. 12.8. Since lignin is hydrophobic in general, it is soluble in organic solvents but is practically insoluble in water. Thus, it was difficult to mix the lignin powder with the NR latex. Among lignin powders, however, a special grade of lignin with ionic groups is now available. Thus, sodium lignosulfonate (Na-LS) was subjected to the preparation of the biocomposite by the soft process [30]. In the case of Na-LS, a basic condition was necessary to produce a lignin aqueous solution. The lignin solution was prepared by stirring Na-LS in an aqueous NaOH solution of 0.1 mol/L. The lignin concentration in the solution was 30 w/v%. In order to prepare a cross-linked film, high ammonia NR latex (60% DRC) of 100 mL was mixed with sulfur cross-linking reagents at r.t. in advance, and followed by mixing with the Na-LS aqueous alkaline solution of 20 mL. Since the NR latex was preserved by NH_4OH, the basic condition of NR latex was preferable for mixing. A film was prepared by casting the liquid mixture on a glass plate at r.t. for several days. The obtained film was subjected to cross-linking at 70°C for 4 h, and was dried at r.t. under a reduced pressure. In the sample code, "L," "number," and "S" mean lignin, a content of lignin in phr, and sulfur cross-linking, respectively. As a reference sample, an unfilled sample (NR-L0-S-soft) was also prepared by the similar method. As expected, the two-phase structure in the lignin/NR biocomposite was clearly observed. Similarly with in situ silica filled NR nanocomposites, the lignin was located around rubber particles. Size of the rubber phase was almost the same as in the case of in situ silica filled NR nanocomposites shown in the TEM photographs of Fig. 12.3.

12.3.3 Conventional mixing method for nanofiller/rubber composite and lignin/natural rubber biocomposite

12.3.3.1 Nanofiller/rubber composites

As mentioned earlier, CB/NR composites were manufactured early in the 1st quarter of 20th century as tires for automobiles and a little later for aircraft. Since then, the manufacturing process of CB/NR composites has still been based on the trial-and-error methodology. In this section, therefore, recipes which are generally used to prepare the composites are given as follows (Table 12.3).

CB/NR composites were prepared by mixing the rubber with cross-linking reagents and various amounts of CB using an open two-roll mill to form compounds, which were subjected to sulfur vulcanization under pressure in a mold at 140°C for 15 min to obtain cross-linked NR samples in sheet form of 1mm thickness. The CB used was HS-HAF from Mitsubishi Chemical Co. (N399), which was dried at 120°C for 2 h before milling.

The recipes for the particulate-silica-filled cross-linked NR examined in this study are shown in Table 12.4. The only ingredient that is changed in this series was the silica loading. The ingredients are mixed with NR on the two-roll mill and

Table 12.3 Recipe for the preparation of sulfur cross-linked carbon black/NR nanocomposites in phr[a]

Sample	Rubber	Stearic acid	Active ZnO	CBS[b]	Sulfur	Carbon black
CB10	100	2.0	1.0	1.0	1.5	10
CB20	100	2.0	1.0	1.0	1.5	20
CB40	100	2.0	1.0	1.0	1.5	40
CB60	100	2.0	1.0	1.0	1.5	60
CB80	100	2.0	1.0	1.0	1.5	80

[a]Parts per one hundred rubber by weight.
[b]*N*-Cyclohexyl-2-benzothiazole sulfenamide.
Source: From Table 17.1 in Kato A, Ikeda Y, Kohjiya S. Carbon black-filled natural rubber composites: physical chemistry and reinforcing mechanism. In: Polymer composites, vol. 1, Thomas S. et al. editors. Wiley-VCH: Weinheim; 2012, Chapter 17, with permission.

Table 12.4 Recipe for the preparation of peroxide cross-linked particulate silica/NR composites in phr[a]

Sample	Rubber	DCP[b]	Silica (VN-3)
VN20	100	1.0	20
VN40	100	1.0	40

[a]Parts per one hundred rubber by weight.
[b]Dicumyl peroxide.
Source: From Table 2 in Tohsan A, Kishi R, Ikeda Y. A model filler network in nanocomposites prepared by in situ silica filling and peroxide cross-linking in natural rubber latex. Colloid Polym Sci 2015;293:2083, with permission.

cross-linked in a mold at 155°C under the pressure of 100–120 kg/cm^2 for 20–30 min to produce cross-linked NR sheets of approximately 1 mm in thickness. The silica that we used was manufactured by a precipitation method (supplied from Nippon Silica Co., Nipsil VN-3), whose specific surface area was around 200 m^2/g.

12.3.3.2 Lignin/natural rubber biocomposite

A lignin/NR biocomposite is prepared by the conventional mixing method for solid NR and lignin powder (Na-LSL) on a two-roll mill at r.t. The recipe in weight is as follows: NR (RSS no. 1) 100, ZnO 1, stearic acid 2, *N*-Cyclohexyl-2-benzothiazole sulfenamide 1, sulfur 1.5, lignin 0 or 10. Each rubber compound was molded into a sheet of 1 mm thickness by heat pressing at 140°C for 12 min, which was the optimal cure time determined by the cure measurement.

12.4 Comparison of mechanical properties of in situ silica and lignin composites from the two methods

In the previous sections, unique morphological features of the nanocomposites which were prepared by soft-processing technique were presented. In this section, attention is paid to their mechanical properties, in comparison with the conventionally-processed nanocomposites.

12.4.1 Mechanical properties of particulate silica composites

12.4.1.1 Tensile properties

Firstly, tensile stress–strain curves of peroxide cross-linked in situ silica/NR nanocomposites (samples in Table 12.2) are focused on [46]. For comparison, the conventionally processed peroxide cross-linked silica/NR composites (samples in Table 12.4) were used and the results are plotted as shown in Fig. 12.9. On comparing with the unfilled sample which was prepared by soft-processing technique, a reinforcement by the in situ silica is clearly recognized in all the nanocomposites. Among them, Sample VN40, which contained the highest amount of silica, exhibited the highest tensile stress for a stretching ratio higher than two. Meanwhile, compared with the others, Sample Si10, which contained 10 phr of in situ silica, showed the lowest tensile stress.

However, as shown in the inset of Fig. 12.9, the stress around the stretching ratio of two was a little higher in Si17 than in VN40, and the same relationship was observed between Si10 and VN20. In general, stress in the initial stage of stretching is known to be related to the presence of filler aggregates and/or agglomerates (most probably, the network-like structure of nanofiller; see Section 12.1) in the rubbery matrix. Therefore, the observed characteristics of the tensile properties of Si10 and Si17 were reasonably compatible with the formation of a nanofiller network structure in the nanocomposites, as suggested in Figs. 12.1–12.3 and 12.7.

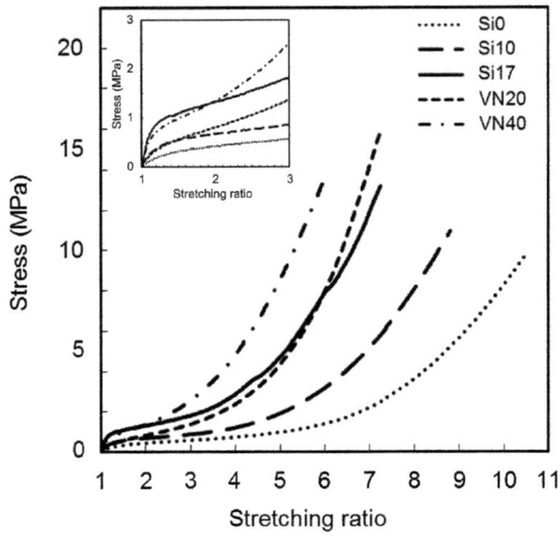

Figure 12.9 Tensile stress–strain curves of peroxide cross-linked nanocomposites and unfilled film. The inset displays the tensile curves at lower stretching ratio.
Source: From Fig. 4 in Tohsan A, Kishi R, Ikeda Y. A model filler network in nanocomposites prepared by in situ silica filling and peroxide cross-linking in natural rubber latex. Colloid Polym Sci 2015;293:2083, with permission.

Moreover, the relationship between the morphological feature and mechanical properties would be of more interest, when strain-induced crystallization (SIC) of NR is taken into account. Recently, we have investigated SIC of the composites by conducting simultaneous tensile and wide-angle X-ray diffraction measurements at SPring-8. It was found that the biphasic structured in situ silica filled composites exhibited unique stepwise SIC behavior [31], which was the first report of this phenomenon as far as we know. For example, the SIC behavior of the cross-linked composite is shown in Fig. 12.10. This unique SIC may be ascribable to the effect of silica filler networks in NR. Up to now, it has been difficult to prepare a model composite composed of a filler network structure in a rubbery matrix. Our samples prepared by in situ silica filling in NR latex, thus may be regarded one of the best model samples to be studied on the reinforcement effect of filler networks [31].

12.4.1.2 Dynamic mechanical properties

Fig. 12.11(A)–(C) shows the master curves of the dynamic mechanical properties, i.e., dynamic storage and loss moduli, and loss tangent of the peroxide cross-linked nanocomposites. They have been constructed using horizontal shifting from the temperature–frequency dependence of the storage modulus (E′), loss modulus (E″), and tan δ value of the samples in Tables 12.2 and 12.4. Effect of difference in the manufacturing methods on the E′ was detected very clearly by comparing Si17 and VN20 whose contents of silica are comparable. It was also clearly seen that E′ in the plateau region of Si17 is much higher than that of VN20. These results

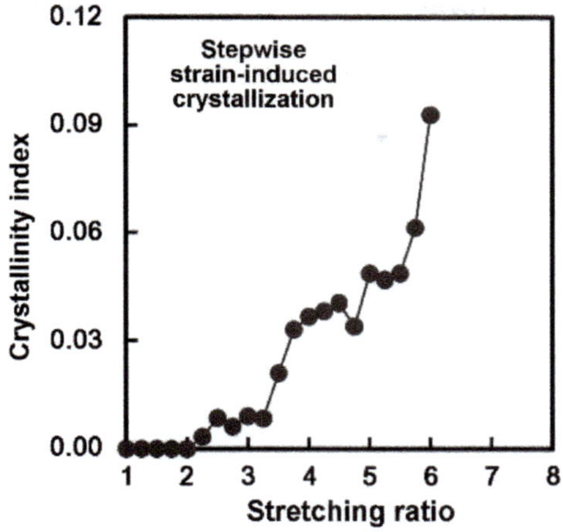

Figure 12.10 Stepwise SIC of cross-linked in situ silica filled composite prepared from NR latex by soft processing.
Source: From Fig. 6.18 in Tohsan A, Ikeda Y. Generating particulate silica fillers in situ to improve the mechanical properties of natural rubber. In: Chemistry, manufacture and applications of natural rubber. Kohjiya S, Ikeda Y. editors. Woodhead/Elsevier: Cambridge; 2014, Chapter 6., with permission.

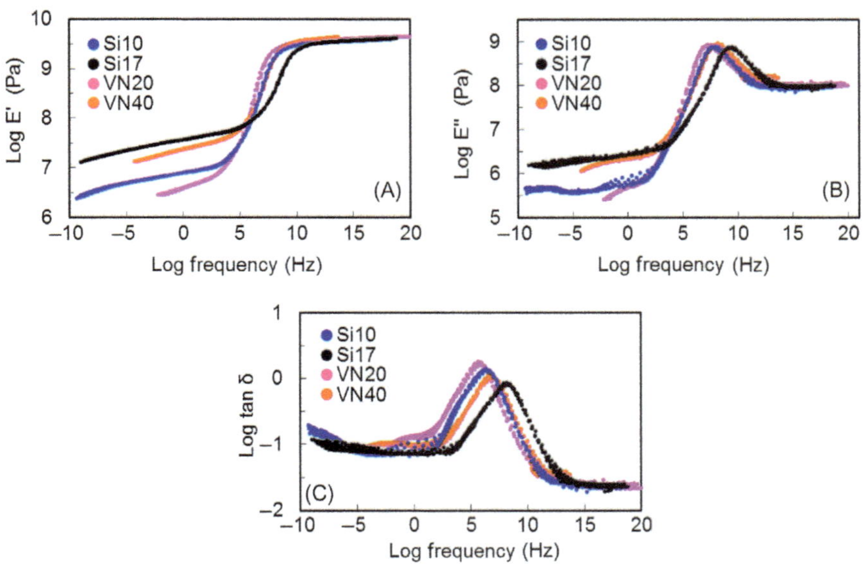

Figure 12.11 Temperature-frequency dispersion of (A) Storage modulus (E′). (B) Loss modulus (E″) and (C) tan δ of the nanocomposites.
Source: From Fig. 5 in Tohsan A, Kishi R, Ikeda Y. A model filler network in nanocomposites prepared by in situ silica filling and peroxide cross-linking in natural rubber latex. Colloid Polym Sci 2015;293:2083, with permission.

suggested that the filler network of in situ silica brought about rigidness to the rubbery matrix, which hindered the Brownian thermal movement of rubber chains upon deformation.

By careful examination of E′ at log frequencies <4 Hz, it was found that the rate of decrease in E′ was slower in the in situ silica filled samples than in the VN-3 filled samples. In general, a decrease in E′ in the low frequency region is ascribable to a slippage of untrapped entanglements in the network structure and/or to a molecular slippage associated with the disruption of pseudo network in the case of filler networks [53]. On the other hand, in a system that has a higher degree of cross-linking, any disruption of the entanglements would be more difficult, thereby extending the plateau region of E′. The in situ silica network is presumed to provide an increase in stiffness to the nanocomposites compared with the conventional silica. As a result, the decrease in E′ was slower and the plateau region was extended in the in situ silica filled samples, Si10 and Si17, compared with the conventional VN20 and VN40.

This behavior becomes pronounce when we consider the variation of E″ at log frequencies <4 Hz, as shown in Fig. 12.11(B), where the similar tendency of E′ was also observed. In general, a lower degree of cross-linking results in a higher loss modulus because the possibility of molecular slippages is higher in samples with a lower degree of cross-linking [53]. Subsequently, in the case of filled samples, the loss modulus has been reported to increase with increasing filler content, because of the high possibility of disruption of the filler networks under dynamic deformation [54]. As a result, the effect of increasing filler content on E″ can be seen very clearly in the plateau region: it was found in this study that E″ increased when the amounts of silica were higher, i.e., the E″ of Si17 and VN40 was higher than Si10 and VN20, respectively.

In terms of E″, by comparing Si17 and VN20, it is seen that E″ in the plateau region of VN20 was lower than that of Si17. It is worth noting that the lower E″ value of the VN-3 nanocomposites was attributable to the higher filler-to-rubber interaction than the in situ silica nanocomposites. Higher filler-to-rubber interaction in the VN-3 samples was predicted from their dispersive morphology, as previously reported [35], compared with the in situ silica networks observed in this study. Therefore, the higher filler-to-rubber interaction in the VN-3 samples may function like cross-linking points do, and restrict the micromovement of rubber molecules.

On the height of the tan δ peak (tan δ_{max}) as shown in Fig. 12.11(C), it was observed that tan δ_{max} decreased with increasing the silica content, i.e., VN40 and Si17 had lower tan δ_{max} than VN20 and Si10, respectively. The tan δ peak has been reported to be influenced by several factors, e.g., cross-linking degree, the fraction of filler and rubber, filler–rubber interaction, and the mobility of the rubber matrix [53]. Therefore, the lower tan δ_{max} of the higher silica filled samples can be explained by the lower mobility of the rubber matrix because of the presence of higher filler contents. Next, when comparing the two filling systems, it was also observed that Si17 had a lower tan δ_{max} than VN20. Similarly to the previous discussion on the effect of filler content on tan δ_{max}, it can be suggested that the lower tan δ_{max} of the in situ silica filled samples was explained by the lower mobility of the rubber matrix because of the presence of a clearer filler network structure compared with the conventional silica samples.

The tan δ peaks of nanocomposites as shown in Fig. 12.11(C) are further considered. It was found that tan δ peaks have shifted to higher frequencies with the increase of silica, i.e., tan δ peaks of Si17 and VN40 were at the higher frequencies than Si10 and VN20, respectively. In the case of VN-3 samples, where precipitated silica was used, it is well known that acidity due to the presence of silanol groups has a significant effect on peroxide cross-linking [55]. As a result, when the higher VN-3 content was used as in VN40, the lower cross-linking degree was expected as suggested by the shift of tan δ peaks to the higher frequencies than that of VN20. On the other hand, in the case of in situ silica samples, the lower cross-linking degree was suggested to be due to the cross-linking activator, i.e., TEPA might work partially as a catalyst for sol–gel reaction before the cross-linking reaction was preceded. Especially, in Si17, which had higher content of TEOS, a larger amount of TEPA was supposed to be used during the sol–gel reaction comparing to Si10. As a result, the cross-linking reaction of the in situ silica filled samples may have been disturbed and it seemed to be larger than that of the effect of acidity in the VN-3 silica. Therefore, when comparing between the two systems, the in situ silica filled nanocomposites seemed to have the lower cross-linking degree comparing with conventional silica filled nanocomposites as suggested by the shift of tan δ to the lower frequencies.

12.4.2 Mechanical properties of lignin biocomposites

The lignin/NR biocomposites prepared by the liquid-state mixing (soft) method and by the conventional mixing method, the sample codes of which are NR-L10-S-soft and NR-L10-S-mill, respectively, were subjected to the tensile measurement at r.t. Stress–strain curves of the biocomposites are illustrated in Fig. 12.12(A) with those of unfilled samples (NR-L0-S-soft and NR-L0-S-mill). It was clearly detected that a presence of lignin enhanced the increase of tensile properties in both cases: the lignin-filled NR composites (NR-L10-S-soft and NR-L10-S-mill) exhibited higher tensile stresses than lignin-unfilled NR composites (NR-L0-S-soft and NR-L0-S-mill), respectively [56]. It is worth noting that the tensile properties of NR-L10-S-soft were much higher than those of NR-L10-S-mill. This result suggests that the specific morphology of NR-L10-S-soft as shown in Fig. 12.12(B) may be a factor to increase the stress up to the high strain, which was in good agreement with that detected in the in situ silica filled nanocomposites, as described in the Subsection 12.4.1. The filler network in NR-L10-S-soft much improved the mechanical properties, even when the filler content was only 10 phr. The role of filler network was observed in the lignin system, i.e., the organic filler system, even though a small amount of filler has been often known to decrease the tensile stress. Therefore, these results may be supportive for the reinforcement effect of filler network for rubbers, which has been more or less recognized for a long time but has been difficult to confirm [56].

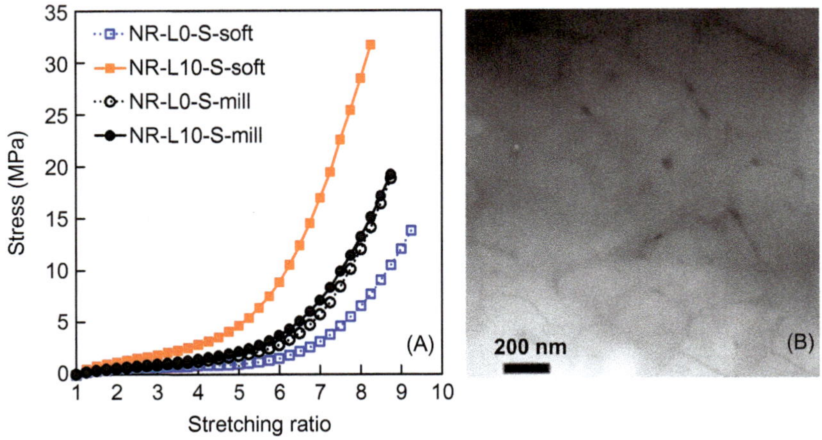

Figure 12.12 (A) Tensile stress–strain curves of the lignin/NR biocomposites prepared by the liquid-state mixing method (NR-L0-S-soft and NR-L10-S-soft) and the conventional mixing method (NR-L0-S-mill and NR-L10-S-mill) and (B) TEM photograph of NR-L10-S-soft. *Source*: From Phakkeeree T, Ikeda Y, Yokohama H, Phinyocheep P, Kitano R, Kato A. Network-like structure of lignin in natural rubber matrix to form high performance elastomeric bio-composite. J Fiber Sci Technol 2016;72:160, with permission.

It is also interesting to observe the larger increase of stress up to the higher strain in the lignin doped biocomposite than in the in situ silica filled nanocomposite prepared by soft processing, when the sample with 10 phr fillers was compared. The softer filler network of the lignin phase than the silica network may have influenced the deformation behavior of the composite. More detailed discussions will be reported in the near future.

12.5 Visualization of nanofiller dispersion in the three-dimensional space

The three-dimensional transmission electron microscopy (3D-TEM, "electron tomography" is interchangeably used.) observation on rubber/nanofiller composites is explained here briefly. Readers who are interested in the theoretical background of this technique are recommended to refer to a book [32], and several review articles [33–37,57] are for more general readers. In conventional TEM observations, diffraction images aligned with the crystal axis have been obtained by tilting crystal specimens [58]. A report had attempted to use this specific technique for obtaining a 3D image from the conventional TEM images of graphitized CB at two tilting angles of $+45°$ and $-45°$ as well as the conventional 0° angle [59], but the obtained 3D image from only three slice images did not afford a sufficient resolution. However, once enough resolution was obtained, it is correctly noted that on the constructed 3D image, any cross-sections could, in principle, be obtained to afford all the 2D information, leading to complete structural information in the 3D space.

In our investigations described here, a Technai G2F20 TEM (FEI Co.) was used to measure unstained specimens by a bright-field imaging technique at an accelerating voltage of 200 kV. The specimens showed good contrast between the inorganic nanofillers and the organic rubbery matrix without electron staining: The brightness of the images is divided into 256 levels from white to black. The 256 means the 8th power of 2, in the binary system. In other words, the images were checked by mapping in 256 levels in terms of intensity. After the checking, the intensity level of 190–255 was assigned to inorganic particles. The transition was sharp enough, and it was not necessary to consider intermediate layers between inorganics and organics. This implies that electron beam did not directly recognize the difference between the bound rubber and the rubbery matrix. Although electron staining is usually an effective way to obtain much higher contrast [60–62], it invariably induces volumetric changes of the stained component, often followed by nanodeformation in the specimen [58]. So, the staining is to be avoided for 3D-TEM observation. Standard preparation method of TEM sample of soft materials is used: An ultramicrotome was used to cut a thin sample from the rubber frozen at the temperature of liquid nitrogen (ca. −198°C). The sample dimensions were ca. 500 nm long, 500 nm wide, and 100–200 nm thick. The thin samples were processed in a gold colloidal solution in order to align the position of each image using the colloidal gold particles on the surface as a fiducial marker. The spherical gold colloidal particles used were a diameter of 10 nm.

Here, the outline of the 3D-TEM process on rubbery samples is explained, using Fig. 12.13 [63,64]. The samples were tilted over a range of angles from −70° to +70°, and image data (tilted images) were continuously obtained in 2° increments as shown in Fig. 12.13. A total of 71 tilted images were automatically

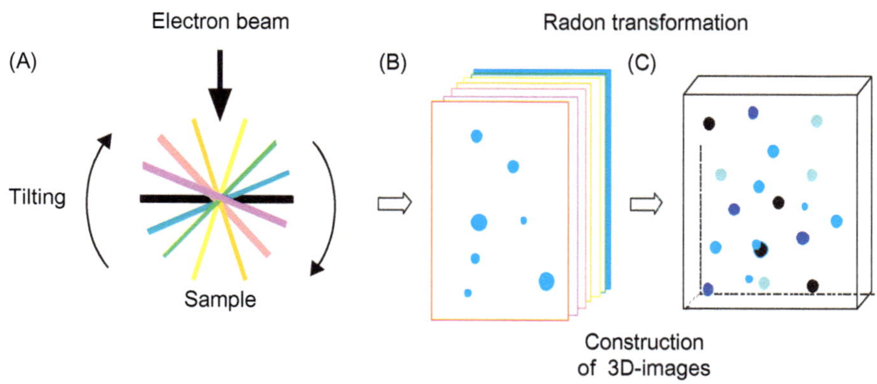

Figure 12.13 3D-TEM observation of filled rubber sample. (A) Sample size: ca. 600 × 600 × 200 nm. Tilting angle: −70° to +70° in 2° steps. (B) 2D-projections (sequential image slices). (C) Reconstructed 3D-image.
Source: From Fig. 2 in Kohjiya S, Kato A, Ikeda Y. Visualization of nanostructure of soft matter by 3D-TEM: nanoparticles in a natural rubber matrix. Prog Polym Sci 2008;33:979, with permission.

uploaded to the computer. The positions of the tilted images were aligned at that time using the colloidal particles. These tilted images were not simply 2D image slices, but rather 2D projected images of the mass-density distribution of the samples. Since tilting the specimen may cause changes of focus and the thickness, it is crucial to obtain a 3D image of high quality. The next step is computerized tomography [32], which is in principle the same as a medical X-ray CT scan. The sequence of tilted images is subject to reconstruction of a 3D image. The IMOD [65] and Amira [66] programs were utilized as the basic softwares.

An important pretreatment of rubber samples which were obtained by sulfur vulcanization has to be mentioned: Zinc atoms in the sample, which were compounded as an activator of zinc oxide for sulfur vulcanization to the rubber, usually together with stearic acid [67], scattered an electron beam to result in the obscure TEM images. Therefore, the residual zinc atoms in the sample should have been removed for the observation. Fortunately, the removal procedure was established, and reported in Ref. [68]. Since sulfur cross-linked rubbers, which usually contain zinc atom as one component after the vulcanization, this pretreatment is mandatory before subjecting them to the 3D-TEM observation.

12.6 Dispersion of silica and optical transparency of silica filled rubber

12.6.1 3D-TEM images of in situ silica and conventional silica

The 3D-TEM (or the electron tomography) images of soft nanocomposites were first reported in 2004 on in situ silica and conventional silica filled natural rubber vulcanizates [11]. NR-mix-V is a conventional silica (Nipsil VN-3) filled one, and NR-in situ-V is in situ silica filled one. The preparation of in situ silica filled sample is described in Section 12.3.1. The compounding recipe for sulfur vulcanization is shown in Table 12.5. The ingredients were mixed on a two-roll mill, and hot-pressed at 150°C for 20 min. Diethylene glycol was not used in the in situ silica stock, which may be one of the features of in situ silica stock. All the vulcanizates were subjected to the pretreatment described in the last paragraph of Section 12.5, in order to remove residual zinc compounds in the vulcanizates. (For the details of this procedure, see Refs. [35,68].)

Fig. 12.14 shows the bright-field 3D-TEM images of silica filled samples in which the neighboring aggregates are shown in different colors to make the silica aggregate to see easily [11]. It is recognized in the images that a few or several silica particles are in contact with each other to form an aggregate, which is the case in both conventional and in situ silica. Thus, the aggregates (see Fig. 12.1B in Section 12.1) are observed clearly by 3D-TEM. In both the figures, only a limited number of the isolated primary silica particles were observed. From these images, we calculated various structural parameters in 3D-space. For example, the sizes of

Table 12.5 Silica filled NR samples[a] for 3D-TEM observation

Sample	MR-mix-V	NR-in situ-V
Ingredient (phr[b]):		
NR (RSS#1)	100	0
Stearic acid (ST)	1	1
ZnO	5	5
Sulfur	2	2
CZ-G[c]	1	1
Diethylene glycol (DG)	2	0
Commercial silica (VN-3[d])	33	0
NR with in situ silica	0	133

[a]Curing conditions: 150°C, 20 mins under pressure.
[b]Parts per one hundred rubber by weight.
[c]Cyclohexylbenzothiazyl-Sulfenamide.
[d]Nipsil VN-3 from Nippon Silica Ind. Co.
Source: From Table 1 in Ikeda Y, Kato A, Shimanuki J, Kohjiya S. Nano-structure observation of in situ silica in natural rubber matrix by three dimensional transmission electron microscopy. Macromol Rapid Commun 2004;25: 1186, with permission.

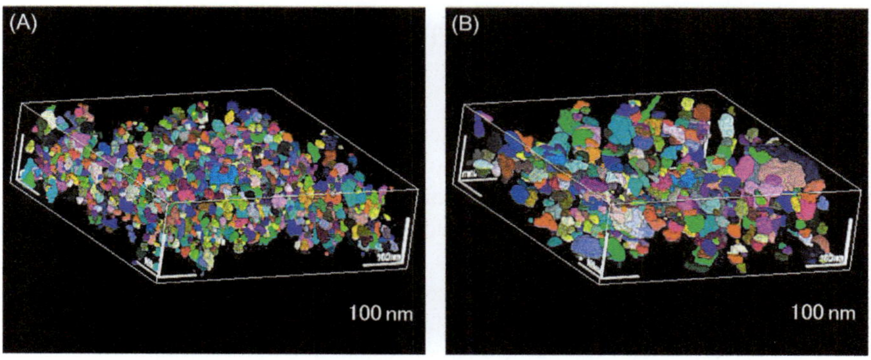

Figure 12.14 Colored 3D-TEM images of silica filled NR samples, where the neighboring particles (aggregates) are shown in different colors. (A) NR-mix-V. (B) NR-in situ-V. *Source*: From Fig. 1B in Ikeda Y, Kato A, Shimanuki J, Kohjiya S. Nano-structure observation of in situ silica in natural rubber matrix by three dimensional transmission electron microscopy. Macromol Rapid Commun 2004;25:1186, with permission.

the species, i.e., the primary particles and the aggregates in these images were measured, and the dispersion in terms of radius was reported in Ref. [11]. The 3D images provided useful structural parameters by the calculation. The further necessity is only the availability of a good software enabling the calculation of specific structural parameters from the 3D image.

12.6.2 Optical transparency and silica network structure

Generally, rubbers are not optically transparent, and only degradation by light has been minded by rubber chemists and technologists among rubber properties related to light. Actually, a development of antiphoto degradation reagents is an important issue for rubber products which are exposed to sunlight outside. So far, scientific reports on the optics of rubbers have been quite limited in number: refractive index (n_D^{20} ; sodium D-ray at 20°C) of silicone elastomers was reported to be changeable between 1.36 and 1.69 by introducing various substituents into the side group [69]. Many studies on the electro-optical properties have been devoted for liquid crystalline elastomers (again mainly silicone elastomers). One of the examples was reported in Ref. [70]. A few papers reported photochromic and photomechanical properties of elastomeric poly(oxyteteramethylene) containing viologen groups [71,72].

In terms of color of rubber products, there has been a problem to be solved: Most of the popular rubber products are black due to the predominant utilization of CB. For example, even if people want yellow-colored automobile tires to be equipped with their car, it is impossible to find them in the tire market. However, a year of 2015 was designated as "International Year of Light and Light-based Technologies (IYL2015)", and nowadays it is predicted that the yellow-colored tires may become available within a few or several years. Silica may play a role for coloring of rubber products. In fact, "Green Tires" were already produced by using silica as a reinforcing filler for tire rubber [73]. In accord with a social trend for sustainable development of our society, the use of silica for tires will be more accelerated and followed by the appearance of colored tires in the near future.

In this section, characteristics of the optical transparency of silica filled NR is shown, the results of which may be useful for the development of colored tires [51,74,75]. In this study, surface-modified hydrophobic silica was used as well as conventional silica, Nipsil VN-3, which is hydrophilic due to the presence of silanol groups on the surface. Peroxide cross-linked NR sheets of 1 mm thickness were prepared in which the loadings of hydrophilic and hydrophobic silica were varied in a range of 0–80 phr (see Table 12.6). The hydrophobic and hydrophilic silica filled cross-linked NRs are designated as RX# sample and VN# sample, respectively. Here, the # symbol indicates the amount of silica loading in phr (parts per one hundred rubber by weight). Fig. 12.15 shows images of the optical transparency of the RX- and VN-filled NR samples obtained by placing them on white section papers. The images of (A), (B), and (C) show that the rubber sheets were transparency, but the degree of transparency decreased a little with the increasing the amount of the hydrophobic silica filler. This indicates that the amount of loading of the hydrophobic silica filler had not much influence on the optical transparency of the NR composites. In contrast, the images of (D), (E), and (F) in Fig. 12.15 show that the visibility of graph paper behind the rubber decreased in an order of VN10 > VN80 > VN30. At the silica content of 30 phr, the section paper was invisible. When the silica loading was further increased to 80 phr, the pattern of the graph

Table 12.6 Compounding recipe for cross-linking of hydrophobic and hydrophilic silica loaded NR samples[a]

Sample	RX0	RX10	RX20	RX30	RX40	RX60	RX80
NR	100	100	100	100	100	100	100
DCP[b] (phr[c])	1	1	1	1	1	1	1
Silca RX[d] (phr)	0	10	20	30	40	60	80
Sample	**VN0**	**VN10**	**VN20**	**VN30**	**VN40**	**VN60**	**VN80**
NR	100	100	100	100	100	100	100
DCP[b] (phr[c])	1	1	1	1	1	1	1
Silica VN3[e] (phr)	0	10	20	30	40	60	80

[a]Cross-linking conditions: 30 min at 155°C under 100–150 kg/cm^2.
[b]Dicumyl peroxide.
[c]Parts per one hundred rubber by weight.
[d]AEROSIL RX200 (trimethyl silyl group treated silica, average radius = ca. 12 nm) from Evonik Degussa Japan Co., Ltd.
[e]Nipsil VN-3 (average primary diameter = ca. 16 nm) from Tosoh Silica Corporation.
Source: From Table 1 in Kato A, Ikeda Y. Nippon Gomu Kyokaishi 2014; 87: 351, with permission.

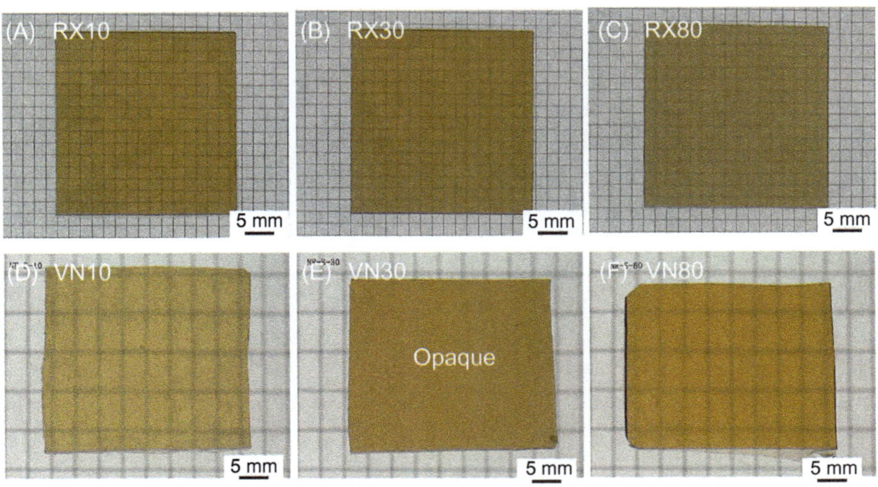

Figure 12.15 Optical transparency of hydrophobic (RX) and hydrophilic (VN) silica loaded NR samples.
Source: From Fig. 1 in Kato A, Ikeda Y. Nippon Gomu Kyokaishi 2014;87:351, with permission.

paper unexpectedly became visible again. It was reported, however, that the increase of amount of silica or glass fiber, which was compounded into transparent polymers, generally tends to reduce the optical transparency of the composite materials [76], which seems to be very acceptable. Thus, a reason for the observed

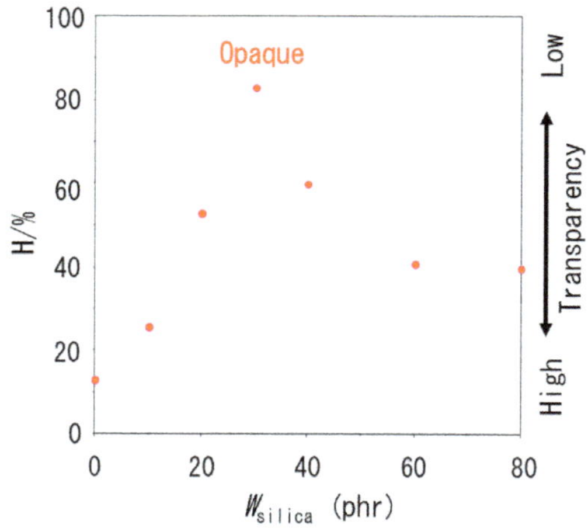

Figure 12.16 Dependence of haze (H) on silica loading (W_{silica}) of hydrophobic and hydrophilic silica filled NR samples.
Source: Modified from Fig. 4 in Kato A, Ikeda Y. Nippon Gomu Kyokaishi 2014;87:351, with permission.

maximum opacity at 30 phr hydrophilic silica content is discussed using a few optical properties relating to transparency.

At first, dependence of the haze (H) of silica filled NR samples on the silica loading was investigated, and results are shown in Fig. 12.16. The H of the hydrophobic silica filled NR samples did not show much dependence on the silica loading. Contrastingly and interestingly, however, the H significantly increased to the silica loading of 30 phr, and it declined steeply to a satulation at 60 phr. The maximum of H at the silica loading was 30 phr for the VN-3 silica filled sample, which was in good agreement with the result in Fig. 12.15, where VN30 was opaque. This optical transparency anomaly observed at 30 phr of hydrophilic silica filled sample was also confirmed by total transmittance obtained by ultraviolet/visible light measurements [51,75].

Fig. 12.17 shows a definition of the closest distance, designated as d_p, between the particles of two silica aggregates. It is defined as the closest distance between the surfaces of two silica aggregates along a straight line connecting their centers of gravity [51,74,75]. The dependency of d_p and the standard deviation of d_p on the silica loading shows that the both values tended to decrease as the loading of the silica was increased and saturated over 40 phr loading, both for the hydrophobic and hydrophilic silica loaded samples. This result indicates that the dispersion of the silica aggregates became more homogeneous as the silica loading was increased. It is especially notable that d_p of the hydrophobic silica filled samples was markedly larger than that of the conventional VN-3 loaded ones in the low loading region. These results suggest that it was easy for the silica aggregates to coalesce in hydrophilic silica probably owing to the effect of filler-to-filler interaction originating from the hydrogen bonding between the silanol groups and others on the silica

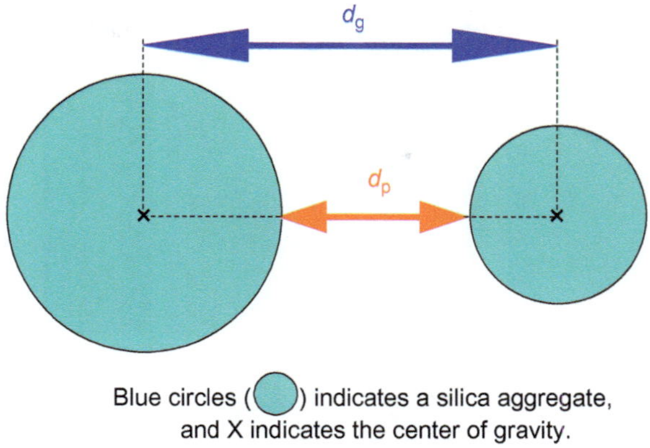

Figure 12.17 Definition of distances between silica aggregates (d_p and d_g) as 3D-structure parameters.
Source: From Fig. 3 in Kato A, Ikeda Y, Kasahara Y, Shimanuki J, Suda T, Hasegawa T, et al. Optical transparency and silica network structure in cross-linked natural rubber as revealed by spectroscopic and three-dimensional transmission electron microscopy techniques. J Opt Soc Am B 2008;25:1602, with permission.

surface. In the case of hydrophobic silica, it was rather difficult for the silica aggregates to coalesce in the low loading region owing to the effect of the filler-to-rubber interaction originating in hydrophobic trimethylsilyl groups on the surface. The group was introduced to make the surface hydrophobic. However, in the high loading region where d_p was shorter, silica aggregates of both types of silica further coalesced to form the agglomerate owing to the van der Waals force (dispersion force). Consequently, d_p is saturated at the same value of loading of 40 phr, and there is almost a constant d_p value of ca. 1.3 nm regardless of whether the surface functional groups were hydrophobic or hydrophilic. The value of ca. 1.3 nm indicated that there was a rubber layer, i.e., a bound rubber layer around the silica aggregates that directly prevented them from contacting each other. It is inferred that there is a network-like structure of silica aggregates, i.e., an agglomerate formed *via* this bound rubber.

In cross-linked NRs with a high CB filler content, d_p was already reported to saturate at ca. 3 nm [35–37,63], which implies that the value of d_p in cross-linked rubber having a high nanofiller content varies according to the type of filler and rubber involved. Also, it is notable that d_p may show the thickness of bound rubber, and silica's d_p was 1.3 nm and CB's was 3 nm. In terms of filler-to-rubber interaction, CB has been supposed to have higher interactions than silica, which are now quantified as 3 nm, compared with 1.3 nm for silica. The values are qualitatively in accordance with the knowledge that many rubber chemists and technologists have had from the huge experiences of the past decades.

The value of $d_p = 1.3$ nm was used here to visualize the structure of the silica aggregate network. Specifically, diagrams were obtained that connected the centers of gravity of nearest-neighbor silica aggregates at this distance. Fig. 12.18 shows the

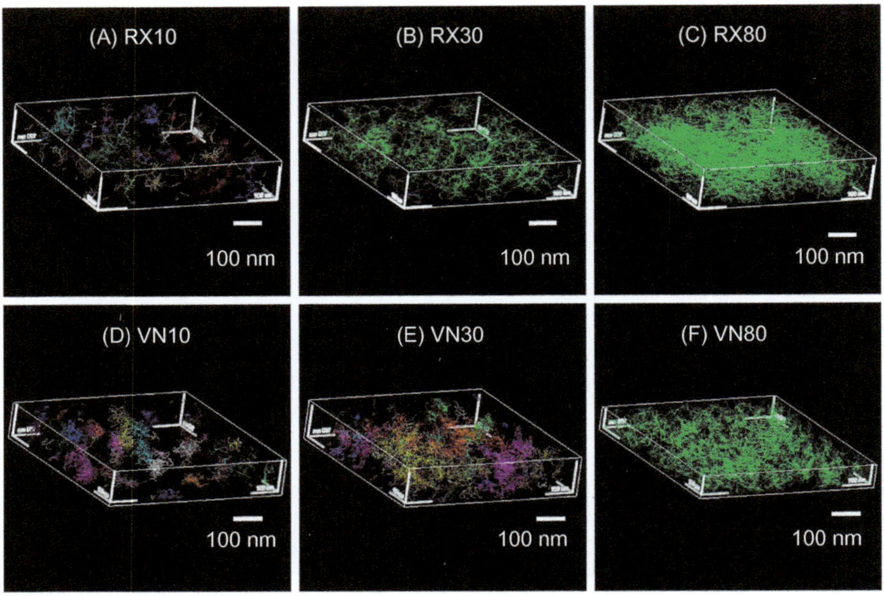

Figure 12.18 Network structure of hydrophobic and hydrophilic silica aggregates in cross-linked NR for silica filler loadings of 10, 30, and 80 phr.
Source: From Fig. 9 in Kato A, Ikeda Y. Nippon Gomu Kyokaishi 2014;87:351, with permission.

network structure of hydrophobic and hydrophilic silica aggregates in cross-linked NR samples for silica filler loadings of 10, 30, and 80 phr. Networks that were not connected are shown in different colors in the figure. For the hydrophobic silica, the networks were nearly all connected at loadings of 30 phr or higher, whereas for the hydrophilic silica, local unconnected networks are seen here and there even at a filler loading of 30 phr, and connections among all the aggregates were not completed until a loading of 80 phr. These results also suggest the effect of the strong filler-to-filler interaction in hydrophilic silica compared with hydrophobic silica. A more detailed observation for Fig. 12.18 reveals that the aggregate network structure consists of three kinds of chains, i.e., cross-linked chains (cross-linking points), branched chains (branching points), and isolated chains that were not connected to the network.

Fig. 12.19 is a schematic diagram of the silica aggregate network structure. The colored particle is a modeling of the silica aggregates and the arrows indicate the direction of the connections to the network. We have previously posited that the cross-linked, branched and isolated chains of the hydrophilic silica aggregate network cause light to scatter [51,74,75]. Based on a multiple scattering equation for a scattering medium using the Monte Carlo method as reported by Kandidov et al. [77], we found the product of the number density of the three types of chains and the square of their average

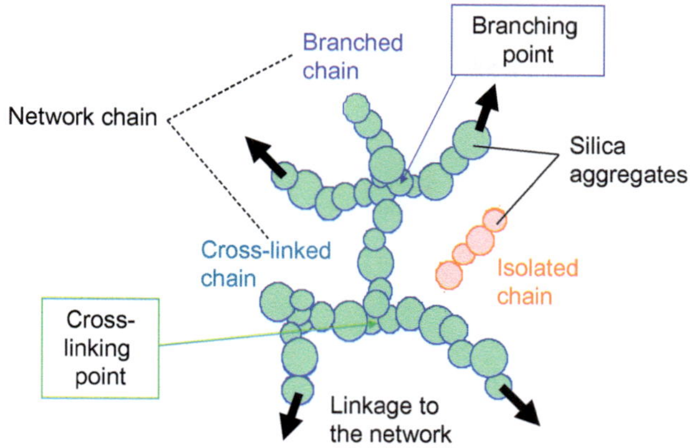

Figure 12.19 Schematic diagram of silica aggregate network structure.
Source: From Fig. 4 in Kato A, Ikeda Y, Kasahara Y, Shimanuki J, Suda T, Hasegawa T, et al. Optical transparency and silica network structure in cross-linked natural rubber as revealed by spectroscopic and three-dimensional transmission electron microscopy techniques. J Opt Soc Am B 2008;25:1602, with permission.

length and used the resultant values to investigate the relationship with haze. The equations are as follows:

$$H \sim \mu_s(=\pi K_p n_{iso} r_{iso}^2) \sim (\pi K_p/4) n_{iso} L_{iso}^2 \quad (12.1)$$

From this proportionality,

$$\begin{aligned} H &= C n_{iso} L_{iso}^2 + D(H) \\ L_{iso} &= 2 r_{iso} \\ C &= \pi K_p/4 \end{aligned} \quad (12.2)$$

where H denotes haze, μ_s the scattering coefficient (m^{-1}), π the circular constant, n_{iso} the number density of the isolated chains (m^{-3}), r_{iso} one-half of the mean length of the isolated chains (L_{iso}), and D (H) is an additional constant only to the cross-linked rubber. As a result, it was found that the multiple scattering of light by the isolated chains in the hydrophilic silica aggregate network caused optical transparency to decline. Product of the number density and square of the length ($n_{iso}L_{iso}^2$) of the isolated chains in the silica aggregate network depended on the hydrophobic and hydrophilic silica loading. For the hydrophilic silica, $n_{iso}L_{iso}^2$ decreased with increasing silica loading, whereas it showed its maximum value near a loading of 30 phr for the hydrophobic silica.

Fig. 12.20 shows the dependence of haze (H) on $n_{iso}L_{iso}^2$ for the hydrophobic and hydrophilic silica filled cross-linked NR. For the hydrophobic silica, H did not display any dependence on $n_{iso}L_{iso}^2$, whereas the hydrophilic silica showed a

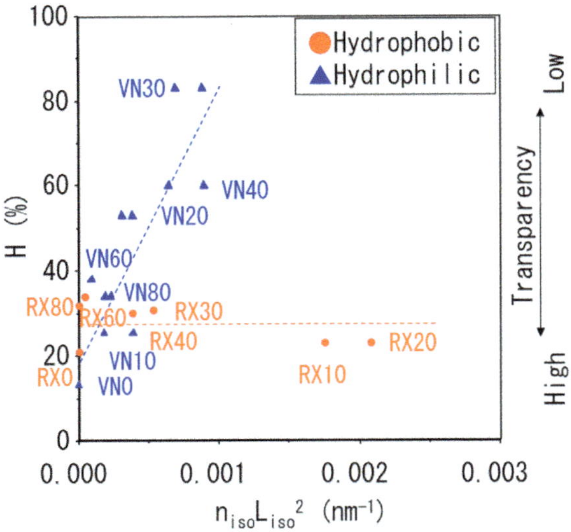

Figure 12.20 Dependence of haze (H) on $n_{\mathrm{iso}}L_{\mathrm{iso}}^2$ for hydrophobic and hydrophilic loaded NR. *Source*: From PLATE XIV in Kato A, Kokubo Y, Tsushi R, and Ikeda Y. Comparative study of hydrophobic and hydrophilic silica in natural rubber: optical transparency and network structure of particulate silica. In: Chemistry, manufacture and applications of natural rubber, Kohjiya S. Ikeda Y. editors. Cambridge: Woodhead/Elsevier; 2014, Chapter 9, with permission.

linear relationship for the dependence of H on $n_{\mathrm{iso}}L_{\mathrm{iso}}^2$, as indicated in Eq. (12.1). Simultaneously, the decrease of the transparency observed in Fig. 12.15 was qualitatively reproduced as VN0 > VN10 > VN60 ≒ VN80 > VN20 ≒ VN40 > VN30. These results suggest that the multiple scattering of light by the isolated chains in the silica aggregate network of the hydrophilic silica filled cross-linked NR caused optical transparency to decline. The observations so far indicate that at filler loadings lower than 30 phr, the matrix rubber probably governs the optical characteristics because few local hydrophilic silica aggregates and isolated chains were formed. While the optical characteristics due to the cross-linked and branched network chains were uniform in the thickness direction of the specimens, the isolated chains probably acted as nonuniformly distributed light scattering bodies. In addition, it is inferred that the optical characteristics of the network chains were dominant at filler loadings higher than 30 phr because isolated chains were markedly scarce. Still, the exact reason why isolated chains noticeably increased at a hydrophilic silica loading of 30 phr was not clear. It is hypothesized that this increase in isolated chains may be dependent on strong filler-to-filler interaction in hydrophilic silica and the cross-linking conditions. These points make it necessary to conduct more detailed investigation in the future work.

12.7 Carbon black network structure and rubber reinforcement

12.7.1 Carbon black loaded nanocomposites

The discussion on rubber reinforcement by nanofillers is not complete without a consideration on CB, because the past studies on reinforcement have been mainly focused on CB as a reinforcer. In addition, the reinforcement is of the utmost importance for describing characteristics of soft rubbery nanocomposites. Therefore, the last section of this chapter is to explain a recent study on the reinforcement of rubber by CB, even though the CB loaded nanocomposites is simply manufactured by the traditional mechanical mixing method. The behavior of a traditional CB would be well compatible with our comment described in Section 12.1.

As CB for this study, a high abrasion furnace (HAF) Grade N330 of the primary particle diameter of 26–30 nm was used, which was supplied by Tokai Carbon Co. HAF is known to be the most popular filler for various rubber products. The amount of CB was changed from 0 to 80 phr in NR in order to elucidate the morphological characteritics of CB in the rubbery matrix. A number in the sample code indicates the amount of loaded CB. Other ingredients for vulcanization of natural rubber (NR, RSS no. 1) were as follows: 1.5 phr sulfur, 1.0 phr CBS (*N*-cyclohexyl-2-benzothiazyl sulfenamide), 1.0 phr zinc oxide, and 2.0 phr stearic acid. The vulcanization was carried out at 140°C for 15 min [78].

12.7.2 Structural variation with the increase of carbon black

Fig. 12.21 presents 3D-TEM images of CB-filled NR vulcanizates. Even in the low CB loadings ,i.e., CB5 (not shown) and CB10, a number of aggregates consisting of several CB particles are observed. This observation is reasonable since most of the CB already coagulated during its manufacturing process (See chapter: Rubber blend nanocomposites, in Ref. [9]). It was seen that the CB aggregates increased in size as well as in number with the increase of CB loading. Moreover, for the CB80 sample with the highest CB loading, it is observed that many large aggregates were positioned closely together.

The dependence of d_p between CB aggregates on the CB loading (W_{CB}) is shown in Fig. 12.22(A). It was observed that d_p declined sharply in the samples with 30 phr or less W_{CB}, whereas it tended to converge to a nearly constant value in the samples with 40 phr or higher W_{CB}. This trend suggested that the CB aggregates in the samples with 40 phr or higher W_{CB} were at a distance of ca. 3 nm from each other, which was the closest they were able to approach one another. Fig. 12.22(B) shows the dependence of the volume resistivity (ρ_v) of CB-filled NR vulcanizates on W_{CB}. For the samples with 30 phr or less W_{CB}, ρ_v declined sharply as W_{CB} was increased, whereas it decreased gradually in the samples having 40 phr or more W_{CB}. This behavior was referred to electrical percolation. It was especially noteworthy that the trend of ρ_v converged to a nearly constant value for CB loadings of

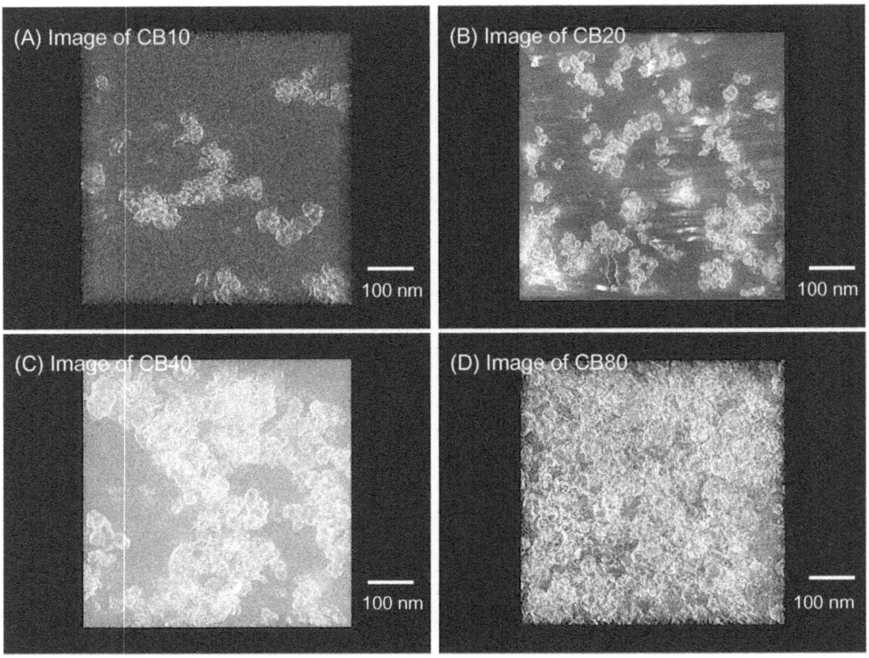

Figure 12.21 3D-TEM images of CB-filled NR vulcanizates.
Source: From Fig. 3 in Kato A, Shimanuki J, Kohjiya S, Ikeda Y. Three-dimensional morphology of carbon black in NR vulcanizates as revealed by 3D-TEM and dielectric measurements. Rubber Chem Technol 2006;79:653, with permission.

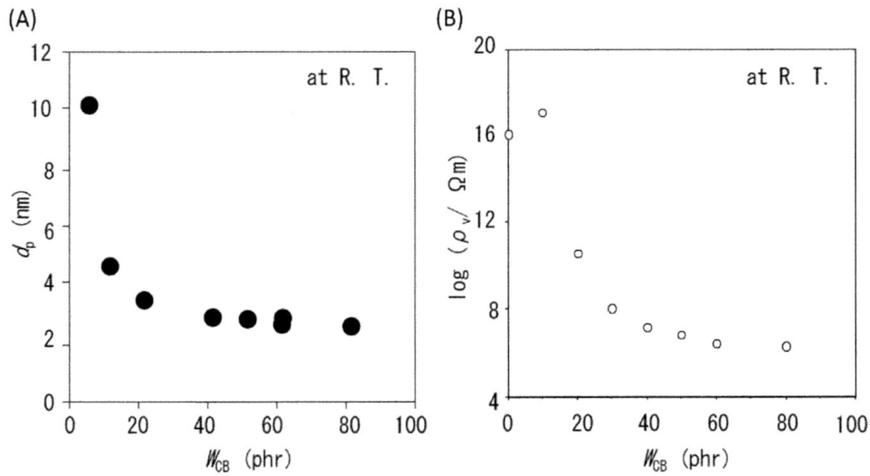

Figure 12.22 Dependence of d_p and ρ_v on CB loading (W_{CB}). (A) d_p versus W_{CB} (B) ρ_v versus W_{CB}.
Source: From Figs. 5 and 6 in Kato A, Shimanuki J, Kohjiya S, Ikeda Y. Three-dimensional morphology of carbon black in NR vulcanizates as revealed by 3D-TEM and dielectric measurements. Rubber Chem Technol 2006;79:653, with permission.

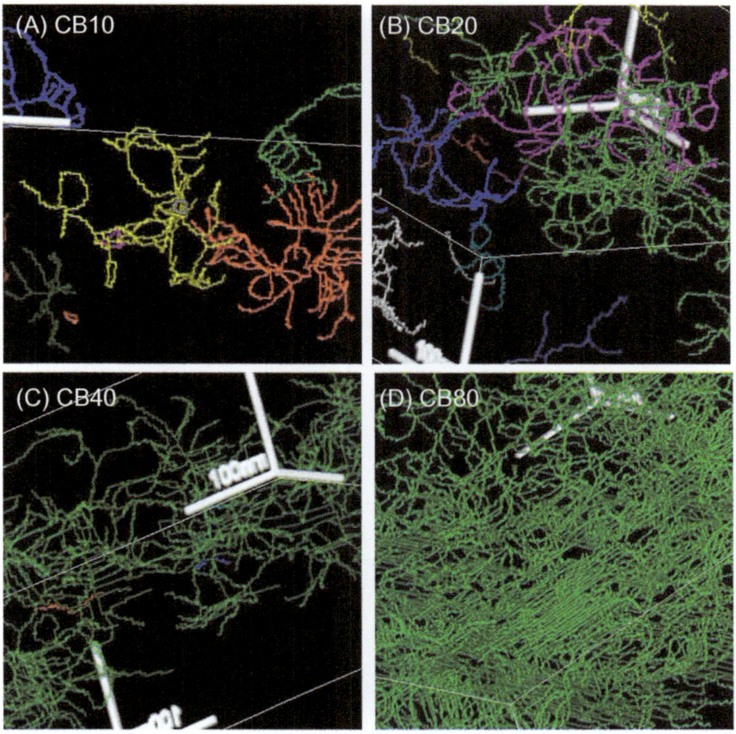

Figure 12.23 Diagrams of 3D network structure of CB aggregates in CB10, CB20, CB40, and CB80.
Source: From Fig. 4.5 in Kato A, Isono Y, Nagata K, Asano A, and Ikeda Y. Study on polymeric nano-composites by 3D-TEM and related techniques. In: Characterization tools for nanoscience & nanotechnology. Kumar CSSR. editor. Berlin: Springer; 2014, pp. 139–193.

more than 40 phr was corresponding to that of d_p. In addition, ρ_v in a temperature range of 23°C–90°C decreased with increasing temperature (not shown).

Fig. 12.23 presents the diagrams of the 3D network structures of the CB aggregates in the CB10, 20, 40, and 80 NR vulcanizates [68]. The structure of the CB aggregate network was visualized by drawing lines connecting the centers of gravity of the closest CB aggregates at a distance of ca. 3 nm, which d_p value of the aggregates saturated in the high CB loading region [36]. Two types of chains were defined in the CB aggregate network: cross-linked chains were ones that connect between two CB aggregates at cross-linked points and branched chains were anothers that extended outward from CB aggregates and had free ends. It was found that cross-linked chains and branched chains of CB aggregates were presented in all the examined samples and there were no isolated chains, i.e., ones unconnected to the CB network. While isolated networks were observed in samples having 20 phr

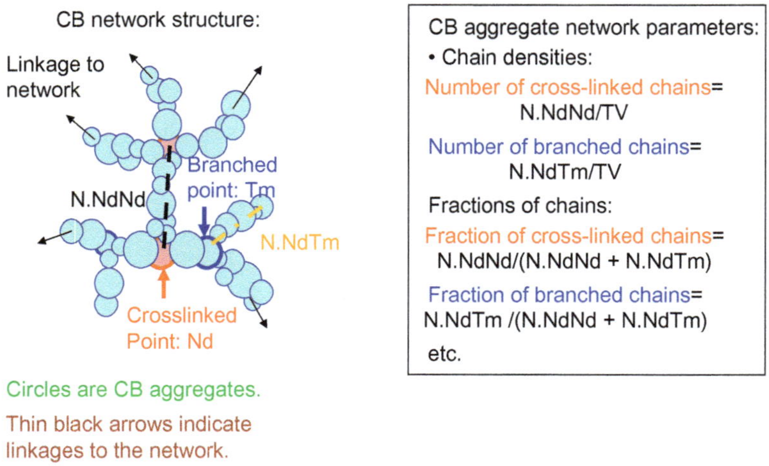

Figure 12.24 Schematic model of CB network structure.
Source: Modified from Fig. 10 in Kato A, Shimanuki J, Kohjiya S, Ikeda Y. Three-dimensional morphology of carbon black in NR vulcanizates as revealed by 3D-TEM and dielectric measurements. Rubber Chem Technol 2006;79:653.

or less W_{CB}, it was found that the CB aggregate network structure was connected and extended throughout the entire vulcanizates with 40 phr or higher W_{CB}. Accordingly, it was assumed that the high conductivity seen in the samples having 40 phr or higher W_{CB} can be attributed to the movement of electrons through such a network.

Fig. 12.24 presents a schematic model of the CB network structure having relatively strong interactions as well as the definitions of the network parameters obtained from the model [78]. The circles in the figure represent CB aggregates and the black arrows indicate that the aggregates are linked to the surrounding network structures. The red and blue arrows represent a cross-linked point and a branched point in the network, respectively. The network parameters are the total volume in the view of the 3D-TEM image (TV), the number of cross-linked points (N.Nd), the number of branched points (N.Tm), the density of the cross-linked chains (N.NdNd/TV), the density of the branched chains (N.NdTm/TV), the number of cross-linked chains (N.NdNd), the number of branched chains (N.NdTm), the fraction of each chain type (N.NdNd/(N.NdNd + N.NdTm, N.NdTm(N.NdNd +N.NdTm), and the fractal dimension. In addition, a novel analysis for characterization of the CB network is proposed on the basis of gelation mechanism of polymer network, where structural parameters of cross-linked and branched chains of CB aggregates were determined by 3D-TEM/electron tomography and were utilized for the characterization.

Fig. 12.25 shows the fractions (N.NdNd/(N.NdNd + N.NdTm, N.NdTm/(N.NdNd + N.NdTm) of the cross-linked and branched chains of the CB network as a function of W_{CB} [78]. The fraction of cross-linked chains increased almost

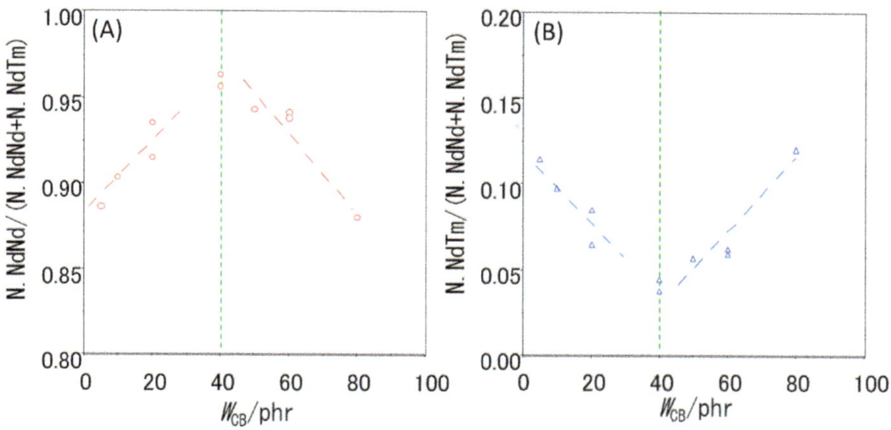

Figure 12.25 Fractions of cross-linked (N.NdNd/(N.NdNd + N.NdTm) and branched chains (N.NdTm/(N.NdNd + N.NdTm) of CB aggregates as a function of CB loading (W_{CB}). (A) Fraction of cross-linked chains. (B) Fraction of branched chains.
Source: From Fig. 13 in Kato A, Shimanuki J, Kohjiya S, Ikeda Y. Three-dimensional morphology of carbon black in NR vulcanizates as revealed by 3D-TEM and dielectric measurements. Rubber Chem Technol 2006;79:653, with permission.

linearly until around $W_{CB} = 40$ phr and then declined. It was noted that the region where W_{CB} was less than 40 phr, was the region where in good agreement with the Charlesby's gelation theory [79–81] (see Ref. [82] for the details). In the region where W_{CB} was greater than 40 phr, the share of the cross-linked chains in the network decreased almost linearly. In contrast, the fraction of branched chains displayed the almost opposite trend to that of the cross-linked chains. In short, the fraction of branched chains decreased almost linearly in the region where W_{CB} was less than 40 phr, whereas it increased almost linearly in the region of 40 phr or greater W_{CB}. These results suggest that the structure of the CB network changed markedly around 40 phr W_{CB}. In other words, in the region where W_{CB} was less than 40 phr, the branched chains linked to form cross-linked chains, whereas in the region of 40 phr or greater W_{CB}, many branched chains were produced without leading to the formation of cross-linked chains owing to a three-dimensional defect in the resultant CB network.

The following discussion presents a hypothesis about the structure of the CB/NR interaction layer. This hypothesis is based on the assumption that the rubber molecular chains are physically constrained in a layer several nm distance from the CB surface as the bound rubber around CB [83]. Nakajima et al. [84] used an atomic force microscopy (AFM) to measure the elastic modulus of bound rubber in CB-filled NR vulcanizates and its thickness from the CB surface. They reported that the bound rubber had a two-layer structure consisting of a glassy layer of about 2 nm in thickness and an uncross-linked rubber layer of about 8 nm in thickness. Kato et al. [35,68,78,82] reported that the closest distance between neighboring particles (d_p) of HAF aggregates is ca. 3 nm, which was estimated from the results of 3D-TEM observations and volume resistivity measurements of HAF-filled NR vulcanizates. It is reasonable to assume that the interfacial layer between two nearest CB aggregates is between 1.5 and 3 nm in thickness from the surface of each CB aggregate [83,84]. From these considerations, d_p value of 3 nm is estimated to correspond to the thickness of the glassy NR layer around CB as found by Nakajima et al. [84] using AFM measurements.

Fig. 12.26 presents 3D images of the CB/NR interaction layer (CNIL) in the CB10,20,40, and 80 samples, assuming that the thickness (d_p) of the immobilized rubber interfacial layer was ca. 3 nm. In order to make the CNIL clearer in this figure, the rubber matrix, CB particles, and CB aggregates were rendered in black, and the CNIL at distances greater than d_p = ca. 3 nm was shown in different color gradations. It was noted that the size of the CNIL in the CB20 sample was made to coincide with the size of the CNIL (e.g., red region) having a few hundred nm, and was made to coincide with the range in a high elastic modulus region in the AFM phase image [85].

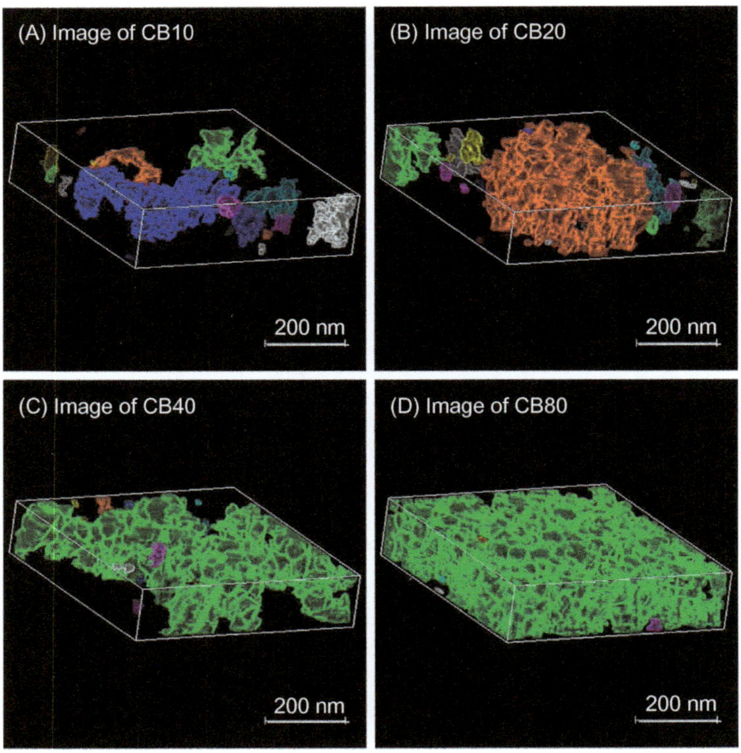

Figure 12.26 3Dimages of CB/NR interaction layer (CNIL) of CB10, 20, 40, and 80.
Source: From Fig. 4 in Kato A, Ikeda Y, Tsushi R, Kokubo Y, Kojima N. A new approach to visualizing the carbon black/natural rubber interaction layer in carbon black-filled natural rubber vulcanizates and to elucidating the dependence of mechanical properties on quantitative parameters. Colloid Polym Sci 2013;291:2101, with permission.

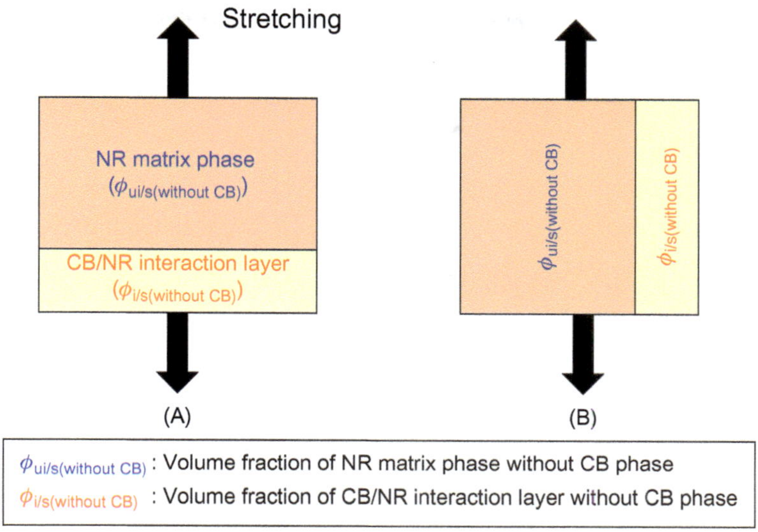

Figure 12.27 Series and parallel mechanical models of mixing law for two phases (NR matrix phase and CB/NR interaction layer) excluding CB phase. (A) Series model. (B) Parallel model.
Source: From Fig. 10 in Kato A, Ikeda Y, Kohjiya S. Nippon Gomu Kyokaishi 2015; 88: 3, with permission.

12.7.3 Mixing law for a mechanistic elucidation of rubber reinforcement

It is well known that a mixing law, such as a series or a parallel mechanical model of the constituent components, can generally be applied to predict the mechanical properties of composites. Firstly, we consider three components, namely, the NR matrix phase, CNIL, and CB phase. Secondly, to reduce the number of components to two, we consider two components system, the NR matrix and the CNIL. While CB is involved in the formation of the CNIL, CB is assumed not directly to affect the mechanical properties. In other words, the mixing law of two-phase mechanical models as shown in Fig. 12.27 is assumed [86].

The following equations are derived for the two-phase series and parallel mechanical models.

$$\phi_{ui/s(without\ CB)} = 100(V_s - V_i - V_{CB}/V_s - V_{CB}) \tag{12.3}$$

$$\phi_{i/s(without\ CB)} = 100(V_i/V_s - V_{CB}) \tag{12.4}$$

$$\phi_{ui/s(without\ CB)} + \phi_{i/s(without\ CB)} = 1 \tag{12.5}$$

Letting G', G'_{ui}, and G'_{i} represent the respective modulus of elasticity of the CB-filled NR and NR matrix and the CNIL, the following Eqs. (12.6) and (12.7) are obtained for the two-phase series and parallel mechanical models, respectively.

$$1/G' = \phi_{\mathrm{ui/s(without\ CB)}}(1/G'_{\mathrm{ui}}) + \phi_{\mathrm{i/s(without\ CB)}}(1/G'_i) \tag{12.6}$$

$$G' = \phi_{\mathrm{ui/s(without\ CB)}}G'_{\mathrm{ui}} + \phi_{\mathrm{i/s(without\ CB)}}G'_i \tag{12.7}$$

Fig. 12.27(A) presents two-phase series mechanical models where the inverse of G' at 293K as a function of $\phi_{i/s(\mathrm{without\ CB})}$ is shown. In the region where 0 phr ≦ CB loading ≦ 20 phr, the increase of G' shows nearly a linear relationship with each volume fraction, where the details are not discussed. However, we found that this series model could not be used for analyzing the G' behavior, and in this CB loading region, the hydrodynamic interaction occurs between the CB aggregate and the NR matrix when a sample is deformed. In contrast, Fig. 12.27(B) presents two-phase parallel models showing G' at 293K as a function of $\phi_{\mathrm{i/s\ (without\ CB)}}$. In the region of CB loading < 30 phr, G' shows nearly a linear relationship with each volume fraction, indicating that a parallel mechanical model is applicable in this region. This result implies that the network structure formed in the high CB loading region of 30 phr or more consisted of CB particles and a layer of rubber molecular chains, i.e., a rubber layer on the outside of the glassy rubber layer attached to the CB particle surface. Consequently, the mechanical properties of the samples seem to follow a parallel mechanical model.

We will now examine whether a two-phase parallel mechanical model is actually valid for CB-filled NR vulcanizates in the high CB loading region. The equations for approximating a straight line in the region of CB loading ≧ 40 phr in Fig. 12.28 (B) are given below.

$$\begin{aligned} G' &= -6.92 + 141\phi_{\mathrm{i/s(without\ CB)}} \\ R^2 &= 0.903 \end{aligned} \tag{12.8}$$

Since R^2 (the square of the correlation coefficient (R) of the regression line) of the models is about 0.9, it is clear that a good linear correlation holds true in both models. Eq. (12.8) is then arranged so as to correspond to the forms of Eq. (12.7) by using the intersection point of the approximated line in Fig. 12.29(B) and the x axis, i.e., when $G' = 0$, $\phi_{\mathrm{i/s\ (without\ CB)}} = 0.0491$. As a result, the following equation is obtained.

$$G' = (1.05G'_{\mathrm{ui}} - 0.0491G'_{\mathrm{i}}) + (G'_{\mathrm{i}} - G'_{\mathrm{ui}})\phi_{\mathrm{i/s(without\ CB)}} \tag{12.9}$$

Comparing the respective values obtained with Eq. (12.8) and Eq. (12.9) yields the following equations about the two-phase parallel mechanical model.

$$1.05G'_{\mathrm{ui}} - 0.0491G'_{\mathrm{i}} = -6.92 \tag{12.10}$$

$$G'_{\mathrm{i}} - G'_{\mathrm{ui}} = 141 \tag{12.11}$$

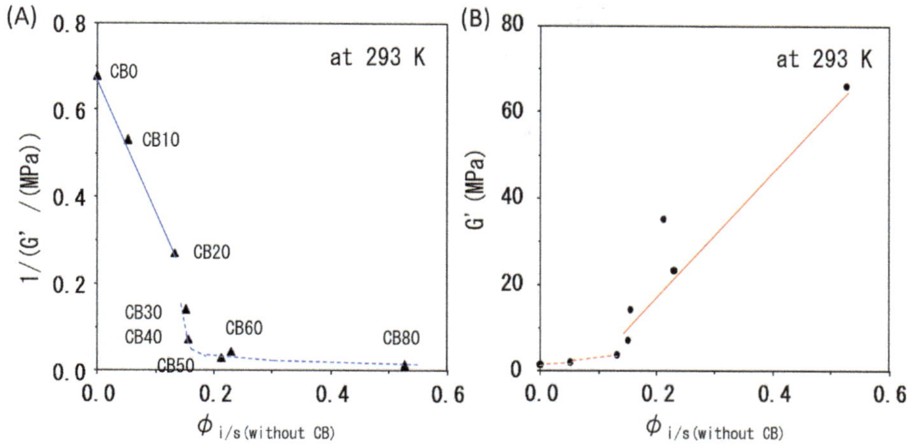

Figure 12.28 Dependence of $1/G'$ and G' on $\phi_{i/s}$ and $\phi_{i/s(without\ CB)}$. (A) Series model for two phases: $1/G'$ versus $\phi_{i/s(without\ CB)}$. (B) Parallel model for two phases: G' versus $\phi_{i/s(without\ CB)}$.
Source: From Figs. 11(B) and 12(B) in Kato A, Ikeda Y, Kohjiya S. Nippon Gomu Kyokaishi 2015;88:3, with permission.

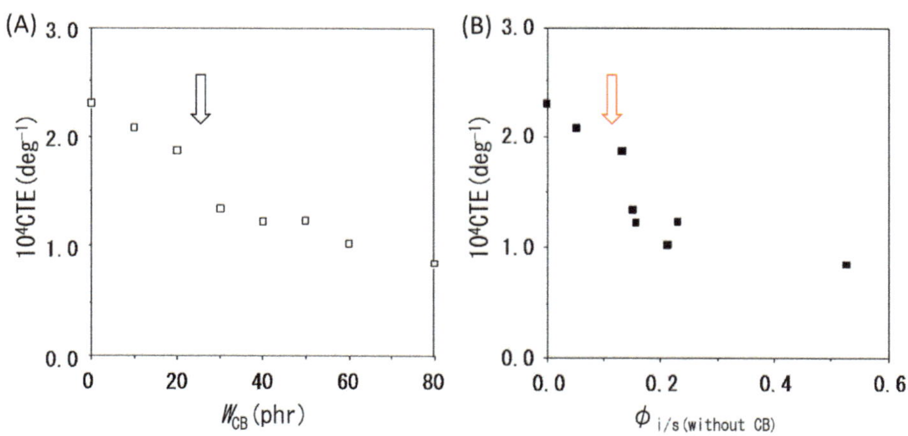

Figure 12.29 Dependence of coefficient of thermal expansion (CTE) on CB loading and volume fraction ($\phi_{i/s(without\ CB)}$) of CNIL in CB-filled NR vulcanizates. (A) CTE versus W_{CB}. (B) CTE versus $\phi_{i/s(without\ CB)}$.
Source: From Fig. 4.23 in Kato A, Isono Y, Nagata K, Asano A, and Ikeda Y. Study on polymeric nano-composites by 3D-TEM and related techniques. In: Characterization tools for nanoscience & nanotechnology. Kumar CSSR. editor. Berlin: Springer; 2014, pp. 139–193, with permission.

Solving the simultaneous expressions in Eqs. (12.10) and (12.11) has resulted in $G'_i = 140$ MPa and $G'_{ui} = 1$ MPa. It is noted that the values of G'_{ui} and G'_i obtained by the abovementioned two-phase mechanical model are positive, indicating that a two-phase mechanical model is valid. This value of G'_i may correspond to the modulus of elasticity between the high elasticity layer surrounding CB and the outer layer, reported by Nakajima et al. [84]. They reported values of 1 GPa, 60 MPa, and 8 MPa for the modulus of elasticity of the high elasticity layer in NR and around CB, the outer layer (combined thickness of both layers = ca. 10 nm), and the rubber matrix, based on AFM measurements of CB-filled NR vulcanizates. In contrast to those values, the value of G'_i in this work is 140 MPa from the aforementioned two-phase parallel mechanical model. This value suggests that it is to be compared with the modulus of elasticity between the high elasticity layer around CB and the outer layer by Nakajima, 60 MPa. A three components system, i.e., NR matrix, the interaction layer, and CB layer, both in parallel and series, has been investigated as well, but it has failed to show a better fitting.

Based on TEM observation results for CB-filled rubber vulcanizates, Medalia [87] reported that the aspect ratio (l/d) of CB aggregates was 1.7–1.9. Halpin et al. [88] found that the significant reinforcement effect of nylon-fiber-reinforced rubber vulcanizates occurred at l/d of 10 or more. These findings suggest that CB aggregates having a l/d value much smaller than 10 may not exert a direct reinforcing effect on the rubber matrix. Rather, as we have made clear in this work, the CNIL has to have an important role for the reinforcing effect. In other words, it is possible to explain the viscoelastic behavior of CB-filled rubber vulcanizates by applying the mixing law concerning the two-phase, the rubbery matrix and CNIL, that forms around the CB aggregates to make CB compatible with rubber.

Moreover, as shown in Fig. 12.29(A), the authors also investigated the relationship between the coefficient of thermal expansion (CTE) of CB-filled NR vulcanizates and the CB loading [68,78,89,90]. Fig. 12.29(A) shows an unclear steplike change of a CTE, as indicated by the arrow, in a CB loading region of 20–30 phr. In contrast to that, by using $\phi_{i/s(\text{without CB})}$ instead of the CB loading, as shown in Fig. 12.29(B), we were able to observe a distinct steplike change regarding the formation of the CB network in comparison with Fig. 12.29(A) in the range of $\phi_{i/s(\text{without CB})} = 0.132–0.151$ (CB loading = 20–30 phr), as indicated by the arrow. These results reveal that the steplike change implies a structural phase transition, and occurs when CB aggregates form a CB network structure in the CB-filled NR vulcanizates.

Concluding Remarks

In this chapter, two new topics are presented. At first, a novel processing technique, the "soft" processing, is introduced for the preparation of "soft" nanocomposites. While the former "soft" means liquid states, e. g., a utilization of natural rubber latex and/or in situ chemical reactions of tetraethoxysilane to produce silica, the latter one means composites based on rubbery matrix. Both are found of much use in realizing network-like structure of nanofillers in the rubbery matrix to result in the formation of high performance soft composites. Secondly, the clustering of nanofillers in the rubbery matrix is studied by a few modern structural techniques, e.g., 3D-TEM, and the detailed nanostructure and morphology are elucidated. The structural researches have brought about an important conclusion to the reinforcement of rubber by nanofillers: "Network-like structure" of the nanofiller via "bound rubber" is of much importance in elucidating the reinforcement mechanism of the rubbery matrix. It is notable, therefore, that the coexisting of two networks, i.e., of rubber networks and of nanofiller networks is the general characteristics of soft rubbery nanocomposites in terms of rubber reinforcement in the soft nanocomposites.

Acknowledgement

The authors acknowledge the kind cooperation of all the coworkers of the studies described here. In particular, cooperation by the researchers at NISSAN ARC to the measurement of 3D-TEM is greatly appreciated.

References

[1] Kelly A. An introduction to composite materials. In: Kelly A, editor. Concise encyclopedia of composite materials. revised ed. London: Pergamon; 1994.
[2] Mulhall D. Our molecular future: How nanotechnology, robotics, genetics, and artificial intelligence will transform our world. New York: Prometheus Books; 2002.
[3] Drexler KE. Molecular engineering: an approach to the development of general capabilities for molecular manipulation. Proc Natl Acad Sci USA 1981;78:5275.
[4] Drexler KE. Nanotechnology: from fineman to funding. Bull Sci Technol Soc 2004;24(1):21.
[5] Smalley RE. Of chemistry, love and nanorobots-how soon will we see the nanometer-scale robots envisaged by K. Evic Drexler and other molecular nanotechnologies? The simple answer is never. Sci Am 2001;September:68.
[6] Kulinowski K. Nanotechnology from "wow" to "yuck"? Bull Sci Technol Soc 2004;24(1):13.
[7] Hirata Y, Kondo H, Ozawa Y. Natural rubber for the tyre industry. In: Kohjiya S, Ikeda Y, editors. Chemistry, manufacture and applications of natural rubber. Amsterdam: Woodhead/Elsevier; 2014.
[8] Kohjiya S. Natural rubber: from the odyssey of the Hevea tree to the age of transportation. Shrewsbury: Smithers RAPRA; 2015.
[9] Kraus G, editor. Reinforcement of elastomers. New York: Interscience; 1956.
[10] Park S-J, Seo M-K. Carbon fiber-reinforced polymer composites: preparation, properties, and applications. In: Thomas S, et al., editors. Polymer composites, vol. 1. Weinheim: Wiley-VCH; 2012.
[11] Ikeda Y, Kato A, Shimanuki J, Kohjiya S. Nano-structure observation of in situ silica in natural rubber matrix by three dimensional transmission electron microscopy. Macromol Rapid Commun 2004;25:1186.
[12] Kohjiya S, Yajima A, Yoon JR, Ikeda Y. In situ silica reinforcement of styrene-butadiene rubber by sol-gel method. Nippon Gomu Kyokaishi (J Soc Rubber Ind, Japan) 1994;67:859.

[13] Ikeda Y, Tanaka A, Kohjiya S. Reinforcement of styrene-butadiene rubber vulcanizate by in situ silica prepared by the sol–gel reaction of tetraethoxsilane. J Mater Chem 1997;7:1497.
[14] Ikeda Y, Kohjiya S. In situ formed silica particles in rubber vulcanizate by the sol-gel method. Polymer 1997;38:4417.
[15] Hashim AS, Azahari B, Ikeda Y, Kohjiya S. Effect of bis(3-triethoxysilylpropyl) tetrasulfide on silica reinforcement of styrene-butadiene rubber. Rubber Chem Technol 1998;71:289.
[16] Kohjiya S, Ikeda Y. Reinforcement of general-purpose grade rubbers by silica generated in situ. Rubber Chem Technol 2000;73:534.
[17] Kohjiya S, Murakami K, Iio S, Tanahashi T, Ikeda Y. In situ filling of silica onto "green" natural rubber by the sol-gel process. Rubber Chem Technol 2001;74:16.
[18] Murakami K, Iio S, Tanahashi T, Kohjiya S, Kajiwara K, Ikeda Y. Reinforcement of NR by silica generated in situ: comparison with carbon black stock. Kaut Gummi Kunst 2001;54:668.
[19] Tohsan A, Ikeda Y. Generating particulate silica fillers *in situ* to improve the mechanical properties of natural rubber. In: Kohjiya S, Ikeda Y, editors. Chemistry, manufacture and applications of natural rubber. Cambridge: Woodhead/Elsevier; 2014.
[20] MacGregor EA, Greenwood CT. Rubber and lignin. Polymer in nature. Chichester: John Wiley & Sons; 1980.
[21] Glasser WG, Sarkanen S. editors. *Lignin: properties and materials*, ACS Symposium Series 397, 1989, American Chemical Society, Washington, DC.
[22] Garver TM, Sarkanen S. Kraft lignin: a new perspective. In: Carraher Jr. CE, Sperling LH, editors. Renewable-resource materials: new polymer sources. New York: Plenum Press; 1986. p. 287–303.
[23] Lora JH, Glasser WG. Recent industrial applications of lignin: a sustainable alternative to nonrenewable materials. J Polym Environ 2002;10:39.
[24] Keilen JJ, Pollak A. Lignin for reinforcing rubber. Ind Eng Chem 1947;39:480.
[25] Keilen JJ, Dougherty WK, Cook WR. Lignin-reinforced nitrile, neoprene, and natural rubbers. Ind Eng Chem 1952;44:163.
[26] Griffith TR, MacGregor DW. Aids in vulcanization of lignin–natural rubber coprecipitates. Ind Eng Chem 1953;45:380.
[27] Tibenham FJ, Grace NS. Compounding natural rubber with lignin and humic acid. Ind Eng Chem 1954;46:824.
[28] Anon. Lignin filler for tires. Chem Eng News 1957;June 3:28.
[29] Yamashita S, Kohjiya S. Use of lignin as a reinforcing filler for rubber. In: Kennedy JF, et al., editors. Wood processing and utilization. Chichester: Ellis Horwood; 1989.
[30] Phakkeeree T, Kittipoom S, Kohjiya S, Phinyocheep P, Ikeda Y. Preparation, structure, and properties of natural rubber/lignin nanocomposites, in *PACCON2009 Abstracts* (Phitsanulok, Thailand; Jan. 14–16, 2009), Lecture no. S10-OR-8 (p. 93).
[31] Tohsan A, Phinyocheep P, Kittipoom S, Pattanasiriwisawa W, Ikeda Y. Novel biphasic structured composite prepared by in situ silica filling in natural rubber latex. Polym Adv Technol 2012;23:1335.
[32] Frank J, editor. Electron tomography: three-dimensional imaging with the transmission electron microscope. New York: Plenum Press; 1992.
[33] Dejong KP, Koster AJ. Three-dimensional electron microscopy of mesoporous materials-recent strides towards spatial imaging at the nanometer scale. Chem Phys Chem 2002;3:776.
[34] Weyland M, Midgley PA. Electron tomography. Mater Today 2004;7(12):32.

[35] Kohjiya S, Kato A, Ikeda Y. Visualization of nanostructure of soft matter by 3D-TEM: nanoparticles in a natural rubber matrix. Prog Polym Sci 2008;33:979.
[36] Kohjiya S, Kato A, Suda T, Shimanuki J, Ikeda Y. Visualisation of carbon black networks in rubbery matrix by skeletonisation of 3D-TEM image. Polymer 2006;47:3298.
[37] Kato A, Ikeda Y, Kohjiya S. Carbon black-filled natural rubber composites: physical chemistry and reinforcing mechanism. In: Thomas S, et al., editors. Polymer composites, vol. 1. Weinheim: Wiley-VCH; 2012.
[38] Koga T, Hashimoto T, Takenaka M, Aizawa K, Amino N, Nakamura M, et al. New insight into hierarchical structures of carbon black dispersed in polymer matrices: a combined small-angle scattering study. Macromolecules 2008;41:453.
[39] Guth E, Gold O. On the hydrodynamic theory of the viscosity of suspensions. Phys. Rev. 1938;53:322.
[40] Guth E. Theory of filler reinforcement. J Appl Phys 1945;16:20.
[41] Mullins L, Tobin NR. Stress softening in rubber vulcanizates. Part I. Use of a strain amplification factor to describe elastic behavior of filler-reinforced vulcanized rubber. J Appl Polym Sci 1965;9:2993.
[42] Kato A, Kokubo Y, Tsushi R, Ikeda Y. Comparative study of hydrophobic and hydrophilic silica in natural rubber: optical transparency and network structure of particulate silica. In: Kohjiya S, Ikeda Y, editors. Chemistry, manufacture and applications of natural rubber. Cambridge: Woodhead/Elsevier; 2014.
[43] Heinrich G, Klüppel M. Recent advances in the theory of filler networking in elastomers. Adv Polym Sci 2002;160:1.
[44] Klüppel M. The role of disorder in filler reinforcement of elastomers on various length scales. Adv Polym Sci 2003;164:1.
[45] Palosuo T. Recent research on natural rubber latex allergy. In: Kohjiya S, Ikeda Y, editors. Chemistry, manufacturing and applications of natural rubber. Cambridge: Woodhead/Elsevier; 2014.
[46] Tohsan A, Kishi R, Ikeda Y. A model filler network in nanocomposites prepared by in situ silica filling and peroxide cross-linking in natural rubber latex. Colloid Polym Sci 2015;293:2083.
[47] Scherer GW, Luong JC. Glasses from colloids. J Non Cryst Solids 1984;63:163–72.
[48] Brinker CJ, Scherer GW. Sol–gel science: the physics and chemistry of sol-gel processing. New York: Academic Press; 1990.
[49] Tohsan A. Characterization of biphasic structured natural rubber/in situ silica composites (PhD Thesis at Kyoto Institute of Technology), 2013.
[50] Poompradub S, Kohjiya S, Ikeda Y. Natural rubber/in situ silica nanocomposite of a high silica content. Chem Lett 2005;34(5):672–3.
[51] Ikeda Y, Poompradub S, Morita Y, Kohjiya S. Preparation of high performance nanocomposite elastomer: effect of reaction conditions on in situ silica generation of high content in natural rubber. J Sol–gel Sci Technol 2008;45(3):299–306.
[52] Miloskovska E, Hansen MR, Friedrich C, Hristova-Bogaerds D, Duin MV, With G. In situ silica nanoparticle formationin a rubber matrix monitored via real-time SAXS and solid-state NMR spectroscopy. Macromolecules 2014;47(15):5174–85.
[53] Ferry JD. Viscoelastic properties of polymers. 3rd ed. New York: John-Wiley; 1980.
[54] Mujtaba A, Keller M, Ilisch S, Radusch HJ, Thurn-Albrecht T, Saalwächter K, et al. Mechanical properties and cross-link density of styrene–butadiene model composites containing fillers with bimodal particle size distribution. Macromolecules 2012;45(16): 6504–651.
[55] Wanger MP. Reinforcing silicas and silicates. Rubber Chem Technol 1976;76:703–74.

[56] Phakkeeree T, Ikeda Y, Yokohama H, Phinyocheep P, Kitano R, Kato A. Network-like structure of lignin in natural rubber matrix to form high performance elastomeric bio-composite, J Fiber Sci Technol **72**, 2016, 160.
[57] Ersen O, Florea I, Hirlimann C, Phan-Huu C. Mater Today 2015;18:395.
[58] Tsuji M, Kohjiya S. Structural studies on crystalline polymer solids by high-resolution electron microscopy. Prog Polym Sci 1995;20:259.
[59] Gruber TC, Zerda TW, Gerspacher M. 3D morphological characterization of carbon-black aggregates using transmission electron microscopy. Rubber Chem Technol 1994;67:280.
[60] Palade GE. A study of fixation for electron microscopy. J Exp Med 1952;95:285.
[61] Kato K. The osmium tetroxide procedure for light and electron microscopy of ABS plastics. Polym Eng Sci 1976;7:38.
[62] Trent JS. Ruthenium tetraoxide staining of polymers: new preparative methods for electron microscopy. Macromolecules 1984;17:2930.
[63] Kohjiya S, Katoh A, Shimanuki J, Hasegawa T, Ikeda Y. Three-dimensional nanostructure of in situ silica in natural rubber as revealed by 3D-TEM/electro tomography. Polymer 2005;46:4440.
[64] Kohjiya S, Katoh A, Shimanuki J, Hasegawa T, Ikeda Y. Nano-structural observation of carbon black dispersion in natural rubber matrix by three-dimensional transmission electron microscopy. J Mater Sci 2005;40:2553.
[65] Kremer JR, Mastronarde DN, McIntosh JR. Computer visualization of three dimensional image data using IMOD. J Struct Biol 1996;116:71–6.
[66] Stalling D, Westerhoff M, Hege H-C. Chapter 38 Amira: a highly interactive system for visual data analysis. In: Hansen CD, Johnson CR, editors. The visualization handbook. San Diego, San Francisco, New York, Boston, London, Sydney and Tokyo: Academic Press; 2005. p. 749–67.
[67] Coran AY. In: Mark JE, Erman B, Eirich FR, editors. Science and technology of rubber. 2nd ed. New York: Academic Press; 1994.
[68] Kato A, Isono Y, Nagata K, Asano A, Ikeda Y. Study on polymeric nano-composites by 3D-TEM and related techniques. In: Kumar CSSR, editor. Characterization tools for nanoscience & nanotechnology. Berlin: Springer; 2014. p. 139–93.
[69] Kohjiya S, Maeda K, Yamashita S, Shibata Y. Chemical modification of silicone elastomers for optics. J Mater Sci 1990;25:3368.
[70] Kohjiya S, Ono A, Kishimoto T, Yamashita S, Yanase H, Asada T. Electrooptic effects on liquid crystalline polysiloxanes with negative dielectric anisotropy. Mol Cryst Liq Cryst 1990;185:183.
[71] Kohjiya S, Hashimoto T, Yamashita S, Irie M. Photomechanical behaviour of elastomeric ionene containing viologen units. Makromol Chem, Rapid Commun 1989;10:9.
[72] Hashimoto T, Kohjiya S, Yamashita S, Irie M. Photochromic and photomechanical ionene elastomer containing poly(tetrahydrofuran) segments and viologen units. J Polym Sci, Part A: Polym Chem 1991;29:651.
[73] Rauline R. (To Compagnie Generale des Establissements Michelin), *European Patent* 0501227 (1993); *US Patent* 5,227,425 (1993).
[74] Kato A, Ikeda Y, Kasahara Y, Shimanuki J, Suda T, Hasegawa T, et al. Optical transparency and silica network structure in cross-linked natural rubber as revealed by spectroscopic and three-dimensional transmission electron microscopy techniques. J Opt Soc Am B 2008;25:1602.
[75] Kato A, Ikeda Y. Structure and property relationship in nano-filler loaded cross-linked rubbers 3. relationship between optical transparency and filled silica networks in natural rubber vulcanizates. Nippon Gomu Kyokaishi 2014;87:351.

[76] Matsumura K, Kagawa Y. Evaluation in "transparency" of glass particle-dispersed optical composites by pico-second order pulse profile. J Mater Sci Lett 2001;20:2101.
[77] Kandidov VP, Militsin VO, Bykov AV, Priezzhev AV. Application of corpuscular and wave Monte-Carlo methods in optics of dispersive media. Quantum Electron 2006;36:1003.
[78] Kato A, Shimanuki J, Kohjiya S, Ikeda Y. Three-dimensional morphology of carbon black in NR vulcanizates as revealed by 3D-TEM and dielectric measurements. Rubber Chem Technol 2006;79:653.
[79] Charlesby A. Solubility and molecular size distribution of crosslinked polystyrene. J. Polym. Sci. 1953;11:513.
[80] Charlesby A. Gel formation and molecular weight distribution in long-chain polymers. Proc. Roy. Soc. 1954;A22:542.
[81] Charlesby A. Effect of molecular weight distribution on gel formation by high energy radiation. J Polym Sci 1954;14:547.
[82] Kato A, Ikeda Y. Structure and property relationship in nano-filler loaded cross-linked rubbers 2. visualization of carbon black network in rubber vulcanizates and its relationship with electric properties. Nippon Gomu Kyokaishi 2014;87:252.
[83] Wang Z, Liu J, Wu S, Wang W, Zhang L. Novel percolation phenomena and mechanism of strengthening elastomers by nanofillers. Phys Chem Chem Phys 2010;12:3014.
[84] Nakajima K, Nishi T. Recent developments in rubber research using atomic force microscopy. In: Bhomik AK, editor. Current topics in elastomers research. Boca Raton: CRC Press; 2008.
[85] Kato A, Ikeda Y, Tsushi R, Kokubo Y, Kojima N. A new approach to visualizing the carbon black/natural rubber interaction layer in carbon black-filled natural rubber vulcanizates and to elucidating the dependence of mechanical properties on quantitative parameters. Colloid Polym. Sci. 2013;291:2101.
[86] Kato A, Ikeda Y, Kohjiya S. Structure and property relationship in nano-filler loaded cross-linked rubbers 5. filler networks and viscoelastic properties of filler loaded cross-linked natural rubber: new proposal for carbon black/natural rubber interaction. Nippon Gomu Kyokaishi 2015;88:3.
[87] Medalia AI. Effective degree of immobilization of rubber occluded within carbon black aggregates. Rubber Chem Technol 1972;45:1171.
[88] Halpin JC, Kardos JL. Moduli of crystalline polymers employing composite theory. J Appl Phys 1972;43:2235.
[89] Kohjiya S, Ikeda Y, Kato A. Visualization of nano-filler dispersion and morphology in rubbery matrix by 3D-TEM. In: Bhowmick AK, editor. Current topics in elastomers research. Boca Raton: CRC Press; 2008.
[90] Kato A, Ikeda Y, Kohjiya S. Chapter 17: carbon black-filled natural rubber composites: physical chemistry and reinforcing mechanism. In: Thomas S, Joseph K, Malhotra SK, Goda K, Streekala MS, editors. Polymer composites volume 1: macro- and microcomposites. Wiley-VCH Verlag & Co.; 2012. p. 515–43.

Rubber nanocomposites with nanocellulose

13

R.C.R. Nunes
Universidade Federal do Rio de Janeiro, Rio de Janeiro, Brazil

13.1 Introduction: the importance of rubber/nanocellulose composites

Cellulose is a fascinating and almost inexhaustible and sustainable natural polymeric raw material characterized by exciting properties such as hydrophilicity, chirality, biodegradability, broad chemical-modifying capacity, and the formation of different semicrystalline fiber morphologies [1]. As chemical characteristic cellulose has a straight chain of carbohydrate polymer consisting of several 1000 β1−4 glucopyranose units; it is the world most abundant biopolymer [2].

Elastomers can also be classed as fascinating materials in view of their elastic extensibility resulting from their molecular structure. These are amorphous, low modulus materials and for technological applications fillers are added for improving the mechanical performance [3].

Over the past decade, the use of wood and other cellulosic fibers as fillers or reinforcements to replace carbon black (CB) and silica in the production of rubber composites has attracted much attention [4−6].

Following the technical and scientific development related to nanotechnology, several publications report the obtention of nanocellulose by different techniques [7−10].

These developments lead to differentiated species of nanocellulose and as a result, to the obtention of nanocomposites of specific characteristics [1,6,11,12].

This chapter aims at covering studies directed to the use of various nanocellulose fibers as reinforcement for different rubbers, as a component in polyblend nanocomposites as well as the preparation of these polyblends and their major characteristics.

The two main groups of nanosized cellulose used as filler in polymer nanocomposites are cellulose nanocrystals (CNCs) and nanofibrillated cellulose. Most of the recent papers employ terms like monocrystals, cellulose crystallites, nanocrystalline cellulose (NCC), whiskers, etc. The terms used by different researchers will be respected in this chapter.

According to Zhou et al. [6], Oksman [9], Siro and Plackett [11], and other researchers in this area the main challenges when producing cellulose-based nanocomposites are related to the following facts: (1) the surface properties of natural

Progress in Rubber Nanocomposites. DOI: http://dx.doi.org/10.1016/B978-0-08-100409-8.00013-9

cellulose are crucial for the successful preparation of the cellulose–rubber composites; (2) the curing characteristics of the composites are complex but closely related to not only the interfacial properties, but also to the compositions and other processing parameters; (3) the nature of hydrophilic cellulose and hydrophobic matrix rubber requires an accurate introduction of a coupling agent, so that one end of the nanocomposite structure should be compatible with a hydrophilic moiety while the other one should be compatible with a hydrophobic moiety.

To overcome some of these problems, our research group developed different procedures to disperse cellulose in the nanoscale and introduce it into elastomeric composites [12–14].

Cellulose nanocomposites find application in many areas such as packaging, optoelectric, barrier, electronic, biomedical, paper, textiles, military, etc. [2,4,15,16].

Considering the advantage of cellulosic filler over the inorganic fillers, such as biodegradability, renewability, and improved mechanical properties, nanocellulose-reinforced rubber nanocomposites are expected to gain a rapid advance in the near future [6].

13.2 Different nanocellulose materials

There are four different polymorphs of cellulose: cellulose I, II, III, and IV. Naturally occurring cellulose is known as cellulose I, which exists in parallel strands without intersheet hydrogen bonding. Cellulose II is thermodynamically more stable and exists in antiparallel strains with intersheet hydrogen bonding. The difference in properties of cellulose I and II arises due to changes in crystal structure. Cellulose III is amorphous and obtained by treatment of cellulose I or II with amines. Cellulose IV is obtained after treatment of cellulose III with glycerol at very high temperatures [7,17,18].

Within the framework of this chapter only cellulose I and cellulose II will be considered.

Cellulose is available in an almost pure state in the seed hairs of cotton and in the case of wood cellulose forms a native composite with lignin and other carbohydrates (hemicelluloses) from which it is isolated by large-scale chemical pulping, separation, and purification processes.

Although wood is certainly the most important industrial source of cellulose, the dominant pathway of this material is its production from plant fibers and crops such as flax, hemp, sisal, and others, especially from by-products of these different plants, which has become of increasing interest. Plant fibers include bast (or stem or soft sclerenchyma) fibers, leaf or hard fibers, seed, fruit, wood, cereal straw, and other grass fibers. These nonwood plants generally contain less lignin than wood and therefore bleaching processes are less demanding [1,11].

Some years ago, a number of researchers were involved in investigating the exploitation of natural fibers as load bearing constituents in composite materials [5,19,20].

Different types of short fibers as reinforcement for rubber have been studied and the most important cellulosic fibers used in composites are: jute [21]; bagasse [22]; bamboo [23]; coir fiber [24]; sisal and oil palm fibers [25].

The primary effects of short fiber reinforcement on the mechanical properties of rubber composites include increased modulus, decreased elongation at failure, greatly improved creep resistance over particulate-filled rubber, increased hardness, and a substantial improvement in cut, tear, and puncture resistance. Hysteresis and fatigue strength are also improved [6].

The main disadvantage of the lignocellulosic fiber-reinforced rubber composite is the poor compatibility between the hydrophilic wood flour or other natural fibers and the hydrophobic rubber matrix. Various approaches, including physical and chemical treatments, have been attempted for modifying the surface of cellulosic fibers and improving the compatibility or interface properties. They can be grouped as follows: radiation treatment [26], electron beam [27], plasma [105], alkali treatment [28], acetylation [29], etc.

The interesting properties resulting from several studies in the field of fibers, including the renewable character and biodegradability among others, aroused the interest of many researchers who focused on reducing the size of these fibers.

According to Lavoine et al. [30] the structure of native cellulosic fibers results in two families of cellulosic nanoparticles and the way to distinguish between these cellulose nanoparticles is to consider the steps involved in their preparation: (1) cellulose nanocrystals (chemical treatment); and (2) cellulose microfibrils (mechanical treatment). There is yet another source of nanocellulose, bacterial cellulose, which is the subject of investigation in areas that are beyond the goal of the present chapter and therefore this is not included for discussion.

Although Herrick et al. [31] and Turbak et al. [32] pioneered and introduced the preparation of cellulosic nanofibers (CNFs) from wood over 20 years ago, numerous investigations have examined different isolation processes and the use of CNFs as a reinforcement in various polymer matrices.

Currently there are different ways to produce fibers with controlled sizes, constituting thereby a new generation material: nanocellulose [6,7]. A recent study [33] shows the influence of sources on the dimensions and structure of CNCs.

Cellulose nanofibers can be extracted from various plant sources and, although the mechanical separation of plant fibers into smaller elementary constituents has typically required high energy input, chemical and/or enzymatic fiber pretreatments have been developed to overcome this drawback [11].

So, the main process in the preparation of CNC is based on strong acid hydrolysis under strictly controlled conditions of temperature, agitation, and time [30].

Different sources of cellulose were studied and used for producing cellulose CNC nanocrystal: tunicin [34]; cotton [35]; wood pulp [36]; microcrystalline cellulose (MCC) [37]; sugar-beet pulp [38], etc.

Further examples of agricultural by-products which might be used to derive nanocellulose include those obtained from the cultivation of corn, wheat, rice, sorghum, barley, sugar cane, pineapple, bananas, and coconut crops [11].

In recent years enormous efforts have been made to develop and characterize nanocellulose materials owing to their biodegradability, biocompatibility, and environmental friendliness [2,6,12,39].

13.3 Preparation of rubber/nanocellulose composites

The nanofillers of choice in polymer nanotechnology are the usual ones. They exhibit different shapes and on that basis these can broadly be categorized into the following classes:

- Spherical/cubical (e.g., nanosilica, polyhedral oligomeric silsesquioxanes, nano-$CaCO_3$, metal oxides, etc.)
- Rod/fiber (e.g., synthetic whiskers, carbon nanotubes, carbon nanofibers, boehmite, sepiolite, nano-$CaCO_3$, etc.)
- Sheet/platelet (e.g., layered silicates such as the smectite group clays, synthetic mica, etc.) [16].

The reinforcement of rubber requires rigid entities, such as CB, and the inclusion of nanofillers or aids to rubber formulations results in optimization of properties to meet a given service application or sets of performance parameters [40,41].

This chapter will show different procedures for obtaining elastomeric nanocomposites with nanocellulose.

Elastomer nanocomposites are currently prepared in the following ways: [16]

- Solution blending
- Melt intercalation
- In situ polymerization
- Latex compounding

In the solution blending method, the elastomer is solubilized in a proper solvent and then the filler dispersion is added into it. After the nanofillers get dispersed, the solvent is evaporated usually under vacuum. Curatives may be added before solvent removal [42].

From the industrial standpoint, the most economic mixing and environmentally friendly technique used for preparing elastomer/nanocomposites is by melt compounding. This technique involves equipment like internal mixers and open two roll mills. To manufacture a nanocomposite, mixing temperatures and shear rates as well as modification of elastomers and the chemistry involved during compounding and curing should be considered [16].

Concerning in situ polymerization, monomers are added and mixed with fillers in a proper solvent medium. Then they are polymerized using some initiator to get the in situ nanocomposite. There are a few elastomer nanocomposites prepared in this way.

Most elastomers are available in the form of latex, which is nothing but an aqueous dispersion of elastomer particles in the submicron to micron range.

A survey of the literature has shown that nowadays studies on CNF-reinforced latex composites have been mostly explored by research groups in France, India, China, and Brazil [6,12,18,43].

It is well-known that emulsion polymerization can lead to materials that are easily processed, either by film casting (water evaporation) or by freeze drying followed by a classical extrusion process. In addition it is possible to blend different types of latexes with different glass–rubber transition temperatures. More generally, it is also possible to mix types of water suspensions including polymer latexes and organic and inorganic suspensions [6].

In summary, the preparation of nanocomposites with latex by the most common process generally involves two steps: the nanocellulose suspension is firstly mixed with a rubber latex for a period of time to obtain a uniform dispersion of nanocellulose. The aqueous suspension is then cast in a formulation mold and dried to form the nanocomposite films [6].

Favier et al. [44] describe the technique used to process a nanocomposite material with a copolymer of styrene and butyl acrylate latex reinforced with a suspension of cellulose whiskers (mantles of tunicates).

Materials having a nanocomposite structure were obtained by casting films from a mixture of latex and a small percentage of cellulose whiskers from tunicin. The improvement in mechanical properties was ascribed to the presence of strong interactions among whiskers, such as hydrogen bonds, which lead to the formation of a rigid network governed by the percolation threshold.

Visakh et al. [45] investigated the transport behavior of three aromatic organic solvents in natural rubber (NR) nanocomposite membranes containing CNFs and cellulose nanowhiskers (CNWs) isolated from bamboo pulp. Both nanocelluloses were found to decrease the uptake of aromatic solvents in nanocomposite membranes, but the effect was more significant in the case of nanofibers compared with nanowhiskers.

In the case of bagasse fiber made from a by-product of sugar cane industry, Jarnthong et al. [46] found that the addition of this nanofiber into a NR matrix improved tensile strength and barrier properties and increased the rate of degradation of rubber in soil.

Abraham et al. [47] developed nanocomposite materials by casting and evaporating a mixture of NR latex and aqueous suspension of cellulose nanofibrils obtained from banana fiber by steam explosion.

The increase in cellulose nanofiber content in the NR matrix causes drastic effects to the mechanical properties of the composite, increasing the Young's modulus and tensile strength of materials, but decreasing the characteristic rubber elongation. A mechanism is suggested for the introduction of the Zn–cellulose complex and its three-dimensional network as a result of the reaction between cellulose and metallic Zinc which is originated during the composite formation.

Latex and melt compounding were used to explore NCC as a green alternative to CB in NR/butadiene rubber/styrene-butadiene rubber blends [48].

As reported in the referred to paper [48] the master batch technique was adopted to fill NR/BR/SBR (natural rubber/polybutadiene rubber/styrene-butadiene rubber) with NCC, which might be also acceptable for other rubber blends. The coagulation method was the approach used to get NR/NCC master batches, which was clean and simple to carry out and could ensure good dispersion of NCC in rubber. Then, the resulting master batches were used for mixing and reinforcing BR and SBR. In order to improve the interaction between rubber and cellulose, resorcinol and relative humidity (RH) was added to modify the NCC/matrix interface, which was widely used as coupling agent to improve the rubber–filler interaction. Results of partial replacement of CB by NCC through the stated method led to good performance of NR/BR/SBR blends according to the investigated aspects, along with environmental benefits and economic interests.

There is another work using nanocellulose as partial filler in nanocomposites of NR. Xu et al. [49] modified nanocrystalline cellulose (MNCC) to investigate the partial replacement of silica in NR. The results were compared with nonmodified nanocellulose (NCC). The rubber—cellulose nanocomposite was obtained by co-coagulation of NCC suspension or MNCC suspension mixed with natural rubber latex. Through the comparison of vulcanization characteristics MNCC was proved to be more efficient than NCC to partially replace silica in NR composites. According to the authors a number of advantages such as being renewable, biodegradable, easily obtained, of low density, low-cost, and environmentally friendly render NCC a new energy-saving filler to partly replace silica.

Bitinis et al. [50] reported another method to prepare cellulose—rubber nanocomposites using direct extrusion. The lyophilized nanocellulose produced from MCC was firstly pulverized in a grinder and the obtained nanocellulose powder was then mixed with NR in an extruder to form the biocomposite.

Modifications of surface properties of CNCs by physical methods have also been studied. Cao et al. [51] investigated the effect of surface modification of CNCs by a mechanical method via vulcanization on the rheological properties of CNC—nitrile butadiene rubber (NBR) composites. NBR latex was mixed with a CNC suspension at a specific ratio by strong mechanical stirring at room temperature before being compressed and vulcanized at 160°C. In this case, cross-link density can be the result of the combination of chemical cross-links and physical adsorption such as chain entanglements.

According to Krawczak [52] the high potential of nanocomposites has already been demonstrated at the lab scale. It is now time to bridge the gap between scientific challenges and industrial stakes.

13.4 Major techniques used for characterizing rubber nanocomposites

Numerous methods have been reported in the literature to correlate the appropriate structure—property of nanocomposites. The two most commonly used methods to determine the structure of nanocomposites are wide angle X-ray scattering and transmission electron microscopy (TEM) [16].

For example, in the work of Xu et al. [49] on the effects of partial replacement of silica with MNCC on surface properties of nanocomposites of NR, both TEM and scanning electron microscopy (SEM) are quite useful for the analysis of the final results. By using TEM initially the silica and NCC suspensions are analyzed. The results show that NCC aggregates can be easily separated due to the following two factors: geometrically, the rodlike NCC aggregates are more easily dispersed than the spherical particles; chemically, these are recognized as exhibiting low hydroxyl density at the surface as compared with silica. The morphology of NCC in the nanocomposites was studied by SEM for the stretch broken surface of different nanocomposites. These results indicated that the addition of MNCC reinforced the bonding between rubber matrix and filler and led to more firm bonds. Therefore, a very good comprehensive performance was reflected in the macroproperties.

In order to analyze the crystalline structure of lignocellulosic materials, X-ray diffraction (XRD) is generally used. It is widely recognized that the cellulose molecular structure is partly crystalline and partly amorphous.

The X-ray technique is most often used to analyze cellulose itself, but not when it is part of a nanocomposite. However the results thus obtained help in the understanding of the final performance of the nanocomposites.

XRD patterns were measured by Pasquini et al. [53] for both cassava bagasse and cellulose whiskers obtained under hydrolysis. The degree of crystallinity was obtained by comparing the area under crystalline peaks and the amorphous curve. Each diffractogram displays the characteristic cellulose peaks. The acidic treatment induces a narrowing and an increase in the magnitude of both peaks most probably because of the higher crystallinity of the hydrolyzed cellulosic residue as compared with the original fibers present in the cassava bagasse. The degree of crystallinity was determined and found to be 43.7% and 54.1% for the cassava bagasse and hydrolyzed residue obtained by hydrolysis condition, respectively.

Jonoobi et al. [43] present an overview of recent research in the area of cellulose nanomaterial production from different sources as well as the methods used for the isolation of nanomaterials and characterization by X-rays including nanocrystals and nanofibers from various sources.

Comparison of MCC with NCC was carried out using the X-ray technique by Chen et al. [48]. According to the XRD pattern, MCC and NCC had similar XRD curves. In addition, the crystallinities of MCC and NCC were calculated to be 87.12% and 87.96%, respectively, which were almost the same. Thus, the acid hydrolysis hardly reduced crystallinity. A slight difference may be in the shapes of the peaks at 38.1° and 44.3°. The explanation was that the acid hydrolysis mainly took place at the amorphous regions and imperfect crystalline regions. Hence, the NCC large-angle peaks tended to fade away while those at small-angle were sharpened. The retention of the crystalline area was able to preserve the mechanical properties of cellulose, which led to a good reinforcing effect.

Not only the morphology but also the dynamic viscoelastic behavior, apparent cross-link density, mechanical performance, and dynamic mechanical properties of NR/BR/SBR/NCC blends were discussed in detail. The results showed that NCC was uniformly dispersed in composites and resorcinol and hexamethylene tetramine (RH) could enhance the adhesion of NCC and matrix. According to the dynamic mechanical analysis (DMA), NCC performed a reinforcing effect comparable with that of CB, and the modulus was improved with modification by RH. Mechanical tests showed that the replacement of CB by NCC in the blends did not deteriorate the mechanical properties of composites. Lastly, the structure, morphology, and performance of NCC loaded NR/BR/SBR blends were investigated by SEM, rubber process analyzer (RPA), apparent cross-link density, DMA, and mechanical properties tests.

Other interesting properties analyzed in relation to elastomeric nanocomposites with nanocellulose are related to gas barrier properties and biodegradability. Bras et al. [22] studied these properties using dynamic mechanical thermal analysis (DMTA) and differential scanning calorimetry (DSC) of the bagasse nanocomposite with NR. The presence of bagasse whiskers resulted in an increase in moisture sorption of rubber films up to 5% whiskers loading while at higher whiskers loading the moisture sorption tended to decrease. Barrier properties to water vapor decreased on increasing cellulose whiskers up to 7.5% whiskers loadings then increased with further increases in whiskers loading. The presence of cellulose whiskers increased the rate of degradation of rubber in soil. DMTA and DSC results showed no change in glass transition temperature (T_g) of the rubber matrix upon addition of cellulose whiskers, but cellulose whiskers have shown a reinforcing effect on the softening of rubber.

The transport behavior of aromatic organic solvents (benzene, toluene, and *p*-xylene) in NR nanocomposite membranes containing CNFs and CNWs isolated from bamboo pulp was investigated by Visakh et al. [45]. The solvent molecules act as molecular probes to study the diffusion, sorption, and permeation through the nanocomposites and provide information on the nanocomposite structure and matrix–filler interactions.

In a review article by Zhou et al. [6], wood flour and other lignocellulosic fibers filled rubber composites including cellulose–rubber composites and nanocellulose-based rubber nanocomposites were analyzed. Regarding the nanocellulose composites, the results show the nonlinear mechanical properties of rubber nanocomposites at different nanocellulose contents. The tensile strength and Young's modulus of the composites increased steadily upon nanocellulose addition, while the elongation at break decreased. The significant reinforcing effect could be ascribed to the mechanical percolation phenomenon of rodlike nanocellulose which helps to form a stiff continuous network of cellulosic nanoparticles linked through hydrogen bonding.

Biodegradation of rubber nanocomposites was also analyzed. It usually took place in soil or compost, in which there is a high microbial diversity for 1–3 months, then the materials are decomposed into water and CO_2 owing to the biological activity which induces chemical structure changes to the exposed material. The review by Zhou et al. [6] reports results on the biodegradation of bagasse nanocellulose/rubber and jute fiber nanocellulose/rubber nanocomposites in soil and compost, respectively. The weight loss of these two composites was the result of loss of both rubber and the reinforced nanocellulose. It is clear that the neat rubber matrix degraded slowly with relatively low weight loss, but the addition of nanocellulose pronouncedly accelerated the biodegradation of the composites. Furthermore, the results revealed that compared with cross-linked composites, the noncross-linked nanocomposites suggested better degradation. The strong interconnected network in the cross-linked samples provided less accessibility of cellulose molecules to degradation and also the diminishment of microorganisms' reactivity. In the meantime, the internal bonding between rubber and nanocellulose formed during cross-linking inhibited the degradation of the composites.

Ching et al. [39] investigated the rheological properties of CNC–polymer composites.

Generally, the addition of nanofillers to a polymer modifies the rheological properties of the resulting composites since they embed themselves among the polymer chains. Furthermore, rheology also provides critical insights into the processing characteristics of these materials and could help predict the effect of nanofillers on the final composite materials.

The viscoelastic behavior of nanocomposite materials can be evaluated by measuring the viscoelastic components, elastic components, G', and viscous components, G'', in which the viscoelastic behavior is strongly dependent on the nanofiber content.

Various studies have shown that the viscoelastic properties of nanocomposites depend strongly on the CNC content. Even at low CNC loading, storage modulus significantly increased compared with the neat polymer matrix. In addition to fiber concentration and shear frequencies, the rheological behavior of CNC-based nanocomposites depends on other factors such as fiber shape and size, surface area, aspect ratio, flexibility, surface chemistry of fiber, rheological properties of matrix materials, and temperature of the system [39].

13.5 Structure and properties of different rubber nanocomposites with nanocellulose

According to Oksman, the main challenge when producing cellulose-based nanocomposites is to disperse the nanoreinforcements into the polymer matrix without degradation of the polymer or the reinforcing phase and also to develop composites based on nanocellulose using suitable large-scale processing technologies [9].

Klemm and collaborators pointed out that the process for obtaining nanocellulose is laborious since it involves several steps, namely, purification, bleaching, fibrillation, and hydrolysis [1].

Another difficulty is associated with the lack of compatibility of hydrophobic polymers using nanocellulose in composites, and various chemical modification methods have been explored in order to address this hurdle [11].

To overcome the difficulty of obtaining a homogeneous dispersion by the incorporation of nanocellulose into the elastomer compound, we carried out the mixing process in liquid phase: the fibrous material was reduced to a viscous, alkaline aqueous solution of sodium cellulose xanthate, while the elastomer was used as an acid-sensitive aqueous rubber emulsion. The apparently homogeneous, viscous, milky mixture was slowly coagulated by addition of a 1:1 molar mixture of sulfuric acid and zinc sulfate with stirring at room temperature [54].

Our research group developed different procedures to disperse cellulose in the nanoscale and introduce it into elastomeric composites [13,14], these procedures being based on the Kalb and Manley [55] work.

Kalb and Manley [55] have demonstrated that cellulose fibers can be formed by precipitation from dimethyl sulfoxide/p-formaldehyde solutions by two methods that involve different mechanisms of fiber formation, i.e., (a) precipitation of cellulose by addition of a nonsolvent to the stirred cellulose solution, and (b) precipitation of cellulose by coagulation of droplets of cellulose solution in a stirred precipitant. Both processes yield fibers with properties depending on the stirring speed and coagulant strength. The two techniques (a) and (b) described above have been used by our research group, based on mixtures of elastomer latices with cellulose xanthate solution [13,14].

Industrially, cellulose from the xanthate process, named regenerated cellulose or cellulose II is prepared by pretreating cellulose with sodium hydroxide to produce alkali cellulose which reacts readily with carbon disulfide at room temperature to form an alkaline soluble cellulose xanthate. The cellulose xanthate solution is forced through spinnerets and the filaments produced are coagulated and converted to cellulose by immersion in an aqueous solution of sulfuric acid and sodium sulfate. During xanthation, secondary reactions take place and the by-products are also decomposed in the spin bath [56]. Although not an industrial process, fiber formation by precipitating polymers in shear fields has been also described in the literature [17,55].

Compared with the melt or solution methods, the approach of co-coagulating rubber latex and filler aqueous suspension is promising due to the simplicity of the preparation process [57].

So, the structure and properties of different nanocomposites of elastomers and nanocellulose are discussed below on the basis of two systems: S1 System, [14], coprecipitation through the addition of aqueous solution of sulfuric acid and zinc sulfate to stirred cellulose xanthate-rubber latex, at room temperature and constant rate; and S2 System, [13], coprecipitation through the addition of cellulose xanthate-rubber latex into aqueous solution of sulfuric acid and zinc sulfate, at room temperature and constant rate.

13.5.1 Natural rubber

The market for products obtained from renewable resources has steadily increased over the recent years. For the development of lightweight materials and the materials with anisotropic properties fiber–polymer composites are of great interest [58,59].

Although NR is an amorphous material it is a very strong matrix due to its strain-induced crystallization. Reinforcing fillers from renewable resources, such as natural fibers, for the production of biosustainable composite materials can be thought of as a potential alternative [19,60].

In a recent report, results obtained by our research group for nanocomposites with nanocellulose were delivered focusing on NR matrix [12]. Further complementary results will be provided related to the development of renewable source, clear materials, nanocellulose nanocomposites having NR as elastomeric matrix. Different nanocomposites were obtained based on the two referred to kinds of co-coagulation Systems for cellulose xanthate and elastomeric latex [13,14].

The results of some properties obtained by comparing the two systems have shown that the S2 System offers the best possibilities of interaction between the two macromolecules, poly-1,4-cis isoprene and cellulose II (or regenerated cellulose or nanocellulose), leading to the best performance [61].

Mechanical properties of NR nanocomposites were determined and the results classified cellulose II as a reinforcing filler. The best performance was achieved by the composite with 15 phr of cellulosic filler [62,63].

The dynamic properties of the nanocomposites studied are presented in Table 13.1 [62].

The crescent incorporation of cellulose II lowers tan δ values as a consequence of higher elastic moduli. In addition, values of T_g for composites with filler content above 20 phr are lower as compared with unfilled NR, while the NR T_g is not altered for other filled composites. Lower values of T_g are indicative of good dispersion of the filler in the elastomeric matrix, and the unchanged T_g of the composite with 15 phr of filler suggests the presence of strong rubber–filler interaction.

Table 13.1 Dynamic properties of NR/Cell II nanocomposites

NR/Cell II		Dynamic properties		
Nanocomposites		Log E′(Pa)	T_g (°C)	Tan δ
Cell II (phr)	0	5.2562	−49.2	2.1663
	10	5.7771	−48.1	1.6369
	15	5.7383	−49.3	1.6053
	20	6.3845	−49.8	1.3639
	30	6.5750	−50.8	1.1854

Source: Martins AF. Nanocompósitos de borracha natural e celulose II. PhD. Universidade Federal do Rio de Janeiro, Instituto de Macromoleculas Professora Eloisa Mano, Rio de Janeiro, Brazil; 2002.

NR/Cell II nanocomposites were examined using ^{13}C high resolution solid-state nuclear magnetic resonance (NMR) with cross-polarization (CP) and magic angle spinning. The 1H relaxation times in the rotating frame ($T_{1}\rho$) were measured using the standard procedure based on a delayed contact time spin lock period preceding a constant CP time. Twelve points were acquired in the range of 800–25 ms [62,64].

Fig. 13.1 shows the plotting of $T_{1}\rho^{H}$ values at 127 ppm versus stress at break for the nanocomposites studied. It can be seen that at 15 phr of cellulose II the composite presents a more rigid molecular dynamics as compared with unfilled NR, and also the highest stress at break. This finding is consistent with the excellent dispersion of the cellulosic filler throughout the elastomeric matrix, and is evidence for a strong rubber–filler interaction. Solid state ^{13}C NMR showed the occurrence of a physical rubber–filler interaction which correlated very well with the mechanical results [62].

The transport of carbon dioxide, oxygen, and nitrogen in vulcanized NR with regenerated cellulose was evaluated.

The values of the permeability coefficient of carbon dioxide, oxygen, and nitrogen for pure gum and composites with 15 phr and 25 phr of nanocellulose were measured at 25°C and 15 cmHg of pressure [65,66].

The apparent solubility coefficient S values are shown in Table 13.2 [65].

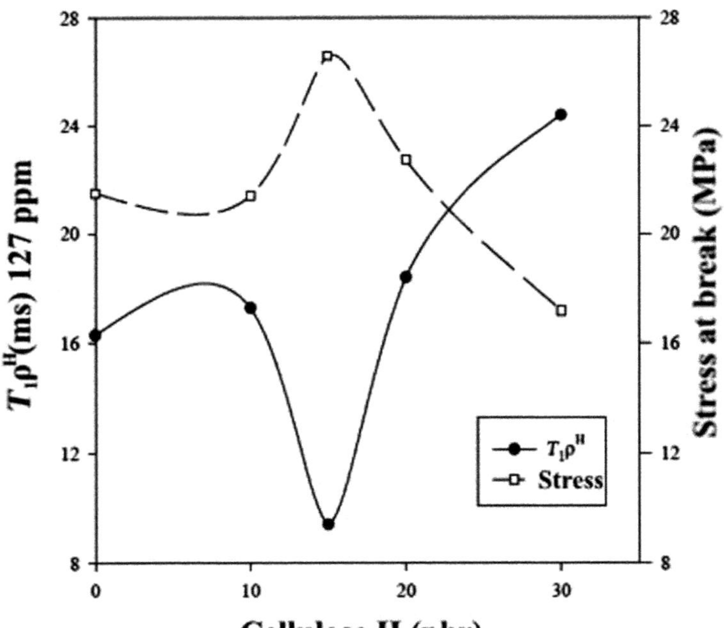

Figure 13.1 $T_{1}\rho^{H}$ *versus* stress at break for NR/Cell II nanocomposites.
Source: Martins AF. Nanocompósitos de borracha natural e celulose II. PhD. Universidade Federal do Rio de Janeiro, Instituto de Macromoleculas Professora Eloisa Mano, Rio de Janeiro, Brazil; 2002.

Table 13.2 Apparent solubility coefficient (S) of CO_2, O_2, and N_2 for the NR, NR15, and NR25 membranes, at 25°C, under upstream chamber pressure of 15 cmHg

Gas	S[a] (NR)	S[a] (NR15)	S[a] (NR25)
CO_2	0.0250	0.0190	0.0170
O_2	0.0030	0.0031	0.0032
N_2	0.0023	0.0020	0.0014

[a]The apparent solubility coefficient is given in cm^3 gas (STP/cm^3) polymer cmHg.
Source: Nunes RCR, Compañ V, Riande E. Gas transport in Vulcanized natural rubber-cellulose II composites. J Polym Sci, Polym Phys 2000; 38: 393–402.

The relatively high boiling point of CO_2 is mainly responsible for the high solubility of this gas in the films in comparison with that of oxygen or nitrogen (see Table 13.2). In the same way, oxygen is more condensable than nitrogen and, hence, the former gas is more permeable than the latter.

The analysis of the permeability characteristics of the membranes in terms of the free-volume theory suggests that gas transport is severely hindered in both the cellulose phase and the cellulose–rubber interphase of the composites.

The values of the permeability coefficient of carbon dioxide, oxygen, and nitrogen in the composites with 25 phr of cellulose, measured at 25°C and 15 cmHg of pressure are roughly one-third of those measured under the same conditions for these gases in a pure gum NR composite [65,66].

Another work reports sorption processes of oxygen, carbon dioxide, methane, ethylene, and propylene in films of vulcanized NR pure gum and with 15 phr and 30 phr of regenerated cellulose (or cellulose II). These studies have shed light on the compatibility of NR–organic filler. Actually, the critical interpretation of the experimental permeation parameters in terms of the free-volume theory suggests that gas transport is severely hindered in the cellulose–rubber interface presumably as a consequence of fair NR–cellulose adhesion. Moreover, the results indicate that permeability and diffusion coefficients decrease with increasing pressure, a behavior attributed to changes in free volume as a result of hydrostatic pressure [67].

13.5.2 Natural rubber and epoxidized natural rubber

NR has many advantages derived from its excellent tensile and tear strengths, besides its dynamical properties. However, NR has poor solvent, oil, and ozone resistance, which can be improved by epoxidation. Epoxidation is an efficient method for introducing reactive groups and polarity onto the polyisoprene backbone. It can be carried out in solution or latex forms but the latter is more economical and is environmentally safe [68–71].

The study of the influence of epoxidized natural rubber (ENR) on nanocomposites with cellulose was conducted by adding ENR into NR master batches with nanocellulose obtained by the S2 System as developed by our research group [13]. The cellulose content was set at 20 phr and the level of NR/ENR was varied

Table 13.3 **Mechanical properties of NR/ENR/Cell II nanocomposites**

NR/ENR/Cell II composites (phr)	Elongation at break (%)	Tensile strength (MPa)	Tear strength (kN/m)
100/0/0	535.4	21.7	58.3
100/0/20	567.4	23.4	57.0
75/25/20	619.6	23.1	56.6
50/50/20	629.2	22.1	50.8
25/75/20	626.9	17.5	39.4

Source: Fernandes RMB, Visconte LLY, Nunes RCR. Curing characteristics, mechanical properties and solvent resistance after ageing of natural rubber/epoxidized natural rubber and cellulose II. Int J Polym Mater Polym Biomater 2011; 60(5): 351–64.

from 100/0, 75/25, 50/50, and 25/75. A pure gum composite (NR without ENR and cellulose II) was also studied to compare the properties. The epoxide content was 25 mol% [71]. The mechanical properties are presented in Table 13.3 [71]. As can be seen, the elongation at break increases with the addition of ENR into the NR/ENR/Cell II nanocomposites. A slight increase in tensile strength with addition of cellulose is noticed, but these values do not undergo large variations with incorporation of ENR until the 50/50/20 blend, as both rubbers crystallize under tension and have similar behavior [72]. There is an excess of ENR in the 25/75/20 blend, decreasing the interaction between the two rubbers and, hence, decreasing the tensile strength. Tensile strength is maximal for the blend containing 75 phr of NR and 25 phr of ENR. Beyond this content, tensile strength decreases due to phase inversion taking place in the system. Tear strength decreases with incorporation of ENR and this could be due to the weak interaction between NR and ENR.

Dynamic mechanical properties were studied and the variation of loss factor (tan δ) as a function of temperature for NR/ENR/cell II compositions is shown in Fig. 13.2 [71]. The incorporation of cellulose increases the T_g of the NR composite, suggesting smaller chain flexibility. The T_g of the 100/0/0 and 100/0/20 blends appear at about $-45°C$. With incorporation of ENR the appearance of two peaks is noticed, one at about $-45°C$ and the second (on the T_g of ENR) between $-20°C$ and $-25°C$. This suggests that the system studied is a physical mixture and that the miscibility between the rubbers is low.

Fig. 13.3 displays the storage modulus (E′) as a function of temperature for NR/ENR/Cell II blends. The storage modulus decreases with increasing temperature. This behavior is observed with the increase in mobility of chains and as it approaches the transition and rubbery regions. It is observed that the curves of storage modulus are different after the addition of ENR, indicating two distinct phases [71].

The best properties were obtained for compositions with 25 phr of ENR and it is noticed that this is the limit for incorporation of ENR. The results of 50/50/20 nanocomposites allow the conclusion that there is separation of the rubber phases, which causes low miscibility.

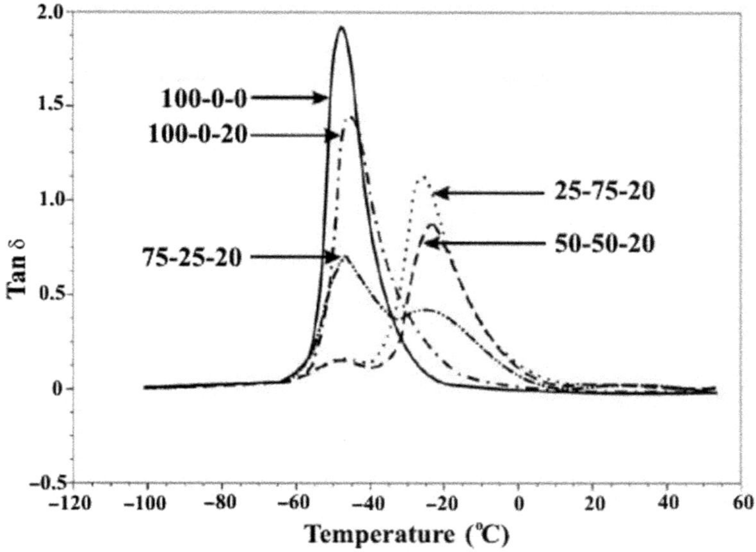

Figure 13.2 Tan δ versus temperature compositions of NR/ENR/Cell II.
Source: Fernandes RMB, Visconte LLY, Nunes RCR. Curing characteristics, mechanical properties and solvent resistance after ageing of natural rubber/epoxidized natural rubber and cellulose II. Int J Polym Mater Polym Biomater 2011; 60(5):351–64.

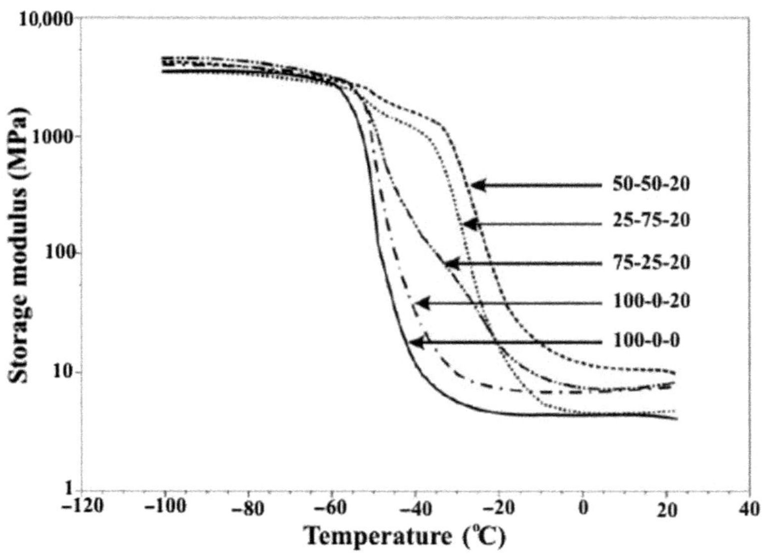

Figure 13.3 Storage modulus versus temperature for NR/ENR/Cell II compositions.
Source: Fernandes RMB, Visconte LLY, Nunes RCR. Curing characteristics, mechanical properties and solvent resistance after ageing of natural rubber/epoxidized natural rubber and cellulose II. Int J Polym Mater Polym Biomater 2011; 60(5):351–64.

The dynamic-mechanical tests are in agreement with the evaluated mechanical properties. Nanocomposites with ENR show two peaks, each one corresponding to the specific rubber, showing that the system studied is a physical mixture.

13.5.3 Natural rubber and polybutadiene rubber

Blending of two or more rubbers is carried out for three main goals: improvement in technical properties, better processing and lower compound cost. Many products in the rubber industry are based on blends. Tires are typical examples of products of large-scale volume production where NR/BR blends are extensively used [73].

Mechanical properties of elastomer blends depend on their morphology, chemical nature, crystallinity, processing, testing procedures, elastomers ratio, and cure system.

It was verified that both systems to incorporate cellulose into elastomers compounds discussed above (S1 and S2) were efficient in producing good performance composites [61,74–76].

The results shown in these items refer to the procedure employing the S1 System for the incorporation of cellulose into a mixture of elastomeric latices of NR and BR [14].

Different ratios of the NR and BR elastomers in latex form were admixed (NR/BR 75/25, 50/50, 25/75) and the nanocellulose content in the blends was varied up to 25 phr. Concerning the mechanical properties of NR/BR/Cell II mixtures under study, the composites with higher NR contents display better tensile strength, tear strength, and abrasion resistance properties, which are affected by filler quantity. These results characterize a positive influence of filler on the composites and the examination of the data confirms the reinforcing action of cellulose in NR/BR compounds. Resilience and compression set are functions of the material elasticity, the obtained values for these two properties are the highest for those compositions where BR predominates and cellulose II is in a lower ratio. The optimum cellulose content depends on each of the analyzed properties and varies from 15 phr to 25 phr [77,78].

In accordance with the literature data this behavior is presumed to be due to the different ratios with which mechanical properties are developed in rubber vulcanizates [79].

The interaction between cellulose and the elastomeric matrix in NR/BR blends of different ratios was also investigated by equilibrium swelling using the Kraus and Cunneen and Russel equations [78,80,81]. It was observed that elastomers–cellulose II compositions obey the Kraus and Cunneen and Russel equations up to a certain volume fraction of cellulose. There is a deviation from linearity at higher filler loadings also observed in other works [82–84].

In our case, the observation of the various compositions having cellulose as filler has shown that this material effectively participates in cross-linking. The degree of polymer–filler interaction follows the order (NR/BR): 25/75 $>$ 50/50 $>$ 25/75 $>$ 100/0, suggesting that higher elastomers–cellulose attachment takes place in systems with higher BR contents.

13.5.4 Natural rubber and ethylene propylene diene methylene rubber

There is an ever increasing technological interest in the use of blends of dissimilar rubbers to improve specific properties of vulcanizates. Ethylene propylene diene methylene rubbers (EPDMs) are relatively inexpensive elastomers and differ from most general purpose elastomers like NR by having saturated hydrocarbon backbones which impart unusually good weathering oxidation and chemical resistance. Blending EPDM with NR and other dienes gives rise to compounds with good ozone and chemical resistance and reduced compression set. NR is often added to EPDM to improve the building of tack properties [85].

A nanocomposite of NR with 15 phr of nanocellulose obtained through the S1 System was mixed with EPDM [14]. Two types of EPDM were included in this study [86]. One of them (EP1) has an ethylene content of 64.6 mol %, which is indicative of a low level of potentially crystallizable sequences, and the other one (EP2) has 74.2 mol% ethylene, which corresponds to approximately 22% crystallinity [87].

Based on the NR-15 Cell II master batch, the following blends of EPDM/NR were obtained: 25/75, 50/50, 75/25, where the concentrations of cellulose were, respectively, (in phr), 11.5, 5.5, and 3.75. Two types of EPDM, respectively, EP1 and EP2 were used in all compositions.

The best mechanical results were given by those blends with 75/25 EPDM/NR which might suggest the existence of some compatibility between the two polymers. As for the two different EPDM's, it was observed that the higher ethylene content, EP2 elastomer, contributes to better performance.

13.5.5 Styrene-butadiene rubber with nanocellulose and clay

SBR represents half of all synthetic rubber production, and is much consumed in tires, where it competes with and complements NR [69].

By using the S1 System described above SBR nanocomposites with nanocellulose were obtained. Results were compared with the CB nanocomposites obtained by the conventional procedure with roll mill. CB exhibits the expected behavior and the presence of nanocellulose does not affect substantially the mechanical properties of the SBR pure gum vulcanizates [14,54].

In the effort to develop a clear, low-cost product and which additionally could be industrially manufactured, hybrid SBR nanocomposites with SBR and nanocellulose were obtained by the addition of pristine clay.

Nonorganophilic pristine clay is an inexpensive natural mineral and can be dispersed in water, which acts as a swelling agent via hydration of the intergallery cations. Admixture of latex with pristine clay followed by coagulation is therefore an interesting way to produce rubber nanocomposites [88].

Based on the S2 System [13] nanocomposites of SBR with nanocellulose and clay were prepared by co-coagulating SBR latex, cellulose xanthate, and clay aqueous suspension mixtures. The incorporated amount of cellulose was 15 phr, the clay content varied from 0 phr to 7 phr and the influence of clay fillers was investigated [89,90].

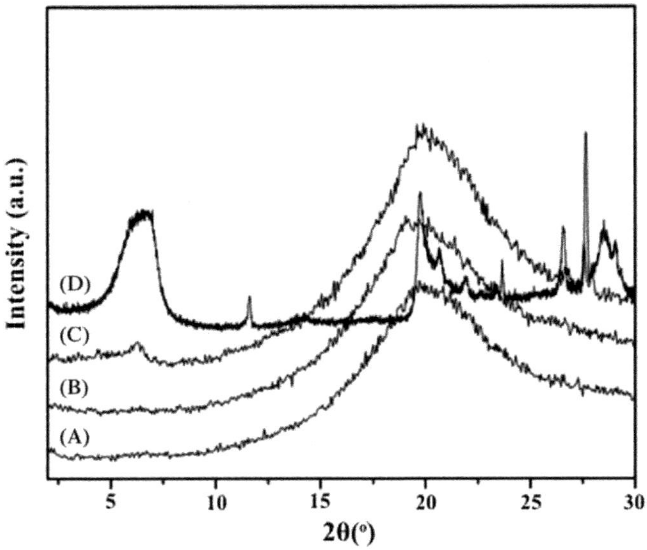

Figure 13.4 XRD patterns of SBR/cellulose/clay cured nanocomposites with different clay contents and pristine clay: (A) 3 phr clay; (B) 5 phr clay; (C) 7 phr clay; (D) pristine clay. *Source*: Zine CLG, da Conceição AJ, Visconte LLY, Ito EN, Nunes RCR. Styrene-butadiene rubber/cellulose II/clay nanocomposites prepared by cocoagulation-mechanical properties. J Appl Polym Sci 2011;120: 1468–74.

XRD is a conventional method to determine the interlayer spacing of clay layers and was used to investigate the original clay and the nanocomposites (Fig. 13.4) [89].

Results in Fig. 13.4 suggest that in the nanocomposites prepared with 3 phr and 5 phr of clay silicate layers are completely exfoliated. Bentonite clay showed a characteristic diffraction peak at $2\theta \approx 6.5°$, which corresponded to an intergallery distance of 1.35 nm. The peak of the nanocomposite with 7 phr of clay appeared at $2\theta \approx 6.25°$, which corresponded to an intergallery distance of 1.41 nm. This value is larger than the initial one of bentonite clay (1.35 nm), which seems to indicate that rubber molecules are intercalated in between the clay interlayers [90].

The values of all the mechanical properties evaluated are shown in Table 13.4 [89]. It is clear that the tensile strength property was improved upon the addition of cellulose II to the rubber matrix and reached about 300% that of the SBR gum. With an increase in the clay amount from 3 phr to 5 phr, the tensile strength of the nanocomposite increases slightly and reaches about 396% compared to SBR gum. Further clay addition led to a decrease in the property, but the value was still higher than that for the unfilled rubber. From these results, it can be said that 5 phr of clay is enough to achieve the best tensile properties concerning the SBR/cellulose II/clay nanocomposites. The energy at break was higher in comparison with SBR gum as the combination of cellulose II and clay improved the SBR matrix toughness. The incorporation of cellulose II into the rubber caused a significant increase in the

Table 13.4 Mechanical properties of SBR and nanocomposites with cellulose II and different clay contents

SBR/cellulose II/clay					
Properties	**100/0/0**	**100/15/0**	**100/15/3**	**100/15/5**	**100/15/7**
300% modulus	1.5 ± 0.16	4.9 ± 0.22	5.0 ± 0.26	5.8 ± 0.09	5.6 ± 0.31
Tensile strength (MPa)	2.17 ± 0.16	7.35 ± 0.54	7.66 ± 0.16	8.59 ± 0.08	8.09 ± 0.31
Elongation at break (%)	368 ± 45	422 ± 18	447 ± 14	468 ± 9	450 ± 27
Tear strength (kN/m)	12.30 ± 0.99	27.03 ± 2.89	32.94 ± 2.31	34.25 ± 2.24	38.58 ± 1.67
Hardness (shore A)	38 ± 0.6	51 ± 0.6	52 ± 0.8	57 ± 1.0	56 ± 1.0
Energy at break (Nmm)	857 ± 144	2803 ± 279	3197 ± 90	4213 ± 221	3780 ± 294

Source: Zine CLG, da Conceição AJ, Visconte LLY, Ito EN, Nunes RCR. Styrene-butadiene rubber/cellulose II/clay nanocomposites prepared by cocoagulation-mechanical properties. J Appl Polym Sci 2011; 120: 1468–74.

hardness of the nanocomposite, as expected, which is indicative of higher stiffness. The hardness also increased with the addition of clay, but the increase was not significant due to the low clay loading. The tear strength of the nanocomposites was excellent in comparison with the gum vulcanizate and further increased with the amount of clay [89].

The swelling measurements for the vulcanized rubbers in contact with organic solvents was used to determine the cross-linking density, which can be related to changes in physical properties (Flory–Rehner method) [91]. The results are compared with data on cure. With an increase in clay amount to 5 phr the cross-linking density increases. Further clay addition leads to a decrease in the property. These results agree with the difference between maximum and minimum torques, which is reflected in an increase in the elastomers strength [92]. In fact, the combined cellulose II and clay behave as an effective reinforcing agent for SBR rubber.

13.5.6 Acrylonitrile butadiene rubber and carboxylated acrylonitrile butadiene rubber

Nitrile rubber is used for its resistance to various organic oils, fuels, and chemicals in seals, gaskets, and hoses. The grades offered differ in acrylonitrile percentage (ACN) in the polymer chain as well as in the overall viscosity of the polymer [93].

Nanocomposites with three types of NBR, respectively, with 26, 33, and 45% acrylonitrile (AN) and nanocellulose content up to 25 phr were obtained [94,95]. As virtually all nitrile elastomers are manufactured by emulsion polymerization, the process used to produce nanocomposites, the results of which are shown in Figs. 13.5 and 13.6 [94], was the coprecipitation through the addition of aqueous

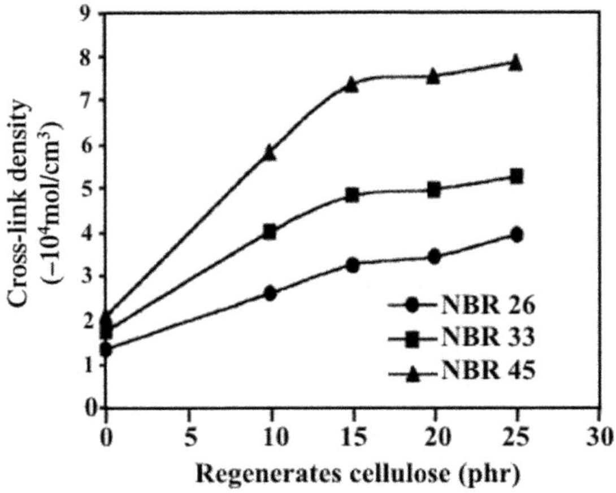

Figure 13.5 Cross-link density for NBR nanocomposites in THF.
Source: Nunes RCR, Affonso JES. Interaction of NBR Rubber with Regenerated Cellulose. Kautschuk Gummi Kunststoffe 1999; 52: 787–93.

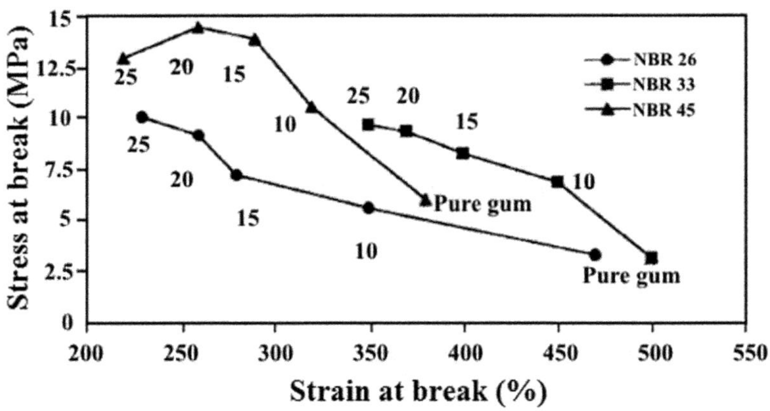

Figure 13.6 Tensile strength of NBR nanocomposites.
Source: Nunes RCR, Affonso JES. Interaction of NBR Rubber with Regenerated Cellulose. Kautschuk Gummi Kunststoffe 1999; 52: 787–93.

solution of sulfuric acid and zinc sulfate to the stirred xanthate–NBR latex cellulose, at room temperature and constant rate, called S1 System [14].

According to the work of Harris [56], during the decomposition of cellulose xanthate, the production of CS_2, H_2S, and CO_2 as gases takes place.

In the literature Singha et al. [96] carried out experiments on the chemical resistance of NBR to acid and gases. It was pointed out that hydrogen sulfide reacts with the carbon–carbon double bond leading to cross-linking. In the case of NBR,

Table 13.5 Tensile strength of NBR with 33% acrylonitrile (AN) and nanofiller content up to 30 phr cellulose

NBR/ cell II	Stress at break (MPa)	Strain at break (%)	Modulus at 100%(MPa)	Tear strength (kN/m)
100/00	3.4 ± 0.76	600 ± 26.5	1.07 ± 0.67	15.24 ± 1.08
100/10	6.1 ± 0.48	675 ± 25.0	1.30 ± 0.73	19.55 ± 2.83
100/20	6.8 ± 0.79	700 ± 50.0	1.73 ± 0.89	24.37 ± 3.06
100/30	5.5 ± 0.61	450 ± 28.9	2.78 ± 0.59	27.48 ± 2.46

Source: Lapa VLC, Suarez JCM, Visconte LLY, Nunes RCR. Fracture behavior of nitrile rubber-cellulose II nanocomposite. J Mater Sci 2007; 42, 9934–39.

H_2S attacks -CN groups to form thioanilide which in turn can react with another nitrile group to form cross-links.

These studies justify the excellent results found by Nunes and Affonso [94] for nanocellulose in NBR with 43% AN as shown in Figs. 13.5 and 13.6.

The S2 System [13] was also used with NBR latex and cellulose xanthate [97,98]. The tensile strength results are shown in Table 13.5 [97] which can be compared with the process used by [94].

Brandt et al. [59] developed a new technique to co-coagulate rubber latices and cellulose xanthate, namely dynamical co-coagulation [59,99]. The viscoelastic properties of the dynamically processed NBR Cell nanocomposites exhibit a nonlinearity behavior similar to that of CB-filled systems. The percolation threshold in NBR is already crossed at a Cell II content of 20 phr. If compared with the CB-filled NBR system the percolation threshold appears above 20 phr but below 50 phr CB. This means that stiffness can be reached with a smaller amount of Cell II than with CB.

The nanocomposite character of NBR nanocellulose prepared by Lapa et al. [97] and Brandt et al. [59] are shown in Fig. 13.7(A) and (B), respectively, by TEM.

The TEM pictures show that cellulose has nanometric dimensions, it is quite well distributed, and that molecules are not touching each other even at the low concentrations of 20 phr and 10 phr.

Another nanocompound with cellulose was obtained with carboxylated nitrile butadiene rubber (XNBR).

By using both S1 and S2 Systems, nanocomposites of XNBR with 33% of ACN content were obtained with nanocellulose and with CB. The results of mechanical properties are shown in Table 13.6 [100].

Data for tensile and tear strength allow the classification of nanocellulose as a reinforcement filler for XNBR due to its improved performance as compared with that of CB. A further advantage is the cellulose clear color. The best performance was with 15 phr cellulose. Among the procedures employed for obtaining nanocomposites the S2 System shows the best results with the consequent better polymer–filler interaction.

A similar study using the S1 and S2 Systems with NR and nanocellulose was carried out, corroborating the results obtained for XNBR [12,61].

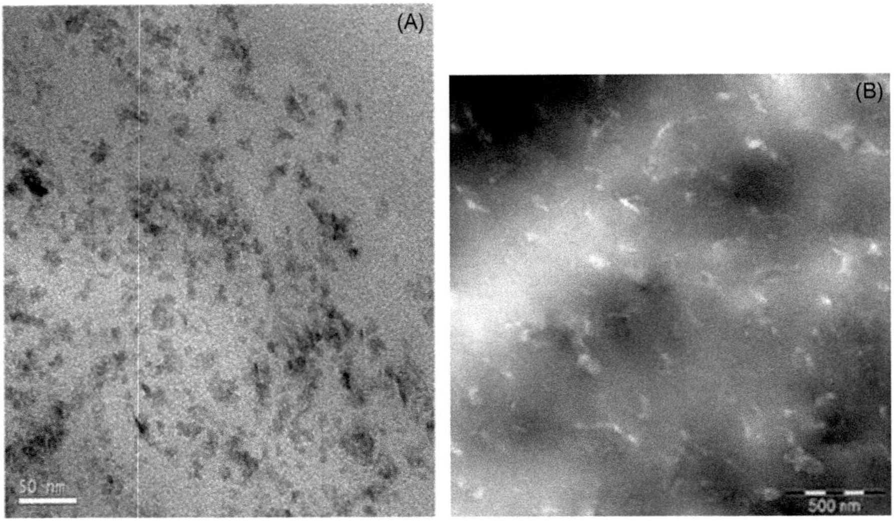

Figure 13.7 (A) and (B) Transmission electron microscopy of NBR nanocomposites. (A) NBR 20 phr of nanocellulose. (B) NBR 10 phr nanocellulose.
Source: (A) Lapa VLC, Suarez JCM, Visconte LLY, Nunes RCR. Fracture behavior of nitrile rubber-cellulose II nanocomposite. J Mater Sci 2007; 42: 9934–39. (B) Brandt K, Schuster RH, Nunes RCR. Influence of process parameters on the viscoelastic behavior of NBR-cell II nanocomposite mixtures. Kautschuk Gummi Kunststoff 2006; 10: 511–15.

Another research with XNBR was also carried out to develop hybrid composites reinforced with cellulose and montmorillonite.

XNBR, cellulose, and montmorillonite were prepared by co-coagulation of XNBR latex, cellulose xanthate, and clay aqueous suspension mixtures (S2 System). Since in previous works the 15 phr nanocellulose content led to the best performance of the XNBR nanocomposites, this figure was kept constant for the montmorillonite hybrids [13,100].

Hybrid nanocomposites have been prepared by dispersing different clay loads (0–7 phr) into water with vigorous stirring, followed by addition of XNBR latex. After homogenizing the mixture, cellulose xanthate (with 15 phr of cellulose II content) was added under stirring and the mixture was coagulated with an acidic buffer solution (S2 System) [13]. The results of elastic torque are shown in Table 13.7 [101].

According to Zhao et al. [102] it could be seen that apparently there was no synergistic effect between the reinforcement of XNBR and cross-linking on elastic torque.

A direct consequence of the incorporation of filler into rubbers is the significant change in the dynamic mechanical properties. It is accepted that the striking reinforcement effects observed in filled rubbers can be attributed to the formation and breakdown of physical bonds between the networking filler particles.

This dependence of the dynamic properties on the strain amplitude was studied by Payne [103] and is reported in the literature as the "Payne effect."

Table 13.6 Mechanical properties of NBR nanocomposites with cellulose II (cell) or carbon black (CB) obtained by S1 and S2 systems

Nanocomposites	Filler (phr)	Tensile strength (MPa)	Strain at break (%)	Modulus at 100% (MPa)	Modulus at 300% (MPa)	Tear strength (kN/m)	Hardness (shore A)
XNBR (S2 System)	0	10.5	500	1.3	3.5	26	59
	10	13.0	475	2.0	6.2	37	68
	15	15.6	575	3.2	8.9	54	76
	20	13.8	450	3.2	9.5	55	78
	25	11.9	450	7.3	10.6	58	84
	30	9.8	450	5.2	9.2	51	84
S1 system–cell	0	5.3	500	0.5	1.8	25	48
S1 system–cell	15	7.9	575	1.4	3.7	40	55
S2 system–CB	20	15.9	450	3.4	10.2	51	70
S1 system–CB	20	10.2	475	2.4	5.8	41	63

Source: Peres ACC. Processo de obtenção de compósitos de borracha nitrílica carboxilada e celulose regenerada. MSc. Universidade Federal do Rio de Janeiro, Instituto de Macromoléculas Professora Eloisa Mano; 2001.

Table 13.7 Elastic shear torque (S′) of XNBR/cellulose/clay nanocomposites

Samples	S′ min (dNm)	S′ max (dNm)	[S′ max−S′ min] (dNm)
100/0/0	3.54	9.63	6.09
100/15/0	5.07	14.64	9.57
100/15/1	5.32	15.37	10.05
100/15/3	5.26	15.90	10.63
100/15/5	5.13	15.33	10.20
100/15/7	4.88	14.24	9.36

Source: Cosme JGL. Nanocompósitos de XNBR com celulose II e argila: preparação e caracterização. MSc. Universidade Federal do Rio de Janeiro, Instituto de Macromoléculas Professora Eloisa Mano; 2012.

The influence of both reinforcement and cure on the dynamic properties of cellulose and clay XNBR nanocomposites was investigated with the aid of a RPA (Fig. 13.8) [101].

The unfilled rubber shows no indication of nonlinearity in the green (Fig. 13.8A) as well as in the vulcanized compound (Fig. 13.8B). With increasing addition of filler, the low strain modulus G_0 rises more significantly than the high strain modulus G_{100}, resulting in a nonlinear viscoelastic behavior known as the Payne effect (G_0-G_{100}).

This can be due to the fact that the interaggregate distances become smaller with rising filler content, and therefore the probability of filler network formation increases. Filler aggregates must have been reformed during the vulcanization process, which is possible in view of the low viscosity of the polymer matrix at high temperatures.

Fig. 13.9 shows the comparison of ($G'_0-G'_{100}$) for XNBR nanocomposites before and after cure. It was observed that for uncured XNBR nanocomposites the addition of cellulose caused an increase in ($G'_0-G'_{100}$) but there was no significant change in this property with clay addition from 1 phr to 7 phr. The cross-link contributions can be analyzed by comparing XNBR cured with the same filler load. The reason for the higher ($G'_0-G'_{100}$) values may be related to rubber chains attached to the surface of cellulose as a consequence of the filler reinforcement, especially for XNBR with 3 phr of clay, indicating a synergistic effect between vulcanization and reinforcement for this clay content.

The different processes involved led to the production of nanocomposites with clear nanocellulose, good mechanical performance, easy preparation, and having properties comparable to those obtained from CB.

13.6 Conclusion

Cellulose and elastomers are polymers of quite distinct properties, which render them attractive in the search for synergistic properties resulting from their combination.

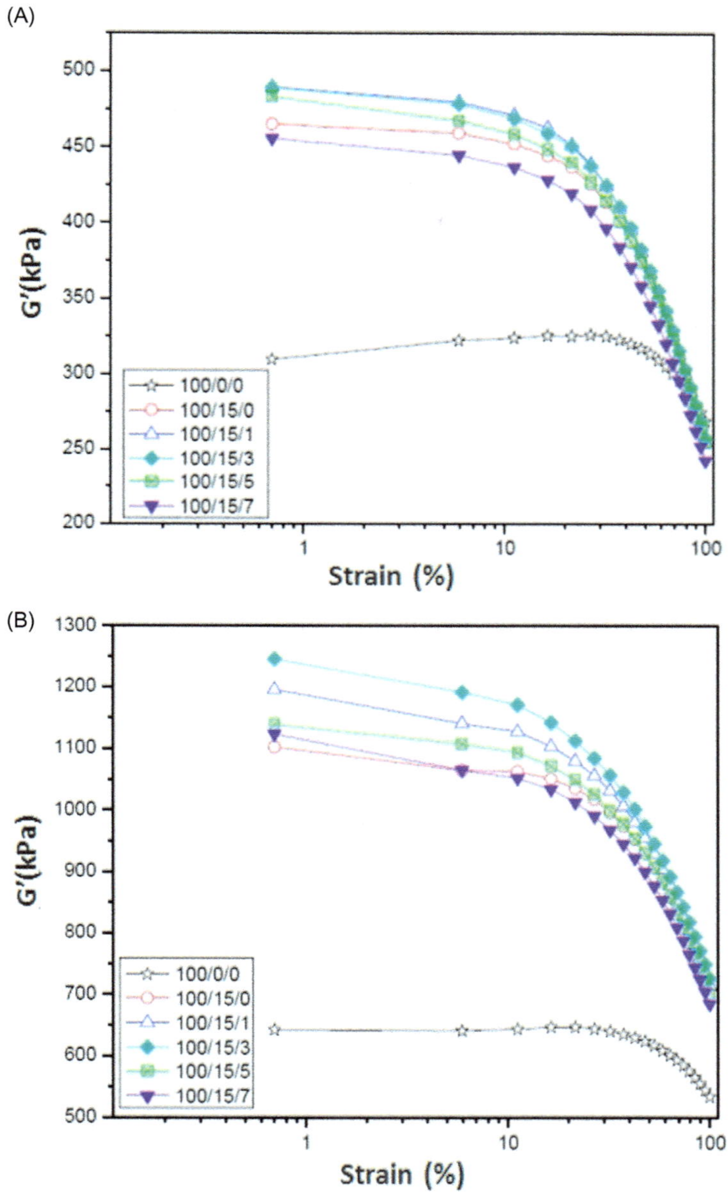

Figure 13.8 The Payne effect before (A) and after (B) cure of XNBR/cellulose/clay nanocomposites.
Source: (A–B) Nanocompósitos de XNBR com celulose II e argila: preparação e caracterização. MSc. Universidade Federal do Rio de Janeiro, Instituto de Macromoléculas Professora Eloisa Mano.

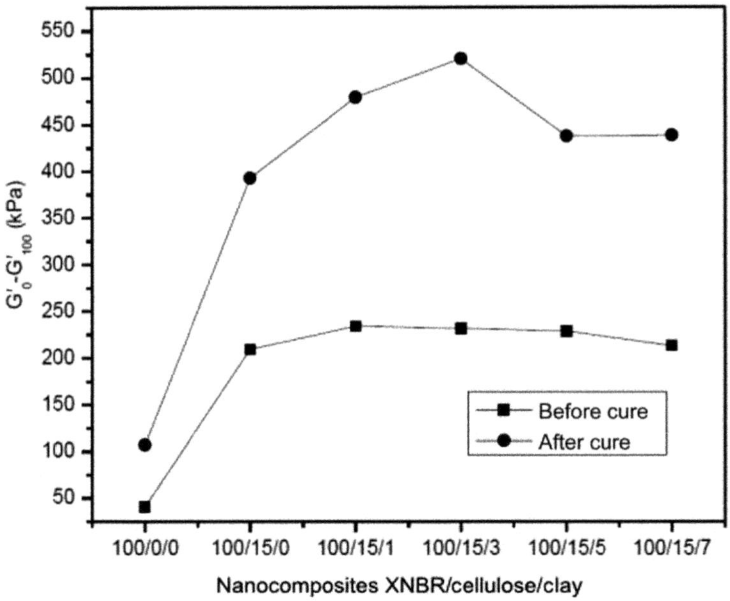

Figure 13.9 Comparison of ($G'_0-G'_{100}$) for uncured and cured NBR/cellulose/clay nanocomposites.
Source: Nanocompósitos de XNBR com celulose II e argila: preparação e caracterizaçáo. MSc. Universidade Federal do Rio de Janeiro, Instituto de Macromoléculas Professora Eloisa Mano.

Elastomers are low modulus, amorphous polymers that require the addition of fillers in order to improve their performance.

Different preparation methods and properties of nanostructured cellulose from various natural resources and residues provide comprehensive possibilities regarding their application in several fields of Polymer Science and Technology as nanomaterials particles.

Based on the studies outlined in this Chapter in Sections 13.2–13.4, a few conclusions can be drawn:

- Nanocellulose is expected to have great potential since cellulose materials are the most abundant biological raw materials and can self-assemble into well-defined architectures at multiple scales, from nano- to micro-size [43]. Furthermore, cellulose is not only renewable but also a multifunctional raw material. It is expected to be able to replace many nonrenewable materials [104].
- These natural fiber-reinforced composites possess acceptable mechanical properties and also offer a number of well-known advantages including light weight, low density, low cost, availability, biodegradability, high specific stiffness and mechanical strength, low hardness, renewability, and recyclability, besides being nonhazardous [39].
- There is a growing interest on biopolymer-based nanocomposites in developed and developing countries, especially if these new products are based on renewable raw materials.

Different cellulose nanoparticles and techniques for their preparation lead to cellulose nanocomposites with various interesting thermal, mechanical, and other properties. However not enough research has been done on the durability of these composites. The improvement in compatibility between cellulose nanoparticles and hydrophobic polymer matrices requires chemical modification of the former [2].
- The research on nanocellulose-reinforced rubber nanocomposites is at its infancy as regards the laboratory stage. Further research efforts shall be devoted to this scientifically promising but demanding field, and furthermore, to bridge the gap between scientific challenges and industrial applications. Novel and advanced technologies for the formulation of cellulosic rubber based materials are and continue to be of great significance to the sustainable development of natural cellulosic, used tire and reused plastics, and many other end-use industrial sectors [6].

As regards the specific properties of regenerated cellulose nanocomposites (Section 13.5) the main conclusions are as follows:

- The different processes involved led to the production of nanocomposites with clear nanocellulose, good mechanical performance, easy preparation, and having properties comparable to those obtained from CB.
- It is a promising industrial system since cellulose xanthate solution and some technologically important rubbers are industrially manufactured by emulsion polymerization.
- Rubber-regenerated cellulose, or cellulose II, or nanocellulose is a very interesting nanomaterial that does not aggregate and can be utilized as a particulate rubber master batch.
- In hybrid nanocomposites the cellulosic filler is the main ingredient responsible for reinforcement, which in combination with clay improves the material properties.
- The cellulose–rubber interaction is of a physical nature and influences the rubber viscoelastic properties without altering the rheological behavior.

References

[1] Klemm D, Schumann D, Kramer F, Heßler N, Hornung M, Schmauder HP, et al. Nanocelluloses as innovative polymers in research and application. Adv Polym Sci 2006;205:49–96.
[2] Feldman D. Cellulose nanocomposites. J Macromol Sci, Part A: Pure Appl Chem 2015;52:322–9.
[3] Treloar LRG. The physics of rubber elasticity. Oxford: Oxford University Press; 1975.
[4] Thomas S, Stephen R. Rubber nanocomposites: preparation, properties and applications. Singapore: John Wiley & Sons; 2010.
[5] Nunes RCR, Visconte LLY. Natural fibers/elastomeric composites. In: Frollini E, Leão AL, Mattoso LHC, editors. Natural polymers and agrofibers based composites. São Carlos: Embrapa Instrumentação Agropecuária; 2000. p. 135–57.
[6] Zhou Y, Fan M, Chen L, Zhuang J. Lignocellulosic fibre mediated rubber composites: an overview. Composites, Part B 2015;76:180–91.
[7] Klemm D, Kramer F, Moritz S, Lindstr T, Ankerfors M, Gray D, et al. Nanocelluloses: a new family of nature-based materials. Angew Chem Int Ed 2011;50:5438–66.
[8] Seppälä JV. Nanocellulose—a renewable polymer of bright future. Express Polym Lett 2012;6(4):257.
[9] Oksman K. Nanocelluloses and their use in composite materials. Express Polym Lett 2012;6(9):687.

[10] Chinga-Carrasco G. Cellulose fibres, nanofibrils and microfibrils: the morphological sequence of MFC components from a plant physiology and fibre technology point of view. Nanoscale Res Lett 2011;6:417–24.

[11] Siro I, Plackett D. Microfibrillated cellulose and new nanocomposite materials: a review. Cellulose 2010;17:459–94.

[12] Nunes RCR. Natural rubber (NR) composites using cellulosic fiber reinforcements. In: Kohjiya S, Ikeda Y, editors. Chemistry, manufacture and applications of natural rubber. Cambridge: Woodhead Publishing; 2014. p. 284–302.

[13] Peres ACC, Nunes RCR, Visconte LY. Processo de obtenção de composições homogêneas de celulose, seus derivados e elastômeros, e suas composições. Brazilian Patent PI0105116-4; 2001.

[14] Mano EB, Nunes RCR, Lima LCOC. Processos para a fabricação de dispersões homogêneas de celulose e seus derivados em elastômeros. Brazilian Patent PI7502614; 1975.

[15] Thomas S, Pothan LA. Natural fibre reinforced polymer composites: from macro to nanoscale. Philadelphia: Editions des archives contemporaines, Old City Publishing; 2008.

[16] Maiti M, Bhattacharya M, Bhowmick AK. Elastomer nanocomposites. Rubber Chem Technol 2008;81(3):384–469.

[17] Woodings C. Regenerated cellulose fibres. Cambridge: Woodhead Publishing Limited; 2001.

[18] John MJ, Thomas S. Cellulosic fibril–rubber nanocomposites. In: Thomas S, Stephen R, editors. Rubber nanocomposites: preparation, properties and applications. Singapore: John Wiley & Sons; 2010. p. 197–208.

[19] Jacob M, Anandjiwala RD, Thomas S. Lignocellulosic fiber reinforced rubber composites. In: Thomas S, Pothan LA, editors. Natural fibre reinforced polymer composites: from macro to nanoscale. Philadelphia: Editions des archives contemporaines; 2008. p. 252–69.

[20] Frollini E, Leão AL, Mattoso LHC. Natural polymers and agrofibers based composites. São Carlos: Embrapa Instrumentação Agropecuária; 2000.

[21] Das D, Datta M, Chavan RB, Datta SK. Coating of jute with natural rubber. J Appl Polym Sci 2005;98:484–9.

[22] Bras J, Hassan ML, Bruzesse C, Hassan EA, El-Wakil NA, Dufresne A. Mechanical, barrier, and biodegradability properties of bagasse cellulose whiskers reinforced natural rubber nanocomposites. Ind Crops Prod 2010;32:627–33.

[23] Ismail H, Edyham MR, Wirjosentono B. Bamboo fibre filled natural rubber composites: the effects of filler loading and bonding agent. Polym Test 2002;21:139–44.

[24] Geethamma VG, Kalaprasad G, Groeninck G, Thomas S. Dynamic mechanical behavior of short coir fiber reinforced natural rubber composites. Composites, Part A: Appl Sci Manuf 2005;36:1499–506.

[25] Jacob M, Francis B, Varughese KT, Thomas S. Dynamical mechanical analysis of sisal/oil palm hybrid fiber-reinforced natural rubber composites. Polym Compos 2006; 27:671–80.

[26] Haque N, Al-Kafi A, Alam MN, Abedin MZ. Jute reinforced polymer composite by gamma radiation: effect of surface treatment with UV radiation. Polym Plast Technol Eng 2006;45(5):607–13.

[27] Chong EL, Ahmad I, Dahlan HM, Abdullah I. Reinforcement of natural rubber/ high density polyethylene blends with electron beam irradiated liquid natural rubber-coated rice husk. Radiat Phys Chem 2010;79(8):906–11.

[28] Mohanty AK, Khan MA, Hinrichsen G. Surface modification of jute and its influence on performance of biodegradable jute-fabric/Biopol composite. Compos Sci Technol 2000;60(7):1115–24.
[29] Haseena AP, Unnikrishnan G, Kalaprasad G. Dielectric properties of short sisal/coir hybrid fibre reinforced natural rubber composites. Compos Interfaces 2007;14(7e9): 763–86.
[30] Lavoine N, Desloges I, Dufresne A, Bras J. Microfibrillated cellulose—its barrier properties and applications in cellulosic materials: a review. Carbohydr Polym 2012;90:735–64.
[31] Herrick FW, Casebier RL, Hamilton JK, Sandberg KR. Microfibrillated cellulose: morphology and accessibility. J Appl Polym Sci Appl Polym Symp 1983;37:797–813.
[32] Turbak AF, Snyder FW, Sandberg KR. Microfibrillated cellulose, a new cellulose product: properties, uses, and commercial potential. J Appl Polym Sci Appl Polym Symp 1983;37:815–27.
[33] Chauve G, Bras J. Industrial point of view of nanocellulose materials and their possible applications. In: Oksman K, Mathew AP, Bismarck A, Rojas O, Sain M, editors. Handbook of green materials: processing technologies, properties and applications, vol. 1. Singapore: World Scientific Publishing; 2014. p. 233–52.
[34] Sturcova A, Davies GR, Eichhorn SJ. Elastic modulus and stress-transfer properties of tunicate cellulose whiskers. Biomacromolecules 2005;6:1055–61.
[35] Wu Q, Meng Y, Wang S, Li Y, Fu S, Ma L, et al. Rheological behavior of cellulose nanocrystal suspension: influence of concentration and aspect ratio. J Appl Polym Sci 2014;131:405–25.
[36] Boluk Y, Lahiji R, Zhao L, McDermott MT. Suspension viscosities and shape parameter of cellulose nanocrystals (CNC). Colloids Surf A 2011;377:297–303.
[37] Bondeson D, Mathew A, Oksman K. Optimization of the isolation of nanocrystals from microcrystalline cellulose by acid hydrolysis. Cellulose 2006;13:171–80.
[38] Azizi Samir MAS, Alloin F, Sanchez JY, El Kissi N, Dufresne A. Preparation of cellulose whiskers reinforced nanocomposites from an organic medium suspension. Macromolecules 2004;37:1386–93.
[39] Ching YC, Ali Md E, Abdullah LC, Choo KW, Kuan YC, Julaihi SJ, et al. Rheological properties of cellulose nanocrystal-embedded polymer composites: a review. Cellulose 2016;23:1011–30.
[40] Kohjiya S, Ikeda Y. Reinforcement of general-purpose grade rubbers by silica generated in situ. Rubber Chem Technol 2000;73(3):534–50.
[41] Medalia AI, Kraus G. Reinforcement. In: Mark E, Erman B, Eirich FR, editors. Science and technology of rubber. San Diego: Academic Press; 1994.
[42] Pramanik M, Srivastava SK, Samantaray B, Bhowmick AK. Rubber–clay nanocomposite by solution blending. J Appl Polym Sci 2003;87:2216–20.
[43] Jonoobi M, Oladi R, Davoudpour Y, Oksman K, Dufresne A, Hamzeh Y, et al. Different preparation methods and properties of nanostructured cellulose from various natural resources and residues: a review. Cellulose 2015;22:935–69.
[44] Favier V, Canova G, Cavaille J, Chanzy H, Dufresne A, Gauthier C. Nanocomposite materials from latex and cellulose whiskers. Polym Adv Technol 1995;6:351–5.
[45] Visakh PM, Thomas S, Oksman K, Mathew AP. Cellulose nanofibres and cellulose nanowhiskers based natural rubber composites: diffusion, sorption, and permeation of aromatic organic solvents. J Appl Polym Sci 2012;124(2):1614–23.
[46] Jarnthong M, Wang F, Wei XY, Wang R, Li JH. Preparation and properties of biocomposite based on natural rubber and bagasse nanocellulose. MATEC Web of Conferences 2015;26:01005.

[47] Abraham E, Elbi PA, Deepa B, Jyotishkumar P, Pothen LA, Narine SS. X-ray diffraction and biodegradation analysis of green composites of natural rubber/nanocomposites. Polym Degrad Stab 2012;97(11):2378–87.
[48] Chen WJ, Gu J, Xu SH. Exploring nanocrystalline cellulose as a green alternative of carbon black in natural rubber/butadiene rubber/styrene-butadiene rubber blends. Express Polym Lett 2014;8(9):659–68.
[49] Xu SH, Gu J, Luo YF, Jia DM. Effects of partial replacement of silica with surface modified nanocrystalline cellulose on properties of natural rubber nanocomposites. Express Polym Lett 2012;6(1):14–25.
[50] Bitinis B, Verdejo R, Cassagnau P, Lopez-Manchado MA. Structure and properties of polylactide/natural rubber blends. Mater Chem Phys 2011;129(3):823–31.
[51] Cao X, Xu C, Wang Y, Liu Y, Liu Y, Chen Y. New nanocomposite materials reinforced with cellulose nanocrystals in nitrile rubber. Polym Test 2013;32(5):819–26.
[52] Krawczak P. Compounding and processing of polymer nanocomposites: from scientific challenges to industrial stakes. Express Polym Lett 2007;1(4):188.
[53] Pasquini D, Teixeira EM, Curvelo AAS, Belgacem MN, Dufresne A. Extraction of cellulose whiskers from cassava bagasse and their applications as reinforcing agent in natural rubber. Industrial Crops and Products 2010;32:486–90.
[54] Mano EB, Nunes RCR. Regenerated cellulose in elastomer compounds. Eur Polym J 1983;19(10/11):919–21.
[55] Kalb B, Manley RSTJ. Hydrodynamically induced formation of cellulose fibers. I. Observations on the formation of cellulose fibers from stirred solutions. J Polym Sci, Polym Phys 1980;18(4):707–21.
[56] Harris GH. Xanthate. In: Kirk-Othmer, editor. *Encyclopedia of chemistry and technology*. New York: John Wiley & Sons; 1970. p. 419–25.
[57] Wu YP, Wang YQ, Zhang HF, Wang YZ, Yu DS, Zhang LQ, et al. Rubber–pristine clay nanocomposites prepared by co-coagulating rubber latex and clay aqueous suspension. Compos Sci Technol 2005;65(7–8):1195–202.
[58] Kaplan DL. Biopolymers from renewable resources. Springer Verlag; 1998. p. 47.
[59] Brandt K, Schuster RH, Nunes RCR. Influence of process parameters on the viscoelastic behavior of NBR-cell II nanocomposite mixtures. Kautsch Gummi Kunstst 2006;10: 511–15.
[60] Ibarra LR, Chamorro AC. Short fiber elastomer composites. J Appl Polym Sci 1991;2: 1805–19.
[61] Nunes RCR, Martins AF, Visconte LLY, Pereira RA, Perez C, Mano EB. Nanocomposites of natural rubber and cellulose II: comparison between different processes of preparation. J Rubber Res 2004;71:01–13.
[62] Martins AF. Nanocompósitos de borracha natural e celulose II. PhD Universidade Federal do Rio de Janeiro, Instituto de Macromoleculas Professora Eloisa Mano, Rio de Janeiro, Brazil; 2002.
[63] Martins AF, Schuster R, Boller F, Visconte LLY, Nunes RCR. Ageing effect on dynamic and mechanical properties of NR/Cell II nanocomposites. Kautsch Gummi Kunstst 2004;57:446–51.
[64] Nunes RCR, Martins AF, Visconte LLY. Interaction between natural rubber and cellulose II. 9th International Seminar on Elastomer (ISE), Kioto, Japão; 2003. pp. 109–110.
[65] Nunes RCR, Compañ V, Riande E. Gas transport in Vulcanized natural rubber-cellulose II composites. J Polym Sci, Polym Phys 2000;38:393–402.

[66] Andrio A, Compañ V, Nunes RCR, Lópes ML, Riande E. Influence of cellulose reinforcers on gas transport through natural rubber. J Memb Sci 2000;178:65–74.
[67] Nunes RCR, Gonzalez MML, Riande E. Basic Studies on gas solubility in natural rubber-cellulose composites. J Polym Sci, Polym Phys 2005;43:2131–40.
[68] Derouet D, Mulder-Houdayer S, Brosse JC. Chemical modification of polydienes in latex medium: study of epoxidation and ring opening of oxiranes. J Appl Polym Sci 2005;95(1):39–52.
[69] Mark JE, Erman B, Eirich FR. Science and technology of rubber. New York: Academic Press Inc.; 1994.
[70] Fernandes RMB. Otimização do desempenho de nanocompósitos de borracha natural e celulose II. PhD Universidade Federal do Rio de Janeiro, Instituto de Macromoléculas Professora Eloisa Mano; 2010.
[71] Fernandes RMB, Visconte LLY, Nunes RCR. Curing characteristics, mechanical properties and solvent resistance after ageing of natural rubber/epoxidized natural rubber and cellulose II. Int J Polym Mater Polym Biomater 2011;60(5):351–64.
[72] Poh BT, Ismail H, Tan KS. Effect of filler loading on tensile and tear properties of SMR L/ENR 25 and SMR L/SBR blends cured via a semi-efficient vulcanization system. Polym Test 2002;21:801–6.
[73] Zanzig DJ, Magnus FL, Halasa AF, Testa ME. IBR block copolymers as compatibilizers in NR/BR blends. Rubber Chem Technol 1993;66(4):538–49.
[74] Mariano RM, Picciani PH de S, Nunes RCR, Visconte LLY. Preparation, structure, and properties of montmorillonite/cellulose II/natural rubber nanocomposites. J Appl Polym Sci 2011;120:458–65.
[75] Martins AF, Visconte LLY, Nunes RCR. Evaluation of natural rubber and cellulose II compositions by curing and mechanical properties. Kautsch Gummi Kunstst 2002; 55:637–41.
[76] Martins AF, Suarez JCM, Visconte LLY, Nunes RCR. Mechanical and fractographic behavior of natural rubber-cellulose II composites. J Mater Sci 2003;38:2415–22.
[77] Vieira A, Nunes RCR, Visconti LLY. Mechanical properties of NR/BR/cellulose II composites. Polym Bull 1996;36:759–66.
[78] Vieira A, Nunes RCR, Costa DMR. Rheometrical behavior and equilibrium swelling In NR/BR/Cell Ii composites. Polym Bull 1997;39:117–24.
[79] Cheremisinoff NP. Polymer mixing and extrusion technology. New York: Marcel Dekker; 1987.
[80] Kraus G. Swelling of filler-reinforced vulcanizates. J Appl Polym Sci 1963;7(3): 861–71.
[81] Cunneen JI, Russel RM. Occurrence and prevention of changes in chemical structure of natural rubber tire tread vulcanizates during service. J Rubber Res Inst Malays 1969;22(3):300.
[82] Porter M. Structural characterization of filled vulcanizates part 1. Determination of the concentration of chemical crosslinks in natural rubber vulcanizates containing high abrasion furnace black. Rubber Chem Technol 1967;40(3):866–82.
[83] Mukhopadhyay R, De SK. Effect of vulcanization temperature and different fillers on the properties of efficiently vulcanized natural rubber. Rubber Chem Technol 1979;52 (2):263–77.
[84] Boonstra BB, Taylor GL. Swelling of filled rubber vulcanizates. Rubber Chem Technol 1965;38(4):943–60.
[85] Blackley DC. Synthetic rubbers: their chemistry and technology. London: Applied Science Publishers; 1983.

[86] Nunes RCR, Costa VG. Mechanical properties of blends of EPDM with NR-cellulose II system. Eur Polym J 1994;30:1025–8.
[87] Scholtens BJR. Linear thermoviscoelasticity and characterization of noncrystalline EPDM rubber networks. J Polym Sci, Polym Phys 1984;22(3):317–44.
[88] Zhang LQ, Wang YZ, Wang YQ, Suy Y, Yu DS. Morphology and mechanical properties of clay/styrene butadiene rubber nanocomposites. J Appl Polym Sci 2000; 78:1873–8.
[89] Zine CLG, Da Conceição AJ, Visconte LLY, Ito EN, Nunes RCR. Styrene-butadiene rubber/cellulose II/clay nanocomposites prepared by cocoagulation-mechanical properties. J Appl Polym Sci 2011;120:1468–74.
[90] Zine CLG. Nanocompósitos de SBR com celulose II e carga mineral. PhD Universidade Federal do Rio de Janeiro, Instituto de Macromoléculas Professora Eloisa Mano; 2011.
[91] Flory PJ. Principles of polymer chemistry. New York: Cornell University; 1953.
[92] López-Manchado MA, Arroyo M, Herrero B, Biagiotti J. Vulcanization kinetics of natural rubber–organoclay nanocomposites. J Appl Polym Sci 2003;15:1–15.
[93] Ciesielski A. An introduction to rubber technology. Shawbury: Rapra Technology Limited; 1999.
[94] Nunes RCR, Affonso JES. Interaction of NBR rubber with regenerated cellulose. Kautsch Gummi Kunstst 1999;52:787–93.
[95] Affonso JES, Nunes RCR. Influence of the filler and monomer quantities in rheometrical behaviour and crosslink density of the NBR-cellulose II composites. Polym Bull 1995;34:669–75.
[96] Singha NK, Bhattacharjee S, Sivaram S. Hydrogenation of diene elastomers, their properties and applications: a critical review. Rubber Chem Technol 1997;70(3):309–67.
[97] Lapa VLC, Suarez JCM, Visconte LLY, Nunes RCR. Fracture behavior of nitrile rubber-cellulose II nanocomposite. J Mater Sci 2007;42:9934–9.
[98] Lapa VLC, Oliveira PD, Visconte LLY, Nunes RCR. Investigation of NBR-cellulose II nanocomposites by rheometric and equilibrium swelling properties. Polym Bull 2008;60:281–90.
[99] German Patent Announcement 10 2006 016 979.4, Deutsches Institut für Kautschuktechnologie.
[100] Peres ACC. Processo de obtenção de compósitos de borracha nitrílica carboxilada e celulose regenerada. MSc Universidade Federal do Rio de Janeiro, Instituto de Macromoléculas Professora Eloisa Mano; 2001.
[101] Cosme JGL. Nanocompósitos de XNBR com celulose II e argila: preparação e caracterização. MSc. Universidade Federal do Rio de Janeiro, Instituto de Macromoléculas Professora Eloisa Mano; 2012.
[102] Zhao F, Shi X, Chen X, Zhao S. Interaction of vulcanization and reinforcement of CB on dynamic property of NR characterized by RPA2000. J Appl Polym Sci 2010;117:1168–72.
[103] Payne AR. The dynamic properties of carbon black loaded natural rubber vulcanizates. Part I. J Appl Polym Sci 1962;6(19):57–63.
[104] Wegner TH, Jones PE. Advancing cellulose-based nanotechnology. Cellulose 2006;13 (2):115–18.
[105] Vladkova TG, Dineff PD, Gospodinova DN. Wood flour: a new filler for the rubber processing industry. II. Cure characteristic and mechanical properties of NBR compounds filled with corona-treated wood floor. J Appl Polym Sci 2004;91(2):883–9.

Thermal conductivity and dielectric properties of silicone rubber nanocomposites

14

B. Du
Tianjin University, Tianjin, China

14.1 Introduction

Silicone rubber (SiR), as a basic insulating material, has been widely used to coat porcelain and glass insulators. The high voltage direct current (HVDC) technology is considered as the most efficient and economical solution to the high voltage, large capacity, and long distance transmission and the power network interconnection [1]. Several ± 800 kV UHVDC transmission lines have been put into operation in China [2]. Due to better contamination resistance, temperature resistance, electrical insulation, and elasticity, SiR is extensively used in insulators and cable accessories for the HVDC transmission lines [3].

Despite all the advantages, SiR suffers from long-term operation and environment effects. In HVDC transmission lines, corona discharge could occur even on well-designed insulators, which can inject charge into the insulator surface and significantly damage the insulators [4]. It is well known that charge injection is predominantly dependent on the initial external field distribution [5,6]. Under DC voltage, the charges are more likely to accumulate on the insulator surface due to the constant electrostatic field, compared to that under AC voltage. The charges may remain on it over a certain period determined by the efficiency of decay process. The existence of surface charges causes early insulation failure and plays an important role during the development of the surface flashover [7]. It also has been reported that with the same field conditions the accumulation of pollutants under DC voltage is 1.2−1.5 times of that under AC voltage [8]. It is necessary to investigate the performance of SiR insulators under DC voltage. When exposed to long-term humidity and severe contamination, the hydrophobicity of SiR can be lost for sustained periods, which will lead to the development of a conductive film on the surface [9]. Thus, dry band arcing may occur and a large amount of heat will be generated. As the thermal conductivity of SiR is very low, the heat is accumulated in the discharge area and cannot spread quickly, which will gradually cause the degradation of SiR and may further induce tracking and erosion [10]. Especially under DC voltage, with more contaminants, conductivity and leakage current are higher, which can result in more severe degradation of SiR. Inclined plane tests on

Progress in Rubber Nanocomposites. DOI: http://dx.doi.org/10.1016/B978-0-08-100409-8.00014-0

polymeric high temperature vulcanized SiR insulators have shown that the tracking and erosion are more severe under positive DC than under AC voltage [11]. Besides, the electrical field inside the accessories is not as uniform as that inside the power cable for its complicated physical structure, and some defects brought during the manufacturing process, such as nonuniform electrical field and defects, may cause dielectric failure inside the accessories [12]. The electrical tree is initiated from the enhanced point of the electric field which may be caused by the void, impurities, or irregular shapes [13]. It is a serious threat to the insulation and it can even cause insulation breakdown.

In order to improve the physical, chemical, mechanical, and electrical properties of SiR, the nanocomposite has recently drawn considerable attention. Venkatesulu and Thomas have investigated good performance on tracking and erosion resistance of nanocomposites due to the interaction between the host dielectric and the nanoparticles [14,15]. Previous studies showed that nanoparticles have a great effect on surface charge behavior [7]. Fleming et al. have presented the space charge profile data of Low Density Polyethylene (LDPE) in which different nanoparticles were incorporated [16]. Kumara et al. have observed that the flashover voltage level varied linearly with the amount of deposited charge for both positive and negative charging [5,6]. Many researchers have also investigated the thermal problems and the resistance to tracking and erosion of SiR. The results have shown that the material degradation is a function of the leakage current magnitude and the time for which dry band arcing exists in a particular spot [17,18]. According to the results in the field, thermal depolymerization activated by electrical discharge is the main degradation factor for SiR insulators exposed to a coastal environment [19]. A correlation study has shown that the resistance to erosion of SiR composites, filled with alumina trihydrate (ATH) or silica, is strongly correlated to the thermal conductivity of the composite [20,21]. In the field of electrical treeing, Chen et al. investigated the tree initiation time of pure epoxy resins and found that nanoparticles were able to increase tree initiation time [22]. Tanaka et al. found that alumina nanofillers were effective in suppressing both tree initiation and propagation [23].

On the bases of all the previous research activities, this chapter discusses three typical SiR nanocomposites and their dielectric properties. In Section 14.2, nano-Boron Nitride (BN) particles are mixed into RTV SiR to obtain SiR/BN nanocomposites. The tracking and erosion processes of SiR/BN nanocomposites were studied by employing a standard inclined plane test, with an exception that the supplied voltage was DC. In Section 14.3, SiR/SiO_2 nanocomposites are studied and the research focuses on the effects of fluorination time and mass fraction of nanoparticles on surface charge and DC flashover characteristics of SiR/SiO_2 nanocomposites. In Section 14.4, AC voltage with the frequency of 50 Hz was applied on the SiR/SiO_2 nanocomposites to initiate electrical trees with the temperature ranging from $-30°C$ to $-90°C$. Both the structure and the growth speed of trees were observed by a digital microscope system, and the treeing proportion was introduced to describe the electrical tree propagation characteristics.

14.2 Thermal conductivity of silicone rubber nanocomposites

14.2.1 Experiment procedure

RTV SiR used as the matrix in this study was supplied by Shenzhen Anpin Silicone Material Co., Ltd., China and the BN particle with a diameter of 100 nm was produced by Nanjing Haitai Nano Material Co., Ltd., China. Appropriate quantities of RTV SiR were kept under vacuum condition for at least 24 h to remove air and moisture from the material. Fillers were kept under high temperature (180°C) for 10 h to remove the moisture. The required quantity of fillers was weighed using a precision digital balance with readability of 0.1 mg [24]. The weight percent of nano-BN is respectively 1, 2.5, 3, 5, and 7%. Specimens with 9 wt% nano-BN particles were tried to prepare and the results were unsatisfactory. Obvious aggregation of nanofiller and small voids were observed in the specimen, which may be caused by the limitation of the preparation method. Therefore, only up to 7 wt% samples were shown in this chapter. Specimens of undoped SiR were also prepared for the purpose of comparison.

The thermogravimetric analysis (TGA) curves of specimens with different contents of BN are shown in Fig. 14.1. It can be seen that the temperature of 20% weight loss for 0, 1, and 7 wt% SiR-BN composites increases with the content of BN, which are 456°C, 460°C, and 463°C, respectively. Using the TGA data, the Differential Thermal Gravity (DTG) lines indicating the rate of weight loss or

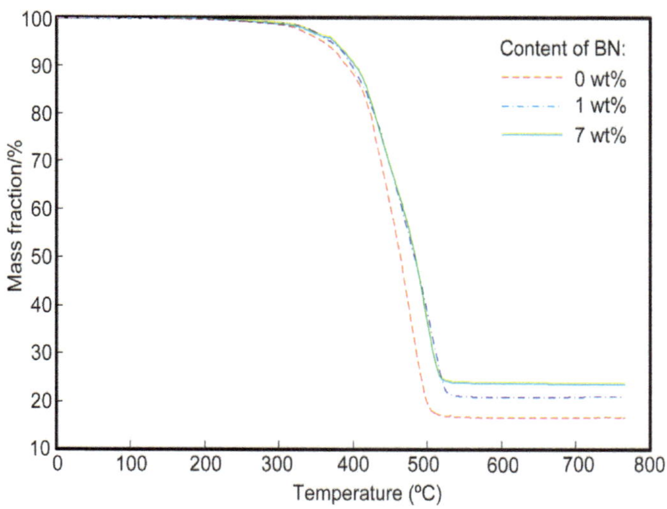

Figure 14.1 TGA curves of specimens with different contents of BN.
Source: Reproduced with permission from Du BX, Xu H. Effects of thermal conductivity on DC resistance to erosion of silicone rubber/BN nanocomposites. *IEEE Trans Dielectr Electr Insul* 2014; **21**(2): 511–18. Copyright 2014, IEEE.

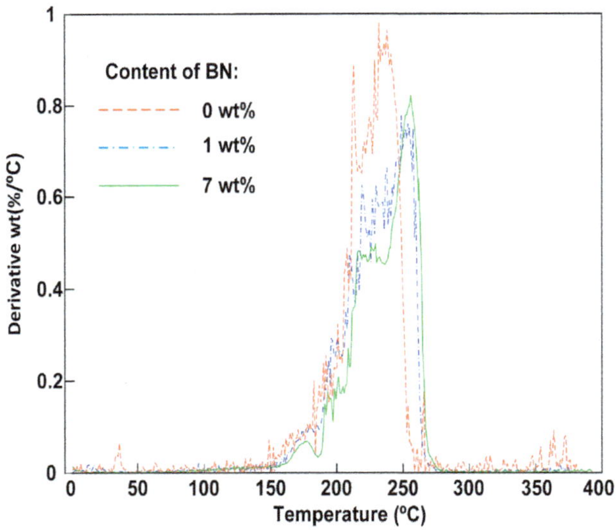

Figure 14.2 DTG curves of specimens with different contents of BN.

degradation of the material with temperature are shown in Fig. 14.2. These slopes indicate the deterioration of unfilled specimen is much more rapid. With the addition of BN particles, the thermal stability of SiR composite becomes better. This is probably due to the increase in physical and chemical cross-linking points created by the interactions between nanoparticles and SiR [25]. Fig. 14.3 shows the distribution of nano-BN particles in 1% BN filled SiR. The distribution is satisfactory as the particles are isolated and scattered uniformly in the matrix.

The test setup, as shown in Fig. 14.4, conformed to the IEC 60587 inclined plane tracking and erosion test, with an exception that a DC power source was used. A pair of stainless steel electrodes was attached to the surface of a flat, rectangular specimen. The upper electrode was connected to the positive pole that generated a high voltage DC power through a series of protective resistors, while the other electrode was connected to the ground. The specimen was mounted at an angle of 45° to the horizontal with the specimen surface on the underside. Filter papers clamped beneath the top electrode were used to introduce the fluid contaminant so that a steady filament flowing down the surface of the specimen between the electrodes was established. The leakage current was recorded by using the sampling resistance connected to an analog-to-digital (A/D) converter. The sampling frequency of the converter was 10 kHz.

14.2.2 Effect on erosion resistance

To predict the relationship between the thermal conductivity and the content of BN particles, the theoretical models of geometric mean were used. The measured thermal conductivity of nano-BN particles used in the tests is 150 W/m K. The

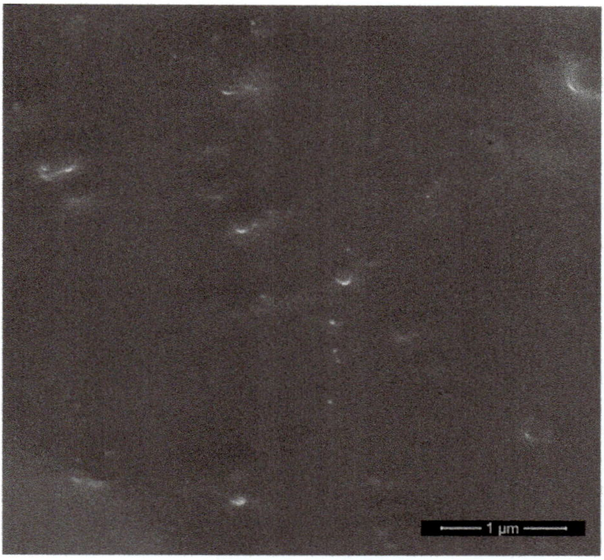

Figure 14.3 Scanning electron micrographs of 1% nano-BN filled silicone rubber at 25,000 × magnification.
Source: Reproduced with permission from Du BX, Xu H. Effects of thermal conductivity on DC resistance to erosion of silicone rubber/BN nanocomposites. *IEEE Trans Dielectr Electr Insul* 2014;**21**(2): 511–18. Copyright 2014, IEEE.

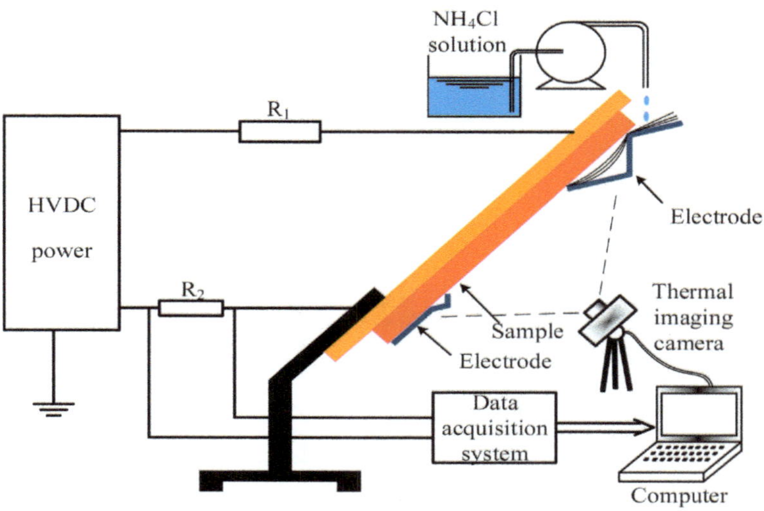

Figure 14.4 Schematic diagram of test system.
Source: Reproduced with permission from Du BX, Xu H. Effects of thermal conductivity on DC resistance to erosion of silicone rubber/BN nanocomposites. *IEEE Trans Dielectr Electr Insul* 2014;**21**(2): 511–18. Copyright 2014, IEEE.

calculated thermal conductivities of samples with weight proportions of 0, 1, 2.5, 3, 5, and 7% are respectively 0.165, 0.168, 0.176, 0.177, 0.186, and 0.194 W/m K. The geometric mean model is based on the amount of fillers and does not take into account the size and geometry of the nanoparticles. Compared to the experimental data shown in Fig. 14.5, the calculated thermal conductivities are a little lower. Fig. 14.5 shows the relationship between the measured thermal conductivity and the content. The measured thermal conductivities of all the SiR/BN specimens are higher than those of the unfilled specimens.

During the inclined plane test, the two electrodes were bridged with a filament of fluid contaminant, which resulted in conduction and Joule heating of the leakage current on the surface. Dry bands formed in the conductive filament and an arc began when the voltage applied to the dry band was over the breakdown voltage. Fig. 14.6 shows the discharge phenomena on the specimen during the test.

As the test went on, the phenomena observed on the surface of the specimen kept varying at different stages of degradation. At the initial stage, dry band arcing formed at random along the conductive filament. The heat was transferred from the arc to the SiR surface, which could cause thermal degradation of specimens, and a continuous, black and solid hydrophilic line was gradually formed between the electrodes, as shown in Fig. 14.6. It was observed that the carbon tracking became wider as the test continued, and the position of the discharges developed down toward the bottom electrode. This was because the hydrophobicity along the filament was lost due to continuous discharges, and the water filament was becoming

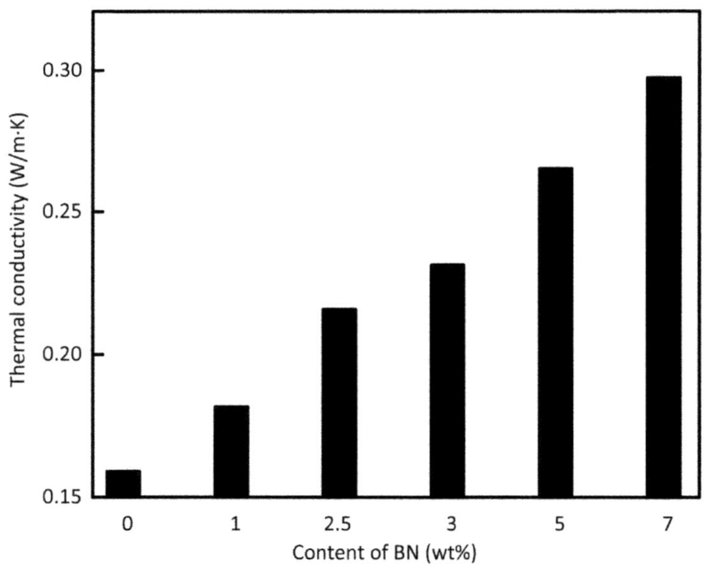

Figure 14.5 Relationship between the thermal conductivity and the content of BN. *Source*: Reproduced with permission from Du BX, Xu H. Effects of thermal conductivity on DC resistance to erosion of silicone rubber/BN nanocomposites. *IEEE Trans Dielectr Electr Insul* 2014;**21**(2): 511–18. Copyright 2014, IEEE.

wider and developed into a water film. Dry bands were more likely to form at the bottom of the water film, so the dry band arcing mainly concentrated near the bottom electrode, which was the main discharge process of the test. Deep erosive material decomposition also took place near the bottom electrode.

When enough energy was transferred to the specimen surface, erosive material decomposition took place, which resulted in weight loss from the specimen. The weight loss of specimen is shown in Fig. 14.7A. Specimens filled with higher BN content show less weight loss than the specimens with lower content. A similar tendency was found in the erosion depth, as shown in Fig. 14.7B.

When the filler loading was 5 and 7 wt%, the depth of erosion was remarkably reduced. It is observed from the results shown in Fig. 14.7 that the tracking resistance and erosion resistance of SiR nanocomposites are improved by increasing filler content.

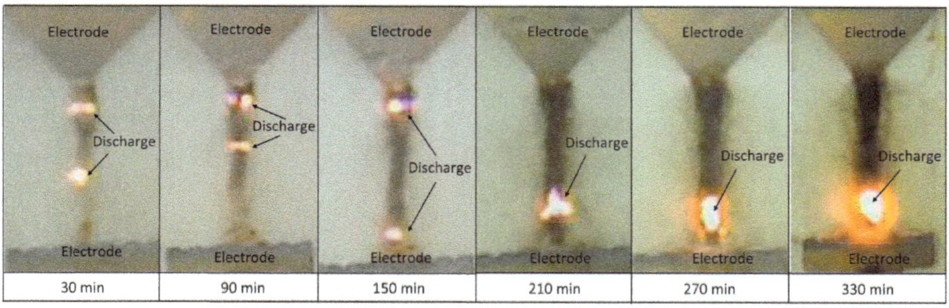

Figure 14.6 Photographs of specimen during the erosion test.
Source: Reproduced with permission from Du BX, Xu H. Effects of thermal conductivity on DC resistance to erosion of silicone rubber/BN nanocomposites. *IEEE Trans Dielectr Electr Insul* 2014;**21**(2): 511–18. Copyright 2014, IEEE.

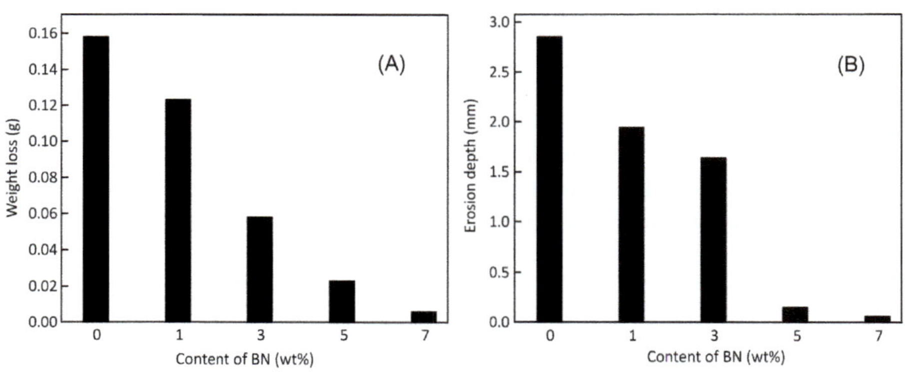

Figure 14.7 Relationship between the content of BN and (A) weight loss, (B) erosion depth.
Source: Reproduced with permission from Du BX, Xu H. Effects of thermal conductivity on DC resistance to erosion of silicone rubber/BN nanocomposites. *IEEE Trans Dielectr Electr Insul* 2014; **21**(2): 511–18. Copyright 2014, IEEE.

While the Joule heating effect of leakage current can result in a slow increase in specimen temperature, dry band arcing can increase the local temperature rapidly. The heat generated by discharges could bring the temperature on the specimen surface to 500°C or more, which is sufficient to cause the SiR backbone to break into low chain polymers and induce tracking and erosion [27]. The temperature rise and heat dissipation have strong effects on the erosion. Adding nano-BN particles improves the thermal conductivity of SiR nanocomposites, which may inhibit the heat accumulation around the discharge area and promote heat dissipation, thus improving the tracking and erosion resistance.

14.2.3 Effect on thermal dissipation

The thermal information on the specimen surface was recorded with an infrared thermal imager during the inclined plane tests. It was observed that dry band arcing mainly occurred along the filament, which increased the local temperature significantly. And the temperature in the surrounding area also rose because of heat diffusion. The effect of thermal conductivity on tracking and erosion is mainly reflected by temperature distribution.

Fig. 14.8 shows the thermographs of SiR/BN nanocomposites with different contents. On the thermographs, a brighter color indicates a higher temperature and a darker color corresponds to a lower temperature. In this chapter, the heat source region is defined, which is close to the discharge area, and has an obvious higher temperature than other regions. It was found that for the same test time, the temperature in the heat source region decreased with the increased content of BN, while the temperature around the heat source region increased with the BN content.

The temperature distribution during the inclined plane test contains some important information reflecting the inhibitory effect of BN particles on tracking and erosion. The isotherm distribution and the maximum temperature after applying the voltage for 10 min are shown in Fig. 14.9. The background temperature on the isotherm is below 45°C. With the increasing content of nano-BN from 0 wt% to 7 wt%, the maximum temperature on the specimens decreases from 113°C to 81°C. The temperature of the unfilled specimen is the highest and the isotherm distribution is the most intensive. On the same test conditions, the intensive distribution indicates that more heat accumulates in the discharge area and cannot be dissipated to the surrounding areas quickly. Relatively sparse distribution is observed in specimens with high content. It is inferred that the BN particles can improve the thermal dissipation in the composites. This indicates that the heat on specimens with higher content was conducted out more quickly and the heat in specimens with lower content accumulated at the discharge area. The heat dissipation correlates with the maximum temperature. When the heat accumulated in the discharge area was dissipated, the maximum temperature decreased.

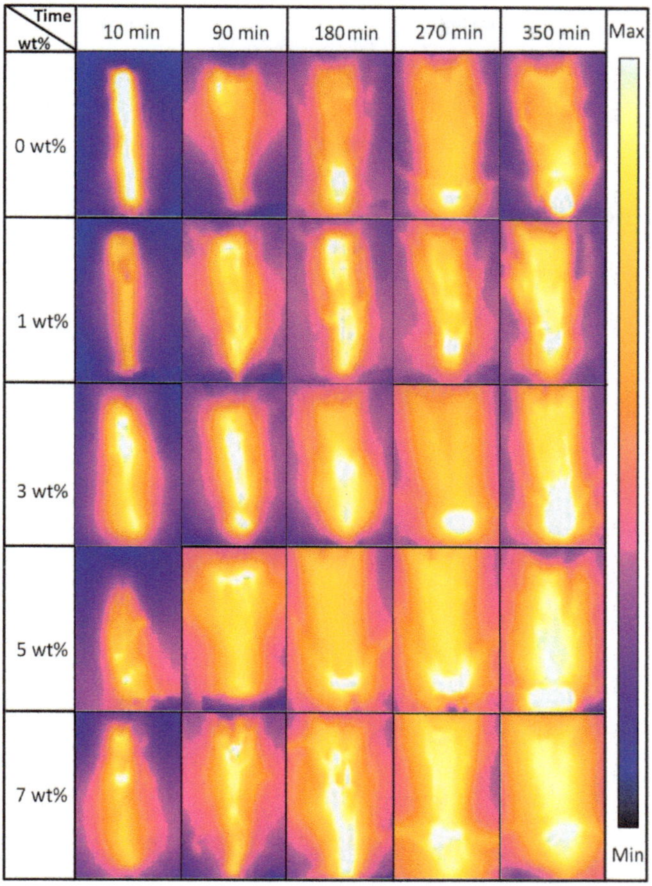

Figure 14.8 Infrared images during the test.
Source: Reproduced with permission from Du BX, Xu H. Effects of thermal conductivity on DC resistance to erosion of silicone rubber/BN nanocomposites. *IEEE Trans Dielectr Electr Insul* 2014;**21**(2): 511–18. Copyright 2014, IEEE.

In practice, thermal accumulation and dissipation often occur simultaneously. In order to investigate the thermal dissipation further without the influence of thermal accumulation, the maximum temperature changes of each specimen were also recorded after turning off the power.

A schematic illustration of mechanisms for thermal diffusion is shown in Fig. 14.10. It is observed that heat would spread to a larger region of the specimen with a higher thermal conductivity. Therefore, specimens with higher thermal conductivity present a relatively lower temperature in the discharge area and a higher temperature in the surrounding area. However, the high temperature in the surrounding areas is not high enough to cause thermal decomposition of specimens. With increasing content of BN particles, the heat dissipation in the specimen is improved and the resistance to erosion increases.

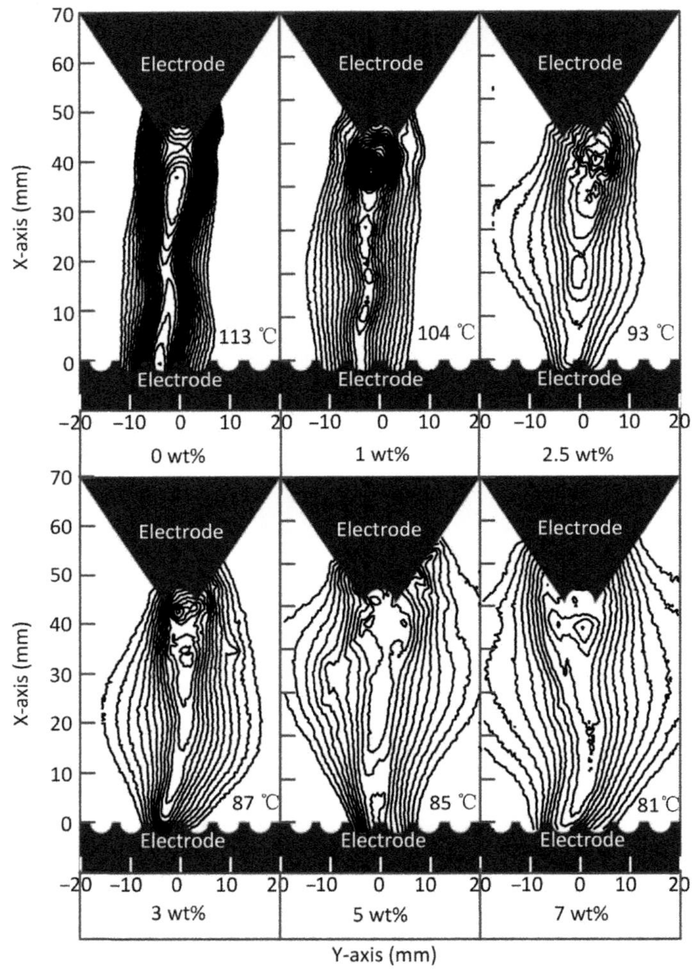

Figure 14.9 Relationship between the isotherm distribution and the content at 10 min after the voltage applied.
Source: Reproduced with permission from Du BX, Xu H. Effects of thermal conductivity on DC resistance to erosion of silicone rubber/BN nanocomposites. *IEEE Trans Dielectr Electr Insul* 2014;**21**(2): 511–18. Copyright 2014, IEEE.

14.3 Surface charge of direct-fluorinated silicone rubber nanocomposites and its effect on DC flashover characteristics

14.3.1 Experiment procedure

One-component RTV SiR without any solid filler, was employed in this experiment. The nano-SiO_2 particles, supplied by Nanjing Haitai Nanomaterials Co.

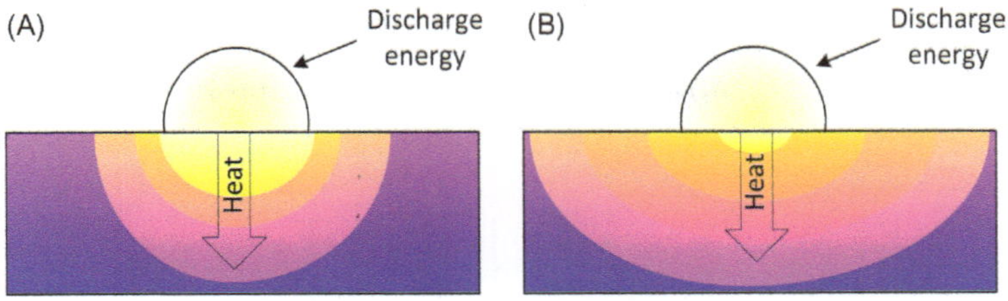

Figure 14.10 Schematic illustration of mechanisms for thermal diffusion. (A) undoped SiR. (B) SiR nanocomposites.
Source: Reproduced with permission from Du BX, Xu H. Effects of thermal conductivity on DC resistance to erosion of silicone rubber/BN nanocomposites. *IEEE Trans Dielectr Electr Insul* 2014;**21**(2): 511–18. Copyright 2014, IEEE.

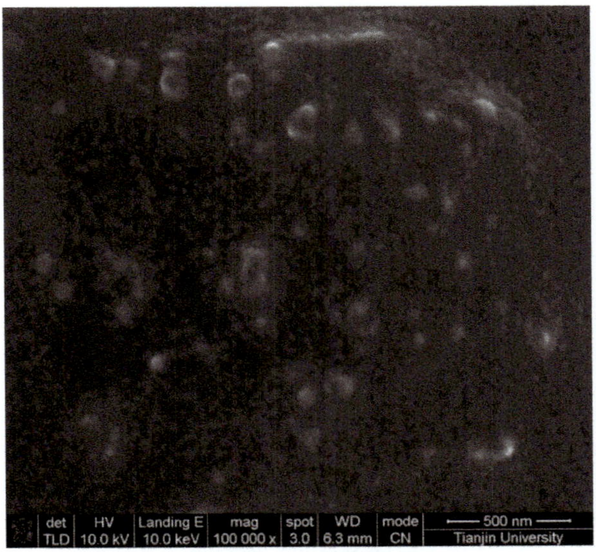

Figure 14.11 Scanning electron micrographs of SiR/SiO_2 nanocomposites.
Source: Reproduced with permission from Du BX, Li ZL. Surface charge and DC flashover characteristics of direct-fluorinated SiR/SiO_2 nanocomposites. *IEEE Trans Dielectr Electr Insul* 2014; **21**(6): 2602–10. Copyright 2014, IEEE.

Ltd., were spherically shaped with average diameters of around 50 nm and also had treated surfaces to avoid agglomeration. Samples, with dimensions of $35 \times 35 \times 2.5\ mm^3$, were prepared by incorporating nano-SiO_2 into SiR matrix with the fraction of 1, 3, and 5 wt% respectively. The undoped samples were employed as a comparison. The distribution of nanoparticles is shown in Fig. 14.11. It can be found that almost all of the nanoparticles are well-dispersed in the RTV SiR matrix.

Before fluorination, the reaction vessel was purified with nitrogen gas by more than three times, in order to exhaust the air in it. Then the F_2/N_2 mixture gas with 12% F_2 by volume was injected slowly into the reaction vessel until the pressure in

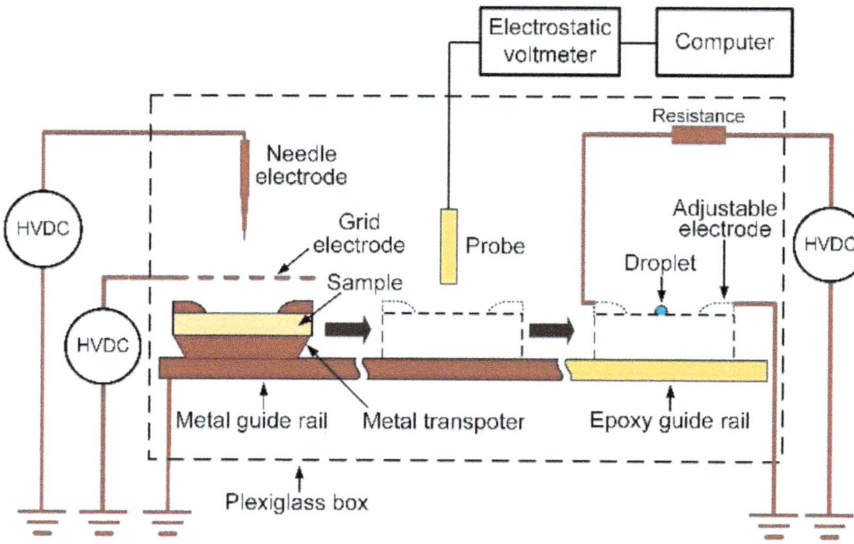

Figure 14.12 Schematic illustration of the experimental setup for corona discharge tests, surface charge measurement, and DC flashover tests.
Source: Reproduced with permission from Du BX, Li ZL. Surface charge and DC flashover characteristics of direct-fluorinated SiR/SiO_2 nanocomposites. *IEEE Trans Dielectr Electr Insul* 2014; **21**(6): 2602–10. Copyright 2014, IEEE.

it reached 0.05 MPa. The samples were treated at 25°C for 5, 10, 15, and 20 min, respectively. After the treatment, the residual mixture gas was removed from the vessel with nitrogen gas by more than five times.

The schematic diagram of surface discharge test, surface charge measurement and DC flashover experimental set-up is shown in Fig. 14.12. A voltage of ± 6 kV was applied between the needle and the grounded electrodes. Meanwhile the grid electrode was set at ± 4 kV. The corona discharge test was performed at room temperature at a relative humidity of $\sim 40\%$. The charging time was set to 10 min.

After the corona discharge test, the sample was rapidly transferred to the underside of a noncontacting probe to determine the surface potential. The sample was immediately transferred to the middle of the adjustable electrodes after the corona discharge. The flashover tests were conducted under DC stress, of which the polarity was identical with the corona discharge. A conductivity of 2.0 ± 0.1 mS/cm of the saline solution was previously obtained by adding NaCl to deionized distilled water which had a conductivity of 5.5 μS/cm. A water droplet with a volume of 10 μL was dropped to the middle of the electrodes. The interval between the corona discharge and the flashover test was controlled within a few seconds. Then, the DC voltage was increased with a rate of 1 kV/s until the flashover via the droplet occurred.

14.3.2 Effect on chemical composition

Surface chemical compositions of the fluorinated samples, with the original samples as reference, were characterized by attenuated total reflection infrared (ATR-IR) analysis.

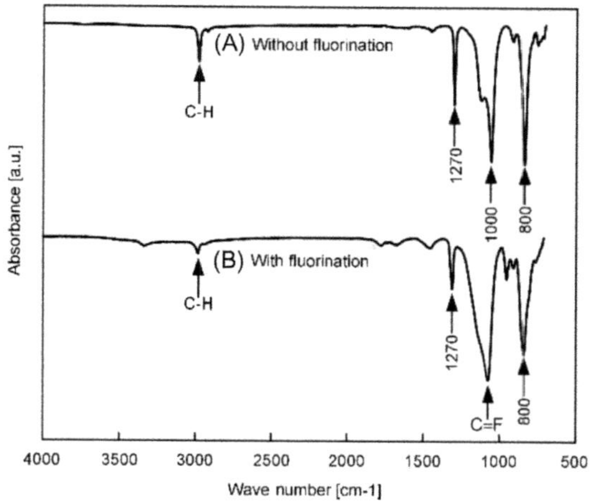

Figure 14.13 ATR-IR spectra of the (A) original sample and (B) fluorinated sample. *Source*: Reproduced with permission from Du BX, Li ZL. Surface charge and DC flashover characteristics of direct-fluorinated SiR/SiO_2 nanocomposites. *IEEE Trans Dielectr Electr Insul* 2014; **21**(6): 2602–10. Copyright 2014, IEEE.

It is well known that the C–F bond is the strongest single bond in organic chemistry and has a bond dissociation energy of 544 kJ/mol, much higher than that of the C–H bond (414 kJ/mol) [29]. As a consequence, the C–H bond would be disrupted by fluorination, followed by a formation of C–F, C–F_2, and C–F_3 groups [30,31].

Fig. 14.13 shows the ATR-IR spectra of the original and fluorinated samples. It is clearly indicated that the direct fluorination gives rise to an obvious change in the chemical component of the RTV SiR surface layer. In the spectrum (a) of the original sample, a strong absorption peak at 1050 cm^{-1} belongs to the SiOSi chain [32]. The absorption peaks at 1270 and 820 cm^{-1} are also observed, which are related to Si-CH_3 and $Si(CH_3)_2$ groups, respectively [33]. In addition, absorption bands observed in the regions of 2830–3000 and 1430–1470 cm^{-1} are corresponding to the C–H stretching vibration and the bending vibrations of CH_2 and CH_3, respectively [30,31]. Spectrum (b) of the fluorinated sample shows that the C-H bond is significantly reduced in strength by the direct fluorination. The absorption band of the Si-O-Si bond is covered by a broad strong absorption band appearing in the 940–1340 cm^{-1} region, which is associated with C–F, CF_2, and CF_3 groups [30,31]. That is because that the absorption intensities of C–F, CF_2, and CF_3 groups are stronger than those of the Si-O-Si chain. In addition, the absorption peaks of Si-CH_3 and $Si(CH_3)_2$ bonds decrease. The results show a clear evidence that some C–H bonds disappear and a fluorine element is incorporated into the surface layer of SiR by fluorination.

14.3.3 Surface charge

14.3.3.1 Effect of fluorination time

The relationship between the normalized surface potential and the decay time of samples with different fluorination times is shown in Fig. 14.14. It is observed that

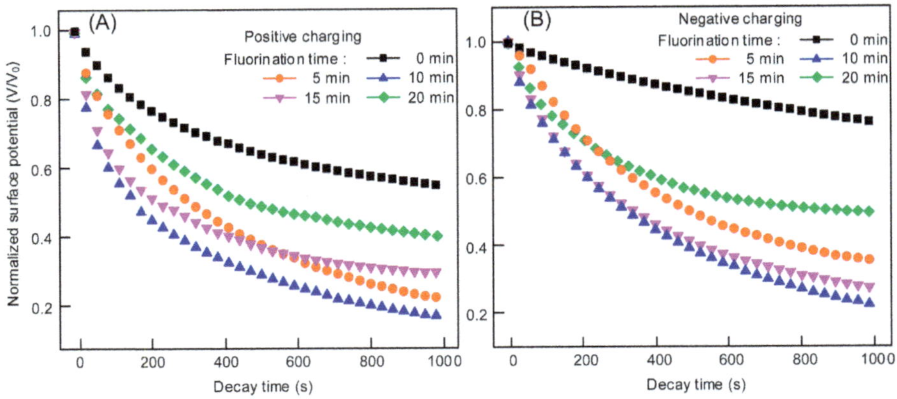

Figure 14.14 Relationship between the normalized surface potential and the decay time with mass fraction of nano-SiO_2 1 wt%.
Source: Reproduced with permission from Du BX, Li ZL. Surface charge and DC flashover characteristics of direct-fluorinated SiR/SiO_2 nanocomposites. *IEEE Trans Dielectr Electr Insul* 2014; **21**(6): 2602–10. Copyright 2014, IEEE.

the surface potential of the fluorinated samples declines faster than that of original samples, regardless of the charge polarity. The descending speed of surface potential becomes faster with the increasing fluorination time from 0 to 10 min, and becomes slower from 10 to 20 min. The result indicates that the fluorination can significantly suppress the surface charge accumulation on SiR. It is mainly because the fluorine element has the largest electro negativity and makes the C–F bond highly polarized. Therefore, the fluorinated layer would capture charges and form a shield. As the charges captured in this shielding layer generate an internal field opposing the applied field which decreases the effective interface field, further charge injection into the bulk could be blocked. However, when the fluorination time exceeds 10 min, the molecular structure of the sample surface may be destroyed by fluorination. It is speculated that the introduced chemical defects would provide trapping centers and make the charge accumulate easily.

14.3.3.2 Effect of mass fraction of nanoparticle

Fig. 14.15 shows a dependence of the normalized surface potential of samples fluorinated for 15 min on the mass fraction of nanoparticles. The decay curves of surface potential have the same decreasing tendency for both positively and negatively charged samples. It can be found that the potential decay of nanocomposites is obviously slower than that of the undoped samples. The slowest descending speed of surface potential appears at a mass fraction of 1 wt% for both positive and negative charges. And with the increase in fraction from 1 to 5 wt%, the descending speed increases.

For further analysis, the average decay rate in the first 5 min of decay time was calculated. The relationship between the average decay rate and the mass fraction of the samples with and without fluorination is shown in Fig. 14.16. The decay rate decreases with the increase in fraction from 0 to 1 wt%, and increases from 1 to

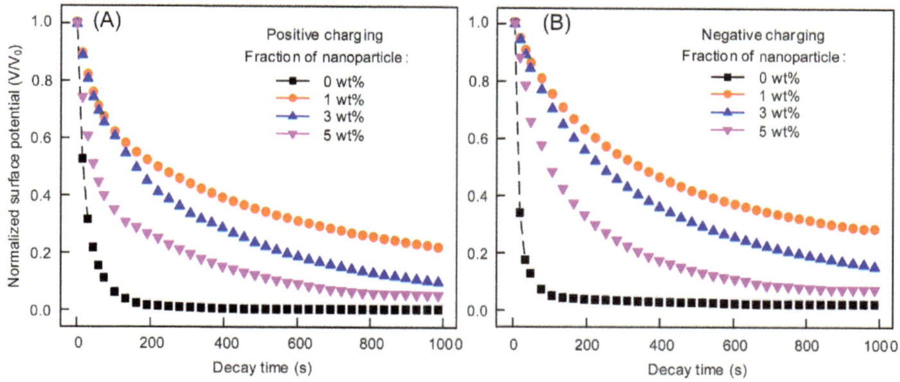

Figure 14.15 Relationship between the normalized surface potential and the decay time with the fluorination time 15 min.
Source: Reproduced with permission from Du BX, Li ZL. Surface charge and DC flashover characteristics of direct-fluorinated SiR/SiO_2 nanocomposites. *IEEE Trans Dielectr Electr Insul* 2014; **21**(6): 2602–10. Copyright 2014, IEEE.

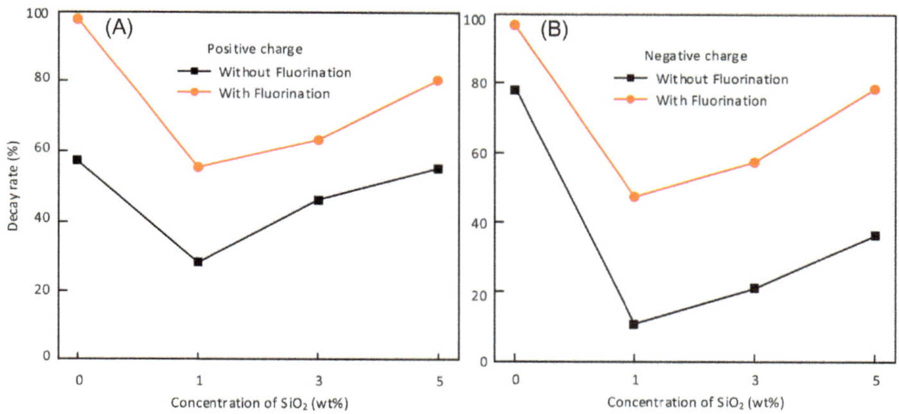

Figure 14.16 Relationship between the decay rate in the first 5 min and the mass fraction of nanoparticle. (The fluorinated samples were treated for 15 min.).
Source: Reproduced with permission from Du BX, Li ZL. Surface charge and DC flashover characteristics of direct-fluorinated SiR/SiO_2 nanocomposites. *IEEE Trans Dielectr Electr Insul* 2014; **21**(6): 2602–10. Copyright 2014, IEEE.

5 wt%, regardless of the charge polarity. The result indicates that the well-dispersed nanoparticles in the matrix could restrain the charge transport through the polymer bulk and influence the surface charge decay rate. These influences are usually considered to be related to the interaction zones introduced by nanoparticles. However, increasing the mass fraction may lead to the overlap of interaction zones surrounding nanoparticles. It was reported that the overlapping layers caused by the interaction zones surrounding nanoparticles will be more reactive than those of low mass fraction [34].

When a lot of interaction zones overlap, many conductive paths form through the overlap of the transition region in the bulk of nanocomposites [35], thus reducing the charge trapping. Therefore, the decay rate increases as the fraction increases from 1 to 5 wt%.

The trap energy distributions shown in Fig. 14.17. It is noted that samples with 1 wt% nano-SiO_2 have the highest energy level for both original and fluorinated samples. It is also observed that the energy level of the fluorinated samples is evidently lower than that of the original samples, and the minimum value in trap depth is ~0.72 eV, appearing at the undoped sample with fluorination. It has been reported that the charges could easily move out from the shallow traps, which are associated with the low energy level [36]. The fluorinated layer may lower the energy level. With this hypothesis, the surface charge of fluorinated samples decays faster than that of the original samples. In contrast, the dispersed nanoparticles would restrain the charge transported through the polymer bulk.

14.3.4 Flashover voltage

Previous researches have explained the flashover mechanism [37]. Once the droplet dropped on the charged surface, the droplet would be electrified. The electrified droplet will be flattened and elongated by the Coulomb force generated by the external DC field. With the increase of applied voltage, the Coulomb force becomes large enough to move the electrified droplet to the grounded electrode, forming a conductive path between the electrodes. This path could reduce the distance between the electrodes and increase the electric field strength at the edges of droplet. With the strengthening of the electric field, the arcs may travel above the wet area and extend in a shape of thin filament until one of the arcs bridges the

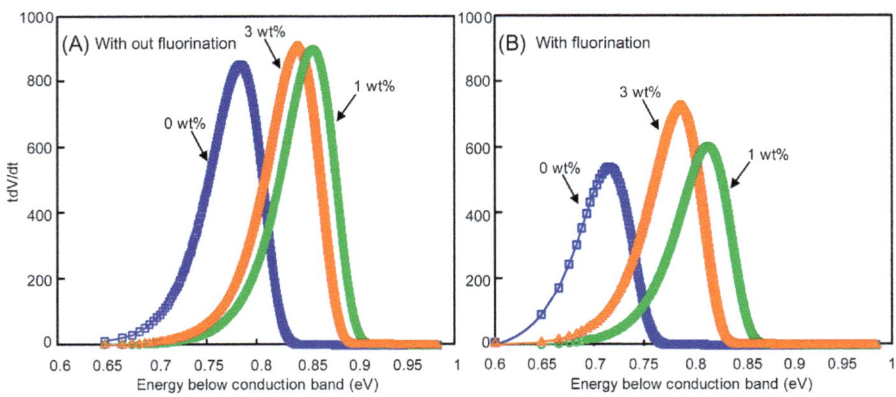

Figure 14.17 The trap distribution at various mass fraction of nanoparticle after negative charging for 10 min. (The fluorinated samples were treated for 15 min).
Source: Reproduced with permission from Du BX, Li ZL. Surface charge and DC flashover characteristics of direct-fluorinated SiR/SiO_2 nanocomposites. *IEEE Trans Dielectr Electr Insul* 2014; **21**(6): 2602–10. Copyright 2014, IEEE.

electrodes and flashover occurs. It can be speculated that the higher surface potential would cause more charge moving into the droplet on the sample surface. In the constant electrostatic field, the droplet with more charge may withstand a larger Coulomb force, which accelerates the elongation process of the electrified droplet and makes the flashover voltage lower.

The DC flashover tests of samples with different mass fraction and fluorination time were conducted after corona discharge finished immediately, in order to study the effect of the mass fraction on the DC flashover voltage. The relationship between the flashover voltage and the mass fraction of nano-SiO_2 is shown in Fig. 14.18. It is observed that the flashover voltage of nanocomposites is lower than that of the undoped samples, regardless of the power polarity. In accordance with the decay rate shown in Fig. 14.16, the DC flashover voltage decreases with increasing the fraction from 0 to 1 wt%, and increases from 1 to 5 wt%. It has been proved that the charge accumulating on the samples could reduce the DC flashover voltage. The incorporated nanoparticles in the matrix restrain the charge transport through the bulk and slow down the charge decay rate, thus decreasing the DC flashover voltage. When the mass fraction exceeds 1 wt%, more conductive paths form through the overlap of the transition region in the bulk, therefore increasing the decay rate of surface charge and raising DC flashover voltage.

14.4 Tree characteristics in silicone rubber/SiO_2 nanocomposites

14.4.1 Experiment procedure

RTV SiR (produced by Shenzhen Anpin Silicone Material Co., Ltd.) was employed for the present study. The nanoparticles used in the study were spherically shaped SiO_2 particles with average particle diameters of 20 nm supplied by Nanomaterials Co., Ltd. Nanjing Haitai. The surface of nanoparticles had been treated in order to avoid the agglomeration. The weight percentage of nanoparticle was set to 0, 0.5, 1, 1.5, and 2 wt% respectively. It was found that agglomeration occurred when the content was increased to 3.0 wt%.

Fig. 14.19 shows the schematic diagram of experimental setup. The temperature in the thermotank for experiment was controlled by the temperature controller, which ranged from 30°C to −196°C. Liquid nitrogen was employed to cool down the temperature inside. The experiment temperatures were set to −30, −60, and −90°C to simulate the low operating temperatures of the power cables in some cold areas. After the samples were fixed between the needle-plate electrodes inside the thermotank, the temperature was decreased to the setting experiment temperature at a rate of 10°C per minute. Subsequently, an AC voltage was applied on the needle electrode to initiate electrical trees. Its frequency was 50 Hz and root-mean-square value was 8 kV. The tree process was observed by using a digital microscope monitoring system (SDK-2000 TianMin). An average of 30 samples were tested under each condition, in which the number of treeing samples was 15. All

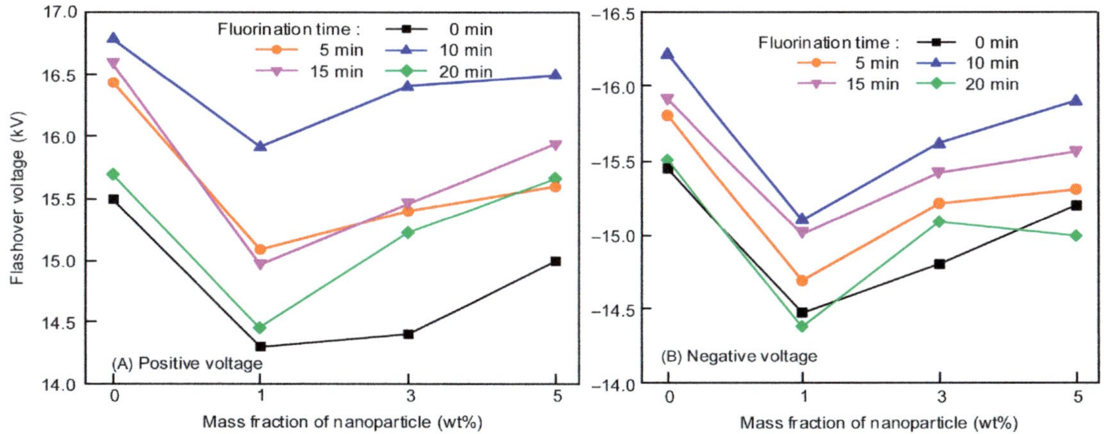

Figure 14.18 Relationship between the flashover voltage and the mass fraction with different fluorination time.
Source: Reproduced with permission from Du BX, Li ZL. Surface charge and DC flashover characteristics of direct-fluorinated SiR/SiO_2 nanocomposites. *IEEE Trans Dielectr Electr Insul* 2014; **21**(6): 2602–10. Copyright 2014, IEEE.

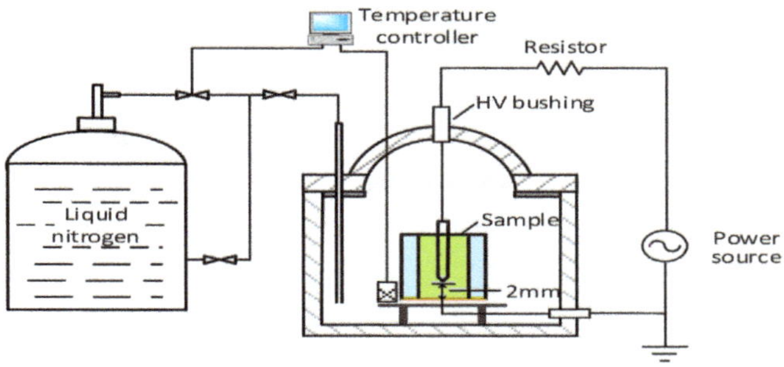

Figure 14.19 Schematic diagram of experimental setup.
Source: Reproduced with permission from Du BX, Han T, Su JG. Tree characteristics in silicone rubber/SiO_2 nanocomposites under low temperature. *IEEE Trans Dielectr Electr Insul* 2014; **21**(2): 503–10. Copyright 2014, IEEE.

the trees in the samples were observed and the tree length was measured with the precision of 10 μm. The results are presented in the following sections.

14.4.2 DSC analysis

Differential scanning calorimetry (DSC) was carried out to define the characteristic temperatures of the samples, including the temperatures of crystallization and melting. The DSC was performed with a 200-F3 Maia differential scanning calorimeter. There were three processes in the test. The cooling process was the first and the cooling rate was set to 10°C per minute, with the temperature changing from 0°C to −100°C. When the temperature reached −100°C, there would be a thermostatic process lasting for 3 min. Then the heating process took place between −100°C and 0°C with the same heating rate as previously. Samples of 5.2 mg and 4.2 mg weight were used for the samples with 0.0 wt% and 1.0 wt%, respectively. As shown in Fig. 14.20, an exothermal peak in the heat flow–temperature curve for the two samples is observed during cooling, the exothermal peak of samples with 0.0 wt% and 1.0 wt% are located at −67.3°C and −64.5°C, respectively. The exothermal peak corresponds to the crystallization of some of the polymer chains. Meanwhile, the crystallization begins when the exotherm rises. During heating, an endothermal peak is observed. The endothermal peak of samples with 0.0 wt% and 1.0 wt% are located at −42.5°C and −41.1°C, respectively. The endothermal peak corresponds to the crystal grain melting [39].

The temperatures at which crystallization and crystallites melting occur are denoted respectively as T_c and T_m in the following. The T_c of samples with content of 0.5, 1.5, and 2 were also measured, and they were all higher than that of 0.0 wt %. These tests do not allow us to determine the glass transition temperature for both materials, which occurs at temperature lower than −100°C. It can be found that the crystallization process occurred when the experiment temperature was set to −60°C and −90°C. At the temperature of 30°C and –30 °C above T_m, there was no crystal region in the sample, and the structure of the molecular chain was

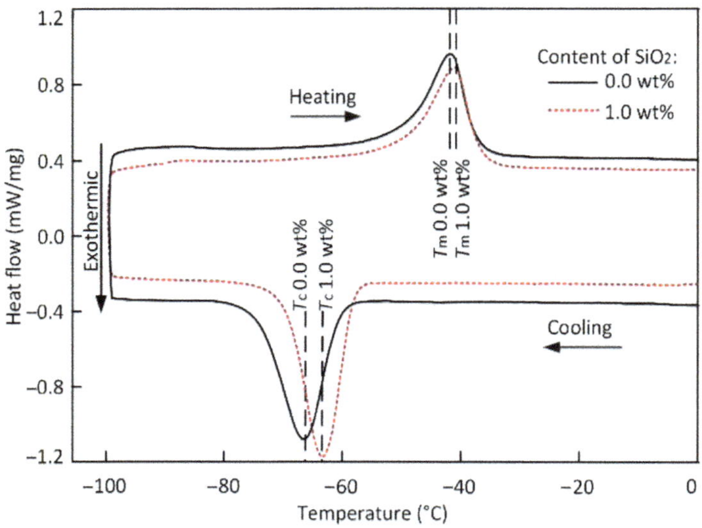

Figure 14.20 Differential scanning calorimetry analysis.
Source: Reproduced with permission from Du BX, Han T, Su JG. Tree characteristics in silicone rubber/SiO_2 nanocomposites under low temperature. *IEEE Trans Dielectr Electr Insul* 2014; **21**(2): 503–10. Copyright 2014, IEEE.

disordered. It is also observed that T_c was changed by the mixed nanoparticles, and this is maybe because the nanoparticles form crystal nuclei during the crystallization process, which accelerates the crystallization [40].

14.4.3 Tree structure and growth characteristics

Three typical profiles of electrical trees namely branch tree, bush tree, and pine branch tree, were observed in samples at different temperatures, as shown in Fig. 14.21A–C, respectively.

The feature of a branch tree is the obviously different size of the tree channel, which also presents in a pine branch tree with a great deal of pine channels. As for the bush tree, the channels propagate in many directions toward the grounded electrode and cross with each other, leaving the area occupied by bush-like channels, and the density of branches is much larger than that of the branch tree and pine branch tree. Moreover, the bush tree shows as a sphere that can uniform the electric field. The main branches of the pine branch tree are fewer than those of the other two types, but there are a large number of small branches generated from the main branches. During the experiment, it is observed that most of the tree channels propagated along the directions of the electric field lines of the electrodes, whereas some of the branch channels propagate vertical to the electric field direction or even in the opposite direction, as shown in Fig. 14.21C. This is maybe the result of high pressure generated by the discharge inside the tree channels during the treeing process.

Fig. 14.22 shows the growth process of a branch tree. In the first 10 s, there are only two main branches which are about 700 and 800 μm. When the time is 30 s, the

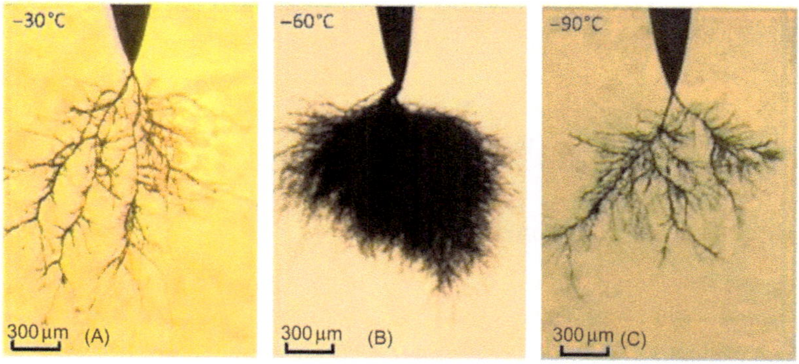

Figure 14.21 Typical profiles of electrical tree. (A) Branch tree. (B) Bush tree. (C) Pine branch tree.
Source: Reproduced with permission from Du BX, Han T, Su JG. Tree characteristics in silicone rubber/SiO_2 nanocomposites under low temperature. *IEEE Trans Dielectr Electr Insul* 2014; **21**(2): 503–10. Copyright 2014, IEEE.

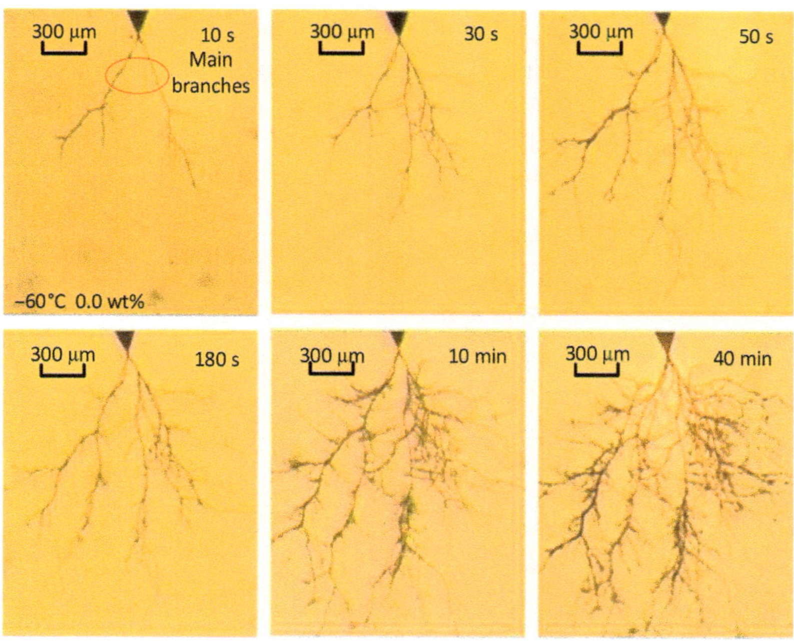

Figure 14.22 Growth process of branch tree.
Source: Reproduced with permission from Du BX, Han T, Su JG. Tree characteristics in silicone rubber/SiO_2 nanocomposites under low temperature. *IEEE Trans Dielectr Electr Insul* 2014; **21**(2): 503–10. Copyright 2014, IEEE.

number of main branches increases and they become a little longer. In the following 20 s, the main branches grow fast. 10 min after the inception, the main branches grow very slowly and there are many small branches generated from the main branches. It should be emphasized that the color of the tree channel is white rather than black, and it is supposed that the channel is composed of silicone compounds.

14.4.4 Effect of nanoparticle

It was found that the tree grew fast in the first 15 min and then grew at a slow speed in the next 100 min during the experiment. In order to observe the effect of nanoparticle content on the tree growth, the average length of the trees at 5, 15, and 110 min in each kind of sample were calculated and are shown in Fig. 14.23.

As shown in Fig. 14.23A, when the temperature is −30°C, the tree length of 5 min decreases rapidly with the content increasing from 0.0 to 0.5 wt% and then decreases slowly until the 1.5 wt% to reach the shortest length. From 1.5 to 2.0 wt %, the length increases obviously. The lengths of 15 and 110 min have the similar tendency. As shown in Fig. 14.23B, when the temperature is −60°C, the length of 5 min decreases slowly with the content increasing from 0.0 to 1.5 wt% and then increases from 1.5 to 2.0 wt%. The lengths of 15 and 110 min have a more rapidly decreasing process. As for the temperature of −90°C, the tendency is more complex, which is shown in Fig. 14.23C. At 5 min, the shortest length is still at 1.5 wt %, but there is an increase from 0.5 to 1.0 wt%. At 15 min, the length decreases from 0.0 to 1.5 wt% and reaches the shortest length. When the time is 110 min, the length increases from 0.0 to 0.5 wt% and then decreases from 0.5 to 1.5 wt%, afterward, there is a fast increase from 1.5 to 2.0 wt%.

It can be assumed that the nanoparticle is the main reason for the phenomena above. The permittivity of SiO_2 is 3.9, which is higher than that of SiR (~3.0). The electric field can be intensified in matrices adjacent to the surface of nanoparticle with high permittivity, and then will be lowered in matrices between neighboring nanoparticles in the AC condition [41], and in the case of positively charged nanoparticles, the electron will be attracted by them; therefore, the tree will grow toward

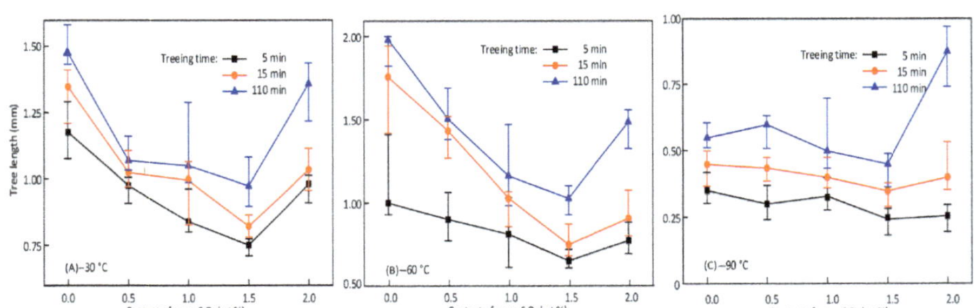

Figure 14.23 Relation between the tree length and nano-SiO_2 contents.
Source: Reproduced with permission from Du BX, Han T, Su JG. Tree characteristics in silicone rubber/SiO_2 nanocomposites under low temperature. *IEEE Trans Dielectr Electr Insul* 2014; **21**(2): 503–10. Copyright 2014, IEEE.

the nanoparticles [42]. Because the nanoparticles have high permittivity and high tolerance to heat and high pressure generated by the partial discharge, the growing process of the tree will be hindered, therefore, the tree is more complex and the tree propagation in nanocomposites takes much more time. All the effects mentioned above certainly lead to the shorter tree length in nanocomposites.

The inner-particle distance starts to reduce with the increase of nanoparticle content, which can result in an overlap of the chemical bond region around the nanoparticles [43]. Then the overlapped nano-SiO_2 particles are presented as the defect, which is due to the form of the clustering station [44]. Thus, the chemical bond in the nanocomposite is weakened by the overlapping SiO_2 nanoparticles, which will lead to the fast growth of the electrical tree at 2.0 wt%. When the experiment temperature is −90°C, all the trees grow at a low speed. This is maybe because the effect of temperature on the treeing process is more obvious than that of the nanoparticle at such a low temperature.

14.4.5 Effect of temperature

The change of inception probability with experiment temperature implies that when the temperature is above T_c, it is easier for the electrical tree to incept in SiR cable accessories at lower temperature than that at higher temperature. The cables in cold areas should be kept over an appropriate temperature to ensure their safe operation.

When the temperature is −60°C, it can be found that the crystallization has begun according to the DSC analysis. But the crystallization rates are different, supposedly higher at 1.0 wt% and lower at 0.0 wt%, and this may be the reason for the different probabilities. The position of the needle tip is important in the tree inception process. When the tip is in the crystalline region, the tree inception will be limited by the mechanical stress from crystal grains, and this will lead to the low inception probability. At the temperature of −90°C, the density of crystal grains may be much higher and the tree inception is more difficult.

To observe the tree inception in much lower temperatures, the treeing experiment was carried out in liquid nitrogen where the temperature could reach −196°C. The result shows that there are no trees at 5 min and the tree inception time can reach more than 100 min in all the samples.

In order to observe the effect of temperature during the treeing process, the distributions of tree structures at different temperature were also calculated, which are shown in Fig. 14.24. Two types of samples with the content of 0.0 and 1.0 wt% are researched and the tree structure at 30°C is employed as a comparison. Fig. 14.24A shows the tree structures in 0.0 wt% samples. At the temperatures from 30°C to −60°C, the distribution of the tree structure is similar, and the main structure is the branch tree. When the temperature is −90°C, there is only the pine branch tree. Fig. 14.24B shows the tree structure in a 1.0 wt% sample. Compared with the 0.0 wt% sample, the proportion of branch tree in the 1.0 wt% sample is lower at the temperatures from 30°C to −60°C, and the bush tree accounts for near 60% at −60°C. When the temperature is −90°C, there is still only the pine branch tree.

In the 0.0 wt% samples, the molecular chains are in the same structure without crystallization at 30°C and −30°C, which means the treeing process is similar. This

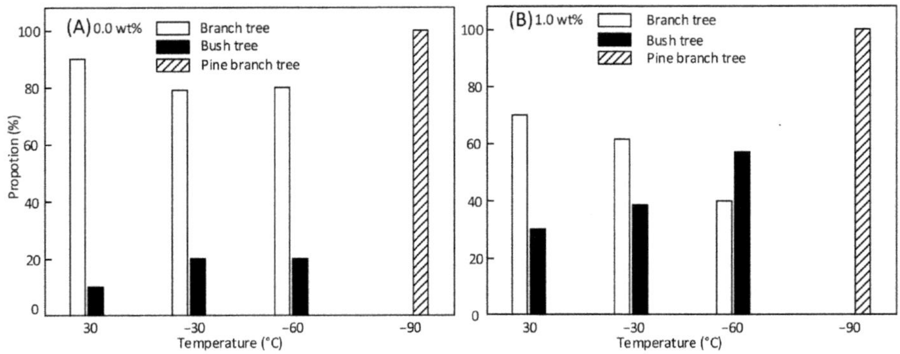

Figure 14.24 Relation between the tree proportion and the temperature.
Source: Reproduced with permission from Du BX, Han T, Su JG. Tree characteristics in silicone rubber/SiO_2 nanocomposites under low temperature. *IEEE Trans Dielectr Electr Insul* 2014; **21**(2): 503–10. Copyright 2014, IEEE.

may lead to the similar distribution of tree structures. When the temperature is −60°C, the inhibitory effect of crystallization on treeing is not obvious for the low density of crystal grains. As the temperature decreases to −90°C, the density of crystal grains is higher, and it is more difficult for the discharge to break the molecular chains of SiR and to propagate new main branches. But a number of tiny branches will generate from the early main branches as a result of the partial discharge in the branch channels. With the growing of these small branches, the structure of the tree appears to be a short pine branch tree.

As in the 1.0 wt% samples, the different distribution is caused by the nanoparticle and temperature. It can be found that the mixed nanoparticles increase the proportion of bush tree at 30°C and −30°C. This might be caused by the existence of the nanoparticle which makes the tree channel complex. When the tree reaches the nanoparticle, it will grow transversely to bypass the nanoparticle, as a result, the channels will cross with each other and the bush tree will be formed. When the temperature is −60°C, the density of crystal grains in a 1.0 wt% sample is higher than that in a 0.0 wt% sample, the tree channels will propagate along the interface of the crystal grains and this is maybe the main reason for the obvious increase of the bush tree proportion.

14.5 Conclusion

This work has shown the effects of mass fraction of nanoparticles on the surface charge accumulation and decay, thermal conductivity, and treeing characteristics of nanocomposites. The main conclusions can be summarized as follows.

1. The thermal conductivity of the SiR/BN specimens increases with the increase in content of BN. From the thermal distribution, it can be seen that the BN particles have a significant inhibitory effect on thermal accumulation, especially at higher content. The thermal

dissipation is accelerated by BN particles because of the increase in thermal conductivity. The discharge quantity during the inclined plane test decreases with the increase in content of BN. Because of the improvement in thermal conductivity, the maximum temperature on the specimens with higher content is reduced, thus leading to fewer dry bands and discharges. The resistance to tracking and erosion of SiR is enhanced by the addition of nano-BN particles. The results indicate that the weight loss and erosion depth decrease with increasing BN content. Because the heat is conducted out quickly on specimens with higher content, the temperature rise in the discharge area is limited, and the erosion resistance is improved.

2. ATR-IR spectra of the original and fluorinated samples show clear evidence that some C-H bonds disappear and the fluorine element is incorporated into the SiR surface layer by the fluorination. The charge decay rate of the fluorinated samples is considerably faster than that of the original samples, which indicates that the fluorinated layer on the sample surface resulting from the appropriate fluorination could significantly improve the decay rate of surface charge. The decay for both positive and negative charge is suppressed by the nanoparticles. The result indicates that the dispersed nanoparticles restrain the charge transport through the bulk and decrease the surface charge decay rate of the nanocomposites. The DC flashover voltage is influenced by the charge on the sample surface. The incorporation of nanoparticles may introduce deep traps into the polymers, which results in the lower flashover voltage. In contrast, the DC flashover voltage could be raised by the direct fluorination.
3. Three typical profiles of electrical trees, namely branch, bush, and pine branch tree, can be observed. The occurrence of different tree structures changes with the temperature. The branch tree and bush tree exist at the temperatures of 30°C, −30°C, and −60°C. When the temperature is −90°C, there is only the pine branch tree. The T_c of a SiR/SiO_2 nanocomposite was analyzed by DSC. It is found that the T_c increases with the addition of nano-SiO_2. The tree growth is restrained by the nano-SiO_2, and the present optimum content with the slowest tree growth speed is 1.5 wt%. Meanwhile, it is found that the nano-SiO_2 can increase the proportion of the bush tree. The tree inception probability is high at 30°C, −30°C, and −60°C. When the temperature is −90°C, it decreases obviously. The tree inception time can reach more than 100 min at −196°C. The tree growth speed is affected by the crystallization. The interface between the crystalline region and noncrystalline region may be the weak region which is easy for the tree growth. Because of the high crystallization ratio at −90°C, the tree growth speed is the lowest.

References

[1] Bruce GP, Rowland SM, Krivda A. Performance of silicone rubber in DC inclined plane tracking tests. IEEE Trans Dilectr Electr Insul 2010;17(2):521−32.

[2] Yang WK, Yin YH, Zeng NC, Wang HW, Wang MX, Zhang J. Analysis on commissioning test of ± 800 kV UHVDC transmission project from Xiangjiaba to Shanghai. Power Syst Technol 2011;35(7):19−23 in Chinese.

[3] Jia ZD, Fang S, Gao HF, Guan ZC, Wang LM, Xu ZH. Development of RTV silicone coatings in China: overview and bibliography. IEEE Electr Insul Mag 2008;24(2): 28−41.

[4] Du BX, Liu Y, Liu HJ, Yang YJ. Recurrent plot analysis of leakage current for monitoring outdoor insulator performance. IEEE Trans Dielec Electr Insul 2009;16(1):139−46.

[5] Kumara S, Ma B, Serdyuk YV, Gubanski SM. Surface charge decay on HTV silicone rubber: effect of material treatment by corona discharges. IEEE Trans Dielectr Electr Insul 2012;19(6):2189–95.
[6] Kumara S, Alam S, Hoque IR, Serdyuk YV, Gubanski SM. DC flashover characteristics of a polymeric insulator polymeric materials under different neutralization surface charges. IEEE Trans Dielectr Electr Insul 2012;19(3):1084–90.
[7] Du BX, Zhang JW, Gao Y. Effects of TiO_2 particles on surface charge of epoxy nanocomposites. IEEE Trans Dielec Electr Insul 2012;19(3):755–62.
[8] Cheng TC, Wu CT. Contamination flashover of HVDC insulators, EL-2016, Research Project 848-1, Final report, EPRI, 1981.
[9] Kim J, Chaudhury MK, Owen MJ. Hydrophobicity loss and recovery of silicone HV insulation. IEEE Trans Dielectr Electr Insul 1999;6(5):695–702.
[10] Kim SH, Cherney EA, Hackam R, Rutherford KG. Chemical changes at the surface of RTV silicone rubber coatings on insulators during dry-band arcing. IEEE Trans Dielectr Electr Insul 1994;1(1):106–23.
[11] Moreno VM, Gorur RS. AC and DC performance of polymeric housing material for HV outdoor insulators. IEEE Trans Dielectr Electr Insul 1999;6(3):342–50.
[12] Nakanishi Y, Fujimori A, Fukunaga S, Tanabe T, Kobayashi M, Shiseki N, et al. Development of prefabricated joint for 275 kV XLPE cable. IEEE Trans Power Deliv 1995;10(3):1139–47.
[13] Laurent C, Mayoux C. Analysis of the propagation of electrical treeing using optical and electrical methods. IEEE Trans Dielectr Electr Insul 1980;15(1):33–42.
[14] Venkatesulu B, Thomas MJ. Corona aging studies on silicone rubber nanocomposites. IEEE Trans Dielectr Electr Insul 2010;17(2):625–34.
[15] Venkatesulu B, Thomas MJ. Erosion resistance of alumina-filled silicone rubber nanocomposites. IEEE Trans Dielectr Electr Insul 2010;17(2):615–24.
[16] Fleming RJ, Pawlowski T, Ammala A, Casey PS, Lawrence KA. Electrical conductivity and space charge in LDPE containing TiO_2 nanoparticles. IEEE Trans Dielectr Electr Insul 2005;12(4):745–53.
[17] Gorur RS, Cherney EA, Hackam R, Orbeck T. The electrical performance of polymeric insulating materials under accelerated aging in a fog chamber. IEEE Trans Power Deliv 1988;3(3):1157–64.
[18] Gorur RS, Cherney EA, Hackam R. The AC and DC performance of polymeric insulating materials under accelerated aging in a fog chamber. IEEE Trans Power Deliv 1988;3(4):1892–902.
[19] Gustavsson TG, Gubanski SM, Hillborg H, Karlsson S, Gedde UW. Aging of silicone rubber under AC or DC voltages in a coastal environment. IEEE Trans Dielectr Electr Insul 2001;8(6):1029–39.
[20] Meyer LH, Jayaram SH, Cherney EA. Correlation of damage, dry band arcing energy, and temperature in inclined plane testing of silicone rubber for outdoor insulation. IEEE Trans Dielectr Electr Insul 2004;11(3):424–32.
[21] Meyer L, Jayaram S, Cherney EA. Thermal conductivity of filled silicone rubber and its relationship to erosion resistance in the inclined plane test. IEEE Trans Dielectr Electr Insul 2004;11(4):620–30.
[22] Chen Y, Imai T, Ohki Y, Tanaka T. Tree initiation phenomena in nanostructured epoxy composites. IEEE Trans Dielectr Electr Insul 2010;17(5):1509–15.
[23] Tanaka T, Matsunawa A, Ohki Y, Kozako M, Kohtoh M, Okabe S. Treeing phenomena in epoxy/alumina nanocomposite and interpretation by a multi-core model. IEEJ Trans FM 2006;126(11):1128–35.

[24] Du BX, Li J. Electrified droplet on corona-charged surface of silicone rubber/SiO_2 nanocomposite. IEEE Trans Dielectr Electr Insul 2012;19(6):2073–80.
[25] Kong SM, Mariatti M, Busfield JJC. Effects of types of fillers and filler loading on the properties of silicone rubber composites. J Reinf Plast Compos 2011;30(13):1087–96.
[26] Du BX, Xu H. Effects of thermal conductivity on DC resistance to erosion of silicone rubber/BN nanocomposites. IEEE Trans Dilectr Electr Insul 2014;21(2):511–18.
[27] Gorur RS, Montesinos J, Varadadesikan L, Simmons S, Shah M. A laboratory test for tracking and erosion resistance of HV outdoor insulation. IEEE Trans Dielectr Electr Insul 1997;4(6):767–74.
[28] Du BX, Li ZL. Surface charge and DC flashover characteristics of direct-fluorinated SiR/SiO_2 nanocomposites. IEEE Trans Dilectr Electr Insul 2014;21(6):2602–10.
[29] Liu YQ, An ZL, Cang J, Zhang YW, Zheng FH. Significant suppression of surface charge accumulation on epoxy resin by direct fluorination. IEEE Trans Dielectr Electr Insul 2012;19(4):1143–50.
[30] Du BX, Li J, Du W. Surface charge accumulation and decay on directfluorinated polyimide/Al_2O_3 nanocomposites. IEEE Trans Dielectr Electr Insul 2013;20(5): 1764–71.
[31] Du BX, Li J, Du W. Dynamic behavior of surface charge on direct-fluorinated polyimide films. IEEE Trans Dielectr Electr Insul 2013;20(3):947–54.
[32] Rajini V, Udayakumar K. Degradation of silicone rubber under AC or DC voltages in radiation environment. IEEE Trans Dielectr Electr Insul 2009;16(3):834–41.
[33] Lu NQ, Deng ZH. Practical infrared spectrum analysis. Electronics Industry Press; 1989.
[34] Zou C, Fothergill JC, Rowe SW. The effect of water absorption on the dielectric properties of epoxy nanocomposites, IEEE International Conference Solid Dielectric, Potsdam, Germany; 2010. p. 1–4.
[35] Molinié P. Measuring and modeling transient insulator response to charging: the contribution of surface potential studies. IEEE Trans Dielectr Electr Insul 2005;12 (5):939–50.
[36] Chen G, Xu Z, Zhang LX. Measurement of surface potential decay of corona charged polymer films using the pulsed Electroacoustic method. Meas Sci Tech 2007;18 (5):1453–8.
[37] Karady GG, Shah M, Brown RL. Flashover mechanism of silicone rubber insulators used for outdoor insulation-I. IEEE Trans Power Deliv 1995;10(4):1965–71.
[38] Du BX, Han T, Su JG. Tree characteristics in silicone rubber/SiO_2 nanocomposites under low temperature. IEEE Trans Dilectr Electr Insul 2014;21(2):503–10.
[39] Rey T, Chagnon G, Le Cam J-B, Favier D. Influence of the temperature on the mechanical behavior of filled and unfilled silicone rubbers. Polym Test 2013;32 (3):492–501.
[40] Tanaka T, Montanari GC, Mulhaupt R. Polymer nanocomposites as dielectrics and electrical insulation-perspectives for processing technologies, material characterization and future applications. IEEE Trans Dielectr Electr Insul 2004;11 (5):763–84.
[41] Tanaka T. Similarity between treeing lifetime and PD resistance in aging mechanisms for epoxy nanocomposites, IEEE International Conference Properties and Applications of Dielectric Materials (ICPADM) 2009. p. 741–44.
[42] Wu JD, Iizuka T, Monden K, Tanaka T. Characteristics of initial trees of 30 to 60 μm length in epoxy/silica nanocomposite. IEEE Trans Dielectr Electr Insul 2012;19(1): 312–20.

[43] Raetzke S, Kindersberger J. The effect of interphase structures in nanodielectrics. IEEJ Trans. Fundam Mater 2006;126(11):1044–9.

[44] Amekura H, Umeda N, Okubo N, Kishimoto N. Ion-induced frequency shift of $\sim$1100 cm^{-1} IR vibration in implanted SiO_2: compaction versus bond-breaking. Nucl Instrum Methods Phys Res B 2003;206:1101–5.

Computational simulation in elastomer nanocomposites

15

J. Liu and L. Zhang
Beijing University of Chemical Technology, Beijing, China

Based on our group's achievements in the study of elastomer nanocomposites (ENCs) via molecular dynamics (MD) simulation and the latest literature, in this chapter we aim to summarize the main research progress of ENCs, hoping to provide some solid basis and scientific guidelines for fabricating high performance and multifunctional ENCs. For better illustration, some important theoretical and experimental results are also covered. This chapter is divided into the following sections: (1) Computer simulation techniques; (2) Dispersion of nanoparticles (NPs): structure and phase behavior; (3) Interfacial chain structure and dynamics between polymer and NPs; (4) Static and dynamic mechanics; (5) Thermal and electrical conductivity; (6) Future simulation opportunities and challenges. Emphasis is placed on introducing three typical simulation methods, such as lattice Monte Carlo (MC), off-lattice MC, and MD, analyzing the spatial arrangement and organization of NPs with different shapes in the polymer chains, discussing the possible change of the chain conformation and glass transition induced by NPs, the interfacial chain structure and dynamics, the interfacial bonding strength and the filler network formed at high loading of NPs, investigating the static mechanical reinforcement and toughness, and the structural evolution under the quiescent or dynamic shear flow, evaluating the key structural parameters influencing the thermal and electrical conductivity. Generally, this chapter is expected to provide an overall picture of the structure-property relation achieved through computer simulation.

15.1 Computer simulation techniques

Before introducing the simulation research of ENCs, we give a brief analysis about the main computational simulation techniques that are adopted to investigate polymeric materials. The first one is the lattice MC simulation by employing the bond fluctuation lattice model, which explicitly considers the excluded volume interaction among the monomers of chains by forbidding each lattice vertex to be occupied simultaneously by two monomers, and also make the crossing between chemical bonds impossible [1]. This simulation method has been utilized to efficiently investigate the phase behavior of polymer blends, polymer solutions,

Progress in Rubber Nanocomposites. DOI: http://dx.doi.org/10.1016/B978-0-08-100409-8.00015-2

and so on, and is much less demanding in terms of computer resources compared to the corresponding off-lattice models discussed later. The second method is the nonlattice MC simulation, which always adopts methods such as configurational bias (CB) reptation, CB rearrangement of chain ends, and rebridging CB of inner sections to promote the equilibrium process of long entangled polymer chains [2]. In fact, few simulation investigations of ENCs have been performed using lattice MC [3] or nonlattice MC technique [4]. The third widely used method is the continuum MD simulation through the bead-spring model, which was first proposed by Kremer and Grest [5] to study the dynamics of entangled linear polymer melts, indicating that good agreement exists among simulated results, theoretical predictions such as the Rouse and Reptation model, and experimental data using neutron spin echo measurement. Since polymer chains possess different length and time scales, ranging from the bond length (1 A), persistence length (10 A), to the gyration radius of the coil (100 A), as depicted in Fig. 15.1A, this model is capable of efficiently simulating dense entangled polymer melts while maintaining its generic physical fundaments through the coarse-graining process from the fully atomistic chain, as displayed in Fig. 15.1C [6]. Presently, most simulation studies of ENCs adopt this coarse-grained polymer model. For the forcefield of ENCs, the interaction between all monomers is always described with the truncated and shifted Lennard-Jones (TSLJ) potential as follows [7]:

$$U_{ij} = \begin{cases} 4\varepsilon\left[\left(\dfrac{\sigma}{r_{ij}}\right)^{12} - \left(\dfrac{\sigma}{r_{ij}}\right)^{6} + C\right] & r_{ij} \leq 2^{*}2^{1/6}\sigma \\ 0, & r_{ij} \geq 2^{*}2^{1/6}\sigma \end{cases} \tag{15.1}$$

where r_{ij} is the distance between monomer i and j. The constant $C = 0.00775$ shifts the potential such that it vanishes at the cutoff distance. The simulated results are in reduced units by setting $\sigma = \varepsilon = 1$. Meanwhile, the connected monomers in every chain are maintained through the finitely extensible nonlinear elastic (FENE) potential as shown below:

$$U_{ij}^{ch} = \begin{cases} -0.5kR_0^2\ln[1-(r_{ij}/R_0)^2], & r_{ij} \leq R_0 \\ \infty, & r_{ij} \geq R_0 \end{cases} \tag{15.2}$$

where the parameters $R_0 = 1.5\sigma$ and $k = 30\varepsilon/\sigma^2$. It is noted that this spring constant is strong enough to prevent the crossing between any bond. The TSLJ and FENE potential are plotted in Fig. 15.1B as a function of the distance between monomers. Note that the distance, corresponding to the minimum of the bond potential (0.96σ) is smaller than that of the LJ potential (1.13σ), and this incompatibility prevents the crystallization of polymer melts. For the interaction U_{mn} between monomers and spherical NPs, and U_{nn} between NPs and NPs, the modified Lennard-Jones potential

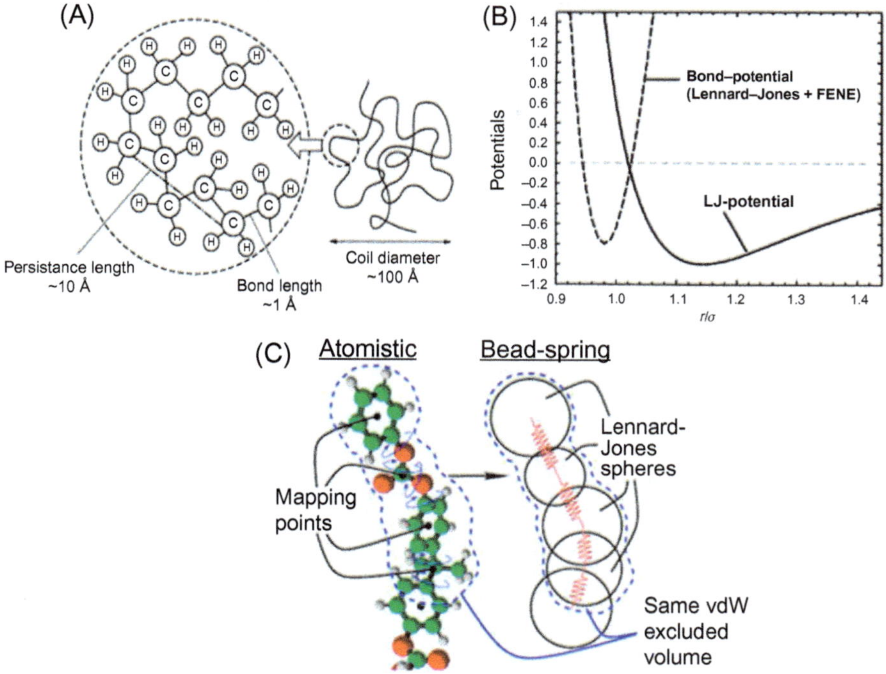

Figure 15.1 (A) Length scales characterizing the structure of a long flexible polymer chain, using polyethylene (PE) as an example. Note that the persistence length characterizes the decay of orientational correlations along the chain backbone. (B) The effective bond potential between connected monomers including LJ and FENE is denoted by the broken curve, which was shifted by −20. LJ potential between monomers in different chains is represented by the full curve. It is noted that the distance corresponding to the minimum of the bond potential (0.96) is smaller than that of the LJ potential (1.13), and this incompatibility maintains an amorphous structure of polymer melts. (C) A schematic representation of the 4:1 mapping scheme for coarse-graining of bisphenol-A-polycarbonate (BPA-PC). A chain ending repeat unit is shown.
Source: For (A) Adapted from Binder K, Baschnagel J, Paul W. Glass transition of polymer melts: test of theoretical concepts by computer simulation. Prog Polym Sci 2003;28:115−72. For (B) Adapted from Binder K, Baschnagel J, Paul W. Glass transition of polymer melts: test of theoretical concepts by computer simulation. Prog Polym Sci 2003;28:115−72. For (C) Adapted from Abrams C, Delle Site L, Kremer K. In: Bridging time scales: molecular simulations for the next decade. Berlin: SPRINGER-VERLAG BERLIN; 2002. pp. 143−64.

is used by offsetting the interaction range by R_{ev} to account for the size effect of NPs:

$$U_{mn} = \begin{cases} 4\varepsilon\left[\left(\frac{\sigma}{r_{ij}-R_{ev}}\right)^{12} - \left(\frac{\sigma}{r_{ij}-R_{ev}}\right)^{6} + C\right], & r_{ij} \le 2^{*}2^{1/6}\sigma + R_{ev}\left(R_{ev} = R_n - \frac{\sigma}{2}\right) \\ 0, & r_{ij} \ge 2^{*}2^{1/6}\sigma + R_{ev}\left(R_{ev} = R_n - \frac{\sigma}{2}\right) \end{cases} \tag{15.3a}$$

$$U_{nn} = \begin{cases} 4\varepsilon\left[\left(\frac{\sigma}{r_{ij}-R_{ev}}\right)^{12} - \left(\frac{\sigma}{r_{ij}-R_{ev}}\right)^{6} + C\right], & r_{ij} \le 2^{*}2^{1/6}\sigma + R_{ev}(R_{ev} = 2R_n - \sigma) \\ 0, & r_{ij} \ge 2^{*}2^{1/6}\sigma + R_{ev}(R_{ev} = 2R_n - \sigma) \end{cases} \tag{15.3b}$$

where R_n represents the radius of the NP, and it is noted that here the cutoff distance r_{cutoff} is set to $2^{*}2^{1/6}\sigma$ for all interactions in order for consistence. In the practical simulation of ENCs, the interaction strength and range can be tuned, and the bond potential energy (Eq. (15.2)) can also be used to model the cross-linking between polymer chains, and the interfacial chemical coupling between polymer chains and NPs. In comparison with this coarse-grained molecular dynamics simulation (CGMDS), the fully atomistic simulation possesses a more accurate force-field, however, the simulated polymer chain length is relatively short, and the system is small. Nevertheless, some advanced simulation techniques are being developed to cover a broad range of length and time scales of specific systems, such as multiscale or hierarchical modeling and simulation by mapping/back-mapping between atomistic and mesoscopic scales, and the adaptive resolution method to allow various length scales within one simulation system [8–10].

15.2 Dispersion of NPs: structure and phase behavior

To successfully prepare ENCs, one key issue is to understand how the NPs are dispersed in polymer matrices, namely, its structure and phase behavior. It is well known that there exist strong attractive van der Waals interaction between NPs, and the affinity between polymer chains and NPs is always bad, if there are is no surface modification of NPs. Moreover, polymer chains proximity to the large surface of NPs always experience the entropy penalty. The first important and basic study comes from Schweizer et al. [11], who used microscopic liquid state theory to calculate the potential of mean force (POMF) between a pair of spherical NPs in the context of homopolymer melt, by varying the diameter ratio between NPs to monomer, chain molecular weight (MW), strength and spatial range between

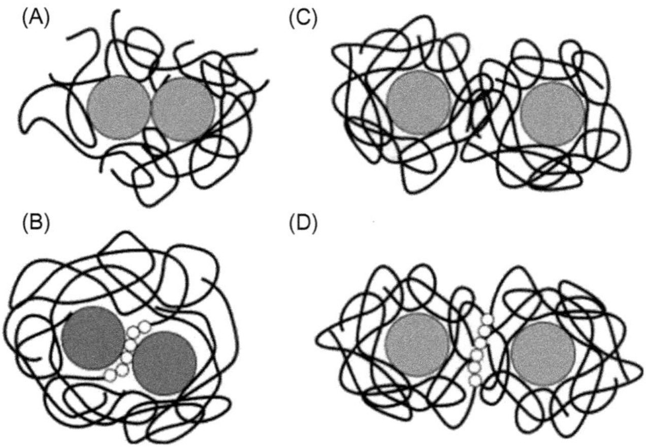

Figure 15.2 Schematic illustration of four states of the spatial organization of NPs in a homopolymer melt: (A) depletion-induced direct contact aggregation; (B) good dispersion due to steric stabilization; (C) short-ranged polymer bridging-induced aggregation; and (D) long-ranged polymer bridging-induced aggregation.
Source: Adapted from Hooper JB, Schweizer KS. Contact aggregation, bridging, and steric stabilization in dense polymer-particle mixtures. Macromolecules 2005;38:8858–69.

NPs–monomer interaction, and interaction between NPs–NPs. In their work, four various spatial organizations of NPs are found: (1) contact aggregation due to depletion attraction; (2) steric stabilization due to thermally stable "bound polymer layers"; (3) segment level tight particle bridging; and (4) "tele-bridging" with longer range bridging contributed by adsorbed polymer layers around neighboring NPs, as schematically displayed in Fig. 15.2.

These four different organization ways can be quantitatively determined by the POMF between NPs, such as Fig. 15.3, by simply varying the strength and spatial range between NPs–monomer interaction [12].

Since a negative value of $W_{cc}(r)$ represents attractive interaction between NPs, and it is easily understood that NPs exhibit direct aggregation for the state of (i). For the state of (ii), the first minimum and negative value occurs at the interparticle distance $r = D + d$, where D and d stand for the diameter of the NP and polymer segment, respectively. This rationalizes the aggregation of NPs bridged via a layer of the polymer segment. For the state of (iv), the first minimum and negative value of $W_{cc}(r)$ appears at much larger distance, such as around $r = D + 3d$, indicating the aggregation of NPs via long-ranged polymer bridging. While for the state of (iii), no negative value of $W_{cc}(r)$ appears as a function of the distance, indicating a good dispersion of NPs without any aggregation. However, the above theoretical studies focus on the ideal situation, namely the infinite dilute case with only two NPs, while in real polymer nanocomposites a finite concentration of NPs exist. It is observed that the multiple NPs case is different from the case of two NPs due to the "many-body effect." Later Schweizer et al. [2] found that the increase of the volume fraction

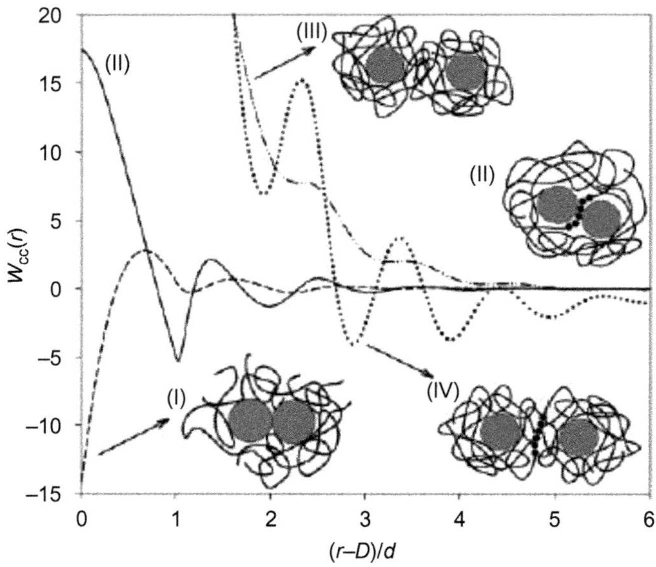

Figure 15.3 Plot of the potential of mean force (POMF), as denoted by $W_{cc}(r)$ as a function of the interparticle distance for these four different states, and it is noted that D and d stand for the diameter of the NP and polymer segment, respectively.
Source: Adapted from Hooper JB, Schweizer KS. Theory of phase separation in polymer nanocomposites. Macromolecules 2006;39:5133−42.

of NPs induces a transition from a highly bridged or sterically stabilized spatial organization to direct clustering of NPs. Schematic illustration of the phase diagram describing the volume fraction of NPs ϕ as a function of the interaction strength ε_{np} between polymer and NPs is displayed in Fig. 15.4 for the multiple NPs.

The miscibility window located in the middle narrows at high volume fraction of NPs, which shares the same effect by increasing the polymer chain length, direct van der Waals interaction between NPs and NPs−monomer size asymmetry ratio in the two NPs case [12]. To validate these observations, Liu et al. [13] have adopted CGMDS to study the structure and phase behavior of NPs in polymer chains. Note that the volume fraction of NPs is $\phi = 7.28\%$, and the diameter of the NPs and bead of polymer chains is 4σ and σ, respectively. The radial distribution function of NPs for various polymer−NPs interaction strength ε_{np} is shown in Fig. 15.5A. At low $\varepsilon_{np} = 0.1$ and 1.0, direct contact aggregation of NPs appears, indicated by the peak located at around $r = 4\sigma$. And at high $\varepsilon_{np} = 5.0$ and 12.0, the values of the peaks located at around $r = 5\sigma$ and 6σ begin to increase, indicating the effect of the chain bridging-induced aggregation of NPs. While at moderate $\varepsilon_{np} = 2.0$, no obvious peaks occur, indicating a uniform dispersion of NPs. And these observations are further validated by the snapshots in Fig. 15.5B. For quantitative analysis, the number of neighbor fillers and the second virial coefficient B_2 as a function of ε_{np} is shown in Fig. 15.5C. The smallest value of the number of neighbor fillers and the positive value of B_2 with $\varepsilon_{np} = 2.0$ just verifies the appearance of a good

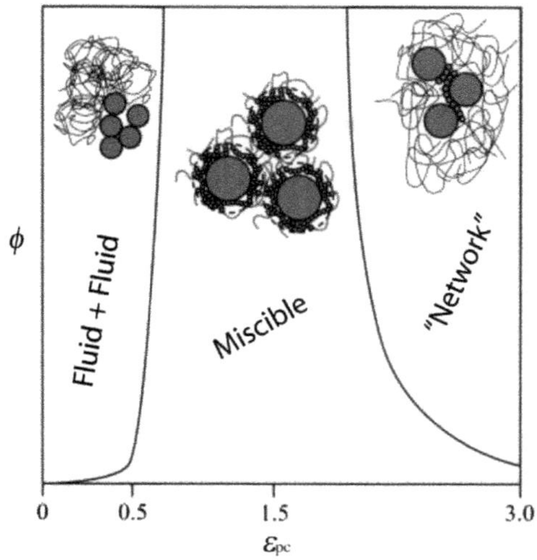

Figure 15.4 Schematics of the phase diagram by plotting the volume fraction of the NPs ϕ as a function of the interaction strength between polymer chains and NPs ε_{pc}. *Source*: Adapted from Hooper JB, Schweizer KS. Theory of phase separation in polymer nanocomposites. Macromolecules 2006;39:5133–42.

dispersion. The POMF, calculated from the radial distribution function, is also shown in Fig. 15.5D.

This trend is also valid for nonspherical NPs, i.e., the details of depletion, stabilization, and bridging behavior are only qualitatively shape-dependent. For instance, Patra et al. [14] adopted CGMDS to study the aggregation and dispersion of the anisotropic NPs, such as cube and tetrahedron, in polymer matrix. The phase diagrams in the $\varepsilon_{np} - \phi$ plane are shown in Fig. 15.6A and B. For the tetrahedrons, at low ε_{np}, the NPs form clusters, whose shape is ordered sphere-like with ϕ smaller than 2.3%, beyond which disordered sheet-like structures form. While with the increase of ε_{np}, the cubic NPs become dispersed, and when ε_{np} becomes larger than 2.0, the NPs exhibit bridging-induced aggregation. In the case of cubic NPs, at low ε_{np} three different ordered structures are formed, such as 3D cubic lattice, 1D square column, and 2D sheet structure. Similar to the case of tetrahedrons, dispersed state and aggregated state of NPs induced by chain bridging occurs at moderate and high ε_{np}, respectively.

In particular, through integration equation theory, Tripathy et al. [15] also found that for nanorod filled polymer nanocomposites, the same kinds of depletion, dispersion, bridging, and telebridging occur, which critically depend on the aspect ratio, surface roughness and nanorod thickness, as schematically shown in Fig. 15.7. Furthermore, the miscible region located between depletion-driven and bridging-driven phase separations quickly narrows with the increase of the aspect ratio of nanorods and attractive interaction between nanorods, which becomes unchanged when the radius of gyration of nanorods exceeds that of the polymer chains.

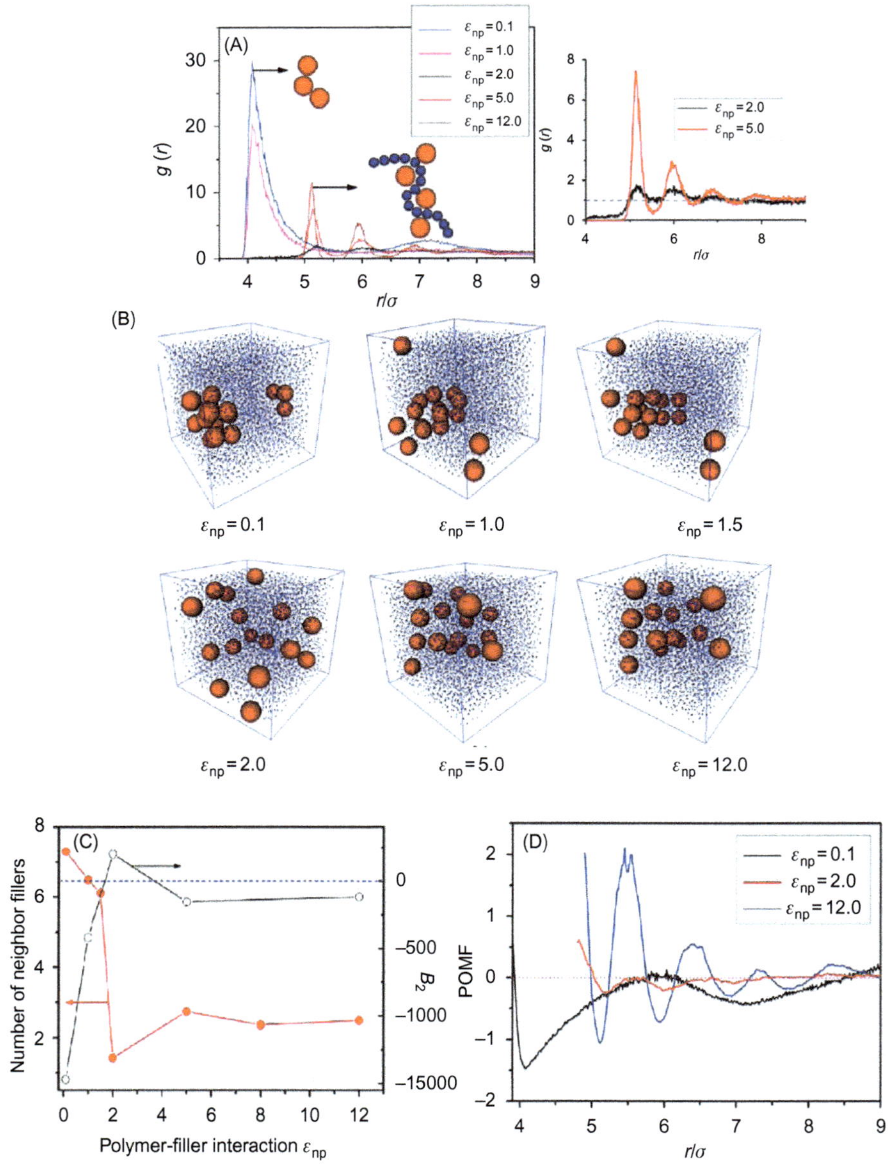

Figure 15.5 (A) The radial distribution function of NPs for various polymer−NPs interaction strength, and inset shows the cases with $\varepsilon_{np} = 2.0$ and 5.0. (B) Snapshots of the dispersion of NPs for different ε_{np}. Note that the red spheres represent NPs, and the blue points stand for polymer chains. (C) The number of neighbor fillers and the second virial coefficient B_2 as a function of ε_{np}. (D) Potential of mean force (POMF) for different ε_{np}. *Source*: Adapted from Liu J, Gao YG, Cao DP, Zhang LQ, Guo ZH. Nanoparticle dispersion and aggregation in polymer nanocomposites: insights from molecular dynamics simulation. Langmuir 2011;27:7926–33.

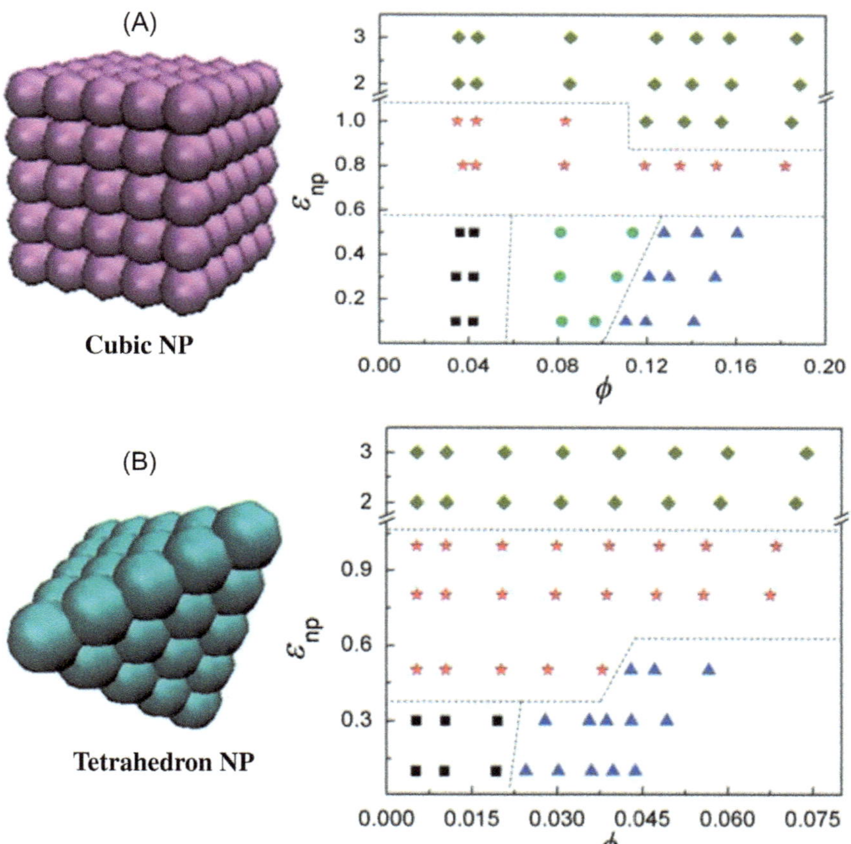

Figure 15.6 (A) Model of cubic NP composed of 98 spherical beads, and its phase diagram standing for all possible spatial organization is shown on the right. ■: cubic array assembly; ●: square column assembly; ▲: sheet-like square array assembly; ★: dispersion; ◆: bridging of NPs via polymer layer. (B) Model of tetrahedrons consisting of 34 spherical beads with its phase diagram representing all possible spatial organization is displayed in the right. ■: Sphere-like ordered structure; ▲: sheet-like disordered structure; ★: dispersion; ◆: bridging of NPs via polymer layer. Note that the dashed lines are used only to guide the eye, and are not thermodynamics boundaries.
Source: Adapted from Patra TK, Singh JK. Polymer directed aggregation and dispersion of anisotropic nanoparticles. Soft Matter 2014;10:1823–30.

In addition, Schweizer et al. [16] adopted polymer reference interaction site model (PRISM) liquid state theory to study the structure and phase behavior of pseudo one-, two-, and three-dimensional particles, such as rod, disk, and cube, as shown in Fig. 15.8. Similarly, the entropic depletion attraction between NPs is dominant for weakly adsorbing polymer chains, and strongly adsorbing chains induce a

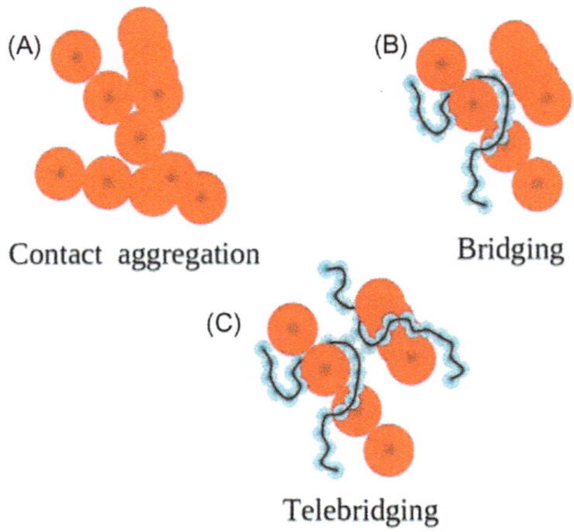

Figure 15.7 Schematics of the spatial organization of nanorods: (A) Contact aggregation; (B) bridging; and (C) telebridging.
Source: Adapted from Sankar UK, Tripathy M. Dispersion, depletion, and bridging of athermal and attractive nanorods in polymer melt. Macromolecules 2015;48:432–42.

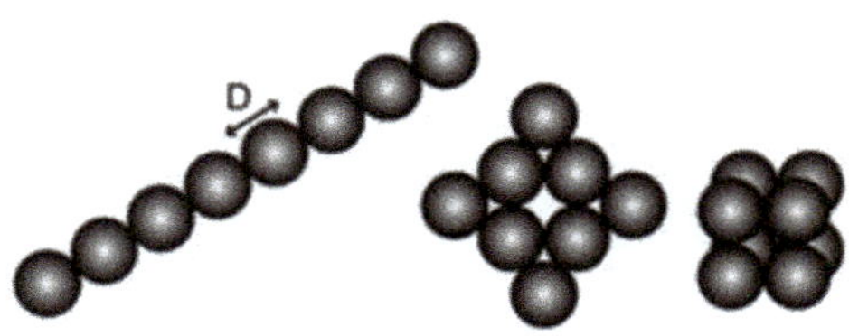

Figure 15.8 Schematics of rod, disk, and cube.
Source: Adapted from Hall LM, Schweizer KS. Structure, scattering patterns and phase behavior of polymer nanocomposites with nonspherical fillers. Soft Matter 2010;6:1015–25.

bridging attraction, while intermediate interfacial cohesion results in the formation of a steric stabilizing adsorbed polymer layer around each nanoparticle.

15.3 Interfacial chain structure and dynamics between elastomer and NPs

Besides the spatial distribution or organization of NPs in polymer matrices, the interfacial interaction between polymer chains and NPs is another key issue, and it has been established that the structure and dynamics of the polymer chains in the vicinity of NPs is very different from those of the bulk polymer chains, such as the thickness and the mobility of the interphase, critically influencing the overall mechanical performance of ENCs. For instance, Padmanabhan [17] found that the mechanical reinforcement could result from the percolation of the high-density regions of polymer chains surrounding the NPs.

As for the chain structure and dynamics induced by the incorporation of NPs, the first interesting issue is that how the conformation of polymer chains such as its radius of gyration R_g and end-to-end distance R_{eed} will be perturbed away from the Gaussian coil in the presence of NPs, which is a complicated function of the size and concentration of fillers, chain length, polymer–filler interaction, the spatial organization of fillers, and so on. But up to now no consistent results have been reached. For example, by employing MC simulation Vacatello et al. [18] observed the reduced size of polymer chains compared with the unfilled polymer melt, attributed to the sequences of surface segments, totally located in the polymer–filler interfacial regions, and of bridge segments connecting different fillers. This, however, is contrary to the experimental results obtained through neutron scattering results, finding that the occurrence of the chain expansion with $R_g > R_n$, where R_g and R_n are the root-mean squared radius of gyration and the nanoparticle radius, respectively, and a decrease of the chain size for $R_g \leq R_n$. This was further validated by Frischknecht et al. [19] through the self-consistent (SC/PRISM) for small NPs compared with R_g and attractive polymer–filler interaction, as shown in Fig. 15.9A, indicating that for all these six systems, the expansion of the chain size occurs as a function of the volume fraction of the fillers with $R_g > R_n$. In addition, Karatrantos et al. [20] also found that polymer chains are unperturbed in the presence of repulsive NPs, while polymer chains can be perturbed by the presence of attractive NPs when $R_g > R_n$, as displayed in Fig. 15.9B. However, Ozmusul et al. [3] have focused on the statistics of bridges, dangling ends, loops, and trains through lattice MC simulation, which indicates that the overall chain statistics remain Gaussian regardless of the filler loading (up to 27 vol %).

Meanwhile, the effect of the incorporated NPs on the segmental dynamics, such as the glass transition temperature T_g, also receives considerable scientific interest. Since for ENCs always with a high loading of NPs, polymer chains are under strong nanoscale confinement exerted by the presence of the dispersed nano-sized fillers, leading to the interfacial chain behavior becoming significantly different from that of the corresponding bulk state. Besides the nanoconfinement effect, the interfacial interaction on the nanoscale is also a key factor to influence T_g, and is also of great importance for static and dynamic mechanical properties, which motivates the investigation of the polymer–filler interface and interphase. Moreover, the understanding of T_g of ENCs becomes more complicated based on the fact that the unified theory or model accounting for the glass transition of pure polymer melts is still lacking, but some efforts are still devoted. For instance, by designing polymer films confined between silica slides, Rittigstein et al. [21] examined the T_g and physical aging by tuning the interlayer distance to mimic the effect of the content of silica NPs in real PNCs. Berriot et al. [22] pointed out that there exists a gradient of T_g in the vicinity of particles, which is used to elucidate the nonlinear mechanical behavior of filled elastomers because of the strain-softening of the glassy polymer shell surrounding the NP surfaces [23]. Moreover, a model based on the presence of the glassy layers around NPs is established, inferring that the strongest reinforcement could be

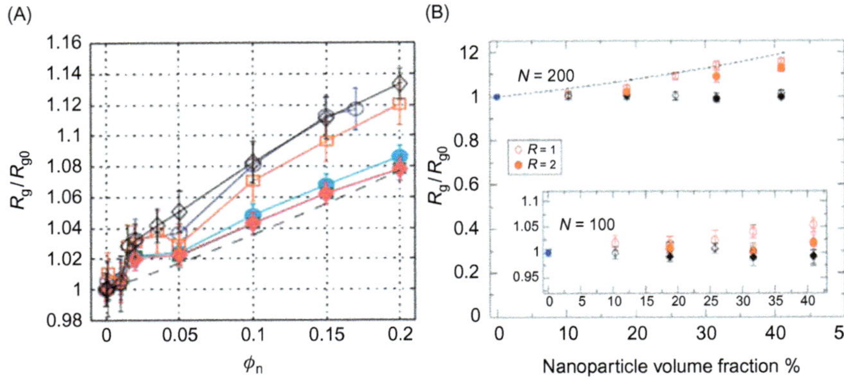

Figure 15.9 (A) Change of the chain dimension normalized by that of the pure system as a function of the volume fraction of the fillers. The chain length $N = 40$, $\sigma_n = 3\sigma$ (filled cyan circles); $N = 40$, $\sigma_n = 5\sigma$ (filled green squares, under the pink diamonds); $N = 40$, $\sigma_n = 6.2\sigma$ (filled pink diamonds); $N = 80$, $\sigma_n = 3\sigma$ (open blue circles); $N = 80$, $\sigma_n = 5\sigma$ (open red squares); $N = 80$, $\sigma_n = 6.2\sigma$ (open black diamonds). The black dashed line shows $R_g/R_{g0} = (1-\phi_n)^{-1/3}$. (B) Radius of gyration of polymers in melt with NPs of radius $R = 1, 2$ normalized with its value in the bulk for $N = 200$ and $N = 100$ (inset): (i) polymer melt (blue filled circles); (ii) nanocomposite: attractive monomer–nanoparticle ($R = 2$) interactions (red filled circles); (iii) nanocomposite: repulsive monomer–nanoparticle ($R = 2$) interactions (black filled diamonds); (iv) nanocomposite: attractive monomer–nanoparticle ($R = 1$) interactions (red open circles); (v) nanocomposite: repulsive monomer–nanoparticle ($R = 1$) interactions (black open diamonds). The black dashed line shows $R_g/R_{g0} = (1-\phi)^{-1/3}$.
Source: For (A) Adapted from Frischknecht AL, McGarrity ES, Mackay, ME. Expanded chain dimensions in polymer melts with nanoparticle fillers. J Chem Phys 2010;132:204901. For (B) Adapted from Karatrantos A, Clarke N, Composto RJ, Winey KI. Polymer conformations in polymer nanocomposites containing spherical nanoparticles. Soft Matter 2015;11:382–8.

achieved when glassy layers around every filler become overlapped [24]. However, contrary to these viewpoints, Ronald et al. [25] have studied the nano-sized silica filled poly(vinyl acetate), which indicates that there exists no difference between the local segmental dynamics of polymer chains in the vicinity of silica particles and that of the bulk chains. And in a later study it is further pointed out that there may exist some stiffening of polymer chains adjacent to the filler particles, which, however, does not translate into a prominent effect on the chain segmental dynamics [26]. By introducing the coupling agents, it is found that the polymer–filler interfacial interaction had a strong effect on the development of the filler network, but the segmental dynamics of polymer chains proximate to filler particles remain unchanged [27]. To gain a better understanding about this confusing issue, Robertson et al. [28] have recently reviewed the effects of fillers on the nearby local segmental dynamics of polymer chains investigated through different experimental approaches such as dynamic mechanical spectroscopy, nuclear magnetic resonance, dielectric spectroscopy, calorimetry, dilatometry, and neutron scattering, and they put forward that the experimental discrepancy stems from the ambiguous data analysis and various polymer–filler interactions for different systems.

Up to now, a large number of computational simulation works have been carried out to elucidate the chain structure and dynamics in the polymer–filler interfacial region, and also the change of T_g triggered by NPs. For instance, for the three-dimensional spherical NPs, Starr et al. [29] firstly used CGMDS simulation to address the structure and dynamics of polymer chains adjacent to the NPs, as shown in Fig. 15.10A and B, respectively presenting the polymer melts surrounding the filler particle and some representative polymer chains nearby the filler. From the figures, it can be seen that the chains become increasingly elongated and flattened when approaching the filler surface, as indicated by the changes of R_g^2 and $R_g^{\perp 2}$ (component perpendicular to the filler surface) in Fig. 15.10C. In order to examine the shift of T_g, the intermediate scattering function $F(t)$ is used to characterize the segmental relaxation of polymer chains, the characteristic relaxation time τ when the value of $F(t)$ decays to 0.2. The temperature dependence of τ for pure polymer melts, filled with attractive and nonattractive systems is plotted in Fig. 15.10D. To reflect the change of T_g, they used the Vogel-Fulcher-Tammann (VFT) equation ($\tau \approx Ae^{B/(T-T_0)}$) to fit the data, since T_o is typically quite close to the experimental T_g values. And it is observed that the T_g of PNCs can be shifted to be higher or lower by changing the interfacial interaction to become attractive or repulsive, compared to that of the unfilled system. At the same time, the interfacial polymer chains are found to orientate parallel to the NP surface through MD simulation of Brown et al. [30]. And in a later study by constructing an atomistic model of silica NP, which are further immersed in a polymer melt, as shown in Fig. 15.11A, the effect of the silica size on the surrounded polymer chains, including its molecular orientation, fraction of trans conformers, and dynamics have been explored [31]. Recently through CGMDS, we have also characterized the interfacial chain structure and dynamics [32]. For different polymer–filler interactions $\varepsilon_{np} = 2.0$ and 10.0, the adsorption state of a single flexible polymer chain is shown in Fig. 15.11B, indicating that great ε_{np} induces the wrapping of the polymer chain around the filler, while the chain is perpendicularly adsorbed onto the filler surface in the case of small ε_{np}. We use the second-order Legendre polynomial $<P_2(\cos\theta)>$ to characterize the bond and segmental orientations as a function of the distance away from the filler center, as given by

$$<P_2(\cos\theta)> = \frac{1}{2}(3\cos^2\theta - 1) \tag{15.4}$$

where θ denotes the angle between the bond or segmental vector and the radial direction connecting the center of the bond or segmental and the filler center. Obviously, $<P_2(\cos\theta)>$ is zero for random orientation, and the orientation is tangential to the filler surface for $<P_2(\cos\theta)> = -0.5$. As shown in Fig. 15.11C, it can be seen that on both bond and segmental length scale, the alignment prefers tangent to the filler surface with stronger polymer–filler interaction ε_{np}, and for

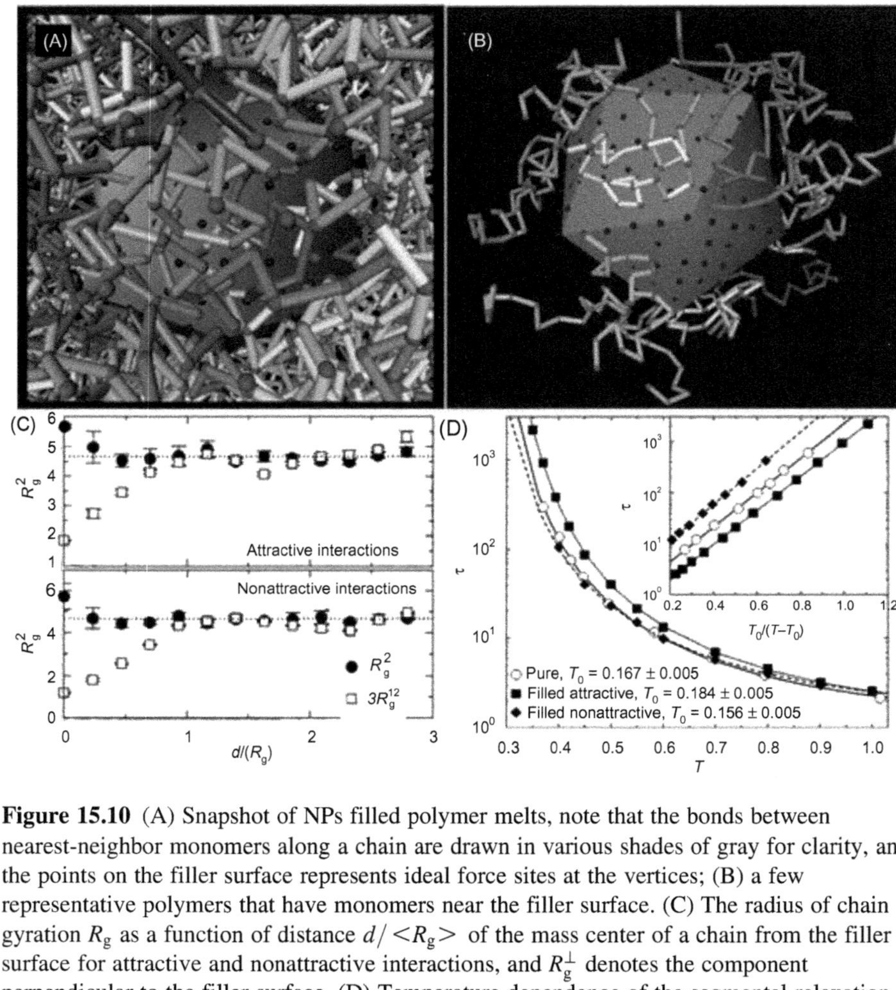

Figure 15.10 (A) Snapshot of NPs filled polymer melts, note that the bonds between nearest-neighbor monomers along a chain are drawn in various shades of gray for clarity, and the points on the filler surface represents ideal force sites at the vertices; (B) a few representative polymers that have monomers near the filler surface. (C) The radius of chain gyration R_g as a function of distance $d/<R_g>$ of the mass center of a chain from the filler surface for attractive and nonattractive interactions, and $R_g^{\perp}$ denotes the component perpendicular to the filler surface. (D) Temperature dependence of the segmental relaxation time τ of the intermediate scattering function of polymer chains for pure polymer melts, filled with attractive and nonattractive cases. The lines are a fit to the VFT from $\tau \approx Ae^{B/(T-T_0)}$, and the inset shows the same data plotted against reduced temperature $T_0/(T-T_0)$ to show the quality of the VFT fit.
Source: Adapted from Starr FW, Schroder TB, Glotzer SC. Molecular dynamics simulation of a polymer melt with a nanoscopic particle. Macromolecules 2002;35:4481–92.

purely repulsive ε_{np} the bond orientates perpendicular to the filler surface. And beyond the distance equal to nearly $R_g = 2.75\sigma$ the bulk behavior is recovered. A similar chain structure near the NPs is also observed in the simulation of gold NPs immersed in polystyrene melts [33]. In response to the intriguing interfacial chain dynamics, our simulated results indicate that it is impossible to form a polymer glassy layer around the spherical filler particles, and in the case of the strongly attractive polymer–filler interaction, polymer chains adjacent to the filler particles

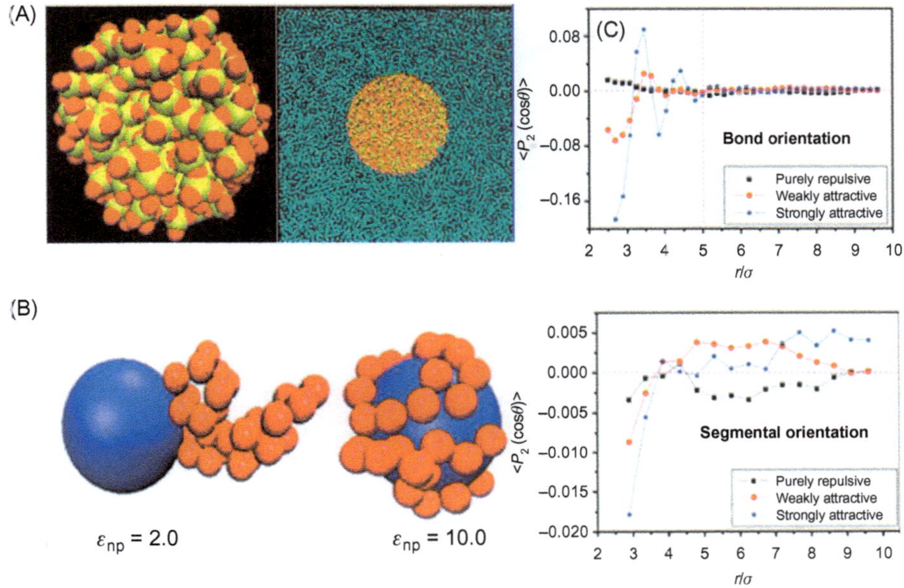

Figure 15.11 (A) A close-up 3D view of silica NP with its radius equal to 1.5 nm (left) and snapshot of the model polymer chains surrounding the silica NP (right), note that oxygen atoms are shown in red and silicon atoms in yellow. (B) The adsorption state of a single polymer chain on the filler surface with two typical polymer–filler interaction $\varepsilon_{np} = 2.0$ and 10.0. (C) The bond and segmental orientation as a function of its distance from the filler center.
Source: For (A) Adapted from Brown D, Marcadon V, Mele P, Alberola, ND. Effect of filler particle size on the properties of model nanocomposites. Macromolecules 2008;41:1499–511, with permission. For (C) Adapted from Liu J, et al. Polymer-nanoparticle interfacial behavior revisited: a molecular dynamics study. Phys Chem Chem Phys 2011;13:13058–69.

still exhibit significantly reduced mobility compared to the bulk case, rather than the glassy behavior.

Very recently through atomistic MC simulation, Pandey et al. [4] observed that the polymer–filler interaction is strongly dependent on the curvature of the NP, such as the thickness of the polymer bound layer. To understand the interaction between elastomer and carbon black (CB), the knowledge of the surface characteristics of CB is also essential, and recently a molecular model of primary particles of porous CB was established using the hexagonal graphite sheets as building units [34], which are shown in Fig. 15.12A, and nonequilibrium MD simulation was performed to generate the CB with different size of building units, as shown in Fig. 15.12B and C for the slice views, and the surface morphology in Fig. 15.12D and E. It is indicated that the CB particle contains amorphous cores and crystalline shells. The continuous grain pattern, as denoted by the dark lines in Fig. 15.12D and E, is formed by the joint edges of graphite layers, which

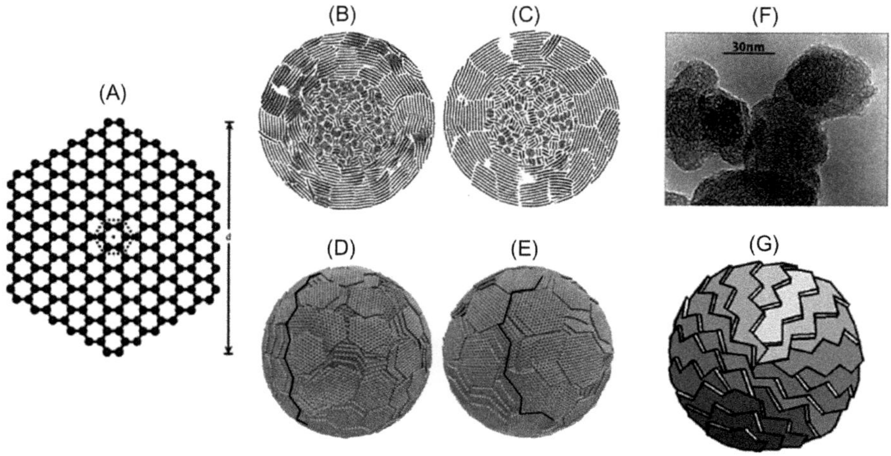

Figure 15.12 (A) Model of the hexagonal graphite unit with $d = 2.7$ nm as building units of CB. The dark dot represents the center of the graphite unit. The slice views are (B) for the carbon black particle with building units $d = 2.7$ nm and (C) $d = 3.7$ nm. The snapshots of the surface morphology are (D) for the carbon black particle with building units $d = 2.7$ nm and (E) $d = 3.7$ nm. The dark lines are drawn to highlight the grain pattern on the CB surface. The radii of the CB particle are 20 nm with its densities 2 g/cm^3. (F) ATEM observation of the turbostratic of a CB particles. (G) A model of the structure of CB primary particle.
Source: For (E) Adapted from Ban SA, Malek K, Huang C, Liu ZS. A molecular model for carbon black primary particles with internal nanoporosity. Carbon NY 2011;49:3362–70. For (F) Adapted from Donnet JB. Nano and microcomposites of polymers elastomers and their reinforcement. Compos Sci Technol 2003;63:1085–8.

possibly exhibits adsorption sites for rubbery macromolecular chains. The simulated structure in fact is perfectly consistent with the structure of CB particles observed through transmission electron microscopy (TEM) and a theoretically proposed model of CB, as shown in (F) and (G) [35]. Hence, it is proposed that in the future by employing this well constructed model of CB, it will be possible to accurately model the interaction between CB and various kinds of rubbery polymer chains.

For the two-dimensional silicate–polymer interfacial interaction, in practical situations organophilic cations such as alkylammoniums are always used to replace the natural metal cations to increase the miscibility between the polymer matrix (always hydrophobic) and the pristine silicate (highly hydrophilic). To understand the interfacial interaction between the polymer chains and the silicate sheet, some MD researches have also been done to study the chain structure, conformation, and dynamics in the vicinity of a solid surface. For example, in the case of poly(ethylene oxide) (PEO) and TiO_2, highly dense polymer layers persist up to 1.5 nm from the solid surface, and the electrostatic interaction between PEO and TiO_2

dramatically slows down the interfacial chain structural relaxation [36]. By using a coarse-grained bead-spring model, the chain conformation near the solid surface is composed of trains, loops, and tails, and a kinetic model is also established to describe the interfacial chain desorption kinetics, which could provide some guidance to the response of the interface to high-frequency oscillations [37]. Moreover, the effects of the flat and structured surfaces on the interfacial chain dynamics and structural heterogeneity are also compared [38].

Considering the fact that the interfacial binding strength critically determines the effectiveness of the interfacial load transfer and therefore the mechanical property of polymer–clay nanocomposites, Sikdar et al. [39] firstly used MD simulation to quantify the molecular interactions between constituents in a polymer–organically modified montmorillonite system, providing important insight into the mechanism of the intercalation process.

For the one-dimensional carbon nanotubes (CNTs), the study of the interfacial chain structure, conformation, and dynamics between polymer and CNTs also receives great attention. For example, the chain structure and conformation in the polymer/CNT interfacial region have been simulated, indicating that the local structure of polymer chains around and nearby single walled carbon nanotubes (SWCNT) becomes more ordered. However, the overall chain configuration characterized by the radius of chain gyration R_g is not altered [40]. In addition, Wei et al. [41] also found that the chain molecules within the adsorption layers nearby CNT prefer to align parallel to the tube axis, contributing to the mechanical modulus of the composite, later the conformation of linear polymer chains influenced by the chirality and radius of CNTs in the interfacial region is also studied through MD simulation [42]. Two different types of polymer chains such as heterocyclic conjugated polymer chains [43] and flexible polymer chains [44] with SWCNTs have also been simulated. For instance, the change of the single poly(caprolactone) chain around the SWCNT as a function of the simulation time is shown in Fig. 15.13, finding that flexible polymers prefer to wrap around the SWCNT.

Attributed to their excellent mechanical properties, CNTs are good candidates to reinforce polymer matrices. However, in order to achieve this goal, one key issue is to guarantee a strong interfacial interaction between polymers and CNTs. For example, Frankland [45] firstly adopted a MD simulation to study the effect of the chemical cross-links between SWCNT and polymer chains on the interfacial shear strength. The results concluded that the interfacial shear strength is enhanced by over an order of magnitude by the formation of cross-links involving less than 1% of the CNT atoms in the interfacial region, and the tensile modulus of the CNT is negligibly modified by the surface chemical modification. Furthermore, the polymer chain wrapping and filling SWCNTs have also been simulated, and in order to characterize the interfacial binding characteristics between SWCNTs chemically modified with $-COOH$, $-CONH_2$, $-C_6H_{11}$, or $-C_6H_5$ groups as shown in Fig. 15.14A and PE chains, the pullout test has been carried out to examine the interfacial shear stress for different chemical surface modifications, as displayed in

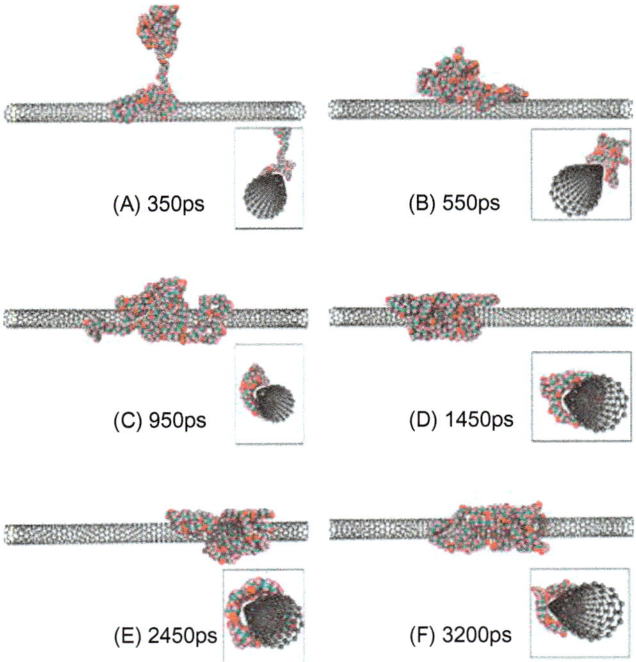

Figure 15.13 MD snapshots of PCL (poly(caprolactone)) interacting with SWCNTs at different simulation times: (A) 350; (B) 550; (C) 950; (D) 1450; (E) 2450; and (F) 3200 ps. The colors used for the polymer represent the following atoms: carbon is aqua; hydrogen is pink; and oxygen is red.
Source: Adapted from Tallury SS, Pasquinelli MA. Molecular dynamics simulations of flexible polymer chains wrapping single-walled carbon nanotubes. J Phys Chem B 2010;114:4122–9.

Fig. 15.14B [46]. The interfacial shear stress, τ_i can be derived from the pullout energy, as shown below:

$$E_{\text{pullout}} = \int_{x=0}^{x=l} 2\pi r(l-x)\tau_i dx = \pi r \tau_i l^2 \Rightarrow \tau_i = \frac{E_{\text{pullout}}}{\pi r l^2} \tag{15.5}$$

where r and l are the outer radius and length of the SWCNT, respectively, and x denotes the coordinate along the longitudinal tube axis. The interfacial bonding energy is equal to the ratio of the interfacial energy to two times of the interfacial contact area. The interfacial interaction energy, interfacial bonding energy, and the shear stress are compared for different chemical functionalizations and the unmodified system, as presented in Fig. 15.14C. The results indicate that the functionalization of the aromatic ring ($-C_6H_5$) has the greatest interaction energy, interfacial bonding energy, and shear stress, such as the increase of the shear stress by about 1700%, which is possibly attributed to more contact area with the polymer matrix for its special ring structure.

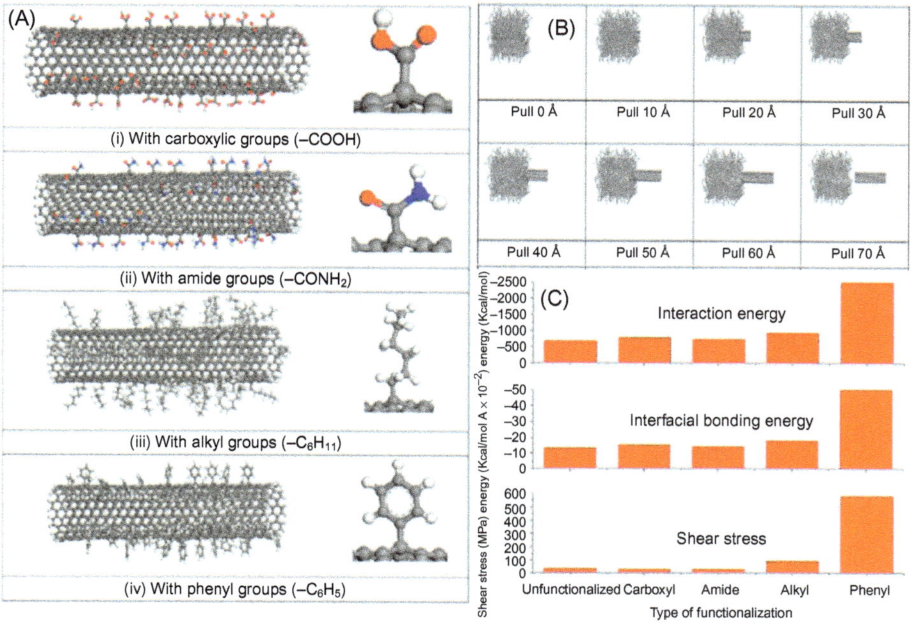

Figure 15.14 (A) The 5% surface chemical modification of SWCNT with (i) carboxylic groups (–COOH), (ii) amide groups (–$CONH_2$), (iii) alkyl groups (–C_6H_{11}), and (iv) phenyl groups (–C_6H_5). (B) Snapshots of the pullout process of the SWCNT from the polyethylene as a function of the displacement. (C) Comparisons of the interaction energy, interfacial bonding energy, and shear stress for the four different surface chemical modifications of SWCNT.
Source: Adapted from Zheng QB, et al. Computational analysis of effect of modification on the interfacial characteristics of a carbon nanotube-polyethylene composite system. Appl Surf Sci 2009;255:3534–43.

In most practical situations, a high loading of fillers are always introduced to the elastomer matrices, which leads to the formation of a three-dimensional filler network via direct contact aggregation or bridged by polymer chains in ENCs. And the breakdown of this filler network under the dynamic strain leads to the famous "Payne effect," namely the nonlinear decrease of the dynamic storage modulus as a function of the shear amplitude. In the case of spherical NPs filled unentangled polymer chains, Gersappe et al. [47] observed the formation of the network structure between polymers and NPs at a critical concentration of fillers through MD simulation in Fig. 15.15. At low filler volume fraction $\phi = 2\%$, no evidence of a percolating network is observed. While for $\phi = 5\%$ and 10%, the presence of major cluster is found, spanning the entire simulation system. In the future more simulation work should be performed, by focusing on the effects of the filler size, shape, polymer–filler interaction, and chain length on the formation of the filler network. Obviously, fillers with different shapes, such as spherical CB, clay sheets, and CNTs, will exhibit totally different filler networks and structures.

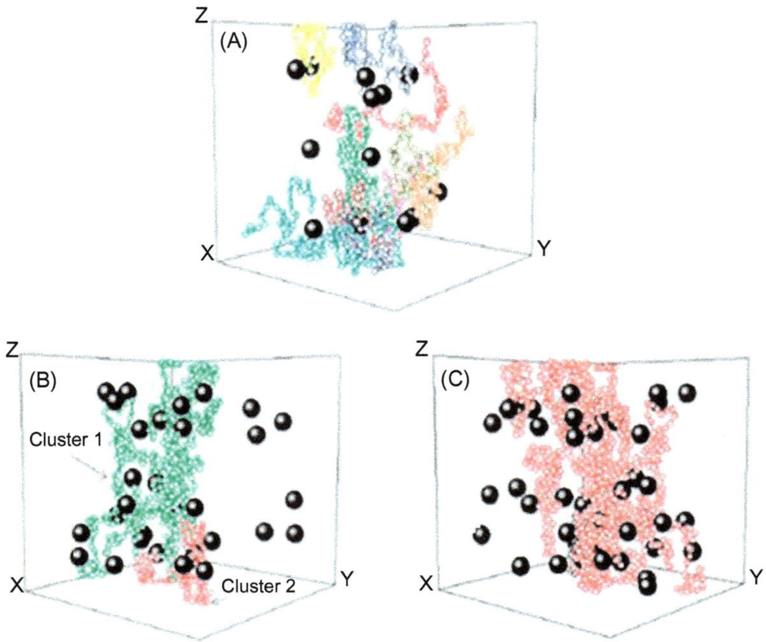

Figure 15.15 (A) Snapshot of a fully equilibrated state in the polymer melt with the volume fraction of the nanofillers equal to $\phi = 2\%$. Note that connecting different color hollow spheres represent different polymer chains belonging to different clusters. (B) $\phi = 5\%$, two clusters are formed and cluster 1 is clearly dominating the system, and (C) $\phi = 10\%$, only one cluster spanning the entire simulation system is formed.
Source: Adapted from Jaber E, Luo HB, Li WT, Gersappe D. Network formation in polymer nanocomposites under shear. Soft Matter 2011;7:3852–60.

Besides focusing on the possible change of the chain size induced by the presence of the incorporated nanofillers, the altered chain structure, and segmental dynamics and relaxation adjacent to the NPs because of nanoscale confinement combined with polymer–filler physical and chemical interactions, which will lead to the ambiguous change of T_g as discussed above, the interfacial binding strength determining the stress transfer from the polymer matrix to the NPs, and the formation of the filler network and its structure, some work has also been carried out to simulate the chain cross-linking process, since the cross-linking process is indispensable to ENCs. For instance, Li et al. [48] recently have adopted CGMDS to study the polymer chain packing and primitive path (PP) network of cross-linked cis-polyisoprene (PI) polymer, indicating that the intrachain packing of PI polymers is enhanced while the interchain packing is weakened by the cross-linking effect. Moreover, the effects of the MW and the cross-linking density on the PP are also examined. To model the practical vulcanization process, such as NPs with different shapes filled elastomers, this approach can be extended to study the cross-linking process of various ENCs, focusing on the influence of NPs on the cross-linking density and distribution. However, up to date no relevant simulation work have been performed.

15.4 Static and dynamic mechanics of ENCs

In the last part we focus on discussing the simulation of the static and dynamic mechanics of ENCs, such as its static (mechanical reinforcement) and dynamic (viscoelasticity) properties. Recently we have investigated the effects of the filler volume fraction, rubber–filler physical and chemical interaction strength and range, and filler size on the stress–strain behavior through CGMDS [49]. The uniaxial tension of spherical NPs filled polymer networks is shown in Fig. 15.16A. For clarity only several polymer chains are shown, and the filler volume fraction is $\phi = 18.81\%$. In the case of strong polymer–filler interaction $\varepsilon_{np} = 5.0$, the stress–strain behaviors for different filler volume fractions $\phi = 3.77\%$, 7.28%, 13.43%, 18.81%, 23.45%, 27.42%, 31.49%, and 34.86% are shown in Fig. 15.16B and C. It is observed that above nearly $\phi = 27.42\%$, the stress–strain behavior begins to decrease. In order to elucidate this observation, the segmental orientation is characterized through $<P_2>$ (see eq 15.4), as shown in (D) and (E). Evidently, the segmental orientation increases monotonically, ranging from $\phi = 0$ to 27.42% at the same strain, while at much higher filler volume fraction it begins to decrease, as shown in (E). In this case, the change of the segmental orientation just reflects the variation trend of the stress–strain behaviors. Generally in this work it is firstly demonstrated via simulation that the mechanical reinforcement mechanism originates from the filler-induced chain orientation and alignment, together with the limited extensibility of chain bridges formed between neighboring fillers at large deformation. The former is controlled by the filler volume fraction and size, and filler–rubber interaction, while the latter will be strengthened by enhancing the physical and chemical interactions between rubber and fillers. Meanwhile, the enhanced orientation of polymer chains induced by the spherical NPs during uniaxial tension is also observed compared to that of the pure polymer network.

Meanwhile, through MC simulation it is also found that attractive (repulsive) polymer–filler interaction will increase (decrease) the average shear and Young's modulus of the nanocomposites, resulting from a glassy polymer layer formed in the vicinity of the attractive fillers [50]. Through MD simulation Sun et al. [51] also observed that the modulus enhancement is attributed to the stiffer layer of polymer of higher densities around the fillers. Meanwhile, in a separate MD simulation study it is revealed that the mechanical reinforcement of spherical NPs filled polymer melts stems from the formation of a long-lived transient polymer–filler network, which can only exist when the polymer–filler interaction strength is strong enough, but with its range in the short distance [52]. Recently CGMDS has also been adopted to study the mechanical reinforcement of nanofilled cross-linked rubber, by focusing on the effects of the cross-linking and the polymer–filler interaction [53].

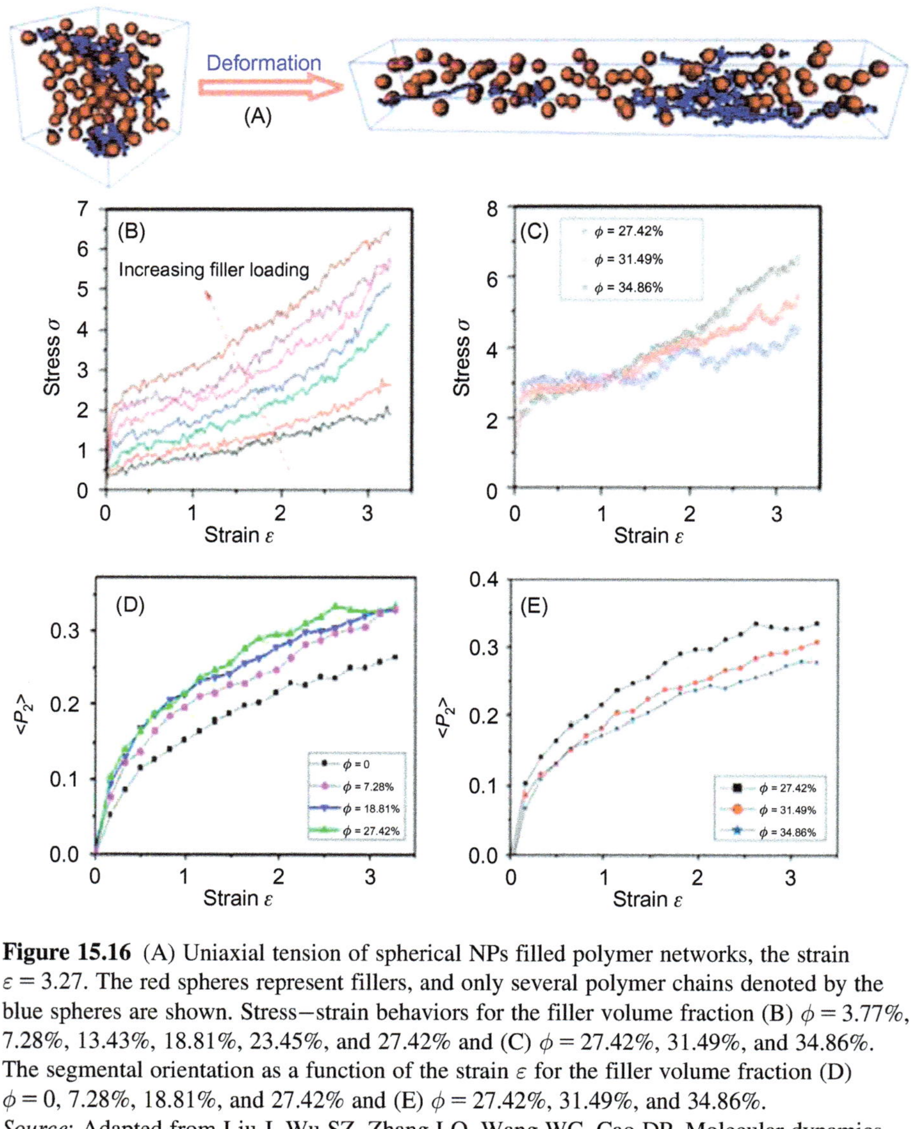

Figure 15.16 (A) Uniaxial tension of spherical NPs filled polymer networks, the strain $\varepsilon = 3.27$. The red spheres represent fillers, and only several polymer chains denoted by the blue spheres are shown. Stress–strain behaviors for the filler volume fraction (B) $\phi = 3.77\%$, 7.28%, 13.43%, 18.81%, 23.45%, and 27.42% and (C) $\phi = 27.42\%$, 31.49%, and 34.86%. The segmental orientation as a function of the strain ε for the filler volume fraction (D) $\phi = 0$, 7.28%, 18.81%, and 27.42% and (E) $\phi = 27.42\%$, 31.49%, and 34.86%.
Source: Adapted from Liu J, Wu SZ, Zhang LQ, Wang WC, Cao DP. Molecular dynamics simulation for insight into microscopic mechanism of polymer reinforcement. Phys Chem Chem Phys 2011;13:518–29.

For the two-dimensional clay sheets, Song et al. [54] have used MD simulation to study the fracture of polymer-exfoliated layered silicate nanocomposites, and the interaction between the surface of the nanoplatelets and the polymer segments has significant influence on the fracture strength. The toughening effectiveness is exhibited more prominently for rubbery polymers with the T_g below the room temperature, in comparison with glassy polymers with the T_g above the room temperature.

Very recently we simulated the fracture behavior of graphene filled polymer nanocomposites, and we observed that the yield point in the stress–strain behavior occurs at a smaller strain for attractive NPs compared to the pure case, attributed to the more mechanically heterogeneous environment. In the case of the same attractive interfacial interaction and same loading, in Fig. 15.17C the toughening efficiency exhibits the following order: graphene > CNT > C_{60}. Furthermore, our simulated results as well indicate polymer matrices with high flexibility and mobility ($>T_g$) are better toughened than that of polymer matrices with low flexibility and mobility ($<T_g$) [55], by changing the volume fraction of graphene sheets, and the interaction strength between graphene and polymer, as shown in Fig. 15.17A and B respectively.

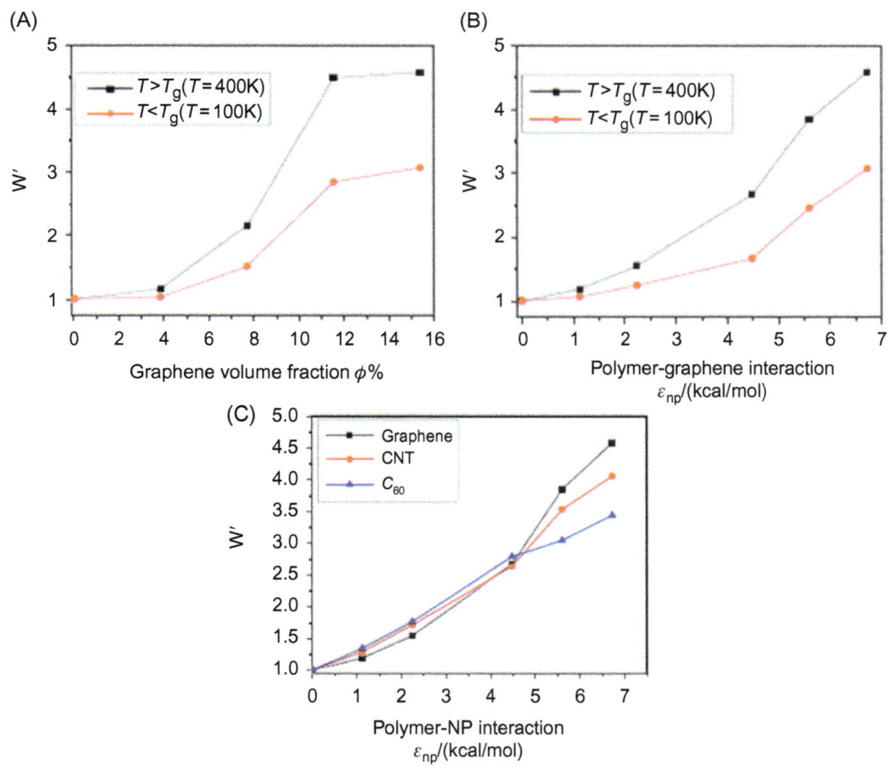

Figure 15.17 The work needed to produce the fracture (A) as a function of the graphene volume fraction in the cases of $T > T_g$ and $T < T_g$, (B) as a function of the polymer–graphene interaction. (C) The work needed to produce the fracture at the same loading as a function of the polymer–NP interaction for C_{60}, CNT, and graphene. *Source*: Adapted from Liu J, Shen JX, Zheng ZJ, Wu YP, Zhang LQ. Revealing the toughening mechanism of graphene-polymer nanocomposite through molecular dynamics simulation. Nanotechnology 2015;26:291003.

While for the one-dimensional CNT, Frankland et al. [56] firstly employed MD simulation to investigate the stress–strain behaviors of PE–nanotube composites, and the mechanical response in the direction of and transverse to the nanotube axis is examined. The elastic moduli of PE–SWCNT was also simulated for infinite and finite CNT to be compared with the macroscopic rule-of-mixtures. Recently from a micromechanics point, Brinson et al. [57] compared the mechanical reinforcing efficiencies of nanotube and nanoplatelet, finding that with the same volume fraction and identical degree of dispersion and extent of influence on the surrounding polymer molecules, the nanotubes produce a significantly larger amount of interphase and therefore higher stiffness than nanoplatelets.

The viscoelastic behavior is always greatly influenced by the introduction of the NPs, Smith et al. [58] conducted the first MD simulation to study the viscoelastic property of polymer chains induced by spherical NPs, focusing on the effects of the loading of NPs and the nature of the polymer–filler interaction on the dynamic shear modulus and the viscosity, later they also developed a multiscale modeling approach to simulate the viscoelasticity of PNCs [59]. The shear viscosity, zero-shear viscosity, and the extent of shear thinning of PNCs were found to increase with the filler content [60]. For the case of the regimes where the NP size and the interparticle distance become comparable to the polymer sizes, Ganesan et al. [61] found that at low NP concentration, the shear rheology of PNCs is dominated by the shear thinning of the polymer chains, while for higher NP loading the shear rheology is dominated by the NPs. Besides the filler size, concentration, and interfacial interaction, Kumar et al. [52] have explored the effect of the filler network on the nonlinear viscoelasticity of PNCs, finding that the occurrence of the stress overshoots is attributed to the interparticle polymer network formed through NPs acting as nodes, compared to that of purely entangled polymers resulting from purely physical entanglements.

In fact, to resolve the mystery of the nonlinear viscoelasticity of ENCs, Raos et al. [62] have simulated the effect of the interparticle interaction on the dynamic storage modulus by tuning the amplitude and frequency of the oscillatory shear deformation by fixing the filler volume fraction $\phi = 20\%$, indicating that the origin of the Payne effect does not seem to entirely and directly result from the particle–particle interaction, in contrast to the traditional assumption. However, in a later study with $\phi = 30\%$, a significant nonlinearity is observed to be associated with the disruption of the filler–filler interactions, as opposed to the existence of the glassy rubber layers around the fillers [63].

In order to characterize the change and contribution of the filler network under the external shear flow, Gersappe et al. [47] used MD simulation to examine the ratio of the number of the polymer chains forming the percolated filler cluster or network to the total polymer chains (N_c/N_Total) as a function of the simulation time for two different shear rates, as shown in Fig. 15.18A and B. Since at $\phi = 2\%$ no percolating filler network is formed (see Fig. 15.15A), therefore the value of N_c/N_Total maintains nearly constant for two shear rates $V_w = 0.03$ and 0.48. Furthermore, the 5% system retains the filler network structure at low shear rate, which, however, breaks down at high shear rate. But the 10% system still maintains some kind of filler cluster structure even at high shear rate. To further analyze the gradual change process of the filler network as a function of the shear rate, the 5% system is chosen and shown in Fig. 15.18C–F. A cartoon representation of the evolution of the filler network formed by polymers and fillers under shear is presented in Fig. 15.18C, respectively corresponding to the snapshots in (D) representing a filler network with the shear rate $V_w = 0.03$, (E) representing different separate filler clusters with $V_w = 0.24$, and (F) representing the complete dispersion of fillers with $V_w = 0.48$. Since all current simulation works are focused on the viscoelastic

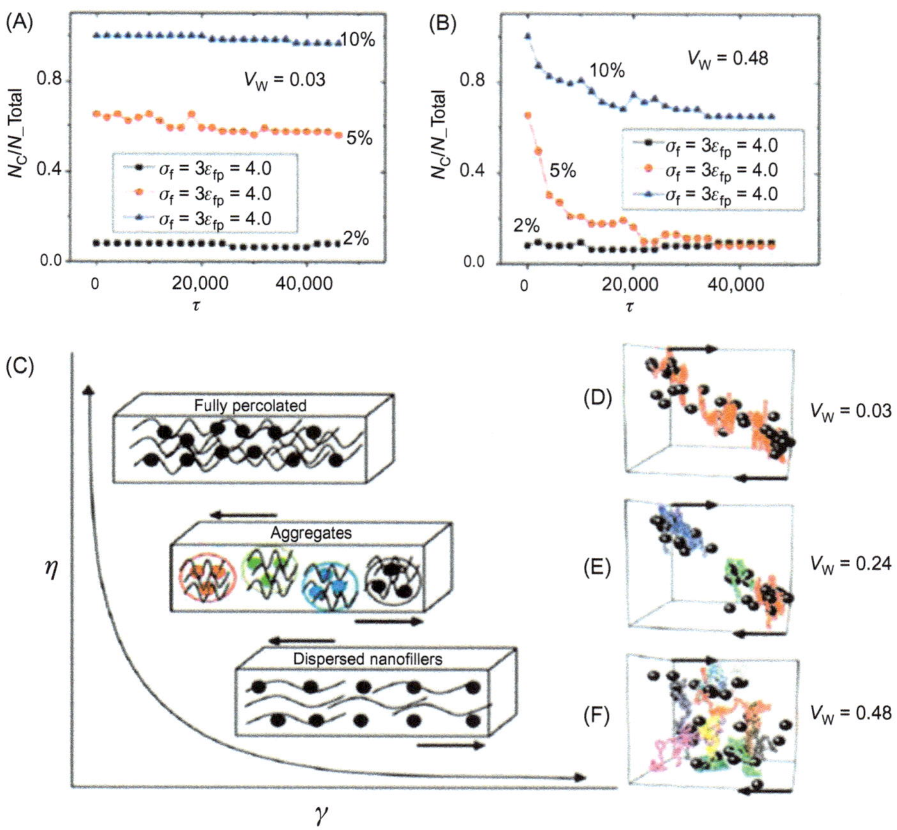

Figure 15.18 The ratio of the number of the polymer chains forming the percolated filler cluster or network to the total polymer chains (N_c/N_Total) versus the shear time for the filler volume fraction 2%, 5%, and 10% at the shear rate (A) $V_w = 0.03$, (B) $V_w = 0.48$. (C) A cartoon representation if the evolution of the filler network under shear for the filler volume fraction 5%. The corresponding snapshots at shear rate (D) $V_w = 0.03$, (E) $V_w = 0.24$, and (F) $V_w = 0.48$. The arrows indicate the direction of the shear.
Source: Adapted from Jaber E, Luo HB, Li WT, Gersappe D. Network formation in polymer nanocomposites under shear. Soft Matter 2011;7:3852–60.

property of spherical NPs filled polymer melts, the shape effect of NPs such as nanorod and clay sheet should be accounted for in the future.

In addition, to elucidate the Payne effect, we also adopted CGMDS to study the tensile and shear deformation of spherical NPs filled elastomer, and we find that the disintegration of the NP network formed through direct contactor the destruction of the NP network formed through the bridging of adsorbed polymer segments among the NPs both contribute to the occurrence of the Payne effect. In Fig. 15.19A the NPs directly aggregate with each other to form a network, in which case the increase of the interaction strength between NPs and NPs ε_{nn} leads to the gradual increase of the nonlinear behavior, while the interaction strength between polymer and NPs ε_{np} does not alter the nonlinear behavior. In Fig. 15.19B The NPs

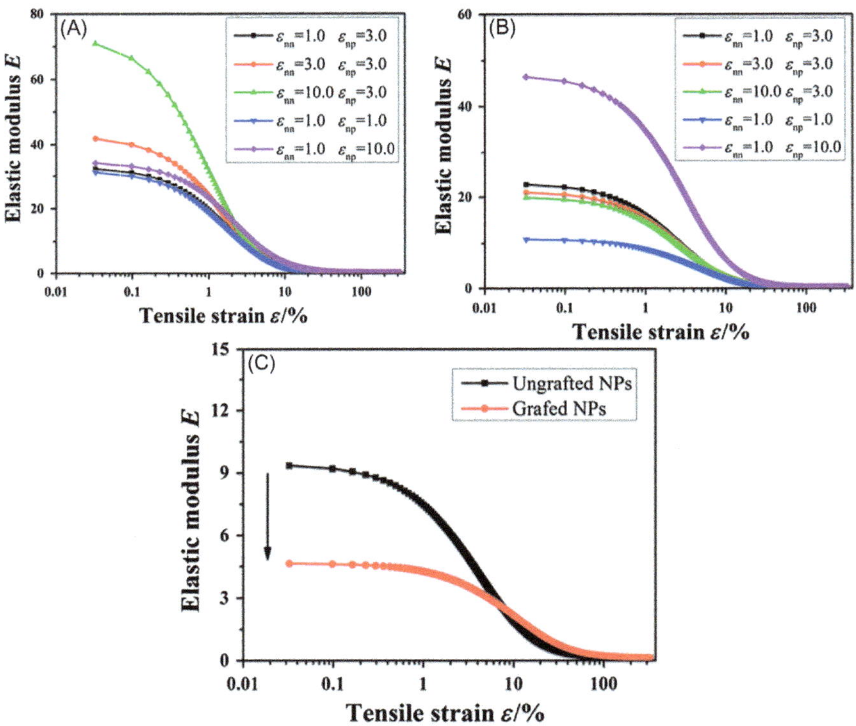

Figure 15.19 (A) In the case of direct aggregation between NPs to form the network, the change of the elastic modulus as a function of the tensile strain for various interaction strength between NPs and NPs, polymer and NPs. (B) In the case of the aggregation of NPs bridged via polymer chains to form the network, the change of the elastic modulus as a function of the tensile strain for various interaction strength between NPs and NPs, polymer and NPs. (C) Comparison of the Payne effect between ungrafted and grafted NPs system. *Source*: Adapted from Shen JX, Liu J, Gao YY, Li XL, Zhang LQ. Elucidating and tuning the strain-induced non-linear behavior of polymer nanocomposites: a detailed molecular dynamics simulation study. Soft Matter 2014;10:5099–113.

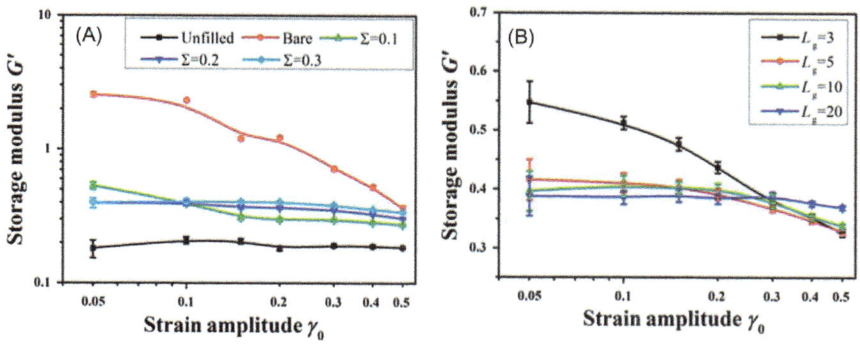

Figure 15.20 Change of the storage modulus as a function of the strain amplitude for (A) various grafting densities and (B) various grafted chain length.
Source: Adapted from Shen JX, et al. Molecular dynamics simulations of the structural, mechanical and visco-elastic properties of polymer nanocomposites filled with grafted nanoparticles. Phys Chem Chem Phys 2015;17:7196–207.

are bridged via polymer chains to form a network, and in this case the interaction strength between NPs and polymer ε_{np} significantly enhances the nonlinear behavior, which is not affected by the change of the interaction strength between NPs and NPs ε_{nn}. As shown in Fig. 15.19C, the Payne effect can be further reduced by grafting NPs with polymer chains to decrease the physical network density [64]. Furthermore, by varying the grafting density (Fig. 15.20A) and the grafted chain length (Fig. 15.20B), the Payne effect is significantly decreased [65].

15.5 Thermal and electrical conductivity of ENCs

Considering some physical properties like thermal and electrical conductivity, which are of great importance in the practical applications, we briefly describe the recent achievements through simulation techniques. For the thermal conductivity of ENCs, which critically determines the service life in the practical applications of ENCs, in the case of spherical NPs, there always exist the interface between the NP and polymer matrix, and the formation of highly densified adsorbed polymer layer around the NP, respectively corresponding to the Kapitza interface and effective interphase, as sketched in Fig. 15.21A, which are the main factors for controlling the thermal transfer between the polymer matrix and NPs. Suyoung et al. [66] used the micromechanics model to examine the effects of the thermal conductivity and thickness of the effective interphase and Kapitza interface on the overall thermal conductivity of the nanocomposites, as shown in Fig. 15.21B and C, by fixing other parameters such as the NP size and volume fraction. The results indicate that the overall thermal conductivity increases with the thickness and conductivity of the effective interphase, which can be further enhanced via the formation of more adhesive pairs between the polymer matrix and NPs. Moreover, for the conductivity of

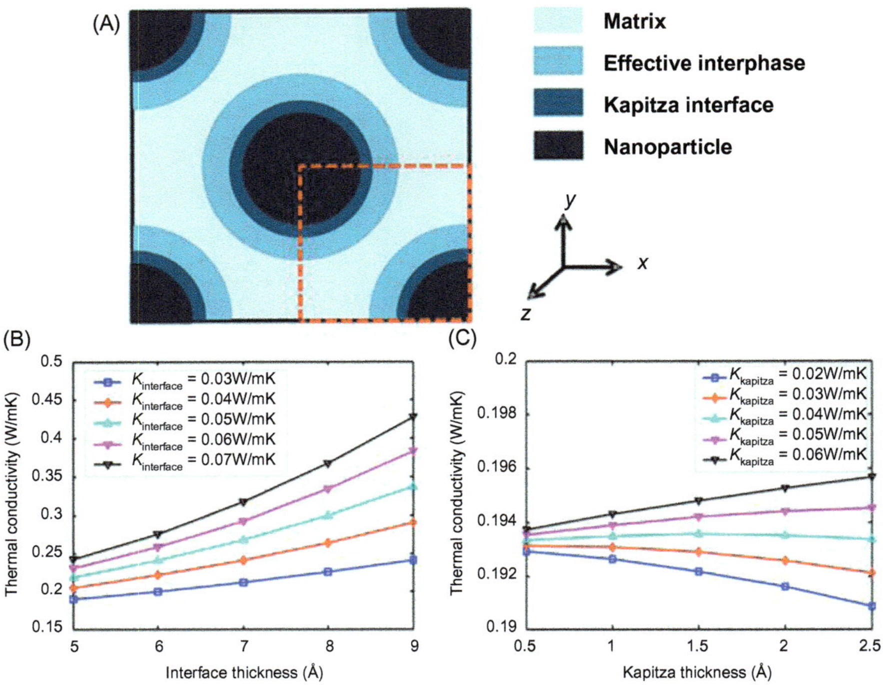

Figure 15.21 (A) Model with four-phase for the thermal conductivity of spherical NPs filled polymers. Change of the overall thermal conductivity of nanocomposites (B) at various thickness and conductivity of the effective interphase and (C) at various thickness and conductivity of the Kapitza interface.
Source: Adapted from Yu S, Yang S, Cho M. Multiscale modeling of cross-linked epoxy nanocomposites to characterize the effect of particle size on thermal conductivity. J Appl Phys 2011;110.

the Kapitza interface greater than 0.05 W/mK, the overall thermal conductivity begins to increase with respect to the Kapitza thickness, and one effective approach to enhance the conductivity of the Kapitza interface is to chemically link the matrix and NPs. Although this theoretical research is focused on the epoxy system, the conclusions can be as well applied to a spherical NPs filled elastomer system.

In particular, CNTs possess excellent thermal conductivity, exceeding the best-conducting metals. Therefore, it is expected that the incorporation of CNTs will significantly improve the thermal conductivity of polymers, however, this is also hindered by the high interfacial thermal resistance between the polymers and nanotubes. Moreover, the anisotropic structure of the polymer phase induced by the presence of the nanotubes leads to a slight increase of the polymer thermal conductivity parallel to the nanotubes, whereas it leads to a decrease perpendicular to it [67]. Clancy et al. [68] have simulated the effect of the functionalization of CNTs on the thermal conductivity of PNCs, finding that the grafting of linear hydrocarbon chains to the surface of SWCNT significantly increases the thermal conductivity

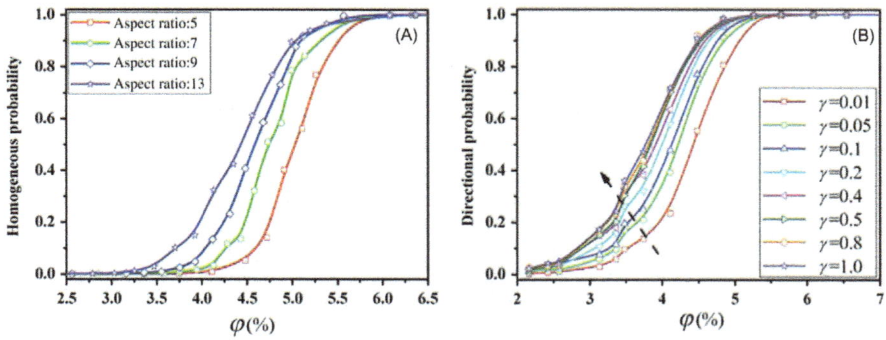

Figure 15.22 (A) The change of the homogeneous probability as a function of the volume fraction φ for various aspect ratio of nanorods. (B) The change of the directional probability versus the volume fraction φ by changing the shear rate.
Source: Adapted from Gao YY, et al. Molecular dynamics simulation of the conductivity mechanism of nanorod filled polymer nanocomposites. Phys Chem Chem Phys 2015;17:22959–68.

attributed to the decrease of the interfacial thermal resistance. Meanwhile, they also probed the effect of the grafting density of oligomers on the surface of the SWCNT on the interfacial thermal resistance through MD simulation in another study [69]. For the electrical conductivity of ENCs, it is directly related to the percolation behavior resulting from the formation of pathways of touching NPs, Ghazavizadeh et al. [70] simulated the effect of the mechanical loading on the electrical percolation threshold of CNT reinforced polymers, finding that a percolating nanocomposite becomes nonpercolating by applying a unidirectional tensile stress. In the meantime, we adopted CGMDS to study the percolation threshold of the electrical conductivity of ENCs filled with nanorods, by changing the effects of the interfacial interaction, the aspect ratio of the nanorods, the external shear field, filler–filler interaction, and temperature [71]. As is shown in Fig. 15.22A, the percolation threshold gradually decreases with the increase of the aspect ratio by fixing the volume fraction of nanorods. By changing the shear rate, we checked the directional probability along the shear direction in Fig. 15.22B. Evidently, the directional percolation threshold gradually decreases with the increase of the shear rate, contributed to by the greater tendency to form the conductivity network because of the alignment and orientation of nanorods along the shear direction.

15.6 Future simulation opportunities and challenges

In this chapter we aim to provide the latest research achievements of ENCs studied via computer simulation technique, by focusing on the dispersion structure and phase behavior of NPs, interfacial chain structure and dynamics, static and dynamic mechanics, and thermal and electrical conductivity. For good illustration, some important theoretical and experimental studies are also included. To provide more rational guidelines in the preparation of ENCs with excellent performance, three future simulation opportunities and challenges are listed below:

1. Presently almost all computer simulation is carried out at one single level, and simulating the system approaching the macroscopic scale from the molecular level is not currently possible because of the limited computational capability. Therefore, developing a multiscale simulation technique is extremely desirable, by covering the chemical characteristics to the continuum level.
2. Since most products made of ENCs are working in the dynamic situation, the resistance against crack propagation is of utmost importance to determine the service life. Computer simulation has the unique advantage to clearly monitor the initialization and propagation of the cavity, by solely changing one physical parameter while fixing others. Hence, future simulation work should be directed to systematically examine the crack resistance mechanism.
3. Through the accurate computer simulation, it is much cheaper to produce a series of data about the influence of the structural parameters, and finally establishing a quantitative structure-property relation is critical as well as challenging for designing and preparing novel ENCs in the future.

Acknowledgment

The authors acknowledge financial supports from the National 973 Basic Research Program of China 2015CB654700(2015CB654704), the Foundation for Innovative Research Groups of the NSF of China (51221002), the National Natural Science Foundation of China (51403015). The supercalculation center of "Tianhe number one" and the cloud calculation platform of Beijing University of Chemical Technology are both greatly appreciated.

References

[1] Binder K, Baschnagel J, Paul W. Glass transition of polymer melts: test of theoretical concepts by computer simulation. Prog Polym Sci 2003;28:115–72.
[2] Banaszak BJ, de Pablo JJ. A new double-rebridging technique for linear polyethylene. J Chem Phys 2003;119:2456–62.
[3] Ozmusul MS, Picu CR, Sternstein SS, Kumar SK. Lattice Monte Carlo simulations of chain conformations in polymer nanocomposites. Macromolecules 2005;38:4495–500.
[4] Pandey YN, Doxastakis M. Detailed atomistic Monte Carlo simulations of a polymer melt on a solid surface and around a nanoparticle. J Chem Phys 2012;136.
[5] Kremer K, Grest GS. J Chem Phys 1990;92:5057.
[6] Abrams C, Delle Site L, Kremer K. Bridging time scales: molecular simulations for the next decade. Berlin: SPRINGER-VERLAG BERLIN; 2002. pp. 143–64.
[7] Liu J, Cao DP, Zhang LQ. Molecular dynamics study on nanoparticle diffusion in polymer melts: a test of the stokes-einstein law. J Phys Chem C 2008;112:6653–61.
[8] Peter C, Kremer K. Multiscale simulation of soft matter systems. Faraday Discuss 2010;144:9–24.

[9] Kremer K, Muller-Plathe F. Multiscale problems in polymer science: simulation approaches. Mrs Bulletin 2001;26:205–10.
[10] Praprotnik M, Delle Site L, Kremer K. Annual review of physical chemistry. Palo Alto: ANNUAL REVIEWS; 2008. pp. 545–71.
[11] Hooper JB, Schweizer KS. Contact aggregation, bridging, and steric stabilization in dense polymer-particle mixtures. Macromolecules 2005;38:8858–69.
[12] Hooper JB, Schweizer KS. Theory of phase separation in polymer nanocomposites. Macromolecules 2006;39:5133–42.
[13] Liu J, Gao YG, Cao DP, Zhang LQ, Guo ZH. Nanoparticle dispersion and aggregation in polymer nanocomposites: insights from molecular dynamics simulation. Langmuir 2011;27:7926–33.
[14] Patra TK, Singh JK. Polymer directed aggregation and dispersion of anisotropic nanoparticles. Soft Matter 2014;10:1823–30.
[15] Sankar UK, Tripathy M. Dispersion, depletion, and bridging of athermal and attractive nanorods in polymer melt. Macromolecules 2015;48:432–42.
[16] Hall LM, Schweizer KS. Structure, scattering patterns and phase behavior of polymer nanocomposites with nonspherical fillers. Soft Matter 2010;6:1015–25.
[17] Padmanabhan V. Percolation of high-density polymer regions in nanocomposites: the underlying property for mechanical reinforcement. J Chem Phys 2013;139:144904.
[18] Vacatello M. Chain dimensions in filled polymers: an intriguing problem. Macromolecules 2002;35:8191–3.
[19] Frischknecht AL, McGarrity ES, Mackay ME. Expanded chain dimensions in polymer melts with nanoparticle fillers. J Chem Phys 2010;132:204901.
[20] Karatrantos A, Clarke N, Composto RJ, Winey KI. Polymer conformations in polymer nanocomposites containing spherical nanoparticles. Soft Matter 2015;11:382–8.
[21] Rittigstein P, Priestley RD, Broadbelt LJ, Torkelson JM. Model polymer nanocomposites provide an understanding of confinement effects in real nanocomposites. Nat Mater 2007;6:278–82.
[22] Berriot J, Montes H, Lequeux F, Long D, Sotta P. Evidence for the shift of the glass transition near the particles in silica-filled elastomers. Macromolecules 2002;35:9756–62.
[23] Montes H, Lequeux F, Berriot J. Influence of the glass transition temperature gradient on the nonlinear viscoelastic behavior in reinforced elastomers. Macromolecules 2003;36:8107–18.
[24] Merabia S, Sotta P, Long DR. A microscopic model for the reinforcement and the nonlinear behavior of filled elastomers and thermoplastic elastomers (payne and mullins effects). Macromolecules 2008;41:8252–66.
[25] Bogoslovov RB, Roland CM, Ellis AR, Randall AM, Robertson CG. Effect of silica nanoparticles on the local segmental dynamics in poly(vinyl acetate). Macromolecules 2008;41:1289–96.
[26] Robertson CG, Lin CJ, Rackaitis M, Roland CM. Influence of particle size and polymer-filler coupling on viscoelastic glass transition of particle-reinforced polymers. Macromolecules 2008;41:2727–31.
[27] Robertson CG, et al. Flocculation, reinforcement, and glass transition effects in silica-filled styrene-butadiene rubber. Rubber Chem Technol 2011;84:507–19.
[28] Robertson CG, Roland CM. Glass transition and interfacial segmental dynamics in polymer-particle composites. Rubber Chem Technol 2008;81:506–22.
[29] Starr FW, Schroder TB, Glotzer SC. Molecular dynamics simulation of a polymer melt with a nanoscopic particle. Macromolecules 2002;35:4481–92.

[30] Brown D, Mele P, Marceau S, Alberola ND. A molecular dynamics study of a model nanoparticle embedded in a polymer matrix. Macromolecules 2003;36:1395–406.
[31] Brown D, Marcadon V, Mele P, Alberola ND. Effect of filler particle size on the properties of model nanocomposites. Macromolecules 2008;41:1499–511.
[32] Liu J, et al. Polymer-nanoparticle interfacial behavior revisited: a molecular dynamics study. Phys Chem Chem Phys 2011;13:13058–69.
[33] Milano G, Santangelo G, Ragone F, Cavallo L, Di Matteo A. Gold nanoparticle/polymer interfaces: all atom structures from molecular dynamics simulations. J Phys Chem C 2011;115:15154–63.
[34] Ban SA, Malek K, Huang C, Liu ZS. A molecular model for carbon black primary particles with internal nanoporosity. Carbon NY 2011;49:3362–70.
[35] Donnet JB. Nano and microcomposites of polymers elastomers and their reinforcement. Compos Sci Technol 2003;63:1085–8.
[36] Borodin O, Smith GD, Bandyopadhyaya R, Byutner E. Molecular dynamics study of the influence of solid interfaces on poly(ethylene oxide) structure and dynamics. Macromolecules 2003;36:7873–83.
[37] Smith KA, Vladkov M, Barrat JL. Polymer melt near a solid surface: a molecular dynamics study of chain conformations and desorption dynamics. Macromolecules 2005;38:571–80.
[38] Smith GD, Bedrov D, Borodin O. Structural relaxation and dynamic heterogeneity in a polymer melt at attractive surfaces. Phys Rev Lett 2003;90.
[39] Sikdar D, Katti DR, Katti KS, Bhowmik R. Insight into molecular interactions between constituents in polymer clay nanocomposites. Polymer (Guildf) 2006;47:5196–205.
[40] Karatrantos A, Composto RJ, Winey KI, Clarke N. Structure and conformations of polymer/SWCNT nanocomposites. Macromolecules 2011;44:9830–8.
[41] Wei CY, Srivastava D, Cho K. Structural ordering in nanotube polymer composites. Nano Lett 2004;4:1949–52.
[42] Wei CY. Radius and chirality dependent conformation of polymer molecule at nanotube interface. Nano Lett 2006;6:1627–31.
[43] Foroutan M, Nasrabadi AT. Investigation of the interfacial binding between single-walled carbon nanotubes and heterocyclic conjugated polymers. J Phys Chem B 2010;114:5320–6.
[44] Tallury SS, Pasquinelli MA. Molecular dynamics simulations of flexible polymer chains wrapping single-walled carbon nanotubes. J Phys Chem B 2010;114:4122–9.
[45] Frankland SJV, Caglar A, Brenner DW, Griebel M. Molecular simulation of the influence of chemical cross-links on the shear strength of carbon nanotube-polymer interfaces. J Phys Chem B 2002;106:3046–8.
[46] Zheng QB, et al. Computational analysis of effect of modification on the interfacial characteristics of a carbon nanotube-polyethylene composite system. Appl Surf Sci 2009;255:3534–43.
[47] Jaber E, Luo HB, Li WT, Gersappe D. Network formation in polymer nanocomposites under shear. Soft Matter 2011;7:3852–60.
[48] Li Y, Kroger M, Liu WK. Primitive chain network study on uncrosslinked and crosslinked cis-polyisoprene polymers. Polymer (Guildf) 2011;52:5867–78.
[49] Liu J, Wu SZ, Zhang LQ, Wang WC, Cao DP. Molecular dynamics simulation for insight into microscopic mechanism of polymer reinforcement. Phys Chem Chem Phys 2011;13:518–29.
[50] Papakonstantopoulos GJ, Yoshimoto K, Doxastakis M, Nealey PF, de Pablo JJ. Local mechanical properties of polymeric nanocomposites. Phys Rev E 2005;72.

[51] Cho J, Sun CT. A molecular dynamics simulation study of inclusion size effect on polymeric nanocomposites. Comput Mater Sci 2007;41:54–62.
[52] Sen S, Thomin JD, Kumar SK, Keblinski P. Molecular underpinnings of the mechanical reinforcement in polymer nanocomposites. Macromolecules 2007;40:4059–67.
[53] Yagyu H, Utsumi T. Coarse-grained molecular dynamics simulation of nanofilled crosslinked rubber. Comput Mater Sci 2009;46:286–92.
[54] Song M, Chen L. Molecular dynamics simulation of the fracture in polymer-exfoliated layered silicate nanocomposites. Macromol Theory Simul 2006;15:238–45.
[55] Liu J, Shen JX, Zheng ZJ, Wu YP, Zhang LQ. Revealing the toughening mechanism of graphene-polymer nanocomposite through molecular dynamics simulation. Nanotechnology 2015;26:291003.
[56] Frankland SJV, Harik VM, Odegard GM, Brenner DW, Gates TS. The stress-strain behavior of polymer-nanotube composites from molecular dynamics simulation. Compos Sci Technol 2003;63:1655–61.
[57] Liu H, Brinson LC. Reinforcing efficiency of nanoparticles: a simple comparison for polymer nanocomposites. Compos Sci Technol 2008;68:1502–12.
[58] Smith GD, Bedrov D, Li LW, Byutner O. A molecular dynamics simulation study of the viscoelastic properties of polymer nanocomposites. J Chem Phys 2002;117:9478–89.
[59] Borodin O, Bedrov D, Smith GD, Nairn J, Bardenhagen S. Multiscale modeling of viscoelastic properties of polymer nanocomposites. J Polym Sci Part B Polym Phys 2005;43:1005–13.
[60] Kairn T, Daivis PJ, Ivanov I, Bhattacharya SN. Molecular-dynamics simulation of model polymer nanocomposite rheology and comparison with experiment. J Chem Phys 2005;123.
[61] Pryamitsyn V, Ganesan V. Mechanisms of steady-shear rheology in polymernanoparticle composites. J Rheol 2006;50:655–83.
[62] Raos G, Moreno M, Elli S. Computational experiments on filled rubber viscoelasticity: what is the role of particle-particle interactions? Macromolecules 2006;39:6744–51.
[63] Raos G, Casalegno M. Nonequilibrium simulations of filled polymer networks: searching for the origins of reinforcement and nonlinearity. J Cheml Phys 2011;134:054902.
[64] Shen JX, Liu J, Gao YY, Li XL, Zhang LQ. Elucidating and tuning the strain-induced non-linear behavior of polymer nanocomposites: a detailed molecular dynamics simulation study. Soft Matter 2014;10:5099–113.
[65] Shen JX, et al. Molecular dynamics simulations of the structural, mechanical and viscoelastic properties of polymer nanocomposites filled with grafted nanoparticles. Phys Chem Chem Phys 2015;17:7196–207.
[66] Yu S, Yang S, Cho M. Multiscale modeling of cross-linked epoxy nanocomposites to characterize the effect of particle size on thermal conductivity. J Appl Phys 2011;110.
[67] Alaghemandi M, Muller-Plathe F, Bohm MC. Thermal conductivity of carbon nanotube-polyamide-6,6 nanocomposites: reverse non-equilibrium molecular dynamics simulations. J Chem Phys 2011;135.
[68] Clancy TC, Gates TS. Modeling of interfacial modification effects on thermal conductivity of carbon nanotube composites. Polymer (Guildf) 2006;47:5990–6.
[69] Clancy TC, Frankland SJV, Hinkley JA, Gates TS. Multiscale modeling of thermal conductivity of polymer/carbon nanocomposites. Int J Thermal Sci 2010;49:1555–60.
[70] Ghazavizadeh A, et al. Evaluating the effect of mechanical loading on the electrical percolation threshold of carbon nanotube reinforced polymers: a 3D Monte-Carlo study. J Comput Theorl Nanosci 2011;8:2087–99.
[71] Gao YY, et al. Molecular dynamics simulation of the conductivity mechanism of nanorod filled polymer nanocomposites. Phys Chem Chem Phys 2015;17:22959–68.

Index

Note: Page numbers followed by "*f*" and "*t*" refer to figures and tables, respectively.

E

H

T

U

V

W

X

Y

Z